International Handbook of Earthquake Engineering

International Handbook of Earthquake Engineering

Codes, Programs, and Examples

edited by Mario Paz

Professor of Civil Engineering
University of Louisville, Kentucky

CHAPMAN & HALL

I(T)P   An International Thomson Publishing Company

New York • Albany • Bonn • Boston • Cincinnati • Detroit • London • Madrid • Melbourne • Mexico City
Pacific Grove • Paris • San Francisco • Singapore • Tokyo • Toronto • Washington

The computer programs developed to implement the seismic codes for the various countries included in this handbook are available, separately directly from the contributors, or as a complete set from MICROTEXT, P.O. Box 35101, Louisville, Kentucky 40232, USA. Technical questions, corrections, and requests for additional information should be directed to this address or directly to the corresponding contributor.

Extreme care has been taken in preparing the programs used in this handbook. Extensive testing and checking have been performed to ensure the accuracy and effectiveness of the computer solutions. However, neither the editor, nor the authors, nor the publisher shall be held responsible or liable for any damage arising from the use of any of the programs in this handbook.

Cover photo: Earthquake in Guatemala City, M.Sgt. Carl D. Martin, US Army
Cover design: Trudi Gershenov

First published in 1994 by
Chapman & Hall
One Penn Plaza
New York, NY 10119

Published in Great Britain by
Chapman & Hall
2–6 Boundary Row
London SE1 8HN

© 1994 Chapman & Hall, Inc.

Printed in the United States of America

Library of Congress Cataloging-in-Publication Data

Paz, Mario.
    International handbook of earthquake engineering : codes,
programs, and examples / Mario Paz.
        p.   cm.
    Includes bibliographical references and index.
    ISBN 0-412-98211-0
    1. Earthquake resistant design—Handbooks, manuals, etc.
2. Earthquake engineering—Handbooks, manuals, etc.   I. Title.
TA658.44.P395   1994
624.1′762—dc20

94-18577
CIP

British Library Cataloguing in Publication Data available

Please send your order for this or any other Chapman & Hall book to **Chapman & Hall, 29 West 35th Street, New York, NY 10001, Attn: Customer Service Department.**
You may also call our Order Department at 1-212-244-3336 or fax your purchase order to 1-800-248-4724.

For a complete listing of Chapman & Hall's titles, send your request to **Chapman & Hall, Dept. BC, One Penn Plaza, New York, NY 10119.**

to her eyes
that are filled
with dewsong
and light
of new morning

to her laughter
which sings sweet notes
of moonsong

to her hands
that spill the art
of lines and curves
and all of

God's colors . . .

to Jean

Sylvia T. Weinberg

Contents

Note: For more detailed information, see individual chapter outlines at chapter opening.

Foreword

The subject of earthquake engineering has been the focus of my teaching and research for many years. Thus, when Mario Paz, the editor of this handbook, asked me to write a Foreword, I was interested and honored by his request.

Worldwide, people are beginning to understand the severity of the danger to present and future generations caused by the destruction of the environment. Earthquakes pose a similar threat; thus, the proper use of methods for earthquake-resistant design and construction is vitally important for countries that are at high risk of being subjected to strong-motion earthquakes. Most seismic activity is the result of tectonic earthquakes. Tectonic earthquakes are very special events in that, although they occur frequently, their probability of becoming natural hazards for a specific urban area is very small. When a severe earthquake does occur near an urban area, however, its consequences are very large in terms of structural destruction and human suffering.

On the average, more than 10,000 earthquakes are recorded each year, of which about 60 are significant or potentially destructive. Between 1890 and 1993, more than 10,000 people were killed and 500,000 were left homeless by earthquakes, per year (on a yearly average). Furthermore, the psychological effect on the millions of people who live through major earthquakes is enormous; the experience inspires a complex fear that lasts for many years. It is, therefore, very important to investigate the reasons for earthquake disasters and to undertake the necessary steps to eliminate or mitigate their potentially catastrophic consequences.

Most of the human and economic losses resulting from a severe earthquake are due to failures of human-made facilities such as buildings and lifelines (dams, bridges, transportation systems, etc.). Although we cannot avert the occurrence of earthquakes, we can avoid their catastrophic effects.

Four conditions determine the occurrence of an earthquake disaster in a region. The first is the magnitude of the earthquake: a small earthquake will not induce groundshaking severe enough to produce extensive damage. The second is the distance between the earthquake source and the urban areas of the region; at large distances the groundshaking is attenuated to a level that cannot cause serious damage.[1] The third is the size and distribution of the population and the level of economic development. The fourth condition is the degree of earthquake preparedness, particularly regarding the methods of design and construction. Clearly, earthquake hazard involves not only the seismicity of the region, but also the population density, the economic development, and the degree of preparedness. Obviously, the potential disaster to an urban center increases when the earthquake magnitude is larger, the earthquake source is closer, the population is larger, the economic development is greater, and the earthquake preparedness program and its implementation are poorer.

---

[1]Under special conditions, however, earthquake damage can occur at distances greater than 240 km from the source (which has usually been assumed as the maximum distance to produce damage). For example, the source distance was 400 km in the 1957 and 1985 Mexico earthquakes and over 900 km in the 1972 Caucete, Argentina earthquake.

Even though seismicity remains constant, the rapid and, in most cases, uncontrolled increases in population, urbanization, and economic development in urban areas are not being counterbalanced by an adequate increase in preparedness.[2] There is an urgent need to reduce the seismic risk to urban areas. The solution is to regulate the built environment because this allows control of the earthquake hazards, which are consequences of the interaction of seismic activity (which cannot be controlled) with the vulnerability of the human-made environment (which can be controlled).

An effective approach to mitigate the destructive effects of earthquakes is the proper enforcement of the knowledge that is currently available for designing, constructing, and maintaining new earthquake-resistant structures and upgrading existing seismically hazardous structures. In practice, the design and construction of most earthquake-resistant structures generally follow seismic code provisions. Specific seismic code regulations for design and construction are different in each country, even though the problems that are created by earthquake ground motions and the application of basic concepts that govern earthquake-resistant design of structure are the same around the world. The differences in the seismic codes occur because basic concepts have to be specifically applied and quantified in each country according to its seismic activity, the education and experience of its professionals, its level of building technology, and its socio-economic conditions.

Modern seismic codes attempt to implement in relatively simple ways the most recent technological advances; however, they may actually obscure the limitations implicit in their regulations. It is, therefore, of utmost importance that structural engineers receive sufficient background in structural dynamics, which governs the response of structures to earthquake excitations, so that they can judge the limitations of their local code specifications. However, currently most structural engineers have limited their education in earthquake engineering to the practice of applying a code in the design of ordinary or standard structures, which compose the bulk of modern construction.

The *International Handbook of Earthquake Engineering: Codes, Programs, and Examples*, edited by Mario Paz, will be of great interest both to engineers practicing in the field of earthquake engineering and to students and teachers of courses on earthquake-resistant design, because it presents the basic theory of structural dynamics applied to earthquake-resistant analysis and design. Of equal importance, the information presented about the seismic codes that are enforced in a large number of countries can be a source for propagating better engineering methods around the world.

The handbook is well conceived and organized. Part I recapitulates the main theoretical topics in structural dynamics and earthquake engineering. In Part II, the reader will find 34 chapters corresponding to 34 countries located in regions with strong seismic activity. In addition to describing and illustrating the uses of the current seismic codes for each country, each chapter contains illustrative examples of building design that use computer programs specifically developed to implement the code regulations for that country.

I believe that this handbook will be enthusiastically received by professionals, educators, and students in the field of earthquake engineering. I congratulate the editor and the authors of the various chapters for their valuable achievement.

Vitelmo Bertero
Berkeley, California, U.S.A.

---

[2]For example, in terms of population and economic development, the disaster potential in California is now at least ten times what it was at the time of the 1906 San Francisco earthquake.

Preface

As a result of my consulting experience with engineers and industry in seismically active areas around the globe, I learned that much effort has to be devoted to understand, interpret, and properly apply the provisions in a building code. In some cases, the effort is a result of unfamiliarity with the language in which the code is written; in other cases, the difficulty is caused by either lack of clarity or lack of sufficient explanations of the code. These experiences led me to plan a handbook that would describe and implement the provisions of the seismic code for earthquake-resistant design of buildings for countries in active seismic regions. I conducted a survey of professional engineers to obtain their opinion of the value of such a handbook. The results of my informal survey encouraged me to proceed with the preparation of this handbook.

From its inception, I decided that the handbook should contain a chapter for each selected country. Therefore, my first task was to locate qualified contributors for each country that was to be included in the handbook. Of the contributors, a few were friends or acquaintances, some were prospective contributors that I contacted after I had examined papers published in technical journals or proceedings from international conferences on earthquake engineering, and others were located indirectly through contacts with universities or engineering societies.

The contributors were given general guidelines for the preparation of their chapters. I indicated that the chapter for each country should contain an introductory paragraph about the pertinent geography and the geology of the country, a brief regional history of recent strong-motion earthquakes, and a recapitulation of the main developments in the seismic code. The body of the chapter was to be devoted to a presentation of the provisions of the code in simple and clear language. This presentation was to include an example of a numerical problem for a building of four to six stories. Also, I requested that a computer program be developed by the contributor to apply the provisions of the seismic code in the solution of a sample numerical problem for a larger building.

My next step was to prepare and send a model chapter to the prospective authors so that they could follow a format that would provide a satisfactory degree of consistency throughout the chapters. However, the reader of this handbook will notice that diversity exists among the chapters written by the contributing experts. Some chapters are brief and concise, while others are extended and detailed. There is no doubt that such variations resulted from differences in the interpretation of the objectives that I sent to each author; I accepted the variations as an indication of the diversity of viewpoints. My decision was based on the perception that the contributors of the handbook are highly qualified professionals in their respective countries. The authors are engineers who are active in seismic-resistant design, many are affiliated with universities; some are also members of committees responsible for revising or updating their country's seismic code for earthquake-resistant design.

This handbook contains two parts: Part I has four chapters that recapitulate the subjects of structural dynamics and earthquake engineering; Part II contains 34 chapters that present the design of structures according to the provisions of the seismic code for each of the 34 countries included in the handbook. Most of the material presented in Part I can be found in other publications, including the textbook on structural dynamics written by the editor of this handbook. Part I gives the reader basic information about structural dynamics and earthquake engineering and provides cross-references to the various chapters of the handbook. In Part II of the handbook, I have tried to maintain a uniform nomenclature of symbols commonly used in earthquake engineering with the exception of cases where a differing author's decision was accepted as final.

This handbook includes an appendix on Magnitude and Intensity of Earthquakes. This appendix provides a readily available reference to the various definitions currently in use in different countries for measuring the magnitude of an earthquake and the intensity at a given locality.

The computer programs developed to implement the seismic codes for the various countries included in this handbook, are available, separately, directly from their respective contributors. These programs are also available from the editor, as a set, which includes all the programs in an interactive menu-driven package. A convenient form to order the complete set of programs is provided at the end of this handbook.

The editor believes the reader will find that this handbook serves as a major source for seismic-resistant design for the countries included.

Acknowledgments

I am indebted to many people for their help. First, I extend my thanks to each of the authors, who not only wrote a chapter and developed a computer program, but also endured my constant pressure to revise, check, and complete the chapter. I consider each of the authors a close friend even though I personally have not met some of them. One of these friends, Dr. Farzad Naeim, editor of an excellent handbook on earthquake engineering, joined me as coauthor of Chapter 2, Seismic Response and Design Spectra. Another friend, Dr. Alex H. Barbat of the Universidad Politécnica de Cataluña, joined me as coauthor of the chapter on Spain. I am most grateful for their contributions. A special thanks to Dr. Auguste Boissonnade who, in addition to preparing his chapter on France, took the time to express his philosophical view of the handbook during the preparatory phase of the task. Also, I give special thanks to Dr. Arturo Cifuentes who prepared the chapter on Chile and provided me with valuable information about potential authors in a number of Latin American countries. I am indebted to Dean Alberto Sarria from the Universidad de Los Andes, in Bogota, Colombia. Dean Sarria kindly authorized me to translate material from his excellent book, *Ingeniería Sísmica*, for inclusion in Appendix, Magnitude and Intensity of Earthquakes. I also wish to thank Professor Luis García, author of the chapter on Colombia, and Professor William Lobo-Quintero, author of the chapter on Venezuela, who recently invited me to conduct seminars on seismic resistant design in their respective countries.

I am most grateful to Mr. Robert D. Anderson and Mr. Joseph P. Colaco, consulting engineers in the states of California and Texas, respectively, who most diligently reviewed my proposal to the publisher for the preparation of this handbook; their comments and suggestions were very useful to me.

The work of compiling and editing the numerous chapters of this handbook would not have been possible without the active support of officials and staff of the University of Louisville. I am grateful to the university president, Dr. Donald Swain, for instituting a special group of secretaries to assist faculty members in preparing academic materials; and to Dr. Thomas R. Hanley, dean of the Speed Scientific School, and Dr. Louis F. Cohn, chairman of the Department of Civil Engineering, for their encouragement and approval of my sabbatical leave so that I could devote sufficient time to the work of editing and completing this handbook. I am also grateful to the provost of the university, Dr. Wallace Mann, who formally authorized my sabbatical leave. I also extend my thanks to my colleague, Dr. Michael Cassaro, for his constructive discussion on a variety of topics in this handbook; and my special thanks to Mrs. Debbie Jones for her competent typing and patience when revisions required retyping.

Finally, I thank my wife, Jean, who not only helped me during the process of checking the structure of the chapters in the handbook, but also most graciously allowed me to devote to this project much of the time that we normally enjoy together. This book is duly dedicated to her.

SPECIAL ACKNOWLEDGMENT

My task of verifying and compiling the contributions of the international group of experts in seismic-resistant engineering would have been very difficult without the collaboration of my friend, Dr. Edwin A. Tuttle, emeritus professor of Education, who most kindly helped me from the very beginning of this project. His primary objective was to achieve an optimal level of readability in each chapter of the handbook; he examined each contribution and revised the text to improve the clarity of the presentation. Dr. Tuttle met regularly with me to discuss proposed modifications. Those modifications that increased clarity without diminishing accuracy were adopted.

Mario Paz

Contributors

Vladimir Nickolayevich Alekhin, Ph.D.
Associate Professor
The Urals State Technical University – UPI
Ekaterinburg, Russia 620002
FAX: 7-343-2-441-624
(Chapter 35: Former USSR)

Celso S. Alfaro,
Professor
Central American University – UCA
Apatado (01) 168
Autopista Sur, San Salvador, El Salvador
TEL.: 503-79-2550 FAX: 503-78-5153
(Chapter 15: EL SALVADOR)

Ricardo D. Ambrosini, M.Eng.
Assistant Professor
Facultad de Ciencias Exactas Tecnología
Universidad Nacional de Tucumán
Lola Mora 380
4000 San Miguel de Tucumán, Argentina
TEL.: 54-81-307403 FAX: 54-81-241338
(Chapter 6: ARGENTINA)

Joao Azevedo, Ph.D.
Associate Professor
Civil Engineering Department
Instituto Superior Técnico
Av. Rovisco Pais
1096 Lisboa Codex, Portugal
TEL.: 351-1-8473457 FAX: 351-1-8497650
(Chapter 28: PORTUGAL)

Alex H. Barbat, Ph.D.
Escuela Técnica Superior de Ingenieros de Caminos,
   Canales y Puertos de Barcelona
Universidad Politécnica de Cataluña 08034 Barcelona,
   Spain
TEL.: 34-3-401-64 96 FAX: 34-3-401-65 17
(Chapter 31: SPAIN)

José Luis Barzuna de Oña
Arquitectura e Ingeniería, S.A.
P.O. Box 4755-1000
San José, Costa Rica
TEL.: 506-21-1932 FAX: 506-22-5397
(Chapter 13: COSTA RICA)

Gianmario Benzoni, D.E.
Assistant Professor
Politecnico Di Milano
Piazza Leonardo da Vinci 32
20133 Milano, Italy
TEL.: 39-2-2399-4228 FAX: 39-2-2399-4220
(Chapter 23: ITALY)

Auguste Boissonnade, Ph.D.
Project Engineer
Jack R. Benjamin & Associates, Inc.
444 Castro Street, Suite 501
Mountain View, CA 94041
TEL.: 1-415-969-8212 FAX: 1-415-969-6671
(Chapter 16: FRANCE)

Athol James Carr, Ph.D.
Professor
Department of Civil Engineering
University of Canterbury
Christchurch, New Zealand
TEL.: 03-366-7001 FAX: 03-364-2758
(Chapter 26: NEW ZEALAND)

Brijesh Chandra, Ph.D.
Professor
Department of Earthquake Engineering
University Roorkee
Roorkee 247667, India
(Chapter 19: INDIA)

Yohchia Chen, Ph.D.
Assistant Professor

School of Science, Engineering Technology
Pennsylvania State University – Harrisburg
Middletown, PA 17057
TEL.: 717-948-6146 FAX: 717-938-6401
(Chapter 32: TAIWAN)

Arturo Cifuentes, Ph.D.
Modeling Systems
IBM T.J. Watson Research Center
P.O. Box 218
Yorktown Heights, NY 10598
TEL.: 1-914-945-4131 FAX: 914-945-4203
(Chapter 10: CHILE)

David B. Crawley, M.A.
Senior Lecturer
Civil Engineering Department
University of Adelaide
Adelaide, South Australia 5000
TEL.: 8-272-084 FAX: 8-303-4359
(Chapter 7: AUSTRALIA)

Mokhtar Daoudi, Civil Engineer
Bureau d'étude Sechaud et Metz
63, Boulevard Reune
75014 Paris, France
FAX: 33-14-370-1674
(Chapter 5: ALGERIA)

Turan Durgunoğlu, Ph.D.
Professor
Civil Engineering Department
Bogazici University
PK2 Bebek Istanbul 80815, Turkey
TEL.: 90-1-2658488 FAX: 90-1-2661034
(Chapter 34: TURKEY)

Fouad H. Fouad, Ph.D.
Professor
Civil Engineering Department
University of Alabama
Birmingham, AL 35294
TEL.: 205-934-8430 FAX: 205-934-8437
(Chapter 14: EGYPT)

Luis E. García, M.S.
Professor
Civil Engineering Department
Universidad de Los Andes
Carrera 1a, No. 18A-10
Bogota, Colombia
TEL.: 51-1-2350155 FAX: 57-1-212-3507
(Chapter 12: COLOMBIA)

Predrag Gavrilovic, Ph.D.
Professor
Institute of Earthquake Engineering and Engineering
  Seismology
University "St. Cyril and Methodius"
Skopje, Republic of Macedonia
FAX: 389-91-112-183
(Chapter 38: Former YUGOSLAVIA)

Carmelo Gentile, D.E.
Assistant Professor
Politecnico Di Milano
Piazza Leonardo da Vinci 32
20133 Milano, Italy
TEL.: 39-2-2399-4242 FAX: 39-2-2399-4220
(Chapter 23: ITALY)

Jacob Gluck, D.Sc.
Professor
Department of Civil Engineering
Technion – Israel Institute of Technology
Technion City, Haifa 32000, Israel
FAX: 972-4-293044
(Chaper 22: ISRAEL)

Michael C. Griffith, Ph.D.
Senior Lecturer
Civil Engineering Department
University of Adelaide
Adelaide, South Australia 5000
TEL.: 8-303-5451 FAX: 8-303-4359
(Chapter 7: AUSTRALIA)

J. L. Humar, Ph.D.
Professor
Department of Civil and Environmental Engineering
Carleton University
Ottawa, Ontario K1S 5B6, Canada
TEL.: 613-788-5784 FAX: 613-788-3951
(Chapter 9: CANADA)

Sudhir K. Jain, Ph.D.
Associate Professor
Indian Institute of Technology Kanpur
Kanpur 208016, India
TEL.: 91-512-259583 FAX: 91-512-250260
(Chapter 19: INDIA)

Dimitar Jurukovski, Ph.D.
Professor
Institute of Earthquake Engineering and Engineering
  Seismology
University "St. Cyril and Methodius"
Skopje, Republic of Macedonia
FAX: 38-91-112-183
(Chapter 38: Former YUGOSLAVIA)

Yoshikazu Kitagawa, Ph.D.
Building Research Institute
Ministry of Construction
1 Tatekara, Tsukuba-shi
Ibraki-ken 305, Japan
FAX: 81-0298-64-2989
(Chapter 24: JAPAN)

David T. Lau, Ph.D.
Associate Professor
Department of Civil and Environmental Engineering
Carleton University
Ottawa, Ontario K1S 5B6, Canada
TEL.: 613-788-5784 FAX: 613-788-3951
(Chapter 9: CANADA)

William Lobo-Quintero, M.Eng.
Professor, Universidad de Los Andes
Mérida, Venezuela
TEL.: 74-441042 FAX: 74-402329
(Chapter 37: VENEZUELA)

Bibiana M. Luccioni, Ph.D.
Assistant Professor
Facultad de Ciencias Exactas Tecnología
Universidad Nacional de Tucumán
Juan B. Terán 375
4107 Yerba Buena, Tucumán, Argentina
TEL.: 54-81-352896 FAX: 54-81-241338
(Chapter 6: ARGENTINA)

Panitan Lukkunaprasit, Ph.D.
Professor
Department of Civil Engineering
Chulalonghorn University
Bangkok 10330, Thailand
TEL.: 66-2-218-6571 FAX: 66-2-252-1513
(Chapter 33: THAILAND)

George C. Manos, Ph.D.
Associate Professor
Department of Civil Engineering
Aristotle University of Thessaloniki
Director of Institute of Engineering Seismology and
    Earthquake Engineering (ITSAK)
Thessaloniki 54006, Greece
TEL.: 3031-992604 FAX: 30-31-99-5769
(Chapter 17: GREECE)

J. P. Mohsen, Ph.D.
Associate Professor
Civil Engineering Department
Speed Scientific School
University of Louisville
Louisville, KY 40292
TEL.: 502-852-6276 FAX: 502-852-8851
(Chapter 21: IRAN)

Farzad Naeim, Ph.D
John A. Martin and Associates
1800 Wilshire Blvd.
Los Angeles, CA 90057
TEL.: 213-483-6490
(Chapter 2: SEISMIC RESPONSE AND DESIGN
    SPECTRA)

Gelu Onu, Ph.D.
Specialist Engineer
IPTANA-Sa
B-dul Dinicu Golescu 38
Bucharest, Romania
TEL.: 400-638-5595 FAX: 400-312-14-16
(Chapter 30: ROMANIA)

Gianfranco Ottazzi, Professor
Pontificia Universidad Católica del Perú
Departamento de Ingeniería
Apartado 1761
Pando, Lima 100, Perú
FAX: 51-14-611785
(Chapter 27: PERU)

D. K. Paul, Ph.D.
Professor
Department of Earthquake Engineering
University Roorkee
Roorkee 247667, India
(Chapter 19: INDIA)

Thomas Paulay, Ph.D.
Professor emeritus
Department of Civil Engineering
University of Canterbury
Christchurch, New Zealand
TEL.: 03-366-7001 FAX: 03-364-2758
(Chapter 26: NEW ZEALAND)

Mario Paz, Ph.D.
Professor
Civil Engineering Department
Speed Scientific School
University of Louisville
Louisville, KY 40292
TEL.: 502-852-6276 FAX: 502-852-8851
(Chapter 31: SPAIN and Chapter 36: USA)

Daniel Quiun, Professor
Pontificia Universidad Católica del Perú
Departamento de Ingeniería
Apartado 1761
Pando, Lima 100, Perú
FAX: 51-14-611785
(Chapter 27: PERU)

Fernando A. M. Reyna, Ph.D.
Professor, Laboratorio de Estructuras
Facultad de Ciencias Exactas y Tecnología
Universidad Nacional de Tucumán
4000, San Miguel de Tucumán, Argentina
TEL.: 54-51-245831
(Chapter 6: ARGENTINA)

Abdenour Salhi, Civil Engineer
Bureau d'Etudes Sechaud et Metz
55 Boulevard de Charonne
75011 Paris, France
FAX: 33-14-370-1674
(Chapter 5: ALGERIA)

Alberto Sarria M., Dean
Facultad de Ingeniería
Universidad de Los Andes
Bogota, Colombia
TEL.: 57-1-271-7693 FAX: 1-57-284-1570
(Appendix: MAGNITUDE AND INTENSITY OF
EARTHQUAKES)

Luis E. Suarez, Ph.D.
Associate Professor
General Engineering Department
University of Puerto Rico
Mayaguez, Puerto Rico 00680
TEL.: 809-265-3816 FAX: 809-832-0119
(Chapter 29: PUERTO RICO)

Suradjin Sutjipto, Ing., M.S.
Lecturer of Earthquake Engineering
Department of Civil Engineering
Trisakti University
Jakarta 11440, Indonesia
TEL.: 62-21-566-3232 FAX: 62-21-549-3895
(Chapter 20: INDONESIA)

Fumio Takino, President
Kozo System Inc.
4-4-7-105 Honcho
Shibuya-ku, Tokyo, Japan
(Chapter 24: JAPAN)

Edward D. Thomson, M.Eng.
Professor, Universidad de Los Andes
Mérida, Venezuela
TEL.: 74-441539 FAX: 74-402947
(Chapter 37: VENEZUELA)

Ludmil Tzenov, Ph.D., Dr.Sc.
Professor and Chairman,
Earthquake Engineering Department

Central Laboratory for Seismic Mechanics and
Earthquake Engineering,
Bulgarian Academy of Sciences, G. Bonchev str.
B1.3
Sofia 1113, Bulgaria
TEL.: 359-02-713-3338 FAX: 359-02-703-107
(Chapter 8: BULGARIA)

Elena Vasseva, Ph.D.
Associate Professor
Earthquake Engineering Department
Central Laboratory for Seismic Mechanics and
Earthquake Engineering,
Bulgarian Academy of Sciences, G. Bonchev str.
B1.3
Sofia 1113, Bulgaria
TEL.: 359-02-713-3338 FAX: 359-02-703-107
(Chapter 8: BULGARIA)

Gyorgy Vértes, Ph.D.
Professor
Faculty of Civil Engineering
Department of Mechanics
Technical University of Budapest
H-1521, Budapest XI, Muegyetem
rkp. 3.k mf.35, Hungary
TEL.: 36-1-650-199 FAX: 36-1-650-199
(Chapter 18: HUNGARY)

Roberto Villaverde, Ph.D.
Associate Professor
Department of Civil Engineering
University of California – Irvine
Irvine, CA 92717
TEL.: 714-856-5482 FAX: 714-725-2117
(Chapter 25: MEXICO)

Julius P. Wong, Ph.D.
Professor
Department of Mechanical Engineering
University of Louisville
Louisville, KY 40292
TEL.: 502-852-6335
(Chapter 32: TAIWAN)

Ye Yaoxian, Ph.D.
Professor
China Building Technology Development Center
19 Che Gong Zhuang Street
Beijing, China 100044
FAX: 01-8328832
(Chapter 11: CHINA)

International Handbook of Earthquake Engineering

I
Introduction to Structural Dynamics and Earthquake Engineering

1

Structures Modeled as Single-Degree-of-Freedom Systems

Mario Paz

1.1 INTRODUCTION

In general, the analysis and design of buildings and other structures to resist the effect produced by earthquakes requires conceptual idealizations and simplifying assumptions through which the physical system is represented by a new idealized system known as the *mathematical model*. In the mathematical model, the number of independent coordinates used to specify the position or configuration of the model at any time is referred to as the number of *degrees of freedom*. In principle, structures, being continuous systems, have an infinite number of degrees of freedom. However, the process of idealization or selection of an appropriate model permits the reduction of the number of degrees of freedom to a discrete number and in some cases, to just a single degree of freedom. Fig. 1.1 shows a one-story building which may be modeled with one degree of freedom. The model represented in this figure contains the following elements: (1) the concentrated mass m, (2) the lateral stiffness indicated by the coefficient k, (3) the damping in the system represented by coefficient c, and (4) the external force

$F(t)$ (considered to be a function of time); the response is indicated by the lateral displacement $y(t)$ of the mass m.

The structural model shown in Fig. 1.2 is assumed to be excited by a horizontal acceleration $\ddot{y}_s(t)$ at its base. In this case, it is convenient to express the response by the relative motion $u(t)$ between the

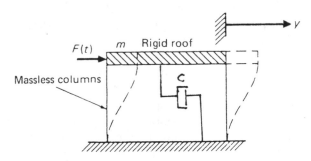

Fig. 1.1. Mathematical model for one-story structure excited by an external force

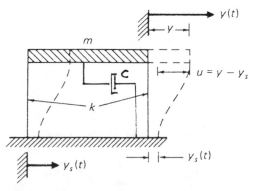

Fig. 1.2. Mathematical model for a one-story structure excited at its base

3

displacement $y(t)$ of the mass m and the displacement of the base $y_s(t)$, that is

$$u(t) = y(t) - y_s(t) \qquad (1.1)$$

In the next section, it will be shown that the structure shown in Fig. 1.1 or 1.2 subjected to two different excitations, a force applied to the mass, or a vibratory motion at the base, results in the same mathematical formulation.

1.2 SINGLE-DEGREE-OF-FREEDOM SYSTEM

A single-degree-of-freedom system such as the structure in Fig. 1.1 can conveniently be described with the simple oscillator shown in Fig. 1.3, which has the following elements: (1) a mass element m representing the mass and inertial properties of the structure, (2) a spring element k representing the elastic restoring force and potential energy capacity of the structure, (3) a damping element c representing the frictional characteristics and energy losses of the structure, and (4) an excitation force $F(t)$ representing the external forces acting on the structure.

In adopting the model shown in Fig. 1.3, it is further assumed that the system is linear. That is, the force-displacement relationship of the restoring force F_s of the spring and the force-velocity relationship of the damper force F_D are linear functions, respectively, of the displacement y and of the velocity $\dot{y}$ as shown in Fig. 1.4. Similarly, the simple oscillator may also be used to represent a structure excited by a motion at its base. Fig. 1.5 shows the oscillator excited at its base by a displacement function $y_s(t)$, and its response indicated by the relative displacement $u(t)$ of the mass.

The equation of motion for a one-degree-of-freedom system represented by the simple oscillator in Fig. 1.3 or 1.5 may be obtained by the application of Newton's law by simply equating to zero the sum of the forces which includes the inertial force $m\ddot{y}$ in the corresponding free-body diagrams shown in these figures. Thus, from Fig. 1.3(b)

$$m\ddot{y} + c\dot{y} + ky = F(t) \qquad (1.2)$$

and from Fig. 1.5(b)

$$m\ddot{y} + c(\dot{y} - \dot{y}_s) + k(y - y_s) = 0 \qquad (1.3)$$

The substitution in eq.(1.3) of $u = y - y_s$ and its derivatives yields

$$m\ddot{u} + c\dot{u} + ku = F_{eff}(t) \qquad (1.4)$$

where $F_{eff} = -m\ddot{y}_s(t)$ is the effective exciting force.

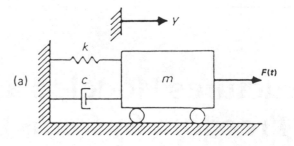

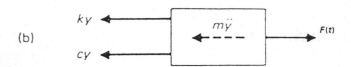

Fig. 1.3.   (a) Simple oscillator excited by an external force, (b) Free-body diagram

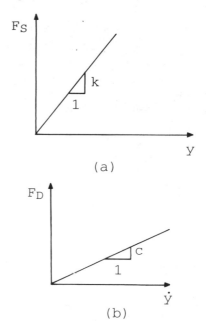

Fig. 1.4.   (a) Linear force-displacement relationship, (b) Linear damping force-velocity relationship

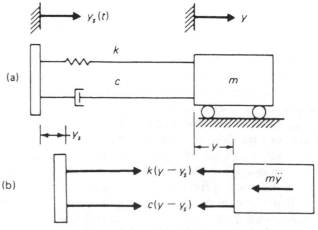

Fig. 1.5.   (a) Simple oscillator excited at its foundation, (b) Free-body diagram

It can be seen that eqs.(1.2) and (1.4) are mathematically equivalent. Equation (1.2) gives the response of the system acted on by the external force $F(t)$ in terms of the absolute displacement $y(t)$, while eq.(1.4) gives the response in terms of the relative motion $u(t)$ between the mass and the base for the case in which the excitation is due to the motion applied to the base of the system.

1.3 FREE VIBRATION

In this section the solution of the differential equation of motion is presented for the case of free vibration; that is, for the case in which the structure vibrates freely under the effect of the initial conditions with no external excitation applied to the system. The following cases of free vibration are considered: undamped free vibration and damped free vibration.

1.3.1 Undamped Free Vibration

Differential equation:

$$m\ddot{y} + ky = 0 \qquad (1.5)$$

Solution:

$$y = C_1 \cos \omega t + C_2 \sin \omega t \qquad (1.6)$$

where $\omega = \sqrt{k/m}$ is the natural frequency in rad/sec and C_1, C_2 are constants of integration. These constants are determined from knowledge of the initial conditions, i.e., the initial displacement, y_0 and the initial velocity, v_0, at time $t = 0$. The introduction of initial conditions into eq.(1.6) yields

$$y = y_0 \cos \omega t + \frac{v_0}{\omega} \sin \omega t \qquad (1.7)$$

which may be written alternatively as

$$y = C \sin (\omega t + \alpha) \qquad (1.8)$$

or

$$y = C \cos (\omega t - \beta) \qquad (1.9)$$

where

$$C = \sqrt{y_0^2 + (v_0/\omega)^2} \quad \text{(amplitude of motion)} \quad (1.10)$$

and

$$\alpha = \tan^{-1} \frac{y_0}{v_0/\omega} , \ \beta = \tan^{-1} \frac{v_0/\omega}{y_0} \quad \text{(phase)} \quad (1.11)$$

1.3.2 Damped Free Vibration (Fig. 1.6)

Differential equation:

$$m\ddot{y} + c\dot{y} + ky = 0 \qquad (1.12)$$

Solution:

$$y = e^{-\xi\omega t}[C_1 \cos \omega_D t + C_2 \sin \omega_D t] \qquad (1.13)$$

or alternatively,

$$y = Ce^{-\xi\omega t} \cos (\omega_D t - \alpha) \qquad (1.14)$$

where

$$\xi = \frac{c}{c_{cr}} \quad \text{(damping ratio)}, \qquad (1.15)$$

$$c_{cr} = 2\sqrt{km} \quad \text{(critical damping coefficient)} \quad (1.16)$$

and

$$\omega_D = \omega\sqrt{1 - \xi^2} \quad \text{(damped frequency)} \qquad (1.17)$$

The constants of integration C and α are evaluated from the initial conditions as

$$C = \sqrt{y_0^2 + \left(\frac{v_0 + y_0\xi\omega}{\omega_D}\right)^2} \qquad (1.18)$$

and

$$\alpha = \tan^{-1}\left(\frac{v_0 + y_0\xi\omega}{\omega_D y_0}\right) \qquad (1.19)$$

Alternatively, the constants of integration in

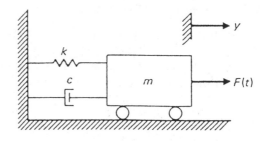

(a)

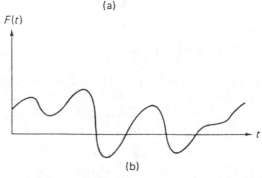

(b)

Fig. 1.6. Oscillator excited by external force $F(t)$ (a) Mathematical model, (b) Load function

Table 1.1.   Recommended Damping Values [Newmark and Hall (1973)]

Stress Level, Working Stress	Type and Condition of Structure	Percentage Damping
≤½ yield point	Vital piping	1–2
	Welded steel, prestressed concrete, reinforced concrete (slight cracking)	2–3
	Reinforced concrete (much cracking)	3–5
	Bolted or riveted steel or wood structures with nailed or bolted joints	5–7
At just below yield point	Vital piping	2–3
	Welded steel, prestressed concrete (no complete loss in prestress)	5–7
	Prestressed concrete (no prestress left)	7–10
	Reinforced concrete	7–10
	Bolted or riveted steel or wood structures with nailed or bolted joints	10–15
	Wood structures with nailed joints	15-20

eq.(1.13) may be determined from initial conditions and introduced into eq.(1.13) to give

$$y(t) = e^{-\xi\omega t}\left(y_0 \cos\omega_D t + \frac{v_0 + y_0\xi\omega}{\omega_D}\sin\omega_D t\right) \qquad (1.20)$$

A structural system will vibrate with an oscillatory motion when initially disturbed if the value of its damping coefficient is less than the value of the critical damping $c_{cr} = 2\sqrt{km}$, that is, the damping ratio, $\xi = c/c_{cr} < 1$. Normally, structures have values for the damping ratio well below the limiting value of $\xi = 1$. Depending on the material employed and the construction of the structure, the range of the damping ratio in structural systems lies between 2% and 20% of the critical damping. A summary of recommended damping values compiled by Newmark and Hall (1973) for different types of structures and different stress levels is given in Table 1.1 for reference.

1.3.3 Frequency and Period

An examination of eq.(1.6) or (1.7) shows that the motion described by these equations is harmonic (sine or cosine function), therefore periodic. It follows that the period T of the motion may be determined from

$$\omega T = 2\pi$$

or

$$T = \frac{2\pi}{\omega} \qquad (1.21)$$

since both functions, sine and cosine, have a period of 2π.

The period is usually expressed in seconds per cycle, or simply in seconds, with the tacit understanding that it is "per cycle." The reciprocal of the period is the natural frequency, i.e.,

$$f = \frac{1}{T} = \frac{\omega}{2\pi} \qquad (1.22)$$

The natural frequency f is usually expressed in hertz

(Hz) or cycles per second (cps). The quantity ω is also referred to as the natural frequency because it differs from the natural frequency f only by the constant factor 2π. To distinguish between these two expressions for the natural frequency, ω may be called the circular or angular natural frequency. Usually, the distinction is understood from the context or from the units. As indicated, the natural frequency f is measured in cps, while the circular frequency ω is measured in radians per second (rad/sec).

1.4 RESPONSE BY DIRECT INTEGRATION

The solution of the differential equation of motion for a structure modeled by the simple oscillator may be obtained in closed form only for some simple excitation functions. For a general excitation function, it is necessary to resort to numerical methods of integration. These numerical methods generally involve the use of approximations in the solution.

The method of solution of the differential equation of motion presented in this section is exact for an excitation function described by linear segments between points defining the excitation function. For convenience, the excitation function is calculated at equal time intervals Δt by linear interpolation between points defining the excitation. Thus, the time duration of the excitation, including a suitable extension of time after cessation of the excitation, is divided into N equal time intervals of duration Δt. For each interval Δt, the response is calculated by considering the initial conditions at the beginning of that time interval and the linear excitation during the interval. In this case, the initial conditions are the displacement and the velocity at the end of the preceding time interval. When the excitation function $F(t)$ is approximated by a piecewise linear function as shown in Fig. 1.7, it may be expressed as

$$F(t) = \left(1 - \frac{t - t_i}{\Delta t}\right)F_i + \left(\frac{t - t_i}{\Delta t}\right)F_{i+1} \quad t_i \leq t \leq t_{i+1} \qquad (1.23)$$

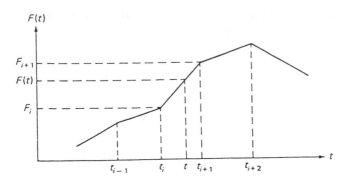

Fig. 1.7. Piecewise linear excitation function

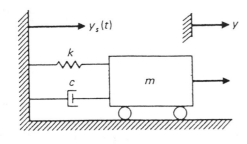

(a)

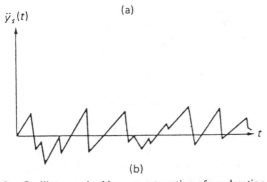

(b)

Fig. 1.8. Oscillator excited by support motion of acceleration $\ddot{y}(t)$

in which $t_i = i \cdot \Delta t$ (for equal intervals of duration Δt), for $i = 1, 2, 3, \ldots, N-1$. The differential equation of motion for the system shown in Fig. 1.8, from eq.(1.4), is then given by

$$m\ddot{u} + c\dot{u} + ku = \left(1 - \frac{t - t_i}{\Delta t}\right)F_i$$

$$+ \left(\frac{t - t_i}{\Delta t}\right)F_{i+1} \quad t_i \leqslant t \leqslant t_{i+1} \quad (1.24)$$

in which F_i is the effective force $-m\ddot{y}_s(t_i)$ evaluated at time t_i. The solution of eq.(1.24), in terms of the displacement and the velocity, is given by

$$u = e^{-\xi\omega(t - t_i)}[C_i \cos \omega_D(t - t_i)$$

$$+ D_i \sin \omega_D(t - t_i)] + B_i + A_i(t - t_i) \quad (1.25)$$

and

$$\dot{u} = e^{-\xi\omega(t - t_i)}[\omega_D D_i - \xi\omega C_i) \cos \omega_D(t - t_i)$$

$$- (\omega_D C_i + \xi\omega D_i) \sin \omega_D(t - t_i)] + A_i \quad (1.26)$$

where

$$\omega_D = \omega\sqrt{1 - \xi^2}, \ \xi = c/c_{cr}, \ \omega = \sqrt{k/m}, \ c_{cr} = 2\sqrt{km}$$

and A_i, B_i, C_i, and D_i are constants of integration.

The constants of integration A_i and B_i (corresponding to the steady-state solution) are given by

$$A_i = \frac{F_{i+1} - F_i}{k\Delta t} \quad (1.27)$$

$$B_i = \frac{F_i - cA_i}{k} \quad (1.28)$$

Knowing the displacement u_i and the velocity $\dot{u}_i$ at the initiation of interval i, we find the constants C_i and D_i after substituting $t = t_i$ in eqs.(1.25) and (1.26)

$$C_i = u_i - B_i \quad (1.29)$$

and

$$D_i = \frac{\dot{u}_i - A_i + \xi\omega C_i}{\omega_D} \quad (1.30)$$

The evaluation of eqs.(1.25) and (1.26) at time $t_{i+1} = t_i + \Delta t$ results in the displacement and the velocity at time t_{i+1}. Namely,

$$u_{i+1} = e^{-\xi\omega\Delta t}[C_i \cos \omega_D \Delta t + D_i \sin \omega_D \Delta t] + B_i + A_i\Delta t \quad (1.31)$$

and

$$\dot{u}_{i+1} = e^{-\xi\omega\Delta t}[D_i(\omega_D \cos \omega_D \Delta t - \xi\omega \sin \omega_D \Delta t)$$

$$- C_i(\xi\omega \cos \omega_D \Delta t + \omega_D \sin \omega_D \Delta t)] + A_i \quad (1.32)$$

Finally, the acceleration at time $t_{i+1} = t_i + \Delta t$ is obtained directly after substituting u_{i+1} and $\dot{u}_{i+1}$ from eqs.(1.31) and (1.32) into the differential eq.(1.24) and letting $t = t_i + \Delta t$. Specifically

$$\ddot{u}_{i+1} = \frac{1}{m}(F_{i+1} - c\dot{u}_{i+1} - ku_{i+1}) \quad (1.33)$$

Example 1.1

The tower shown in Fig. 1.9 is subjected to a constant impulsive acceleration of magnitude $\ddot{y}_s = 0.5 \, g$ applied at its foundation during 0.1 second. Determine the response of the tower in terms of the displacement and the velocity relative to the motion of the foundation. Also determine its maximum absolute acceleration. Assume 20% of the critical damping.

Solution:

The differential equation of motion of this structure (modeled by the simple oscillator) is given, in terms of the relative motion, $u = y - y_s$, by eq.(1.4) as

$$\ddot{u} + c\dot{u} + ku = F_{eff}(t)$$

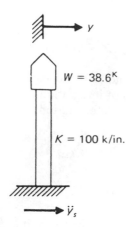

Fig. 1.9. Idealized structure for Example 1.1

where the effective force is

$$F_{eff}(t) = -m\ddot{y}_s = -100 \times 0.5 \times 386 = -19,300 \text{ lb}$$

For this system, the natural frequency is

$$\omega = \sqrt{k/m} = \sqrt{1000,000/100} = 31.62 \text{ (rad/sec)}$$

Hence the natural period is

$$T = \frac{2\pi}{\omega} = \frac{2\pi}{31.62} \approx 0.20 \text{ sec}$$

Recommended practice is to select $\Delta t \leqslant T/10$. Specifically, we select $\Delta t = 0.02$ sec and calculate the following constants:

$$c = c_{cr}\,\xi = 2\sqrt{km}\,\xi$$

$$= 2\sqrt{100,000 \times 100} \times 0.20 = 1265 \text{ lb·sec/in.}$$

$$\omega_D = \omega\sqrt{1-\xi^2} = 31.62\sqrt{1-0.2^2} = 30.99 \text{ rad/sec}$$

With initial conditions $u_0 = 0$, $\dot{u}_0 = 0$, and the use of eq.(1.33), $\ddot{u}_0 = F_{eff}(0)/m = -193$ in./sec². Next the coefficients A_0, B_0, C_0, D_0 are determined using

eqs.(1.27) through (1.32). Finally, the relative motion at time $t_1 = 0.02$ sec is as follows:

$$u_1 = -0.009 \text{ in.}$$
$$\dot{u}_1 = -1.783 \text{ in./sec}$$
$$\ddot{u}_1 = -161.267 \text{ in./sec}^2$$

and the absolute acceleration $\ddot{y}_1$ at time $t = 0.02$ sec is

$$\ddot{y}_1 = \ddot{u}_1 + \ddot{y}_s$$
$$\ddot{y}_1 = -161.267 + 193$$
$$\ddot{y}_1 = 31.733 \text{ in./sec}^2$$

The continuation of this process results in the response shown, up to 0.10 sec, in Table 1.2.

1.5 RESPONSE BY DUHAMEL'S INTEGRAL

The response to earthquakes for structures modeled as one-degree-of-freedom systems may be obtained directly from Duhamel's integral. The effective force $F_{eff} = -m\ddot{y}_s(\tau)$, on an oscillator [Fig. 1.10(a)] of mass m excited by an acceleration function $\ddot{y}_s(\tau)$ at its base, is shown in Fig. 1.10(b). The differential equation of motion, in terms of the relative displacement $u = y - y_s$, is given from eq.(1.4) as

$$m\ddot{u} + c\dot{u} + ku = F_{eff}(\tau) \tag{1.34}$$

where $F_{eff}(\tau) = -m\ddot{y}_s(\tau)$ is the effective force. The impulsive force $F_{eff}(\tau)d\tau$ at time τ [Fig. 1.10(b)] applied to the mass m of the oscillator results in a change of velocity dv. This change in the velocity can be determined from Newton's law of motion, namely

$$m\frac{dv}{d\tau} = F_{eff}(\tau)$$

or

$$dv = \frac{F_{eff}\,d\tau}{m} \tag{1.35}$$

Table 1.2. Calculation of the Response for Example 1.1

t_i (sec)	u_i (in.)	$\dot{u}_i$ (in./sec)	$\ddot{u}_i$ (in./sec²)	$\ddot{y}_i$ (in./sec²)	F_{eff} (lb)	B_i	A_i	C_i	D_i
0	0	0	−193.000	0	−19300	0	−0.19300	0.19300	0.03940
0.01	−0.009	−1.783	−161.267	31.733	−19300	0	−0.19300	0.19300	0.03940
0.02	−0.034	−3.188	−188.283	74.717	−19300	0	−0.19300	0.18382	−0.02002
0.03	−0.071	−4.129	−69.388	122.612	−19300	0	−0.19300	0.15861	−0.07051
0.04	−0.115	−4.573	−19.839	173.162	−19300	0	−0.19300	0.12162	−0.10843
0.05	−0.161	−4.539	25.671	218.671	−19300	0	−0.19300	0.07769	25.67084
0.06	−0.205	−4.086	63.381	256.381	−19300	0	−0.19300	0.03175	63.38117
0.07	−0.242	−3.306	90.706	283.706	−19300	0	−0.19300	−0.01691	90.70636
0.08	−0.270	−2.311	106.335	299.335	−19300	0	−0.19300	−0.04888	−0.11669
0.09	−0.288	−1.219	110.201	303.201	−19300	0	−0.19300	−0.07099	−0.09033
0.10	−0.295	−0.143	103.342	296.342	−19300	0	−0.19300	−0.09478	−0.05869

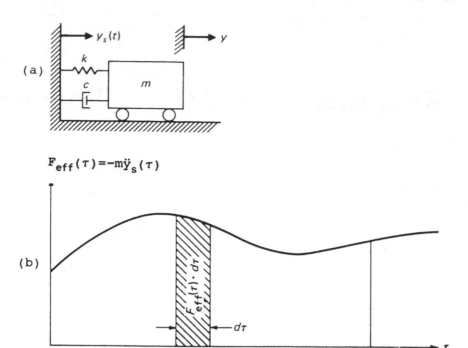

Fig. 1.10. Simple oscillator excited at its support. (a) Mathematical model, (b) Excitation in terms of the effective force $F_{eff}(\tau) = -m\ddot{y}_s(\tau)$

The change in velocity dv may be considered to be an initial velocity of the mass at time τ. This change in velocity is then introduced in eq.(1.20) as the initial velocity v_0 together with the initial displacement $u_0 = 0$ at time τ resulting in a change of displacement $du(t)$, at a later time t, given by

$$du(t) = e^{-\omega(t-\tau)} \frac{F_{eff}(\tau)d\tau}{m\omega_D} \sin \omega_D(t-\tau) \qquad (1.36)$$

The exciting function $F_{eff}(\tau)$ may be regarded as a series of short impulses at successive incremental times $d\tau$, each producing its own differential response at time t in the form given by eq.(1.36). Therefore, the total displacement $u(t)$ at time t, due to the continuous action of the effective force $F_{eff}(\tau)$, is given by the sum

or integral of the differential displacements $du(t)$ from time $\tau = 0$ to time t, that is

$$u(t) = \frac{1}{m\omega_D} \int_0^t F_{eff}(\tau)e^{-\xi\omega(t-\tau)} \sin \omega_D(t-\tau)d\tau \qquad (1.37)$$

The integral in equation (1.37) is known as *Duhamel's integral*. For seismic motion, the excitation function is known only from experimentally recorded data and the response must be evaluated by a numerical method (Paz, 1991, p. 70–75).

REFERENCE

PAZ, M. (1991) *Structural Dynamics: Theory and Computation*, 3rd Ed. Chapman & Hall, New York.

2

Seismic Response and Design Spectra

Farzad Naeim and Mario Paz

2.1 INTRODUCTION

Response spectrum analysis is the dominant contemporary method for dynamic analysis of building structures under seismic loading. The main reasons for the widespread use of this method are: its relative simplicity, its inherent conservatism, and its applicability to elastic analysis of complex systems. In this chapter the concept of *response spectrum* is introduced and it is differentiated from the closely related but different concept of *design spectrum*. Spectral entities and their "pseudo" counterparts such as spectral acceleration and spectral pseudo-acceleration are defined in Section 2.2. Response spectra and the common tripartite representation are introduced in Section 2.3. The concept of design spectra and several techniques for construction of elastic design spectra are presented in Sections 2.4 and 2.5. The influence of local soil conditions is treated in Section 2.6 and corresponding

methods for the construction of spectra are presented in Sections 2.7 and 2.8. Section 2.9 introduces recent advances in the construction of response spectra using predictive attenuation relationships. Finally, techniques intended to extend the response spectrum to inelastic response are introduced in Sections 2.10 and 2.11. Several numerical examples are also provided throughout the chapter.

2.2 SPECTRAL ENTITIES

As explained in Section 1.5 of Chapter 1, the response of a single-degree-of-freedom (SDOF) system to general excitation may be obtained through the application of Duhamel's integral. Consider in Fig. 2.1 a typical SDOF system subjected to base excitation. Let $u(t)$ be the time-dependent displacement of the mass (m) relative to the ground, $y(t)$ be the absolute displacement of the mass (m) with respect to a Newtonian "fixed" coordinate system, and $y_s(t)$ be the absolute displacement of the ground with respect to the same "fixed" reference. The equation of motion for the mass is given by eq.(1.4) and may be written as

$$\ddot{u} + 2\omega\xi\dot{u} + \omega^2 u = -\ddot{y}_s(t) \tag{2.1}$$

where $\omega = \sqrt{k/m}$ is the undamped natural frequency of the system and $\xi = c/2m\omega$ is the fraction of critical damping. From Duhamel's integral [eq.(1.37)], if the system starts from rest (that is if $\dot{u}_0 = u_0 = 0$),

$$u(t) = -\frac{1}{\omega_D} \int_0^t \ddot{y}_s e^{-\xi\omega(t-\tau)} \sin \omega_D(t-\tau) d\tau \tag{2.2}$$

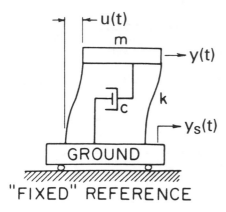

Fig. 2.1. Reference coordinates for a typical SDOF system

in which the damped natural frequency is $\omega_D = \omega\sqrt{(1-\xi^2)}$ and the effective force is $F_{eff} = -m\ddot{y}_s$.

Similar expressions for velocity and acceleration may be obtained by using the following differentiation rule:

$$\frac{\partial}{\partial t}\left[\int_0^t f(\tau, t)\,d\tau\right] = \int_0^t \frac{\partial}{\partial t} f(\tau, t)\,d\tau + [f(\tau, t)]_{\tau=t} \quad (2.3)$$

The resulting expressions for relative velocity and absolute acceleration are:

$$\dot{u}(t) = -\int_0^t \ddot{y}_{(\tau)} e^{-\xi\omega(t-\tau)} \cos\omega_D(t-\tau)\,d\tau$$
$$+ \int_0^t \frac{\ddot{y}_s(\tau)\xi}{\sqrt{1-\xi^2}} e^{-\xi\omega(t-\tau)} \sin\omega_D(t-\tau)\,d\tau \quad (2.4)$$

and

$$\ddot{y}(t) = 2\xi\omega \int_0^t \ddot{y}_s(\tau) e^{-\xi\omega(t-\tau)} \cos\omega_D(t-\tau)\,d\tau$$
$$+ \int_0^t \frac{\ddot{y}_s(\tau)\omega(1-2\xi^2)}{\sqrt{1-\xi^2}} e^{-\omega\xi(t-\tau)} \sin\omega_D(t-\tau)\,d\tau \quad (2.5)$$

It should be noted that eqs.(2.2) and (2.4), give, respectively, the relative displacement and the relative velocity of the mass with respect to the foundation while eq.(2.5) gives the absolute acceleration of the mass. These three quantities, the absolute acceleration, the relative displacement, and the relative velocity are the important quantities. The inertial forces are proportional to absolute acceleration, according to Newton's second law; the member forces are proportional to relative displacements of the member, from Hooke's law; and the damping forces are directly proportional to the relative velocity, for viscous damped systems.

The damping ratio ξ is usually small for structural systems ($\xi \ll 1$); therefore $1-\xi^2 \approx 1$, $\sqrt{1-2\xi^2} \approx 1$,

and $\omega_D \approx \omega$. In this case, since the second term in eq.(2.4) and the first term in eq.(2.5) are directly proportional to ξ, they would be small and therefore may be ignored. Consequently, with small damping as the only simplifying assumption, eqs.(2.4) and (2.5) may be written as

$$\dot{u}(t) = -\int_0^t \ddot{y}_s(\tau) e^{-\omega\xi(t-\tau)} \cos\omega_D(t-\tau)\,d\tau \quad (2.6)$$

and

$$\ddot{y}(t) = \omega \int_0^t \ddot{y}_s(\tau) e^{-\omega\xi(t-\tau)} \sin\omega_D(t-\tau)\,d\tau \quad (2.7)$$

It may be seen from eqs.(2.2) and (2.7) that in the absence of damping, the absolute acceleration $\ddot{y}(t)$ is equal to $-\omega^2 u(t)$. The absolute value of this quantity $|\omega^2 u(t)|$ which is a good approximation of the absolute acceleration when damping is small, is called *pseudo-acceleration*. If eq.(2.6) contained the sine term instead of $\cos\omega_d(t-\tau)$, a similar deduction could have been made about the velocity. This is not the case; nevertheless, for convenience, the relative pseudo-velocity is defined as $|\omega u(t)|$.

The absolute values for the maximum responses are defined as spectral values. Hence, the spectral displacement S_D, spectral velocity S_V, and spectral acceleration S_A are:

$$\begin{aligned}
S_D &= |u(t)|_{\max} \quad \text{(maximum relative displacement)} \\
S_V &= |\dot{u}(t)|_{\max} \quad \text{(maximum relative velocity)} \quad (2.8) \\
S_A &= |\ddot{y}(t)|_{\max} \quad \text{(maximum absolute acceleration)}
\end{aligned}$$

For convenience, both the spectral acceleration S_A and the spectral velocity S_v are represented, respectively, by their counterparts, the pseudo-acceleration S_{pA} and pseudo-velocity S_{pV}; that is,

$$S_{pV} = \omega S_D = \frac{2\pi}{T} S_D$$

(spectral pseudo-velocity) (2.9)

$$S_{pA} = \omega S_{pV} = \omega^2 S_D = \left(\frac{2\pi}{T}\right)^2 S_D$$

(spectral pseudo-acceleration) (2.10)

It should be obvious from this derivation that the spectral pseudo-acceleration S_{pA} is a good approximation for the spectral acceleration S_A, but the spectral pseudo-velocity S_{pV} is not a precise estimation of the spectral velocity S_V. However, for small values of damping, S_{pV} is usually an acceptable approximation of S_V. A comparison between these two spectral functions for a typical earthquake excitation is shown in Fig. 2.2. It may be seen in this figure that at a low

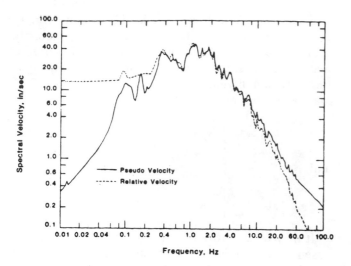

Fig. 2.2. Comparison of S_V and S_{pV} for a typical earthquake record (from Gupta 1990)

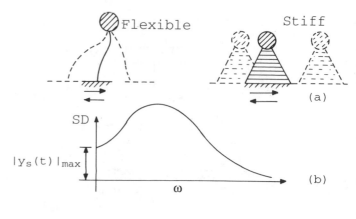

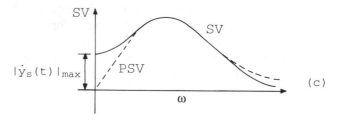

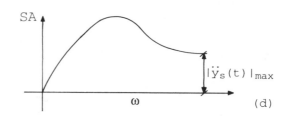

Fig. 2.3. Limiting states of spectral entities

frequency if S_{pV} were used instead of S_V, a large error would be introduced.

An examination of the spectral functions for extreme frequency values reveals the following: for low values of the frequency, $\omega \to 0$ (or $T \to \infty$), *the system is very flexible and the mass will remain essentially stationary while the base is vibrating as simulated in Fig. 2.3(a). Consequently, the maximum relative displacement or spectral displacement S_D would equal the* maximum absolute ground displacement, $|\ddot{y}_s(t)|_{max}$. Also, the maximum relative velocity or spectral velocity S_v would reach the maximum absolute ground velocity $|\dot{y}_s(t)|_{max}$. However, the spectral pseudo-velocity S_{pV}, is proportional to ω and would equal zero, and not $|\dot{y}_s(t)|_{max}$, as shown in Fig. 2.3(c). On the other hand, the maximum absolute acceleration S_A for this virtually stationary mass would be zero. Since the pseudo-acceleration is proportional to ω^2, it would be zero as shown for $\omega = 0$ in Fig. 2.3(d). Therefore, in the limiting case when $\omega \to 0$, the spectral pseudo-acceleration and the spectral acceleration tend to the same limiting value.

When the frequency of the system is large, $\omega \to \infty$ (or $T \to 0$), the system is very rigid and tends to move with the ground as sketched in Fig. 2.3(a). The relative displacement S_D and relative velocity S_V tend towards zero. Also, the pseudo-velocity S_{pV} tends towards zero as a result of its proportionality to S_D. In this case, when $\omega \to \infty$, the spectral acceleration and the pseudo spectral-acceleration reach the maximum ground acceleration $|\ddot{y}_s(t)|_{max}$ as indicated in Fig. 2.3(d). In passing, it is pertinent to share Hudson's observation (Hudson 1979) that: ". . . the widespread adoption of the prefix 'pseudo' in connection with the approximate spectrum representations is in one sense misleading. The literal meaning of 'pseudo' (false) is not really appropriate, since we are dealing with

approximations rather than with concepts that are in any sense incorrect or inappropriate."

2.3 RESPONSE SPECTRA

In the previous section, the maximum response of an SDOF system excited at its base by a time acceleration function $\ddot{y}_s(t)$, was expressed in terms of only two parameters: (1) the natural frequency ω of the system, and (2) the amount of damping ξ. Charts may be prepared depicting the maximum response values for any SDOF system within the frequency (or period) range of interest. Each one of these charts is called a response spectrum; a collection of them (i.e., for various damping levels) is termed response spectra. An example of such a chart for spectral velocity is Fig. 2.2. Another example for spectral acceleration for SDOF systems with $\xi = 0.05$ subjected to the N–S component of the famous 1940 El Centro Earthquake

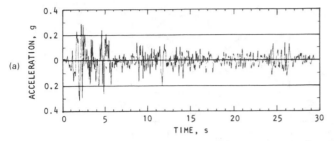

ACCELEROGRAM, EL CENTRO, CALIFORNIA EARTHQUAKE, MAY 18, 1940
(N-S COMPONENT)

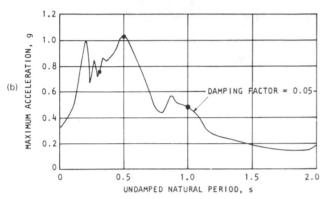

ACCELERATION RESPONSE SPECTRUM, EL CENTRO GROUND MOTIONS

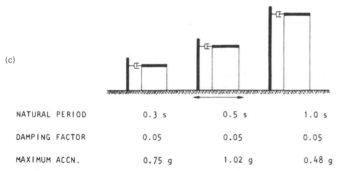

NATURAL PERIOD	0.3 s	0.5 s	1.0 s
DAMPING FACTOR	0.05	0.05	0.05
MAXIMUM ACCN.	0.75 g	1.02 g	0.48 g

Fig. 2.4. Application of acceleration response spectrum (after Seed and Idriss 1982)

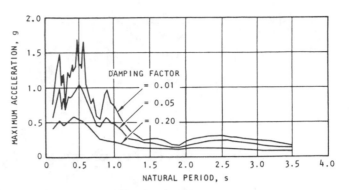

Fig. 2.5. Acceleration response spectra for El Centro (1940) earthquake (from Seed and Idriss 1982)

The relationships among the spectral entities S_D, S_{pV}, and S_{pA} provided by eqs.(2.9) and (2.10) make it possible to present all three of these spectral values in a single chart using logarithmic scales. These charts are called *tripartite logarithmic response spectra*. A spectrum response chart of this type is shown in Fig. 2.6. In this chart, values of the spectral pseudo-velocity, S_{pV} are read on the vertical axis while those of the spectral displacement S_D and spectral pseudo-acceleration S_{pA} are obtained from the diagonal axes. Figure 2.7 shows the same response spectra but now plotted as a function of the natural frequency instead of the natural period. Diagonal axes for displacement and acceleration are switched in Fig. 2.7 relative to their location in Fig. 2.6.

To demonstrate the construction of a tripartite chart such as the one in Fig. 2.7, eq.(2.9) is written in terms of the natural frequency f in cycles per second (cps) or hertz (Hz) as

$$S_{pV} = \omega S_D = 2\pi f S_D$$

Then, taking logarithms of the terms

$$\log S_{pV} = \log f + \log(2\pi S_D) \tag{2.11}$$

For constant values of S_D, eq.(2.11) is the equation of a straight line of $\log S_{pV}$ versus $\log f$ with a slope of 45°. Analogously, from eq.(2.10)

$$S_{pV} = \frac{S_{pA}}{\omega} = \frac{S_{pA}}{2\pi f}$$

$$\log S_{pV} = -\log f + \log \frac{S_{pA}}{2\pi} \tag{2.12}$$

For a constant value of S_{pA}, eq.(2.12) is the equation of a straight line of $\log S_{pV}$ versus $\log f$ with a slope of 135°. In closing, it should be remarked that although the duration of earthquake ground motion is used in the computation of the spectral values, this important aspect of the ground motion is not explicitly represented in response spectra.

(Fig. 2.4(a)) is shown in Fig. 2.4(b). It should be noted that the maximum acceleration experienced by any SDOF system with damping ratio $\xi = 0.05$ can be read directly from this response spectrum chart. Figure 2.4(c) shows three structures of periods 0.3 sec, 0.5 sec, and 1.0 sec for which the maximum acceleration responses are directly obtained from the chart (b) of this figure. The dots in the chart indicate the maximum responses corresponding to these three structures.

Spectral charts are usually prepared as a family of curves corresponding to different values of damping values as shown in Fig. 2.5. The curves in these charts may be plotted either as functions of the natural frequency or of the period. Either type of plot presents the same information. However, the graph in one plot is the mirror image of the graph in the other plot.

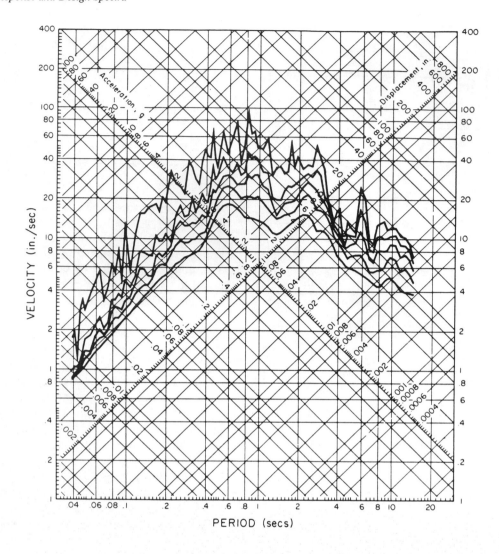

Fig. 2.6. Response spectra for Imperial Valley earthquake of May 18, 1940 plotted on tripartite logarithmic paper as function of the natural period. (Damping values are 0, 2, 5, 10 and 20 percent of critical)

2.4 ELASTIC DESIGN SPECTRA

A response spectral chart can be constructed for each earthquake record. Earthquakes show many common characteristics, but each tremor shows unique attributes as well. Response spectra of earthquake records contain many sharp peaks and valleys as can be seen in Fig. 2.6, for example. It would not be reasonable to expect that the spectra for future earthquakes will exhibit exactly the same peaks and valleys. Furthermore, the natural periods and mode shapes of building structures cannot be predicted exactly. Many other uncertainties are present in design for seismic

resistance. Some of these uncertainties are: (a) unavoidable variations in the mass and stiffness properties of the building from those used in design, (b) difficulties involved in establishing values of parameters representative of site soil conditions, and (c) inelastic response that tends to lengthen the natural period of the structure. For these reasons, it is more rational for design purposes to use average curves obtained from a number of earthquake records. These average curves, which do not reflect the sharp peaks and valleys of individual records, are known also as smoothed response spectra, or more accurately as *design spectra*. While a response spectrum is an

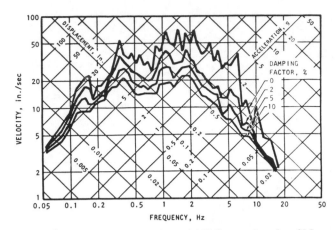

Fig. 2.7. Response spectra for Imperial Valley earthquake of May 18, 1940 plotted on tripartite logarithmic paper as function of the natural frequency

attribute of a particular ground motion, a design spectrum is not; the design spectrum reflects a defined set of criteria for structural analysis and design.

2.5 CONSTRUCTION OF ELASTIC SPECTRA

In this section, the fundamental techniques for construction of design spectra are introduced. Developing a design spectrum for a specific construction project is a complex task. For example, the development of a design spectrum for a typical site in the Los Angeles area involves a detailed evaluation of possible seismic events on more than 100 active and semi-active faults, and a historical study of the past 200 years of local seismic activity. A discussion of this matter with examples is given in Naeim and Lew (1991).

The first design spectrum, based on the horizontal components of three earthquakes, two in the state of California and one in Washington state, was developed by Housner (Housner 1959, 1970). Plots of Housner spectra normalized to 20% of the acceleration of gravity g, that is, to $0.20\,g$ are shown in Fig. 2.8. Spectral charts for other values of peak ground acceleration may be obtained by multiplying the spectral ordinates in Fig. 2.8 by the ratio of the specified ground acceleration to $0.20\,g$.

Typical earthquake response spectra plotted on a tripartite chart as the one shown in Fig. 2.9 reveal three distinct regions: (a) a low period range exhibiting somewhat constant acceleration, (b) a mid-period range exhibiting a somewhat constant velocity, and (c) a long period range exhibiting a somewhat constant displacement. These three regions are commonly referred to, respectively, as the acceleration, velocity, and displacement regions of response spectra. This observation led Newmark and Hall to suggest a modified trilinear shape for design spectra (Newmark and Hall 1973, 1982). They constructed smoothed

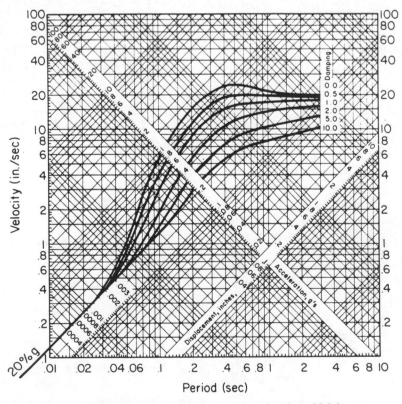

Fig. 2.8. Housner design spectra scaled to 0.2g at zero period (or peak ground acceleration of 0.2g) from Housner (1970)

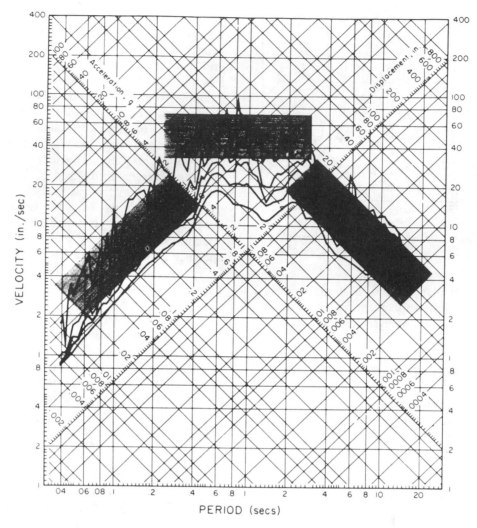

Fig. 2.9.   Trilinear idealization of response spectra

response spectra of idealized ground motion by amplifying the ground motion by factors dependent on
the damping in the system. In general, for any given
site, estimates must be made of the maximum ground
acceleration, maximum ground velocity, and maximum ground displacement. The lines representing
these maximum values are drawn on a tripartite
logarithmic paper as sketched in Fig. 2.10. Newmark's
and Hall's original estimates of constant amplification
factors were based on 28 accelerograms recorded on
alluvial sites (Newmark and Hall 1973). These amplification factors were refined later using data
obtained from the 1971 San Fernando earthquake
(Newmark and Hall 1982).

The primary scaling parameter used in the construction of a Newmark–Hall spectrum is the design maximum ground acceleration. Newmark and Hall (1982)
recommend that, lacking other information, the ratio
of maximum ground velocity to maximum ground
acceleration (v/a) be taken as 48 in./sec/g (122 cm/sec/
g) for competent soil conditions and 36 in./sec/g

(92 cm/sec/g) for rock sites. Furthermore, to ensure
that the spectrum represents a frequency bandwidth
adequate to incorporate a range of earthquakes, they
recommend that the ratio ad/v^2 be taken as about 6.0,
where d is the maximum ground displacement.

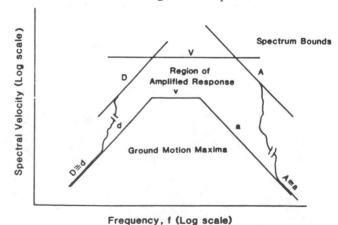

Fig. 2.10.   General shape of a modified trilinear response spectra
(from Gupta 1990)

Table 2.1. Spectral Amplification Factors (Newmark and Hall 1982)

Damping ratio (%)	84th percentile			Median		
	F_a	F_v	F_d	F_a	F_v	F_d
0.5	5.10	3.84	3.04	3.68	2.59	2.01
1	4.38	3.38	2.73	3.21	2.31	1.82
2	3.66	2.92	2.42	2.74	2.03	1.63
3	3.24	2.64	2.24	2.46	1.86	1.52
5	2.71	2.30	2.01	2.12	1.65	1.39
7	2.36	2.08	1.85	1.89	1.51	1.29
10	1.99	1.84	1.69	1.64	1.37	1.20
20	1.26	1.37	1.38	1.17	1.08	1.01

The following is a step-by-step procedure for constructing a Newmark–Hall spectrum on a tripartite logarithmic chart [Dunbar and Charlwood (1991)]:

Step 1. Estimate the maximum design ground motion parameters, a, v, and d for the site.

Step 2. Plot a, v and d by drawing a horizontal line through v, a line parallel to the displacement axis through a, and a line parallel to the acceleration axis through d. Connect all these lines.

Step 3. Select a confidence level, median (50th percentile), or median plus one standard deviation (84.1th percentile).

Step 4. Obtain amplification factors F_a, F_v, F_d from Table 2.1 corresponding to the selected confidence level and specified damping ratio.

Step 5. Calculate the following: $A = a \times F_a$, $V = v \times F_v$, and $D = d \times F_D$.

Step 6. Plot on the tripartite logarithmic paper A, V, and D by drawing a horizontal line through V, a line parallel to the displacement axis through A, and a line parallel to the acceleration axis through D. Connect all these lines.

Step 7. Draw a line between the A ordinate at 0.125 sec period (8 Hz) and the line a at 0.03 second period (33 Hz).

Example 2.1   (Construction of a Newmark–Hall Design Spectrum)

Construct a median Newmark–Hall elastic design spectrum for a maximum design ground acceleration of $0.5g$ on an alluvial site. Assume 5% damping.

- Determine ground motion parameters:

$$a = (1.0)(0.5g) = (0.5)(386.4)$$
$$= 193.2 \text{ in./sec}^2 \ (491 \text{ cm/sec}^2)$$
$$v = (48.0)(0.5) = 24 \text{ in./sec} \ (30.5 \text{ cm/sec})$$
[based on $v/a = 48$ in./sec/g]

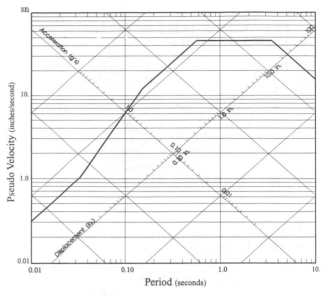

Fig. 2.11.   Newmark–Hall design spectrum of Example 2.1

$$d = (6.0)v^2/a = 17.9 \text{ in.} \ (45.4 \text{ cm})$$
[based on $ad/v^2 = 6$]

- Read amplification factors for median spectrum (5% damping) from Table 2.1:

$$F_a = 2.12 \quad F_v = 1.65 \quad F_d = 1.39$$

- Determine amplified response parameters:

$$A = a \times F_a = (0.5)(2.12) = 1.06g$$
$$V = v \times F_v = (24.0)(1.65) = 39.6 \text{ in./sec}$$
$$(100.6 \text{ cm/sec})$$
$$D = d \times F_d = (17.9)(1.39) = 24.9 \text{ in./sec} \ (63.2 \text{ cm})$$

- Follow steps 2, 6, and 7 of the step-by-step procedure to plot the Newmark–Hall design spectrum. The final spectrum is shown in Fig. 2.11.

2.6 INFLUENCE OF LOCAL SOIL CONDITIONS

Before the San Fernando earthquake of 1971, earthquake accelerograms were limited in number, and the majority had been recorded on alluvium. Therefore, it is only natural that the design spectra based on those data, such as those suggested by Housner (1959) and Newmark–Hall (1973), mainly represent alluvial sites. Since 1973, the wealth of information obtained from earthquakes worldwide and from subsequent studies have shown the very significant effect that the local site conditions have on spectral shapes (Figs. 2.12, 2.13, and 2.14) [Hayashi et al. (1971); Kuribayashi et al. (1972); Mohraz et al. (1972); Hall et al. (1975); Mohraz (1976); Seed and Idriss (1982), Idriss (1985, 1987); Mohraz and Elghadamsi (1989)].

An example of how a conservative design spectrum may be constructed from these spectral shapes is shown in Fig. 2.12. This figure shows four spectral acceleration curves representing the average of nor-

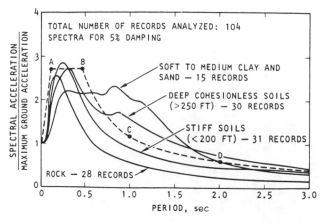

Fig. 2.12. Average acceleration spectra for different soil conditions (after Seed et al. 1976; from Seed and Idriss 1982)

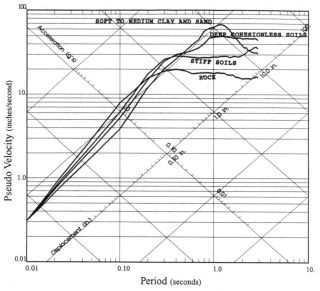

Fig. 2.13. Average spectra of Fig. 2.12 plotted on a tripartite logarithmic chart

malized spectral values corresponding to several sets of earthquake records registered on four types of soils. The dashed line through the points A, B, C, and D defines a possible conservative design spectrum for rock and stiff soil sites. Normalized design spectral shapes, as those included in the new editions of the Uniform Building Code (ICBO 1988, 1991) (Fig. 2.15) are based on such simplifications. The UBC spectral shapes become trilinear (Fig. 2.16), when drawn on a tripartite logarithmic chart, similar in shape to a Newmark–Hall spectrum. UBC spectral shapes (ICBO 1991) can be defined[1] by the following rather simple formulas:

Soil Type I (Rock and Stiff Soils):

$$
\begin{array}{llll}
S_A = 1 + 10T & \text{for} & 0 < T \leqslant 0.15 \sec & \\
S_A = 2.5 & \text{for} & 0.15 < T \leqslant 0.39 \sec & (2.13) \\
S_A = 0.975/T & \text{for} & T > 0.39 \sec &
\end{array}
$$

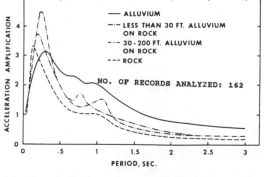

Fig. 2.14. Average horizontal accelerations for 2% damping for four soil categories (after Mohraz 1976; from Mohraz and Elghadamsi 1989)

Soil Type II (Deep Cohesionless or Stiff Clay Soils):

$$
\begin{array}{llll}
S_A = 1 + 10\,T & \text{for} & 0 < T \leqslant 0.15 \sec & \\
S_A = 2.5 & \text{for} & 0.15 < T \leqslant 0.585 \sec & (2.14) \\
S_A = 1.463/T & \text{for} & T > 0.585 \sec &
\end{array}
$$

Soil Type III (Soft to Medium Clays and Sands):

$$
\begin{array}{llll}
S_A = 1 + 75\,T & \text{for} & 0 < T \leqslant 0.2 \sec & \\
S_A = 2.5 & \text{for} & 0.2 < T \leqslant 0.915 \sec & (2.15) \\
S_A = 2.288/T & \text{for} & T > 0.915 \sec &
\end{array}
$$

where S_A is the spectral acceleration for 5% damping normalized to a peak ground acceleration of one g, and T is the fundamental period of the building. It should be noted that values obtained from the UBC spectral chart of Fig. 2.15, or alternatively, calculated with eqs.(2.13)–(2.15), are too conservative. In actual design practice, these values are scaled down by the structural factor R_w with values between 4 and 12, depending on the type of building. However, the UBC establishes limitations for the resultant base shear force obtained by the dynamic method relative to the base shear given by the static method of analysis, as explained by Paz (1991, p. 577).

2.7 MOHRAZ DESIGN SPECTRUM

Mohraz (1976) studied a total of 162 earthquake accelerograms obtained from 16 seismic events to arrive at his suggestions for construction of design spectra. He considered four different soil categories and divided the response spectra of each soil category into three sets: (1) the horizontal components with the larger peak ground accelerations, (2) the horizontal components with the smaller peak ground accelerations, and (3) vertical components. Instead of using a fixed relationship among v, d, and a, as suggested earlier by Newmark and Hall, he used ratios for v/a and ad/v^2 obtained from a statistical study. These

---

[1]From Recommended Lateral Force Requirements and Tentative Commentary SEAOC-90, p. 36-c.

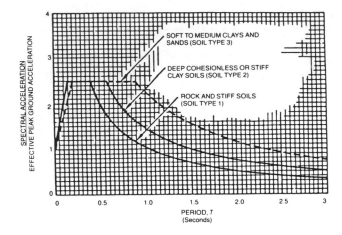

Fig. 2.15. Normalized design spectra shapes contained in Uniform Building Code (ICBO 1988, 1991)

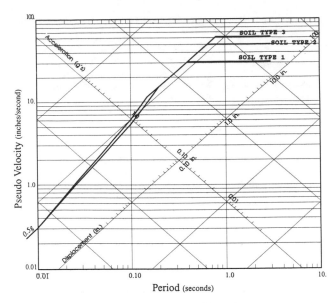

Fig. 2.16. Uniform Building Code's spectral shapes normalized to 0.5g ground acceleration and plotted on tripartite logarithmic chart

ratios are presented in Table 2.2. Mohraz obtained design amplification factors for the three frequency regions (displacement, velocity, and acceleration) by computing the mean of amplification factors plus one standard deviation for different frequencies and averaging them for each frequency region. The resulting average values of the amplification factors were applied to the ground motion ratios in Table 2.2 to compute spectral ordinates and to construct design spectra, normalized to one g peak ground acceleration. The site-dependent spectral ordinates A, V, D corresponding to the three frequency regions for different values of damping are given in Table 2.3 as Acc., Vcl., and Dis. Construction of the Mohraz

design spectrum is completed using values of A, V, and D from Table 2.3 in the following steps:

Step 1. Read from Table 2.2 mean values (50th percentile) of v/a, ad/v^2, and d/a for soil category and earthquake component (Group) selected. Determine the peak ground velocity v and the peak ground displacement d using the design peak ground acceleration.

Step 2. Read from Table 2.3 the spectral values D, V, and A for the site category, for specified damp-

Table 2.2. Summary of v/a and ad/v^2 Ratios (after Mohraz 1976; reproduced from Mohraz and Elghadamsi 1989)

Soil category	Group*	v/a, (in./sec)/g		ad/v^2		d/a, in./g
		Percentile		Percentile		Percentile
		50	84.1	50	84.1	50
Rock	L	24	38	5.3	11.0	8
	S	27	44	5.2	11.2	10
	V	28	45	6.1	11.8	12
<30 ft of alluvium underlain by rock	L	30	57	4.5	7.7	11
	S	39	62	4.2	8.2	17
	V	33	53	6.8	13.3	19
30–200 ft of alluvium underlain by rock	L	30	46	5.1	7.8	12
	S	36	58	3.8	6.4	13
	V	30	46	7.6	13.7	18
Alluvium	L	48	69	3.9	6.0	23
	S	57	85	3.5	4.9	29
	V	48	70	4.6	7.0	27

*L: Horizontal components with the larger of the two peak accelerations, S: Horizontal components with the smaller of the two peak accelerations, V: Vertical components.

Table 2.3. Spectral Ordinates for Unit Ground Acceleration (after Mohraz 1976; reproduced from Mohraz and Elghadamsi 1989)

Site category	Group*	Damping = 0%† Dis., in.	Vel., in./sec	Acc., g	2%† Dis., in.	Vel., in./sec	Acc., g	5%† Dis., in.	Vel., in./sec	Acc., g	10%† Dis., in.	Vel., in./sec	Acc., g	20%† Dis., in.	Vel., in./sec	Acc., g
Rock	L	54	97	7.02	43	66	3.80	35	51	2.82	28	40	2.11	21	30	1.54
	S	71	118	8.14	53	75	4.16	44	58	3.04	36	46	2.29	27	34	1.65
	V	87	115	8.15	67	74	3.81	54	57	2.70	44	45	2.03	30	35	1.65
<30 ft of	L	93	174	10.42	75	106	4.76	59	77	3.38	46	58	2.54	32	42	1.81
alluvium	S	127	158	10.17	106	101	4.73	87	77	3.37	68	59	2.49	49	44	1.73
underlain by rock	V	151	140	10.55	123	88	4.28	101	68	2.93	81	52	2.09	61	37	1.46
30–200 ft of	L	68	167	8.48	53	96	4.13	44	72	2.94	35	54	2.19	26	39	1.60
alluvium	S	80	201	8.85	62	117	4.11	50	89	2.88	41	68	2.17	31	49	1.62
underlain by rock	V	110	198	10.04	84	110	4.22	68	80	2.95	53	59	2.18	38	41	1.60
Alluvium	L	132	242	7.52	99	141	3.55	81	106	2.58	63	81	1.99	47	59	1.53
	S	160	288	9.18	122	169	4.00	99	128	2.86	78	98	2.19	58	70	1.65
	V	146	246	10.91	111	143	4.43	92	109	3.08	73	81	2.32	55	57	1.69

*L: horizontal components with the larger of the two peak ground accelerations; S: horizontal components with the smaller of the two peak ground accelerations; V: vertical components.

†Of critical damping.

ing and for selected earthquake component (Group).

Step 3. Plot a, v, and d on a tripartite logarithmic paper by drawing a horizontal line through the value of v, drawing a line parallel to the inclined axis for displacements through the value of a, and drawing a line parallel to the acceleration axis through the value of d.

Step 4. Plot lines A, V, and D parallel to the lines a, v, and d, respectively.

Step 5. Draw a line between the A ordinate at the period 0.125 sec (8 Hz) and the line a at the period 0.03 sec (33 Hz).

Example 2.2 (Construction of Mohraz Design Spectrum)

Construct a median Mohraz elastic design spectrum for a maximum design horizontal ground acceleration of 0.5g for each of the following sites: (a) rock, (b) 20 ft (6 m) of alluvium underlain by bedrock, (c) 120 ft (37 m) of alluvium underlain by bedrock, and (d) 300 ft (91 m) of alluvium underlain by bedrock. Assume 5% damping.

- Determine ground motion parameters:
  (a) Rock site:
    from Table 2.2: $v/a = 24$(in./sec)/g, $ad/v^2 = 5.3$, $d/a = 8$(in./g) for $a = 0.5g$, $v = (0.5)(24) = 12.0$ in./sec (30.5 cm/sec), and

$$d = \frac{(5.3)(12.0)^2}{(0.5g \times 386.4)} = 3.95 \text{ inches (10 cm)}$$

    or from $d/a = 8$, $d = (8)(0.5) = 4.0 \approx 3.95$ in.

(b) 20 ft of alluvium underlain by rock:
    from Table 2.2: $v/a = 30$(in./sec)/g, $ad/v^2 = 4.5$, $d/a = 11$(in./g) for $a = 0.5g$, $v = (0.5)(30) = 15.0$ in./sec (38.1 cm/sec), and

$$d = \frac{(94.5)(15.0)^2}{(0.5g \times 386.4)} = 5.2 \text{ inches (13.3 cm)}$$

    or from $d/a = 11$, $d = (11)(0.5) = 5.5$ in. ≈ 5.2 in.

(c) 120 ft of alluvium underlain by rock:
    from Table 2.2: $v/a = 30$(in./sec)/g, $ad/v^2 = 5.1$, $d/a = 12$ (in./g) for $a = 0.5g$, $v = (0.5)(30) = 15.0$ in./sec (38.1 cm/sec), and

$$d = \frac{(5.1)(15.0)^2}{(0.5g \times 386.4)} = 5.94 \text{ inches (15.1 cm)}$$

    or from $d/a = 12$, $d = (12)(0.5) = 6 \approx 5.94$ in.

(d) 300 ft of alluvium underlain by rock:
    from Table 2.2: $v/a = 48$ (in./sec)/g, $ad/v^2 = 3.9$, $d/a = 23$ (in./g) for $a = 0.5g$, $v = (0.5)(48) = 24.0$ in./sec (61 cm/sec), and

$$d = \frac{(3.9)(24.0)^2}{(0.5g \times 386.4)} = 11.6 \text{ inches. (29.5 cm)}$$

    or from $d/a = 23$, $d = (23)(0.5) = 11.5 \approx 11.6$ in.

- Calculate spectral ordinates by multiplying spectral ordinates for unit ground accelerations for 5% damping, from Table 2.3, by the value of design ground acceleration (0.5g):
  (a) Rock site:

$D = (35 \text{ in./g})(0.5g) = 27 \text{ in. (68.6 cm)}$
$V = (51 \text{ in./sec/g})(0.5g) = 25.5 \text{ in./sec}$ (64.8 cm/sec)
$A = (2.82)(0.5g) = 1.41g$

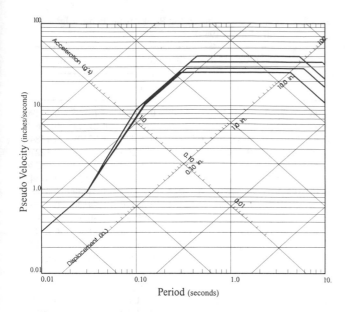

Fig. 2.17. Design spectra of Example 2.2

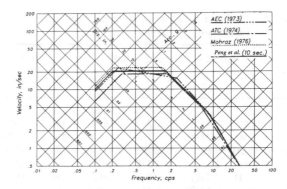

Fig. 2.18. Comparison of various design spectra for 0.2g ground acceleration and 5% damping on alluvium (after Peng et al. 1989)

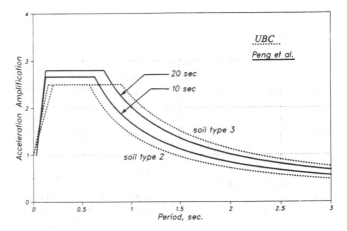

Fig. 2.19. Comparison of UBC and Peng et al. spectral shapes for 5% damping on alluvium (after Peng et al. 1989; from Mohraz and Elghadamsi 1989)

(b) 20 ft of alluvium underlain by rock:
$D = (59 \text{ in.}/g)(0.5g) = 29.5 \text{ in. (75 cm)}$
$V = (77 \text{ in.}/\text{sec}/g)(0.5g) = 38.5 \text{ in./sec}$
(98 cm/sec)
$A = (3.38)(0.5g) = 1.69g$

(c) 120 ft of alluvium underlain by rock:
$D = (44 \text{ in.}/g)(0.5g) = 22 \text{ in. (56 cm)}$
$V = (72 \text{ in.}/\text{sec}/g)(0.5g) = 36 \text{ in./sec (91 cm/}$
$\text{sec})$
$A = (2.94)(0.5g) = 1.47g$

(d) 300 ft of alluvium underlain by rock:
$D = (81 \text{ in.}/g)(0.5g) = 40.5 \text{ in. (103 cm)}$
$V = (106 \text{ in.}/\text{sec}/g)(0.5g) = 53 \text{ in./sec}$
(135 cm/sec)
$A = (2.58)(0.5g) = 1.29g$

• Construct design spectra following the various steps indicated for the construction of the Mohraz design spectra. These spectra are shown in Fig. 2.17.

2.8 PENG, ELGHADAMSI, AND MOHRAZ PROBABILISTIC DESIGN SPECTRA

Peng, Elghadamsi, and Mohraz (1989) introduced a simplified procedure for computing the probabilistic maximum response of an elastic SDOF system subjected to earthquake ground motion. This procedure incorporates the effects of the duration of strong ground motion and of the site soil conditions as parameters in computing response spectra. The resulting spectra are in good agreement with Mohraz design spectra for similar situations, as can be observed by examining Figs. 2.18 and 2.19. The reader is referred to Peng et al. (1989) for a detailed discussion of the theoretical background information. A step-by-step procedure for construction of Peng et al. spectra is as follows:

Step 1. Given the peak ground acceleration a expressed as a fraction of the acceleration of gravity, compute the root-mean-square (rms) acceleration as

$$a_{rms} = 220a^{0.87} \tag{2.16}$$

Step 2. Select a natural frequency value f in the range of interest and obtain the corresponding values of f'_c (the adjusted central frequency) and σ_I (the rms response to a unit *rms* ground acceleration) from Table 2.4 for alluvial sites or from Table 2.5 for rock sites.

Step 3. Compute the number of response maxima n as a function of f'_c and of the duration t_d of the strong ground motion as follows:

$$n = 2t_d f'_c \tag{2.17}$$

Step 4. Compute the peak factor r as

$r = 1.0$ for $n < 2$
$r = 1.0 + (0.7312 + 0.2276m)L_n(n/2)$
for $2 \leqslant n < 20$ (2.18)

Table 2.4. Values of f'_c and σ_I for Alluvium (from Peng, Eghadamsi, and Mohraz 1989)

Freq. (Hz)	$\xi = 0.02$		$\xi = 0.05$		$\xi = 0.10$	
	f'_c	σ_I	f'_c	σ_I	f'_c	σ_I
0.06	0.06474	1.15363	0.09095	0.72580	0.12910	0.50927
0.08	0.08632	0.89659	0.11841	0.56986	0.16273	0.40556
0.10	0.10687	0.78168	0.14124	0.49846	0.18733	0.35598
0.20	0.20320	0.66148	0.23260	0.40740	0.27410	0.27645
0.40	0.40349	0.23036	0.46378	0.14532	0.54381	0.10249
0.60	0.60641	0.14815	0.68868	0.09145	0.79793	0.06334
0.80	0.80105	0.10671	0.89117	0.06709	1.00997	0.04655
1.00	1.00017	0.07778	1.10615	0.04902	1.24176	0.03404
1.20	1.19563	0.05631	1.32424	0.03575	1.48218	0.02526
1.40	1.39926	0.04518	1.54721	0.02837	1.72703	0.01992
1.60	1.59244	0.03710	1.75239	0.02359	1.94771	0.01652
1.80	1.78828	0.03177	1.96166	0.01966	2.17362	0.01368
2.00	1.98471	0.02480	2.18229	0.01593	2.41086	0.01137
2.20	2.18444	0.02183	2.39717	0.01387	2.64078	0.00980
2.40	2.38037	0.01994	2.59984	0.01233	2.85616	0.00858
2.60	2.57033	0.01685	2.80408	0.01065	3.07088	0.00748
2.80	2.76483	0.01462	3.01348	0.00927	3.28905	0.00654
3.00	2.95500	0.01294	3.21693	0.00816	3.50429	0.00575
3.20	3.14448	0.01126	3.42195	0.00711	3.72056	0.00504
3.40	3.32866	0.00969	3.62739	0.00616	3.94152	0.00442
3.60	3.51565	0.00811	3.84436	0.00533	4.17078	0.00389
3.80	3.71280	0.00741	4.06056	0.00477	4.40138	0.00347
4.00	3.90202	0.00663	4.27161	0.00428	4.62820	0.00312
4.20	4.09187	0.00600	4.48118	0.00386	4.85189	0.00281
4.40	4.27957	0.00539	4.68959	0.00349	5.07274	0.00255
4.60	4.46771	0.00493	4.89538	0.00317	5.29059	0.00231
4.80	4.64728	0.00442	5.09777	0.00286	5.50649	0.00210
5.00	4.82364	0.00390	5.30403	0.00258	5.72299	0.00192
5.50	5.28463	0.00314	5.81606	0.00206	6.25655	0.00154
6.00	5.71194	0.00244	6.31217	0.00164	6.77637	0.00125
7.00	6.57255	0.00162	7.31018	0.00111	7.79978	0.00086
8.00	7.40192	0.00113	8.25448	0.00078	8.73670	0.00062
9.00	8.12205	0.00079	9.08332	0.00056	9.53758	0.00046
10.00	8.81181	0.00058	9.84157	0.00042	10.21724	0.00035
12.00	9.89494	0.00033	11.01220	0.00026	11.17459	0.00022
14.00	9.93013	0.00019	11.10312	0.00016	11.28842	0.00015
16.00	9.85680	0.00013	11.02267	0.00012	11.20111	0.00011
18.00	10.01688	0.00010	11.13593	0.00009	11.21517	0.00009
20.00	9.93569	0.00007	11.15453	0.00007	11.22890	0.00007
25.00	9.58973	0.00004	10.45702	0.00004	10.54691	0.00004

Table 2.5. Values of f'_c and σ_I for Rock (from Peng, Elghadamsi, and Mohraz 1989)

Freq. (Hz)	$\xi = 0.02$		$\xi = 0.05$		$\xi = 0.10$	
	f'_c	σ_I	f'_c	σ_I	f'_c	σ_I
0.06	0.07149	0.40416	0.11949	0.25701	0.18746	0.18296
0.08	0.09121	0.36757	0.14099	0.23375	0.21111	0.16639
0.10	0.10986	0.35540	0.15814	0.22473	0.22708	0.15841
0.20	0.20653	0.29980	0.26120	0.18295	0.34634	0.12270
0.40	0.42270	0.13985	0.54727	0.08660	0.71173	0.05986
0.60	0.61442	0.10671	0.71049	0.06428	0.84525	0.04343
0.80	0.80918	0.08104	0.91022	0.05035	1.04554	0.03467
1.00	1.00175	0.06930	1.10303	0.04199	1.24530	0.02799
1.20	1.19674	0.04268	1.33306	0.02759	1.50454	0.01957
1.40	1.40735	0.02884	1.59370	0.01934	1.81787	0.01418
1.60	1.60381	0.02478	1.82419	0.01602	2.09823	0.01159
1.80	1.82574	0.02028	2.07833	0.01365	2.36256	0.01010
2.00	2.01677	0.02320	2.24826	0.01403	2.53920	0.00968
2.20	2.22048	0.02150	2.43643	0.01344	2.70554	0.00920
2.40	2.40329	0.02124	2.60213	0.01308	2.86142	0.00864
2.60	2.58474	0.01946	2.77825	0.01162	3.03679	0.00768
2.80	2.76706	0.01504	2.98068	0.00940	3.25112	0.00650
3.00	2.95309	0.01121	3.21320	0.00748	3.50392	0.00546
3.20	3.16755	0.00998	3.45828	0.00647	3.77203	0.00473
3.40	3.36051	0.00880	3.68737	0.00580	4.02776	0.00425
3.60	3.58386	0.00863	3.91202	0.00549	4.26087	0.00392
3.80	3.77000	0.00873	4.09380	0.00530	4.46564	0.00365
4.00	3.95578	0.00762	4.29207	0.00481	4.66968	0.00336
4.20	4.14939	0.00692	4.49955	0.00438	4.87893	0.00309
4.40	4.34789	0.00642	4.70518	0.00406	5.08486	0.00285
4.60	4.54276	0.00617	4.89873	0.00381	5.28180	0.00263
4.80	4.71909	0.00569	5.08203	0.00350	5.47303	0.00241
5.00	4.89838	0.00499	5.27185	0.00313	5.66712	0.00220
5.50	5.32589	0.00355	5.76818	0.00231	6.19307	0.00171
6.00	5.78981	0.00264	6.33109	0.00179	6.78985	0.00136
7.00	6.77719	0.00213	7.41072	0.00137	7.94317	0.00100
8.00	7.66229	0.00147	8.39486	0.00097	8.94789	0.00073
9.00	8.47803	0.00101	9.33627	0.00070	9.89345	0.00054
10.00	9.21327	0.00074	10.18974	0.00051	10.75184	0.00041
12.00	10.77582	0.00043	12.02588	0.00031	12.51935	0.00026
14.00	11.96416	0.00027	13.39429	0.00021	13.79632	0.00018
16.00	12.86042	0.00018	14.24520	0.00014	14.41590	0.00012
18.00	13.21881	0.00012	14.65880	0.00010	14.73869	0.00009
20.00	13.19055	0.00009	14.72889	0.00008	14.72369	0.00007
25.00	12.07801	0.00005	12.92741	0.00004	13.22180	0.00004

$$r = \sqrt{2L_n(n)} + \frac{0.5772 + 1.28m}{\sqrt{2L_n(n)}} \quad \text{for } n \geqslant 20$$

in which $m = 0$ for median spectra or $m = 0.92$ for 84.1 percentile spectra (mean + one standard deviation for a normal distribution).

Step 5. Compute spectral ordinates:

$$\begin{aligned} S_D &= ra_{rms}\sigma_I \\ S_{pV} &= \omega S_D \\ S_{pA} &= \omega^2 S_D \end{aligned} \quad (2.19)$$

Step 6. Repeat Steps 2 through 5 for other values of natural frequencies and connect the points obtained to plot the spectrum.

Example 2.3 (Construction of Peng Spectra)

Construct Peng et al. mean elastic spectra for a peak design horizontal ground acceleration of 0.5g on an alluvial site for durations of 10, 20, 30, and 40 seconds of strong ground motion. Assume 5% damping.

Sample calculations of spectra for $T = 2.5$ sec ($f = 0.4$ Hz) and strong motion duration of $t_d = 10$ seconds are shown below. Computation of spectral ordinates for other natural frequencies, or other values for the duration of strong ground motion, are similar.

$$a_{rms} = 220a^{0.87} = 220(0.5)^{0.87} = 120.4 \text{ cm/sec}^2$$

eq.(2.16)

$f'_c = 0.46378$ and $\sigma_I = 0.14532$ (from Table 2.4)

$$n = 2t_d f'_c = 2(10)(0.46378) = 9.28 \qquad \text{eq.(2.17)}$$

For mean spectral response $m = 0$ and from eq.(2.18):

$$r = 1.0 + (0.7312)L_n \frac{n}{2}$$

$$= 1.0 + 0.7312 L_n \frac{9.28}{2} = 2.12 \qquad \text{eq.(2.18)}$$

The spectral ordinates from eq.(2.19) are:

$$S_D = r a_{rms} \sigma_I = (2.12)(120.4)(0.14532)$$
$$= 37.09 \text{ cm} = 14.6 \text{ in.}$$

$$\omega = 2\pi f = 2\pi(0.40) = 2.51 \text{ Hz}$$

$$S_{pV} = \omega S_D = (2.51)(37.09) = 93.1 \text{ cm/sec}$$
$$= 36.7 \text{ in./sec}$$

$$S_{pA} = \omega^2 S_D = (2.51)^2(37.09) = 234 \text{ cm/sec}^2$$
$$= 92 \text{ in./sec}^2 = 0.24g$$

The completed spectra for this example are shown in Fig.2.20. It should be noted that increasing the duration of the ground motion increases spectral ordinates for larger periods.

2.9 PREDICTIVE ATTENUATION RELATIONSHIPS

A new trend in the generation of design spectra is the development of comprehensive attenuation relationships to predict spectral ordinates. The advantage of this approach lies in the fact that the predicted spectral response includes the effects of both the magnitude of the earthquake and the site distance from the quake source. In using these proposed relationships, the reader should keep in mind the observations made by Joyner and Boore (1988) that,

> . . . data are sparse or nonexistent for important ranges of the predictive variables. . . . A key feature of the data set for shallow earthquakes is the scarcity of data points for distances less than about 20 km and magnitudes greater than 7.0. Confident predictions can simply not be made in that range of magnitude and distance, which is, unfortunately, where predictions are most needed.

Predictive relationships for spectral ordinates have been suggested by Joyner and Boore (1988), Sadigh (1987), and Crouse (1987) for shallow-focus earthquakes (earthquakes with depth of focus less than 70 km) and by Crouse et al. (1988) and Kawashima et al. (1984) for subduction-zone earthquakes. This section is limited to the presentation of Sadigh's (1987) relationship as a representative example of available

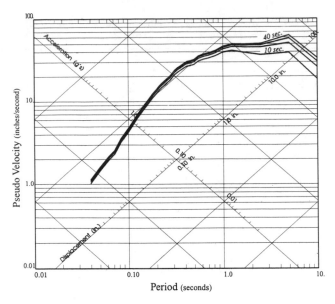

Fig. 2.20. Design spectra of Example 2.3

procedures. Readers are referred to Joyner and Boore (1988) for a comprehensive state-of-the-art review of the subject. A very informative comparison of the various empirical methods for prediction of design spectra may be found in Dunbar and Charlwood (1991).

Sadigh's relationship for predicting the peak ground acceleration and pseudo-acceleration response for $\xi = 0.05$ was developed using data from the western United States, supplemented by significant recordings of shallow-focus earthquakes from other parts of the world. Both horizontal components of recorded earthquakes were used. The resulting relationship proposed by Sadigh may be expressed as:

$$L_n(Y) = a + bM + c_1(8.5 - M)^{c_2} + dL_n[r + h_1 e^{(h_2 M)}] \quad (2.20)$$

where L_n is the Napierian logarithm, Y is the predicted acceleration in units of gravitational acceleration (g), M is the magnitude of the earthquake, and r is the closest distance to the rupture surface in km. Values of other parameters are given in Table 2.6 where the standard deviation of an individual prediction of $L_n(Y)$ is given as $\sigma_{L_n(Y)}$. Equation (2.20) is intended for strike-slip events. For reverse-slip earthquakes Sadigh recommends that values of Y obtained from eq.(2.20) be multiplied by 1.20.

Example 2.4 (Spectral Ordinates Using Sadigh's Relationship)

Using Sadigh's relationship, eq.(2.20), estimate the peak ground acceleration at the site and the pseudo-acceleration response of an elastic SDOF system with a natural period of 0.50 second ($f = 2$ Hz) subjected to an earthquake of magnitude $M = 7.5$ with the

Table 2.6. Parameters in the Predictive Equations of Sadigh (1987) for the Randomly Oriented Horizontal Component of Pseudo-Acceleration Response (g) at 5% Damping and of Peak Acceleration (reproduced from Joyner and Boore 1988)

Period (s)	a	b	c_1	c_2	d	$M < 6.5$			$M \geqslant 6.5$		
						h_1	h_2	$\sigma_{\ln y}$	h_1	h_2	$\sigma_{\ln y}$
					Pseudo-acceleration response at soil sites						
0.1	−2.024	1.1	0.007	2.5	−1.75	0.8217	0.4814	1.332–0.148M	0.3157	0.6286	0.37
0.2	−1.696	1.1	0.0	2.5	−1.75	0.8217	0.4814	1.453–0.162M	0.3157	0.6286	0.40
0.3	−1.638	1.1	−0.008	2.5	−1.75	0.8217	0.4814	1.486–0.164M	0.3157	0.6286	0.42
0.5	−1.659	1.1	−0.025	2.5	−1.75	0.8217	0.4814	1.584–0.176M	0.3157	0.6286	0.44
1.0	−1.975	1.1	−0.060	2.5	−1.75	0.8217	0.4814	1.62–0.18M	0.3157	0.6286	0.45
2.0	−2.414	1.1	−0.105	2.5	−1.75	0.8217	0.4814	1.62–0.18M	0.3157	0.6286	0.45
4.0	−3.068	1.1	−0.160	2.5	−1.75	0.8217	0.4814	1.62–0.18M	0.3157	0.6286	0.45
					Peak acceleration at soil sites						
	−2.611	1.1	0.0	2.5	−1.75	0.8217	0.4184	1.26–0.14M	0.3157	0.6286	0.35
					Pseudo-acceleration response at rock sites						
0.1	−0.688	1.1	0.007	2.5	−2.05	1.353	0.406	1.332–0.148M	0.579	0.537	0.37
0.2	−0.479	1.1	−0.008	2.5	−2.05	1.353	0.406	1.453–0.162M	0.579	0.537	0.40
0.3	−0.543	1.1	−0.018	2.5	−2.05	1.353	0.406	1.486–0.164M	0.579	0.537	0.42
0.5	−0.793	1.1	−0.036	2.5	−2.05	1.353	0.406	1.584–0.176M	0.579	0.537	0.44
1.0	−1.376	1.1	−0.065	2.5	−2.05	1.353	0.406	1.62–0.18M	0.579	0.537	0.45
2.0	−2.142	1.1	−0.100	2.5	−2.05	1.353	0.406	1.62–0.18M	0.579	0.537	0.45
4.0	−3.177	1.1	−0.150	2.5	−2.05	1.353	0.406	1.62–0.18M	0.579	0.537	0.45
					Peak acceleration at rock sites						
	−1.406	1.1	0.0	2.5	−2.05	1.353	0.406	1.26–0.14M	0.579	0.537	0.35

closest distance to a rupture surface of $r = 20$ km to the site. Assume a deep soil site condition, reverse faulting, and 5% damping.

- Estimate peak acceleration at the site:
  Substitute for various parameters in eq.(2.20) the values read from Table 2.6 for peak acceleration at deep soil sites:

$$L_n(a) = -2.611 + 1.1(7.5) \\ + 0 + (-1.75)L_n[20 + 0.3157\,e^{(0.6286 \times 7.5)}] = -1.381$$

therefore,

$a = e^{-1.381} = 0.25g$ for strike-slip faulting

and

$a = (1.20)(0.25) = 0.30g$ for reverse-faulting

- Estimate S_{pA} at $T = 0.5$ sec.:
  Substitute for various parameters in eq.(2.20) the values read from Table 2.6:
  At $T = 0.5$ sec on deep soil sites:

$$L_n S_{pA} = -1.659 + (1.1)(7.5) + (-0.025)(8.5 - 7.5)^{2.5} + \\ (-1.75)L_n[20 + 0.3157\,e^{(0.628 \times 7.5)}] = -0.454$$

$S_{pA} = e^{-0.454} = 0.64g$ for strike-slip faulting

$S_{pA} = (1.20)(0.64) = 0.76g$ for reverse-faulting

Example 2.5 (Effect of Earthquake Magnitude on Spectral Shape)

It is known that earthquake magnitude has a significant effect on the shape of response spectra. Stronger earthquakes release a larger percentage of their energy in longer periods of vibration. Using eq.(2.20), compare spectral amplification at $T = 2.0$ sec for earthquakes of $M = 5.5$ and 7.5. Assume a deep soil site, $r = 10$ km, strike-slip faulting, and 5% damping.

- Estimate peak ground acceleration for each event:
  for $M = 5.50$

$$L_n(a) = -2.611 + (1.1)(5.5) + 0 \\ + (-1.75)L_n[10 + 0.8217\,e^{(0.6286 \times 7.5)}] = -1.64$$
$$a_{5.5} = e^{-1.639} = 0.19g$$

Similarly for $M = 7.50$

$$L_n(a) = -2.611 + (1.1)(7.5) + 0 \\ + (-1.75)L_n[10 + 0.3157\,e^{(0.6286 \times 7.5)}] = -1.03$$
$$a_{7.5} = e^{-1.03} = 0.36g$$

- Estimate S_{pA} at $T = 2.0$ sec for each event:
  for $M = 5.50$

$$L_n s_{pA} = -2.414 + (1.1)(5.5) + (-0.105)(8.5 - 5.5)^{2.5} + \\ (-1.75)L_n[10 + 0.8217\,e^{(0.4814 \times 5.5)}] = -3.378$$

$(S_{pA})_{5.5} = e^{-3.378} = 0.034g$
for $M = 7.50$

$L_n S_{pA} = -2.414 + (1.1)(7.5) + (-0.105)(8.5 - 7.5)2.5 +$
$\quad (-1.75)L_n[10 + 0.3157 e^{(0.6286 \times 7.5)}] = -0.939$
$(S_{pA})_{7.5} = e^{-0.939} = 0.391g$

- Compare pseudo-acceleration amplifications:

$$\text{for } M = 5.5, \quad \frac{S_{pA}}{a} = \frac{0.034}{0.19} = 0.18$$

$$\text{for } M = 7.5, \quad \frac{S_{pA}}{a} = \frac{0.391}{0.36} = 1.09$$

The amplification at $T = 2.0$ sec for $M = 7.5$ is about six times $(1.09/0.18 \approx 6)$ that for $M = 5.5$.

2.10 RESPONSE SPECTRA FOR INELASTIC SYSTEMS

Current seismic design criteria do not rely on building systems to remain elastic during major earthquakes. The most common design approach (ATC 1978; FEMA 1988; SEAOC 1990; ICBO 1991) is formulated around a two-level seismic design concept:

1. Buildings should resist moderate earthquakes with essentially no structural damage (i.e. essentially elastic behavior).
2. Buildings should resist catastrophic earthquakes with some structural damage but without collapse and major injuries or loss of life (i.e., inelastic response within acceptable limits).

It is generally accepted that it is neither economically feasible, nor realistically possible, to design structures to remain elastic under very severe earthquake ground motions when there is a small probability of occurrence during the life of the structure. The inelastic response spectrum is an attempt to extend the application of response spectrum analysis beyond the linear elastic range. Generally, an elastoplastic idealization is used. Such a force–displacement relationship is shown in Fig. 2.21. The elastoplastic behavior assumes that the material will return along its loading line to the origin, if the force is removed prior to the occurrence of yielding. However, when yielding occurs at a displacement y_t, the restoring force remains constant at a magnitude R_t. If the displacement is not reversed, the displacement may reach a maximum value y_{max}. If, however, the displacement is reversed, the elastic recovery follows along a line parallel to the initial line and the recovery proceeds elastically until a negative yield value R_c is reached in the opposite direction. Construction of response spectra for inelastic systems is more difficult than for elastic systems. However, inelastic response spectra for several important earth-

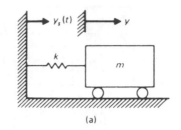

(a)

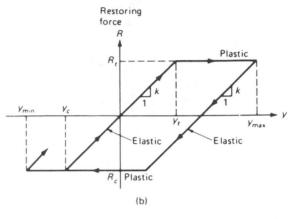

(b)

Fig. 2.21 Force-displacement relationship for an elastoplastic SDOF system (from Paz 1991)

quake records have been prepared. These spectra are usually plotted as a series of curves corresponding to definite values of the ductility ratio μ. The ductility ratio μ is defined as the ratio of the maximum displacement y_{max} of the structure in the inelastic range to the displacement corresponding to the yield point y_y, that is,

$$\mu = \frac{y_{max}}{y_y} \tag{2.21}$$

The response spectra for an SDOF system with 10% critical damping subjected to the N–S component of the 1940 El Centro earthquake record is shown in Fig. 2.22 for several values of the ductility ratio μ.

2.11 INELASTIC DESIGN SPECTRA

A number of procedures have been suggested for construction of inelastic design spectra (ATC 1974; Newmark and Hall 1973; Riddel and Newmark 1979; Lai and Biggs 1980; Newmark and Riddell 1980; Elghadamsi and Mohraz 1987). Newmark and Hall suggested that an inelastic spectrum be constructed by modifying an elastic design spectrum to reflect a specified ductility ratio μ. Two distinct inelastic spectra are derived from each elastic spectrum: (1) the inelastic acceleration spectrum that is used in the calculation of inelastic forces, and (2) the inelastic displacement spectrum. The theoretical basis for such

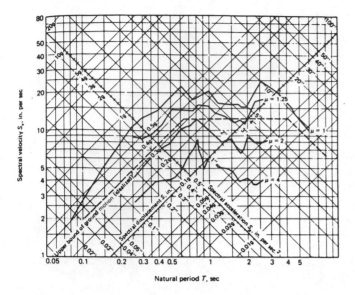

Fig. 2.22. Response spectra for elastoplastic systems with 10% critical damping for the 1940 El Centro earthquake (from Blume et al. 1961)

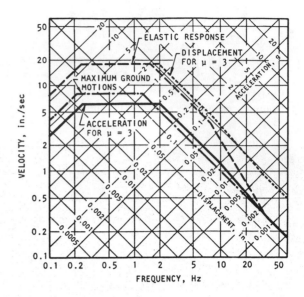

Fig. 2.24. Inelastic design spectra of Example 2.6 (from Newmark and Hall 1982)

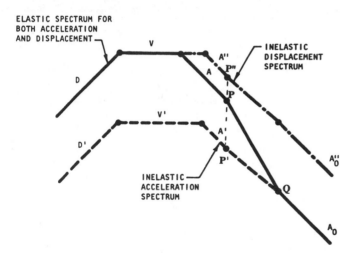

Fig. 2.23. Modifications of Newmark–Hall design spectrum for inelastic response (modified from Newmark and Hall 1982)

Step 2. Obtain lines D' and V' parallel to lines D and V by dividing the ordinates of D and V by the specified ductility μ.

Step 3. Divide the ordinate of point P on the elastic spectrum by $\sqrt{(2\mu - 1)}$ to locate point P'.

Step 4. Draw from the newly located point P', line A' at 45° until it intersects line V'.

Step 5. Join points P' and Q to complete the spectrum for accelerations.

Step 6. Draw segments A'' and A_0'' obtained from the ordinates of the corresponding segments A' and A_0 multiplied by the ductility ratio μ.

Following this step-by-step procedure, Paz (1991, pp. 189–191) has prepared inelastic spectral charts for ductility ratios $\mu = 0$, 2, 5, and 10, and for damping ratios $\xi = 0\%$, 5%, and 10%.

Example 2.6 (Application of Newmark–Hall Inelastic Design Spectrum)

Use the elastic and inelastic design spectra of Fig. 2.24 to estimate the maximum acceleration and displacement of an elastoplastic SDOF system with a natural period of $T = 2$ sec. Assume $\mu = 3$. For $T = 2.0$ sec, $f = 1/T = 0.5$ Hz.

The inelastic acceleration curve corresponding to $f = 0.5$ Hz provides $S_{pA} = 0.05g$. Similarly, from the curve corresponding to inelastic displacement, $S_D = 6$ in. (15 cm). If the system had remained elastic, it would have experienced an acceleration in excess of $0.2g$, more than four times that experienced by the inelastic system.

modification is Newmark's observation that three equivalences can be made between elastic and inelastic response: (1) at low frequencies displacements are equal, (2) at intermediate frequencies absorbed energies are the same, and (3) at high frequencies forces (or accelerations) are equal.

A step-by-step procedure for construction of a Newmark and Hall inelastic design spectrum with reference to Fig. 2.23 is as follows:

Step 1. Draw on tripartite logarithmic paper the elastic design spectrum developed for the stipulated ground motion parameters and the amount of damping specified (line D-V-A-A_0 in Fig. 2.23).

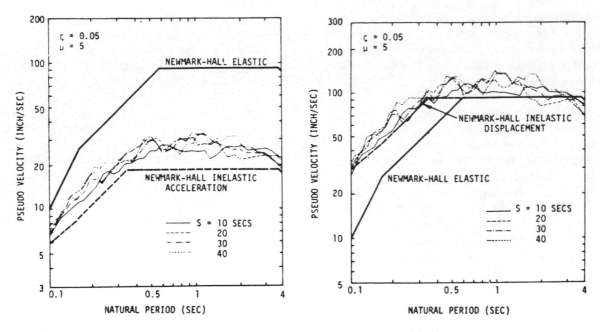

Fig. 2.25. Mean inelastic acceleration and displacement response for different strong ground motion durations (from Lai and Biggs 1980)

Lai and Biggs (1980) used a set of artificial ground motions of various durations to study the validity of the Newmark–Hall procedure for constructing inelastic design spectra. The artificial motions used by Lai and Biggs were generated to have elastic response matching the Newmark–Hall target elastic spectrum. They concluded that the Newmark–Hall procedure may be unconservative for 5% damping elastoplastic systems (Fig. 2.25) and suggested a new procedure to eliminate this concern.

A Lai–Biggs inelastic spectrum is constructed by dividing the ordinates of the corresponding elastic design spectrum by a set of factors called "inelastic

response ratios." These ratios for a set of control points and various ductility ratios are shown in Table 2.7. Figure 2.26 compares Lai–Biggs inelastic spectra for $\mu = 4$ and $\xi = 0.05$ with those of Newmark–Hall.

Table 2.7. Inelastic Response Ratios at Control Periods (after Lai and Biggs 1980)

Ductility Ratio	Inelastic Acceleration Response Ratios				Inelastic Displacement Response Ratios			
	0.10	0.50	0.70	4.0	0.10	0.50	0.70	4.0
		(sec)				(sec)		
Damping: 5%								
2	1.35	1.58	1.80	2.00	0.72	0.90	1.00	1.00
3	1.50	2.01	2.38	2.88	0.54	0.79	0.89	1.00
4	1.60	2.34	2.84	3.55	0.44	0.71	0.81	1.00
5	1.68	2.68	3.24	4.11	0.37	0.65	0.75	1.00
Damping: 2%								
2	1.53	2.00	2.22	2.45	0.79	1.08	1.20	1.20
3	1.76	2.85	3.22	3.72	0.61	0.98	1.10	1.20
4	1.94	3.50	4.00	4.68	0.51	0.91	1.03	1.20
5	2.07	4.10	4.70	5.56	0.43	0.86	0.98	1.20

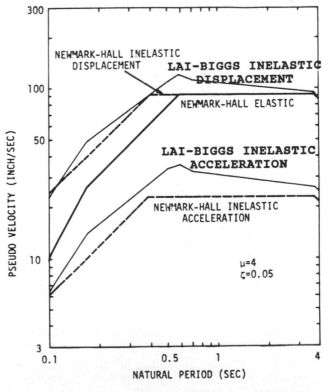

Fig. 2.26. Comparison of Lai–Biggs and Newmark–Hall inelastic spectra (from Lai and Biggs 1980)

Notice for example, that for $T = 0.6$ scc, thc displacement response indicated by the Lai–Biggs procedure is about 30% larger than that obtained from the Newmark–Hall procedure. The same ratio for inelastic acceleration response is 52%.

In their evaluation of inelastic design spectra, Mahin and Bertero (1981) concluded that inelastic design spectra constructed by modifying elastic design spectra tend to overestimate the effect of viscous damping on inelastic response. Generally, care should be taken in application of inelastic spectra in structural design of complex structures. Many studies (Mahin and Bertero 1981; Anderson and Naeim 1984; Anderson and Bertero 1987; Elghadamsi and Mohraz 1987; and Bertero et al. 1991) have consistently shown that there are many parameters not considered in construction of such spectra that can dominate nonlinear response of multidegree-of-freedom (MDOF) systems. A general consensus on methods for application of inelastic design spectra in analysis of MDOF systems has not yet been achieved.

REFERENCES

ANDERSON, J. C., and NAEIM, F. (1984) "Design Criteria and Ground Motion Effectson Seismic Response of Multistory Buildings." In *Critical Aspects of Earthquake Ground Motion and Building Damage Potential*, ATC-10-1. Applied Technology Council, San Francisco, CA.

ANDERSON, J. C., and BERTERO, V. V. (1987) "Uncertainties in Establishing Design Earthquakes." *J. of Structural Engineering, ASCE 113(8)*: 1709–1724.

ATC (Applied Technology Council) (1974) *An Evaluation of A Response Spectrum Approach to Seismic Design of Buildings, ATC-2.* Applied Technology Council, San Francisco, CA.

——— (1978) *Tentative Provisions for the Development of Seismic Regulations for Buildings, ATC-3-06.* National Bureau of Standards, Spec. Publ. 510.

BERTERO, V. V.; ANDERSON, J. C.; KRAWINKLER, H.; and MIRANDA, E. (1991) *Design guidelines for ductility and drift limits: Review of state-of-the-practice and of-the-art on ductility and drift-based earthquake-resistant design of buildings.* A Cure-Kajima Research Report, Berkeley, CA.

BLUME, J. A.; NEWMARK, N. M.; and CORNING, L. H. (1961) *Design of Multistory Reinforced Concrete Buildings for Earthquake Motions.* Portland Cement Association, Skokie, IL.

CROUSE, C. B. (1987) Written communication referred to in Joyner and Boore (1988).

CROUSE, C. B.; VYES, Y. K.; and SCHELL, B. A. (1988) "Ground Motions from Subduction-zone Earthquakes." *Bull. Seism. Soc. Am.* 78: 1–25.

DUNBAR, W. S., and CHARLWOOD, R. G. (1991) "Empirical Methods for the Prediction of Response Spectra."

Earthquake Spectra. Earthquake Engineering Research Institute, 7(3), pp. 333–353. Oakland, CA.

ELGHADAMSI, F., and MOHRAZ, B. (1987) "Inelastic Earthquake Spectra." *J. Earthquake Engineering and Structural Dynamics* 15: 91–104.

FEMA (Federal Emergency Management Agency) (1988) *NEHRP Recommended Provisions for the Development of Seismic Regulations for Buildings, FEMA-97.* Washington, D.C.

GUPTA, A. K. (1990) *Response Spectrum Method in Seismic Analysis and Design of Structures.* Blackwell Scientific Publications, Boston, MA.

HALL, W. J.; MOHRAZ, B.; and NEWMARK, N. M. (1975) *Statistical Studies of Vertical and Horizontal Earthquake Spectra.* Nathan M. Newmark Consulting Engineering Services, Urbana, IL.

HAYASHI, S.; TSUCHIDA, H.; and KURATA, E. (1971) "Average Response Spectra for Various Subsoil Conditions." Third Joint Meeting, U.S.-Japan Panel on Wind and Seismic Effects, UJNR, Tokyo, Japan.

HOUSNER, G. W. (1959) "Behavior of Structures During Earthquakes." *J. Eng. Mech. Div., ASCE* 85(EM4): 109–129.

——— (1970) "Design Spectrum." Chapter 5 in *Earthquake Engineering*, ed. R. L. Wiegel, 93–106. Prentice-Hall, Englewood Cliffs, NJ.

HUDSON, D. E. (1979) *Reading and Interpreting Strong Motion Accelerograms.* Earthquake Engineering Research Institute, Berkeley, CA.

ICBO (International Conference of Building Officials) (1988, 1991) *Uniform Building Code.* Whittier, CA.

IDRISS, I. M. (1985) "Evaluating Seismic Risk in Engineering Practice." Proc. 11th International Conf. on Soil Mechanics and Foundation Engineering 1, 255–320. August 12–16, 1985, San Francisco, CA.

——— (1987) "Earthquake Ground Motions." Lecture Notes, Course on Strong Ground Motion, Earthquake Engineering Research Institute, Pasadena, CA.

JOYNER, W. B., and BOORE, D. M. (1988) "Measurement, Characterization, and Prediction of Strong Ground Motion." In *Earthquake Engineering and Soil Dynamics, II*, J. L. Von Thun, ed. ASCE Geotechnical Special Publication No. 20, 43–102. Park City, UT.

KAWASHIMA, K. K.; AIZAWA, K.; and TAKAHASHI, K. (1984) "Attenuation of Peak Ground Motion and Absolute Acceleration Response Spectra." Proceedings of 8th World Conf. on Earthquake Engineering. 2, 257–264. San Francisco, CA.

KURIBAYASHI, E.; IWASAKI, T.; IIDA, Y.; and TUJI, K. (1972) "Effects of Seismic and Subsoil Conditions on Earthquake Response Spectra." 499–512. Proceedings of International Conference on Microzonation, Seattle, WA.

LAI, S. P., and BIGGS, J. M. (1980) "Inelastic Response Spectra for Aseismic Building Design." *J. Structural Div., ASCE* 106(ST6): 1295–1310.

MAHIN, S. A., and BERTERO, V. V. (1981) "An Evaluation of Inelastic Seismic Response Spectra." *J. Structural Div., ASCE* 107(ST9): 1777–1795.

Mohraz, B. (1976) "A Study of Earthquake Response Spectra for Different Geological Conditions." *Bull. Seism. Soc. Am.* 66: 915–935.

Mohraz, B., and Elghadamsi, F. (1989) "Earthquake Ground Motion and Response Spectra." Chapter 2 of *The Seismic Design Handbook*, F. Naeim, ed., Van Nostrand Reinhold, New York, NY.

Mohraz, B.; Hall, W. J.; and Newmark, N. M. (1972) *A Study of Vertical and Horizontal Earthquake Spectra.* AEC Report WASH-1255. Nathan M. Newmark Consulting Engineering Services, Urbana, IL.

Naeim, Farzad, ed. (1989) *The Seismic Design Handbook.* Van Nostrand Reinhold, New York, NY.

Naeim, F., and Lew, M. (1991) "Scaling Design Spectra and Seismic Safety of Tall Buildings in Greater Los Angeles." Proceedings of the Second Conference on Tall Buildings in Seismic Regions, Organized by Council on Tall Buildings and Urban Habitat and Los Angeles Tall Buildings Structural Design Council, Los Angeles, CA.

Newmark, N. M., and Hall, W. J. (1973) *Procedures and Criteria for Earthquake Resistant Design.* Building Practices for Disaster Mitigation. Building Sciences Series 46, 1, 209–236. National Bureau of Standards, Washington, D.C.

——— (1982) *Earthquake Spectra and Design.* Earthquake Engineering Research Institute, Berkeley, CA.

Newmark, N. M., and Riddell, R. (1980) "Inelastic Spectra for Seismic Design." Proc. 7th World Conf. on Earthquake Engineering. 4, 129–136. Istanbul, Turkey.

Paz, Mario (1991) *Structural Dynamics: Theory and Computation.* 3d ed. Van Nostrand Reinhold, New York, NY.

Peng, M. H.; Elghadamsi, F.; and Mohraz, B. (1989) "A Simplified Procedure for Constructing Probabilistic Response Spectra." *Earthquake Spectra.* Earthquake Engineering Research Institute, 5(2), Oakland, CA.

Riddell, R., and Newmark, N. M. (1979) *Statistical Analysis of the Response of Nonlinear Systems Subjected to Earthquakes.* Civil Engineering Studies, Structural Research Series, 468. Dept. of Civil Eng., Univ. of Illinois, Urbana, IL.

Sadigh, K. (1987) Written communication referred to in Joyner and Boore (1988).

SEAOC (Structural Engineers Association of California) (1990) *Recommended Lateral Force Requirements and Commentary.* Seismology Committee, Sacramento, CA.

Seed, H. B., and Idriss, I. M. (1982) *Ground Motions and Soil Liquefaction During Earthquakes.* Earthquake Engineering Research Institute, Berkeley, CA.

Seed, H. B.; Ugas, C.; and Lysmer, J. (1976) "Site-dependent Spectra for Earthquake Resistant Design." *Bull. Seism. Soc. Am.* 66: 221–243.

3

Structures Modeled by Generalized Coordinates

Mario Paz

3.1 INTRODUCTION

Coordinates of linear or angular position in the structure are the ones most frequently used to specify the configuration of a dynamic system. Coordinates do not necessarily have to correspond directly to displacements; they may be independent quantities, sufficient in number to specify the position of all parts of the system. These coordinates are usually called *generalized coordinates*. Their number is equal to the number of degrees of freedom in the system. When the system has only one degree of freedom, the configuration of the structure is determined in terms of a single generalized coordinate. For example, a four-story building represented by the frame shown in Fig. 3.1(a) and by the discrete model in Fig. 3.1(b), still may be modeled as a single-degree-of-freedom system if its lateral displacement $y(x, t)$ at any level could be expressed as the product of the *shape function* $\phi(x)$ [such as the one shown in Fig. 3.1(c)] and the generalized coordinate $Y(t)$, that is

$$y(x, t) = \phi(x) Y(t) \qquad (3.1)$$

The generalized coordinate $Y(t)$ will be the lateral displacement at the top level of the building provided

that the shape function is assigned a unit value at that level, that is, provided that $\phi(H) = 1.0$.

3.2 GENERALIZED EQUATION OF MOTION

The equation of motion for the generalized single-degree-of-freedom system in Fig. 3.1(d) is obtained by equating the sum of the forces to zero in the corresponding free-body diagram [Fig. 3.1(e)]. That is

$$m^* \ddot{Y} + c^* \dot{Y} + k^* Y = F^*(t) \qquad (3.2)$$

where m^*, c^*, k^*, and $F^*(t)$ are referred to, respectively, as the generalized mass, generalized damping, generalized stiffness, and generalized force, and are given for discrete systems by

$$
\begin{aligned}
m^* &= \Sigma \, m_i \phi_i^2 \\
c^* &= \Sigma \, c_i \Delta \phi_i^2 \\
k^* &= \Sigma \, k_i \Delta \phi_i^2 \\
F^* &= \Sigma \, F_i \phi_i
\end{aligned}
\qquad (3.3)
$$

In eqs.(3.3), the relative displacement $\Delta \phi_i$ between two consecutive levels of the building is given by

$$\Delta \phi_i = \phi_i - \phi_{i-1} \qquad (3.4)$$

with $\phi_0 = 0$ at the ground level. As shown in Fig. 3.1(b), m_i and F_i are, respectively, the mass and the force at level i of the building, while k_i and c_i are,

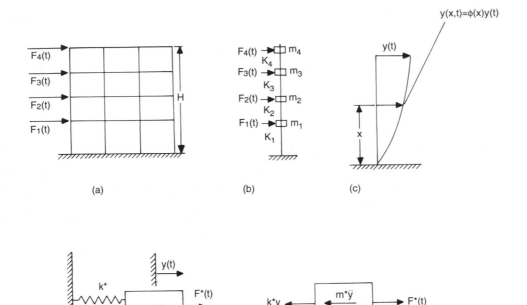

Fig. 3.1. Development of generalized single-degree-of-freedom system; (a) Actual frame, (b) Discretized system, (c) Assumed shape, (d) Generalized single-degree-of-freedom system, (e) Free-body diagram

respectively, the stiffness and damping coefficients corresponding to the *i*th story.

The various expressions in eq.(3.3) are obtained by equating the kinetic energy, potential energy, damping work, and the work done by the external force in the actual system, with the corresponding expressions for the generalized single-degree-of-freedom system.

3.3 PARTICIPATION FACTOR

For a time-dependent base acceleration $[\ddot{y}_s(t)]$ of the building, the generalized force becomes

$$F^*(t) = \Gamma^* \ddot{y}_s(t) \tag{3.5}$$

in which Γ^* is the *participation factor* given by

$$\Gamma^* = \frac{\Sigma m_i \phi_i}{\Sigma m_i \phi_i^2} \tag{3.6}$$

3.4 DAMPING RATIO

It is convenient to express the generalized damping coefficient c^* [eq.(3.3)] in terms of the damping ratio as

$$c^* = \Sigma c_i \Delta\phi_i^2 = 2\xi m^* \omega \tag{3.7}$$

in which ω is the natural frequency of the generalized system

$$\omega = \sqrt{\frac{k^*}{m^*}} \tag{3.8}$$

and ξ is the damping ratio, defined as

$$\xi = \frac{c^*}{c_{cr}} \tag{3.9}$$

where c_{cr}, the critical damping of the generalized system, is given by

$$c_{cr} = 2\sqrt{k^* m^*} \tag{3.10}$$

3.5 SHAPE FUNCTION

The use of generalized coordinates transforms a multidegree-of-freedom system into an equivalent single-degree-of-freedom system. The shape function describing the deformed structure could be any arbitrary function that satisfies the boundary conditions. However, in practical applications, the success of this approach will depend on how close the assumed shape function approximates the displacements of the actual dynamic system. For structural buildings, selection of the shape function is most appropriate by considering the *aspect ratio* of the structure, which is defined as the

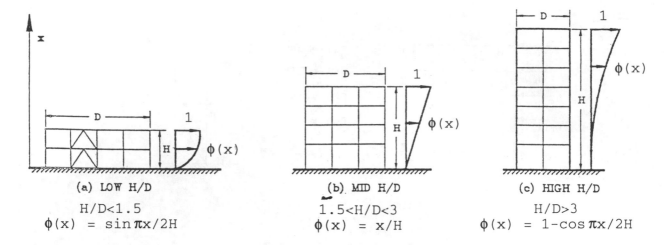

(a) LOW H/D

H/D<1.5

$\phi(x) = \sin \pi x/2H$

(b) MID H/D

1.5<H/D<3

$\phi(x) = x/H$

(c) HIGH H/D

H/D>3

$\phi(x) = 1 - \cos \pi x/2H$

Fig. 3.2.   Possible shape functions based on aspect ratio (Naeim 1989, p. 100)

ratio of the building height to the dimension of the base. The recommended shape functions for high-rise, mid-rise, and low-rise buildings are summarized in Fig. 3.2. Most seismic building codes use the straight-line shape which is shown for the mid-rise building. The displacements in the structure are calculated using eq.(3.1) after the dynamic response is obtained in terms of the generalized coordinate.

Example 3.1

A four-story reinforced concrete framed building has the dimensions shown in Fig. 3.3. The sizes of the exterior columns (nine each on lines A and C) are 12 in. × 20 in., and the interior columns (nine on line B) are 12 in. × 24 in. for the bottom two stories, and, respectively, 12 in. × 16 in. and 12 in. × 20 in. for the highest two stories. The height between floors is 12 ft. The dead load per unit area of the floor (floor slab, beam, half the weight of columns above and below the floor, partition walls, etc.), is estimated to be 140 pounds per square foot (psf). The design live load is taken as 25% of an assumed live load of 125 psf. Determine the generalized mass, generalized stiffness, generalized damping for damping ratio $\xi = 0.1$, and the fundamental period for lateral vibration perpendicular to the long axis of the building. Assume the following shape functions: (a) $\phi(x) = x/H$, and (b) $\phi(x) = \sin(\pi x/2H)$ where H is the height of the building.

Solution

1. Effective weight at various floors. No live load needs to be considered on the roof. Hence, the effective weight at all floors, except at the roof, will be $140 + 0.25 \times 125 = 171.25$ psf, and the effective weight for the roof will be 140 psf. The plan area is 48

ft × 96 ft = 4,608 ft^2. Hence, the weights of various levels are:

$$W_1 = W_2 = W_3 = 4,608 \times 0.17125 = 789.1 \text{ kips}$$

$$W_4 = 4,608 \times 0.140 = 645.1 \text{ kips}$$

The total seismic design weight of the building is then

$$W = 789.1 \times 3 + 645.1 = 3,012.4 \text{ kips}$$

2. Story lateral stiffness. It will be assumed that horizontal beam-and-floor diaphragms are rigid compared to the columns of the building in order to simplify the hand calculation. In this case, the stiffness between two consecutive levels is given by

$$k = \frac{12EI}{L^3}$$

where

$L = 12$ ft (distance between two floors)
$E = 3 \times 10^3$ kip/square inch (ksi) (modulus of elasticity of concrete)
$I = \dfrac{1}{12} 12 \times 20^3 = 8,000$ in.^{4} (moment of inertia for the concrete section for columns 12 in. × 20 in.)

Therefore, for these columns,

$$k = \frac{12 \times 3 \times 10^3 \times 8,000}{144^3} = 96.450 \text{ kip/in.}$$

Similarly, for columns 12 in. × 24 in.,

$$I = 13,824 \text{ in.}^4, \ k = 166.667 \text{ kip/in.}$$

The total stiffness for the first and second stories is then

$$K_1 = K_2 = 18 \times 96.45 + 9 \times 166.67 = 3,236 \text{ kip/in.}$$

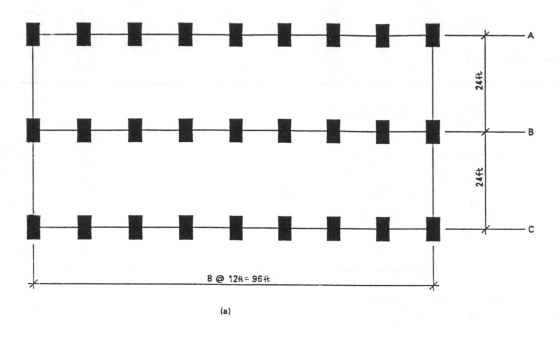

(a)

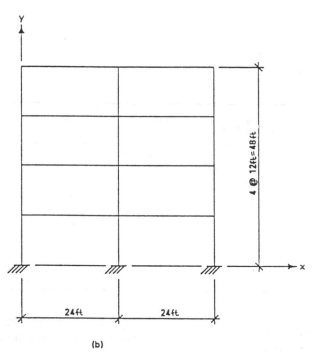

(b)

Fig. 3.3.   Plan and elevation for a four-story building for Example 3.1; (a) Plan, (b) Elevation

Similarly, for the top and third stories,

$$I = \frac{1}{12}12 \times 16^3 = 4{,}096 \text{ in.}^4,$$

$$k = \frac{12 \times 3 \times 10^3 \times 4{,}096}{144^3} = 49.4 \text{ kip/in.}$$

$$I = \frac{1}{12}12 \times 20^3 = 8{,}000 \text{ in.}^4,$$

$$k = \frac{12 \times 3 \times 10^3 \times 8{,}000}{144^6} = 96.5 \text{ kip/in.}$$

Hence, total stiffness for the third or fourth stories is

$$K_3 = K_4 = 18 \times 49.4 + 9 \times 96.5 = 1{,}757.7 \text{ kip/in.}$$

3. Generalized mass and stiffness

(a) Assuming $\phi(x) = x/H$:

Level	k_i (kip/in)	m_i (kip·sec²/in.)	ϕ_i	$\Delta\phi_i$	$m_i\phi_i^2$	$k_i\Delta\phi_i^2$
4		1.671	1.000		1.671	
	1,758			0.250		109.875
3		2.044	0.750		1.150	
	1,758			0.250		109.875
2		2.044	0.500		0.610	
	3,236			0.250		202.250
1		2.044	0.250		0.128	
	3,236			0.250		202.250
					$m^* = 3.559$	$k^* = 625.250$

$$\omega = \sqrt{\frac{k^*}{m^*}} = \sqrt{\frac{625.250}{3.559}} = 13.25 \text{ rad/sec and } T_a = 0.47 \text{ sec.}$$

$$c_{cr} = 2\sqrt{k^* m^*} = 2\sqrt{(625.250)(3.559)} = 94.345 \text{ (lb.sec/in.)}$$

$$c^* = \xi c_{cr} = (0.1)(94.345) = 9.43 \text{ (lb.sec/in.)}$$

(b) Assuming $\phi(x) = \sin(\pi x/2H)$:

Level	k_i (kip/in.)	m_i (kip·sec²/in.)	ϕ_i	$\Delta\phi_i$	$m_i\phi_i^2$	$k_i\Delta\phi_i^2$
4		1.671	1.000		1.671	
	1,758			0.076		10.154
3		2.044	0.924		1.745	
	1,758			0.217		82.782
2		2.044	0.707		1.022	
	3,236			0.324		339.702
1		2.044	0.383		0.300	
	3,236			0.383		476.686
					$m^* = 4.738$	$k^* = 909.324$

$$\omega = \sqrt{\frac{k^*}{m^*}} = \sqrt{\frac{909.324}{4.738}} = 13.85 \text{ rad/sec and } T_b = 0.45 \text{ sec.}$$

$$c_{cr} = 2\sqrt{k^* m^*} = 2\sqrt{(909.324)(4.738)} = 131.276 \text{ (lb.sec/in.)}$$

$$c^* = \xi c_{cr} = (0.1)(131.276) = 13.13 \text{ (lb.sec/in.)}$$

For this example, either of the two assumed shape functions results in essentially the same value for the fundamental period; $\phi(x) = x/H$ is a slightly better approximation[1] to the true deflected shape than is $\phi(x) = \sin(\pi x/2H)$ because $T_a > T_b$.

3.6 RAYLEIGH'S METHOD

The Rayleigh method is a procedure to determine the natural frequency of a vibrating system. It is based on the law of conservation of energy. The success of the method for determining an approximate value for the fundamental frequency has been recognized in most

---

[1]For an assumed displacement shape closer to the actual displacement shape, the structure will vibrate closer to the free condition with less imposed constraints; thus with the stiffness reduced and a longer period.

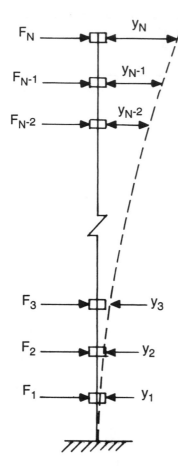

Fig. 3.4. Building modeled as a column with discrete masses subjected to lateral forces

seismic building codes. Such building codes include the Rayleigh method as an alternative for estimating the fundamental period of vibration.

For undamped elastic systems, the maximum potential energy may be expressed in terms of the external work done by the applied forces. Thus, for the multi-story building modeled by a column with discrete masses shown in Fig. 3.4, the maximum potential energy $(PE)_{max}$ is

$$(PE)_{max} = \tfrac{1}{2}\Sigma F_i y_i \qquad (3.11)$$

where y_i is the lateral displacement at coordinate i and F_i the external force at this coordinate. Similarly, the maximum kinetic energy $(KE)_{max}$ can be expressed in terms of the natural frequency of vibration ω in (rad/sec) as

$$(KE)_{max} = \frac{\omega^2}{2}\sum_i m_i y_i^2 \qquad (3.12)$$

The principle of conservation of energy requires that these two maximum values (for the potential energy and for the kinetic energy) be equal to each other and equal to the total energy of the system. Hence,

equating eq.(3.11) with eq.(3.12) produces the following expression for the natural frequency of the system:

$$\omega = \sqrt{\frac{\displaystyle\sum_i F_i y_i}{\displaystyle\sum_i m_i y_i^2}} \qquad (3.13)$$

The natural period T then is given conveniently in terms of the weights $W_i = m_i g$ as

$$T = 2\pi \sqrt{\frac{\displaystyle\sum_i W_i y_i^2}{g \displaystyle\sum_i F_i y_i}} \qquad (3.14)$$

which is the formula found in most seismic building codes.

The period T may also be expressed in terms of the generalized coordinates Y after substitution of eq.(3.1) into eq.(3.14). Hence

$$T = 2\pi \sqrt{\frac{\displaystyle\sum_i m_i \phi_i^2 Y}{\displaystyle\sum_i F_i \phi_i}} \qquad (3.15)$$

in which $m_i = W_i/g$.

The forces that must be applied laterally to obtain either the shape function ϕ_i or the lateral displacement y_i are inertial forces, which are products of the mass times the acceleration.

If the acceleration is assumed to vary linearly over the height of a building with uniform weight distribution, the distribution of inertial force is shaped like an inverted triangle, maximum at the top and zero at the base of the building. This distribution is similar to that of the base shear force adopted in most seismic building codes; it is an appropriate distribution to use in the application of Rayleigh's method. The resulting lateral deflections can be used directly in eq.(3.14) to estimate the fundamental period of vibration or they can be normalized, in terms of the generalized coordinates, to obtain the shape function for use in the generalized-coordinate method.

Example 3.2

Use Rayleigh's method to determine the fundamental period of vibration in the lateral direction and to estimate the shape function for the reinforced concrete building of Example 3.1. Model the structure as a shear building and assume a linear distribution of lateral inertial forces.

Level	k_i (kip/in.)	m_i $\left(\dfrac{\text{kip·sec}^2}{\text{in.}}\right)$	F_i (kip)	V_i† (kip)	$\Delta = V_i/k_i$ in.	y_i in.	ϕ_i	$m_i\phi_i^2$	$F_i\phi_i$
4		1.671	400	400	0.228	1.213	1.000	1.671	400.0
	1,758								
3		2.044	300	700	0.398	0.985	0.812	1.348	243.6
	1,758								
2		2.044	200	900	0.278	0.587	0.484	0.479	96.8
	3,236								
1		2.044	100	1000	0.309	0.309	0.255	0.133	25.5
	3,236								
							$\Sigma =$	3.631	765.9

†V_i is the story shear force.

Solution

Assume that the inertial forces vary linearly from the base to the roof of the building as shown in Fig. 3.5. It should be noted that the absolute magnitude of these forces is not relevant and numerical values were chosen for ease of computation. The above table shows the necessary computation for application of eq.(3.15). Substitution into eq.(3.15) of values obtained in this table yields:

$$T = 2\pi \sqrt{\frac{(3.631)(1.213)}{765.9}} = 0.48 \text{ sec}$$

Since $T = 0.48$ sec is greater than either of the periods calculated in Example 3.1, the deflected shape that results by applying the lateral loads is a better approximation to the true shape function than the two previous shapes.

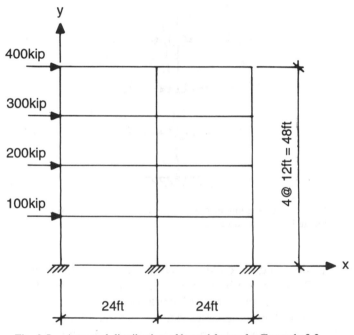

Fig. 3.5. Assumed distribution of lateral forces for Example 3.2

4

Structures Modeled as Multidegree-of-Freedom Systems

Mario Paz

4.1 THE SHEAR BUILDING MODEL

Buildings as well as other structures may be idealized as an assembly of *elements* connected at joints or nodal points. These elements can be unidirectional such as beams or rod elements, two dimensional like plates and shell elements, and three dimensional such as solid elements. The structure may be modeled as a *shear building* when the horizontal diaphragms at the floor levels of a multistory building are assumed to be rigid. In such a model, it is assumed that: (1) the total mass of the structure is concentrated at the levels of the floors, (2) the horizontal diaphragms at the floor levels are plane rigid, and (3) the deformation of the structure is independent of the axial force present in the columns. These assumptions transform the problem from a system with an infinite number of degrees of freedom (due to the distributed mass) to a system that has only as many degrees of freedom as it has lumped masses at the floor levels. A three-story building modeled as a shear building will have only three degrees of freedom, the three horizontal displacements shown in Fig. 4.1(a).

It should be noted that, for convenience, the building is presented solely in terms of a single bay rather than many bays. Actually, the building can be idealized further as a single column [Fig. 4.2(a)], having concentrated masses at the floor levels, with the understanding that only horizontal displacements of these masses are possible. Another alternative is to represent the shear buildings by adopting a multiple

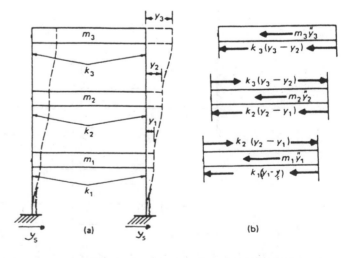

Fig. 4.1. Single-bay model representation of a shear building

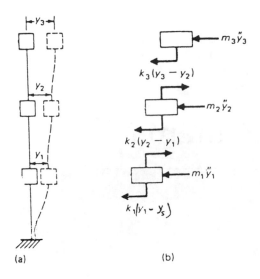

Fig. 4.2. Single-column model representation of a shear building

mass-spring system as shown in Fig. 4.3(a). In any of the three representations depicted in these figures, the stiffness coefficient or spring constant k_i shown between any two consecutive masses is the force required to produce a unit relative displacement of the two adjacent floor levels. For a uniform column with the two ends fixed against rotation, the spring constant is given by

$$k = \frac{12EI}{L^3} \qquad (4.1)$$

and for a column with one end fixed and the other end pinned, by

$$k = \frac{3EI}{L^3} \qquad (4.2)$$

as shown in Fig. 4.4(a) and (c), where E is the material modulus of elasticity, I the cross-sectional moment of inertia, and L the length of the column.

The lateral stiffness of a cantilever structural wall [Fig. 4.4(d)] which includes the effect of shear deformation is obtained from the expression for its lateral displacement.

$$\Delta = \frac{PL^3}{3EI} + \frac{\alpha PL}{AG} \qquad (4.3)$$

in which the shape constant $\alpha = 1.2$ for rectangular sections and A is the area. Solving for $k = P/\Delta$ and substituting $G = E/2(1 + \nu)$ where ν is Poisson's ratio, yields

$$k = \frac{3EI}{L^3[1 + 0.6(1 + \nu)(d/L)^2]} \qquad (4.4$$

Finally, as shown in Fig. 4.4(b), the stiffness of a diagonal bracing element is

$$k = \frac{AE}{L} \cos^2 \theta \qquad (4.5)$$

where A is the cross-sectional area of the bracing element.

4.2 EQUATIONS OF MOTION FOR BASE EXCITATION

The equations of motion for a shear building caused by motion at its base or foundation are conveniently expressed in terms of the lateral displacements at the various levels relative to the base. For an N-story shear building [Fig. 4.5(a)] subjected to excitation motion at its foundation, the equations of motion obtained by equating to zero the sum of the forces in

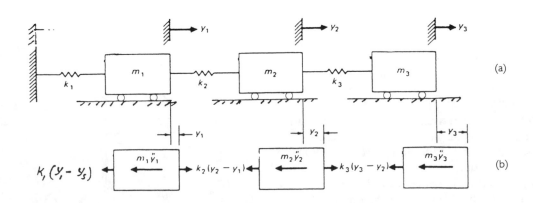

Fig. 4.3. Multimass-spring model representation of a shear building

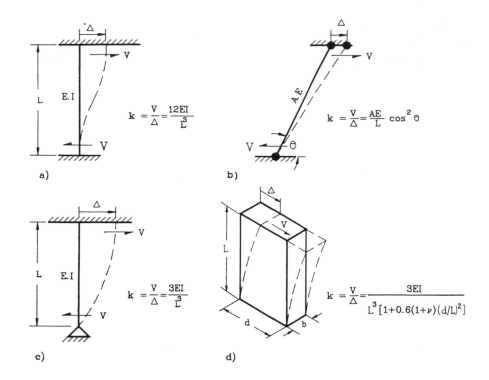

Fig. 4.4. Expressions for the lateral stiffness coefficients; (a) Fixed-fixed column, (b) Diagonal brace, (c) Fixed-pinned column, (d) Cantilever wall. (From Farzad Naeim, Seismic Design Handbook, Van Nostrand Reinhold, 1989)

the corresponding free-body diagrams [Fig. 4.5(b)] are:

$$m_1 \ddot{y}_1 + k_1(y_1 - y_s) - k_2(y_2 - y_1) = 0$$

$$m_2 \ddot{y}_2 + k_2(y_2 - y_1) - k_3(y_3 - y_2) = 0$$

$$\cdots$$

$$m_{N-1} \ddot{y}_{N-1} + k_{N-1}(y_{N-1} - y_{N-2}) - k_N(y_N - y_{N-1}) = 0$$

$$m_N \ddot{y}_N + k_N(y_N - y_{N-1}) = 0 \tag{4.6}$$

Introducing the notation

$$u_i = y_i - y_s \quad (i = 1, 2, \ldots, N) \tag{4.7}$$

yields

$$m_1 \ddot{u}_1 + k_1 u_1 - k_2(u_2 - u_1) = -m_1 \ddot{y}_s$$

$$m_2 \ddot{u}_2 + k_2 u_2 - k_3(u_3 - u_2) = -m_2 \ddot{y}_s$$

$$\cdots$$

$$m_{N-1} \ddot{u}_{N-1} + k_{N-1}(u_{N-1} - u_{N-2}) - k_N(u_N - u_{N-1}) = -m_{N-1} \ddot{y}_s$$

$$m_N \ddot{u}_N + k_N(u_N - u_{N-1}) = -m_N \ddot{y}_s \tag{4.8}$$

where $\ddot{y}_s = \ddot{y}_s(t)$ is the acceleration function exciting the base of the structure.

Equation (4.8) may be written conveniently in matrix notation as

$$[M]\{\ddot{u}\} + [K]\{u\} = -[M]\{1\}\ddot{y}_s(t) \tag{4.9}$$

in which $[M]$, the mass matrix, is a diagonal matrix; $[K]$, the stiffness matrix, is a symmetric matrix; $\{1\}$ is a vector with all its elements equal to 1; $\ddot{y}_s = \ddot{y}_s(t)$ is the applied acceleration at the foundation of the building; and $\{u\}$ and $\{\ddot{u}\}$ are, respectively, the displacement and acceleration vectors relative to the motion of the foundation.

4.3 NATURAL FREQUENCIES AND NORMAL MODES

When free vibration is being considered, the structure is not subjected to any external excitation (force or support motion) and its motion is governed only by initial conditions. Only infrequently are there circumstances for which it is necessary to determine the motion of the structure under conditions of free vibration. Nevertheless, the most important dynamic properties of the structure, the natural frequencies, and corresponding modal shapes, are obtained from the analysis of the structure in free vibration.

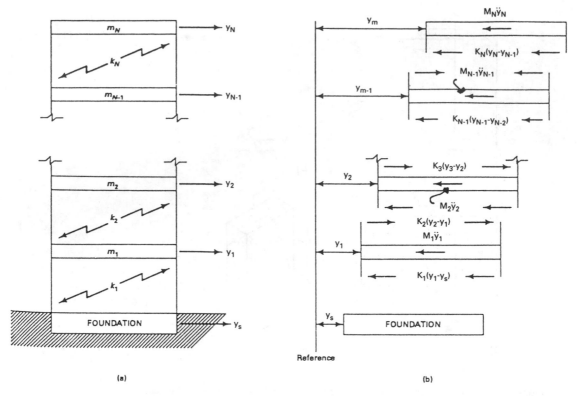

(a)   (b)

Fig. 4.5. Multistory shear building excited at the foundation; (a) Structural model, (b) Free-body diagrams

The free vibration requires that $\ddot{y}_s$ in eq.(4.9) be equal to zero. Hence,

$$[M]\{\ddot{y}\} + [K]\{y\} = \{0\} \tag{4.10}$$

The solution of eq.(4.10) is in the form

$$\{y\} = \{a\}\sin(\omega t - \alpha) \tag{4.11}$$

where $\{a\}$ is the vector of the amplitude of motion. The substitution of eq.(4.11) into eq.(4.10) gives, after rearranging terms,

$$[[K] - \omega^2[M]]\{a\} = \{0\} \tag{4.12}$$

which, for the general case, is a set of N homogeneous (right-hand side equal to zero) algebraic linear equations with N unknown displacements a_i and an unknown parameter ω^2. The formulation of eq.(4.12) is an important mathematical problem known as an eigenproblem. Its nontrivial solution, that is, the solution for which not all $a_i = 0$, requires that the determinant of the matrix factor of $\{a\}$ be equal to zero; in this case,

$$|[K] - \omega^2[M]| = 0 \tag{4.13}$$

In general, eq.(4.13) yields a polynomial equation of degree n in ω^2 which should be satisfied for n values of

ω^2. This polynomial is known as the *characteristic equation* of the system. Equation (4.12) can be solved for $a_1, a_2, \ldots, a_N$ in terms of an arbitrary constant for each of these values of ω^2 satisfying the characteristic eq.(4.13).

Example 4.1

Determine the natural frequencies and corresponding mode shapes for the shear building shown in Fig. 4.6.

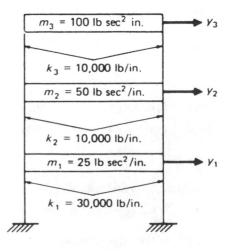

Fig. 4.6. Shear building for Example 4.1

Solution

The equation of motion in free vibration for this structure is given by eq.(4.10) as

$$[M]\{\ddot{y}\} + [K]\{y\} = \{0\} \qquad (a)$$

in which $[M]$ and $[K]$ are, respectively, the mass and stiffness matrices of the system. The substitution of numerical values into eq.(a) yields

$$\begin{bmatrix} 25 & 0 & 0 \\ 0 & 50 & 0 \\ 0 & 0 & 100 \end{bmatrix} \begin{Bmatrix} \ddot{y}_1 \\ \ddot{y}_2 \\ \ddot{y}_3 \end{Bmatrix}$$
$$+ \begin{bmatrix} 40{,}000 & -10{,}000 & 0 \\ -10{,}000 & 20{,}000 & -10{,}000 \\ 0 & -10{,}000 & 10{,}000 \end{bmatrix} \begin{Bmatrix} y_1 \\ y_2 \\ y_3 \end{Bmatrix} = \begin{Bmatrix} 0 \\ 0 \\ 0 \end{Bmatrix}$$

Upon substituting $y_i = a_i \sin \omega t$ and canceling the factor $\sin \omega t$, we obtain

$$\begin{bmatrix} 40{,}000 - 25\omega^2 & -10{,}000 & 0 \\ -10{,}000 & 20{,}000 - 50\omega^2 & -10{,}000 \\ 0 & -10{,}000 & 10{,}000 - 100\omega^2 \end{bmatrix} \qquad (b)$$
$$\begin{Bmatrix} a_1 \\ a_2 \\ a_3 \end{Bmatrix} = \begin{Bmatrix} 0 \\ 0 \\ 0 \end{Bmatrix}$$

for which a nontrivial solution requires that the determinant of the coefficients be equal to zero, that is,

$$\begin{vmatrix} 40{,}000 - 25\omega^2 & -10{,}000 & 0 \\ -10{,}000 & 20{,}000 - 50\omega^2 & -10{,}000 \\ 0 & -10{,}000 & 10{,}000 - 100\omega^2 \end{vmatrix} = 0$$

The expansion of this determinant results in a third-degree equation in terms of ω^2 having the following roots:

$$\omega_1^2 = 36.1, \ \omega_2^2 = 400.0, \ \omega_3^2 = 1{,}664.0 \qquad (c)$$

The natural frequencies are calculated as $f = \omega/2\pi$, so that

$$f_1 = 0.96 \text{ cps}, \ f_2 = 3.18 \text{ cps}, \ f_3 = 264.8 \text{ cps} \qquad (d)$$

The modal shapes are then determined by substituting, in turn, each of the values for the natural frequencies into eq.(b): deleting a redundant equation, and solving the remaining two equations for two of the unknowns in terms of the third. In solving for these unknowns, it is expedient to set the first nonzero unknown equal to one. Performing these operations,

we obtain from eqs.(b) and (c) the following values for the mode shapes:

$$\begin{aligned} a_{11} &= 1.00, & a_{12} &= 1.00, & a_{13} &= 1.00 \\ a_{21} &= 3.91, & a_{22} &= 3.00, & a_{23} &= 3.338 \\ a_{31} &= 6.11, & a_{32} &= -1.00, & a_{33} &= -2.025 \end{aligned} \qquad (e)$$

Because the mode shapes are known only in proportional numbers, any mode may be normalized by an arbitrary factor. The following is a particularly convenient normalization for a general system:

$$\phi_{ij} = \frac{a_{ij}}{\sqrt{\{a\}_j^T [M]\{a\}_j}} \qquad (4.14)$$

which, for a system having a diagonal mass matrix, may be written as

$$\phi_{ij} = \frac{a_{ij}}{\sqrt{\sum_{k=1}^{N} m_k a_{kj}^2}} \qquad (4.15)$$

in which ϕ_{ij} is the normalized i component of the j modal vector.

An important orthogonality condition is satisfied by the mode shapes. For normalized mode shapes this condition may be expressed as

$$\begin{aligned} \{\phi\}_i^T [M]\{\phi\}_j &= 0 \quad \text{for } i \neq j \\ &= 1 \quad \text{for } i = j \end{aligned} \qquad (4.16)$$

Another orthogonality condition relates normal modes:

$$\begin{aligned} \{\phi\}_i^T [K]\{\phi\}_j &= 0 \quad \text{for } i \neq j \\ &= \omega^2 \quad \text{for } i = j \end{aligned} \qquad (4.17)$$

Example 4.2

For the three-degree-of-freedom system of Example 4.1, determine the normalized modal shapes.

Solution

For the first mode the normalization factor, from eq.(4.15), is

$$\sqrt{\sum_k m_k a_{k1}^2} = \sqrt{25(1.00)^2 + 50(3.91)^2 + 100(6.11)^2} = 67.25$$

Consequently, the normalized components of the first modal shapes are

$$\phi_{11} = \frac{1.00}{67.25} = 0.01487, \quad \phi_{21} = \frac{3.91}{67.25} = 0.05814,$$

$$\phi_{31} = \frac{6.11}{67.25} = 0.09085$$

Similarly, for the second and third normalized modes

$$\phi_{12} = \frac{1.00}{23.98} = 0.04170, \quad \phi_{22} = \frac{3.00}{23.98} = 0.12511,$$

$$\phi_{32} = \frac{-1}{23.98} = -0.04170$$

$$\phi_{13} = \frac{1.00}{31.50} = 0.03174, \quad \phi_{23} = \frac{3.338}{31.50} = 0.10597,$$

$$\phi_{33} = \frac{-2.025}{31.50} = -0.06429$$

The modal shapes usually are presented as the columns of a matrix known as the *modal matrix*. For this example the modal matrix $[\Phi]$ is given by

$$[\Phi] = \begin{bmatrix} 0.01487 & 0.04170 & 0.03174 \\ 0.05814 & 0.12511 & 0.10597 \\ 0.09085 & -0.04170 & -0.06429 \end{bmatrix}$$

4.4 STATIC CONDENSATION

In the process of modeling buildings and other structures, often it is necessary to divide the structure into a large number of elements because of changes in geometry, loading, or material properties. For such structures, the total number of unknown displacements (that is, the number of degrees of freedom) may be quite large. As a consequence, the stiffness and mass matrices will be of large dimensions, and the solution of the corresponding eigenproblem, to determine natural frequencies and modal shapes, will be difficult and expensive. In such cases, it is desirable to reduce the size of these matrices in order to make the solution of the eigenproblem more manageable and economical. Such reduction is termed condensation.

A popular method of reduction is the *static condensation method*. This method, although simple to apply, is only approximate and may produce relatively large errors when applied to dynamic problems. An improved method for the condensation of dynamic problems that gives virtually exact results has recently been proposed. This new method, called the *dynamic condensation method*, is presented in the next section of this chapter.

A practical procedure to implement static condensation is to separate the primary or independent coordinates $\{y_p\}$ from the secondary or dependent coordinates $\{y_s\}$ to be condensed and to write the static stiffness equation in partition matrices as follows:

$$\begin{bmatrix} [K_{ss}] & \vdots & [K_{sp}] \\ \cdots & \vdots & \cdots \\ [K_{ps}] & \vdots & [K_{pp}] \end{bmatrix} \begin{Bmatrix} \{y_s\} \\ \cdots \\ \{y_p\} \end{Bmatrix} = \begin{Bmatrix} \{0\} \\ \cdots \\ \{F_p\} \end{Bmatrix} \quad (4.18)$$

In eq.(4.18), it has been assumed that the external forces were zero at the dependent (i.e., secondary) degrees of freedom; this assumption is not mandatory (Gallagher 1975), but serves to simplify explanations without affecting the final results. A simple multiplication of the matrices on the left side of eq.(4.18) expands this equation into the following two matrix equations:

$$[K_{ss}]\{y_s\} + [K_{sp}]\{y_p\} = \{0\} \quad (4.19)$$

$$[K_{ps}]\{y_s\} + [K_{pp}]\{y_p\} = \{F_p\} \quad (4.20)$$

Equation (4.19) is equivalent to

$$\{y_s\} = [\bar{T}]\{y_p\} \quad (4.21)$$

where $[\bar{T}]$ is the transformation matrix given by

$$[\bar{T}] = -[K_{ss}]^{-1}[K_{sp}] \quad (4.22)$$

Substituting eq.(4.21) and using eq.(4.22) in eq.(4.20) yields the reduced stiffness equation relating forces and displacements at the primary coordinates; that is,

$$[\bar{K}]\{y_p\} = \{F_p\} \quad (4.23)$$

where $[\bar{K}]$ is the reduced stiffness matrix given by

$$[\bar{K}] = [K_{pp}] - [K_{ps}][K_{ss}]^{-1}[K_{sp}] \quad (4.24)$$

Equation (4.21), which expresses the static relation between the secondary coordinates $\{y_s\}$ and primary coordinates $\{y_p\}$, may also be written using the identity $\{y_p\} = [I]\{y_p\}$ as

$$\begin{Bmatrix} \{y_s\} \\ \cdots \\ \{y_p\} \end{Bmatrix} = \begin{bmatrix} [\bar{T}] \\ \cdots \\ [I] \end{bmatrix} \{y_p\} \quad (4.25)$$

or

$$\{y\} = [T]\{y_p\} \quad (4.26)$$

where

$$\{y\} = \begin{Bmatrix} \{y_s\} \\ \cdots \\ \{y_p\} \end{Bmatrix} \quad \text{and} \quad [T] = \begin{bmatrix} [\bar{T}] \\ \cdots \\ [I] \end{bmatrix} \quad (4.27)$$

Substituting eqs.(4.26) and (4.27) into eq.(4.18) and pre-multiplying by the transpose of $[T]$ gives

$$[T]^T[K][T]\{y_p\} = [[\bar{T}]^T[I]] \begin{Bmatrix} \{0\} \\ \{F_p\} \end{Bmatrix}$$

or

$$[T]^T[K][T]\{y_p\} = \{F_p\}$$

and using eq.(4.23)

$$[\bar{K}] = [T]^T[K][T] \qquad (4.28)$$

thus showing that the reduced stiffness matrix $[\bar{K}]$ can be expressed as a transformation of the system stiffness matrix $[K]$.

It may appear that the calculation of the reduced stiffness matrix $[\bar{K}]$ given by eq.(4.24) requires the inconvenient calculation of the inverse matrix $[K_{ss}]^{-1}$. However, the practical application of the static condensation method does not require a matrix inversion. Instead, the standard Gauss-Jordan elimination process is applied systematically on the system stiffness matrix $[K]$ to eliminate the secondary coordinates $\{y_s\}$. At this stage of the elimination process, the stiffness equation (4.18) has been reduced to

$$\left[\begin{array}{c:c} [I] & -[\bar{T}] \\ \hdashline [0] & [\bar{K}] \end{array}\right] \left\{\begin{array}{c} \{y_s\} \\ \cdots \\ \{y_p\} \end{array}\right\} = \left\{\begin{array}{c} \{0\} \\ \cdots \\ \{F_p\} \end{array}\right\} \qquad (4.29)$$

It may be seen by expanding eq.(4.29) that the partition matrices $[\bar{T}]$ and $[\bar{K}]$ indicated in this equation are precisely the transformation matrix and the reduced stiffness matrix defined by eqs.(4.22) and (4.24), respectively. In this way, the Gauss-Jordan elimination process yields both the transformation matrix $[\bar{T}]$ and the reduced stiffness matrix $[\bar{K}]$. There is no need to calculate $[K_{ss}]^{-1}$ in order to reduce the secondary coordinates of the system. To reduce the mass and damping matrices, it is assumed that the same static relation between the secondary and primary degrees of freedom remains valid in the dynamic problem. Hence, the same transformation based on static condensation for the reduction of the stiffness matrix is used to reduce the mass and damping matrices. In general, this method of reducing the dynamic problem introduces errors in the results because it is not an exact method. The magnitude of these errors depends on the relative number of degrees of freedom reduced and the specific selection of these degrees of freedom for a given structure.

In the case in which the process of modeling the structure has left a number of massless degrees of freedom, it is only necessary to carry out the static condensation of the stiffness matrix and to delete from the mass matrix the rows and columns corresponding to the massless degrees of freedom. In this case, the static condensation method does not alter the original problem. Thus, it yields an equivalent eigenproblem without introducing any errors.

In the general case in which mass has been allocated to every degree of freedom, the reduced mass and the reduced damping matrices are obtained using a transformation analogous to eq.(4.28). Specifically,

if $[M]$ is the mass matrix of the system, then the *reduced mass matrix* is given by

$$[\bar{M}] = [T]^T[M][T] \qquad (4.30)$$

where $[T]$ is the transformation matrix defined by eq.(4.27). Analogously, for a damped system, the reduced damping matrix is given by

$$[\bar{C}] = [T]^T[C][T] \qquad (4.31)$$

where $[C]$ is the damping matrix of the system.

This method of reducing the mass and damping matrices may be justified as follows. The potential elastic energy V and the kinetic energy KE of the structure may be written, respectively, as

$$V = \frac{1}{2}\{y\}^T[K]\{y\} \qquad (4.32)$$

$$KE = \frac{1}{2}\{\dot{y}\}^T[M]\{\dot{y}\} \qquad (4.33)$$

Analogously, the virtual work δW_d done by the damping forces $F_d = [C]\{\dot{y}\}$ corresponding to the virtual displacement $\{\delta y\}$ may be expressed as

$$\delta W_d = \{\delta y\}^T[C]\{\dot{y}\} \qquad (4.34)$$

Introduction of the transformation eq.(4.26) in the above equations produces

$$V = \frac{1}{2}\{y_p\}^T[T]^T[K][T]\{y_p\} \qquad (4.35)$$

$$KE = \frac{1}{2}\{\dot{y}_p\}^T[T]^T[M][T]\{\dot{y}_p\} \qquad (4.36)$$

$$\delta W_d = \{\delta y_p\}^T[T]^T[C][T]\{\dot{y}_p\} \qquad (4.37)$$

The respective substitution of $[\bar{K}]$, $[\bar{M}]$, and $[\bar{C}]$ from eqs.(4.28), (4.30), and (4.31) for the products of the three central matrices in eqs.(4.35), (4.36), and (4.37) yields

$$V = \frac{1}{2}\{y_p\}^T[\bar{K}]\{y_p\}$$

$$KE = \frac{1}{2}\{\dot{y}_p\}^T[\bar{M}]\{\dot{y}_p\}$$

$$\delta W_d = \{\delta y_p\}^T[\bar{C}]\{\dot{y}_p\}$$

These last three equations express the potential energy, the kinetic energy, and the virtual work of the damping forces in terms of the independent coordinates $\{y_p\}$. Hence, the matrices $[\bar{K}]$, $[\bar{M}]$, and $[\bar{C}]$ may be interpreted, respectively, as the stiffness, mass, and

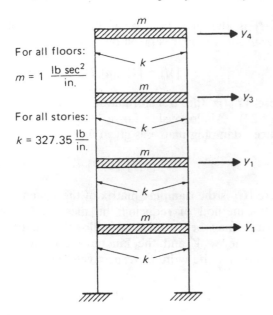

Fig. 4.7. Uniform shear building for Example 4.3

damping matrices of the structure corresponding to the independent degrees of freedom $\{y_p\}$.

Example 4.3

Figure 4.7 shows a uniform four-story building. For this structure, determine the following: (a) the natural frequencies and corresponding modal shapes as a four-degree-of-freedom system, (b) the natural frequencies and modal shapes after static condensation of coordinates y_1 and y_3.

Solution

(a) Natural Frequencies and Modal Shapes as a Four-Degree-of-Freedom System: The stiffness and mass matrices for this structure are, respectively,

$$[K] = 327.35 \begin{bmatrix} 2 & -1 & 0 & 0 \\ -1 & 2 & -1 & 0 \\ 0 & -1 & 2 & -1 \\ 0 & 0 & -1 & 1 \end{bmatrix} \quad \text{(a)}$$

and

$$[M] = \begin{bmatrix} 1 & 0 & 0 & 0 \\ 0 & 1 & 0 & 0 \\ 0 & 0 & 1 & 0 \\ 0 & 0 & 0 & 1 \end{bmatrix} \quad \text{(b)}$$

Substituting eqs.(a) and (b) into eq.(4.12) and solving the corresponding eigenvalue problem yields

$$\omega_1^2 = 39.48, \ \omega_2^2 = 327.35, \ \omega_3^2 = 768.3, \text{ and } \omega_4^2 = 1156.00$$

corresponding to the natural frequencies $(f = \omega/2\pi)$

$$f_1 = 1.00 \text{ cps}, f_2 = 2.88 \text{ cps}, f_3 = 4.41 \text{ cps}$$
$$\text{and } f_4 = 5.41 \text{ cps} \quad \text{(c)}$$

and the normalized modal matrix

$$[\Phi] = \begin{bmatrix} 0.2280 & 0.5774 & -0.6565 & 0.4285 \\ 0.4285 & 0.5774 & 0.2280 & -0.6565 \\ 0.5774 & 0 & 0.5774 & 0.5774 \\ 0.6565 & -0.5774 & -0.4285 & -0.2280 \end{bmatrix} \quad \text{(d)}$$

(b) Natural Frequencies and Modal Shapes after Reduction to Two-Degree-of-Freedom System: To reduce coordinates y_1 and y_3, first, for convenience, rearrange the stiffness matrix in eq.(a) to have the coordinates in order y_1, y_3, y_2, y_4:

$$[K] = 327.35 \begin{bmatrix} 2 & 0 & -1 & 0 \\ 0 & 2 & -1 & -1 \\ -1 & -1 & 2 & 0 \\ 0 & -1 & 0 & 1 \end{bmatrix} \quad \text{(e)}$$

Applying Gauss-Jordan elimination to the first two rows of the matrix in eq.(e) produces

$$\begin{bmatrix} 1 & 0 & \vdots & -0.5 & 0 \\ 0 & 1 & \vdots & -0.5 & -0.5 \\ \cdots & \cdots & \vdots & \cdots & \cdots \\ 0 & 0 & \vdots & 327.35 & -163.70 \\ 0 & 0 & \vdots & -163.70 & 163.70 \end{bmatrix} \quad \text{(f)}$$

A comparison of eq.(f) with eq.(4.29) reveals that

$$[\bar{T}] = \begin{bmatrix} 0.5 & 0 \\ 0.5 & 0.5 \end{bmatrix} \quad \text{(g)}$$

and

$$[\bar{K}] = \begin{bmatrix} 327.35 & -163.70 \\ -163.70 & 163.70 \end{bmatrix} \quad \text{(h)}$$

Use of eq.(4.27) gives

$$[T] = \begin{bmatrix} 0.5 & 0 \\ 0.5 & 0.5 \\ \cdots & \cdots \\ 1 & 0 \\ 0 & 1 \end{bmatrix} \quad \text{(i)}$$

The reduced mass matrix now can be calculated by eq.(4.30) as

$$[\bar{M}] = [T]^T[M][T] = \begin{bmatrix} 1.5 & 0.25 \\ 0.25 & 1.25 \end{bmatrix} \quad \text{(i)}$$

The condensed eigenproblem is

$$\begin{bmatrix} 327.35 - 1.5\omega^2 & -163.70 - 0.25\omega^2 \\ -163.70 - 0.25\omega^2 & 163.70 - 1.25\omega^2 \end{bmatrix} \begin{Bmatrix} a_2 \\ a_4 \end{Bmatrix} = \begin{Bmatrix} 0 \\ 0 \end{Bmatrix} \quad \text{(j)}$$

and its solution is

$$\omega_1^2 = 40.39, \quad \omega_2^2 = 365.98 \qquad \text{(k)}$$

$$[a]_p = \begin{bmatrix} 0.4380 & 0.7056 \\ 0.6723 & -0.6128 \end{bmatrix} \qquad \text{(l)}$$

where $[a]_p$ is the modal matrix corresponding to the primary degrees of freedom. The eigenvectors for the four-degree-of-freedom system are calculated for the first mode by eq.(4.26) as

$$\begin{Bmatrix} a_1 \\ a_3 \\ a_2 \\ a_4 \end{Bmatrix}_1 = \begin{bmatrix} 0.5 & 0 \\ 0.5 & 0.5 \\ 1 & 0 \\ 0 & 1 \end{bmatrix} \begin{Bmatrix} 0.4380 \\ 0.6723 \end{Bmatrix} = \begin{Bmatrix} 0.2190 \\ 0.5552 \\ 0.4380 \\ 0.6723 \end{Bmatrix}$$

or

$$\begin{Bmatrix} a_1 \\ a_2 \\ a_3 \\ a_4 \end{Bmatrix}_1 = \begin{Bmatrix} 0.2190 \\ 0.4380 \\ 0.5552 \\ 0.6723 \end{Bmatrix} \qquad \text{(m)}$$

and, analogously, for the second mode

$$\begin{Bmatrix} a_1 \\ a_2 \\ a_3 \\ a_4 \end{Bmatrix}_2 = \begin{Bmatrix} 0.3528 \\ 0.7056 \\ 0.0464 \\ -0.6128 \end{Bmatrix} \qquad \text{(n)}$$

4.5 DYNAMIC CONDENSATION

A method of reduction that may be considered an extension of the static condensation method has been proposed (Paz 1984a,b). The algorithm for the proposed method is developed by assigning an approximate value (e.g., zero) to the first eigenvalue ω_1^2, applying condensation to the dynamic matrix of the system $[D_1] = [K] - \omega_1^2[M]$, and then solving the reduced eigenproblem to determine the first and second eigenvalues ω_1^2 and ω_2^2. Next, condensation is applied to the dynamic matrix $[D_2] = [K] - \omega_2^2[M]$ to reduce the problem and calculate the second and third eigenvalues, ω_2^2 and ω_3^2. The process continues with one virtually exact eigenvalue and an approximation of the next order eigenvalue calculated at each step.

The dynamic condensation method requires neither matrix inversion nor series expansion. Consider the eigenvalue problem of a discrete structural system for which it is desired to reduce the secondary degrees of freedom $\{y_s\}$ and retain the primary degrees of

freedom $\{y_p\}$. In this case, the equations of free motion may be written in partitioned matrix form as

$$\begin{bmatrix} [M_{ss}] & \vdots & [M_{sp}] \} \\ \cdots & \cdots & \cdots \\ [M_{ps}] & \vdots & [M_{pp}] \end{bmatrix} \begin{Bmatrix} \{\ddot{y}_s\} \\ \cdots \\ \{\ddot{y}_p\} \end{Bmatrix}$$

$$+ \begin{bmatrix} [K_{ss}] & \vdots & [K_{sp}] \\ \cdots & \cdots & \cdots \\ [K_{ps}] & \vdots & [K_{pp}] \end{bmatrix} \begin{Bmatrix} \{y_s\} \\ \cdots \\ \{y_p\} \end{Bmatrix} = \begin{Bmatrix} \{0\} \\ \cdots \\ \{0\} \end{Bmatrix} \quad (4.38)$$

The substitution of $\{y\} = \{a\} \sin \omega_i t$ in eq.(4.38) produces the generalized eigenproblem

$$\begin{bmatrix} [K_{ss}] - \omega_i^2[M_{ss}] & \vdots & [K_{sp}] - \omega_i^2[M_{sp}] \\ \cdots & \cdots & \cdots \\ [K_{ps}] - \omega_i^2[M_{ps}] & \vdots & [K_{pp}] - \omega_i^2[M_{pp}] \end{bmatrix} \begin{Bmatrix} \{a_s\} \\ \cdots \\ \{a_p\} \end{Bmatrix} = \begin{Bmatrix} \{0\} \\ \cdots \\ \{0\} \end{Bmatrix}$$

$$(4.39)$$

where ω_i^2 is the approximation of the ith eigenvalue which was calculated in the preceding step of the process. To start the process one takes an approximate or zero value for the first eigenvalue ω_1^2. The following three steps are executed to calculate the ith eigenvalue ω_i^2 and the corresponding eigenvector $\{a\}_i$ as well as an approximation of the eigenvalue of the next order ω_{i+1}^2.

Step 1. The approximation of ω_i^2 is introduced in eq.(4.39); Gauss-Jordan elimination of the secondary coordinates $\{a_s\}$ is then used to reduce eq.(4.39) to

$$\begin{bmatrix} [I] & \vdots & -[\bar{T}_i] \\ \cdots & \cdots & \cdots \\ [O] & \vdots & [\bar{D}_i] \end{bmatrix} \begin{Bmatrix} \{a_s\} \\ \cdots \\ \{a_p\} \end{Bmatrix} = \begin{Bmatrix} \{0\} \\ \cdots \\ \{0\} \end{Bmatrix} \quad (4.40)$$

The first equation in eq.(4.40) can be written as

$$\{a_s\} = [\bar{T}_i]\{a_p\} \qquad (4.41)$$

Consequently, the ith modal shape $\{a_i\}$ can be expressed as

$$\{a\}_i = [T_i]\{a_p\} \qquad (4.42)$$

where

$$[T_i] = \begin{bmatrix} [\bar{T}_i] \\ \cdots \\ [I] \end{bmatrix} \quad \text{and} \quad \{a\}_i = \begin{Bmatrix} \{a_s\} \\ \cdots \\ \{a_p\} \end{Bmatrix} \quad (4.43)$$

Step 2. The reduced mass matrix $[\bar{M}_i]$ and the reduced stiffness matrix $[\bar{K}_i]$ are calculated as

$$[\bar{M}_i] = [T_i]^T[M][T] \qquad (4.44)$$

and

$$[\bar{K}_i] = [\bar{D}_i] + \omega_i^2[\bar{M}_i] \qquad (4.45)$$

where the transformation matrix $[T_i]$ is given by eq.(4.43) and the reduced dynamic matrix $[\bar{D}_i]$ is defined in eq.(4.40).

Step 3.   The reduced eigenproblem

$$[[\bar{K}_i] - \omega^2[\bar{M}_i]]\{a_p\} = \{0\} \qquad (4.46)$$

is solved to obtain an improved eigenvalue ω_i^2, its corresponding eigenvector $\{a_p\}_i$, and also an approximation for the next order eigenvalue ω_{i+1}^2.

This three-step process may be applied iteratively. That is, the value of ω_i^2, obtained in Step 3, may be used as an improved approximate value in Step 1, thereby obtaining an improved value of ω_i^2. Experience has shown that one or two such iterations will produce virtually exact eigensolutions. Once an eigenvector $\{a_p\}_i$ for the reduced system is found, the ith modal shape of the system is determined as $\{a\}_i = [T_i] \{a_p\}_i$ using eq.(4.42).

Example 4.4

Repeat Example 4.3 of Section 4.4 using the dynamic condensation method.

Solution

The stiffness matrix and the mass matrix with the coordinates in the order y_1, y_3, y_2, y_4 are given, respectively, by eqs.(a) and (b) of Example 4.3. Substitution of these matrices into eq.(4.39) produces the dynamic matrix for the system:

$$[D] = \begin{bmatrix} 654.70 - \omega_i^2 & 0 & -327.35 & 0 \\ 0 & 654.70 - \omega_i^2 & -327.35 & -327.35 \\ -327.35 & -327.35 & 654.70 - \omega_i^2 & 0 \\ 0 & -327.35 & 0 & 327.35 - \omega_i^2 \end{bmatrix} \quad (a)$$

Step 1.   Assuming no initial approximation of ω_i^2, start Step 1 by setting $\omega_1^2 = 0$ and substituting this value into eq.(a):

$$[D_1] = \begin{bmatrix} 654.70 & 0 & -327.35 & 0 \\ 0 & 654.70 & -327.35 & -327.35 \\ -327.35 & -327.35 & 654.70 & 0 \\ 0 & -327.35 & 0 & 327.35 \end{bmatrix} \quad (b)$$

Application of the Gauss-Jordan elimination process to the first two rows gives

$$\begin{bmatrix} 1 & 0 & -0.5 & 0.0 \\ 0 & 1 & -0.5 & -0.5 \\ 0 & 0 & 327.35 & -163.67 \\ 0 & 0 & -163.67 & 163.67 \end{bmatrix}$$

from which, by eqs.(4.40) and (4.43)

$$[T_i] = \begin{bmatrix} 0.5 & 0.0 \\ 0.5 & 0.5 \\ 1 & 0 \\ 0 & 1 \end{bmatrix} \quad (c)$$

and

$$[\bar{D}_i] = \begin{bmatrix} 327.35 & -163.67 \\ -163.67 & 163.67 \end{bmatrix}$$

Step 2.   The reduced mass and stiffness matrices, eqs.(4.44) and (4.45), are

$$[\bar{M}_1] = [T_1]^T[M][T_1] = \begin{bmatrix} 1.5 & 0.25 \\ 0.25 & 1.25 \end{bmatrix}$$

and

$$[\bar{K}_1] = [\bar{D}_1] + \omega_1^2[\bar{M}_1] = \begin{bmatrix} 327.35 & -163.67 \\ -163.67 & 163.67 \end{bmatrix}$$

Step 3.   The solution of the reduced eigenproblem $[[\bar{K}_1] - \omega^2[\bar{M}_1]]\{Y_p\} = \{0\}$ yields

$$\omega_1^2 = 40.39 \quad \text{and} \quad \omega_2^2 = 365.98 \qquad (d)$$

The value for ω_1^2 may be improved by iteration, that is, by introducing $\omega_1^2 = 40.39$ into eq.(a). This substitution yields

$$[D_1] = \begin{bmatrix} 614.31 & 0 & -327.35 & 0 \\ 0 & 614.31 & -327.35 & -327.35 \\ -327.35 & -327.35 & 614.31 & 0 \\ 0 & -327.35 & 0 & 286.96 \end{bmatrix}$$

Application of the Gauss-Jordan elimination process to the first two rows gives

$$\begin{bmatrix} 1 & 0 & \vdots & -0.533 & 0.0 \\ 0 & 1 & \vdots & -0.533 & -0.533 \\ 0 & 0 & \vdots & 265.44 & -174.44 \\ 0 & 0 & \vdots & -174.44 & 112.53 \end{bmatrix}$$

from which

$$[T_1] = \begin{bmatrix} 0.533 & 0.0 \\ 0.533 & 0.533 \\ 1 & 0 \\ 0 & 1 \end{bmatrix} \qquad (e)$$

and

$$[\bar{D}_1] = \begin{bmatrix} 265.44 & -174.44 \\ -174.44 & 112.53 \end{bmatrix}$$

The reduced mass and stiffness matrices are

$$[\bar{M}_1] = [T_1]^T[M][T_1] = \begin{bmatrix} 1.568 & 0.284 \\ 0.284 & 1.284 \end{bmatrix}$$

and

$$[\bar{K}_1] = [\bar{D}_1] + \omega_1^2[\bar{M}_1] = \begin{bmatrix} 328.76 & -162.97 \\ -162.67 & 164.39 \end{bmatrix}$$

The solution of the reduced eigenproblem

$$[[\bar{K}_1] - \omega^2[M_1]]\{a_p\} = \{0\}$$

yields the eigenvalues

$$\omega_1^2 = 39.48 \quad \text{and} \quad \omega_2^2 = 360.21$$

and the corresponding eigenvectors

$$\{a_p\}_1 = \begin{bmatrix} 0.4283 \\ 0.6562 \end{bmatrix}, \quad \{a_p\}_2 = \begin{bmatrix} 0.6935 \\ -0.6171 \end{bmatrix}$$

The first modal shape for the system is calculated by substituting $\{a_p\}_1$ and $[T_1]$ into eq.(4.42) to obtain

$$\{a\}_1 = \begin{Bmatrix} 0.2283 \\ 0.4283 \\ 0.5780 \\ 0.6562 \end{Bmatrix}$$

The same process is applied now to the second mode, starting by substituting into eq.(a) the approximate eigenvalue $\omega_2^2 = 360.21$ to obtain as a result a very close approximation to the second mode:

$$\omega_2^2 = 328.61$$

or

$$f_2 = \frac{\omega_2}{2\pi} = 2.88 \text{ cps}$$

and

$$\{a\}_2 = \begin{Bmatrix} 0.5789 \\ 0.5760 \\ 0.0 \\ -0.5766 \end{Bmatrix}$$

4.6 MODIFIED DYNAMIC CONDENSATION

The dynamic condensation method essentially requires the application of elementary operations, as is routinely done to solve a linear system of algebraic equations, using the Gauss-Jordan elimination process. The elementary operations are required to transform eq.(4.39) to the form given by eq.(4.40). However, the method also requires the calculation of the reduced mass matrix by eq.(4.44). This last equation involves the multiplication of three matrices of dimensions equal to the total number of coordinates in the system. Therefore, the calculation of the reduced mass matrix $[M]$ requires a large number of numerical operations for a system defined with many degrees of freedom. A modification of the method (Paz 1989) obviates the large number of numerical operations. This modification consists of calculating the reduced stiffness matrix $[\bar{K}]$ only once by simple elimination of s displacements in eq.(4.39) after setting $\omega^2 = 0$, thus making unnecessary the repeated calculation of $[\bar{K}]$ for

each mode using eq.(4.45). Furthermore, it also eliminates the time consumed in calculating the reduced mass matrix $[\bar{M}]$ using eq.(4.44). In the modified method, the reduced mass matrix for any mode i is calculated from eq.(4.45) as

$$[\bar{M}_1] = \frac{1}{\omega_i^2}[[\bar{K}] - [\bar{D}_i]] \qquad (4.47)$$

where $[\bar{K}]$ = the reduced stiffness matrix, already calculated, and $[\bar{D}_i]$ = the dynamic matrix given in the partitioned matrix of eq.(4.40).

The modified algorithm essentially requires only the application of the Gauss-Jordan process to eliminate s unknowns in a linear system of equations such as the system in eq.(4.39), for each eigenvalue calculated.

Example 4.5

Use the modified dynamic condensation method to solve Example 4.3 and compare the results with those obtained in Examples 4.3 and 4.4.

Solution

Table 4.1 shows the values for the first two natural frequencies obtained by static condensation, by dynamic condensation, and by modified dynamic condensation and the comparison of these values with the exact solution.

4.7 MODAL SUPERPOSITION METHOD

The modal superposition method consists of transforming the system of differential equations of motion for multidegree-of-freedom systems into a set of independent differential equations, solving these independent equations, and superimposing the results to obtain the solution of the original system. Thus, the *modal superposition method* reduces the problem of finding the response of a multidegree-of-freedom system to the determination of the response of single-degree-of-freedom systems.

The equations of motion of a building modeled with

Table 4.1. First Two Natural Frequencies for the Structure Modeled in Fig. 4.7.

Mode	Exact Frequencies (cps)	Frequency/(% Error)		
		Static Condensation	Dynamic Condensation	Modified Dynamic Condensation
1	1.000	1.011 (1.1%)	1.000 (0.0%)	1.000 (0.0%)
2	2.888	3.045 (5.4%)	2.885 (0.1%)	2.844 (1.5%)

lateral displacement coordinates at N levels and subjected to seismic excitation at the base may be written, neglecting damping, from eq.(4.9) as

$$[M]\{\ddot{u}\} + [K]\{u\} = -[M]\{1\}\ddot{y}_s(t) \qquad (4.48)$$

In eq.(4.48), $[M]$ and $[K]$ are, respectively, the mass and stiffness matrices of the system, $\{u\}$ and $\{\ddot{u}\}$ are, respectively, the displacement and acceleration vectors (relative to the motion of the base), $\ddot{y}_s(t)$ is the seismic acceleration at the base of the building, and $\{1\}$ is a vector with all its elements equal to 1. Introducing into eq.(4.48) the linear transformation of coordinates

$$\{u\} = [\Phi]\{Z\} \qquad (4.49)$$

in which $[\Phi]$ is the modal matrix, yields

$$[M][\Phi]\{\ddot{Z}\} + [K][\Phi]\{Z\} = -[M]\{1\}\ddot{y}_s(t) \qquad (4.50)$$

The pre-multiplication of this last equation by the transpose of mth mode shape $\{\phi\}_m^T$, gives

$$\{\phi\}_m^T[M][\Phi]\{\ddot{Z}\} + \{\phi\}_m^T[K][\Phi]\{Z\} =$$
$$-\{\phi\}_m^T[M]\{1\}\ddot{y}_s(t) \qquad (4.51)$$

The orthogonality conditions between normalized modes, eqs.(4.17) and (4.18), imply that

$$\{\phi\}_m^T\{m\}[\Phi] = 1$$

and

$$\{\phi\}_m^T[K][\Phi] = \{\omega\}_m^2$$

Consequently, eq.(4.51) reduces to

$$\ddot{Z}_m + \omega_m^2 Z_m = \Gamma_m \ddot{y}_s(t) \qquad (4.52)$$

in which Γ_m (omitting the inconsequential negative sign), is the participation factor for the mth mode given in terms of the weights W_i at the various levels by

$$\Gamma_m = \frac{\displaystyle\sum_{i=1}^{N} W_i \phi_{im}}{\displaystyle\sum_{i=1}^{N} W_i \phi_{im}^2} \qquad (4.53a)$$

For normalized mode shapes, the participation factor reduces to

$$\Gamma_m = \frac{1}{g} \sum_{i=1}^{N} W_i \phi_{im} \qquad (4.53b)$$

because in this case

$$\sum_{i=1}^{N} \frac{W_i}{g} \phi_{im}^2 = 1 \qquad (4.54)$$

where g is the acceleration due to gravity.

Damping may be introduced in the modal equation (4.52) by simply adding the damping term to this equation, namely,

$$\ddot{Z}_m + 2\xi_m \omega_m \dot{Z}_m + \omega_m^2 Z_m = \Gamma_m \ddot{y}_s(t) \qquad (4.55)$$

where ξ_m is the modal damping ratio. Equation (4.55) can be written, for convenience, with omission of the participation factor as

$$\ddot{q}_m + 2\xi_m \omega_m \dot{q}_m + \omega_m^2 q_m = \ddot{y}_s(t) \qquad (4.56)$$

with the substitution

$$Z_m = \Gamma_m q_m \qquad (4.57)$$

4.7.1 Modal Shear Force

The value of the maximum response in eq.(4.56) for the modal spectral acceleration, $S_{am} = (\ddot{q}_m)_{\max}$, is found from an appropriate response spectral chart, such as the charts in Chapter 2 or the spectral chart provided by the Uniform Building Code, UBC-91 (Fig. 4.8).

From eqs.(4.49) and (4.57), the maximum acceleration a_{xm} of the mth mode at the level x of the building is given by

$$a_{xm} = \Gamma_m \phi_{xm} S_{am} \qquad (4.58)$$

in which S_{am} and a_{xm} are usually expressed as multiples of the gravitational acceleration g.

As stated in Chapter 2, the modal values of the spectral acceleration S_{am}, the spectral velocity S_{vm}, and the spectral displacement S_{dm} are related by an apparent harmonic relationship:

$$S_{am} = \omega_m S_{vm} = \omega_m^2 S_{dm}$$

or in terms of the modal period $T_m = 2\pi/\omega_m$ by

$$S_{am} = \frac{2\pi}{T_m} S_{vm} = \left(\frac{2\pi}{T_m}\right)^2 S_{dm}$$

On the basis of these relations, the spectral acceleration S_{am} in eq.(4.58) may be replaced by the spectral displacement S_{dm} times ω_m^2 or by the spectral velocity S_{vm} times ω_m.

The modal lateral force F_{xm} at the level x of the building is then given by Newton's law as

$$F_{xm} = a_{xm} W_x$$

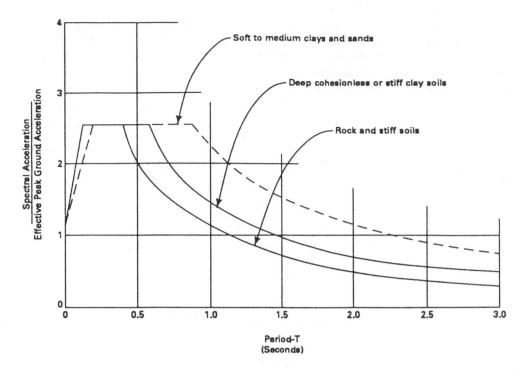

Fig. 4.8. Normalized response spectra shapes. (Reproduced from the *1991 Uniform Building Code*, © 1991, with permission from the publishers, the International Conference of Building Officials)

or by eq.(4.58) as

$$F_{xm} = \Gamma_m \Phi_{xm} S_{am} W_x \tag{4.59}$$

in which S_{am} is the modal spectral acceleration in g units and W_x is the weight attributed to the level x of the building.

The modal shear force V_{xm} at that level is

$$V_{xm} = \sum_{i=x}^{N} F_{im} \tag{4.60}$$

The total modal shear force V_m at the base of the building is then calculated as

$$V_m = \sum_{i=1}^{N} F_{im} \tag{4.61}$$

or using eq.(4.59)

$$V_m = \sum_{i=1}^{N} \Gamma_m \phi_{im} W_i S_{am} \tag{4.62}$$

4.7.2 Effective Modal Weight

The effective modal weight W_m is defined in the equation

$$V_m = W_m S_{am} \tag{4.63}$$

Then, from eq.(4.62), the modal weight is given by

$$W_m = \Gamma_m \sum_{i=1}^{N} \phi_{im} W_i \tag{4.64}$$

Combining eqs.(4.53a) and (4.64) produces the following important expression for the effective modal weight:

$$W_m = \frac{\left[\sum_{i=1}^{N} \phi_{im} W_i \right]^2}{\sum_{i=1}^{N} \phi_{im}^2 W_i} \tag{4.65}$$

It can be proven analytically (Clough and Penzien 1975, pp. 559–560) that the sum of the effective modal weights for all the modes of the building is equal to the total design weight of the building; that is,

$$\sum_{m=1}^{N} W_m = \sum_{i=1}^{N} W_i \tag{4.66}$$

Equation (4.66) is most convenient in assessing the number of significant modes of vibration to consider in the design. Generally, seismic design codes require that all the significant modes of vibration be included when applying the dynamic method of analysis. This requirement can be satisfied by including a sufficient number of modes such that their total effective modal weight is at least 90% of the total design weight of the building. Thus, this requirement can be satisfied by simply adding a sufficient number of effective modal weights [eq.(4.65)] until their total weight is 90% or more of the seismic design weight of the building.

4.7.3 Modal Lateral Force

By combining eq.(4.59) with eqs.(4.63) and (4.64), the modal lateral force F_{xm} may be expressed as

$$F_{xm} = C_{xm} V_m \qquad (4.67)$$

where the modal seismic coefficient C_{xm} at level x is given by

$$C_{xm} = \frac{\phi_{xm} W_x}{\sum\limits_{i=1}^{N} \phi_{im} W_i} \qquad (4.68)$$

4.7.4 Modal Displacement

The modal displacement δ_{xm} at the level x of the building may be expressed, in view of eqs.(4.49) and (4.57), as

$$\delta_{xm} = \Gamma_m \phi_{xm} S_{dm} \qquad (4.69)$$

where Γ_m is the participation factor for the mth mode, ϕ_{xm} is the component of the modal shape at level x of the building, and S_{dm} is the spectral displacement for that mode.

Alternatively, the modal displacement δ_{xm} may be calculated from Newton's law of motion in the form

$$F_{xm} = \frac{W_x}{g} \omega_m^2 \delta_{xm} \qquad (4.70)$$

because the magnitude of the modal acceleration, corresponding to the modal displacement δ_{xm}, is $\omega_m^2 \delta_{xm}$. Hence, from eq.(4.70)

$$\delta_{xm} = \frac{g}{\omega_m^2} \cdot \frac{F_{xm}}{W_x} \qquad (4.71)$$

or substituting $\omega_m = 2\pi/T_m$

$$\delta_{xm} = \frac{g}{4\pi^2} \cdot \frac{T_m^2 F_{xm}}{Wx} \qquad (4.72)$$

where T_m is the mth natural period.

4.7.5 Modal Story Drift

The modal drift Δ_{xm} for the xth story of the building, defined as the relative displacement of two consecutive levels, is given by

$$\Delta_{xm} = \delta_{xm} - \delta_{(x-1)m} \qquad (4.73)$$

with $\delta_{0m} = 0$.

4.7.6 Modal Overturning Moment

The modal overturning moment M_{xm} at the level x of the building, which is calculated as the sum of the moments of the seismic forces F_{xm} above that level, is given by

$$M_{xm} = \sum_{i=x+1}^{N} F_{im}(h_i - h_x) \qquad (4.74)$$

where $x = 0, 1, 2, 3, \ldots, N-1$ and h_i is the height of level i.

The modal overturning moment M_m at the base of the building is given by

$$M_m = \sum_{i=1}^{N} F_{im} h_i \qquad (4.75)$$

4.7.7 Modal Torsional Moment

The modal torsional moment T_{xm} at level x, which is due to eccentricity e_x between the center of the mass of the stories above level x and the center of stiffness the story (normal to the direction considered), is calculated as

$$T_{xm} = e_x V_{xm} \qquad (4.76)$$

where V_{xm} is the modal shear force at level x.

4.8 TOTAL DESIGN VALUES

The design values for the base shear, story shear, lateral displacement, story drift, and overturning and torsional moments are obtained by combining corresponding modal responses. If these modal responses were determined as functions of time (time histories), their combined response would be given by the corresponding transformation such as the one indicated in eq.(4.49) for the displacements at the levels of the building relative to the displacement at its base. However, when response spectral values are used to obtain the maximum response for each mode, the direct superposition of modal maximum responses presents a problem. In general, these modal maximum values will not occur simultaneously as the transformation in eq.(4.49) requires. To overcome this difficulty, it is necessary to use an approximate method. An upper limit for the maximum response may be obtained by adding the absolute values of the maximum modal contributions. Generally, the results obtained by this method will greatly overestimate the maximum response. Another estimate of the maximum response, which is widely accepted and which usually gives a reasonable estimate, is the Square Root of the Sum of the Squared values of the modal contributions (SRSS method). The SRSS estimate of the total response from calculated maximum modal values may be expressed in general form as

$$R = \sqrt{\sum_{m=1}^{N} R_m^2} \qquad (4.77)$$

where R is the estimated response (displacement, force, moment, etc.) at a specified coordinate and R_m is the maximum contribution of mode m at that coordinate.

Application of the SRSS method for combining modal responses generally provides an acceptable estimation of the total maximum response. However, when some of the modes are closely spaced, the use of the SRSS method may either grossly underestimate or overestimate the maximum response. In particular, large errors have been found in the analysis of three-dimensional structures in which torsional effects are significant. The term "closely spaced" may define arbitrarily the case in which the difference between two natural frequencies is within 10% of the smallest of the two frequencies.

A formulation known as the Complete Quadratic Combination (CQC), which is based on the theory of random vibrations, has been proposed by Der Kiureghian (1980) and by Wilson et al. (1981). The CQC method, which may be considered as an extension of the SRSS method, is given by the following equations:

$$R = \sqrt{\sum_{i=1}^{N} \sum_{j=1}^{N} R_i \rho_{ij} R_j} \qquad (4.78)$$

in which the cross-modal coefficient ρ_{ij} may be approximated by

$$\rho_{ij} = \frac{8(\xi_i \xi_j)^{1/2}(\xi_i + r\xi_j)r^{3/2}}{(1-r^2)^2 + 4\xi_i \xi_j r(1+r^2) + 4(\xi_j^2 + \xi_i^2)r^2} \qquad (4.79)$$

where $r = \omega_i/\omega_j$ is the ratio of the natural frequencies of modes i and j, and ξ_i and ξ_j are the corresponding damping ratios for the modes i and j. For constant modal damping ξ eq.(4.79) reduces to

$$\rho_{ij} = \frac{8\xi^2(1+r)r^{3/2}}{(1-r^2)^2 + 4\xi^2 r(1+r)^2} \qquad (4.80)$$

It is important to note that for $i = j$, eq.(4.79) or eq.(4.80) yields $\rho_{ii} = 1$ for any value of the damping ratio, including $\xi = 0$. Thus, for an undamped structure, the CQC estimate [eq.(4.78)] is identical to the SRSS estimate [eq.(4.77)].

4.9 P-DELTA EFFECTS (P-Δ)

The so-called *P-Δ* effect refers to the additional moment produced by the vertical loads and the lateral displacements of columns or other elements of the building resisting lateral forces. Figure 4.9 shows a column supporting an axial compressive force P, a shear force V, and bending moments M_A and M_B at the two ends. Because of this system of loads, the column undergoes a relative lateral displacement, or drift, Δ. This displacement produces a secondary moment, $M_s = P\Delta$, which is resisted by an additional shear force $P\Delta/L$ in the column.

It should be realized that this simple calculation of the *P-Δ* effect involves an approximation because the secondary moment $M_s = P\Delta$ will further increase the drift in the column and, consequently, it will produce an increment of the secondary moment and shear force in the column.

An acceptable estimate of the final drift can be obtained by adding, for each story, the incremental drifts: Δ_x due to the primary overturning moment M_{xp}; $\Delta_x \cdot \Theta_x$ due to the secondary moment $M_{xs} = M_{xp} \cdot \Theta_x$; $\Delta_x \cdot \Theta_x^2$ due to the next incremental moment $M_{xs} \cdot \Theta_x = (M_{xp} \cdot \Theta_x) \cdot \Theta_x$, etc.

Hence, the total story drift Δ_{xTOTAL} including the *P-Δ* effect is given by the geometric series

$$\Delta_{xTOTAL} = \Delta_x + \Delta_x \Theta_x + \Delta_x \Theta_x^2 + \ldots \qquad (4.81)$$

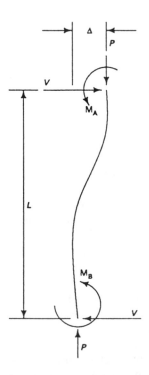

Fig. 4.9. Deflected column showing the *P-Δ* effect

which is equal to

$$\Delta_{xTOTAL} = \Delta_x \left(\frac{1}{1 - \Theta_x} \right) \qquad (4.82)$$

and where $\Theta_x = M_{xs}/M_{xp}$ is the ratio between the secondary and primary moments, M_{xs} and M_{xp}.

Consequently, the *P*-Δ effect may be included by multiplying, for each story, the calculated story drift and the calculated story shear force by the amplification factor $1/(1 - \Theta_x)$ and then recalculating the overturning moments, and other seismic effects, for these amplified story shear forces. Most seismic design codes specify that the resulting member forces and moments, as well as the story drift induced by the *P*-Δ effect, be considered in the evaluation of overall structural frame stability.

The *P*-Δ effect generally need not be considered when the ratio of secondary moment (resulting from the story drift) to the primary moment (due to the seismic lateral forces), for any story, does not exceed 0.10. This ratio may be evaluated at each story as the product of the total dead and live loads above the story, times the seismic drift, divided by the product of the seismic shear force times the height of the story. That is, the ratio Θ_x at level x between the secondary moment M_{xs} produced by the *P*-Δ effect, and the primary moment M_{xp} due to the seismic lateral forces, may be calculated from the following formula:

$$\Theta_x = \frac{M_{xs}}{M_{xp}} = \frac{P_x \Delta_x}{V_x H_x} \qquad (4.83)$$

where

P_x = total weight at level x and above

Δ_x = drift of story x

V_x = shear force of story x

H_x = height of story x

REFERENCES

Structural Dynamics

BATHE, K. J. (1982) "Finite Element Procedures in Engineering Analysis." Prentice-Hall, Englewood Cliffs, NJ.

BERG, GLEN V. (1989) "Elements of Structural Dynamics." Prentice-Hall, Englewood Cliffs, NJ.

BIGGS, J. M. (1964) "Introduction to Structural Dynamics." McGraw-Hill, New York, NY.

BLEVINS, R. D. (1979) "Formulas for Natural Frequency and Mode Shape." Van Nostrand Reinhold, New York, NY.

CHOPRA, A. (1981) "Dynamics of Structures: A Primer, Earthquake Engineering Research Institute." Berkeley, CA.

CLOUGH, R. W., and PENZIEN, J. (1975) "Dynamics of Structures." McGraw-Hill, New York, NY.

DER KIUREGHIAN, A. (1980) "A Response Spectrum Method for Random Vibration." Report No. UCB/EERC-80/15, Earthquake Engineering Research Center, University of California, Berkeley, CA.

GALLAGHER, R. H. (1975) "Finite Element Analysis," p. 115, Prentice-Hall, Englewood Cliffs, NJ.

GUYAN, R. J. (1965) "Reduction of Stiffness and Mass Matrices." AIAA J., 13, 380.

HARRIS, CYRIL M. (1987) "Shock and Vibration Handbook." 3d ed. McGraw-Hill, New York, NY.

NASHIF, A. D., JONES, D. I. C.; and HENDERSON, J. P. (1985) "Vibration Damping." Wiley, New York, NY.

NEWMARK, N. M. (1959) "A Method of Computation for Structural Dynamics." Trans. ASCE, 127, pp. 1406–35.

PAZ, MARIO (1973) "Mathematical Observations in Structural Dynamics." Int. J. Comput. Struct., 3, 385–396.

——— (1983) "Practical Reduction of Structural Problems." J. Struct. Eng., ASCE, 109(111), 2590–2599.

——— (1984a) "Dynamic Condensation." AIAA J., 22(5), 724–727.

——— (1984b) in "Structural Mechanics Software Series." The University Press of Virginia, Charlottesville, Vol. V, pp. 271–286.

——— (1985) "Micro-Computer Aided Engineering: Structural Dynamics." Van Nostrand Reinhold, New York, NY.

——— (1989) "Modified Dynamic Condensation Method." J. Struct. Eng., ASCE, 115(1), 234–238.

——— (1991) "Structural Dynamics: Theory and Computation." Van Nostrand Reinhold, New York, NY.

PAZ, M., and DUNG, L. (1975) "Power Series Expansion of the General Stiffness Matrix for Beam Elements." Int. J. Numer. Methods Eng., 9, 449–459.

WILSON, E. L.; DER KIUREGHIAN, A.; and BAYO, E. P. (1981) "A Replacement for the SRSS Method in Seismic Analysis." Int. J. Earthquake Eng. Struct. Dyn., 9, 187–194.

WILSON, E. L.; FARHOOMAND, I.; and BATHE, K. J. (1973) "Nonlinear Dynamic Analysis of Complex Structures." Int. J. Earthquake and Structural Dynamics, Vol. 1, pp. 241–252.

Earthquake Engineering

BLUME, J. A.; NEWMARK, N. M.; and CORNING, L. (1961) *Design of Multi-story Reinforced Concrete Buildings for Earthquake Motions.* Portland Cement Association, Chicago, IL.

HART, GARY C., and ENGLEKIRK, ROBERT E. (1982) *Earthquake Design of Concrete Masonry Buildings: Response Spectra Analysis and General Earthquake Modeling Considerations.* Prentice-Hall, Englewood Cliffs, NJ.

HOUSNER, G. W. (1970) "Design Spectrum." In *Earthquake Engineering.* R. L. Weigel, ed. Prentice-Hall, Englewood Cliffs, NJ.

HOUSNER, G. W., and JENNINGS, P. C. (1982) *Earthquake Design Criteria*. Earthquake Engineering Institute, Berkeley, CA.

HUDSON, D. E. (1970) "Dynamic Tests of Full Scale Structures." In *Earthquake Engineering*. R. L. Weigel, ed. Prentice-Hall, Englewood Cliffs, NJ.

NAEIM, FARZAD (1989) *The Seismic Design Handbook*. Van Nostrand Reinhold, New York, NY.

NEWMARK, N. M., and HALL, W. J. (1973) *Procedures and Criteria for Earthquake Resistant Design: Building Practices for Disaster Mitigation*. Building Science Series 46, pp. 209–237. National Bureau of Standards, Washington, D.C.

——— (1982) *Earthquake Spectra and Design*. Earthquake Engineering Research Institute, Berkeley, CA.

NEWMARK, N. M., and RIDDELL, R. (1980) "Inelastic Spectra for Seismic Design." Proceedings of Seventh World Conference on Earthquake Engineering, Istanbul, Turkey, Vol. 4, pp. 129–136.

NEWMARK, N. M., and ROSENBLUETH, E. (1971) *Fundamentals of Earthquake Engineering*. Prentice-Hall, Englewood Cliffs, NJ.

POPOV, E. P., and BERTERO, V. V. (1980) "Seismic Analysis of Some Steel Building Frames." *J. Eng. Mech., ASCE* 106: 75–93.

STEINBRUGGE, KARL V. (1970) "Earthquake Damage and Structural Performance in the United States." In *Earthquake Engineering*, R. L. Weigel, ed. Prentice-Hall, Englewood Cliffs, NJ.

WASABAYASHI, MINORU (1986) *Design of Earthquake-Resistant Buildings*. McGraw-Hill, New York, NY.

II
Earthquake-Resistant Design of Buildings: Countries in Seismic Regions

5

Algeria

Abdenour Salhi and Mokhtar Daoudi

5.1 INTRODUCTION

Strong earthquakes with intensities[1] VI to XI have occurred in Algeria almost every year, historically and since records have been kept. The first official regulation for seismic-resistant design of buildings was published by the Dey of Algiers in 1717. This regulation was a response to the devastating earthquake in Algeria in February of 1716, which caused many deaths and much material loss. However, comprehensive regulations for earthquake-resistant design were not issued until 1955; these seismic regulations were prepared after the 1954 earthquake in Chlef, Algeria, which caused the loss of 1,500 lives. Subsequently, the Algerian authorities adapted the French seismic design regulations of 1970.

The current Algerian Seismic Code which is a revised version of the regulations issued in 1981 (RPA-81), was published in 1988 under the title *Regles Parasismiques Algeriennes (RPA-88)*. Both of these regulations contain provisions for the earthquake-resistant design of buildings that are similar to the recommendations of the Structural Engineering Association of California (SEAOC-1988) and to the provisions of the Uniform Building Code (International Conference of Building Officials, 1988); they were prepared at the Blume Center of Stanford University. The RPA-88 version of the Seismic Code includes special provisions for seismic design of masonry buildings. The consideration of nonlinear behavior of structures during strong motion earthquakes is the most significant improvement in the RPA-88 Code. In this behavior, the integrity of the structure is preserved by designing the structure with an appropriate level of ductility, that is, with the capacity to dissipate energy during inelastic deformations.

5.2 EARTHQUAKE-RESISTANT DESIGN METHODS

The Algerian Seismic Code contains detailed provisions for the design of buildings by the static lateral

---

[1]Earthquake intensities are given in the Modified Mercalli Intensity (MMI) scale. (See Appendix on Magnitude and Intensity of Earthquakes.)

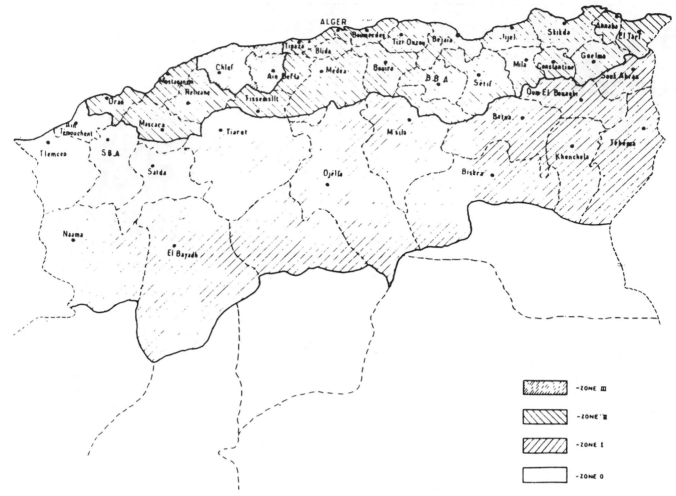

Fig. 5.1. Seismic zoning of Algeria

force method, but only general recommendations for the application of a dynamic method of seismic design. The static lateral force method is applicable to regular structures not higher than 30 m, located in Seismic Zone III (seismic zones for Algeria are shown in Fig. 5.1) and those for structures in Seismic Zones I and II not higher than 60 m. The dynamic method may be used for any structure but *must* be used for irregular structures or for structures that are higher than the limits indicated for the use of the equivalent static lateral force method. Consideration also must be given to foundation support conditions. For example, sites on which soils may liquefy or show resonant behavior with the structure during earthquakes must be examined with detailed dynamic and geotechnical analyses.

The following conditions must be met if a building is to be classified as a regular structure:

(a) The plan dimensions of the building may decrease for higher stories, but by not more than 25% between any two consecutive levels of the building.

(b) The 25% change limitation between consecutive stories also should not be exceeded for the eccentricity in any story of the building (eccentricity is defined as the distance between the mass center and the stiffness center for any level of the building).

(c) The building should have about the same degree of damping for each story.

(d) The building should have a simple and essentially symmetric shape.

Structures that do not conform to the restrictions of regularity should be designed by a dynamic method such as the use of the modal superposition and response spectra analysis, or a time-history analysis with an appropriate geological study of the site.

The equivalent static lateral force method is based on the determination of equivalent forces that are applied statically at the various levels of the building. The evaluation of these equivalent forces and their effect on the building should be performed independently along the two orthogonal main directions of the building. In the Algerian Seismic Code, the magnitude of the equivalent lateral forces are determined for an earthquake of moderate intensity. A structure designed on this basis must possess enough ductility to dissipate large amounts of energy during a major earthquake. Chapter 4 of the Algerian Seismic Code contains requirements for detailing that are intended to provide the ductility implicitly assumed by the Code. Extensive appendices to the Code are provided to indicate requirements for construction techniques, determination of fundamental period, important earthquake events, seismic classification of Algerian cities and towns, and a detailed explanation of the dynamic response spectrum analysis method.

5.3 SEISMIC ZONES

The seismic hazard map of Algeria (Fig. 5.1) divides the territory of the country into four seismic zones as follows:

Zone 0: Negligible seismic effects
Zone I: Low seismicity
Zone II: Average seismicity
Zone III: High seismicity

5.4 BUILDING CATEGORIES

The degree of safety for each type of structure, required by the Code, varies according to the type of building occupancy. Three risk periods, 500, 100, and 50 years, are assigned to structures in the following categories:

Category 1 (seismic design for 500-year return earthquake)
Structures required for emergency medical, economic, social, or safety needs in the event of a major earthquake:

● General hospitals and other medical facilities
● Fire and police stations
● Power plants
● Communication centers
● Airports and railroad stations
● Covered structures occupied by large groups of people (large schools, universities, religious centers, cinemas and auditoriums, etc.)
● Important water towers and reservoirs
● Structures housing, supporting, or containing sufficient quantities of toxic or explosive substances that would be dangerous if released.

Category 2 (seismic design for 100-year return earthquake)
Structures that are of average importance:

● Small school buildings, nursing homes, dispensaries
● Administrative buildings
● Water towers
● Commercial and industrial buildings
● Houses

Category 3 (seismic design for 50-year return earthquake)
The collapse of these structures would have minor consequences for everyday life:

● All structures with occupancies or functions not listed above.

5.5 SEISMIC COEFFICIENT A

The seismic coefficient *A* expressed as a fraction of the acceleration of gravity is a function of both the Seismic Zone and Building Category as given in Table 5.1.

5.6 LOAD COMBINATIONS

In the seismic design of structures, the most unfavorable of the following load combinations should be used:

$$L = G + Q + E \qquad (5.1a)$$
$$L = 0.8G + E \qquad (5.1b)$$

where

G = dead load;
Q = live load; and
E = earthquake load

For columns of a structural frame, the first load combination in eq.(5.1) is replaced by

$$L = G + Q + 1.2E \qquad (5.1c)$$

Table 5.1. Seismic Coefficient *A*

Building Category	Seismic Zones		
	I	II	III
1	0.12	0.25	0.35
2	0.08	0.15	0.25
3	0.05	0.10	0.15

In determining the most unfavorable load combination, earthquake loads should be considered to act both positively and negatively in any two orthogonal directions.

5.7 SEISMIC WEIGHT *W*

The seismic weight *W* of the structure includes the dead weight of the building and permanent equipment, and a fraction of the live load, as indicated in Table 5.2.

5.8 FUNDAMENTAL PERIOD *T*

The Algerian Seismic Code (Annex II) states that the fundamental period *T* of the structure may be calculated using theoretical methods [Rayleigh's formula, for example, see eq.(3.14) of Chapter 3] or estimated by using an empirical formula.

For buildings with frames designed to resist lateral forces without shear walls,

$$T = 0.1N \sec \tag{5.2}$$

where *N* is the number of stories.

For other types of buildings,

$$T = \frac{0.09h}{\sqrt{L}} \tag{5.3}$$

where

h = height of the building (m); and
L = building length in the direction of the seismic action (m)

In the calculation of the fundamental period, the value of *D* (Section 5.10) obtained by using any theoretical formula must not be less than 80% of the value of *D* obtained by the use of eq.(5.2) or eq.(5.3).

5.9 QUALITY FACTOR Q

The quality factor *Q* of a structural lateral force-resistant system is determined by

$$Q = 1 + \sum_{q=1}^{6} P_q \tag{5.4}$$

Table 5.2. Live Load Multiplying Factor

Description	Live Load Multiplying Factor
• Record office • Silo, grain elevator • Water tank • Structures with long-duration live loads	1.0
• Covered structures with public occupancy • Goods warehouses, stores, markets	0.5
• Other buildings (residential buildings, office buildings, schools . . .)	0.2

Table 5.3. Penalty Factor P_q

Criteria	P_q value Criteria Observed	P_q value Criteria not Observed
1 – Minimal requirements of structural design	0	0.05
2 – Shape redundancy in plan	0	0.05
3 – Shape symmetry in plan	0	0.05
4 – Elevation regularity	0	0.05
5 – Materials quality control	0	0.05
6 – Construction quality control	0	0.10

Table 5.4. Mean Dynamic Amplification Factor *D*

Fundamental Period T (sec)	Firm Soil	Soft Soil
≤0.3	2.00	2.00
0.4	1.65	2.00
0.5	1.42	2.00
0.6	1.26	1.77
0.7	1.14	1.59
0.8	1.04	1.46
0.9	0.96	1.35
1.0	0.896	1.26
1.1	0.841	1.18
1.2	0.793	1.12
1.3	0.752	1.06
1.4	0.716	1.00
1.5	0.684	0.96
1.6	0.655	0.921
1.7	0.629	0.880
1.8	0.606	0.852
1.9	0.584	0.821
≥2.0	0.564	0.794

where the penalty factors P_q, depend on the ductility of the structure, the degree of redundancy in the lateral-resistant system, the regularity between stories, the plan symmetry of the structure, and the design and construction quality control. Table 5.3 provides values for the penalty factors P_q.

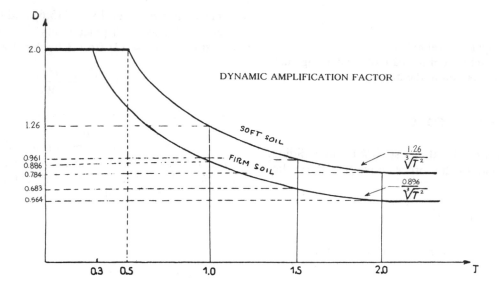

Fig. 5.2. Dynamic Amplification Factor D

5.10 MEAN DYNAMIC AMPLIFICATION FACTOR D

The mean dynamic amplification factor D is obtained from spectral acceleration curves provided by the code (Fig. 5.2) or from values given in Table 5.4. The Algerian regulations give spectral acceleration curves for two types of soils, firm soil and soft soil, and for a damping value that is equal to 10% of critical damping. This damping value is considered appropriate for most buildings in Algeria.

The value of D varies only slightly for large values of T; thus for large values of T, small errors in estimation of T do not result in much of a change in the design seismic forces.

5.11 BEHAVIOR FACTOR B

The behavior factor B depends on the type of lateral-resistant system used in the structure. Values of the behavior factor B are given in Table 5.5

Table 5.5. Behavior Factor B

No.	Structural System	B
1	Reinforced concrete buildings composed of shear walls interacting with frames (dual system)	0.20
2	Steel frame structure	0.20
3	Reinforced concrete frame structure	0.25
4	Columns and beams with walls to resist lateral forces	0.25
5	Reinforced concrete shear walls	0.34
6	Reinforced masonry shear walls	0.67
7	• Inverted pendulum type structures • Water tower supported by columns	0.67
8	Other structures	0.50

5.12 BASE SHEAR FORCE

The total shear force V, at the base of the building, due to seismic action is calculated by

$$V = ADBQW \tag{5.5}$$

where

A = seismic coefficient (Table 5.1)
D = mean dynamic amplification factor (Table 5.4)
B = behavior factor (Table 5.5)
Q = quality factor [eq.(5.4)]
W = seismic weight (Section 5.7)

5.13 EQUIVALENT LATERAL FORCES

The base shear force V induces seismic forces F_k at the various levels of the building according to the following formula:

$$F_k = \frac{(V - F_t)\, W_k h_k}{\sum_{i=1}^{N} W_i h_i} \tag{5.6}$$

with an additional force F_t at the top of the building given by

$$F_t = 0 \qquad\qquad \text{if } T \leq 0.7 \text{ sec, or} \tag{5.7}$$
$$F_t = 0.07TV (\leq 0.25V) \qquad \text{if } T > 0.7 \text{ sec}$$

where

W_k = seismic weight at level k;
h_k = height of level k from the base of the building; and
T = fundamental period of the building

5.14 STORY SHEAR FORCE

The story shear force V_k at any story k of the building, is given by the sum of the forces above that story; that is,

$$V_k = F_t + \sum_{i=k}^{N} F_k \qquad (5.8)$$

where F_t and F_k are given, respectively, by eqs.(5.6) and (5.7).

5.15 OVERTURNING MOMENTS

The Algerian regulations for seismic design of buildings require that the structure be designed to support the overturning moments induced at the various levels by the equivalent lateral forces F_k and F_t. The overturning moment at level k is determined by statics as

$$M_k = F_t(h_N - h_k) + \sum_{i=k+1}^{N} F_i(h_i - h_k) \qquad (5.9)$$

where $k = 0, 1, 2, \ldots, N-1$ and N is the number of levels in the building.

5.16 TORSIONAL MOMENTS

The torsional moment at a given story of the building is caused by the eccentricity between the center of the mass above that story and the center of stiffness of the resisting elements in that story. The Code stipulates a minimum design eccentricity of 5% of the largest plan dimension of the building. Thus, the torsional moment M_{tk}, at level k of the building, is given by

$$M_{tk} = V_k e_k \qquad (5.10)$$

where e_k is the actual value or the design minimum value for the eccentricity at level k, and V_k is the story shear force.

5.17 LATERAL DISPLACEMENTS

The lateral elastic displacements at the various levels of the building may be determined by elastic analysis of the structure, which is acted upon by the seismic

forces F_t and F_k given by eqs.(5.6) and (5.7). Alternatively, the elastic displacements X_{ek} may be determined using eq.(4.72) from Chapter 4 as

$$X_{ek} = \frac{g}{4\pi^2} \frac{T^2 F_k}{W_k} \qquad (5.11)$$

in which the additional force F_t is included in F_k for the top level of the building.

The Algerian regulations for seismic design stipulate that the design lateral displacements X_k be calculated by multiplying the elastic displacements X_{ek} by $1/(2B)$ (B is the behavior factor given in Table 5.5) with the provision that $1/(2B)$ should not be less than 2.0.

5.18 STORY DRIFT

The story drift, or relative displacement, Δ_i, between the upper and lower levels of story i, is calculated by

$$\Delta_i = X_i - X_{i-1} \qquad (5.12)$$

$$\text{with } X_0 = 0$$

where X_i and X_{i-1} are, respectively, the lateral displacements for level i and for level $i-1$. The Code mandates that the story drift be limited to a maximum value of 0.0075 times the interstory height.

Example 5.1

A five-story reinforced concrete building with three longitudinal and five transverse frames, shown in Fig. 5.3, is used to illustrate the application of the Algerian regulations (RPA-88) for seismic-resistant design. The cross-section of every column is 40 cm × 40 cm, that of the beams is 0.30 cm × 0.35 cm. The seismic weight at each level is 840 kN. The building is located in Seismic Zone III on a site with firm soil. Anticipated use of the building is for offices.

Solution

1. Fundamental period T:
   $T = 0.1N = 0.1 \times 5 = 0.5$ sec   [eq.(5.2)]
2. Seismic coefficient A:
   Category 3 (Section 5.4)
   $A = 0.15$ (from Table 5.1)
3. Mean dynamic amplification factor D
   $D = 1.42$ (from Table 5.4 with $T = 0.5$, firm soil)
4. Behavior factor B:
   $B = 0.25$ (from Table 5.5, reinforced concrete frame system)
5. Quality factor Q:

$$Q = 1 + \sum_{q=1}^{6} P_q \qquad \text{[eq.(5.4)]}$$

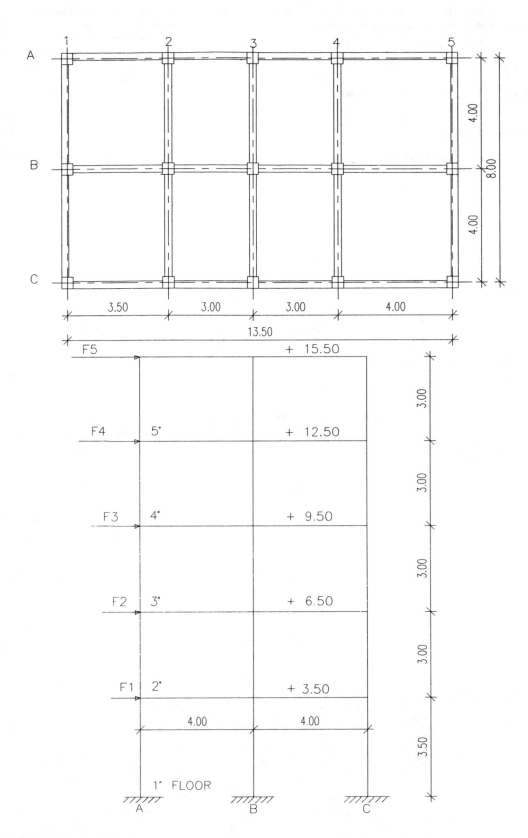

Fig. 5.3. Five-Story Building for Example 5.1

$Q = 1.05$ (using Table 5.3, assumed all criteria observed, except for materials quality control)

6. Base shear force

$V = ADBQW$ [eq.(5.5)]

$= 0.15 \times 1.42 \times 0.25 \times 1.05 \times (5 \times 840)$

$= 235 \text{ kN}$

7. Equivalent lateral forces

$$F_k = \frac{(V - F_t)\,W_k h_k}{\displaystyle\sum_{i=1}^{N} W_i h_i}$$ [eq.(5.6)]

$F_t = 0 \quad \text{for} \quad T = 0.5 < 0.7\,\text{sec}$ [eq.(5.7)]

The calculation of the equivalent lateral forces is shown in Table 5.6.

8. Story shear forces

$$V_k = F_t + \sum_{i=k}^{N} F_k$$ [eq.(5.8)]

The results are in Table 5.6.

9. Overturning moments

$$M_k = F_t(h_N - h_k) + \sum_{i=k+1}^{N} F_i(h_i - h_k)$$ [eq.(5.9)]

where $k = 0, 1, 2, \ldots, N-1$.

Table 5.6. Calculation of Seismic Forces, Story Shear Forces, and Overturning Moments

Level	h_k (m)	$h_k/\Sigma h_i$	F_k (kN)	V_k (kN)	M_k (kN-m)
5	15.5	0.326	76.67	76.67	0
4	12.5	0.263	61.85	138.52	230.01
3	9.5	0.200	47.04	185.56	645.57
2	6.5	0.137	32.22	217.78	1,202.25
1	3.5	0.0737	17.33	235.11	1,855.59
0	0.0	—	—	—	2,678.38

The results are in Table 5.6.

REFERENCES

INTERNATIONAL CONFERENCE OF BUILDING OFFICIALS (1988) *Uniform Building Code (UBC)*. Whittier, CA.

Regles Parasismiques Algeriennes (1981) *RPA-81* (Regulations of Algerian Seismic Code) Publication OPU, Algiers, Algeria.

Regles Parasismiques Algeriennes (1988) *RPA-88* (Regulations of Algerian Seismic Code) Publication OPU, Algiers, Algeria.

STRUCTURAL ENGINEERING ASSOCIATION OF CALIFORNIA (SEAOC) (1988) *Recommended Lateral Force Requirement and Tentative Commentary*. San Francisco, CA.

6

Argentina

Fernando A. M. Reyna, Bibiana M. Luccioni, and Ricardo D. Ambrosini*

6.1 INTRODUCTION

6.1.1 Seismic Activity in Argentina

The Republic of Argentina is located between the Atlantic Ocean and the Andes mountain range along the southeastern coast of South America. The seismic activity in Argentina is concentrated mostly in the western and northwestern parts of the country; this area of the Andes, part of the Pacific Ring, has been subjected to a series of violent earthquakes.

On April 14, 1927, the city of Mendoza suffered severe damage from an earthquake that registered 7.4 on the Richter scale. In the North, three other earthquakes with magnitudes greater than 7.0 occurred in the province of Salta, and another in the province of San Juan. In July 1934, an earthquake destroyed the city of Sampacho, in the province of

*The authors are very grateful to Dr. Rodolfo Danesi, director of the Structures Laboratory of the National University of Tucumán, Argentina, for the support given during the development of this work, and to Dr. Mario Paz, Professor, University of Louisville, Louisville, Kentucky, U.S.A., for the encouragement and information supplied for the preparation of this chapter.

Córdoba. On January 15, 1944, another strong earthquake registering 7.8 on the Richter scale, destroyed the city of San Juan.

May 21, 1960, marked the beginning of a series of earthquakes that occurred south of the 36th parallel in Chile. The earthquake of May 31, 1960 (magnitude 8.5 on the Richter scale), which was strongly felt in the western part of Argentina, has to be considered a consequence of the Chilean earthquakes. In the same year, another strong earthquake (magnitude 7.4 on the Richter scale) caused near total destruction of the city of Caucete, the epicenter of this earthquake, and resulted in extensive damage to the neighboring city of San Juan.

6.1.2 Seismic Codes in Argentina

In 1972, the seismic code CONCAR 70 (*Normas Antisísmicas CONCAR 70* 1972), was adopted in Argentina. A new seismic code, NAA 80 (*Normas Antisísmicas Argentinas – NA* 1980), was adopted in 1981 by a resolution of The National Institute for Seismic Prevention (Instituto Nacional de Prevención Sísmica, INPRES). In the same year, the collaboration between INPRES and the Centro de Investigación de los Reglamentos Nacionales de Seguridad para las Obras Civiles (CIRSOC) resulted in a new seismic code, Reglamento INPRES-CIRSOC 103 (*Normas Argentinas para las Construcciones Sismorresistentes* 1983). This new code incorporated internationally accepted and recommended concepts for earthquake-resistant design.

The INPRES-CIRSOC 103 Code has three parts: Part 1 – Constructions in General (Construcciones en General), Part 2 – Reinforced and Prestressed Concrete Constructions (Construcciones de Hormigón Armado y Pretensado), and Part 3 – Masonry Constructions (Construcciones de Mampostería). This code established the minimum requirements for the design of structures located in seismically active regions. The Code's use is mandatory for all public structures built in the national territory. Recently, a supplement was added to the Code. This supplement contains modifications that facilitate the use of the Code and adapt the norms to the technological and economic conditions in Argentina. This new document (INPRES-CIRSOC 103 1991) is now part of the Sistema Reglamentario Argentino (SIREA).

6.2 SEISMIC CODE (INPRES-CIRSOC 103)

6.2.1 Seismic Zones

The Republic of Argentina is divided into five seismic zones (Fig. 6.1) that are defined according to

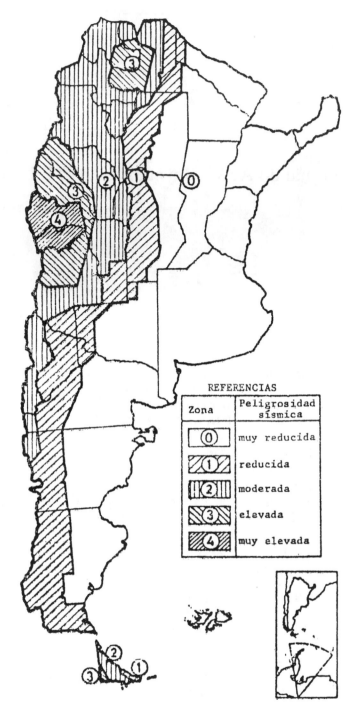

Zona	Peligrosidad sísmica
⓪	muy reducida
①	reducida
②	moderada
③	elevada
④	muy elevada

REFERENCIAS

Fig. 6.1.   Seismic zones of the Republic of Argentina

their seismic risk, as indicated in Table 6.1. The Seismic Code INPRES-CIRSOC 103 of 1983 is mandatory in Seismic Zones 1, 2, 3, and 4.

For Zone 0, the provisions of the Code must be applied to public structures whose failure could produce a catastrophic effect on the population and to those structures that are required for national security. In general, the seismic provisions of the Code are considered to be satisfied if the structure contains

Table 6.1. Seismic Risk Assigned to Seismic Zones in Argentina

Seismic Zone	Seismic Risk
0	very low
1	low
2	moderate
3	high
4	very high

structural elements in vertical planes, in two perpendicular directions, that are resistant to horizontal forces; in addition, these structural elements should conform to a torsional-resistant mechanism.

6.2.2 Building Classification

Buildings are classified according to their function during and after a seismic event:

Group A_o. Includes buildings that play a vital role after a destructive event; they are essential in providing services to the population (for example, military centers, police stations, hospitals, fire stations, communication centers, and power stations);

Group A. Includes buildings that are important in the manufacture of goods and in the maintenance of security and order in the city and those buildings that provide essential services to the population that are not included in Group A_o (for example, governmental centers, military and police buildings, hospitals and health care centers, educational facilities, movie theaters, stadiums, big hotels);

Group B. Includes buildings whose damage would result in losses of intermediate magnitude (for example, private houses and all the public buildings or commercial and industrial centers that are not included in group A);

Group C. Includes buildings that, if damaged, would cause almost no economic loss or could not harm buildings included in previous categories (for example, containers, stables, sheds, silos).

6.2.2.1 Risk factor. Table 6.2 establishes the *risk factor* γ_d, corresponding to the different types of buildings.

It is not required to perform seismic analysis of structures in Group C. However, the Code requires that arrangements and construction details that help provide seismic protection must be considered in the design and construction of structures in this group.

Table 6.2. Risk Factor

Building Type	γ_d
Group A_o	1.4
Group A	1.3
Group B	1.0

6.2.3 Soil Local Conditions

The code classifies soils as dynamically stable and dynamically unstable because the seismic response of a structure is strongly influenced by the local conditions of the soil.

(a) *Dynamically Stable Soils*
Dynamically stable soils are subdivided as follows:
Type I: Very firm and compact (firm rocks or similar formations, and stiff soils above firm rock with a depth less than 50 m).
Type II: Soils with intermediate stiffness (firm soils with a depth greater than 50 m or soils with intermediate characteristics of stiffness with a depth greater than 8 m).
Type III: Soft soils (loose granular soils, soft and medium-stiff cohesive soils).

(b) *Dynamically Unstable Soils*
The Code assigns special measures for soils that could become unstable under a seismic load (for details see the INPRES-CIRSOC 103, 1983).

6.2.4 Seismic Activity and Design Spectra

Seismic actions are basically defined by acceleration or pseudo-acceleration spectra whose shape and magnitude depend on the seismic zone and the type of foundation soil.

6.2.4.1 Horizontal seismic spectra. The Code provides the following analytical expressions for the elastic spectral pseudo-accelerations S_a (in g):

$$S_a = a_s + (b - a_s)\frac{T}{T_1} \quad \text{for} \quad T \leqslant T_1 \qquad (6.1a)$$

$$S_a = b \quad \text{for} \quad T_1 \leqslant T \leqslant T_2 \qquad (6.1b)$$

$$S_a = b\left[\frac{T_2}{T}\right]^{2/3} \quad \text{for} \quad T \geqslant T_2 \qquad (6.1c)$$

where T is the fundamental period of the structure and where the values of the parameters a_s, b, T_1, and T_2 are given in Table 6.3 for the different seismic zones and types of soil. A plot of the pseudo-acceleration spectra for Seismic Zone 1 with 5% damping is shown in Fig. 6.2.

The response spectra defined in eq.(6.1) are based on a damping value of no less than 5% of the critical damping. For structures having a damping value smaller than 5%, the following expressions should be used:

$$S_a = a_s + (f_A b - a_s)\frac{T}{T_1} \quad \text{for} \quad T \leqslant T_1 \qquad (6.2a)$$

$$S_a = f_A b \quad \text{for} \quad T_1 \leqslant T \leqslant T_2 \qquad (6.2b)$$

$$S_a = \left[1 + (f_A - 1)\frac{T_2}{T}\right]\left[b\left[\frac{T_2}{T}\right]^{2/3}\right] \quad \text{for } T \geqslant T_2 \qquad (6.2c)$$

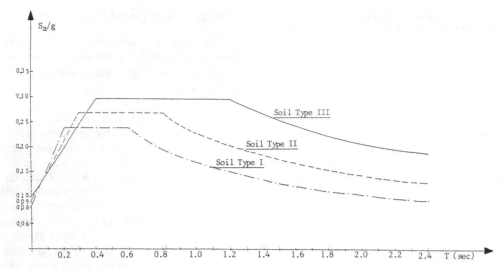

Fig. 6.2. Elastic pseudo-acceleration spectra for Seismic Zone 1 with damping $\xi = 5\%$

where f_A, the amplification factor due to damping, is calculated by

$$f_A = \sqrt{\frac{5}{\xi}} \quad \text{for } 0.5\% \leqslant \xi \leqslant 5\% \qquad (6.3)$$

in which ξ is the relative damping expressed as a percentage of the critical damping.

6.2.4.2 Vertical seismic actions. When it is necessary to consider the vertical seismic actions, the corresponding acceleration spectra S_{av} is obtained from the horizontal spectra acceleration S_a as

$$S_{av} = f_v S_a \qquad (6.4)$$

where the factor f_v is given in Table 6.4.

Table 6.3. Values of a_s, b, T_1, and T_2 for Different Seismic Zones and Types of Soil

Seismic Zone	Soil Type	a_s	b	T_1	T_2
4	I	0.35	1.05	0.20	0.35
	II	0.35	1.05	0.30	0.60
	III	0.35	1.05	0.40	1.00
3	I	0.25	0.75	0.20	0.35
	II	0.25	0.75	0.30	0.60
	III	0.25	0.75	0.40	1.00
2	I	0.16	0.48	0.20	0.50
	II	0.17	0.51	0.30	0.70
	III	0.18	0.54	0.40	1.10
1	I	0.08	0.24	0.20	0.60
	II	0.09	0.27	0.30	0.80
	III	0.10	0.30	0.40	1.20
0	I	0.04	0.12	0.10	1.20
	II	0.04	0.12	0.10	1.40
	III	0.04	0.12	0.10	1.60

6.2.5 Reduction Factor

The capacity of a structure to dissipate energy by inelastic deformations is taken into account by a reduction factor applied to the ordinate of the elastic spectra. The reduction factor R which depends on the global ductility μ of the structure and on its fundamental period T, is given by

$$R = 1 + (\mu - 1)\frac{T}{T_1} \quad \text{for } T \leqslant T_1 \qquad (6.5a)$$

$$R = \mu \quad \text{for } T \geqslant T_1 \qquad (6.5b)$$

The value of the global ductility μ established by the Code corresponds to regular structures in elevation, in which it is assumed that the structure uniformly dissipates energy during inelastic deformations. Values of the global ductility μ are given in Table 6.5 for different types of structures.

6.2.6 Seismic Weight

The seismic weight considered for seismic analysis includes the dead weight of the building and a fraction of the live load. Thus, the seismic weight W_i at each level i of the building is determined as

$$W_i = G_i + \eta L_i \qquad (6.6)$$

Table 6.4. Values of the Factor f_v According to the Seismic Zone

Seismic Zone	f_v
4	0.6
3	0.6
2	0.5
1	0.4
0	0.4

Table 6.5. Values of the Global Ductility (μ)

System	μ
Ductile steel frames; ductile coupled shear walls of reinforced concrete.	6
Ductile reinforced concrete; Ductile reinforced concrete frames combined with shear walls with the frames carrying more than 30% of the shear force.	5
Conventional steel frames; systems of ductile reinforced concrete shear walls joined by beams.	4
Frames-shear wall systems or ductile shear walls of reinforced concrete that do not fulfill the previous conditions; chained reinforced masonry walls with strong bricks; reinforced masonry walls with distributed reinforcement.	3.5
Chained masonry walls with strong bricks; inverted pendulum-type structures with special join and support design details blocks; inverted pendulum-type structures that do not satisfy above requirements; hanging structures; reinforced concrete columns without links in the anlyzed direction.	3
Chained masonry walls with hollowed blocks; inverted pendulum-type structures that do not satisfy above requirements; hanging structures; reinforced concrete columns without links in the analyzed direction.	2
Structures for which an elastic behavior is required during destructive earthquakes.	1

where G_i and L_i are, respectively, the dead load and the design live load for level i, and where the load factor η is given in Table 6.6.

6.2.7 Load States

The values of seismic actions defined in the Code are considered ultimate values. The most unfavorable combination of effects corresponding to the following load combinations must be considered for the seismic analysis and design of buildings:

$$1.3E_w \pm E_s$$

$$0.85E_w \pm E_s$$

where

E_w = actions due to gravitational loads
E_s = seismic actions

6.2.8 Directional Loads

Structures shall be analyzed for seismic horizontal actions acting independently in two orthogonal directions, and for the seismic vertical action when this vertical action is significantly important. Regular buildings must be designed for the most unfavorable combination of the effects produced by gravity loads

Table 6.6. Load Factor η for Different Probabilities of Full Live Load

Probability of Full Live Load	Examples	η
Very low	Roofs	0
Low	Hotels, offices, and residential buildings	0.25
Intermediate	(a) Schools, theaters, movies, public buildings, etc.	0.50
	(b) Load due to accumulation of snow and ice	0.50
High	Warehouses, libraries, etc.	0.75
Very high	(a) Liquid tanks, silos, etc.	1.00
	(b) Critical structural elements such as overhanging elements and balconies	1.00

and one horizontal seismic component. The most unfavorable values obtained by the combination of the effects produced by gravity loads, one horizontal seismic component according to one direction, and 30% of the other horizontal seismic component on the other direction must be used for irregular buildings. When the structure is approximately symmetric with respect to one axis, one of the directions of the seismic analysis must correspond to that axis. One of the following situations should be adopted when the structure has no axes of symmetry:

(a) Select any arbitrary two orthogonal directions and use an increase of 15% of the seismic action for the analysis.

(b) Select two sets of orthogonal directions rotated 45° with respect to each other and consider the most unfavorable situations in relation to the effects stated previously.

6.2.9 Dynamic Characteristics of the Structure

The dynamic characteristics of a structure are determined assuming that the structure behaves in the elastic range of deformations.

6.2.9.1 Fundamental period. The fundamental period of a structure is obtained using conventional procedures of structural dynamics, tests performed on similar structures, or empirical relationships.

6.2.9.1.1 APPROXIMATE FORMULA. When the structure is fixed at its base and when the masses are assumed to be concentrated at the floor levels of the building, the fundamental vibration period T_o may be obtained using the Rayleigh formula as

$$T_o = 2\pi \sqrt{\frac{\sum_{i=1}^{n} W_i u_i^2}{g \sum_{i=1}^{n} \bar{F}_i u_i}} \tag{6.7}$$

where

W_i = gravity load located at level i
g = acceleration of gravity
u_i = displacement at level i, due to the normalized horizontal forces
$\overline{F}_i$ = normalized horizontal force defined by eq.(6.8), applied at level i:

6.2.9.1.2 NORMALIZED HORIZONTAL FORCES. The equivalent seismic forces, known as normalized horizontal forces, applied at the various levels of the building are given by

$$\overline{F}_i = \frac{W_i h_i}{\sum_{i=1}^{n} W_i h_i} \qquad (6.8)$$

where

W_i = seismic weight at level i
h_i = height of level i above the base level[1]
n = number of levels in the building.

For regular buildings with equal distance between levels, the natural period of vibration may be calculated as

$$T_o = 2\pi \sqrt{\frac{W_n u_n}{g \overline{F}_n}} \qquad (6.9)$$

6.2.9.1.3 EMPIRICAL FORMULA. The following empirical formula is suggested in the Code as an alternative procedure to calculate the fundamental period of a structure:

$$T_{oe} = \frac{h_n}{100} \sqrt{\frac{30}{L} + \frac{2}{1 + 30d}} \qquad (6.10)$$

where

T_{oe} = fundamental period of vibration of the structure in the direction considered (in seconds)
h_n = height of the building from the base level (in meters)

L = length of the building in the direction considered (in meters)
d = density of the structural walls in the direction considered (ratio between the cross-sectional area of these walls and the plan area of the building).

The Code establishes that the value for the fundamental period used in the calculations must not be greater than $1.25T_{oe}$ for structures located in Seismic Zones 3 and 4, and not greater than $1.5T_{oe}$ for structures in other seismic zones.

6.2.9.2 Damping value. The damping value ξ, expressed as a percentage of the critical damping, depends on the type of structure. It varies from 1% for steel structures to 5% for common reinforced concrete structures.

6.2.10 Deformations

Deformation analysis in structures subjected to seismic loads is associated with the following problems:

- Damage in non-structural elements
- Problems of stability and strength
- Hammering between adjacent structures.

In order to avoid these potential problems, the following parameters must be controlled:

- Story drift
- P-Δ effects
- Separation of adjacent buildings.

6.2.10.1 Lateral displacement and story drift. The lateral displacements of a building may be determined using a computer program that implements the stiffness method for analysis of plane frames. These displacements[2] are determined by applying to the building the seismic lateral forces F_i [eqs.(6.21) and (6.22)] multiplied by the global ductility μ (Table 6.5).

---

[1]The base level is considered to be the union of the structure and the foundation; otherwise it is the horizontal plane from which considerable displacements are produced due to seismic action.

---

[2]Editor's note: Alternatively, the displacement δ_i at the level i of the building may be determined using eq.(4.72), Chapter 4, as

$$\delta_i = \frac{g}{4\pi^2} \cdot \frac{T^2 F_i}{W_i} \qquad (6.11)$$

where T is the fundamental period; g the acceleration of gravity; F_i the seismic lateral force, [eqs.(6.21) and (6.22)], and W_i the seismic weight at level i.
    The drift Δ_i for story i is then given by

$$\Delta_i = \delta_i - \delta_{(i-1)} \qquad (6.12)$$

Table 6.7. Limited Values for Story Drift

Condition*	Building Group		
	A_o	A	B
D	0.010	0.011	0.014
ND	0.010	0.015	0.019

*Condition D: nonstructural elements may be damaged during structural deformations
Condition ND: nonstructural elements are attached to the structure in a form that they will not be damaged during structural deformation

The story drift or interstory displacement Δ_i for story i is then calculated by

$$\Delta_i = \delta_i - \delta_{i-1} \qquad (6.13)$$

with $\delta_o = 0$.

where

δ_i, δ_{i-1} = horizontal displacements for levels i, $i-1$ of the building.

For the protection of nonstructural elements, the value of the relative story drift (story distortion) Δ_i/H_i (H_i = interstory height) must not be greater than certain limit values established in the Code as shown in Table 6.7 for different building groups.

6.2.10.2 P-Δ effects. The effects due to additional deformations caused by the gravity loads in a structure deformed by an earthquake must be considered in the calculation of effects and deformations when

$$\beta_i = \frac{P_i \Delta_i}{V_i H_i} \geqslant 0.08 \qquad (6.14)$$

where

β_i = P-Δ coefficient for story i
Δ_i = drift for story i
V_i = shear force at story i
H_i = height of story i
P_i = total seismic weight at level i and above.

An approximate procedure to consider P-Δ effects is to amplify the calculated forces and displacements due to seismic action by the amplification factor

$$\psi = \frac{1}{1 - \beta_{max}} \qquad (6.15)$$

in which the P-Δ coefficient β_{max} is the maximum value calculated by eq.(6.14) for all the stories in the building.

6.2.10.3 Building separation. To avoid a possible hammering, the separation Y_i between adjacent structures must fulfill the following conditions at each level:

$$Y_i \geqslant \delta_i + f_s h_i \qquad (6.16a)$$
$$Y_i \geqslant f_o h_i + 1 \text{ cm} \qquad (6.16b)$$
$$Y_i \geqslant 2.5 \text{ cm} \qquad (6.16c)$$

where

f_s = factor depending on the type of foundation soil (Table 6.8a)
f_o = factor depending on the type of foundation soil and the seismic zone (Table 6.8b)
h_i = level height measured from the base of the building.

6.2.11 Methods of Analysis

Seismic effects can be obtained using the following methods:

- Static
- Simplified static
- Spectral modal analysis
- Step-by-step modal superposition
- Step-by-step direct integration

6.2.11.1 Static method. The static method consists of the replacement of seismic action by an equivalent system of static lateral forces, applied at the various levels of the building. The static method is applicable only to regular structures in plan and in elevation. The total height should be no more than the limit established in Table 6.9 for structures in Group A, A_o, and B in Seismic Zones 3 and 4 or 1 and 2. The Code also limits the application of the static method to structures for which $T_o < 3T_2$, where T_o is the fundamental

Table 6.8a. Values of f_s for Different Types of Soil

Type of Soil	I	II	III
f_s	0.0010	0.0015	0.0025

Table 6.8b. Values of f_o for Different Seismic Zones and Types of Soil

Seismic Zone	Values of f_o		
	Soil Type I	Soil Type II	Soil Type III
1 and 2	0.003	0.004	0.006
3 and 4	0.005	0.007	0.010

Table 6.9. Height Limit for Application of Static Method

Seismic Zone	Group A_o	Group A	Group B
3 and 4	12 m	30 m	40 m
1 and 2	16 m	40 m	55 m

period calculated by eq.(6.7), and T_2 is the time parameter given in Table 6.3.

The most important provisions of the static method are presented in the following sections.

(a) Base shear force. The base shear force V_o in the direction considered for the seismic action is given by

$$V_o = CW \qquad (6.17)$$

where C is the *seismic coefficient* and W is the total seismic weight of the building:

$$W = \sum_{i=1}^{n} W_i \qquad (6.18)$$

in which W_i is the seismic weight at level i.

The seismic coefficient C is calculated by

$$C = \frac{S_a \gamma_d}{R} \qquad (6.19)$$

where

S_a = seismic horizontal spectral value [eq.(6.1) or eq.(6.2)]
γ = risk factor (Table 6.2)
R = reduction factor (Section 6.2.5).

(b) Lateral seismic forces. The lateral seismic forces F_i at the various levels of the building are calculated by

$$F_i = \frac{W_i h_i}{\sum_{k=1}^{n} W_k h_k} V_o \qquad (6.20)$$

where

W_i, W_k = seismic weights at level i or k, respectively
h_i, h_k = height of levels i and k measured from the base
V_o = base shear force [eq.(6.17)].

When the value calculated for the fundamental period of the building is greater than $2T_2$ (T_2 given in Table 6.3), the lateral seismic forces are determined as follows.

For all levels, except the top level

$$F_i = \alpha V_o \frac{W_i h_i}{\sum_{k=1}^{n} W_k h_k} \qquad (6.21)$$

and for the top level,

$$F_n = V_o \left(\alpha \frac{W_n h_n}{\sum_{k=1}^{n} W_k h_k} + (1 - \alpha) \right) \qquad (6.22)$$

where α is a coefficient given by

$$\alpha = 1 - \frac{T_o - 2T_2}{10T_2} \quad \text{with } \alpha \le 1.0 \qquad (6.23)$$

in which

T_o = fundamental period
T_2 = parameter of the spectrum given in Table 6.3.

(c) Story shear force. The story shear force V_i at level i, is

$$V_i = \sum_{k=1}^{n} F_k \qquad (6.24)$$

(d) Overturning moment. The overturning moment M_i at any level i of the building is

$$M_i = \alpha \sum_{k=i+1}^{n} F_k(h_k^* - h_i^*) \qquad (6.25)$$

$$(i = 0, 1, 2, 3, \ldots, n - 1)$$

where h_k^*, h_i^* = height at level k, i from the foundation level of the building. In eq.(6.25) the reduction factor $\alpha = 0.9$ for the overturning moment (M_o) at the base of the building and $\alpha = 1.0$ for the moments at all other levels.

(e) Torsional effects. The Code provides the following definitions in relation to torsional effects:

- Center of Mass (C.M.): The centroid, at each level, of the seismic weight attributed to the level.
- Center of Stiffness of the vertical resisting elements of the story (C.S.): The centroid, at a level, of the cross-sectional moments of inertia of the lateral resisting structural elements with respect to the main axis perpendicular to the analyzed direction.
- Center of Stiffness (C.R.): A point at a level where the application of a horizontal force produces only a displacement of the level.
- Geometric eccentricity: The distance between C.M. and C.S. measured in a direction perpendicular to the analyzed direction.
- Structural eccentricity: The distance between C.M. and C.R. measured in a direction perpendicular to the analyzed direction.

The torsional effects are determined according to the type of structure as described in the following cases.

Case (a) Structures with two axes of symmetry

These structures should conform to the following conditions:

- The geometric eccentricity in all levels is smaller than 5% of the longest dimension of the plan normal to the analyzed direction;
- The C.S., at all levels, are aligned approximately in a vertical direction; and
- The C.M., at all levels, are aligned approximately in a vertical direction.

The torsional moment M_{ti} at level i is given by the most unfavorable moment calculated from

$$M_{ti} = (1.5e_1 + 0.10L)V_i \qquad (6.26a)$$

and

$$M_{ti} = (e_1 - 0.10L)V_i \qquad (6.26b)$$

where

e_1 = distance between the C.S., at level i, and the line of action of the shear force, measured perpendicular to the analyzed direction; and

L = maximum dimension in plan measured perpendicular to the direction of the story shear force V_i.

Case (b) Asymmetric structures with seismic resistant vertical planes of similar behavior

Structures in this group must satisfy the following conditions:

- Structural eccentricity at any level must be less than 25% of the longest dimension in plan, normal to the analyzed direction.
- The center of stiffness (C.R.) at any level must be approximately aligned in a vertical direction.
- The C.M. at any level must be approximately aligned in a vertical direction.

The torsional moment M_{ti} at level i, is given by the most unfavorable moment calculated by the following equations:

$$M_{ti} = (1.5e_3 + 0.07L)V_i \qquad (6.27a)$$

and

$$M_{ti} = (e_3 - 0.07L)V_i \qquad (6.27b)$$

where e_3 = distance between the C.R. at level i and the action line of the shear force measured perpendicular to the analyzed direction.

Case (c) Asymmetric structures with seismic resistant vertical planes with different behavior

Structures in this group must satisfy the following conditions:

- Geometric eccentricity in all planes must be less than 25% and greater than 5% of the longest dimension in plan normal to the direction under consideration.
- The C.S. at any level must be approximately aligned in the vertical direction.
- The C.M. at any level must be approximately aligned in the vertical direction.

For this case, a three-dimensional static analysis that couples translation and torsion must be performed.

Case (d) Other structures

For structures that do not conform to the conditions of the above cases, a three-dimensional dynamic analysis considering the coupling of translational and torsional displacements, must be performed.

(f) Vertical seismic forces. With the exception of the three cases listed below, it is not necessary to consider the vertical component of the seismic action.

Case 1 Cantilevers and balconies.
Case 2 Building roofs with large spans; horizontal members of prestressed concrete.
Case 3 Structures with lateral projections.

The components of such exceptional structures must be designed for a vertical force F_v proportional to the weight of the component:

$$F_v = \pm C_v \gamma_d W \qquad (6.28)$$

where

C_v = vertical seismic coefficient (Table 6.10);
γ_d = risk factor (Table 6.2); and
W = weight of the structural component.

Table 6.10. Vertical Seismic Coefficient C_v for Cases 1 and 2*

Seismic Zone	C_v	
	Case 1	Case 2
4	1.20	0.65
3	0.86	0.47
2	0.52	0.28
1	0.24	0.13

*The value of the vertical seismic coefficient C_v for Case 3 structures must be obtained from an appropriate response spectral chart.

6.2.11.2 Simplified static method. The simplified static method is applicable when all of the following conditions are fulfilled:

- The total height of the structure (measured from the base level) is less than 14 m and the number of stories in the building is not more than four.
- The building must be of seismic-resistant reinforced concrete, cast in situ, with a lateral resisting system of structural walls, structural frames, or a combination of both.
- The structural configuration of the building must be regular (i.e., regular distribution of mass and stiffness). Also, the structure must be symmetric with respect to two horizontal orthogonal axes.
- The ratio of the total height of the building must not be greater than 3, with respect to the smallest dimension of the rectangle that circumscribes the plan.
- The ratio of the largest dimension of the base must not be greater than 2.3, with respect to the smallest dimension.
- The floors and roofs must behave as rigid diaphragms in their own planes.

The simplified static method and the static method use the same formulas with the following exceptions:

(a) Seismic coefficient. In the simplified static method, the base shear force V_o is determined as

$$V_o = CW \qquad (6.29)$$

where

$$C = C_n \gamma_d \qquad (6.30)$$

in which C_n is the normalized seismic coefficient, with values given in Table 6.11, and γ_d is the risk factor, given in Table 6.2.

(b) Torsional effects. The torsional moment M_{ti} at level i is given by the most unfavorable value of the moments calculated by

$$M_{ti} = (2.0e_3 + 0.10L)V_i \qquad (6.31a)$$

and

$$M_{ti} = (e_3 - 0.10L)V_i \qquad (6.31b)$$

where L and e_3 are defined in Section 6.2.11.1.

Table 6.11.   Normalized Seismic Coefficient C_n

Seismic Zone	C_n
1	0.10
2	0.18
3	0.25
4	0.35

The Code does not permit a reduction of shear force due to torsion; only an increase of the shear force is permitted.

6.2.11.3 Spectral modal analysis
(a) Modal base shear force. The base shear force V_m, corresponding to the mth mode, is given by

$$V_m = \frac{\gamma_d S_{am} \bar{W}_m}{R} \qquad (6.32)$$

where

S_{am} = elastic horizontal pseudo-acceleration corresponding to the mth mode [eq.(6.1) or eq.(6.2)]
$\bar{W}_m$ = modal effective weight.

The modal effective weight $\bar{W}_m$ is given by

$$\bar{W}_m = \frac{\left[\sum_{i=1}^{n} W_i \phi_{im} \right]^2}{\sum_{i=1}^{n} W_i \phi_{im}^2} \qquad (6.33)$$

where

ϕ_{im} = lateral displacement in level i corresponding to the mth mode
W_i = seismic weight at level i of the building.

(b) Modal lateral force. The modal lateral force F_{im} at level i is given by

$$F_{im} = \frac{W_i \phi_{im}}{\sum_{k=1}^{n} W_k \phi_{km}^2} V_m \qquad (6.34)$$

(c) Modal story shear force. The modal story shear force V_{im} at level i is calculated as

$$V_{im} = \sum_{k=i}^{n} F_{km} \qquad (6.35)$$

(d) Modal overturning moment. The modal overturning moment M_{im} at level i is calculated as

$$M_{im} = \sum_{k=i+1}^{n} F_{km}(h_k^* - h_i^*) \qquad (6.36)$$

$$(i = 0, 1, 2, 3, \ldots, n-1)$$

where

F_{km} = modal lateral force at level k
h_k^*, h_i^* = height at level k, i from the foundation level of the building.

(e) Effective values. All the modes whose contributions to the total response are greater than 5% of the

contribution corresponding to the fundamental mode must be included in the analysis. However, at least three modes must be considered in the analysis, except for structural systems modeled with two degrees of freedom. The effective value E for forces, moments, or displacements is calculated by the combination of the modes included in the analysis as

$$E = \sqrt{\sum_i E_i^2 + \left[\sum_j |E_j| \right]^2} \qquad (6.37)$$

where

E_i = modal contributions for a separated mode
E_j = modal contributions for a nonseparated mode.

A mode is considered separated if its natural period differs from the other natural periods by more than 10%.

The Code requires that, for each direction analyzed, the base shear resulting from the spectral modal method of analysis be no less than 75% of the base shear force calculated by the static method. When the value for the base shear does not fulfill this condition, all the results obtained for the earthquake effects using the spectral modal method must be incrementally increased by the ratio between 75% of the static base shear force and the corresponding value obtained from the spectral modal analysis.

Depending on the symmetry of the structure, the application of the spectral modal method will be implemented as follows:

Case (a) Structures with two axes of symmetry

Vibrational modes in the direction of each axis of symmetry must be determined independently. Torsional effects must be obtained using the static method and algebraically added to the total response in the analyzed direction. The vibratory model of analysis must have at least one translational degree of freedom at each level of the building where concentrated masses are located.

Case (b) Asymmetric structures with seismic resistant vertical planes of similar behavior

These structures may be analyzed using either the above requirements specified for Case (a) or the requirements that follow for Case (c).

Case (c) Asymmetric structures not included in the preceding cases

The vibratory model must have, at each level of the building, at least one translational degree of freedom in each direction of analysis and one rotational degree of freedom about a vertical axis.

6.2.11.4 Step-by-step modal superposition method.
In this method, the seismic actions are defined by accelerogram records of historical or artificial earthquakes, applied at the base of the structure. The acceleration-time history used must observe the following:

- The maximum acceleration a_{max} must be equal to or greater than the product of the risk factor γ_d (Table 6.2) and the ordinate of the spectral acceleration a_s, which depends on the seismic zone and the type of soil (Table 6.3); that is,

$$a_{max} \geqslant \gamma a_s \qquad (6.38)$$

- The elastic spectrum obtained from the acceleration-time history should have an area equal to the area (between 0.05 sec and the fundamental period of the structure) of the spectrum given by the Code amplified by the risk factor γ_d (Section 6.2.4). Also, the spectral ordinates can not be less than 70% of the spectral values established by the Code, amplified by γ_d.

- The results of at least three independent acceleration-time histories must be considered for structures of Groups A and B; and in the case of Group A_o, at least four.

- The values of the seismic actions and the deformations that result from the average of the envelopes obtained from each acceleration-time history must be adopted. This average should not include values that are smaller than 85% of the maximum value found.

- The reduction factor R (Section 6.2.5) can be applied in order to reduce the actions obtained from the elastic analysis. However, the limitations of this criterion must be taken into consideration due to the possible concentration of inelastic deformations in some sections of the structure.

- The shear force at the base must not be smaller than 70% of the force determined by the static method based on the fundamental period.

6.2.11.5 Step-by-step direct integration method.
The acceleration-time history to be used must observe the conditions of the step-by-step modal superposition method. When a nonlinear analysis is performed and the building is located in an epicenter area, long intense pulses must be included in the acceleration-time history.

When a linear elastic analysis is being performed, the values obtained from the seismic action and deformations (that result from the average of the corresponding envelope obtained from the application of each acceleration–time history) must be adopted. Such an average must not include values that are smaller than 80% of the maximum value found. In a nonlinear analysis the average of the envelopes will not be taken.

In this method, the elastic actions could be reduced by R.

6.3 APPLICATION EXAMPLES

6.3.1 Example 6.1

6.3.1.1 Building description. This example consists of a static seismic analysis of a building (see Fig. 6.3) with the Seismic Code INPRES-CIRSOC 103 (1983). The example corresponds to the one described in Example 23.1 of Structural Dynamics: Theory and Computation, Paz (1991). The building has four stories and a reinforced concrete frame. The dimensions of the columns from lines A and C are 0.3048 m by 0.5080 m for the two bottom floors, and 0.3048 m by 0.4064 m for the two top floors. The cross-section for the columns from line B is 0.3048 m by 0.6096 m for the two bottom floors and 0.3048 m by 0.5080 m for the two top floors. The dead load considered on each level is estimated to be 683.40 kg/m^2. This load is calculated by adding the weights of the floor slab, beams, and half of the weight of the columns above and below

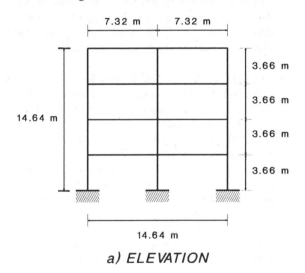

a) ELEVATION

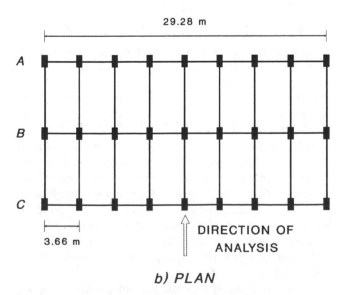

b) PLAN

Fig. 6.3. Plane frame modeling building for Example 6.1

each floor. The building will be used as a warehouse, and the live load is estimated to be 610.12 kg/m^2. The building is supposed to be constructed on rock in Seismic Zone 3. The seismic direction is indicated in Fig. 6.3b.

6.3.1.2 Static method

(a) Seismic weights. The seismic weights W_i at each level are calculated as

$$W_i = G_i + \eta L_i \qquad \text{[eq.(6.6)]}$$

where

$$\eta = 0.75 \text{ for levels 1 to 3 (Table 6.6)}$$
$$\eta = 0 \text{ for level 4}$$
$$W_1 = W_2 = W_3 = 428.66 \text{ m}^2 \times (683.40 + 0.75 \times 610.12)$$
$$\text{kg/m}^2$$
$$= 4.891 \times 10^5 \text{ kg}$$
$$W_4 = 428.66 \text{ m}^2 \times 683.40 \text{ kg/m}^2$$
$$= 2.929 \times 10^5 \text{ kg}$$

The total seismic weight is

$$W = \sum_{i=1}^{4} W_i = 17.602 \times 10^5 \text{ kg}$$

(b) Fundamental period. Because the building is regular, the fundamental period of vibration may be determined using eq.(6.10). Since there is no information about the density of the walls, d is assumed equal to 0. The height of the building above the basal level is $h_n = 14.64$ m, and the length of the building in the analyzed direction is $L = 14.64$ m. Then by eq.(6.10)

$$T_{oe} = \frac{14.64}{100} \sqrt{\frac{30}{14.64} + \frac{2}{1}}$$

$$= 0.3 \text{ sec.}$$

(c) Seismic coefficient. A 5% critical damping is assumed for the reinforced concrete frame building. The elastic spectral ordinate is given by eq.(6.1) with the parameters for Seismic Zone 3 and Soil Type I (rock) obtained from Table 6.3 as

$$a_s = 0.25, \ b = 0.75, \ T_1 = 0.20 \text{ sec}, \ T_2 = 0.35 \text{ sec}$$

In this case $T_1 < T < T_2$, then by eq.(6.1b)

$$S_a = 0.75$$

The seismic coefficient of design is calculated using eq.(6.19) where $\gamma_d = 1$, according to Table 6.2, since the building is a warehouse belonging to Group B. The reduction factor is $R = \mu$ [from eq.(6.5b), for $T > T_1$] and $\mu = 5$ (from Table 6.5) for a reinforced

Table 6.12. Lateral Seismic Forces, Story Shear Forces, Overturning Moments, and Torsional Moments

Level	Lateral Force F_i (10^4 kg)	Shear Force V_i (10^4 kg)	Overturning Moment M_i (10^5 kg-m)	Torsional Moment M_{ti} (10^5 kg-m)
4	7.533	7.533	—	2.2057
3	9.434	16.967	2.76	4.9679
2	6.289	23.256	8.97	6.8094
1	3.145	26.401	17.48	7.7302
0	—	—	24.43	—

concrete frame building. With these values the seismic coefficient is

$$C = \frac{0.75 \times 1}{5} = 0.15 \qquad [\text{eq.}(6.19)]$$

(d) Base shear force

$$
\begin{aligned}
V_o &= CW \\
&= 0.15 \times 17.602 \times 10^5 \text{ kg} \qquad [\text{eq.}(6.17)] \\
&= 2.64 \times 10^5 \text{ kg}
\end{aligned}
$$

(e) Lateral seismic forces and story shear. The lateral seismic forces and story shear forces are given, respectively, by eqs.(6.20) and (6.24). The calculated values are shown in Table 6.12.

(f) Overturning moments. The overturning moments calculated by eq.(6.25) are shown in Table 6.12.

(g) Torsional moments. Equation (6.26a) can be applied for the calculation of the torsional moments because the structure has two axes of symmetry ($e_1 = 0$). Hence,

$$M_{ti} = 0.10 \times L \times V_i$$

where $L = 29.28$ m. Table 6.12 gives the torsional moments at each level of the structure.

(h) Story drifts, lateral displacements, and story distortions. For this structure, it is assumed that the horizontal diaphragms at the floor levels are rigid in their planes. In this case, the stiffness K_i for story i between two consecutive levels is given by

$$K_i = \frac{12EI_i}{H_i^3} \qquad (6.39)$$

where

$E = 2.1 \times 10^9$ kg/m^2 (modulus of elasticity, concrete)
$H_i = 3.66$ m (story height)
$I_i =$ total cross-sectional moment of inertia for story columns.

(1) Moment of inertia of columns
Levels 1 and 2:

$$
\begin{aligned}
I_A &= I_C = (1/12) \times 0.3048 \text{ m} \\
&\quad \times (0.508 \text{ m})^3 = 3.3298 \times 10^{-3} \text{ m}^4 \\
I_B &= (1/12) \times 0.3048 \times (0.6096)^3 \\
&= 5.7539 \times 10^{-3} \text{ m}^4
\end{aligned}
$$

Levels 3 and 4:

$$
\begin{aligned}
I_A &= I_C = (1/12) \times 0.3048 \text{ m} \\
&\quad \times (0.4064 \text{ m})^3 = 1.7048 \times 10^{-3} \text{ m}^4 \\
I_B &= (1/12) \times 0.3048 \times (0.5080)^3 \\
&= 3.3298 \times 10^{-3} \text{ m}^4
\end{aligned}
$$

The total moment of inertia for the story is then given by

$$I_i = 9 \times I_A + 9 \times I_B + 9 \times I_C$$

(2) Story drift

$$\Delta_i = \frac{\mu V_i}{K_i} \qquad (6.40)$$

where μ is the global ductility (Table 6.5).

Table 6.13 shows calculated values for the story drifts and for the relative drift Δ_i/H_i for the various levels of the building.

(3) Lateral displacements
Lateral displacements are calculated using the following equation:

$$\delta_i = \sum_{j=1}^{i} \Delta_j \qquad (6.41)$$

Table 6.13 shows the lateral displacements at each level of the building.

(4) P-Δ coefficient
The P-Δ coefficient β_i is given by

$$\beta_i = \frac{P_i \Delta_i}{V_i H_i} \qquad [\text{eq.}(6.14)]$$

Calculated values of the P-Δ coefficient βi are given in Table 6.13.

Table 6.13. Story Drifts, Lateral Displacements, Story Relative Displacements, and Stability Coefficients

Level i	Story Drift Δ_i (m)	Lateral Displ. δ_i (m)	Relative Drift Δ_i/H_i	Seismic Weight W_i (10^5 kg)	Sum Weight ΣW_i (10^5 kg)	P-Δ Coeff. β_i
4	0.012	0.082	0.0033	2.929	2.929	0.013
3	0.027	0.070	0.0074	4.891	7.820	0.034
2	0.020	0.043	0.0055	4.891	12.710	0.030
1	0.023	0.023	0.0063	4.891	17.602	0.042

The maximum value of the relative story drift from Table 6.13 is 0.0074. This value is well below the maximum value of 0.014 permitted by the Code for building Group B, as indicated in Table 6.7.

(5) Verification of P-Δ effects. In order to determine whether or not the P-Δ effects must be taken into consideration, the P-Δ coefficient β_i defined by eq.(6.14) must be calculated for each level of the building. Calculated values of this coefficient are shown in Table 6.13. The condition established in eq.(6.14), $\beta_i \geqslant 0.08$ is not met at any of the stories; hence, it is not necessary to consider P-Δ effects in this example.

6.3.2 Example 6.2

This example applies the spectral modal analysis to the building of Example 6.1. The structure is modeled as a system with four masses concentrated in each floor level of the building with one translational degree of freedom per level. The horizontal diaphragms at the floor levels are assumed to be rigid.

(a) Mass matrix. The mass matrix for this example is constructed as follows:

$$[M] = \begin{bmatrix} W_1 & 0 & 0 & 0 \\ 0 & W_2 & 0 & 0 \\ 0 & 0 & W_3 & 0 \\ 0 & 0 & 0 & W_4 \end{bmatrix} \frac{1}{g}$$

where g is the gravitational acceleration. Substituting numerical values results in

$$[M] = \begin{bmatrix} 4.986 & 0 & 0 & 0 \\ 0 & 4.986 & 0 & 0 \\ 0 & 0 & 4.986 & 0 \\ 0 & 0 & 0 & 2.986 \end{bmatrix} \times 10^4 \text{ kg sec}^2/\text{m}$$

(b) Stiffness matrix. The stiffness matrix for the idealized structure is

$$[K] = \begin{bmatrix} K_1+K_2 & -K_2 & 0 & 0 \\ -K_2 & K_2+K_3 & -K_3 & 0 \\ 0 & -K_3 & K_3+K_4 & -K_4 \\ 0 & 0 & -K_4 & K_4 \end{bmatrix}$$

where the stiffness of each story is calculated as in the static method; the stiffness matrix results in

$$[K] = \begin{bmatrix} 11.485 & -5.742 & 0 & 0 \\ -5.742 & 8.86 & -3.118 & 0 \\ 0 & -3.118 & 6.235 & -3.118 \\ 0 & 0 & -3.118 & 3.118 \end{bmatrix} 10^7 \text{ kg/m}$$

(c) Natural frequencies and vibration modes. The natural frequencies and the vibration modes are obtained from the nontrivial solutions of the following system of equations:

$$[[K] - \omega^2[M]]\{\phi\} = 0$$

where

$$\omega = \text{natural frequency}$$
$$\{\phi\} = \text{vibration mode}$$

The calculated results for the natural frequencies and periods are given in Table 6.14.

The modal shapes are as follows:

$$[\Phi] = \begin{bmatrix} 1.0 & 1.0 & 1.0 & 1.0 \\ 0.863 & 0.128 & -0.925 & -0.876 \\ 0.529 & -0.931 & 0.122 & 0.309 \\ 0.282 & -0.769 & 0.477 & -0.142 \end{bmatrix}$$

Each column of the modal matrix $[\Phi]$ represents a modal shape, with the first row corresponding to the fourth level.

(d) Spectral ordinates. The seismic coefficient C for each mode is obtained using eq.(6.19) in which $\gamma_d = 1$ and $R = 5$ as in the static method. The elastic spectral ordinates S_a are calculated from eqs.(6.1a) and (6.1b) for each modal period considered. Table 6.15 shows the calculated values for the seismic coefficient.

(e) Modal effective weights. The modal effective weights $\bar{W}_m$ are calculated using eq.(6.33) and are shown in Table 6.15.

$$\bar{W}_m = \frac{\left[\sum_{i=1}^{n} W_i \phi_{im} \right]^2}{\sum_{i=1}^{n} W_i \phi_{im}^2} \qquad \text{[eq.(6.33)]}$$

(f) Modal base shear. The modal base shear force V_m, corresponding to the mth mode, is obtained by using eq.(6.32). Table 6.15 shows the calculated values.

Table 6.14.   Natural Frequencies and Periods

Mode	Frequency ω (rad/sec)	Frequency $f = \omega/2\pi$ (Hz)	Period $T = 1/f$ (sec)
1	11.95	1.90	0.526
2	30.18	4.80	0.208
3	44.83	7.13	0.140
4	57.55	9.16	0.109

Table 6.15. Spectral Ordinates, Effective Weights, and Modal Shears

Mode	Period T (sec)	Spectral Ordinates S_a	$C = \dfrac{S_a \gamma_d}{R}$	Effective Weight $\bar{W}_m$ (10^5 kg)	Modal Shear V_m (10^5 kg)
1	0.526	0.572	0.114	14.837	1.697
2	0.208	0.750	0.150	2.234	0.335
3	0.140	0.600	0.158	0.216	0.034
4	0.109	0.523	0.164	0.315	0.052
				Sum = 17.602	

(g) Modal lateral forces and modal shears forces. The modal lateral force at each level of the building is calculated using eq.(6.34), and the corresponding modal shear force for each story with eq.(6.35).

(h) Modal overturning moments. The modal overturning moments at the base of the building and at all the levels are determined by using eq.(6.36). In this case, the modal forces corresponding to the mode being analyzed are used in eq.(6.36).

(i) Modal torsional moments. The modal torsional moments are calculated using eq.(6.26a) with $e_1 = 0$ as in the static method; the modal shear is used in the calculation.

(j) Modal drifts, displacements, and distortions. The modal drifts, displacements and distortions are calculated in the same way as in the static method; the modal shear is used for their calculations.

(k) Modal superposition. The total effects are calculated by superposing the modal effects according to eq.(6.37) and considering the vibration periods of the four modes that differ from each other by more than 10%. Table 6.16 summarizes the results of superposing the four modes of vibration. Effective values for the story distortion Δ_i/H_i in Table 6.16 are not greater than the limit value of 0.014 established by the Code for buildings in Group B, as given in Table 6.7.

(l) Verification of P-Δ effects. The verification is done as in the static method with the shear values and the floor drifts obtained from the modal superposition. Calculated values of the P-Δ coefficient β_i from eq.(6.14) are shown in Table 6.16. The established condition of inequality (eq.6.14) is verified in all levels; it is not necessary to consider the P-Δ effects.

The base shear force ($V = 1.731 \times 10^5$ kg) obtained by using the spectral modal analysis method is 66% of the value obtained with the static method ($V = 2.64 \times 10^5$ kg). According to the Code, all the effects obtained with the spectral modal analysis must be amplified with a factor, in this case equal to 0.75/0.66 = 1.14.

6.4 COMPUTER PROGRAM

6.4.1 Description

A computer program written in FORTRAN was developed for the seismic analysis of buildings following the requirements in the INPRES-CIRSOC 103 Code. The program can be executed on an IBM or a compatible microcomputer. The building is modeled as a system with rigid diaphragms at the floor levels, where the masses are concentrated. This model has one translational degree of freedom per level. The program can perform both static and dynamic (spectral modal) analyses.

A file that contains the following input data is prepared:

General data:

- Seismic Zone
- Building Group
- Type of foundation soil
- Damping (percentage of critical damping)
- Global ductility
- Acceleration due to gravity
- Case of torsion and distortion according to structural configuration
- Wall density
- Method of analysis (static or dynamic)
- Number of stories

Data for each level:

- Dead load, live load, and simultaneity factor
- Total story stiffness
- Height of story measured from base of the structure
- Eccentricity
- Plan dimension in analyzed direction
- Plan dimension in orthogonal direction

The output of the program includes period, seismic coefficient, base shear force, lateral seismic forces, story shears, overturning moments, torsional moments, story drifts, and lateral displacements.

Table 6.16. Effective Values for Example 6.2

Level i	Story Shear V_i (10^5 kg)	Overturn Moment M_i (10^5 kgm)	Torsional Moment M_{ti} (10^5 kgm)	Story Drift Δ_i (10^{-2} m)	Lat. Displ. δ_i (10^{-2} m)	Distor. Δ_i/H_i (10^{-2})	P-Δ Coeff. β_i
4	0.498	—	1.459	0.799	5.404	0.218	0.013
3	1.112	1.823	3.283	1.798	4.605	0.49	0.034
2	1.492	5.876	4.367	1.300	2.807	0.35	0.030
1	1.731	11.166	5.068	1.507	1.507	0.41	0.042
0	—	17.285	—	—	—	—	—

The program also checks story drifts and the need for considering P-Δ effects in the analysis. If P-Δ effects must be considered, all the resulting effects are multiplied by the amplification factor ψ suggested in the Code [eq.(6.15)].

When a static analysis is requested, the program first checks the conditions of applicability of the method. If these conditions are not fulfilled, the program stops and gives a warning message. For the static analysis, the fundamental period of the building is evaluated with both the empirical formula of eq.(6.10) and the Rayleigh formula [eq.(6.7)]. If the period evaluated with the Rayleigh formula falls below the limits established in the Code, this value is used in the analysis. Otherwise, limit values are taken for the analysis.

For the dynamic analysis, the standard Jacobi diagonalization method is used to calculate the natural periods and modal shapes by solving the corresponding eigenproblem (Bathe and Wilson 1976). A symmetric stiffness matrix and a diagonal mass matrix are assumed. The output includes both modal response values and effective values obtained by modal superposition. The effective values are calculated using eq.(6.37). Only those modes whose contribution to the total response is greater than 5% of the contribution of the fundamental mode are included.

6.4.2 Application Examples

The following examples illustrate the use of the computer program.

6.4.2.1 Example 6.3. The structure in this example is the 15-story steel Kajima International Building, located in Los Angeles, California. The building has four moment-resisting frames in both longitudinal and transverse directions; the general dimensions are indicated in Fig. 6.4. The building is analyzed for seismic motion in the y direction. The building was designed using a fundamental period of 1.5 sec (Housner and Jennings 1982). The design value for the base shear force was 172,000 kg, using the Los Angeles Building Code. Wind response tests, performed after the construction was completed, indicated that the period of the building was about 1.9 sec.

In 1971, during the San Fernando earthquake, the building suffered only nonstructural damage. According to records obtained during the earthquake, the base shear force could have been as high as 350,000 kg.

This example could not be solved with the static method because the building exceeded the high limit established in the Code. The Kajima International Building was analyzed by the spectral modal method

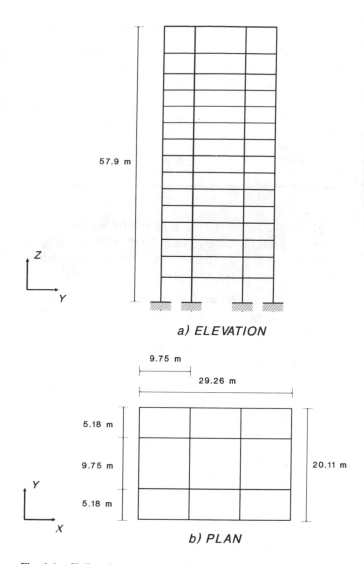

Fig. 6.4. Kajima International Building, Example 6.3

of analysis. A condensed output is presented in this section.

The value obtained for the fundamental period ($T_1 = 1.89$ sec) is very close to the value estimated experimentally from wind effects. Only five modes were included in the modal superposition. The value of the base shear force was 325,137 kg. The P-Δ effects were included in the analysis; all effects were multiplied by the amplification factor ψ. The resulting value for the base shear force was then 400,443 kg. The 350,000 kg value of the base shear force, estimated from records obtained during the San Fernando earthquake of 1971, falls between the calculated results.

6.4.2.2 Example 6.4. This example involves the static analysis of the building shown in Fig. 6.5, which was originally analysed in an internal report (INPRES 1987). The structure is a conventional four-story

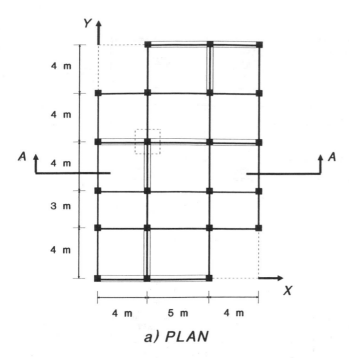

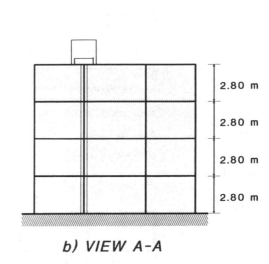

a) PLAN

b) VIEW A-A

Fig. 6.5. Plane frame modeling structure for Example 6.4

```
          INPUT DATA AND OUTPUT RESULTS FOR EXAMPLE 6.3

    INPUT DATA:

    SEISMIC ZONE NUMBER        :     4
    CONSTRUCTION GROUP         :     B
    RISK FACTOR                :     1.0

    TYPE OF SOIL               :     1
    DAMPING                    :     2.0%
    GLOBAL DUCTILITY           :     6.0
    ACCELERATION DUE TO GRAVITY :    9.81

    NUMBER OF DEGREE OF FREEDOM :     15
```

STORY	HEIGHT (M)	SEISMIC WEIGHT (10^6 KG)	STIFFNESS (10^8 KG/M)	ECCENTRICITY (M)
1	4.30	0.2464	0.500	0.0
2	7.90	0.3986	0.500	0.0
3	11.60	0.3932	0.500	0.0
4	15.20	0.3932	0.450	0.0
5	18.90	0.3905	0.450	0.0
6	22.50	0.3905	0.450	0.0
7	26.20	0.3905	0.420	0.0
8	29.90	0.3905	0.420	0.0
9	33.60	0.3905	0.400	0.0
10	37.20	0.3905	0.400	0.0
11	40.80	0.3905	0.400	0.0
12	44.50	0.3905	0.400	0.0
13	48.20	0.3905	0.380	0.0
14	52.10	0.5844	0.360	0.0
15	57.90	0.3624	0.280	0.0

OUTPUT RESULTS (DYNAMIC ANALYSIS):

MODE	FREQ. (CPS)	PERIOD (SEC)	SEISMIC COEFF.	BASE SHEAR (KG)
1	3.32	1.89	0.063	3.078 E5
2	9.70	0.65	0.152	0.892 E5
3	16.10	0.39	0.248	0.472 E5
4	22.41	0.28	0.277	0.241 E5
5	28.47	0.22	0.277	0.117 E5

P-DELTA EFFECT MUST BE CONSIDERED!

ALL RESULTING EFFECTS ARE MULTIPLIED BY THE AMPLIFICATION FACTOR IN THE CODE

EFFECTIVE VALUES:

BASE SHEAR FORCE = $0.400 \ 10^6$ (KG)

STORY	LATERAL FORCE (10^5KG)	STORY SHEAR (10^5KG)	OVERT. MOMENT (10^6KG-M)	TORSION MOMENT (10^6KG-M)	STORY DRIFT (10^{-1}M)	LATERAL DISPL. (M)
15	0.598	0.598	-----	0.175	0.128	0.534
14	0.772	1.344	0.347	0.393	0.224	0.526
13	0.459	1.713	0.866	0.501	0.270	0.509
12	0.472	2.013	1.487	0.589	0.302	0.488
11	0.472	2.284	2.202	0.668	0.363	0.463
10	0.470	2.532	2.978	0.741	0.380	0.433
9	0.468	2.765	3.827	0.809	0.415	0.399
8	0.452	2.979	4.773	0.872	0.426	0.361
7	0.444	3.175	5.789	0.929	0.454	0.321
6	0.444	3.354	6.870	0.981	0.447	0.278
5	0.434	3.519	7.980	1.030	0.469	0.235
4	0.439	3.673	9.175	1.075	0.490	0.189
3	0.446	3.818	10.386	1.117	0.458	0.141
2	0.390	3.954	11.677	1.157	0.474	0.095
1	0.134	4.004	12.974	1.172	0.481	0.048
0	-----	-----	14.564	-----	-----	-----

reinforced concrete building. The structural configuration is one frequently used for residential buildings. The building, founded on a medium-dense granular soil, is located in Godoy Cruz, Mendoza (Seismic Zone 4).

An output of the program corresponding to the seismic analysis in the *x* direction is presented below. The results obtained were identical to those included in the internal report (INPRES 1987).

INPUT DATA AND OUTPUT RESULTS FOR EXAMPLE 6.4

INPUT DATA

SEISMIC ZONE NUMBER	:	4
CONSTRUCTION GROUP	:	B
RISK FACTOR	:	1.0
TYPE OF SOIL	:	2
DAMPING	:	5.0%
GLOBAL DUCTILITY	:	4.0
ACCELERATION DUE TO GRAVITY	:	9.81
NUMBER OF DEGREE OF FREEDOM	:	4

STORY	HEIGHT (M)	SEISMIC WEIGHT (10^4KG)	STIFFNESS (10^7KG/M)	ECC (M)
1	2.80	0.260	0.1429	0.650
2	5.60	0.260	0.1286	0.540
3	8.40	0.260	0.2333	0.390
4	11.20	0.180	0.4000	0.280

OUTPUT RESULTS (STATIC ANALYSIS)

FUNDAMENTAL PERIOD:

```
EMPIRICAL FORMULA      T = 0.206 SEC
USED                   T = 0.224 SEC
SEISMIC COEFFICIENT    C = 0.269
BASE SHEAR FORCE       V = 2586 KG
```

STORY	LATERAL FORCE (10^3KG)	STORY SHEAR (10^3KG)	OVERT. MOMENT (10^4KG-M)	TORSION MT1 (10^4KG-M)	TORSION MT2 (10^4KG-M)	STORY DRIFT (10^{-2}M)	LATERAL DISPL. (10^{-1}M)
4	0.817	0.817	-----	0.143	0.086	0.082	0.181
3	0.885	1.701	0.229	0.326	0.160	0.292	0.173
2	0.590	2.291	0.705	0.490	0.181	0.713	0.144
1	0.295	2.586	1.347	0.596	0.176	0.724	0.072
0	-----	-----	2.213	-----	-----	-----	-----

REFERENCES

BATHE, K. J., and WILSON, E. L. (1976) *Numerical Methods in Finite Element Analysis*. Prentice-Hall, Englewood Cliffs, NJ.

HOUSNER, G. W., and JENNINGS, P. C. (1982) *Earthquake Design Criteria*. Earthquake Engineering Research Institute, Berkeley, CA.

INPRES (1987) *Análisis Estático de Edificios Según el Reglamento INPRES-CIRSOC 103*. Publicación Técnica No. 14, Instituto Nacional de Prevención Sísmica, San Juan, Argentina.

Normas Antisísmicas CONCAR 70 (1972) Instituto Nacional de Prevención Sísmica (INPRES). San Juan, Argentina.

Normas Antisísmicas Argentinas – NAA 80 (1980) Instituto Nacional de Prevención Sísmica (INPRES), San Juan, Argentina.

PAZ, M. (1991) *Structural Dynamics*. 3d ed., Van Nostrand Reinhold, New York, NY.

REGLAMENTO INPRES-CIRSOC 103 (1983) *Normas Argentinas para las Construcciones Sismorresistentes. Parte I: Construcciones en General, Parte II: Construcciones de Hormigón Armado y Hormigón Pretensado, Parte III: Construcciones de Mampostería*, ed. por el INTI (Instituto Nacional de Tecnología Industrial). Buenos Aires, Argentina.

REGLAMENTO INPRES-CIRSOC 103 (1991) *Normas Argentinas para las Construcciones Sismorresistentes*. Nueva Edición, ed. por el INTI (Instituto Nacional de Tecnología Industrial). Buenos Aires, Argentina.

7

Australia

David B. Crawley and Michael C. Griffith

7.1 HISTORICAL BACKGROUND

Prior to the occurrence of the earthquakes in Meckering, West Australia, in 1971 and 1972, the engineering design of most buildings in Australia completely ignored the possible effects produced by seismic ground motion. From the earliest days of the colonies tremors had been noted, but it was not until 1897, upon recommendations from the British Association of Science Seismological Committee, that a network of seismographs was installed. These seismographs were intended primarily for monitoring worldwide seismic activity, and so were not entirely suitable for detecting local events. Subsequently, seismic activity in Australia has been monitored and recorded in the earthquake data file of the Bureau of Mineral Resources (Denham et al. 1975). Most of the events recorded during this period, from 1897 to 1972, corresponded to earthquakes of relatively low magnitude with only a few quakes of magnitude just over five ($M > 5$) on the Richter scale. Following the Meckering earthquakes in 1971 and 1972 ($M > 7$), in which substantial damage occurred to masonry buildings in a small township, the possibility of a similar event occurring in a major city led to the establishment of a National Committee on Earthquake Engineering, with a brief to prepare a National Standard. This activity led to the publication of the Australian Standard 2121 (1979), which is generally known as the *SAA Earthquake Code*. However, only in Adelaide, of the state capitals, was it mandatory to design for earthquake resistance for buildings other than those of unreinforced masonry.

In December 1989, an earthquake of magnitude $M = 5.6$ that occurred near Newcastle, N.S.W. caused for the first time loss of life in Australia due to seismic action. Because this regional center had been zoned previously as having zero risk, urgent reconsideration of risk levels throughout Australia was initiated. The committee already established to revise the existing code was required to expedite the preparation of a new seismic code for earthquake-resistant design. As a result of the committee's work, a revised version of the Australian earthquake code was published in 1993 [Australian Standard 1170.4 (1993)].

This chapter presents a brief recapitulation of the Australian earthquake code of 1979 followed by a more detailed description of the new earthquake code of 1993.

7.2 EARTHQUAKE CODE OF 1979

The first set of Australian provisions for the design of earthquake-resistant structures, Australian Standard 2121 of 1979 [AS 2121-1979], was prepared during the 1970s. The Standard reflected the work in 1977 of the Seismology Committee, the Structural Engineers Association of California [SEAOC], and the 1976 edition of the *Uniform Building Code*, published by the International Conference of Building Officials, California, U.S.A.[1] A zone map for Australia, based on collated historical records for the period up to 1972, was prepared by the Australian Bureau of Mineral Resources. Zones were defined using risk levels similar to those in the United States, but with an additional Zone A for those areas where the expected low seismic activity could be withstood by "reasonably" ductile buildings, designed for wind loadings.

The primary aim of the Australian Standard of 1979 was the prevention of collapse of the structure rather than minimization of damage, although the Standard also had provision for limiting story drift. However, small domestic structures were not included, and special structures such as nuclear power stations, dams, and bridges were excluded specifically. In addition to provisions for estimating seismically induced loads, the Australian Standard included detailing requirements for different types of construction. Increasingly stringent design rules were required in higher-risk-level seismic zones for levels corresponding to those specified in Appendix A of the American Concrete Institute (ACI) *Building Code Requirements for Reinforced Concrete*, ACI 318-76 (1976).

The calculation of the total horizontal base shear force followed the SEAOC method using quasi-static forces, unless a dynamic analysis was performed. The use of dynamic analysis was not encouraged because very little guidance was available for the use of this type of analysis. In determining the base shear force V, consideration was given to a number of factors by use of the following equation:

$$V = ZIKCSW \qquad (7.1)$$

with a minimum value of $V = 0.02W$, for nonzero values of Z where

V = total horizontal force [in kilonewtons (kN)]
Z = zone factor
  $Z = 0.00$ for Zone A (ductile buildings)
  $Z = 0.09$ for Zone A (nonductile buildings)
  $Z = 0.18$ for Zone 1
  $Z = 0.36$ for Zone 2
I = occupancy importance factor
  $I = 1.2$ for essential facilities
  $I = 1.0$ for all other buildings
K = horizontal force factor for the structural system
  (Values of K vary between 0.75 for certain ductile structures to 3.2 for brittle structures.)
C = seismic response factor for base shear

$$C = \frac{1}{15\sqrt{T}} \leqslant 0.12 \qquad (7.2)$$

  (T = fundamental elastic period of vibration of building in seconds)
W = total dead weight of building + 0.25 live load for storage areas (kN), and
S = soil-structure resonance factor
  $S = 1.5$ (unless it is evaluated)

The 1979 Code further stipulated that the product CS need not exceed a value of 0.14.

7.3 EARTHQUAKE CODE OF 1993

The new version of Australian Standard AS 1170.4 (1993), contains a part with provisions for earthquake-resistant design of buildings, including domestic construction. This part contains two sections: (1) minimum load requirements common to all structural materials and (2) design clauses for specific construction materials. The earthquake loading code is complemented by provisions in the structural design codes for specific construction materials. These provisions include requirements for strength, ductility, serviceability, and detailing. Standards are provided in the new code for steel, concrete, masonry, and timber structures with specific requirements for different levels of earthquake hazard. The new version of the

---

[1]The SEAOC recommendations were revised and published in 1990, and the UBC was revised in 1991; see References to Chapter on United States of America.

Australian Standard *Minimum Design Loads on Structures* was published in August 1993 under the title AS 1170.4 as an interim standard to be replaced in due course by a Joint Australian-New Zealand Code. The Australian Standard considers loads on structures and attachments due to earthquake-induced ground motion. Other possible effects of earthquakes such as surface rupture, landslides, liquefaction of foundation, or secondary effects such as fire, loss of service, and loads from tsunamis that may develop after an earthquake are not considered in the new Australian Standard. The main purpose of the Standard is to prevent loss of life, particularly in structures of high occupancy, and to maintain the functional capability of essential facilities after an earthquake. The provisions contained in the standard implicitly accept the notion that structures may be damaged in the event of a severe earthquake; e.g., an event with a specified recurrence interval of 500 years.

Special structures such as nuclear reactors, cryogenic tanks, and industrial facilities of high economic importance are not considered in the Standard provisions. Single-story and two-story domestic structures are included, but the design of such buildings generally must include consideration only of detailing, without analysis for earthquake loads.

The provisions of the Australian earthquake loading standard are based largely on the recommendations of Applied Technology Council ATC 3-06 (1988). Zoning contour maps (see Fig. 7.1) have been developed from the work of Gaull et al. (1990) who used seismic data from the Australian Bureau of Mineral Resources. These data were derived from records of locations and magnitudes of all known Australian earthquakes between the years 1856 and 1990. Every state and territory of Australia is deemed to be subject to earthquakes, although by world standards the seismicity levels are not severe.

Acceleration coefficients were calculated on the basis of a probabilitistic analysis using recorded peak ground velocity as the intensity value for historic earthquakes in Australia. These coefficients vary between a minimum of 0.03 and a maximum of 0.22. They are based on a risk of 10% probability of exceedance in a 50-year period. On this basis, specific values of acceleration coefficients were assigned, as shown in Table 7.1, to the 21 major cities of the country, where over 90% of the population lives. Values of the acceleration coefficients range from a minimum of 0.05 for the city of Hobart to a maximum of 0.11 for Newcastle.

Requirements for analysis and design in the new Australian Standard are related to the seismic design categories specified in ATC 3-06 (1988). These categories reflect the effect of seismic activity and building occupancy, modified to suit Australian conditions (see

Table 7.1. Acceleration Coefficient *a* for Major Cities

City	Acceleration Coefficient *a*
Adelaide	0.10
Albury/Wodonga	0.08
Ballarat	0.08
Bendigo	0.09
Brisbane	0.06
Cairns	0.06
Canberra	0.08
Darwin	0.08
Geelong	0.10
Gold Coast/Tweed Heads	0.06
Hobart	0.05
Latrobe Valley	0.10
Launceston	0.06
Melbourne	0.08
Newcastle	0.11
Perth	0.09
Rockhampton	0.08
Sydney	0.08
Toowoomba	0.06
Townsville	0.07
Wollongong	0.08

Table 7.2). Construction is classified as ductile or nonductile, and building configuration as regular or irregular. On the basis of these classifications and on the height of the buildings, restrictions are imposed on the use of certain structural materials, and additional requirements on detailing are stipulated. Static analysis is acceptable for seismic design in most situations, but for irregular structures in Design Categories D or E (see Table 7.3), dynamic analysis must be used.

Where dynamic analysis is used, ground motion time histories of earthquakes for specific sites will seldom be available; hence the response spectrum method is considered more appropriate. The design spectra should be scaled to the site acceleration coefficient, and factored to account for occupancy and for structure ductility. For buildings of irregular configuration, when the base shear force determined by dynamic analysis is less than 100% of the value calculated by static analysis, the results of the dynamic analysis must be scaled up to the level of the static analysis force. Values calculated for other forces, moments, and deflections must be scaled similarly. For regular buildings, the calculated dynamic base shear force must be scaled up to 90% of the value obtained by static analysis.

In using the static analysis method to calculate the base shear force, the vertical distribution of seismic lateral forces, and the story drift, the approach of ATC 3-06 (1988) is followed, although certain factors have been changed to reflect the local conditions prevalent in Australia.

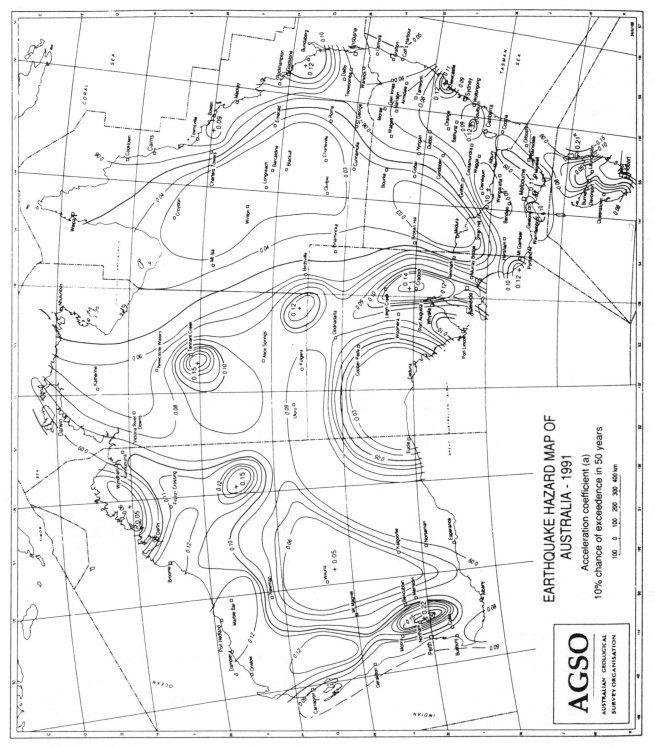

Fig. 7.1. Acceleration coefficient map of Australia. (By kind permission of the Australian Bureau of Mineral Resources, Canberra.)

7.3.1 General Requirements

Australian Standard AS 1170, Parts 1–3 (1989–1990), prescribes loads and combinations of loads for the analysis and design of buildings. Part 1 provides information to determine the dead and live loads in a building, Part 2 applies to wind loads, and Part 3 deals with snow loads. Earthquake loads have been considered in Part 4 [AS 1170.4 (1993)].

There are two main steps in the process of determining earthquake loads: (1) evaluating the seismic base shear force and (2) distributing the base shear force as acting forces laterally applied at the various levels of a building. It is also necessary to determine the torsional moments due to these lateral forces applied eccentrically with respect to the center of rigidity at each story of the building, as these moments modify the force distribution among the resisting structural elements.

The method of analysis for seismic-resistant design of buildings depends on the following:

- Type of building occupancy
- Building configuration
- Value of the acceleration coefficient at the specific location.

The new standard for seismic design provides for two methods of analysis: (1) the equivalent static method, which is generally adequate for areas of moderate and low levels of risk, and (2) the dynamic method, which may be used in any structure but is mandatory for structures in the highest design categories when the building configuration is classified as irregular. Also, it may be desirable to use the dynamic method in the design of post-earthquake recovery facilities and for structures where hazardous materials are manufactured or stored.

7.3.2 Structural Classification

The code provisions classify structures as either domestic or general structures. Within the general structures classification there are three types of building based on occupancy and use.

Type I.   Buildings not classified as Type II or III and including domestic buildings of more than two stories

Type II. Buildings with high occupancy capacity (schools, theaters, auditoriums, etc.)

Type III. Buildings used for essential functions or to store hazardous materials (power stations, hospitals, etc.)

7.3.3 Building Configurations

Buildings are classified as regular or irregular structures. A building is deemed irregular in plan if any of the following applies:

Torsional irregularity. The center of mass and the center of stiffness at any story are separated by a distance greater than one-tenth of the dimension normal to the direction of seismic loading under consideration

Re-entrant corners. Projections greater than $0.15L$ or $0.15B$ where L is the length of the building along the direction considered and B is the normal dimension of the building

Diaphragm discontinuity. Open areas in a horizontal diaphragm are greater than half of the area enclosed by the diaphragm

Nonparallel systems. Resisting elements are not orthogonal to main axes of the building.

A structure is deemed irregular in the vertical direction if any of the following applies:

Stiffness irregularity or soft story. The stiffness of a story is less than 70% of that of the story above.

Gravity load irregularity. The load at any level of the building is greater than 1.5 times the load of either adjacent level, except for the roof load which can be less than two-thirds of the load on the next lower level.

Column or wall offsets. The resisting elements are offset from one floor to the next.

Strength discontinuity or weak story. The strength of a story is less than 80% of the strength of the story above.

7.3.4 Acceleration Coefficient

The acceleration coefficient used to predict the magnitude of the earthquake forces depends on the geographical location and on the earthquake risk as evaluated from historic seismic records and geological data. The code provides a contour map for acceleration coefficient (Fig. 7.1) in addition to Table 7.1, which contains values for seismic acceleration coefficients for the major cities of Australia.

7.3.5 Seismic Design Category

There are five seismic design categories for general structures. These categories determine the level of design and detailing, and also the requirement to use dynamic analysis. For the lowest category (A), it is not deemed necessary to consider earthquake loading.

The seismic design categories are given in Table 7.2; they are determined by the structure classification specified and the product of the acceleration coefficient a from Table 7.1 (or from Fig. 7.1) and the site factor S from Table 7.4.

Table 7.2. Seismic Design Categories*

Product of Acceleration Coefficient and Site Factor aS	Structure Classification			
	III	II	I	Domestic
$aS \geqslant 0.20$	E	D	C	H3
$0.10 \leqslant aS < 0.20$	D	C	B	H2
$aS < 0.10$	C	B	A	H1

*The requirements for analysis corresponding to Categories A–E assigned in this table are summarized in Table 7.3.

Table 7.3. Analysis Requirements for Seismic Design Categories

Design Category	Regular Buildings	Irregular Buildings
A	None	None
B	None (if ductile)	Static analysis
C	Static analysis	Static analysis
D	Static analysis*	Dynamic analysis
E	Static analysis*	Dynamic analysis

*Subject to height limits on certain types of structural systems (Code Sections 2.7.5 and 2.7.6). Only domestic structures in Design Category H3 require static analysis. For detailing requirements, see Section 7.3.16.

7.3.6 Determination of Seismic Forces by Static Analysis

Structures must be designed for seismic forces in all directions. Thus, it is necessary to evaluate actions due to forces in the direction of each major axis of the building. Where individual structural elements contribute to the resistance in both directions, the worst combination of 100% loading in one direction, and 30% loading in the orthogonal direction must be used in their design.

7.3.7 Seismic Base Shear Force

The seismic base shear force V (in kN) is given by

$$V = \frac{ICS}{R_f} G_g \qquad (7.3)$$

subject to the condition that

$$0.01 G_g \leqslant V \leqslant \frac{2.5Ia}{R_f} G_g$$

where

I = importance factor
$\quad I = 1.0$ for structural classification Types I and II
$\quad I = 1.25$ for structural classification Type III
C = earthquake design coefficient

$$C = \frac{1.25a}{T^{2/3}} \qquad (7.4)$$

in which a is the acceleration coefficient for the location and T is the fundamental period of the structure in seconds.

The fundamental period of the structure corresponding to the translational mode under consideration can either be calculated using rigorous dynamic analysis methods or estimated by using one of the following empirical formulas:

$$T = \frac{h_n}{46} \text{ for fundamental translation period in the main direction of the building} \qquad (7.5)$$

$$T = \frac{h_n}{58} \text{ for fundamental translation period in the orthogonal direction in which the building is more rigid}$$

where h_n is the building height above the structural base in meters.

In eq.(7.3), S is the site factor, which depends on the soil profile as established by geotechnical investigations. Table 7.4 gives values of S between 0.67 and 2.0 for a range of conditions; interpolation is permitted. These values reflect the acceleration amplification that develops from foundations on soft soils.

G_g is the gravity load, which includes that part of the live load ψQ used in assessing the design load for the strength limit state. Values of ψ vary from 0.4 to 0.6, depending on the building use [see AS 1170.1 (1992)]. Hence,

$$G_g = G + \psi Q \qquad (7.6)$$

where G and Q are, respectively, the dead and live loads (in kN).

Table 7.4. Site Factor (S)

Soil Profile	Site Factor (S)
1. A profile of rock materials with rock strength class L (Low) or better	0.67
2. A soil profile with either: (a) rock materials Class EL (Extremely Low) or VL (Very Low) characterized by a shear wave velocity >760 m/sec; or (b) not more than 30 m of medium-dense to very dense coarse sands and gravels; firm, stiff or hard clays; or controlled fill	1.0
3. A soil profile with more than 30 m of medium-dense to very dense coarse sands and gravels; firm, stiff or hard clays; or controlled fill.	1.25
4. A soil profile with a total depth of 20 m or more and containing 6 m to 12 m of very soft to soft clays; very loose or loose sands; silts or uncontrolled fill.	1.5
5. A soil profile with more than 12 m of very soft to soft clays; very loose or loose sands; silts; or uncontrolled fill characterized by a shear wave velocity <150 m/sec.	2.0

Table 7.5. Structural Response Modification Factor (R_f) and Deflection Amplification Factor (K_d) for Building Structures (from AS 1170.4)

Type of Structural System	R_f	K_d
Bearing Wall Seismic-Resisting System		
Light-framed walls with shear panels	6.0	4.0
Reinforced concrete shear walls	4.5	4.0
Reinforced masonry shear walls	4.0	3.0
Concentrically braced frames	4.0	3.5
Unreinforced masonry shear walls	1.5	1.25
Building Frame Seismic-Resisting System		
Eccentrically braced steel frames, connections at	7.0	4.0
columns away from link	7.0	4.5
Light-framed walls with shear panels	5.0	4.5
Concentrically braced frames	6.0	5.0
Reinforced concrete shear walls	5.0	4.0
Reinforced masonry shear walls	1.5	1.5
Unreinforced masonry shear walls		
Moment Resisting Frame Seismic-Resisting System		
Special moment frames of steel	8.0	5.5
Special moment frames of reinforced concrete	8.0	5.5
Intermediate moment frames of steel	6.5	4.5
Intermediate moment frames of reinforced concrete	6.0	3.5
Ordinary moment frames of steel	4.5	4.0
Ordinary moment frames of reinforced concrete	4.0	2.0
Dual System with a Special Moment Frame		
Complementary Seismic-Resisting Elements		
Eccentrically braced steel frames	8.0	4.0
Concentrically braced frames	6.5	5.0
Reinforced concrete shear walls	8.0	6.5
Reinforced masonry shear walls	6.5	5.5
Dual System with an Intermediate Moment Frame of		
Reinforced Concrete or an Ordinary Moment Frame of		
Steel and Complementary Seismic-Resisting Elements		
Concentrically braced frames	5.5	4.5
Reinforced concrete shear walls	6.0	5.0
Reinforced masonry shear walls	5.5	4.5

Note: A height limitation of 30 m above the base of the building applies for ordinary and intermediate moment frames, 50 m for bearing wall systems, and 70 m for building frame systems, for Design Category E.

R_f is the structural response factor, which depends on the structural system. This factor takes into account the potential of the structural system to dissipate energy through damping and through ductility. The code values for this factor vary from 1.5 for a brittle system such as unreinforced masonry, to 8 for very ductile reinforced concrete or steel frame structures. Table 7.5 provides values for R_f as given in the latest code, Australian Standard 1170.4 (1993).

7.3.8 Distribution of the Base Shear Force

In contrast to wind loading, which is assumed to be distributed uniformly over the building height, seismic loading is related to the amplitude of the motion, and thus increases from the bottom to the top of the building. In the simplified approach used for static analysis, the lateral displacements are assumed to be a function of the height. On the basis of this assumption, the force F_x (in kN) at a given level x is calculated by

$$F_x = C_{Vx} V \qquad (7.7)$$

where V is equal to the total shear force given by eq.(7.3) and C_{Vx} is the seismic coefficient given by

$$C_{Vx} = \frac{G_{gx} h_x^k}{\sum_{i=1}^{n} G_{gi} h_i^k} \qquad (7.8)$$

where

n = number of levels in the building
G_{gx}, G_{gi} = the gravity loads at levels x, i
h_x, h_i = the heights above base of levels x, i
k = an exponent related to the fundamental period of the building
$k = 1.0$ for $T \leqslant 0.5$ sec
$k = 2.0$ for $T \geqslant 2.5$ sec
$k = 1 + 0.5(T - 0.5)$ for $0.5 < T < 2.5$ sec
(i.e., linear interpolation between 1.0 and 2.0)

Figs 7.2 (a), (b), and (c) illustrate the distribution of the force F_x respectively for $k = 1.0$, $k = 1.5$, and $k = 2.0$; the base shear force V has the same value in all these cases. However, the distribution of the forces F_x is different and therefore the shear forces and overturning moments at the various levels of the building also will be significantly different for different values of the exponent k.

7.3.9 Horizontal Shear Force

The horizontal shear force V_i at any story i of the building is determined simply by statics as the sum of the forces F_x from level i to the roof; that is,

$$V_i = \sum_{x=i}^{n} F_x \qquad (7.9)$$

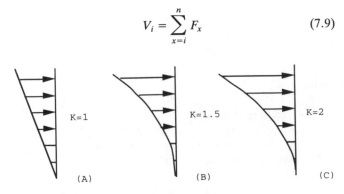

(A)   (B)   (C)

Fig. 7.2. Vertical distribution of seismic forces

where n is the number of levels in the building, including the roof.

The horizontal shear force V_i at story i is resisted by the structural elements on the next lower story according to their relative stiffness values.

7.3.10 Torsional Moments

The seismic code requires that horizontal torsional moments at each story of the building be considered. The most severe of the torsional moments M_{t1x} or M_{t2x} calculated at each story x, for the following two cases should be used in design.

Case 1

$$M_{t1x} = V_x e_{d1} \qquad (7.10)$$

Case 2

$$M_{t2x} = V_x e_{d2}$$

$$e_{d1} = A_1 e_s + 0.05b \qquad (7.11)$$

and

$$e_{d2} = A_2 e_s - 0.05b$$

where

V_x = the story shear force at level x
e_s = the static eccentricity (normal distance from the center of the mass above story x to the stiffness center of that story)
A_1 and A_2 = the dynamic eccentricity factors given by

$$A_1 = 2.6 - 3.6\frac{e_s}{b} \geqslant 1.4$$

and

$$A_2 = 0.5$$

and where b is the maximum dimension of the structure at level i in the direction perpendicular to the line of action of the shear force V_x.

Figs 7.3(a) and 7.3(b) show, respectively, the formulation of the factors e_{d1} and e_{d2} and their values as a function of the ratio e_s/b.

7.3.11 Overturning Moment

The overturning moment M_0 at the base of the building is determined by statics as the sum of the moments due to the lateral seismic forces F_i. However, the code permits a reduction, by a factor $\alpha = 0.75$, in the resulting overturning moment for all structures except inverted pendulum structures. This reduction is consistent with results obtained from dynamic analysis

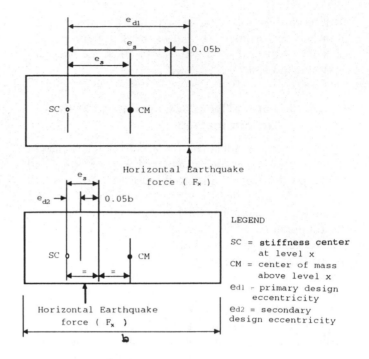

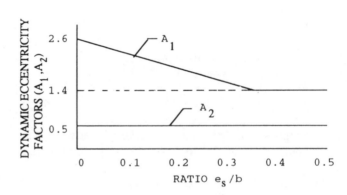

Fig. 7.3.   Maximum and minimum torsional eccentricity

because the design shear envelope based on the simple static analysis method is generally conservative as the story shears do not all attain their maximum values simultaneously. Also any slight uplifting during vibration will result in a reduction in moment. Hence,

$$M_0 = \alpha \sum_{i=1}^{n} F_i h_i \qquad (7.12)$$

where

α = 0.75 for all structures except inverted pendulum structures;
n = number of levels in the building
F_i = lateral seismic force at level i
h_i = height above base of the building to level i.

As noted above, the reduction factor α in eq.(7.12) may not be applied for inverted pendulum structures, or in the design of piers and columns for such structures. For this type of structure the overturning moment is assumed to vary linearly between M_0 (calculated by eq.(7.12) with $\alpha = 1.0$) at the base and $0.5M_0$ at the top.

7.3.12 Lateral Displacements and Drift Determination

All structures must be separated from adjacent structures by a distance calculated as $0.375R_f\delta_x$, and where portions of a building do not act as an integral unit, there must be the same separation to prevent damaging contact. The interstory drift δ_x, including the torsional effects, is limited to a value that ensures structural stability and controls damage (but does not necessarily eliminate damage to architectural components). The permissible value of drift at each story is limited to $0.015h_{sx}$, where h_{sx} is the interstory height.

The code provides for the determination of lateral displacements δ_{xe} at the various levels of the building by means of elastic analysis of the structure subjected to the seismic forces previously calculated. To allow for inelastic effects, these elastic displacements are multiplied by the amplification factor K_d from Table 7.5, and where applicable, increased to allow for P-delta effects as described in Section 7.3.13. Thus, the design lateral displacement δ_x at level x is given by

$$\delta_x = K_d\delta_{xe} \qquad (7.13)$$

The design story drift is the difference between the displacements δ_x at the top and at the bottom of the story.

The alternative method of calculating δ_x, used elsewhere, with modal analysis [see eq.(4.72), Chapter 4], gives:

$$\delta_{xe} = a_i \cdot \left(\frac{T}{2\pi}\right)^2$$

$$\delta_{xe} = \frac{gF_i}{G_{gi}} \cdot \left(\frac{T}{2\pi}\right)^2 \qquad (7.14)$$

where

a_i = the acceleration at level i
F_i = the seismic force at level i
G_{gi} = the seismic weight at level i
T = the fundamental period.

The Australian code does not specify this method of calculating displacements, which is used in some codes for multimode analysis. However, it has been used in the computer program developed for this chapter to provide an estimation of displacements and drifts; a more precise determination would require detailed information on structural member sizes, or on the total stiffness of each story.

7.3.13 P-Delta Effects

To allow for P-delta (P-Δ) effects, the story drift is increased by a factor $0.9/(1 - m) \geqslant 1.0$ (i.e., no effect when $m < 0.1$).

The stability coefficient m is given by:

$$m = \frac{P_x\Delta_x}{V_xh_{sx}K_d} \qquad (7.15)$$

where

P_x = total vertical design load at story x
Δ_x = design story drift
V_x = seismic horizontal shear force at level x
h_{sx} = height of story x
K_d = deflection amplification factor (Table 7.5)

When $m < 0.1$, the P-delta effect may be ignored. When $m \geqslant 0.1$, in addition to increasing the story drift, modifications also must be made in the story shears and overturning moments, which were determined without considering the P-delta effect.

7.3.14 Vertical Component of Ground Motion

Only for buildings of Design Categories D and E are the effects of the vertical seismic motion deemed to be significant. Unless otherwise indicated by specific data, the vertical component may be defined as half of the corresponding horizontal acceleration. In ultimate strength design, this additional vertical load is unlikely to be significant.

7.3.15 Dynamic Analysis

Where dynamic analysis is used, the earthquake actions may be represented by:

(a) Response spectra normalized to the site acceleration coefficient a previously described, modified by the ratio I/R_f to account for structural importance and ductility; or

(b) Design response spectra developed for the specific site; or

(c) Ground motion time histories chosen for the specific site, which must be representative of actual earthquake motions.

Where dynamic analysis is required, a three-dimensional mathematical model must be used that adequately represents the spatial distribution of mass and stiffness of the actual physical structure. A sufficient number of modes must be included in the analysis so that the total effective modal mass is at least 90% of the total mass of the structure. The peak member forces and displacements or story shears, etc., may be determined by any recognized method, such as the SRSS (Square Root of Sum of Squares) rule, for combining the modal responses. However, when modal periods are closely spaced, interaction effects among modes must also be considered.

The base shear for any given direction must be compared with the values given by the static analysis previously described, and must not be less than the following values:

(a) For irregular buildings: 100% of the static analysis base shear

(b) For regular buildings: 90% of the static analysis base shear.

Where calculated base shear values are greater than static analysis results, they *may* be scaled down correspondingly. All other response parameters, including displacements, member forces, and moments also shall be adjusted proportionately.

The effects of torsion also must be accommodated by appropriate adjustments to the model to include the effects of "accidental torsion." Where dynamic analysis is used, the calculated overturning moments may not be less than those calculated by the static analysis procedure in Section 7.3.11.

7.3.16 Additional Requirements: Detailing, Architectural Components, and Services

7.3.16.1 General detailing requirements. Detailing requirements are given in Section 4 of the code for each design category, based on relative structural ductility. For buildings in Design Category C, D, or E, restrictions are placed on the use of unreinforced masonry for buildings of more than two stories. Apart from ductile buildings in categories A and H1, the following detailing requirements must be satisfied:

1. Load paths shall be provided so that earthquake-generated forces from both structural and nonstructural components are transmitted to the foundations.
2. Ties and continuity must be provided, including connections across separation joints, and a positive connection must be provided to resist horizontal forces at beam or truss supports equal to 5% of the gravity load reactions (2.5% for Category A, 7.5% for Category H3). For Categories C, D, and E, portions of the structure must also be tied to the remainder with elements having a strength to resist $0.33(aS)$ times the gravity load of the smaller part.
3. Walls shall be anchored to the roof and to all floors that provide lateral support, with a connection capable of resisting a force of $10(aS)$ kN per meter (only 50% of this for Category A), or a minimum of 0.8 kN per meter.

Additional details are specified for buildings in Design Categories C, D, and E. These details include requirements for floor and roof diaphragms to ensure that such diaphragms function in distributing loads. Also, interconnections of bearing wall elements must be ductile and sufficiently strong to resist shrinkage, temperature, and settlement effects, in addition to seismic forces. Where there are openings in shear walls or diaphragms, local strengthening must be provided at the edges of the openings.

Where footings are supported on piles or caissons, or where spread footings are located on soils with a bearing capacity of less than 250 kPa, restraint in two directions must be provided between the separate footings. This restraint must be capable of resisting a tie force equal to $0.25(aS)$ times the larger gravity load on the footings being tied together.

7.3.16.2 Requirements for architectural, mechanical, and electrical components. These requirements have two objectives:

1. To ensure safety of the occupants and the general public from danger of falling objects.
2. To prevent failure of the mechanical and electrical components.

Emphasis is given to the performance of connections, attachments, and restraining devices, so that the integrity of components is maintained under seismic loading. Seismic coefficients are prescribed for a wide range of components including attachments such as parapets and noload-bearing walls, partitions, ceilings, racks, lifts, heating systems, and air-conditioning plants and ducts.

Connections must be designed to resist a force F_p, applied at the center of gravity, given by:

$$F_p = aSa_c a_x C_{c1} IG_c \leqslant 0.5G_c \qquad (7.16)$$

where

a = acceleration coefficient (see Table 7.1 or Fig. 7.1)
S = site factor (see Table 7.4)
a_c = attachment amplification factor,
a_x = height amplification factor at height x,
C_{c1} = seismic coefficient for components of architectural or other systems
I = importance factor
G_c = weight of component or equipment.

For some mechanical and electrical systems, further multipliers may be required to allow for attachment and height amplification. Values for these coefficients and multipliers may be found in AS 1170.4.

7.4 EXAMPLE 7.1

Fig. 7.4 models a ten-story building with a uniform story height of 4 m. The weight of each floor is 2,900 kN, and the weight of the roof is 800 kN, including the live load. The floor dimensions are 25 m × 20 m, at all levels. The building structural classification is Type II, and the building configuration is regular. The location is Adelaide, where the acceleration coefficient is 0.10. The building has a concrete building frame and shear wall construction, and the centers of mass coincide with the centers of resistance of the horizontal load resisting system. The site conditions are defined by a soil profile consisting of 25 m including 10 m of soft clay.

Solution

From Table 7.2, the seismic design category is C (Type II, $a = 0.10$), and as indicated in Table 7.3, static analysis is adequate.

Using the equivalent static force method, the analysis will determine first the seismic base shear and then the vertical distribution of the seismic forces. From this distribution, the shears, overturning moments, torsional moments, lateral displacements, and drifts will be found.

1. Determination of seismic forces by static analysis. Structures must be designed for seismic forces in all directions. However, those structural elements that

contribute to the lateral resistance in one direction only need to be designed for 100% of the earthquake force in that direction only. Those structural elements that contribute to the lateral resistance in both directions must be considered for the combination of 100% in one direction and 30% in the orthogonal direction.

2. Seismic base shear. The total seismic base shear V (kN) is given by eq.(7.3):

$$V = \frac{ICS}{R_f} G_g; \quad V_{min} = 0.01 G_g, \quad V_{max} = \frac{2.5 I a}{R_f} G_g$$

where

$I = 1.0$ (importance factor for structural classification Type II, Section 7.3.7)
$C =$ seismic design coefficient:

$$C = \frac{1.25a}{T^{2/3}} \qquad [eq.(7.4)]$$

in which

$$a = 0.10$$

$$T_1 = h_n/46 = 40/46 \qquad [eq.(7.5)]$$

$$= 0.87 \, sec$$

$$T_2 = h_n/58 \text{ for orthogonal direction}$$

$$= 40/58 = 0.69 \, sec$$

Hence,

$$C_1 = 0.137 \quad \text{and} \quad C_2 = 0.160$$

The orthogonal direction is that in which the structure is more rigid ($L = 25$ m).

$S = 1.5$ (site factor for the given soil profile, Table 7.4)
$G_g =$ gravity load including the appropriate part of the live load.

Thus, nine floors at 2,900 kN per floor and 800 kN for roof:

$$G_g = 9 \times 2,900 + 800 = 26,900 \, kN \qquad [eq.(7.6)]$$

$R_f =$ structural response modification factor:
$R_f = 6$ and $K_d = 5$ (for reinforced concrete shear wall system, Table 7.5)

Substituting,

$$V_1 = 0.0343 G r g$$

$$= 921 \, kN \text{ for N–S direction.}$$

This value is within the limits $V_{min} = 0.01 G_g$ to $V_{max} = 0.042 G_g$. Similarly,

$$V_2 = 1,076 \, kN \quad \text{for} \quad \text{E–W direction}$$

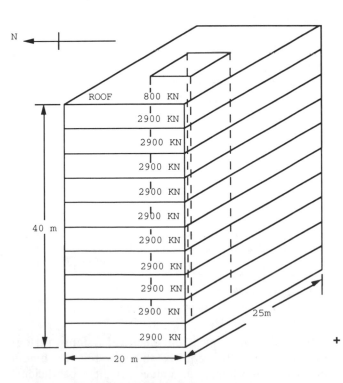

Fig. 7.4. Design Example 7.1, structural model

3. Vertical distribution of horizontal seismic forces. The lateral force F_x (in kN) induced at level x is given by eq.(7.7):

$$F_x = C_{vx} V$$

where

$$C_{Vx} = \frac{G_{g_x} h_x^k}{\sum_{i=1}^{n} G_{g_i} h_i^k} \qquad [\text{eq.}(7.8)]$$

$n = 10$

$G_{gi} = 2{,}900$ kN for $i = 1$ through 9, and $G_{g10} = 800$ kN for the roof

$h_i = 4.0 \times i$

$k = 1 + 0.5(T - 0.5)$ for $0.5 < T < 2.5$ sec (Section 7.3.8)

$\quad k_1 = 1.185$ for the fundamental period

$\quad k_2 = 1.095$ for the orthogonal period

Values of F_{x1} and F_{x2} calculated for the two directions are given in Table 7.6.

4. Horizontal shear at each level. The seismic shear force (horizontal) at each level is determined simply by statics as the sum of the forces F_x from level i to the roof; that is,

$$V_i = \sum_{x=i}^{n} F_x \qquad [\text{eq.}(7.9)]$$

where n is the number of levels in the building, including the roof.

Calculated values of shear for V_{1i} for the N–S direction and V_{2i} for the E–W direction are given in Table 7.6.

5. Overturning moments at each level. The overturning moment M_i at level i is the sum of the moments of all the horizontal forces above level i to the roof, level n, i.e.,

$$M_i = \sum_{x=i}^{n} F_x(h_x - h_i)$$

Calculated values of M_{1i} and M_{2i}, respectively, for the N–S and E–W directions are given in Table 7.6.

6. Torsional effects. The maximum and minimum eccentricities are determined by

$$e_{d1} = A_1 e_s + 0.05b$$
$$e_{d2} = A_2 e_s - 0.05b \qquad [\text{eq.}(7.11)]$$

in which

e_s = static eccentricity = 0 (symmetry of mass and resistance)

A_1 = dynamic eccentricity factor = $2.6 - 3.6 e_s/b \geqslant 1.4 = 2.6$

A_2 = dynamic eccentricity factor = 0.5

b = the maximum dimension of the structure at level i, perpendicular to the line of action of the shear force V_{ix}.

$e_{d1} = -e_{d2} = +0.05 \times 25 = +1.25$ for N–S direction

$e_{d1} = -e_{d2} = +0.05 \times 20 = +1.00$ for E–W direction

Hence, the design torsional moments are given by

$$(M_{ti})_1 = \pm 1.25 V_{1i} \quad \text{for the N–S direction} \qquad [\text{eq.}(7.10)]$$

$$(M_{ti})_2 = \pm 1.00 V_{2i} \quad \text{for the E–W direction}$$

The calculated torsional moments are also given in Table 7.6.

Table 7.6. Results of Calculations for Example 7.1

Level	Forces N–S (kN)	E–W	Shears N–S (kN)	E–W	*Overturning Moments N–S (kN-m)	E–W	Lateral Displacements N–S (mm)	E–W	Torsional Moments N–S (kN-m)	E–W
Roof	58	66	58	66	232	264	68.0	48.8	72	66
9	187	210	245	276	1,212	1,368	60.5	42.8	306	276
8	162	185	407	461	2,840	3,212	52.4	37.7	509	461
7	138	159	545	620	5,020	5,692	44.6	32.4	681	620
6	115	135	660	755	7,660	8,712	37.2	27.5	825	755
5	93	111	753	866	10,672	12,176	30.1	22.6	941	866
4	71	87	824	953	13,968	15,988	23.0	17.7	1,030	953
3	51	63	875	1,016	17,468	20,052	16.5	12.8	1,094	1,016
2	31	41	906	1,057	21,092	24,280	10.0	8.3	1,132	1,057
1	14	19	921	1,076	24,776	28,584	4.5	3.9	1,151	1,076

*Overturning moments are at the base of the column below the level considered, e.g., moment for level 1 applies at ground level. Hence design moments for foundations are equal to moments at foundation reduced by the factor $\alpha = 0.75$, that is $0.75 M_{11} = 18{,}582$ (kN-m) and $0.75 M_{21} = 21{,}438$ (kN-m), respectively, for N–S and E–W directions.

7. Lateral displacements. The alternative method of calculating the elastic lateral displacements δ_{xe} is given by eq.(7.14) as

$$\delta_{xe} = \frac{gF_i}{G_{g_i}} \cdot \left(\frac{T}{2\pi}\right)^2 \qquad \text{[eq.(7.14)]}$$

and the total lateral displacements, including inelastic effects, by

$$\delta_x = K_d \delta_{xe} \qquad \text{[eq.(7.13)]}$$

Substituting for T, and values of F_i and G_{gi} for each level and for both directions, with $K_d = 5$, lateral displacements are calculated and shown in Table 7.6.

The design story drift Δ_x is the difference between the displacements δ_x at the top and at the bottom of each story. Calculated drift values for this example have not been tabulated. The maximum calculated value for story drift was 8 mm, which is less than $0.015 \times 4,000 = 60$ mm, the code limit on story drift.

8. P-Delta effects. P-delta effects must be considered when $m > 0.1$:

$$m = \frac{P_x \Delta_x}{V_x h_{sx} K_d} \qquad \text{[eq.(7.15)]}$$

$$m = \frac{800 \times 7.5}{58 \times 4,000 \times 5} = 0.005 \text{ (for 10th story)}$$

$$m = \frac{24,000 \times}{906 \times 4,000 \times 5} = 0.007 \text{ (for 2nd story)}$$

No correction is necessary for P-delta effects, as $m < 0.1$.

7.5 STRUCTURAL DESIGN AND DETAILING REQUIREMENTS FROM OTHER AUSTRALIAN CODES

7.5.1 Concrete Structures AS 3600 (1988)

The current version of AS 3600, *Concrete Structures*, relates to the 1979 Earthquake Code AS 2121 and will be revised according to the new Standard, AS 1170.4. Therefore, certain requirements that are shown below may be revised in the near future.

Reinforced and prestressed concrete construction can be classified as ductile construction for which the standard requirements of AS 3600 provide adequate detailing, including the use of footing ties when soil conditions are poor. Design for lateral loading is required only in Zones 1 and 2, with a number of special detailing requirements to be satisfied; these requirements are similar to those of ACI 318-76.

For strength design, alternative load combinations of $1.25G + 1.5F_{eq} + \psi_c Q$, and $0.8G + 1.5F_{eq}$ must be considered, where G = dead load, Q = live load, F_{eq} = seismic load, and ψ_c is a live load combination factor (generally 0.4, but 0.6 for storage areas). For serviceability design, the load is F_{eq} and a limit of $h/200$ is given for story drift.

If supported by piles or caissons, footings must be tied together. They must also be tied together whenever the bearing pressure under isolated footings is less than 100 kPa, unless it can be established that soil structure collapse or liquefaction cannot occur under earthquake conditions. These ties must be able to resist a force equal to 10% of the vertical column/wall loads on such footings.

Concrete cladding elements must accommodate relative movements, have connections with strength sufficient to resist local earthquake forces, and have ductility sufficient to preclude brittle failure at welds or in the concrete.

In the case of moment-resisting frames, detailing requirements graded for Zones 1 and 2 also depend on the building height. They may include minimum areas for reinforcement continuity in both faces, and confinement at lapping splices, as well as shear reinforcement over the whole length of the beam. Closer spacing of stirrups or ties adjacent to beam/column connections is also required, particularly on perimeter joints. In shear walls, anchorage of reinforcement must be given special consideration; confinement must be provided for longitudinal steel in the members of a braced frame. For flat slabs, there are specific recommendations for location of the reinforcement to provide moment transfer to the columns.

7.5.2 Steel Structures AS 4100 (1990)

The AS 4100 (1990) *Steel Structures*, also relates to AS 2121 (1979); it will be revised when the new earthquake loading code is published. Although AS 4100 provisions are restricted to steel structures, they include the use of concrete shear walls as the sole system or as part of a combined system for resisting lateral forces. The steel code does not give guidance for the use of steel in conjunction with masonry structural walls.

The provisions of AS 4100 (1990) for steel structures are similar to those of AS 3600 (1988) for concrete structures; they require detailing to ensure that ductility levels and the separation of adjacent buildings to avoid hammering, are appropriate to the seismic risk. Similarly, explicit guidance is given for the support of cladding elements to ensure that they are capable of accommodating movements resulting from horizontal earthquake forces.

For structures in Zone A (see AS 4100 section

3.2.1) sufficient ductility is deemed to be provided by the use of a steel grade for which the specified yield stress is less than 450 megapascals (MPa). For ductile structures in Zone 1, the maximum specified yield stress may not exceed 350 MPa. In addition, the design axial tension force in bracing members is limited to 0.85 times the design capacity, and web stiffeners in beam/column connections must extend the full depth between flanges and be butt-welded to both flanges. For ductile structures in Zone 2, the design of members must comply with the requirements specified for plastic design; all welds must be of category SP with strict quality control.

Also, restrictions are placed on the methods of fabrication for components subject to plastic deformation. This will apply to all steel structures deemed to be ductile, i.e., where the brittle factor ($K = 3.2$) is not used in calculating seismic forces.

7.6 COMPUTER PROGRAM

A computer program has been developed to implement the provisions of Draft Australian Standard 1170.4 (1993) for minimum design loads on structures, due to earthquakes. The program is restricted to the calculation of loads using the equivalent static force method, but a prompt is given to indicate when this method is not valid; i.e., for "irregular" buildings under certain higher risk conditions. The program allows structural and foundation data to be entered interactively from the keyboard or from an existing data file. It includes provisions for the calculation of the fundamental period from the empirical formulas of the code. The program determines the base shear in each principal direction of the building and values of lateral forces, shears, torsional and overturning moments, and lateral displacements for each story level. These values are printed in tabular form for each level of the building. In addition, a table of calculated and allowable story drift values is provided.

7.6.1 Example 7.2

Example 7.1 is solved by the computer program developed for this chapter.

```
INPUT DATA AND OUTPUT RESULTS FOR EXAMPLE 7.2

*********************************************************************

              ***EARTHQUAKE RESISTANT DESIGN***  "

          DAVID CRAWLEY,  UNIVERSITY OF ADELAIDE"

       ***USING  AUSTRALIAN STANDARD 1170.4 ***PROGRAM 7A"

*********************************************************************

          ***DATA FILE INFORMATION***

              1. PREPARE NEW DATA FILE

              2. MODIFY EXISTING DATA FILE

              3. USE EXISTING DATA FILE

          SELECT NUMBER : 3

          DRIVE USED FOR DATA FILES (A:,B:,or C:)  :B

          FILE NAME (OMIT DRIVE LETTER)(SAMPLE 10B)  :10R
```

```
INPUT DATA:

SEISMIC ACCELERATION                                    ACC = 0.1
BASE DIMENSION (EARTHQUAKE DIRECTION)    (M)    B = 20
BASE DIMENSION (ORTHOGONAL DIRECTION)    (M)    L = 25
STRUCTURAL RESPONSE FACTOR                              R = 6
TYPE OF SOIL                                            NST = 4
SITE FACTOR                                             SF = 1.5
NUMBER OF STORIES                                       ND = 10
```

STORY #	STORY HEIGHT METER	STORY WEIGHT kN.
10	4.00	800.00
9	4.00	2900.00
8	4.00	2900.00
7	4.00	2900.00
6	4.00	2900.00
5	4.00	2900.00
4	4.00	2900.00
3	4.00	2900.00
2	4.00	2900.00
1	4.00	2900.00

STORY TORSIONAL ECCENTRICITIES

STORY # NO.	ECCENTRICITY (METER) N - S	E - W
10	0.00	0.00
9	0.00	0.00
8	0.00	0.00
7	0.00	0.00
6	0.00	0.00
5	0.00	0.00
4	0.00	0.00
3	0.00	0.00
2	0.00	0.00
1	0.00	0.00

```
FUNDAMENTAL PERIOD MENU: (BASED ON AS1170.4)

     1. FUNDAMENTAL TRANSLATIONAL PERIOD   : T1=H/46"
        AND PERIOD FOR ORTHOGONAL DIRECTION : T2=H/58"

     2. VALUES PREVIOUSLY DETERMINED      :"

        SELECT A NUMBER ";1

     RESULTS :

     SEISMIC ACCELERATION (a)              = 0.1
     SITE FACTOR (S)                       = 1.5
     IMPORTANCE FACTOR (I)                 = 1
     STRUCTURAL RESPONSE MOD.FACTOR (R)    = 6
```

FACTOR	MAIN DIRECTION	ORTHOGONAL
STRUCTURE PERIOD (T)	0.870	0.690
SEISMIC DESIGN COEFFICIENT (C)	0.137	0.160

LEVEL	SEISMIC FORCES AT EACH LEVEL (Fx) MAIN DIRECTION	ORTHOGONAL	SHEAR FORCE (kN.) MAIN DIRECTION	ORTHOGONAL
10	58.39	65.25	58.39	65.25
9	186.83	210.76	245.23	276.00
8	162.50	185.26	407.73	461.26
7	138.72	160.06	546.45	621.32
6	115.56	135.20	662.01	756.53
5	93.11	110.74	755.13	867.27
4	71.48	86.74	826.61	954.00
3	50.84	63.30	877.44	1017.30
2	31.44	40.61	908.89	1057.91
1	13.83	19.01	922.72	1076.93

LEVEL	OVERTURNING MOMENT (kN.m) MAIN DIRECTION	ORTHOGONAL
9	233.57	260.99
8	1214.48	1365.01
7	2845.38	3210.06
6	5031.17	5695.35
5	7679.22	8721.47
4	10699.72	12190.53
3	14006.15	16006.55
2	17515.92	20075.77
1	21151.47	24307.42
0	24842.34	28615.13

DESIGN MOMENTS FOR FOUNDATIONS ARE:

```
     0     18632.76          21461.35.
```

LEVEL	LATERAL DISPLACEMENTS (mm) MAIN DIRECTION	ORTHOGONAL	REQUIRED CLEARANCE (mm) MAIN DIRECTION	ORTHOGONAL
10	68.50	48.15	154.13	108.33
9	60.46	42.90	136.04	96.53
8	52.59	37.71	118.32	84.85
7	44.89	32.58	101.01	73.31
6	37.40	27.52	84.15	61.92
5	30.13	22.54	67.80	50.72
4	23.13	17.66	52.05	39.73
3	16.45	12.89	37.02	28.99
2	10.10	8.27	22.90	18.60
1	4.48	3.87	10.07	8.71

RELATIVE DISPLACEMENTS BETWEEN LEVELS (mm):

PERMISSIBLE STORY DRIFT (MM) IS .015 x STORY HEIGHT

LEVEL	MAIN DIRECTION	ORTHOGONAL	PERMISSIBLE
10	8.039	5.245	60.00
9	7.875	5.190	60.00
8	7.695	5.129	60.00
7	7.494	5.060	60.00
6	7.265	4.980	60.00
5	7.000	4.886	60.00
4	6.681	4.770	60.00
3	6.275	4.619	60.00
2	5.700	4.396	60.00
1	4.476	3.870	60.00

```
THESE VALUES SHOULD BE CHECKED USING CODE CLAUSE 2.10

   TORSIONAL MOMENTS

        TORSIONAL MOMENT (kN.m):

   LEVEL        MAIN DIRECTION            ORTHOGONAL

   No.    MAXIMUM      MINIMUM        MAXIMUM      MINIMUM

   10      73.0        -73.0          65.2        - 65.2
    9     306.5       -306.5         276.0        -276.0
    8     509.7       -509.7         461.3        -461.3
    7     683.1       -683.1         621.3        - 621.3
    6     827.5       - 827.5        756.5        - 756.5
    5     943.9       - 943.9        867.3        - 867.3
    4    1033.3      -1033.3         954.0        - 954.0
    3    1096.8      -1096.8        1017.3        -1017.3
    2    1136.1      -1136.1        1057.9        -1057.9
    1    1153.4      -1153.4        1076.9        -1076.9
```

7.6.2 Example 7.3

The 25-story ordinary moment resisting steel building shown in Fig. 7.5 has a uniform story height of 3.6 m. This building is modeled as separate two-dimensional frames, one in each principal direction, for the determination of earthquake loads and for analysis. The effective weight of each floor and the roof is 2,900 kN. The plan dimensions are 25 m × 20 m

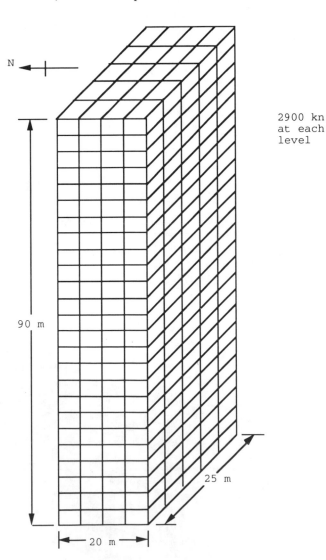

2900 kn at each level

90 m

20 m

25 m

Fig. 7.5. Design Example 7.3, structural model

at all levels. The building structural classification is Type II, and the building configuration is regular. It is located in Adelaide, where the acceleration coefficient is 0.10. The center of mass of the building coincides with the center of stiffness of the horizontal load resisting system at each level. The site conditions are defined by a soil profile consisting of 20 m of clay, including 10 m of soft clay (Soil Profile Type 4). Analyze this 25-story steel frame building, using the program developed for this chapter.

```
       INPUT DATA AND OUTPUT RESULTS FOR EXAMPLE 7.3
   ********************************************************************
          ***EARTHQUAKE RESISTANT DESIGN***   "

          DAVID CRAWLEY,   UNIVERSITY OF ADELAIDE"

          ***USING  AUSTRALIAN STANDARD AS1170.4 ***PROGRAM 7A"
   ******************************************************************** **

          ***DATA FILE INFORMATION***

              1. PREPARE NEW DATA FILE

              2. MODIFY EXISTING DATA FILE

              3. USE EXISTING DATA FILE

          SELECT NUMBER : 3
          DRIVE USED FOR DATA FILES (A:,B:,or C:)  :B
          FILE NAME (OMIT DRIVE LETTER)(SAMPLE 10B) :25R

       INPUT DATA:

   SEISMIC ACCELERATION                         ACC = 0.1
   BASE DIMENSION (EARTHQUAKE DIRECTION) (M)      B = 20
   BASE DIMENSION (ORTHOGONAL DIRECTION) (M)      L = 25
   STRUCTURAL RESPONSE FACTOR                     R = 4.5
   TYPE OF SOIL                                 NST = 4
   SITE FACTOR                                   SF = 1.5
   NUMBER OF STORIES                             ND = 25

   STORY #        STORY HEIGHT        STORY WEIGHT
                     METER                  kN.

      25            3.60               2900.00
      24            3.60               2900.00
   .....            .....             ........
       1            3.60               2900.00

   STORY TORSIONAL ECCENTRICITIES

   STORY #         ECCENTRICITY (METER)
     NO.          N - S           E - W

     25           0.00            0.00
     24           0.00            0.00
   ....           ......          ......
      1           0.00            0.00

       FUNDAMENTAL PERIOD MENU: (BASED ON AS1170.4)

          1. FUNDAMENTAL TRANSLATIONAL PERIOD  : T1=H/46"
             AND PERIOD FOR ORTHOGONAL DIRECTION : T2=H/58"

          2. VALUES PREVIOUSLY DETERMINED       :"

              SELECT A NUMBER : 1

       RESULTS :

   SEISMIC ACCELERATION   (a)      = 0.1
   SITE FACTOR (S)                 = 1.5
   IMPORTANCE FACTOR (I)               = 1
   STRUCTURAL RESPONSE MOD.FACTOR (R)  = 4.5

       FACTOR                 MAIN DIRECTION    ORTHOGONAL

   STRUCTURE PERIOD (T)           1.957           1.552
   SEISMIC DESIGN COEFFICIENT (C) 0.080           0.093

       SEISMIC FORCES AT EACH LEVEL (Fx)          SHEAR FORCE (kN.)

   LEVEL    MAIN DIRECTION   ORTHOGONAL    MAIN DIRECTION   ORTHOGONAL

    25        199.71          216.66         199.71          216.66
    24        186.11          203.57         385.82          420.23
    23        172.91          190.77         550.74          611.00
    22        160.13          178.26         718.86          789.27
    21        147.76          166.05         866.62          955.31
    20        135.81          154.13        1002.43         1109.45
    19        124.29          142.53        1126.71         1251.98
    18        113.20          131.24        1239.91         1383.22
    17        102.55          120.28        1342.46         1503.51
    16         92.35          109.66        1434.81         1613.16
    15         82.60           99.37        1517.42         1712.53
    14         73.32           89.44        1590.73         1801.98
    13         64.50           79.88        1655.24         1881.85
    12         56.17           70.70        1711.41         1952.55
    11         48.33           61.91        1759.74         2014.46
    10         40.99           53.53        1800.73         2067.98
     9         34.17           45.58        1834.89         2113.56
     8         27.87           38.08        1862.76         2151.64
     7         22.13           31.06        1884.89         2182.70
     6         16.95           24.55        1901.85         2207.25
     5         12.37           18.59        1914.22         2225.84
     4          8.41           13.22        1922.63         2239.06
     3          5.12            8.53        1927.75         2247.59
     2          2.54            4.59        1930.29         2252.18
     1          0.77            1.59        1931.05         2253.78
```

OVERTURNING MOMENT (kN.m)

LEVEL	MAIN DIRECTION	ORTHOGONAL
24	718.97	779.96
23	2107.94	2292.79
22	4119.39	4492.40
21	6707.29	7333.76
20	9827.11	10772.89
19	13435.85	14776.90
18	17492.01	19274.03
17	21955.69	24253.63
16	26788.56	29666.26
15	31953.08	35473.64
14	37416.58	41638.76
13	43143.23	48125.87
12	49102.08	54900.55
11	55263.16	61929.73
10	61598.21	69181.76
9	68080.83	76626.50
8	74696.44	84235.31
7	81392.40	91981.21
6	88178.01	99838.93
5	95024.66	107785.0
4	101915.8	115798.1
3	108837.3	123858.7
2	115777.2	131950.0
1	122726.2	136556.4
0	129678.0	148171.5

DESIGN MOMENTS FOR FOUNDATIONS ARE :

0	97258.5.	111128.6

	LATERAL DISPLACEMENTS (mm)		REQUIRED CLEARANCE (mm)	
LEVEL	MAIN DIRECTION	ORTHOGONAL	MAIN DIRECTION	ORTHOGONAL
25	261.75	178.61	441.71	301.41
24	243.92	167.83	411.62	283.21
23	226.62	157.27	382.43	265.40
22	209.87	146.96	354.15	248.00
21	193.65	136.89	326.79	231.00
20	177.99	127.07	300.36	214.43
19	162.89	117.50	274.88	198.29
18	148.36	108.20	250.36	182.58
17	134.41	99.16	226.81	167.34
16	121.04	90.40	204.25	152.55
15	108.26	81.92	182.69	138.24
14	96.09	73.74	162.16	124.43
13	84.54	65.85	142.66	111.13
12	73.62	58.28	124.23	98.35
11	63.34	51.04	106.89	86.12
10	53.72	44.13	90.65	74.47
9	44.78	37.57	75.56	63.41
8	36.53	31.39	61.65	52.98
7	29.00	25.61	48.94	43.21
6	22.22	20.24	37.50	34.15
5	16.21	15.32	27.36	25.86
4	11.03	10.90	18.61	18.40
3	6.71	7.03	11.32	11.86
2	3.33	3.79	5.62	6.39
1	1.00	1.31	1.69	2.22

RELATIVE DISPLACEMENTS BETWEEN LEVELS DRIFT (mm)

PERMISSIBLE STYORY DRIFT (MM) IS 0.015 x STORY HEIGHT

LEVEL	MAIN DIRECTION	ORTHOGONAL	PERMISSIBLE
25	17.83	10.79	54
24	17.30	10.55	54
23	16.76	10.31	54
22	16.21	10.07	54
21	15.66	9.82	54
20	15.10	9.57	54
19	14.53	9.30	54
18	13.95	9.04	54
17	13.37	8.76	54
16	12.77	8.48	54
15	12.17	8.19	54
14	11.55	7.88	54
13	10.92	7.57	54
12	10.28	7.25	54
11	9.62	6.91	54
10	8.94	6.90	54
9	8.25	6.55	54
8	7.52	6.18	54
7	6.78	5.79	54
6	6.01	4.92	54
5	5.19	4.42	54
4	4.32	3.87	54
3	3.38	3.24	54
2	2.32	2.47	54
1	1.00	1.32	54

THESE VALUES SHOULD BE CHECKED USING CODE CLAUSE 2.10

TORSIONAL MOMENTS (kN.m):

LEVEL No.	MAIN DIRECTION MAXIMUM	MINIMUM	ORTHOGONAL MAXIMUM	MINIMUM
25	249.6	-249.6	216.7	-216.7
24	482.3	-482.3	420.2	-420.2
23	698.4	-698.4	611.0	-611.0
22	898.6	-898.6	789.3	-789.3
21	1083.3	-1083.3	955.3	-955.3
20	1253.0	-1253.0	1109.4	-1109.4
19	1408.4	-1408.4	1252.0	-1252.0
18	1549.9	-1549.9	1383.2	-1383.2
17	1670.1	-1670.1	1503.5	-1503.5
16	1793.5	-1793.5	1613.2	-1613.2
15	1896.8	-1896.8	1712.5	-1712.5
14	1988.4	-1988.4	1802.0	-1802.0
13	2069.0	-2069.0	1881.9	-1881.9
12	2139.3	-2139.3	1952.5	-1952.5
11	2199.7	-2199.7	2014.5	-2014.5
10	2250.9	-2250.9	2068.0	-2068.0
9	2293.6	-2293.6	2113.6	-2113.6
8	2328.5	-2328.5	2151.6	-2151.6
7	2356.1	-2356.1	2182.7	-2182.7
6	2377.3	-2377.3	2207.3	-2207.3
5	2392.8	-2392.8	2225.8	-2225.8
4	2403.3	-2403.3	2239.1	-2239.1
3	2409.7	-2409.7	2247.6	-2247.6
2	2412.9	-2412.9	2252.1	-2252.1
1	2413.8	-2413.8	2253.8	-2253.8

REFERENCES

AMERICAN CONCRETE INSTITUTE (1976) [ACI 318-76], *Building Code Requirements for Reinforced Concrete.* Detroit, MI.

APPLIED TECHNOLOGY COUNCIL (1988) [ATC 3-06 (1988)] *Tentative Provisions for the Development of Seismic Regulations for Buildings.* 2d printing, Palo Alto, CA.

AUSTRALIAN STANDARD 2121 (1979) *The Design of Earthquake-Resistant Buildings (SAA Earthquake Code).* Standards Association of Australia, Sydney, Australia.

AUSTRALIAN STANDARD 3600 (1988) [AS 3600] *Concrete Structures.* Standards Association of Australia, Sydney, Australia.

AUSTRALIAN STANDARD 4100 (1990) [AS 4100] *Steel Structures.* Standards Association of Australia, Sydney, Australia.

AUSTRALIAN STANDARD 1170.1–3 (1989–1990) *Minimum Design Loads on Structures; Part 1 – Dead and Live Loads, Part 2 – Wind Loads, Part 3 – Snow Loads.* Standards Association of Australia, Sydney, Australia.

AUSTRALIAN STANDARD 1170.4 (1993) *Minimum Design Loads on Structures; Part 4 – Earthquake Loads.* Standards Association of Australia, Sydney, Australia.

DENHAM, D., SMALL, G. R., CLEARY, J. R., GREGSON, P. J., SUTTON, D. J., and UNDERWOOD, R. (1975) "Australian Earthquakes (1897–1972)." *Search* Vol. 6, 1975, pp. 34–37, Sydney, Australia.

GAULL, B. A.; MICHAEL-LEIBA, M. O.; and RYNN, J. M. W. (1990) "Probabilistic Earthquake Risk Maps of Australia." *Proceedings Australian Institute of Earth Sciences* Vol. 37, 1990, pp. 169–187. Sydney, Australia.

INTERNATIONAL CONFERENCE OF BUILDING OFFICIALS (1991) *Uniform Building Code* [UBC-91]. Whittier, CA.

STRUCTURAL ENGINEERS ASSOCIATION OF CALIFORNIA [SEAOC] (1990) *Recommended Lateral Force Requirements and Commentary.* San Francisco, CA.

8

Bulgaria

Ludmil Tzenov and Elena Vasseva

8.1 INTRODUCTION

Bulgaria is in the seismically active region around the Mediterranean Sea and the Carpathian Mountains, where the tectonics create a strong seismic effect. Severe earthquakes have occurred frequently in the territory of Bulgaria. One of the oldest historic records (first century B.C.) describes how earthquakes destroyed the town of Kavarna and devastated all its surrounding villages. Seismologists have estimated the intensity of this earthquake to be about X on the MSK scale, which is similar to the Modified Mercalli Intensity (MMI) scale. In the month of September A.D. 543, an earthquake with the intensity of about IX occurred near the town of Balchik; the sea flooded the land

producing further devastation throughout the region. In November 1444, a chronicler wrote that in the Balchik region, "The towns completely collapsed, rivers changed their course" (Rijikova 1981). Seismologists have concluded that "The strongest earthquake in Bulgaria in this century," which was also the strongest earthquake in Europe, occurred on April 4, 1904 in the valley of the Struma River, with the epicenter in the vicinity of the Kresna Gorge. The magnitude of this earthquake was 7.8, the intensity was X, and the depth to the focus was about 18 km.

Before 1892, information about earthquakes and their effects was generally in the form of subjective accounts of witnesses. The foundation of the Seismological Service occurred in 1900. Its first Director, Spas Vuzov, published information, collected using scientific methods, yearly on the earthquakes in Bulgaria. Later, the Seismological Service became the main division of the Central Meteorological Institute of Bulgaria. At the present time, the data collected from earthquake activity is registered and processed using the modern facilities of the Geophysical Institute of the Bulgarian Academy of Sciences, Department of Seismology. In 1928 the effects of earthquakes were investigated from an engineering point of view for the first time in Bulgaria. In that year the special Directorate for Support and Restoration of the Seismic Zone was established. A comprehensive report was published (*DIPOZE* 1931) containing the results of extensive investigations and the analysis of numerous photographs conducted from January 1928 to November 1931. This report described and evaluated damage to different types of buildings and structures during the earthquakes that occurred on April 14, 18, and 25,

1928. The information in the report has been used as the basis for the delineation of the seismic regions in Bulgaria.

The first code for design and building in seismic regions in Bulgaria was published in 1949. The provisions of this code were based entirely on equivalent static analysis for earthquake-resistant design. This first code was revised in 1957 to include also the dynamic method of analysis for earthquake resistant design. The 1957 seismic code also introduced a new coefficient to rate the structural characteristics of the buildings. The Bulgarian seismic code was revised in 1961 and 1964. The design provisions of the 1964 seismic code are based entirely on application of dynamic theory to earthquake engineering. In 1983 the Central Laboratory for Seismic Analysis and Earthquake Engineering was founded and placed under the control of the Bulgarian Academy of Sciences. The main objective of this laboratory has been the investigation of practical methods for the seismic design of buildings and other structures. Some of the results of this investigation have been incorporated in the current Bulgarian *Code for Design of Buildings and Structures in Seismic Regions* (1987) as well as in publications by Boncev et al. (1982), Tzenov and Bonneville (1985), and Tzenov (1983).

8.2 GENERAL REQUIREMENTS

Seismic excitation can occur in any direction in space; however, for regularly shaped rectangular buildings, it may be assumed that the seismic loads are applied independently along the two main horizontal axes of the buildings. In determining whether or not a building is regular, the code requirements for uniform distribution of stiffness and mass in plane and along the structure's height should be given primary consideration. For buildings and other structures of complex shapes, the seismic forces should be applied in the most unfavorable direction using a three-dimensional model of the structure (Tzenov 1990). The internal forces in the elements of the structure are calculated by superposition of the effects of the seismic loads and of permanent and temporary loads (of both long and short duration) reduced by the loading coefficients shown in Table 8.1.

The following loads need not be considered in the seismic design of structures: dynamic loads caused by machines and equipment, lateral and longitudinal braking forces caused by crane movements, loads resulting from settlement of foundations, loads on flexible suspensions, and loads due to changes in temperature and climate.

The larger effect produced by wind forces should be used in the design except for mast, towers, chimneys,

Table 8.1. Loading Coefficients for Calculation of Design Seismic Forces

Type of Load	Loading Coefficient
Permanent (dead) loads	1.0
Temporary loads of long duration	
● Live loads for storage facilities, warehouses, and the like	1.0
● All other types of loads of long duration	0.8
Loads of short duration	
● Snow loads	0.8
● Live loads for industrial buildings	0.8
● Live loads for residential and public buildings	0.5
● Live loads for railway bridges	0.7
● Live loads for motorway bridges	0.4
● All other types of loads of short duration	0.5

and other similar structures. In the design of these structures, 30% of the effect due to wind loads is added to those due to seismic loads.

The horizontal seismic forces of overhead cranes are considered only in the direction perpendicular to the crane track. These forces are determined as follows:

1. When the load is suspended on a rope (flexible suspension), only the weight of the crane and the grab are taken into account.
2. When the suspension of the live load is rigid, the weight of the crane, the grab, and the live load (equal to the lifting capacity of the crane) are taken into account.
3. The seismic forces in the columns of multispan industrial buildings with several bridge cranes are calculated considering only the weight of the heaviest crane.

8.3 DESIGN SEISMIC FORCES

The design seismic forces F_{ik} in the direction under consideration, at level k of the building for vibration mode i, are given in the current Bulgarian *Code for Design of Buildings and Structures in Seismic Regions* (1987) by the following formula:

$$F_{ik} = CRK_c\beta_i\eta_{ik}Q_k \qquad (8.1)$$

where

C = the importance coefficient (Table 8.2)
R = the response coefficient (Table 8.3)
K_c = the seismic coefficient representing the ratio of the design peak ground acceleration to gravitational acceleration
β_i = the dynamic coefficient corresponding to the ith mode of vibration of the structure

η_{ik} = the distribution coefficient of the design seismic forces corresponding to the *i*th mode of vibration of the structure at level *k*

Q_k = the seismic weight at level *k* of the building.

8.3.1 Importance Coefficient *C*

Table 8.2 provides the values for the importance coefficient *C* for categories of buildings and structures classified according to the importance of their occupancy and their continuous use after an earthquake.

8.3.2 Response Coefficient *R*

Values of the response coefficient *R* for different types of buildings and structures are shown in Table

Table 8.2. Importance Coefficient (*C*)

Category of Buildings and structures	Description of Buildings and Structures	Coefficient *C*
A	Essential facilities and buildings that must function during and after an earthquake (hospitals, fire stations, electric power plants, etc.).	1.5
	Large buildings and structures (buildings over 20 stories high, bridges with spans larger than 50 m, large tanks and water treatment facilities, etc.).	
	Important state buildings and historical structures (monuments, museums, etc.).	
	Facilities for production or storage of toxic materials.	
	National or regional communications centers.	
	Dams over 80 m high or with large reservoirs.	
B	Large tanks or water treatment plants.	1.0
	Public and residential buildings up to 20 stories high.	
	Buildings for assembly of large numbers of people (schools, universities, hotels, department stores, etc.).	
	Structures not included in Category A, C or D.	
C	Secondary buildings and other structures (warehouses, auxiliary industrial buildings, farm buildings for animals, etc.).	0.75
	Small dams, bridges, power plants, substations.	
D	Temporary buildings and structures.	0

Note: In Building Category D, the value of the coefficient *C* = 0 indicates that this type of structure need not be designed for seismic action.

8.3. The Code provides for modification of the value of the coefficient *R* for the following types of buildings:

1. For frame buildings where the wall filling does not affect wall deformability, the coefficient *R* obtained from Table 8.3 must be multiplied by:
   (a) 1.5 when $h/b \geq 25$
   (b) 1.00 when $h/b \leq 15$
   (c) An interpolated value between 1.00 and 1.50, when $15 < h/b < 25$ where *h* is the story height and *b* is the cross-sectional dimension of the columns in the direction perpendicular to the direction of the seismic forces;

2. For buildings with a soft story (story stiffness is less than half the stiffness of the adjacent stories), the coefficient *R* should be multiplied by 2.0. This factor is applied only to the computation of the internal forces and moments in the soft story;

3. For buildings on pile foundations with a high base grid (more than 1.0 m above ground level), the coefficient R should be multiplied by 1.5.

8.3.3 Seismic Coefficient *K*$_c$

The value of the seismic coefficient K_c is given by arabic numerals next to the value of the corresponding seismic intensity (given by roman numerals) on the Bulgarian seismic zone map (Fig. 8.1) for events with a 1,000-year return period. This seismic map illustrates the high seismic activity in almost all of Bulgaria; over 95% of the country lies in regions for which earthquakes are very dangerous events (Boncev et al. 1982).

According to the Code, the internal forces and moments can be determined by the direct integration method using real or simulated accelerograms that consider the development of deformations. The maximum acceleration of the accelerograms should not be less than $K_c g$, where K_c is the seismic coefficient and *g* is the gravitational acceleration.

8.3.4 Dynamic Coefficient β_i

The dynamic coefficient β_i is a function of both the natural period T_i in the *i*th mode of vibration and the soil conditions at the site of the structure (Table 8.4). The dynamic coefficient β_i corresponding to mode *i* is given by the following formulas:

1. For Soil Category I

$$0.8 \leq \beta_i = \frac{0.9}{T_i} \leq 2.5 \tag{8.2}$$

2. For Soil Category II

$$0.8 \leq \beta_i = \frac{1.2}{T_i} \leq 2.5 \tag{8.3}$$

Table 8.3. Response Coefficient (R)

Type of Structural System	Response Coefficient R	Type of Structural System	Response Coefficient R
REINFORCED CONCRETE STRUCTURES		• Reinforced masonry walls with precast floor panels and cast-in-place reinforced concrete tie beams	0.35
1. MONOLITHIC			
• Frame-beam structures with reinforced concrete shear walls	0.25	• Nonreinforced or partially reinforced masonry walls with cast-in-place concrete slabs and tie beams	0.40
• Frame-beam structures with brick walls participating in seismic force resistance	0.30	• Nonreinforced or partially reinforced masonry walls with prefabricated floor panels and cast-in-place reinforced concrete tie beams	0.50
• Flat plate structures with reinforced concrete shear walls	0.28		
• Flat plate structures with frames	0.25		
• One-span frame structures	0.25	**SPECIALLY DESIGNED STRUCTURES**	
• Frame structures with two or more spans	0.20	• Silos, bunkers, tanks, and other similar rigid structures:	
• Shear wall buildings constructed with large-scale formwork, tunnel formwork, sliding formwork, etc.	0.25	(a) reinforced concrete	0.25
		(b) steel	0.20
2. PREFABRICATED STRUCTURES		• Tall structures such as chimneys, masts, towers, scaffold bridges, and others:	
• Large-panel structures	0.25	(a) reinforced concrete	0.35
• Frame-panel structures	0.22	(b) steel	0.28
• Frame-beam structures with shear walls	0.25	• Retaining walls:	
• Flat plate structures with shear walls and frames	0.28	(a) concrete	0.28
• Single-story and two-story frame structures where seismic force resistance is provided by the columns	0.28	(b) reinforced concrete	0.25
		• Water and intake towers:	
• Multistory frame structures with rigid joints	0.25	(a) reinforced concrete	0.38
		(b) steel	0.30
STEEL STRUCTURES		• Dams:	
• Frames with rigid joints	0.20	(a) earth-fill dams and rock-fill dams	0.25
• Frame structures where seismic force resistance is provided by the columns	0.22	(b) concrete and reinforced concrete dams	0.30
		• Other types of hydrotechnical structures	0.25
• Frame structures with reinforced concrete shear walls or rigid cores	0.25	• Bridges and other types of transportation structures	0.25
		• Underground structures	0.25
BUILDINGS WITH BRICK MASONRY BEARING WALLS			
• Reinforced masonry walls with cast-in-place reinforced concrete slabs and tie beams	0.30		

3. For Soil Category III

$$1.0 \leqslant \beta_i = \frac{1.6}{T_i} \leqslant 2.5 \qquad (8.4)$$

In the case of a heterogeneous soil profile, the dynamic coefficient β_i can be determined as the average value of the coefficients β_i for the soil layers under the foundation, down to the base rock. The term "base rock" also applies to dense soils with a seismic shear wave propagation velocity $V_s \geqslant 750$ m/sec. Values of the propagation velocity V_s for the different soil types are taken from the geotechnical report for the site. If such data are not available in the engineering-geological report, the values of V_s may be taken from published data available for the region.

In the case of a pile foundation, the dynamic coefficient β_i is assumed to be the average value of the coefficients β_i for load-bearing soil layers that are penetrated by the piles. Load-bearing soil layers are those layers for which the shear strength is taken into consideration, including the resistance of soils under piles.

The Code does not permit building construction on soils not included in Table 8.4 (uncompacted fills, very soft or sensitive clays, loose sands, silt, organic soils, and the like) without additional foundation support measures. These measures must be in accordance with "Shallow Foundation Design Regulations" (1983).

The density of sandy soils (except sand with fines) must be determined by dynamic penetration tests. The density of sands with fines and the strength of plastic clays, silt, organic soils, and other similar soft soils may be determined by static penetration tests and field vane shear tests. When data on the consistency of cohesive soils or the density of sandy soils are not available, the necessary data can be taken from regional engineering maps.

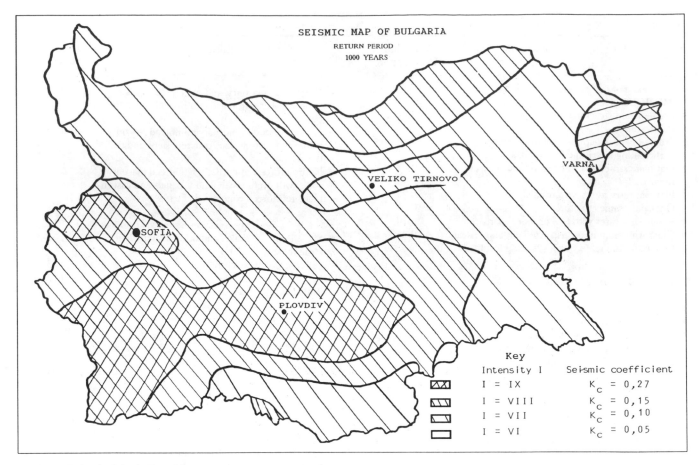

SEISMIC MAP OF BULGARIA
RETURN PERIOD
1000 YEARS

Key		
Intensity I	Seismic coefficient	
I = IX	$K_c = 0,27$	
I = VIII	$K_c = 0,15$	
I = VII	$K_c = 0,10$	
I = VI	$K_c = 0,05$	

Fig. 8.1.   Bulgaria Seismic Zone Map

8.3.5 Distribution Coefficient η_{ik}

The distribution coefficient η_{ik} is determined by the following formula:

$$\eta_{ik} = X_{ik} \frac{\sum_{j=1}^{n} Q_j X_{ij}}{\sum_{j=1}^{n} Q_j X_{ij}^2} \qquad (8.5)$$

where

X_{ik}, X_{ij} = the modal displacement amplitude of the *i*th mode of vibration at level *k* or *j* of the structure

Q_j = the seismic weight assigned to level *j* calculated using the loading coefficients in Table 8.1

n = the number of levels in the building.

For buildings up to five stories in height, when the fundamental period $T_1 \leqslant 0.4$ sec, the coefficient η_{ik} can be determined by the following formula:

$$\eta_{ik} = h_k \frac{\sum_{j=1}^{n} Q_j h_j}{\sum_{j=1}^{n} Q_j h_j^2} \qquad (8.6)$$

where

h_k, h_j = the height above the foundation of level *k* or *j*.

For buildings and structures of irregular shape in plan or in height, with asymmetrically distributed

Table 8.4.   Soil Categories

Soil Category	Description of Soils
I	• All kinds of rock (excluding weathered rock) • Dense gravel • Marl (not weathered) • Stiff clay
II	• Weathered rocks and marls • Gravel sands, coarse and medium sands, dense to medium dense • Fine sand, dense • Clayey sand and sandy clay, stiff, firm, and medium • Stiff plastic clay
III	• Fine sand, medium dense • Sand with fines, dense to medium dense • Clayey sand and sandy clay, medium stiff to soft • Clay, medium stiff to soft • Loess

masses, the Code requires that the analysis be made for the most unfavorable direction of the seismic excitation. In this case the coefficient η_{ik} is determined by

$$\eta_{ik} = X_{ik} \frac{\sum\limits_{j=1}^{n} Q_j X_{ij} \nu_j}{\sum\limits_{j=1}^{n} Q_j X_{ij}^2} \qquad (8.7)$$

where

ν_j = cosine of the angle between the considered direction of the seismic excitation and the jth seismic force.

8.3.6 Seismic Weight Q_k

The seismic weight Q_k at each level k is calculated as factored loads using Table 8.1 for the loads attributed at the various levels of the building.

8.4 VERTICAL SEISMIC RESPONSE

The effects of the vertical components of the seismic excitation should be considered for the following structures:

1. Horizontal or inclined cantilever structures;
2. Superstructures of bridges;
3. Structures such as frames, arches, trusses, and plate roofs with a span equal to or greater than 24 m;
4. Buildings and structures susceptible to overturning and sliding; and
5. Brick or stone structures.

For all these cases, the vertical design seismic forces are determined as a fraction of the horizontal forces calculated by eq.(8.1).

Masonry structures are designed for simultaneous action of both the horizontal and seismic vertical forces, calculated as follows:

1. The horizontal design seismic forces are determined by eq.(8.1).
2. The vertical seismic forces are calculated as a percentage of the seismic weight at each level of the building with 15% for buildings in seismic zones with $K_c \leq 0.15$ and 30% for buildings in seismic zones with $K_c > 0.15$.
3. Horizontal and inclined cantilever elements (balconies, bay windows, etc.) and their anchorages are designed for a vertical seismic force equal to $2CK_cQ$, where Q is the weight of the element.
4. Elements rising above the building (parapets, ornaments, chimneys, watertanks, etc.), as well as their

anchorage, must be designed for a horizontal seismic force equal to $2CK_cQ$, where Q is the weight of the element.
5. Panels, structural, and partition walls, as well as their anchorage, must be designed for a horizontal seismic force equal to $0.6CK_cQ$, that is, applied in the direction normal to the plane of the element, where Q is the weight of the element.

The vertical seismic forces for overhead cranes are determined by considering both the weight of the crane and the grab and the weight of the live load, which is equal to 0.3 of the lifting capacity of the crane.

8.5 HIGHER MODES OF NATURAL VIBRATIONS

When the fundamental period of natural vibration of the structure T_1 is greater than 0.4 sec, the Code requires that the internal forces and moments be determined including, at least, the lowest three modes of vibration. However, for buildings irregular in plan, a greater number of the modes must be included in the seismic analysis.

When more than one mode of vibration is considered, the "effective" design seismic forces or moments N_d are determined by the following formula:

$$N_d = \sqrt{\sum_{i=1}^{m} N_{id}^2} \qquad (8.8)$$

where

N_{id} = internal force or moment in the cross-section corresponding to the ith mode of vibration
m = number of modes of vibration considered.

8.6 DISTRIBUTION OF STORY SHEAR FORCE

The distribution of the horizontal shear force at each level of the building among the vertical bearing elements (shear walls, frames, columns, etc.) is carried out as follows:

1. For buildings with rigid horizontal diaphragms at the floor levels, the story shear force is distributed by equalization of lateral displacement of the structural resisting elements in the story.
2. For buildings with flexible floors or floors that do not behave like rigid horizontal diaphragms (e.g., prefabricated floor systems whose separate elements are not rigidly interconnected), the story shear force is distributed among the structural resisting elements considering the deformability of the floor diaphragm. As an exception, for buildings up to five stories high, the story shear force may be distributed in proportion

to the seismic weights transmitted to the vertical resisting elements of the story.

8.7 TORSIONAL EFFECTS

When designing structures that do not satisfy all the criteria for regular systems, the Code requires taking into account the seismic forces along the two main axes of the building and the torsional moments caused by the eccentricity of the horizontal forces relative to the stiffness centers. Therefore, the eccentricity between the mass center and the stiffness center at each story must be determined. When the eccentricity at any story is less than $0.02B$ (B is the dimension of the building in the direction perpendicular to the seismic forces), the eccentricity must be assumed to be equal to $0.02B$.

For strength and stability calculations, the design characteristics of the materials must be multiplied by a performance coefficient γ_k whose values are presented in Table 8.5.

8.8 PERFORMANCE COEFFICIENT

Grouted connections between separate structural elements of prefabricated and prefabricated-monolithic structures must be designed for internal forces and moments due to the seismic forces. For the design of welded joints of steel and reinforced concrete elements in prefabricated structures, the internal forces and moments due to seismic forces must be increased by 25%. The design compressive strength of the reinforcement in reinforced concrete structures is not allowed to exceed 450 megapascals (MPa).

8.9 ALLOWABLE DEFORMABILITY AND SEISMIC JOINTS

The story drift caused by the seismic forces of eq.(8.1) corresponding to the first mode of vibration must not exceed the following values:

1. For Building Category A (Table 8.2) except for one-story and two-story industrial buildings: $h/400$
2. For one-story and two-story industrial buildings of Category A: $h/250$
3. For one-story and two-story industrial buildings of Category B: $h/180$
4. For all other types of buildings of Category B: $h/250$, where h is the story height.

For all other categories of buildings, the story drift is not limited by the Code. Second-order effects on

Table 8.5.   Values of the Performance Coefficient γ_k

Type of Structure	Coefficient γ_k
1. For strength verification	
Steel and wood structures	1.4
Reinforced concrete structures (without strength verification of inclined sections):	
(a) normal weight concrete with reinforcement class A-I, A-II, A-III, A-Ib	1.2
(b) normal weight concrete with other reinforcement classes	1.1
(c) columns resisting seismic forces (for all types of reinforcement)	1.0
(d) light weight aggregate concrete and reinforcement of all classes	1.1
(e) cellular concrete and reinforcement of all classes	1.0
Reinforced concrete structures (with strength verification of inclined sections):	
(a) beams and girders	1.0
(b) columns	0.8
Concrete, brick, and stone masonry structures	1.0
Welded joints	1.0
Bolts (standard and high strength), rivets, and glued joints	1.1
2. For stability calculation	
Steel structures whose slenderness ratio is greater than 80	1.0
Steel structures whose slenderness ratio does not exceed 20	1.2
Steel structures whose slenderness ratio is between 20 and 80	1.2–1.0 by interpolation

internal forces and moments need not be considered if the following condition is satisfied at every story:

$$\theta = \frac{Q\Delta}{FhR} \leq 0.10 \qquad (8.9)$$

where

θ = deformability index
Q = total gravity load above the considered story
Δ = elastic interstory drift due to the design seismic forces for the first mode of vibration
F = shear force in the considered story
h = story height below the considered level
R = response coefficient from Table 8.3

For values of θ within the range $0.10 < \theta \leq 0.20$, the additional internal forces and moments due to P-delta (P-Δ) effects must be considered. To include P-Δ effects, the seismic forces determined by eq.(8.1) must be multiplied by the factor $1/(1-\theta)$ for each story level. When $\theta > 0.20$ the stiffness of the structure should be increased or the structure must be analyzed by more precise methods.

Buildings and structures must be separated by seismic joints in the following cases:

1. When the building or structure has an irregular shape in plan
2. In frame structures with a difference in the floor levels of the adjacent sections
3. When the difference in height of the adjacent sections of the building is 6 m or more.

For one-story buildings up to 10 m in height, when $K_c < 0.10$, seismic joints need not be included in building design.

The seismic joints must separate the building along its entire height above the foundation. The seismic joint width $J\Delta$ must be determined as follows:

1. For one-story and two-story industrial buildings

$$J\Delta = \frac{1}{2}\left(\frac{D_1}{R_1} + \frac{D_2}{R_2}\right), \text{ but may not be less than 5 cm.}$$

For two adjacent buildings of equal height, D_1 and D_2 are the deflections at the top level of the buildings, whereas for buildings of different height, they are the deflections at the top level of the lower building. D_1 and D_2 must be determined by an elastic analysis for the seismic forces corresponding to the first mode of vibration. R_1 and R_2 are the response coefficients for the respective buildings.

2. For all other types of buildings, $J\Delta = H/250$, but may not be less than 5 cm.

The maximum distance between two seismic joints in concrete and reinforced concrete buildings, or in specially designed structures, is limited as follows:

60 m in seismic zones with $K_c \leqslant 0.27$
50 m in seismic zones with $K_c > 0.27$

If the distance between the expansion joints (in accordance with *Concrete and Reinforced Concrete Structures Design Code* (1988)) is smaller than the above-mentioned distances, the expansion joints shall also be designed as seismic joints.

For steel frame buildings, the maximum distance between the seismic joints is limited to be not greater than 150 m.

For all other types of structures, the maximum distance between the seismic joints is assumed to be equal to the distance between the expansion joints in nonseismic zones, but not greater than:

80 m in seismic zones with $K_c \leqslant 0.27$
60 m in seismic zones with $K_c > 0.27$

8.10 ALLOWABLE NUMBER OF STORIES

The allowable height of the buildings and the corresponding number of stories are given in Table 8.6. The height of the buildings and structures listed in Table 8.6 must be measured from the ground level to the cornice of the building or to the highest level of the structure. The number of stories of hospitals and schools is restricted as follows:

for zones with $K_c \leqslant 0.27$, up to 8 stories
for zones with $K_c > 0.27$, up to 5 stories

8.11 REQUIREMENTS FOR FOUNDATIONS

The Code prescribes that the ultimate limit state method be applied in the design of foundations. The design of the foundation for bearing capacity under static loads combined with seismic loads must be performed in accordance with "Shallow Foundation Design Regulations" (1983). The safety factors for overturning and sliding are given in this document. The safety factor under combined total loads should not be less than 1.20. The bearing capacity of the soil must be calculated separately for each one of the biaxial moments acting in both principal directions, at the foundation level. In the design of the foundation for a specific load combination, partial loss of contact of the footing can be accepted if the following requirements are satisfied:

1. $$e < B/3$$

where

$B =$ width of the foundation perpendicular to direction of the seismic forces
$e =$ eccentricity of the design load

2. $$\alpha \leqslant 4R_1$$

where

$\alpha =$ maximum soil stress at edge of footing
$R_1 =$ allowable soil bearing pressure

For rigid foundations, the edge soil stress must be determined using Navier's hypothesis. The footings must be proportioned and the reinforcing steel must be selected for the internal forces and moments resulting from the increased soil stresses (up to $4R_1$).

The following general requirements must be considered for design of foundations and basement walls:

1. The foundation ties between the footings must be designed in compliance with "Shallow Foundation Design Regulations" (1983).

Table 8.6. Allowable Height of Buildings and Number of Stories for Different Seismic Regions

Type of Building and Structural System	Seismic Regions with					
	$0.05 \leqslant K_c < 0.15$		$0.15 \leqslant K_c < 0.27$		$K_c > 0.27$	
	Height (m)	Number of Stories (N)	Height (m)	Number of Stories (N)	Height (m)	Number of Stories (N)
Residential, public, and industrial buildings with monolithic reinforced concrete or steel frames	No restrictions					
Large-panel building	39	12	30	9	27	8
Prefabricated frame buildings	45	14	39	12	30	9
Package lift-slabs buildings	39	12	30	9	27	8
Residential, public, office, and other buildings of this type with monolithic R.C. floor structures and bearing brick masonry walls framed by R.C. columns and tie beams	16	5	13	4	10	3
Brick masonry buildings with wooden floor structures	7	2	4	1	–	–
Chimneys: – reinforced concrete, steel	No restrictions					
– brick with external steel rings, longitudinal ribs or other type of strengthening	50	–	–	30	–	20
Brick chimneys Unreinforced	Not allowed					

2. The use of more than one basement for tall buildings is suitable when the depth of underground water is more than 4 m.
3. Pile foundations are designed in compliance with "Pile Foundation Design Code" (1981).
4. For construction in loess soils, the special requirements of "Shallow Foundation Design Regulations" (1983) must be considered.

In seismic zones with $K_c \geqslant 0.27$, as well as in zones of complex engineering-geological conditions (severely inclined terrain, saturated soils with $R_1 \leqslant 150 \, kN/m^2$, regions with active geologic processes, etc.), prefabricated concrete and reinforced concrete basement walls and foundations are not permitted. In seismic zones with $K_c \geqslant 0.27$, all footings under a building or under a separate building wing must be designed and built at the same level.

For strip foundations that extend under adjacent sections founded at different levels, the transition zone between levels of the foundations must be designed with steps not higher than 0.60 m and sloped no steeper than 1:2. Strip foundations under double walls should be founded at the same level, at a distance of 1 m on both sides from the seismic joint.

For adjacent footings at different levels, the following requirement must be satisfied:

$$\frac{\Delta h}{d} \leqslant \tan(\phi - \Delta\phi) \qquad (8.10)$$

where

Δh = difference in foundation levels
d = clear distance between footings
ϕ = angle of shearing resistance of soil
$\Delta\phi$ = decrease of angle for seismic excitation assumed as follows:

1. $\Delta\phi = 2°$ for $K_c = 0.10$
2. $\Delta\phi = 3°$ for $K_c = 0.15$
3. $\Delta\phi = 5°$ for $K_c = 0.27$

For intermediate values of K_c, the values of $\Delta\phi$ should be found by interpolation.

As an exception, foundations at different levels of two adjacent building sections are permitted if this arrangement is required for technical reasons and if the seismic safety of both sections is proven.

An example is now provided to illustrate the application of the Bulgarian code for seismic resistant design.

Example 8.1

A six-story residential building is presented as a design example. The site of the building is located in a zone of design intensity VIII (on the MSK scale). The local soil conditions are gravel/sand corresponding to those of Soil Category II of Table 8.4. The building is a frame with reinforced concrete shear walls. Figure 8.2 shows the plan view of the building. The interstory

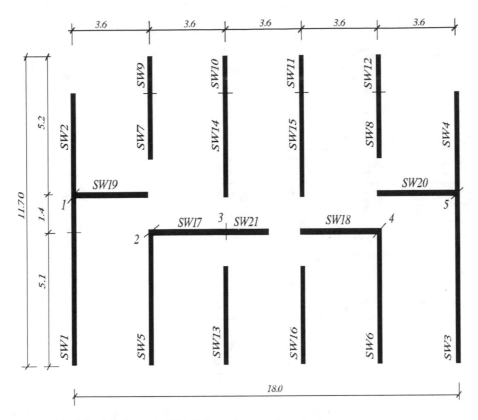

Fig. 8.2. Elevation and plan of building for Example 8.1 (all dimensions are in meters)

height is 2.80 m for all the stories. The calculated seismic weight for all the levels is 235 tonnes except for the top level where it is 367 tonnes.

A computer program developed to implement the provisions of the Bulgarian seismic code is used for the seismic analysis of this building. Input data and output results are shown in the following tables:

INPUT DATA AND OUTPUT RESULTS FOR EXAMPLE 8.1

INPUT DATA:

A) GENERAL DATA:
NUMBER OF STORIES: N = 6
SEISMIC COEFFICIENT: KC = 0.15 (from Seismic Zone Map)
IMPORTANCE COEFFICIENT: C = 1.00 (from Table 8.1, Category B)
STRUCTURAL RESPONSE COEFFICIENT: R = 0.25 (Table 8.3, Concrete frame)
SOIL CATEGORY: S = 2 (from Table 8.4, Gravel/sand)

B) STORY DATA:

LEVEL	MASS CENTER X(M)	Y(M)	HEIGHT (M)	WEIGHT (TONNE)	ROTATIONAL MASS MOMENT OF INERTIA (TONNE-M**2)
6	9.00	5.94	16.80	367.0	14313.0
5	9.00	5.94	14.00	235.0	9165.0
4	9.00	5.94	11.20	235.0	9165.0
3	9.00	5.94	8.40	235.0	9165.0
2	9.00	5.94	5.60	235.0	9165.0
1	9.00	5.94	2.80	235.C	9165.0

C) STRUCTURAL ELEMENT DATA:

ELEMENT	TYPE	CENTROIDAL COORDINATES X(M)	Y(M)	ANGLE BETWEEN X-ELEMENT AND X-FRAME (DEGREE)
1	1	0.0	2.55	90.000
2	1	0.0	7.65	90.000
3	1	18.00	2.55	90.000
4	1	18.00	7.65	90.000
5	2	3.60	2.55	90.000
6	2	14.40	2.55	90.000
7	3	3.60	8.85	90.000
8	3	14.40	8.85	90.000
9	4	3.60	10.95	90.000
10	4	7.20	10.95	90.000
11	4	10.80	10.95	90.000
12	4	14.40	10.95	90.000
13	5	7.20	1.95	90.000
14	5	7.20	8.25	90.000
15	5	10.80	8.25	90.000
16	6	10.80	1.91	90.000
17	7	5.40	5.10	0.0
18	7	12.60	5.10	0.0
19	7	1.80	6.54	0.0
20	7	16.20	6.54	0.0
21	8	8.24	5.10	0.0

D) CONNECTIVITY:

CONNECTION NUMBER	ELEMENT NUMBERS CONNECTED		CONNECTION COORDINATES X(M)	Y(M)	STORIES FROM-TO	ELEMENT STIFFNESS (TON/m**2)
1	2	19	0.00	6.54	1-6	1.0E6
2	5	17	3.60	5.10	1-6	1.0E6
3	17	21	7.20	5.10	1-6	1.0E6
4	18	6	14.40	5.10	1-6	1.0E6
5	20	4	18.00	6.54	1-6	1.0E6

E) ELEMENT DATA:

TYPE NUMBER	NUMBER OF STORIES	DIMENSION CROSS-SECTION A(M)	B(M)
1	6	0.130	5.220
2	6	0.140	4.940
3	6	0.140	2.540
4	6	0.130	1.620
5	6	0.140	3.740
6	6	0.140	3.660
7	6	0.140	3.440
8	6	0.140	1.920

OUTPUT RESULTS:

A) NATURAL PERIODS:

MODE	PERIOD (SEC)
1	0.466
2	0.321
3	0.300

B) SEISMIC FORCES AND TORSIONAL MOMENT:

LEVEL	X-FORCE (TONNE)	Y-FORCE (TONNE)	TORSIONAL MOMENT (T-M*2)
6	45.156	-4.007	-287.077
5	22.726	-1.948	-144.228
4	16.550	-1.356	-104.800
3	10.722	-0.827	- 67.704
2	5.675	-0.400	- 35.694
1	1.909	-0.113	- 11.925

C) DISPLACEMENTS AND CONNECTION SHEAR FORCES:

LEVEL	DIRECTION X(MM)	DIRECTION Y(MM)	ROTATION RADIAN/1000	CONNECTION FROM-TO	SHEAR FORCE (TONNE) CONNECTION (1)	(2)	(3)	(4)	(5)
6	7.046	-1.178	0.065	1-6	-3.96	-7.49	-6.27	-4.15	-3.24
5	5.525	-0.890	0.049	1-6	-10.30	-16.56	-14.89	-10.17	-8.81
4	4.012	-0.617	0.034	1-6	-14.49	-19.48	-19.00	-13.46	-12.95
3	2.590	-0.373	0.020	1-6	-17.59	-21.34	-21.77	-15.78	-16.09
2	1.364	-0.178	0.010	1-6	-19.76	-21.10	-22.96	-17.03	-18.45
1	1.455	-0.049	0.003	1-6	-20.70	-17.16	-22.21	-16.94	-19.81

D) ELEMENT DISPLACEMENT AND INTERNAL FORCES:

EXAMPLE ELEMENT NO. 1 ON FIRST MODE:

LEVEL	DISPLACEMENT (MM)	AXIAL FORCE (TONNE)	SHEAR FORCE (TONNE)	BENDING MOMENT TOP (T-M)	BENDING MOMENT BOTTOM (T-M)
6	-1.178	2.584	-2.584	0.0	-7.235
5	-0.890	-0.269	-2.853	-7.235	-15.224
4	-0.616	-0.111	-2.964	-15.224	-23.522
3	-0.373	0.148	2.816	23.522	31.408
2	-0.178	0.476	-2.340	-31.408	-37.960
1	-0.049	0.726	-1.615	-37.960	-42.481

(Output results continue for other elements and second and third modes.)

REFERENCES

Boncev, E.; Bune, V.; Christoskov, L.; Karagjuleva, J.; Kostadinov, V.; Reisner, G.; Rizhikova, S.; Shebalin, N.; Sholpo, V.; and Sokerova, D. (1982) "A Method for Compilation of Seismic Zoning Prognostic Maps for the Territory of Bulgaria." *Geologic Balcanica* Vol. 12, No. 2. Sofia, Bulgaria.

Code for Design of Buildings and Structures in Seismic Regions (1987) Bulgarian Academy of Science Committee of Territorial and Town System at the Council of Ministers, Sofia, Bulgaria.

Concrete and Reinforced Concrete Structures Design Code (1988) Bulgarian Academy of Science Committee of Territorial and Town System at the Council of Ministers, Sofia, Bulgaria.

DIPOZE (1931) State Printing House, Sofia, Bulgaria.

"Pile Foundation Design Code" (1981) *Bulletin for Structures and Architecture* N 3. Sofia, Bulgaria.

Rijikova, Sn (1981) *Earthquake-Disaster and Source of Knowledge*. "Technika" Publishing House, Sofia, Bulgaria.

"Shallow Foundation Design Regulations" (1983) *Bulletin for Structures and Architecture* N 1–2. Sofia, Bulgaria.

Tzenov, L. (1983) *Reduction of the Seismic Vulnerability of Structures*. Bulgarian Acad. of Sciences, Sofia, Bulgaria.

——— (1990) "Response of Bridges to Strong Ground Motion." *Proc. IX ECEE*. Moscow, CIS (former USSR).

Tzenov, L., and Bonneville, P. (1985) *Calcul des Structures Parasismiques*. N Edition: 1824/84. Enterprise Nationale du Livre, Algiers, Algeria.

9
Canada

David T. Lau and J. L. Humar

9.1 INTRODUCTION

The *National Building Code of Canada* (NBCC) is a national standard of safety for building design. In Canada, building design is governed by regulations of the local (municipal) governments. Usually the local governments incorporate the NBCC in their regulations; the code then becomes a legally binding document.

The first Canadian regulations for earthquake-resistant design of buildings were published in the first edition of the National Building Code, 1941. The seismic design provisions stipulated in the 1941 Code have been revised and updated in each of the subsequent editions of the NBCC to reflect the current practice in earthquake-resistant design in Canada; 1990 is the most current edition.

Canada was divided into four seismic intensity regions in the first official seismic zoning map (NBCC 1953). This map remained in use until 1970 when the first probabilistic seismic zoning map of Canada was introduced. In the 1970 seismic map, the estimated seismic risk was specified in terms of the peak horizontal acceleration on firm ground with a 40% probability of exceedance in 50 years.

A foundation factor was introduced in the calculation of the design base shear, in recognition of the effect of local soil conditions on the intensity of ground motion (NBCC 1965). At the same time, an importance factor was included to account for the social importance of a building based on its intended use. Seismic torsion provisions also were introduced for the first time in 1965.

Based on the observed damage and on lessons learned from a major Venezuelan earthquake in 1967, further refinements of the equivalent static load procedure were incorporated in the 1975 edition of the NBCC. The commentary to the 1975 Code included for the first time an alternative dynamic analysis procedure for the design of buildings with unusual or complex structural configurations. The dynamic procedure employed an average inelastic response spectrum to determine the design earthquake forces. In the 1980 edition of the NBCC, improvements were made in the provisions for torsional loads that related to the determination of eccentricity for the calculation of design torsional moment.

Compared to the previous editions, the NBCC of 1990 represents a major restructuring of the seismic loading provisions, because of the development of the concept of force modification factor. The Code also includes provisions that were developed from studies on the effect of foundation and soil generated by the 1985 earthquake in Mexico.

Over the years, developments in the Canadian and U.S. Code requirements often have followed similar paths because of the close technical cooperation between engineers of the two countries. An overview of the early evolution of the Canadian seismic design provisions has been presented by Uzumeri, Otani, and Collins (1978).

9.2 DESIGN OBJECTIVES AND PHILOSOPHIES

The fundamental aim of earthquake-resistant design of buildings is to provide structures with an acceptable level of safety for public use. The occurrence of a major destructive earthquake is rare in the lifetime of a structure. Therefore, the Code does not require all buildings to have the necessary reserve strength and capacity to withstand the largest possible earthquake without suffering any damage—this is not practical economically. The NBCC aims only to minimize the probability of injury and loss of life by adopting the philosophy that structures should be designed (a) to resist severe major earthquakes without collapse or major failure, (b) to resist moderate earthquakes without significant structural damage but possibly with some nonstructural damage, and (c) to withstand minor earthquakes without any damage. These design objectives are implicit in the Code provisions, which stipulate a balance between the objectives of designing for strength and designing for ductile behavior.

The seismic design provisions of the NBCC of 1990 are based on an equivalent static load approach. The equivalent static lateral loads do not represent the peak dynamic forces that may be exerted on the structure during earthquakes. It is expected that structures designed to resist these equivalent static loads will perform satisfactorily in accordance with the design objectives.

9.3 SEISMIC REGIONALIZATION

The current seismic zoning maps for Canada, first adopted in the NBCC of 1985, are based on results derived from statistical analyses of recent and past earthquakes in Canada (Basham et al. 1985). The seismicity of Canada and adjacent regions was modeled in relation to 32 earthquake source zones, identified on the basis of seismological and geological evidence. A magnitude-recurrence interval relation and an attenuation relation for the ground motion earthquake were obtained for each of the source zones. Numerical integration of the ground motion parameters derived from all relevant source zones was used to obtain the probability of exceedance.

In the current NBCC seismic zoning maps, Canada is divided into seven acceleration-related seismic zones Z_a as shown in Fig. 9.1(a), and seven velocity-related seismic zones Z_v as shown in Fig. 9.1(b). The level of seismic risk is considered to be similar within each zone. The level of seismic risk at any site is specified in terms of both peak horizontal ground acceleration and peak horizontal ground velocity, each with a probability of exceedance of 10% in 50 years or 0.21% per annum. The reference seismic ground motion associated with this lower level of probability is considered appropriate for the level of protection provided by the seismic design provisions in the current code. The use of two ground motion parameters to characterize seismicity is another significant improvement toward a more rational quantification of the seismic risk.

The zonal acceleration ratio a is the ratio of the peak horizontal ground acceleration to the acceleration due to gravity; whereas the zonal velocity ratio v is the ratio of the peak horizontal ground velocity to a velocity of one m/sec. The definition of the seismic zones and the corresponding ranges of peak horizontal ground acceleration and velocity are given in Table 9.1.

The ratio Z_a/Z_v gives an indication of the characteristic of the ground motion expected at a particular site. For sites with a low zonal ratio Z_a/Z_v, the expected ground motion is velocity dominated (typical of distant

Table 9.1.   Definition of Seismic Zones (NBCC 1990)

Seismic Zone Z_a, Z_v	Range of Peak Horizontal Ground Acceleration (g) and Velocity in m/sec	Zonal Acceleration Ratio a; Zonal Velocity Ratio v
0	<0.04	0
1	0.04 to <0.08	0.05
2	0.08 to <0.11	0.10
3	0.11 to <0.16	0.15
4	0.16 to <0.23	0.20
5	0.23 to <0.32	0.30
6	≥0.32	0.40

large earthquakes), whereas at sites with a high zonal ratio, acceleration dominates the ground motion (typical of earthquakes originating at a nearby source). The separate acceleration and velocity seismic maps also provide independent ground motion reference levels for small rigid structures having short fundamental periods and taller, more flexible structures with longer fundamental periods. Details of the technical basis of the current seismic zoning maps in relation to earthquake-resistant design have been presented by Heidebrecht et al. (1983).

9.4 DIRECTION OF SEISMIC MOTIONS

In general, earthquake ground motions are multidirectional. However, the NBCC seismic provisions assume that structures designed to resist seismic forces acting independently along each of the two horizontal principal axes of the building, together with any associated torsional effects, will provide adequate resistance to seismic motion applied in any direction. Furthermore, the gravity load requirements of the Code ensure a reserve of strength sufficient to accommodate the effects of the vertical component of the seismic ground motion.

9.5 LIMITATIONS AND RESTRICTIONS

The NBCC assumes that the deformations of buildings during the severe ground shaking of major earthquakes will include permanent, inelastic displacements. To provide the appropriate level of safety to the occupants of the building, structures designed according to the NBCC equivalent static lateral load method also must be designed and detailed to prevent premature failure due to inadequate member ductility, insufficient strength or resilience at the connections, stress reversals, or excessive lateral displacement or drift. The NBCC seismic design provisions have been developed to cover the design of typical building structures. The seismic load provisions in the Code are not directly applicable for highly irregular or unusually complex structures and industrial facilities. The design of those structures requires special considerations. Dynamic analyses usually are needed to determine the vertical distribution of the seismic lateral load for buildings with complex configurations.

9.6 LATERAL SEISMIC FORCE

The NBCC of 1990 specifies that the minimum lateral seismic shear force V at the base of the structure be calculated from the equivalent elastic base shear V_e, which represents elastic behavior, and a force modification factor R (Section 9.13) as follows:

$$V = (V_e/R)U \qquad (9.1)$$

where $U = 0.6$.

The force modification factor R is a measure of the ductility of the structure, while U is a calibration factor developed from experience on the satisfactory seismic performance of buildings designed for the level of forces implied by the Code. It has been observed that structures possess considerable reserve strength to resist seismic actions exceeding that provided in design. This reserve strength (overstrength) is due to the lateral resistance provided by elements such as stairwells, partitions, and infill walls, as well as the fact that actual material strengths are usually higher than the nominal design values. Thus, a structure designed to resist a base shear of $V_d U$ will, in fact, possess a strength equal to V_d.

The elastic equivalent base shear V_e is determined from the following formula:

$$V_e = \nu SIFW \qquad (9.2)$$

where

ν = zonal velocity ratio
S = seismic response factor
I = importance factor
F = foundation factor
W = dead load plus a portion of the live load

The value of the zonal velocity ratio ν associated with each seismic zone is given in Table 9.1, except that the value of ν shall be taken as 0.05 when the velocity-related zone $Z_v = 0$ and the acceleration-related zone $Z_a > 0$.

9.7 NATURE OF SEISMIC ACTION

In the current seismic design provisions, seismic action is regarded as an "accidental action" which does not require any further "factor of safety" (Rainer 1987). Thus, in limit states design of buildings, the lateral seismic force is assigned a load factor of 1.0, whereas the other source of lateral load, the wind load, is given a load factor of 1.5, because the wind load is considered to be a "frequent variable action." In building design, the factored seismic load is combined with other factored loads (dead load, live load, and load due to temperature changes) to produce the most critical condition for the design and detailing of structural members.

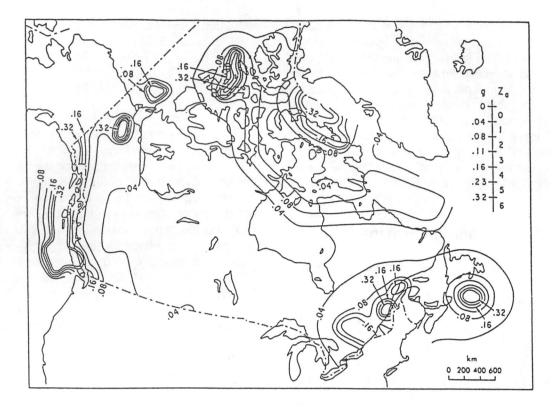

Fig. 9.1(a).   Contours of peak horizontal ground acceleration. Acceleration-related seismic zone map of Canada (NBCC 1990)

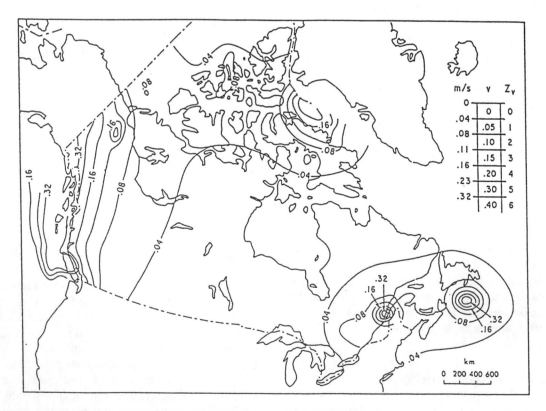

Fig. 9.1(b).   Contours of peak horizontal ground velocity. Velocity-related seismic zone map of Canada (NBCC 1990)

9.8 GRAVITY LOAD W

In the evaluation of the equivalent elastic base shear, eq.(9.2), the gravity load W shall be taken as the sum of the structure dead load plus 25% of the design snow load, 60% of the design live load for areas used for storage (storage load), and the full weight of the contents of any storage tanks. This gravity load W, used in the calculation of the seismic lateral force, is distributed throughout the building height as discrete loads applied at the floor levels; i.e.,

$$W = \sum_{i=1}^{n} W_i \qquad (9.3)$$

where W_i = portion of W assigned to level i.

9.9 SEISMIC RESPONSE FACTOR S

The seismic response factor S in eq.(9.2) is a function of the fundamental period T of the structure, and the seismic zoning ratio Z_a/Z_v of a particular site, as shown in Table 9.2. The seismic response factor represents the idealized elastic response spectrum for a 5% critically damped, single-degree-of-freedom oscillator subjected to the average ground motion implied by the Building Code. The response spectrum values for structures with periods longer than 0.5 sec are controlled by the maximum ground velocity represented by the zonal velocity ratio. The maximum

Table 9.2. Seismic Response Factor S (NBCC 1990)

Fundamental Period T (in sec)	Seismic Zoning Ratio Z_a/Z_v	Seismic Response Factor S
$T \leqslant 0.25$	<1.0	2.1
	1.0	3.0
	>1.0	4.2
$0.25 < T < 0.50$	<1.0	2.1
	1.0	$3.0 - 3.6\,(T - 0.25)$
	>1.0	$4.2 - 8.4\,(T - 0.25)$
$T \geqslant 0.50$	All values	$1.5/\sqrt{T}$

ground velocity enters eq.(9.2) through v. For structures with periods smaller than 0.5 sec, the response spectrum values are dependent on maximum ground acceleration rather than maximum ground velocity. Equation (9.2), which contains the zonal velocity ratio, is still applicable and the dependence on ground acceleration is achieved by suitably adjusting the values of S. For T less than 0.5 sec, the seismic response factor S follows three different branches: one corresponds to $Z_a > Z_v$; another corresponds to $Z_a = Z_v$; and a third corresponds to $Z_a < Z_v$. The variation of the seismic response factor S is shown in Fig. 9.2. To preclude large variations in the seismic lateral forces, the seismic response factors for short period structures have been adjusted. Thus, for $T < 0.25$ sec and $Z_a > Z_v$, the seismic response factor S is limited to 1.4 times the value relative to $Z_a = Z_v$, while for $T < 0.25$ sec and $Z_a < Z_v$, S is taken as 0.71

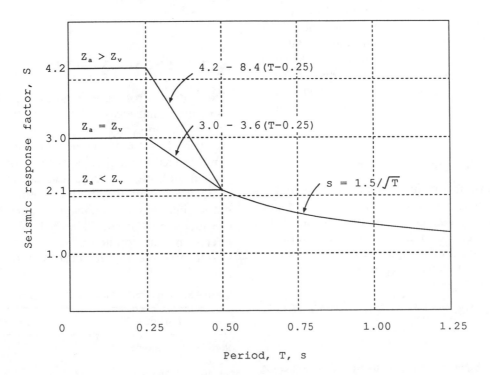

Fig. 9.2. Seismic response factor S (NBCC 1990)

times its value relative to $Z_a = Z_v$. The multipliers 1.4 and 0.71 correspond to a difference of one between the acceleration and velocity-defined seismic zones.

9.10 FUNDAMENTAL PERIOD *T*

The determination of the seismic response factor *S* requires an estimate of the fundamental period *T* of the structure. Depending on the nature of the lateral force-resisting system of the building, the Code permits different empirical formulas to be used to calculate the fundamental period.

9.10.1 Empirical Formulas

For moment-resisting space frames, the fundamental period *T* in seconds may be determined by

$$T = 0.1N \tag{9.4}$$

where N = total number of stories above exterior grade.

Other than moment-resisting space frames, *T* may be obtained for buildings with lateral load-resisting systems using the following formula:

$$T = \frac{0.09h_n}{\sqrt{D_s}} \tag{9.5}$$

In eq.(9.5), h_n is the height of the structure above the base to the uppermost level n, in meters, and D_s, in meters, is the dimension of the main lateral load-resisting system parallel to the direction of the seismic action considered. The Code stipulates that the dimension of the building D, in meters, measured parallel to the applied force, should replace D_s in eq.(9.5) when the length of the lateral load-resisting system is not well defined. The Code ensures that the design seismic force is conservative by using the largest possible value for D_s.

9.10.2 Rayleigh Approximation

The Code permits the fundamental period of the structure to be determined directly from a dynamic analysis or by other established methods of analysis, such as the Rayleigh formula. The Rayleigh formula for determining the fundamental period of the structure is given by the following expression:

$$T = 2\pi \sqrt{ \left(\sum_{i=1}^{n} W_i \delta_i^2 \right) \bigg/ \left(g \sum_{i=1}^{n} W_i \delta_i \right) } \leq 1.2T_a \tag{9.6}$$

where

T_a = the value obtained for the fundamental period using eq.(9.4) or eq.(9.5)

g = the acceleration due to gravity

W_i = the seismic gravity load assigned to level i

δ_i = the elastic lateral displacement at level i due to horizontal forces W_i applied simultaneously at all the levels of the building.

In order to prevent excessive reduction in the seismic lateral forces, which may be nonconservative, the Code establishes an upper limit on the value of the fundamental period T determined from analysis. This limit is 1.2 times the value obtained from eq.(9.4) or eq.(9.5). This implies that the design base shear, obtained with the period calculated by Rayleigh approximation or a more precise dynamic analysis, is no less than 90% of that obtained by using the period given by one of the empirical expressions.

9.11 IMPORTANCE FACTOR *I*

The Code recognizes the social importance of certain buildings in relation to their intended use, and increases the required minimum design base shear by means of the importance factor given in Table 9.3. The highest importance factor, 1.5, is assigned to so-called "post-disaster" buildings such as hospitals, fire and police stations, and other structures related to public safety that must remain functional during and immediately after earthquakes. An importance factor of 1.3 is assigned to schools to reduce the loss of life and to function as post-disaster shelters. Buildings other than those specified are assigned an importance factor of 1.0.

9.12 FOUNDATION FACTOR F

Observations from previous earthquakes clearly demonstrated that local soil conditions can modify significantly the intensity and frequency characteristics of the ground motions. Depending on the nature and depth of local soil deposits, the intensity of the seismic ground motions may be amplified and the frequency characteristics of the motions significantly altered as the seismic waves propagate through the intervening soil media to the site. The NBCC approximates the effects of soil conditions on the magnitude of the seismic base shear by means of the foundation factor **F**

Table 9.3. Importance Factor *I* (NBCC 1990)

Type of Building	Importance Factor *I*
Post-disaster buildings	1.5
Schools	1.3
All other buildings	1.0

Table 9.4. Foundation Factor **F** (NBCC 1990)

Category	Type of Soil	Depth of Soil (m)	F
1	Rock; dense and very dense coarse-grained soils; very stiff and hard fine-grained soils	Any depth	1.0
	Compact coarse-grained soils; Firm and stiff fine-grained soils	0–15	
2	Compact coarse-grained soils; Firm and stiff fine-grained soils	>15	1.3
	Very loose and loose coarse-grained soils; Very soft and soft fine-grained soils	0–15	
3	Very loose and loose coarse-grained soils	>15	1.5
4	Very soft and soft fine-grained soils	>15	2.0

defined in Table 9.4. Factor **F** depends on the nature and depth of the soil deposit measured from the foundation or pile cap level.

The incorporation of Category 4 with **F** = 2.0 in Table 9.4 in the last revision of the Code is based largely on the damage and response behavior observed in Mexico City during the 1985 Michoacán earthquake. In the NBCC seismic design provisions, the soil-structure interaction effect caused by the flexibility of the foundation structure is not considered explicitly; ignoring the interaction effect usually yields conservative designs for most buildings. Furthermore, in applying the foundation factor **F** to the design of buildings located on sites with multiple soil layers of different types, the Code approves the use of a reasonable average value of the foundation factor **F**, weighted in accordance with the relative thicknesses of the different soil layers. However, when much of the soil profile consists of soft clay or when thick layers of soft clay are present, rather than using the averaging technique, the value of **F** should be determined based on the total thickness of the soft clay layers only.

The Code limits the increase in the base shear caused by the local soil condition by specifying the following limits on the product of the foundation factor **F** and the seismic response factor S:

$$\text{for } Z_a \leqslant Z_v : FS \leqslant 3.0$$
$$\text{for } Z_a > Z_v : FS \leqslant 4.2 \qquad (9.7)$$

9.13 FORCE MODIFICATION FACTOR R

The value of the force modification factor R is assigned according to the type of structural system used in the building. Factor R is related to the reduction in the elastic strength demand on the structure, which is expected to behave inelastically

during an earthquake; i.e., R represents the capability of a structure to dissipate the seismic energy through inelastic behavior. In the Code, values of R have been developed for different types of structural systems based on cumulative design and construction experience, as well as on observed performance of different structural systems during major and moderate earthquakes. Consequently, higher values for the factor R are assigned to structural systems that can be designed and detailed to exhibit ductile behavior (which have been observed generally to perform well during earthquakes), whereas lower values are given to structural systems with little ductile capacity to accommodate cyclic inelastic deformations. For example, the types of structural systems to which a value of $R = 1.0$ has been assigned have performed poorly during earthquakes. The values of R assigned to different structural systems are shown in Table 9.5.

For buildings with combinations of different lateral load-resisting systems acting in the same direction, the Code stipulates that the lowest value of R corresponding to any one of these systems should be used in the determination of the seismic design base shear.

Table 9.5. Force Modification Factor R (NBCC 1990)

Case	Type of Lateral Load-Resisting System	R
	Steel Structures (CAN/CSA-S16.1-M)	
1	Ductile moment-resisting space frame	4.0
2	Ductile eccentrically braced frame	3.5
3	Ductile braced frame	3.0
4	Moment-resisting space frame with nominal ductility	3.0
5	Braced frame with nominal ductility	2.0
6	Other steel structural systems not defined in Cases 1–5	1.5
	Reinforced Concrete Structures (CAN3-A23.3-M)	
7	Ductile moment-resisting space frame	4.0
8	Ductile flexural wall	3.5
9	Moment-resisting space frame with nominal ductility	2.0
10	Wall with nominal ductility	2.0
11	Other concrete structural systems not defined in cases 7–10	1.5
	Timber Structures (CAN/CSA-086.1-M)	
12	Nailed shear panel with plywood, waferboard, or strandboard	3.0
13	Concentrically braced heavy timber space frame with ductile connections	2.0
14	Moment-resisting wood space frame with ductile connections	2.0
15	Other timber structural systems not defined in cases 12–14	1.5
	Masonry Structures (CAN3-S304-M)	
16	Reinforced masonry	1.5
17	Unreinforced masonry	1.0
18	Other lateral load-resisting systems not defined in cases 1–17	1.0

9.14 DISTRIBUTION OF LATERAL FORCES

9.14.1 Distribution Throughout Building Height

In the evaluation of seismic lateral force distribution throughout the building height, the building is assumed to respond primarily in the fundamental mode of vibration, which for strong ground motions is assumed to take the shape of a triangle, the lateral displacement increasing linearly with height above ground. For a tall flexible structure, the contribution to seismic forces from higher modes of vibration is significant, and therefore is taken into account approximately by specifying that a concentrated force F_t be applied at the top of the structure, where

$$F_t = 0.07TV \leq 0.25V \quad \text{for } T \geq 0.7 \text{ sec}$$

$$F_t = 0 \quad \text{for } T < 0.7 \text{ sec} \tag{9.8}$$

The remaining base shear is then distributed throughout the entire height of the building as forces at individual floor levels, as shown in Fig. 9.3, according to the formula

$$F_x = (V - F_t) \frac{W_x h_x}{\displaystyle\sum_{i=1}^{n} W_i h_i} \tag{9.9}$$

where h_x, W_x and F_x are the height, gravity load for calculation of seismic forces, and applied seismic lateral force at level x, respectively.

9.14.2 Distribution Among Lateral Load-Resisting Elements

The Code specifies that the seismic lateral shear at each story shall be distributed to the various load-resisting elements in proportion to their stiffnesses. The effect of the torsional moment caused by the eccentricities between the center of rigidity at a story and the centers of mass at and above that story also should be taken into consideration.

9.15 SEISMIC OVERTURNING MOMENT

The seismic lateral forces assigned to the different floor levels of a building cause an overturning moment at the base of the structure. As stated earlier, a concentrated force F_t is applied at the top of the structure to account for the contributions from higher vibration modes for a tall, relatively flexible building. However, contributions of the higher modes to overturning moment are not as significant as those to lateral loads. Thus, the Code stipulates that the base

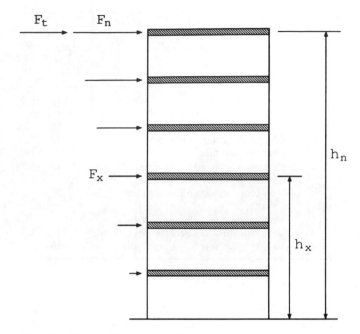

Fig. 9.3. Vertical distribution of lateral forces

moment M and the moment at each floor level M_x shall be reduced by the reduction factors J and J_x, respectively.

The reduction factor J for the overturning moment at the base of the structure is

$$\begin{array}{lll} J = 1 & \text{for} & T < 0.5 \text{ sec} \\ J = (1.1 - 0.2T) & & 0.5 \leq T \leq 1.5 \text{ sec} \\ J = 0.8 & & T > 1.5 \text{ sec} \end{array} \tag{9.10}$$

and the reduction factor J_x at level x is

$$J_x = J + (1 - J)(h_x/h_n)^3 \tag{9.11}$$

where h_n is the full height of the structure (n stories). The Code also specifies that the incremental change of the overturning moment in any floor level under consideration shall be distributed to the various elements resisting lateral loads in the same proportion as in the distribution of the lateral story shear to those elements.

9.16 TORSIONAL MOMENT

Unsymmetrical buildings undergo torsional oscillations during an earthquake in addition to translational vibrations. In such buildings, the centers of mass, through which the inertial forces act, and the centers of rigidity at the various stories of the building do not coincide because of the uneven distributions of mass in the structure and/or lateral load-resisting elements.

Also, direct torsional ground motion may cause torsional effects.

The NBCC seismic provisions require that buildings be designed for both a lateral load and a torsional moment applied to each story. The design torsional moment at any story level is the moment obtained by multiplying the lateral forces (at and above the story under consideration) by the design eccentricity measured normal to the direction of the seismic action. The Code stipulates that the eccentricity used for design be composed of two parts (Tso and Dempsey 1980): the first part is the computed eccentricity adjusted to take into consideration the effect of the dynamic torsional response; the second part, commonly referred to as the accidental eccentricity, accounts for the difference between the actual and computed eccentricity, and the response generated from torsional ground motion. The difference between the actual and computed eccentricity may arise from one or more of the following causes: variations in estimates of the relative rigidities of the lateral load-resisting elements, addition of partition walls after completion of the building, uncertainties in the estimates and distributions of dead and live loads at the floor levels, and redistribution of lateral stiffness due to inelastic action during earthquakes.

The design eccentricity is given by eq.(9.12); the value that produces the worst effect is to be used.

$$e_x = 1.5e + 0.10D_n$$
$$e_x = 0.5e - 0.10D_n$$
(9.12)

In these equations e is the computed eccentricity at the level being considered, which is the perpendicular distance between the line of action of the resultant of all the forces (at and above the level being considered) and the center of rigidity of the floor; D_n is the building dimension normal to the line of action of the forces being considered.

The design torsional moment at each floor is determined using the following formula:

$$M_{tx} = \left(F_t + \sum_{i=x}^{n} F_i \right) e_x = V_x e_x$$
(9.13)

These torsional provisions apply only for buildings in which the centers of mass and the centers of rigidity of different floors lie approximately on vertical lines. For buildings with unusual or complex configurations, full dynamic analyses should be carried out to determine the torsional effects.

9.17 LATERAL DISPLACEMENT AND STORY DRIFT

The maximum lateral displacements of a building during an earthquake should be calculated according to accepted practices. One such practice is to apply the design seismic lateral forces, obtained from the equivalent lateral force procedure, as static loads acting on the building. The total lateral displacements then are equal to the resulting elastic displacements multiplied by the force modification factor R: this accounts for the inelastic deformations that the building may sustain during a major earthquake.

In order to limit damage to nonstructural components of the building during moderate earthquakes, the Code establishes the following limits on the interstory drift

$$\delta_{i+1} - \delta_i \leq 0.01h_s \quad \text{for post-disaster buildings}$$
$$\delta_{i+1} - \delta_i \leq 0.02h_s \quad \text{for all others}$$
(9.14)

in which δ_{i+1} and δ_i are the total lateral displacements at levels $i+1$ and i, respectively, and h_s is the interstory height.

9.18 SEPARATION OF BUILDINGS

During an earthquake, lateral collisions of buildings (structural pounding) occur when the separations between adjacent buildings are insufficient to accommodate the relative motions of the buildings; many instances of such structural failure have been observed. These failures result in anything from minor damage to nonstructural elements, to total collapse of buildings. The NBCC of 1990 attempts to eliminate or minimize this source of structural failure. The Code specifies that either the separation between adjacent buildings equal or exceed the absolute sum of the lateral displacements of the buildings obtained from the equivalent static load procedure, or the adjacent buildings be connected properly to each other. For the latter case, the lowest force modification factor R of the connected buildings should be used to determine the equivalent seismic design base shear.

9.19 SPECIAL PROVISIONS

As stated earlier, the equivalent lateral force procedure has been developed for typical building structures. The use of this procedure to design highly irregular buildings or complex structures may introduce unacceptable errors. For this reason, the Code commentary specifies that, for buildings with complex configurations, dynamic analyses should be carried out to obtain a more reliable and accurate vertical distribution of the seismic design base shear than that given by eqs.(9.8) and (9.9). According to the Code, the purpose of the dynamic analysis is not to replace eq.(9.1) in the determination of the seismic design base shear, but rather to obtain better estimates for the vertical distribution of the equivalent lateral seismic forces.

The Code also stipulates that the equivalent seismic design base shear V be increased by 50% for buildings located in seismic velocity-related zone $Z_v \geq 4$ that are greater than 60 m in height and have a structural system requiring a force modification factor R of 1.5 or 2.0. This stipulation recognizes the need for extra caution in designing buildings with low ductility when such buildings are located in zones of comparatively greater seismicity.

For structural systems with discontinuous lateral load-resisting elements, the NBCC of 1990 requires that the overturning moment at the lowest story of the discontinuous element be carried down to the foundation. The Code also has provisions for the determination of seismic design forces on architectural appendages, portions of buildings, and mechanical and electrical equipment components. However, details of those provisions are not discussed here.

9.20 COMPUTER PROGRAM FOR THE EVALUATION OF SEISMIC DESIGN FORCES

A computer program has been developed to implement the seismic design provisions as stipulated in the NBCC of 1990. The program is written in FORTRAN for execution on microcomputers with DOS operating system. The program may be executed either interactively, with the data supplied and stored in a file during the interactive session, or in a batch mode with the data input from a previously prepared data file.

The program is capable of analyzing structural systems with discontinuous resisting elements. The program algorithm assumes that all lateral load-resisting elements extend the entire height of the building. Structural systems with discontinuous elements may, however, be analyzed by modeling the discontinuous element as a continuous element having zero lateral stiffness in those stories in which it does not exist.

A lateral load-resisting element may be oriented in an arbitrary direction on a floor plan. The orientation of the resisting element is defined by the angle measured counterclockwise from the positive global X-axis to the positive local element x-axis. The lateral stiffnesses of the resisting elements are described in their local element coordinate systems. These coordinates are specified as input data for the analysis.

The program calculates the seismic design base shear and the corresponding lateral seismic forces distributed to the stories. The program also determines the centers of rigidity at each of the stories. The

distribution of the lateral shear among the resisting elements at each story is also determined. The effect of the torsional moments corresponding to the design eccentricities specified in the Code is included. The program also determines the story drift resulting from the lateral story shear, and rotations due to the story torsional moments.

9.21 NUMERICAL EXAMPLES

Example 9.1

A four-story reinforced concrete office building is shown in Fig. 9.4; the lateral load in the N–S direction is resisted primarily by a shear wall system. The dead load attributed to each level is taken to be 4,608 kN, except for the roof where it is 4,800 kN. The accumulated snow load on the roof is estimated to be 2.2 kN/m². The soil at the site is assumed to be stiff clay, less than 15 m deep. The building is located in acceleration-related seismic Zone 4 ($Z_a = 4$) and velocity-related seismic Zone 2 ($Z_v = 2$).

Assuming that the seismic action is in the N–S direction, according to the NBCC of 1990 determine the following: (a) the fundamental period, (b) seismic

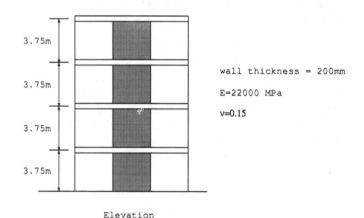

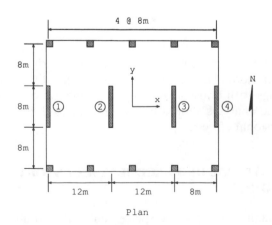

Fig. 9.4.   Building modeled for Example 9.1

response factor, (c) force modification factor, foundation factor, and importance factor, and (d) the equivalent seismic base shear of the building. Also, at each floor level for the story as a whole and for each element of the lateral load-resisting system, determine the (e) assigned lateral force and story shear, (f) overturning moment, (g) torsional moment, and (h) interstory drift and lateral displacement.

Solution

(a) Fundamental period. The fundamental period of a shear wall building is given by

$$T = \frac{0.09 h_n}{\sqrt{D_s}} \qquad [\text{eq.}(9.5)]$$

with

$h_n = 15$ m (height of building)
$D_s = 8$ m (dimension of the main lateral load-resisting system)

$$T = \frac{0.09(15)}{\sqrt{8}} = 0.48 \text{ sec}$$

(b) Seismic response factor. According to Table 9.2, for a seismic zoning ratio greater than one ($Z_a/Z_v = 2$) and a fundamental period T between 0.25 and 0.5 sec, the seismic response factor is given by the formula

$$S = 4.2 - 8.4(T - 0.25)$$

$$= 4.2 - 8.4(0.48 - 0.25)$$

$$= 2.27$$

(c) Force modification factor, foundation factor and importance factor. Assuming that the shear wall lateral load-resisting system in the N–S direction of the building is designed and detailed to exhibit ductile flexural behavior during strong ground motions, the force modification factor for such a structural system, as given in Table 9.5, is

$$R = 3.5$$

The shear wall system is designed to take 100% of the lateral load in the N–S direction.

From Tables 9.4 and 9.3, the foundation factor and the importance factor are as follows:

$$\textbf{F} = 1.0 \text{ (Soil Category 1)}$$

$$I = 1.0 \text{ (not a post-disaster building)}$$

Note that the product **FS** equals 2.27, which is less than 4.2 as specified in eq.(9.7).

(d) Equivalent seismic base shear. The gravity load W required in the calculation of the seismic base shear is determined as follows:
Roof area $= 32 \text{ m} \times 24 \text{ m} = 768 \text{ m}^2$

$$W = \sum_{i=1}^{n} W_i \qquad [\text{eq.}(9.3)]$$

$$= 4{,}608 \text{ kN} \times 3 + 4{,}800 \text{ kN} + 0.25 \times 2.2 \text{ kN/m}^2 \times 768 \text{ m}^2$$

$$= 19{,}046 \text{ kN}$$

The elastic base shear is given by

$$V_e = \nu SIFW \qquad [\text{eq.}(9.2)]$$

with the zonal velocity ratio $\nu = 0.10$ for velocity-related seismic Zone 2 ($Z_v = 2$) as determined from Table 9.1.

$$V_e = 0.10 \times 2.27 \times 1.0 \times 1.0 \times 19{,}046$$

$$= 4{,}323 \text{ kN}$$

The equivalent seismic base shear is then obtained from the elastic base shear and the force modification factor R as follows:

$$V = (V_e/R)U$$

$$= (4{,}323/3.5)0.6 \qquad [\text{eq.}(9.1)]$$

$$= 741.1 \text{ kN}$$

(e) Story lateral force and story shear. Since the fundamental period of the building $T = 0.48 < 0.7$ sec, the lateral force F_t applied at the top level is zero, as specified in eq.(9.8).
The story lateral forces are given by

$$F_x = (V - F_t) \frac{W_x h_x}{\displaystyle\sum_{i=1}^{n} W_i h_i} \qquad [\text{eq.}(9.9)]$$

with

$$F_t = 0 \qquad [\text{eq.}(9.8)]$$

(f) Overturning moment. The products of the story lateral forces and the vertical distances to those forces from a given story yield the overturning moment.
The story lateral forces F_x, the story shear forces V_x, and the story overturning moments M_x are shown in Table 9.6.
According to eq.(9.10), the seismic overturning moment reduction factors J and J_x, for the moments at the base and at level x, both are equal to 1.0. Thus, no

Table 9.6. Story Lateral Forces, Story Shear Forces, and Story Overturning Moments

Level x	h_x (m)	W_x (kN)	$W_x h_x$	F_x (kN)	V_x (kN)	M_x (kN-m)
4	15.00	5,222	78,330	318.9	318.9	1,196
3	11.25	4,608	51,840	211.1	530.0	3,183
2	7.50	4,608	34,560	140.7	670.7	5,699
1	3.75	4,608	17,280	70.4	741.1	8,478
$\Sigma =$			182,010	741.1		

adjustment is needed for the moments calculated in Table 9.6.

(g) Torsional moment. Considering the four shear wall elements in the N–S direction shown in Fig. 9.4, the center of rigidity $(\bar{x}, \bar{y})$ at each floor level is determined as follows:

$$\bar{x} = \frac{\sum_j K_j x_j}{\sum_j K_j} \qquad (9.15)$$

where K_j = lateral stiffness in the N–S direction of wall j

$$\bar{x} = \frac{-16K - 4K + 8K + 16K}{4K}$$

$$= 1\,\text{m}$$

$$\bar{y} = 0\,\text{m (from symmetry)}$$

The computed eccentricity at each story is calculated using the following formula:

$$e = \frac{\sum_{i=x}^{n} F_i e_{ix}}{\sum_{i=x}^{n} F_i} \qquad (9.16)$$

where

F_i = lateral force applied at level i
e_{ix} = distance between the vertical line through the center of mass at level i and the vertical line through the center of rigidity at level x, where both vertical lines are projected onto a vertical plane perpendicular to the lateral force.

Assuming that the centers of mass are located at the centers of the floor slabs at $(x, y) = (0, 0)$ throughout the entire building, the computed eccentricity at each story is simply

$$e = 1\,\text{m}$$

Table 9.7. Story Torsional Moments

Level x	M_{tx} (kN-m)	
	$e_x = 4.70$ m	$e_x = -2.70$ m
4	1,499	−861.0
3	2,491	−1,431
2	3,152	−1,811
1	3,483	−2,001

The design eccentricity is determined from eq.(9.12):

$$e_x = 1.5e + 0.10D_n$$
$$e_x = 0.5e - 0.10D_n$$

Substitution gives

$$e_x = 1.5(1) + 0.10(32) = 4.70\,\text{m}$$
$$e_x = 0.5(1) - 0.10(32) = -2.70\,\text{m}$$

Of these two eccentricities, the one producing the greater force in a member is to be used for the design of that member.

The story torsional moments are given by $M_{tx} = V_x e_x$, with the results shown in Table 9.7.

For the building and seismic action considered in this example, the story shear distributed to each individual wall, including the torsional effect, is obtained as follows:

$$V_{jx} = V_x \frac{K_j}{\sum_i K_i} - M_{tx} \frac{K_j(x_j - \bar{x})}{\sum_i K_i(x_i - \bar{x})^2} \qquad (9.17)$$

where V_{jx} = story shear distributed to wall j at level x.

Story shears distributed to the individual walls for the two cases of design eccentricity are shown in Tables 9.8a and 9.8b; the governing forces for the

Table 9.8a. Distribution of Story Shear to Individual Walls ($e_x = 4.70$ m)

Level x	Wall 1 (kN)	Wall 2 (kN)	Wall 3 (kN)	Wall 4 (kN)
4	123.1*	92.47*	61.89	41.49
3	204.5*	153.7*	102.9	68.95
2	258.8*	194.5*	130.2	87.27
1	286.0*	214.9*	143.8	96.42

Table 9.8b. Distribution of Story Shear to Individual Walls ($e_x = -2.70$ m)

Level x	Wall 1 (kN)	Wall 2 (kN)	Wall 3 (kN)	Wall 4 (kN)
4	54.83	72.40	89.98*	101.7*
3	91.13	120.3	149.5*	169.0*
2	115.3	152.3	189.2*	213.9*
1	127.4	168.3	209.1*	236.3*

Table 9.9. Distribution of Story Overturning Moment to Individual Walls

Level x	M_x (kN-m)	ΔM (kN-m)	Wall 1 (kN-m)	Wall 2 (kN-m)	Wall 3 (kN-m)	Wall 4 (kN-m)
4	1,196	—	299	299	299	299
3	3,183	1,987	796	796	796	796
2	5,699	2,516	1,425	1,425	1,425	1,425
1	8,478	2,779	2,120	2,120	2,120	2,120

design of the resisting elements are marked with asterisks (*).

The story overturning moment also is distributed to the individual walls in proportion to the wall stiffnesses and rigidities, as shown in Table 9.9.

(h) Interstory drift and lateral displacement. The lateral stiffness K_j of an individual wall element may be evaluated as follows: [see eqs.(4.3) and (4.4)].

$$K_j = \frac{1}{(L^3/3EI) + (L/GA_s)} \qquad (9.18)$$

where

    L = interstory height
    E = modulus of elasticity
    I = moment of inertia of the wall section
    A_s = A/1.2 = the effective shear area
    A = wall cross-sectional area
    G = modulus of rigidity = $E/2(1 + \nu)$
    ν = Poisson's ratio

The substitution of numerical values into eq.(9.18) results in

$$K_j = 2.579 \times 10^6 \text{ kN/m}$$

The interstory drift due to the translational story shear force only is given by

$$\delta_x = \frac{V_x}{\sum_j K_j} R \qquad (9.19)$$

where

δ_x = interstory drift along the seismic direction at level x
R = force modification factor (Table 9.5)

Calculated drifts are shown in Table 9.10; the summation of stiffnesses K_j is equal to 1.032×10^7 kN/m. The rotation of the floor slab $\delta_{\theta x}$ due to the torsional moment M_{tx} at level x is obtained from the following:

$$\delta_{\theta x} = \frac{M_{tx}}{K_{tx}} R \qquad (9.20)$$

Table 9.10. Interstory Drifts and Lateral Displacements, N–S

Level x	V_x (kN)	Interstory Drift (mm)	Lateral Displacement (mm)
4	318.9	0.108	0.766
3	530.0	0.180	0.658
2	670.7	0.227	0.478
1	741.1	0.251	0.251

Level x	δ^θ		Interstory Drift at NW Corner (e_x = 4.7 m)	
	e_x - 4.7 m	e_x = −2.7 m	y(mm)(N–S)	x(mm)(E–W)
4	$3.46 \ 10^{-6}$	$-1.99 \ 10^{-6}$	0.0588	0.0416
3	$5.75 \ 10^{-6}$	$-3.30 \ 10^{-6}$	0.0978	0.0690
2	$7.28 \ 10^{-6}$	$-4.18 \ 10^{-6}$	0.124	0.0873
1	$8.04 \ 10^{-6}$	$-4.62 \ 10^{-6}$	0.137	0.0965

where K_{tx} is the story torsional rigidity. Assuming that the floor slab itself is rigid, the story torsional rigidity in this example is given by

$$K_t = \sum_j K_j(x_j - \bar{x})^2 \qquad (9.210)$$

where

    K_j = lateral stiffness of wall j
    x_j = coordinate of centroid of wall j

$$\begin{aligned} K_t &= (17^2 + 5^2 + 7^2 + 15^2)K \\ &= 588 \text{ m}^2(2.579 \times 10^6 \text{ kN/m}) \\ &= 1.516 \times 10^9 \text{ kN-m} \end{aligned}$$

The rotations of the floor slab, and the corresponding interstory drifts at the northwest corner of the building, the farthest point from the center of rigidity, are shown in Table 9.11.

The maximum resultant interstory drift due to lateral shear and torsional moment at the first story can be checked against the allowable drift limit specified in the code, $0.02h_s$:

$$\sqrt{(0.251 + 0.137)^2 + 0.0965^2} = 0.4 \text{ mm} < 0.02h_s = 75 \text{ mm}.$$

Satisfactory.

Example 9.2

For the same building model as in Example 9.1, repeat the analysis considering seismic action in the E–W direction. Use the computer program to carry out the analysis. The dimensions and numbering scheme of the lateral load-resisting elements for the program input are shown in Fig. 9.5.

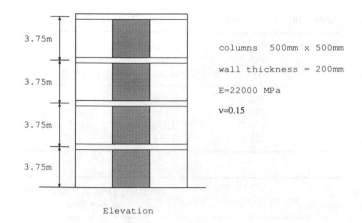

columns 500mm x 500mm

wall thickness = 200mm

E=22000 MPa

v=0.15

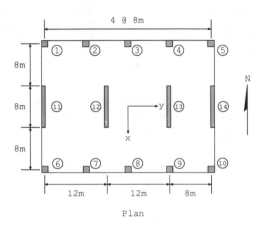

Fig. 9.5. Building modeled for Example 9.2

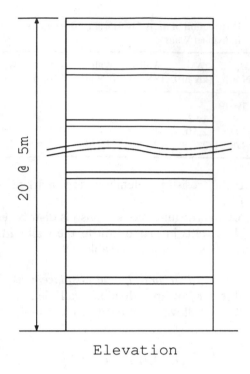

Elevation

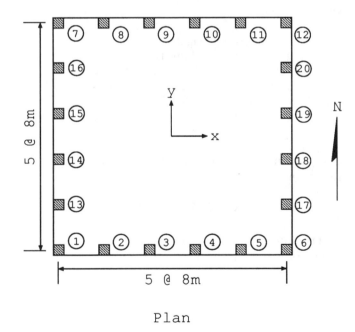

Plan

Fig. 9.6. Building modeled for Example 9.3

Solution

Along the E–W direction, lateral seismic action is resisted by a ductile moment-resisting space frame. The minimal lateral stiffness of the shear walls in the E–W direction is ignored in the determination of the seismic forces. The force modification factor for a ductile steel moment-resisting frame is

$$R = 4$$

This example includes the contributions of the lateral stiffness of the shear walls and columns in the N–S direction in resisting the torsional moment induced by the E–W seismic action.

The procedure for determining the lateral stiffnesses of shear walls in the N–S direction is shown in Example 9.1. Using the same formula as used in the calculation of wall stiffnesses, the lateral stiffness of the square column in both the N–S and E–W directions is determined to be

$$K_j = 6.440 \times 10^3 \text{ kN/m}$$

A computer printing showing the input data and calculated results is given on the following pages.

Example 9.3

A 20-story steel building with a lateral load-resisting system composed of a moment-resisting space frame is shown in Fig. 9.6. A gravity load of 6,000 kN is attributed to each floor level, except the roof where it

```
**************************************************
*                                                *
*            * * * * NBCC90 * * * *              *
*                                                *
*          Earthquake Resistant Design           *
*                                                *
*                 according to                   *
*                                                *
*     National Building Code of Canada (1990)    *
*                                                *
*                                                *
*         D.T. Lau, Carleton University          *
*   Dept. of Civil and Environmental Engineering *
*            Ottawa, Ontario, Canada             *
*           (416) 788-2600 ext.7473              *
*                                                *
**************************************************
```

Version 1.03 May 1993

Program limits:

Maximum number of storeys = 100
Maximum different plans of resisting element layout = 10
Maximum number of resisting element in each layout = 50

Lateral load resisting elements:

* may be oriented in any arbitrary directions
* may be discontinuous along the height of the building

Coordinate System: y-axis is defined as parallel to the direction of seismic action

Units: meter, second, kN

* * * T i t l e * * *

Example 2: Repeat of example 1 with seismic action in E-W direction

Number of storeys (N) = 4

Plan dimension of the main lateral load resisting system (Ds) = .320000D+02
parallel to direction of seismic action (y-direction)

Plan dimension of building normal to (Dn) = .240000D+02
direction of seismic action (x-direction)

Force modification factor R = .400000D+01

Options for fundamental period estimate:

T = 0.09 * hn / sqrt(Ds) (Option 1)
T = 0.1 * N (Option 2)
T = direct input (Option 3)

Fundamental period estimate option 2

Foundation factor (F) = .100000D+01

Importance factor (I) = .100000D+01

Seismic risk Data:

Acceleration-related seismic zone (Za) = 4
Velocity-related seismic zone (Zv) = 2

* * * * Structural Layout Information * * * *

Storey #	Storey height (hs)	Storey weight (Wi)	Center of mass x-coord	Center of mass y-coord
4	.375000D+01	.522240D+04	.000000D+00	.000000D+00
3	.375000D+01	.460800D+04	.000000D+00	.000000D+00
2	.375000D+01	.460800D+04	.000000D+00	.000000D+00
1	.375000D+01	.460800D+04	.000000D+00	.000000D+00

Storey #	Layout
4	1
3	1
2	1
1	1

* * Layout 1 * *

Resisting element #	Centroid x-coord	Centroid y-coord	Orientation angle (c.clockwise deg.)
1	-.120000D+02	-.160000D+02	.000000D+00
2	-.120000D+02	-.800000D+01	.000000D+00
3	-.120000D+02	.000000D+00	.000000D+00
4	-.120000D+02	.800000D+01	.000000D+00
5	-.120000D+02	.160000D+02	.000000D+00
6	.120000D+02	-.160000D+02	.000000D+00
7	.120000D+02	-.800000D+01	.000000D+00
8	.120000D+02	.000000D+00	.000000D+00
9	.120000D+02	.800000D+01	.000000D+00
10	.120000D+02	.160000D+02	.000000D+00
11	.000000D+00	-.160000D+02	.000000D+00
12	.000000D+00	-.400000D+01	.000000D+00
13	.000000D+00	.800000D+01	.000000D+00
14	.000000D+00	.160000D+02	.000000D+00

Resisting Element #	Local Stiffness Kxx	Kyy	Kxy
1	.644000D+04	.644000D+04	.000000D+00
2	.644000D+04	.644000D+04	.000000D+00
3	.644000D+04	.644000D+04	.000000D+00
4	.644000D+04	.644000D+04	.000000D+00
5	.644000D+04	.644000D+04	.000000D+00
6	.644000D+04	.644000D+04	.000000D+00
7	.644000D+04	.644000D+04	.000000D+00
8	.644000D+04	.644000D+04	.000000D+00
9	.644000D+04	.644000D+04	.000000D+00
10	.644000D+04	.644000D+04	.000000D+00
11	.257900D+07	.000000D+00	.000000D+00
12	.257900D+07	.000000D+00	.000000D+00
13	.257900D+07	.000000D+00	.000000D+00
14	.257900D+07	.000000D+00	.000000D+00

* * * * Live Loads Due to Earthquake * * * *

Fundamental period (T) = .400000D+00

Seismic response factor (S) = .294000D+01

Total weight (W) = .190464D+05

Zonal velocity ratio (v) = .100000D+00

Elastic base shear (Ve) = .559964D+04

Design base shear (V) = .839946D+03

Storey #	Storey lateral force (Fx)	Storey shear (Vx)	Moment reduction factor (Jx)	Storey moment (Mx)
4	.361496D+03	.361496D+03	.100000D+01	.135561D+04
3	.239225D+03	.600721D+03	.100000D+01	.360831D+04
2	.159483D+03	.760205D+03	.100000D+01	.645908D+04
1	.797417D+02	.839946D+03	.100000D+01	.960888D+04

Storey #	Center of rigidity x-coord.	Center of rigidity y-coord.
4	.000000D+00	.993796D+00
3	.000000D+00	.993796D+00
2	.000000D+00	.993796D+00
1	.000000D+00	.993796D+00

Storey#	Computed (e) eccent.	Design eccentricity (ex1)	Design eccentricity (ex2)	Torsional moment (Mtx1)	Torsional moment (Mtx2)
4	.000000D+00	.240000D+01	-.240000D+01	.867590D+03	-.867590D+03
3	.000000D+00	.240000D+01	-.240000D+01	.144173D+04	-.144173D+04
2	.000000D+00	.240000D+01	-.240000D+01	.182449D+04	-.182449D+04
1	.000000D+00	.240000D+01	-.240000D+01	.201587D+04	-.201587D+04

Storey #	Storey drift (due to lateral shear only) x	Storey drift y	Lateral deflection x	Lateral deflection y
4	.000000D+00	.224532D-01	.000000D+00	.159153D+00
3	.000000D+00	.373119D-01	.000000D+00	.136700D+00
2	.000000D+00	.472177D-01	.000000D+00	.993882D-01
1	.000000D+00	.521706D-01	.000000D+00	.521706D-01

Storey #	Storey rotation (clockwise rad.) Des. eccent.1 (ex1)	Des. eccent.2 (ex2)
4	.226225D-05	-.226225D-05
3	.375932D-05	-.375932D-05
2	.475737D-05	-.475737D-05
1	.525640D-05	-.525640D-05

* * * * Lateral Shear Distribution to Resisting Elements * * * *

* * * Design eccentricity (ex1) * * *

Storey 4 (** Design eccentricity (ex1) = .240000D+01 **)

Elm#	Lateral shear x-global	y-global	x-local	y-local	Overturning moment Mx-global	My-global	Mx-local	My-local
1	-.6190D-01	.3619D+02	-.6190D-01	.3619D+02	.1357D+03	.2321D+00	.1357D+03	.2321D+00
2	-.3276D-01	.3619D+02	-.3276D-01	.3619D+02	.1357D+03	.1228D+00	.1357D+03	.1228D+00
3	-.3620D-02	.3619D+02	-.3620D-02	.3619D+02	.1357D+03	.1357D-01	.1357D+03	.1357D-01
4	.2552D-01	.3619D+02	.2552D-01	.3619D+02	.1357D+03	-.9569D-01	.1357D+03	-.9569D-01
5	.5466D-01	.3619D+02	.5466D-01	.3619D+02	.1357D+03	-.2050D+00	.1357D+03	-.2050D+00
6	-.6190D-01	.3611D+02	-.6190D-01	.3611D+02	.1354D+03	.2321D+00	.1354D+03	.2321D+00
7	-.3276D-01	.3611D+02	-.3276D-01	.3611D+02	.1354D+03	.1228D+00	.1354D+03	.1228D+00
8	-.3620D-02	.3611D+02	-.3620D-02	.3611D+02	.1354D+03	.1357D-01	.1354D+03	.1357D-01
9	.2552D-01	.3611D+02	.2552D-01	.3611D+02	.1354D+03	-.9569D-01	.1354D+03	-.9569D-01
10	.5466D-01	.3611D+02	.5466D-01	.3611D+02	.1354D+03	-.2050D+00	.1354D+03	-.2050D+00
11	-.2479D+02	.0000D+00	-.2479D+02	.0000D+00	.0000D+00	.9295D+02	.0000D+00	.9295D+02
12	-.7284D+01	.0000D+00	-.7284D+01	.0000D+00	.0000D+00	.2731D+02	.0000D+00	.2731D+02
13	.1022D+02	.0000D+00	.1022D+02	.0000D+00	.0000D+00	-.3832D+02	.0000D+00	-.3832D+02
14	.2189D+02	.0000D+00	.2189D+02	.0000D+00	.0000D+00	-.8208D+02	.0000D+00	-.8208D+02

Storey 3 (** Design eccentricity (ex1) = .240000D+01 **)

Elm#	Lateral shear x-global	y-global	x-local	y-local	Overturning moment Mx-global	My-global	Mx-local	My-local
1	-.1029D+00	.6014D+02	-.1029D+00	.6014D+02	.3613D+03	.6178D+00	.3613D+03	.6178D+00
2	-.5444D-01	.6014D+02	-.5444D-01	.6014D+02	.3613D+03	.3270D+00	.3613D+03	.3270D+00
3	-.6015D-02	.6014D+02	-.6015D-02	.6014D+02	.3613D+03	.3613D-01	.3613D+03	.3613D-01
4	.4241D-01	.6014D+02	.4241D-01	.6014D+02	.3613D+03	-.2547D+00	.3613D+03	-.2547D+00
5	.9083D-01	.6014D+02	.9083D-01	.6014D+02	.3613D+03	-.5456D+00	.3613D+03	-.5456D+00
6	-.1029D+00	.6000D+02	-.1029D+00	.6000D+02	.3604D+03	.6178D+00	.3604D+03	.6178D+00
7	-.5444D-01	.6000D+02	-.5444D-01	.6000D+02	.3604D+03	.3270D+00	.3604D+03	.3270D+00
8	-.6015D-02	.6000D+02	-.6015D-02	.6000D+02	.3604D+03	.3613D-01	.3604D+03	.3613D-01
9	.4241D-01	.6000D+02	.4241D-01	.6000D+02	.3604D+03	-.2547D+00	.3604D+03	-.2547D+00
10	.9083D-01	.6000D+02	.9083D-01	.6000D+02	.3604D+03	-.5456D+00	.3604D+03	-.5456D+00
11	-.4119D+02	.0000D+00	-.4119D+02	.0000D+00	.0000D+00	.2474D+03	.0000D+00	.2474D+03
12	-.1210D+02	.0000D+00	-.1210D+02	.0000D+00	.0000D+00	.7270D+02	.0000D+00	.7270D+02
13	.1698D+02	.0000D+00	.1698D+02	.0000D+00	.0000D+00	-.1020D+03	.0000D+00	-.1020D+03
14	.3637D+02	.0000D+00	.3637D+02	.0000D+00	.0000D+00	-.2185D+03	.0000D+00	-.2185D+03

Storey 2 (** Design eccentricity (ex1) = .240000D+01 **)

Elm#	Lateral shear x-global	y-global	x-local	y-local	Overturning moment Mx-global	My-global	Mx-local	My-local
1	-.1302D+00	.7611D+02	-.1302D+00	.7611D+02	.6467D+03	.1106D+01	.6467D+03	.1106D+01
2	-.6889D-01	.7611D+02	-.6889D-01	.7611D+02	.6467D+03	.5853D+00	.6467D+03	.5853D+00
3	-.7612D-02	.7611D+02	-.7612D-02	.7611D+02	.6467D+03	.6467D-01	.6467D+03	.6467D-01
4	.5366D-01	.7611D+02	.5366D-01	.7611D+02	.6467D+03	-.4559D+00	.6467D+03	-.4559D+00
5	.1149D+00	.7611D+02	.1149D+00	.7611D+02	.6467D+03	-.9766D+00	.6467D+03	-.9766D+00
6	-.1302D+00	.7593D+02	-.1302D+00	.7593D+02	.6451D+03	.1106D+01	.6451D+03	.1106D+01
7	-.6889D-01	.7593D+02	-.6889D-01	.7593D+02	.6451D+03	.5853D+00	.6451D+03	.5853D+00
8	-.7612D-02	.7593D+02	-.7612D-02	.7593D+02	.6451D+03	.6467D-01	.6451D+03	.6467D-01
9	.5366D-01	.7593D+02	.5366D-01	.7593D+02	.6451D+03	-.4559D+00	.6451D+03	-.4559D+00
10	.1149D+00	.7593D+02	.1149D+00	.7593D+02	.6451D+03	-.9766D+00	.6451D+03	-.9766D+00
11	-.5213D+02	.0000D+00	-.5213D+02	.0000D+00	.0000D+00	.4429D+03	.0000D+00	.4429D+03
12	-.1532D+02	.0000D+00	-.1532D+02	.0000D+00	.0000D+00	.1301D+03	.0000D+00	.1301D+03
13	.2149D+02	.0000D+00	.2149D+02	.0000D+00	.0000D+00	-.1826D+03	.0000D+00	-.1826D+03
14	.4603D+02	.0000D+00	.4603D+02	.0000D+00	.0000D+00	-.3911D+03	.0000D+00	-.3911D+03

Storey 1 (** Design eccentricity (ex1) = .240000D+01 **)

Elm#	Lateral shear x-global	y-global	x-local	y-local	Overturning moment Mx-global	My-global	Mx-local	My-local
1	-.1438D+00	.8410D+02	-.1438D+00	.8410D+02	.9620D+03	.1645D+01	.9620D+03	.1645D+01
2	-.7611D-01	.8410D+02	-.7611D-01	.8410D+02	.9620D+03	.8707D+00	.9620D+03	.8707D+00
3	-.8410D-02	.8410D+02	-.8410D-02	.8410D+02	.9620D+03	.9621D-01	.9620D+03	.9621D-01
4	.5929D-01	.8410D+02	.5929D-01	.8410D+02	.9620D+03	-.6783D+00	.9620D+03	-.6783D+00
5	.1270D+00	.8410D+02	.1270D+00	.8410D+02	.9620D+03	-.1453D+01	.9620D+03	-.1453D+01
6	-.1438D+00	.8389D+02	-.1438D+00	.8389D+02	.9597D+03	.1645D+01	.9597D+03	.1645D+01
7	-.7611D-01	.8389D+02	-.7611D-01	.8389D+02	.9597D+03	.8707D+00	.9597D+03	.8707D+00
8	-.8410D-02	.8389D+02	-.8410D-02	.8389D+02	.9597D+03	.9621D-01	.9597D+03	.9621D-01
9	.5929D-01	.8389D+02	.5929D-01	.8389D+02	.9597D+03	-.6783D+00	.9597D+03	-.6783D+00
10	.1270D+00	.8389D+02	.1270D+00	.8389D+02	.9597D+03	-.1453D+01	.9597D+03	-.1453D+01
11	-.5759D+02	.0000D+00	-.5759D+02	.0000D+00	.0000D+00	.6589D+03	.0000D+00	.6589D+03
12	-.1692D+02	.0000D+00	-.1692D+02	.0000D+00	.0000D+00	.1936D+03	.0000D+00	.1936D+03
13	.2374D+02	.0000D+00	.2374D+02	.0000D+00	.0000D+00	-.2716D+03	.0000D+00	-.2716D+03
14	.5086D+02	.0000D+00	.5086D+02	.0000D+00	.0000D+00	-.5818D+03	.0000D+00	-.5818D+03

*** NOTE: The complete computer output also includes results for design eccentricity (ex2) = -.240000D+01, not reproduced here to save space.

is 6,600 kN. The lateral stiffness of each column at all floor levels is 5×10^4 kN/m; it is assumed to be the same in both principal directions. Analyze the building using the computer program. Consider seismic action in the N–S direction.

Solution

The input data and complete results are given by the computer output; however, to save space only selected portion of the results is reproduced for this example.

```
* * * *  Live Loads Due to Earthquake  * * * *
```

Fundamental period	(T) =	.200000D+01
Seismic response factor	(S) =	.212132D+01
Total weight	(W) =	.120600D+06
Zonal velocity ratio	(v) =	.200000D+00
Elastic base shear	(Ve) =	.511662D+05
Design base shear	(V) =	.767494D+04

Storey #	Storey lateral force (Fx)	Storey shear (Vx)	Moment reduction factor (Jx)	Storey moment (Mx)
20	.175944D+04	.175944D+04	.100000D+01	.879722D+04
19	.591549D+03	.235099D+04	.971475D+00	.199659D+05
18	.560415D+03	.291141D+04	.945800D+00	.332063D+05
17	.529281D+03	.344069D+04	.922825D+00	.482754D+05
16	.498147D+03	.393884D+04	.902400D+00	.649790D+05
15	.467013D+03	.440585D+04	.884375D+00	.831632D+05
14	.435878D+03	.484173D+04	.868600D+00	.102707D+06
13	.404744D+03	.524647D+04	.854925D+00	.123517D+06
12	.373610D+03	.562008D+04	.843200D+00	.145517D+06
11	.342476D+03	.596256D+04	.833275D+00	.168647D+06
10	.311342D+03	.627390D+04	.825000D+00	.192852D+06
9	.280208D+03	.655411D+04	.818225D+00	.218082D+06
8	.249073D+03	.680318D+04	.812800D+00	.244284D+06
7	.217939D+03	.702112D+04	.808575D+00	.271400D+06
6	.186805D+03	.720792D+04	.805400D+00	.299360D+06
5	.155671D+03	.736360D+04	.803125D+00	.328084D+06
4	.124537D+03	.748813D+04	.801600D+00	.357474D+06
3	.934025D+02	.758153D+04	.800675D+00	.387413D+06
2	.622684D+02	.764380D+04	.800200D+00	.417766D+06
1	.311342D+02	.767494D+04	.800025D+00	.448375D+06

Storey#	Computed (e) eccent.	Design eccentricity (ex1)	(ex2)	Torsional moment (Mtx1)	(Mtx2)
20	.000000D+00	.400000D+01	-.400000D+01	.703777D+04	-.703777D+04
19	.000000D+00	.400000D+01	-.400000D+01	.940397D+04	-.940397D+04
18	.000000D+00	.400000D+01	-.400000D+01	.116456D+05	-.116456D+05
17	.000000D100	.400000D+01	-.400000U01	.137628D+05	-.137628D+05
16	.000000D+00	.400000D+01	-.400000D+01	.157553D+05	-.157553D+05
15	.000000D+00	.400000D+01	-.400000D+01	.176234D+05	-.176234D+05
14	.000000D+00	.400000D+01	-.400000D+01	.193669D+05	-.193669D+05
13	.000000D+00	.400000D+01	-.400000D+01	.209859D+05	-.209859D+05
12	.000000D+00	.400000D+01	-.400000D+01	.224803D+05	-.224803D+05
11	.000000D+00	.400000D+01	-.400000D+01	.238502D+05	-.238502D+05
10	.000000D+00	.400000D+01	-.400000D+01	.250956D+05	-.250956D+05
9	.000000D+00	.400000D+01	-.400000D+01	.262164D+05	-.262164D+05
8	.000000D+00	.400000D+01	-.400000D+01	.272127D+05	-.272127D+05
7	.000000D+00	.400000D+01	-.400000D+01	.280845D+05	-.280845D+05
6	.000000D+00	.400000D+01	-.400000D+01	.288317D+05	-.288317D+05
5	.000000D+00	.400000D+01	-.400000D+01	.294544D+05	-.294544D+05
4	.000000D+00	.400000D+01	-.400000D+01	.299525D+05	-.299525D+05
3	.000000D+00	.400000D+01	-.400000D+01	.303261D+05	-.303261D+05
2	.000000D+00	.400000D+01	-.400000D+01	.305752D+05	-.305752D+05
1	.000000D+00	.400000D+01	-.400000D+01	.306997D+05	-.306997D+05

Storey #	Storey drift (due to lateral shear only) x	y	Lateral deflection x	y
20	.000000D+00	.703777D-02	.000000D+00	.448361D+00
19	.000000D+00	.940397D-02	.000000D+00	.441323D+00
18	.000000D+00	.116456D-01	.000000D+00	.431919D+00
17	.000000D+00	.137628D-01	.000000D+00	.420274D+00
16	.000000D+00	.157553D-01	.000000D+00	.406511D+00
15	.000000D+00	.176234D-01	.000000D+00	.390756D+00
14	.000000D+00	.193669D-01	.000000D+00	.373132D+00
13	.000000D+00	.209859D-01	.000000D+00	.353765D+00
12	.000000D+00	.224803D-01	.000000D+00	.332779D+00
11	.000000D+00	.238502D-01	.000000D+00	.310299D+00
10	.000000D+00	.250956D-01	.000000D+00	.286449D+00
9	.000000D+00	.262164D-01	.000000D+00	.261353D+00
8	.000000D+00	.272127D-01	.000000D+00	.235137D+00
7	.000000D+00	.280845D-01	.000000D+00	.207924D+00
6	.000000D+00	.288317D-01	.000000D+00	.179840D+00
5	.000000D+00	.294544D-01	.000000D+00	.151008D+00
4	.000000D+00	.299525D-01	.000000D+00	.121554D+00
3	.000000D+00	.303261D-01	.000000D+00	.916011D-01
2	.000000D+00	.305752D-01	.000000D+00	.612750D-01
1	.000000D+00	.306997D-01	.000000D+00	.306997D-01

REFERENCES

BASHAM, P. W.; WEICHERT, D. H.; ANGLIN, F. M.; and BERRY, M. J. (1985) "New Probabilistic Strong Seismic Ground Motion Maps of Canada." *Bulletin of the Seismological Society of America* 75(2): 563–595.

HEIDEBRECHT, A. C.; BASHAM, P. W.; RAINER, J. H.; and BERRY, M. J. (1983) "Engineering Applications of New Probabilistic Seismic Ground-Motion Maps of Canada." *Canadian Journal of Civil Engineering* 10: 670–680.

NBCC (1953, 1970, 1975, 1980, 1985, 1990) *National Building Code of Canada*. National Research Council of Canada, Ottawa, Ontario, Canada.

RAINER, J. H. (1987) "Force Reduction Factors for the NBCC Seismic Provisions." *Canadian Journal of Civil Engineering* 14(4): 447–454.

TSO, W. K., and DEMPSEY, K. M. (1980) "Seismic Torsional Provisions for Dynamic Eccentricity." *Earthquake Engineering and Structural Dynamics* 8: 275–289.

UZUMERI, S. M.; OTANI, S.; and COLLINS, M. P. (1978) "An Overview of Canadian Code Requirements for Earthquake Resistant Concrete Buildings." *Canadian Journal of Civil Engineering* 5(3): 427–441.

10
Chile

Arturo Cifuentes*

10.1 BACKGROUND

Chile, located along the southwestern coast of South America between the Pacific Ocean and the Andes Mountains, has a long history of seismic activity. The country has been struck by a number of severe earthquakes. The most notable quake, in 1960, measured 8.5 on the Richter scale and caused more than 5,000 deaths. Another quake in 1939 destroyed the city of Chillán and caused approximately 30,000 deaths. In addition, at the beginning of this century (1906) the city of Valparaíso also suffered a serious earthquake. Based on the extent of the damage, the intensity of that earthquake in the area close to the epicenter has been estimated as IX on the Modified Mercalli Intensity (MMI) scale.

Back in 1835 the cities of Concepción and Talcahuano were hit by a devastating combination of quake and tsunami, which was also accompanied by volcanic activity. Luckily, the death toll was less than one hundred. This low loss of lives was attributed to the fact that the quake occurred during the day and that most people were accustomed to rush outdoors directly during these events (Moorehead 1969). Charles Darwin, who was traveling aboard the Beagle at that time, was in Valdivia during the quake. A month later he visited Concepción with Captain Robert Fitzroy, and gave in his diary a frightening description of the event. He mentions, among other things, that some

*The author is very grateful to Rodrigo Araya, Paul Burridge, Ventura Charlin, María Ofelia Moroni, Wilfred Iwan, Lucía Ovalle, and Mario Paz for their help during the preparation of this chapter.

people who were in Valparaíso during the dreadful quake in 1822, found the 1835 quake even more powerful (The London Folio Society 1977).

The history of the Chilean seismic regulations goes back to 1928. In that year the city of Talca was affected by a quake and, as a result, in 1929 the government passed a law creating a committee to propose earthquake related regulations. Then, in 1935 a building code known as *Ordenanza General de Construcciones y Urbanizaciones* became official. The Chillán earthquake of 1939 prompted the government to appoint several committees to study the current seismic codes and propose modifications. These modifications went into effect officially in 1949. In 1958 another earthquake rocked the country, prompting a review of the current design practices. A government body known as INDITECNOR, which eventually changed its name to Instituto Nacional de Normalización (INN), started reviewing the design practices of the *Ordenanza*. Finally, in 1972 a seismic code known as *Cálculo Antisísmico de Edificios* (Earthquake Design of Buildings) was officially approved. A second edition of this seismic code was published in 1985 (INN 1985). The most recent version of this code, also known as *Diseño Sísmico de Edificios* (INN 1989), is currently used in Chile and it is the document on which this chapter is based.[1] A more detailed account of the evolution of the Chilean building codes from an earthquake engineering standpoint can be found in Sarrazin (1979).

10.2 OVERVIEW OF THE CODE

In essence, the aim of the Chilean seismic code is to provide the designer with the minimum values for base shear, torsional moments, horizontal forces, and overturning moments that should be used to estimate the response (e.g., stresses, displacements) of a building subjected to an earthquake. The code allows the designer to use one of two alternative approaches: a so-called static method and a dynamic method.

The static method reduces to consideration of a set of constant loads, acting at each floor level, all in the same horizontal direction (and with the same sign). In principle, this set of loads can be acting in any direction. However, it is acceptable to determine the response for two orthogonal (or almost orthogonal) main directions of the building.

The dynamic method is really a modal superposition method; it uses a base acceleration spectrum that depends on the characteristics of both the building and the type of soil at the site. It is assumed that the

ground motion is horizontal and can occur in any direction. Again, it is sufficient to consider two orthogonal (or almost orthogonal) directions for the analysis.

The code states explicitly that for buildings of more than 15 stories (or higher than 45 m) an analysis based on the dynamic method is mandatory, if the structure has a nonregular mass or stiffness distribution in the vertical direction. In addition, the code states that its applicability is limited to those structures that are normally described as buildings; it does not apply to other types of structures.

10.3 CLASSIFICATION OF BUILDINGS

The code classifies buildings, depending on their use or structural configuration, as follows:

Group A. Government buildings, municipal and public service buildings, buildings that are normally occupied by a large number of people, and buildings whose integrity is vital in case of national catastrophe or emergency (e.g., hospitals, churches, railroad terminals, fire department buildings).

Group B. Private residences and public buildings where it is uncommon for large crowds to gather.

Group C. Light or temporary constructions not used as residences (e.g., stables, granaries).

Group D. Buildings in general.

Group E. Buildings whose floors behave as rigid diaphragms.

Group F. Buildings whose floors behave as rigid diaphragms and which resist horizontal forces exclusively by rigid frames of adequate ductility. It is assumed that a rigid frame has adequate ductility if: (a) it is made of steel and it is riveted with high-strength bolts, or welded connections, or (b) it is made of reinforced concrete in a manner that allows the development of plastic hinges.

10.4 STATIC METHOD

10.4.1 Base Shear

The base shear, denoted as Q_0, is computed as

$$Q_0 = K_1 K_2 C W \tag{10.1}$$

where K_1 is a coefficient that depends on the use of the building (see Table 10.1), K_2 is a coefficient that depends on the building configuration (see Table 10.2), and W represents the weight of the structure

---

[1]A copy of the official code can be obtained from INN, P.O. Box 995, Santiago, Chile or Matías Cousiño 64, Santiago, Chile. Telephone: 56-2-696-8144.

Table 10.1. Value of Coefficient K_1

Use of Building	Value of K_1
Group A	1.2
Group B	1.0
Group C	0.8

Table 10.2. Value of Coefficient K_2

Structural Configuration of Building	Value of K_2
Group D	1.2
Group E	1.0
Group F	0.8

Table 10.3. Values of Parameter T_0

Type of Foundation Soil	Value of T_0 (sec)
Rock, dense gravel, dense sandy gravel	0.20
Dense sand, hard or firm cohesive soils	0.30
Granular soils, soft cohesive soils	0.90

plus a percentage of the live load. This percentage must be at least 25% in the case of private residences or buildings where large gatherings of people are not common, and 50% for those structures where large crowds commonly gather.

The factor C in eq.(10.1) depends on the period T (in seconds) of the building, and is determined as follows:

$$C = 0.10 \quad \text{for} \quad T \leqslant T_0 \quad (10.2)$$

or

$$C = \frac{0.2TT_0}{T^2 + T_0^2} \quad \text{for} \quad T > T_0 \quad (10.3)$$

where T_0 is given by Table 10.3 for different types of foundation soils.

In addition, the code states that: (a) under no circumstances can the value of C be less than 0.06, (b) a value of T larger than T_0 must be justified with a theoretical approach, an empirical formula, or experimental data, and (c) in multistory buildings the value of Q_0 must be at least $0.06W$, while in one-story buildings Q_0 must be at least $0.12W$, except for unreinforced masonry structures in which this value must be increased to $0.18W$.

10.4.2 Lateral Distribution of Forces

Let Z_k be the height of floor k measured from the ground level (thus, for the base level, $Z_0 = 0$). The horizontal forces F_k acting at each floor level are calculated as

$$F_k = \frac{W_k A_k Q_0}{\sum\limits_{j=1}^{N} W_j A_j} \quad (10.4)$$

where W_k represents the floor weight, A_k is a coefficient given by eq.(10.5) (it can be taken as Z_k for buildings of no more than five stories and not taller than 16 m), and N denotes the number of stories.[2]

The value of A_k is given by

$$A_k = \sqrt{1 - \frac{Z_{k-1}}{H}} - \sqrt{1 - \frac{Z_k}{H}} \quad (10.5)$$

where H is the total height of the structure measured from the base level, i.e., $H = Z_N$.[3]

10.4.3 Overturning Moments

The overturning moment M_k at a floor k is calculated by applying the horizontal forces specified by eq.(10.4) and considering a reduction factor J_k. This reduction factor applies only to buildings with four or more stories. Thus, the overturning moment M_k is given by

$$M_k = \sum_{i=k+1}^{N} F_i (Z_i - Z_k) J_k \quad (10.6)$$

where

$$J_k = 0.8 + \frac{0.2Z_k}{H} \quad (10.7)$$

and $k = 1, 2, \ldots, N-1$.

---

[2]The reader should be aware that in the original INN (1985) document (NCh.433.Of72, 2d ed.) there is a typographical error in the formula. The correct value for the lower index in the summation is 1, not i, as it is printed in the code.

---

[3]Again, in the original INN (1985) document there is a typographical error in this formula. The second term in the first square root is, as indicated here, Z_{k-1}/H; not $(Z_k - 1)/H$ as the code states.

10.4.4 Shear Forces

The shear Q_k at floor level k is calculated as

$$Q_k = \sum_{i=k}^{N} F_i \qquad (10.8)$$

The code specifies the following in relation to the distribution of shear forces:

(a) For buildings having rigid diaphragms at the levels of the floors, the horizontal shear force at a story should be distributed among the resisting elements in such a manner that equilibrium is satisfied. In addition, the deformations of the resisting elements must be compatible with the assumption of rigidity in the diaphragms. These diaphragms must be shown to behave as rigid bodies; otherwise, their flexibility must be taken into account when distributing the shear forces.

(b) For floors that do not satisfy the rigid diaphragm condition, the resisting elements should be designed using the actual horizontal forces that act on these elements.

(c) Slabs made of prefabricated reinforced concrete, prestressed concrete, or reinforced masonry can be considered as rigid diaphragms if the connections between the slab and the vertical elements are rigid. It must also be verified that the slab is capable of distributing the horizontal forces among the vertical elements connected to it.

(d) If the elongation of columns due to axial loads resulting from the seismic forces is not negligible, this effect must be taken into account when distributing the shear at the floor level.

10.4.5 Torsional Moments

For design purposes, at each floor level, a torsional moment must be considered in addition to the shear forces. The torsional moment at floor level k, denoted as M_{tk}, is computed as follows:

$$M_{tk} = 1.5 Q_k e_k \pm 0.05 I_k \sum_{j=k}^{N} F_j b_j \qquad (10.9)$$

where I_k, a reduction factor for accidental torsional moment, is given by

$$I_k = 0.7 + 0.3 \frac{Z_k}{H} \qquad (10.10)$$

The factor e_k refers to the distance between the center of resistance (rigidity) of the floor under consideration (k) and the line of action of the shear force Q_k; b_j represents the largest plan dimension at floor j in the direction perpendicular to F_j. The code specifies that the building should be designed so that, at each floor level, the force due to torsion acting on any resisting member should not exceed the force due to shear acting on that same member. To this end, a certain direction must be assumed to distribute the shear, such that if the shear Q_k is applied at the center of resistance, it should cause only a translation of the floor in the direction of the element under consideration. The code also specifies that those elements in which the force produced by the torsional moment opposes the force produced by the shear distribution should be designed to withstand the difference between the latter and 50% of the former.

10.4.6 Formulas to Estimate the Fundamental Period of a Building

The application of the static method requires the designer to estimate the fundamental period T of the building. The Chilean code does not endorse the use of any specific formula to compute T. Some formulas that Chilean engineers commonly use to estimate T, depending on the type of building, are given by eqs.(10.11), (10.12), and (10.13).

$$T = 0.1N \qquad (10.11)$$

where N is the number of stories. This is an empirical expression, intended for ductile moment-resisting frames, which has been recommended by the Structural Engineers Association of California and the U.S. Geological Survey (Beles, Ifrim, and Garcia 1975; Paz 1991).

$$T = \frac{0.05H}{\sqrt{B}} \qquad (10.12)$$

where H is the height of the building and B is the plan dimension in the direction for which the period is being estimated. Both H and B must be expressed in feet. The formula in eq.(10.12) is employed in several codes (Newmark and Rosenblueth 1971).

$$T = 0.024 H^{0.21} d^{-0.14} \qquad (10.13)$$

where d is a coefficient obtained by dividing the summation of the lengths of all the resisting walls in all the building stories (in meters) by the summation of areas (in square meters) for all the floors in the building. The formula in eq.(10.13), proposed by Arias and Husid, is recommended for reinforced concrete building with rigid walls (Arias and Husid 1962).

10.5 DYNAMIC METHOD

This approach allows the building to be modeled as a lumped mass system, with the masses concentrated at the floor levels. Unless translational as well as rotational degrees of freedom are considered at the floor levels, the torsional effects should be computed using the formula employed in the static method, eq.(10.9). The specifications given in Section 10.4.5 for the static method are also applicable to the dynamic method.

The dynamic method is simply a modal superposition method in which the modal contributions are combined using

$$S = \frac{1}{2}\left[\sum_{i=1}^{r} |S_i| + \sqrt{\sum_{i=1}^{r} S_i^2} \right]$$ (10.14)

where S denotes the response quantity computed (moment, shear, displacement, etc.), S_i refers to the contribution of the i mode to this quantity, and r represents the number of modes included, which must be at least three. If the base shear computed with eq.(10.14) is less than $0.06K_1K_2W$, then all the response quantities should be scaled so that the base shear equals $0.06K_1K_2W$.

The acceleration spectrum for use in the dynamic method is the following:

$$\frac{a}{g} = 0.10K_1K_2 \quad \text{for} \quad T \leq T_0$$ (10.15a)

or

$$\frac{a}{g} = 0.10K_1K_2 \frac{2TT_0}{T^2 + T_0^2} \quad \text{for } T > T_0$$ (10.15b)

In these expressions g denotes gravitational acceleration; a is the acceleration assumed to be caused by the ground motion; T is the period of the mode under consideration; and K_1, K_2, and T_0 are given, as in the static method, by Tables 10.1, 10.2, and 10.3, respectively.

In addition, the following conditions must be observed:

(a) In determining the masses at each floor level, a percentage increase corresponding to the live load should be added (as in the static method).

(b) The effect of damping is implicitly assumed in the values of a/g given by the spectrum, so no additional reductions due to damping are allowed.

(c) In the determination of the natural modes of vibration of the building, it is not necessary to consider the modal coupling due to damping.

(d) Acceleration spectra other than the one specified by eq.(10.15) may be used in the dynamic method. An alternative dynamic method, such as the time history response method, also may be used. However, the resulting base shear must be compared with

$0.06K_1K_2W$; if it is less than this value, all the results should be scaled to satisfy this minimum base shear requirement.

(e) Participation factors, modal lateral loads, story drifts, etc., for each mode of response can be determined simply using the standard expressions for modal methods. See for example Chopra 1980; Craig 1981; Paz 1991.

(f) As in the static method, paragraphs (a), (c), and (d) of Section 10.4.4 still apply.

10.6 UNITS

The Chilean code does not require the use of any particular system of units in the seismic analysis and design of buildings. However, it must be noted that in Chile it is customary to use the metric system (also called "technical system") of units for most engineering calculations. In this system, distances are expressed in meters (m), forces (and weights) are expressed in kilograms (kg), and time is expressed in seconds (s or sec). Therefore, the mass of a body is obtained by dividing its weight (expressed in kg) by the acceleration of gravity (9.81 m/s^2), which yields $kg \cdot s^2/m$. All the numerical examples presented in this chapter are expressed in these units.[4]

10.7 EXAMPLE 10.1

10.7.1 Building Description

Consider the four-story building depicted in Fig. 10.1. This structure corresponds to the one described in Example 23.1 of *Structural Dynamics: Theory and Computation* (Paz 1991). The structure has a reinforced concrete frame with the dimensions shown in the figure. The cross-section of the A and C columns is 0.3048 m by 0.5080 m for the two bottom floors, and 0.3048 m by 0.4064 m for the top two levels. The cross-section of the B columns is 0.3048 m by 0.6096 m for the two bottom floors, and 0.3048 m by 0.5080 m for the two top floors. The dead load on each floor is equivalent to 683.40 kg/m^2. The normal live load is 610.12 kg/m^2. The foundation is rock. Assume that the building will be used as a warehouse and take the value of E (modulus of elasticity) for concrete as $2.10 \times 10^9 \text{ kg/m}^2$. Determine the loads acting on this structure for a seismic motion having the direction indicated in the figure, using both the static and dynamic methods. Assume that the floors behave as rigid diaphragms.

---

[4]For the reader familiar with the British system, 1 m is equal to 39.37 in., and 1 kg is equal to 2.20 lb. (1 kg is also equivalent to 9.81 newtons (N).)

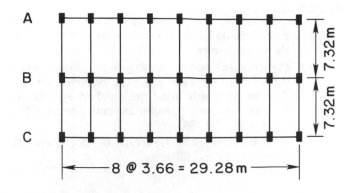

(A) PLAN

DIRECTION
OF
GROUND
MOTION

7.32 m   7.32 m

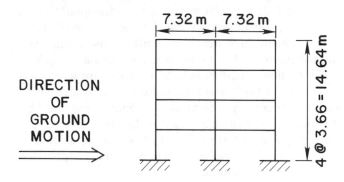

(B) ELEVATION

Fig. 10.1.   Plan and elevation for a four-story building for Example 10.1

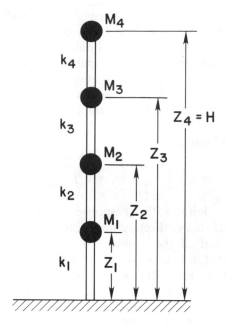

Fig. 10.2.   Four-story building of Example 10.1 modeled as column with concentrated masses

10.7.2 Static Method

The building is modeled using a four-degree-of-freedom system with the masses concentrated at each floor level. Only one degree of freedom per floor is considered (horizontal displacement). The simplified model for this building is shown in Fig. 10.2.

(a) Moment of inertia of columns

For the top two floors:

$$I_{Atop} = I_{Ctop} = 1/12(0.3048)(0.4064)^3 = 1.7048 \times 10^{-3} \, m^4;$$

$$I_{Btop} = 1/12(0.3048)(0.5080)^3 = 3.3298 \times 10^{-3} \, m^4;$$

For the bottom two floors:

$$I_{Abottom} = I_{Cbottom} = 1/12(0.3048)(0.5080)^3 = 3.3298 \times 10^{-3} \, m^4;$$

$$I_{Bbottom} = 1/12(0.3048)(0.6096)^3 = 5.7539 \times 10^{-3} \, m^4.$$

(b) Stiffness for each floor

$$k_1 = k_2 =$$
$$\frac{12(2.1 \times 10^9)(9 \times 5.7539 \times 10^{-3} + 18 \times 3.3298 \times 10^{-3})}{(3.66)^3}$$

$$= 5.7424 \times 10^7 \, kg/m;$$

$$k_3 = k_4 =$$
$$\frac{12(2.1 \times 10^9)(9 \times 3.3298 \times 10^{-3} + 18 \times 1.7048 \times 10^{-3})}{(3.66)^3}$$

$$= 3.1176 \times 10^7 \, kg/m.$$

(c) Weights. The total area of each floor is $3.66 \times 8 \times 7.31 \times 2 = 428.1 \, m^2$. Since large gatherings of people are unlikely in this building, 25% of the live load is added (except for the roof) to compute the total weight on each floor.

$$W_1 = W_2 = W_3 = 428.1(683.40 + 0.25 \times 610.17)$$
$$= 3.5786 \times 10^5 \, kg$$

and

$$W_4 = 428.1 \times 683.40 = 2.9256 \times 10^5 \, kg.$$

(d) Base shear coefficients. K_1 is taken as 1.0 (Table 10.1) because this building is unlikely to house large crowds and it is not a temporary construction. K_2 is taken as 0.8 (Table 10.2), consistent with the building configuration, and T_0 (Table 10.3) is assumed to be 0.2 (rock).

Several empirical formulas could be used to estimate the fundamental period T of a building. (The code does not give any specifications regarding this matter.) Using the empirical formula given by eq.(10.11) yields

$$T = 0.1 \times 4 = 0.40 \, \text{sec}$$

Thus, by eq.(10.3)

$$C = \frac{0.2 \times 0.4 \times 0.2}{0.4 \times 0.4 + 0.2 \times 0.2} = 0.08$$

(e) Base shear

$$Q_0 = 1.0 \times 0.8 \times 0.08(3 \times 3.5786 \times 10^5 + 2.9256 \times 10^5)$$
$$= 87,432 \, \text{kg} \qquad \text{by eq.(10.1)}$$

(f) A coefficients

$Z_0 = 0$, $Z_1 = 3.66 \, \text{m}$, $Z_2 = 7.32 \, \text{m}$, $Z_3 = 10.98 \, \text{m}$, and $Z_4 = H = 14.64 \, \text{m}$.

Then from eq.(10.5),

$A_1 = 0.133$, $A_2 = 0.159$, $A_3 = 0.207$, and $A_4 = 0.501$

(g) Horizontal forces

$$W_1 A_1 + W_2 A_2 + W_3 A_3 + W_4 A_4 = 325,145 \, \text{kg}$$

and by eq.(10.4),

$$F_1 = (3.5786 \times 10^5 \times 0.133 \times 87,432)/325,145 = 12,798 \, \text{kg}$$

$$F_2 = (3.5786 \times 10^5 \times 0.159 \times 87,432)/325,145 = 15,301 \, \text{kg}$$

$$F_3 = (3.5786 \times 10^5 \times 0.207 \times 87,432)/325,145 = 19,919 \, \text{kg}$$

$$F_4 = (2.9256 \times 10^5 \times 0.501 \times 87,432)/325,145 = 39,414 \, \text{kg}$$

(h) Shear. Using eq.(10.8),

$$Q_1 = 87,432 \, \text{kg}, \quad Q_2 = 74,634 \, \text{kg}, \quad Q_3 = 59,333 \, \text{kg},$$
$$Q_4 = 39,414 \, \text{kg}$$

(i) Overturning moment. The reduction factors from eq.(10.7) are

$$J_0 = 0.80, \quad J_1 = 0.85, \quad J_2 = 0.90, \quad \text{and} \quad J_3 = 0.95$$

Then, from eq.(10.6),

$$M_0 = 0.80[(12,798 \times 3.66) + (15,301 \times 7.32)$$
$$+ (19,918 \times 10.98) + (39,414 \times 14.64)] = 763,651 \, \text{kg-m}$$

$$M_1 = 0.85[(15,301 \times 3.66) + (19,919 \times 7.32)$$
$$+ (39,414 \times 10.98)] = 539,385 \, \text{kg-m}$$

$$M_2 = 0.90[(19,919 \times 3.66) + (39,414 \times 7.32)]$$
$$= 325,277 \, \text{kg-m}$$

$$M_3 = 0.95(39,414 \times 3.66) = 137,043 \, \text{kg-m}$$

(j) Torsional moments. The reduction factors for the torsional moment calculated by eq.(10.10) are:

$I_1 = 0.775$, $I_2 = 0.85$, $I_3 = 0.925$, and $I_4 = 1.00$.

The torsional moments are computed by eq.(10.9) noting that, because of the symmetric configuration of the building, $e_1 = e_2 = e_3 = e_4 = 0$; and $b_1 = b_2 = b_3 = b_4 = 29.28 \, \text{m}$. Thus,

$$M_{t1} = 0.05 \times 0.775 \times 29.28 \times 87,432 = 99,200 \, \text{kg-m}$$

$$M_{t2} = 0.05 \times 0.850 \times 29.28 \times 74,634 = 92,874 \, \text{kg-m}$$

$$M_{t3} = 0.05 \times 0.925 \times 29.28 \times 59,333 = 80,347 \, \text{kg-m}$$

$$M_{t4} = 0.05 \times 1.000 \times 29.28 \times 39,414 = 57,702 \, \text{kg-m}$$

(k) Story Drift and Lateral Displacements. The story drift Δ_j can be obtained at each floor by computing the ratio between the shear force Q_j and the corresponding stiffness k_j. In this case,

$$\Delta_1 = 0.0015 \, \text{m}, \quad \Delta_2 = 0.0013 \, \text{m}, \quad \Delta_3 = 0.0019 \, \text{m}, \quad \text{and}$$
$$\Delta_4 = 0.0013 \, \text{m}.$$

The corresponding total lateral displacements are

$$\delta_1 = 0.0015 \, \text{m}, \quad \delta_2 = 0.0028 \, \text{m}, \quad \delta_3 = 0.0047 \, \text{m}, \quad \text{and}$$
$$\delta_4 = 0.0060 \, \text{m}.$$

Table 10.4 summarizes the results of applying the static method to the analysis of this building.

(l) Comparison of results using UBC-85 and UBC-88. A similar analysis for this building using the Uniform Building Code, 1985 (UBC-85) and UBC-88 can be found in Paz (1991). In both cases, it was assumed that the building was located in an area described as Zone 3. (Zone 4 corresponds to an area of the highest seismic risk.) UBC-85 gives a base shear of 72,333 kg (83% of the value obtained with the Chilean code), whereas UBC-88 gives a value of 63,594 kg (73% of the value obtained with the Chilean code). Considering that Chile, like most of the southern California coast, is located in a highly active

Table 10.4. Example 1. Summary of Results Using Static Method

Level	Horiz. Force (kg)	Shear Force (kg)	Overturning Moment (kg-m)	Torsional Moment (kg-m)	Drift (m)	Lateral Disp. (m)
4	39,414	39,414	—	57,702	0.0013	0.0060
3	19,919	59,333	137,043	80,347	0.0019	0.0047
2	15,301	74,634	325,277	92,874	0.0013	0.0028
1	12,798	87,432	539,385	99,200	0.0015	0.0015
0			763,651			

Table 10.5. Lateral Seismic Forces for the UBC and Chilean Code (in kg)

Level	Chilean Code	UBC-85	UBC-88
4	39,414	34,044	29,931
3	19,919	31,250	27,451
2	15,201	20,800	18,321
1	12,798	10,400	9,130
Base Shear	87,432	96,444	84,712

seismic region, a computation for Zone 4 would be more appropriate for comparison purposes. In this case the values for base shear given by UBC-85 and UBC-88 are 96,444 kg and 84,792 kg, respectively. These values are, roughly speaking, in agreement with the values given by the Chilean code. However, the distribution of lateral forces given by the UBC codes follows a slightly different pattern. Forces computed according to UBC-85 and UBC-88 are shown for comparison with the Chilean code forces in Table 10.5.

10.7.3 Dynamic Method

First, the normal modes and natural frequencies (at least for the first three modes) must be determined. The mass matrix $[M]$ and the stiffness matrix $[k]$ corresponding to the model shown in Fig. 10.2 are (only translational degrees of freedom are considered):

$$[M] = \begin{bmatrix} 3.5786 & 0 & 0 & 0 \\ 0 & 3.5786 & 0 & 0 \\ 0 & 0 & 3.5786 & 0 \\ 0 & 0 & 0 & 2.9256 \end{bmatrix} \times \frac{10^5 \, kg}{g}$$

and

$$[k] = \begin{bmatrix} k_1 + k_2 & -k_2 & 0 & 0 \\ -k_2 & k_2 + k_3 & -k_3 & 0 \\ 0 & -k_3 & k_3 + k_4 & -k_4 \\ 0 & 0 & -k_4 & k_4 \end{bmatrix}$$

becomes

$$[k] = \begin{bmatrix} 1.4848 & -5.7424 & 0 & 0 \\ -5.7424 & 8.8600 & -3.1176 & 0 \\ 0 & -3.1176 & 6.2352 & -3.1176 \\ 0 & 0 & -3.1176 & 3.1176 \end{bmatrix}$$

$$\times 10^7 \, kg/m$$

Solution of the eigensystem $[[k] - \omega^2 [M]]\{\phi\} = 0$ (where ω represents the frequency and $\{\phi\}$ the normal mode) gives

$$\omega_1 = 13.144 \, rad/sec \quad \omega_2 = 33.668 \, rad/sec$$
$$\omega_3 = 50.086 \, rad/sec \quad \omega_4 = 67.288 \, rad/sec$$

The corresponding periods T_i are ($T_i = 2\pi/\omega_i$):

$$T_1 = 0.478 \, sec, \, T_2 = 0.187 \, sec, \, T_3 = 0.126 \, sec,$$
$$T_4 = 0.093 \, sec$$

and the modal matrix $[\Phi]$ is

$$[\Phi] = \begin{bmatrix} 1.0000 & 0.7806 & -0.5320 & 1.0000 \\ 0.8348 & 1.0000 & -0.2176 & -0.8714 \\ 0.5010 & 0.0787 & 1.0000 & 0.2912 \\ 0.2650 & -0.9464 & -0.7149 & -0.0874 \end{bmatrix}$$

Let ϕ_{ki} denote the kth component of the ith normal mode. The modal participation factor p_i is determined by eq.(4.53a) from Chapter 4 as follows:

$$p_i = \frac{\sum_{j=1}^{N} W_j \phi_{ji}}{\sum_{j=1}^{N} W_j \phi_{ji}^2} \qquad (10.16)$$

Then,

$$p_1 = 1.317, \, p_2 = 0.462, \, p_3 = 0.191, \text{ and } p_4 = 0.188$$

The modal contributions to the various parameters for the first mode are computed as follows:

(a) Modal ground acceleration. The values for K_1, K_2, and T_0 are the same as used for the static method; because $T_1 = 0.478 \, sec > T_0 = 0.2 \, sec$, eq.(10.15b) yields

$$a/g = 0.05697$$

(b) Horizontal forces. The horizontal force F_{ki}, where the first index refers to the floor level and the second index refers to the mode, is given from eq.(4.59) of Chapter 4 after using (eq.10.15a) by

$$F_{ki} = p_i \phi_{ki} (a/g) W_k \qquad (10.17)$$

Thus,

$$F_{11} = 1.317 \times 0.265 \times 0.05697 \times 3.5786 \times 10^5 = 7,116 \, kg$$
$$F_{21} = 1.317 \times 0.501 \times 0.05697 \times 3.5786 \times 10^5 = 13,451 \, kg$$
$$F_{31} = 1.317 \times 0.835 \times 0.05697 \times 3.5786 \times 10^5 = 22,406 \, kg$$
$$F_{41} = 1.317 \times 1.000 \times 0.05697 \times 2.9256 \times 10^5 = 21,939 \, kg$$

(c) Shear forces. The corresponding shear forces Q_{ki} for mode i are given from eq.(10.8) as

$$Q_{ki} = \sum_{j=k}^{N} F_{ji} \tag{10.18}$$

Thus, for the first mode:

$Q_{11} = 64{,}913$ kg, $Q_{21} = 57{,}797$ kg, $Q_{31} = 44{,}345$ kg,

$Q_{41} = 21{,}939$ kg

(d) Overturning moment. In this case, in contrast with the static analysis, no reduction factors are used. Therefore, the moments are simply the result of multiplying the horizontal forces F_{ki} by the corresponding distances.

$M_{01} = (7{,}116 \times 3.66) + (13{,}451 \times 7.32) + (22{,}406 \times 10.98)$
$\qquad\qquad + (21{,}939 \times 14.64) = 691{,}710$ kg-m

$M_{11} = 13{,}451 \times 3.66 + 22{,}406 \times 7.32 + 21{,}939 \times 10.98$
$\qquad\qquad\qquad = 454{,}137$ kg-m

$M_{21} = 22{,}406 \times 3.66 + 21{,}939 \times 7.32 = 242{,}600$ kg-m

$M_{31} = 21{,}939 \times 3.66 = 80{,}297$ kg-m

(e) Torsion. Torsional moments are computed as in the static method, but using the horizontal loads of the corresponding mode.

$M_{t11} = 0.05 \times 0.775 \times 29.28 \times 64{,}913 = 73{,}650$ kg-m

$M_{t21} = 0.05 \times 0.850 \times 29.28 \times 57{,}797 = 71{,}922$ kg-m

$M_{t31} = 0.05 \times 0.925 \times 29.28 \times 44{,}345 = 60{,}052$ kg-m

$M_{t41} = 0.05 \times 1.000 \times 29.28 \times 21{,}939 = 32{,}118$ kg-m

(f) Lateral displacements and story drift. The modal lateral displacement δ_{ki} can be computed using eq.(4.72)

$$\delta_{ki} = \frac{T_i^2 F_{ki}}{4\pi^2 W_k} g \tag{10.19}$$

Then,

$\delta_{11} = 0.0011$ m, $\delta_{21} = 0.0021$ m, $\delta_{31} = 0.0036$ m,

$\delta_{41} = 0.0043$ m

The corresponding story drifts Δ_{ki} by eq.(4.73) are

$\Delta_{11} = 0.0011$ m, $\Delta_{21} = 0.0010$ m, $\Delta_{31} = 0.0014$ m,

$\Delta_{41} = 0.0007$ m

Table 10.6 summarizes the results of the first mode. The contributions of the second and third mode are determined in a similar fashion, noting that for the

Table 10.6. Summary of Results for First Mode

Level	Horiz. Force (kg)	Shear Force (kg)	Overturning Moment (kg-m)	Torsional Moment (kg-m)	Drift (m)	Lateral Disp. (m)
4	21,939	21,939	—	32,118	0.0007	0.0043
3	22,406	44,345	80,297	60,052	0.0014	0.0036
2	13,451	57,797	242,600	71,992	0.0010	0.0021
1	7,116	64,913	454,137	73,650	0.0011	0.0011
0			691,710			

Table 10.7. Summary of Results for Second Mode

Level	Horiz. Force (kg)	Shear Force (kg)	Overturning Moment (kg-m)	Torsional Moment (kg-m)	Drift (m)	Lateral Disp. (m)
4	−10,241	−10,241	—	14,993	−0.0003	−0.0003
3	1,041	−9,199	−37,484	12,458	−0.0003	0.0000
2	13,230	4,031	−71,155	−5,016	0.0001	0.0003
1	10,331	14,362	−56,400	−16,295	0.0003	0.0003
0			−3,843			

Table 10.8. Summary of Results for Third Mode

Level	Horiz. Force (kg)	Shear Force (kg)	Overturning Moment (kg-m)	Torsional Moment (kg-m)	Drift (m)	Lateral Disp. (m)
4	3,200	3,300		−4,686	0.0001	0.0000
3	−5,471	−2,270	11,715	3,075	−0.0001	−0.0001
2	1,187	−1,083	3,404	1,347	0.0000	0.0000
1	2,912	1,829	−559	−2,075	0.0000	0.0000
0			6,123			

second and third modes, $T < T_0$, and from eq.(10.15a),

$$a/g = 0.08$$

Tables 10.7 and 10.8 show, respectively, the modal contributions obtained for the second and third modes.

The Chilean code states that at least three modes should be included in the dynamic method, regardless of the percentage of vibrating mass to which they correspond. It is nevertheless advisable to make sure that the modes included account for a reasonable fraction—say 90%—of the total mass. In this example three modes are sufficient to account for most of the mass. The modal weights Wm_i associated with the first

Table 10.9. Combined Results for First Three Modes

Level	Horiz. Force (kg)	Shear Force (kg)	Overturning Moment (kg-m)	Torsional Moment (kg-m)	Drift (m)	Lateral Disp. (m)
4	29,902	29,902	—	43,776	0.0010	0.0044
3	26,003	50,581	109,441	68,496	0.0016	0.0036
2	23,388	60,429	285,001	75,199	0.0011	0.0023
1	16,618	73,806	484,362	83,740	0.0013	0.0013
0	—	—	696,712	—	—	—

three modes are calculated using eq.(4.65) from Chapter 4 as

$$Wm_i = \frac{\left(\sum_{q=1}^{N} \phi_{qi} W_q \right)^2}{\sum_{q=1}^{N} \phi_{qi}^2 W_q} \qquad (10.20)$$

yielding

$$Wm_1 = 11.37 \times 10^5 \text{ kg}, \; Wm_2 = 1.79 \times 10^5 \text{ kg},$$

$$Wm_3 = 0.22 \times 10^5 \text{ kg}$$

and

$$Wm_1 + Wm_2 + Wm_3 = 13.38 \times 10^5 \text{ kg}$$

The total weight of the structure is 13.66×10^5 kg; thus, the first three modes account for about 97% of the total weight. Therefore, three modes suffice for a dynamic analysis.

The value for base shear obtained with the dynamic method is about 85% of that computed with the static method. Table 10.9 shows the combined contributions of the first three modes using eq.(10.14).

10.8 COMPUTER PROGRAM

A computer program written in BASIC has been developed to implement the provisions of the Chilean seismic code. The code assumes that the building is modeled as a mass-spring assembly, with the masses concentrated at the floor levels (see Fig. 10.2). The floors are assumed to be rigid, and therefore, one degree of freedom per floor (horizontal displacement) is considered. The program is menu driven and the designer is requested to enter the appropriate data (e.g., weight of each floor, interstory stiffness, height of each floor).

The code outputs the horizontal force at each floor, as well as the shear, overturning moment, torsional moment, story drift, and total lateral displacements.

Both the static and the dynamic methods have been implemented in the program. In the dynamic method, the program computes the normal modes and the natural frequencies solving the actual eigensystem using the Jacobi method (Bathe 1982, Chapter 11). If the eigensystem is larger than 6 by 6, the code automatically performs the Guyan reduction, and reduces the size of the eigensystem to 6 by 6 (Irons 1963, 1965; Guyan 1965). In this case, the analyst must select the "master" nodes (those to be retained). This method is much more reliable than that of estimating the normal modes and natural frequencies using empirical formulas.

Running this program on a microcomputer produces a response almost immediately after the data are entered, except for the eigenvalue computation. Solution of the eigensystem, as expected, is the most costly operation from the computational time standpoint.

Finally, if the "master" nodes are chosen wisely, it is possible to determine fairly accurately the properties of the first four or five modes for a building of 20 to 25 stories [a good discussion of this topic can be found in Section 10.4 of Hughes' book (Hughes 1987)]. This technique is usually sufficient for estimating the dynamic response of a "plane" (two-dimensional) structure. The program does not have a "built-in" system of units; the user can employ any set of units as long as they are consistent.

10.9 USING THE COMPUTER PROGRAM

The following two examples illustrate the use of the computer program. Forces and weights are expressed in kilograms, and distances are expressed in meters.

10.9.1 Steel Building (Example 10.2)

Figure 10.3 shows a modeled 15-story steel building. This idealized model describes, approximately, the Kajima International Building, a moment-resisting frame structure located in downtown Los Angeles, California. The building, which has a height of 57.9 m, was built with four moment-resisting frames in both the traverse and longitudinal directions. Lightweight reinforced concrete slabs act as rigid diaphragms for horizontal motion (Housner and Jennings 1982, Appendix A). The building was designed assuming a period of 1.5 sec (although wind response tests performed after its completion indicated a period of 1.9 sec). The design value for base shear was 172,000 kg, following the 1961 Los Angeles building code.

In 1971, during the San Fernando earthquake, this building suffered nonstructural damage. It has been estimated, from earthquake records, that the base

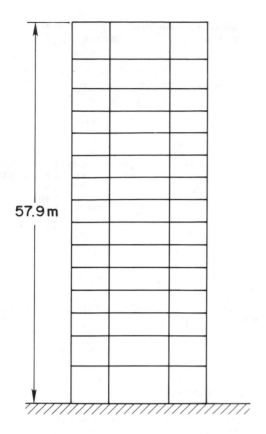

(A) PROFILE

Fig. 10.3. (a) Elevation of the Kajima International Building for Example 10.2

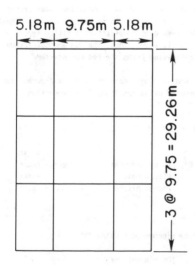

(B) PLAN VIEW

Fig. 10.3. (b) Plan view of the Kajima International Building for Example 10.2

shear during the quake could have been as high as 350,000 kg (Housner and Jennings 1982, Appendix A). The following pages show the result of applying the Chilean code to the analysis of this structure using the computer program. Values were assigned to parameters as follows: 1 for K_1, 0.8 for K_2, and 0.3 for T_0. The weights employed in the computations are listed in the computer printout that follows (they account for both the actual weight and the appropriate fraction of the live load).

From the computer results presented on the following pages, it can be seen that the value obtained for base shear using the static method is 25% higher than that obtained with the dynamic method. (The minimum value for Q_0 for the static method is 0.06 × weight, whereas for the dynamic method the minimum shear is 0.06 × K_1 × K_2 × weight). Since in this case $K_1 = 1$ and $K_2 = 0.8$, the ratio between the two values is simply 1/0.8 = 1.25. It is interesting to observe that the value for the base shear ($Q_0 = 353{,}557$ kg) computed with the static method was very close to the maximum value for base shear estimated from the earthquake records ($Q_0 = 350{,}000$ kg).

For comparison, considering that California is a highly active seismic area, the expression for base shear given by UBC-85 (International Conference of Building Officials 1985) using a set of parameters roughly comparable to those employed in the static method, yields

$$Q_0 = ZIKCSW$$
$$= 1.0 \times 1.0 \times 0.67 \times 0.048 \times 1.2 \times 58.9 \times 10^5$$
$$= 227{,}230 \text{ kg}$$

The expression for base shear used in UBC-88 (International Conference of Building Officials 1988) gives

$$Q_0 = (ZICW)/R_w$$
$$= (0.4 \times 1.0 \times 0.97 \times 58.9 \times 10^5)/12$$
$$= 190{,}443 \text{ kg}$$

10.9.2 Input Data and Output Computer Results for Example 10.2

(a) Static Method:

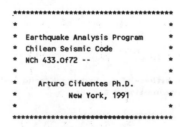

What would you like to do?

1   Prepare a new data file and run an analysis
     using this file

2   Create a new data file by modifying an existing
     one, and then run the corresponding analysis

3   Use an existing data file to run the analysis

     Enter your selection   ? 3
Enter the drive where the file to be read resides
Must be A: , B: , C: or D: ? C:
Now enter the name of the file ? KAJIMA
Data for the building to be analyzed
================================
STORY	WEIGHT	STIFFNESS	Height	Ecc.	b (dist.)
1	2.464E+05	5.0000E+07	4.30	0.00	29.26
2	3.986E+05	5.0000E+07	7.90	0.00	29.26
3	3.932E+05	5.0000E+07	11.60	0.00	29.26
4	3.932E+05	4.5000E+07	15.20	0.00	29.26
5	3.905E+05	4.5000E+07	18.90	0.00	29.26
6	3.905E+05	4.5000E+07	22.50	0.00	29.26
7	3.905E+05	4.2000E+07	26.20	0.00	29.26
8	3.905E+05	4.2000E+07	29.90	0.00	29.26
9	3.905E+05	4.0000E+07	33.60	0.00	29.26
10	3.905E+05	4.0000E+07	37.20	0.00	29.26
11	3.905E+05	4.0000E+07	40.80	0.00	29.26
12	3.905E+05	4.0000E+07	44.50	0.00	29.26
13	3.905E+05	3.8000E+07	48.20	0.00	29.26
14	5.844E+05	3.6000E+07	52.10	0.00	29.26
15	3.624E+05	2.8000E+07	57.90	0.00	29.26

****

Enter the value of K1
It must be 0.8 , 1.0 or 1.2
 K1 ? 1

Enter the value of K2
It must be 0.8 , 1.0 or 1.2
 K2 ? .8

Enter value of T0 (in seconds)
It must be 0.2 , 0.3 or 0.9
depending on the foundation soil

T0 ? .3

Which type of analysis do you want to perform?
1 Static method
2 Dynamic method
              enter your selection

Analysis Code ? 1

Enter an estimate (seconds) of fundamental period
T1 ? 1.9
Results of Static Method
****

K1   1
K2   .8
T0   .3
T1 in seconds   1.9

   T1 is the estimate of the fundamental period used
   in the determination of C

C   .06

Minimum Value for Base Shear
(0.06 * Weight)   353557.5

    Q0, base shear   353557.5

STORY	F HORI.	SHEAR	MOMENT	TORSION	DRIFT	DELTA
15	1.028E+05	1.028E+05	0.000E+00	1.5035E+05	3.67E-03	8.93E-02
14	4.859E+04	1.514E+05	5.841E+05	2.1479E+05	4.20E-03	8.57E-02
13	2.511E+04	1.765E+05	1.147E+06	2.4520E+05	4.64E-03	8.15E-02
12	2.182E+04	1.983E+05	1.754E+06	2.6996E+05	4.96E-03	7.68E-02
11	1.906E+04	2.174E+05	2.421E+06	2.8982E+05	5.43E-03	7.19E-02
10	1.746E+04	2.348E+05	3.116E+06	3.0669E+05	5.87E-03	6.64E-02
9	1.664E+04	2.515E+05	3.848E+06	3.2157E+05	6.29E-03	6.06E-02
8	1.558E+04	2.670E+05	4.635E+06	3.3400E+05	6.36E-03	5.43E-02
7	1.469E+04	2.817E+05	5.449E+06	3.4447E+05	6.71E-03	4.79E-02
6	1.357E+04	2.953E+05	6.286E+06	3.5279E+05	6.56E-03	4.12E-02
5	1.331E+04	3.086E+05	7.117E+06	3.6027E+05	6.86E-03	3.46E-02
4	1.250E+04	3.211E+05	7.985E+06	3.6585E+05	7.14E-03	2.78E-02
3	1.235E+04	3.335E+05	8.840E+06	3.7082E+05	6.67E-03	2.06E-02
2	1.174E+04	3.452E+05	9.726E+06	3.7419E+05	6.90E-03	1.40E-02
1	8.357E+03	3.536E+05	1.059E+07	3.7360E+05	7.07E-03	7.07E-03
0			1.162E+07			

(b) Dynamic Method:

Which type of analysis do you want to perform?
1 Static method
2 Dynamic method
              enter your selection

Analysis Code ? 2
Enter the value of the gravity acceleration (g)
g ? 9.8
Static Reduction is Needed Before Solving The EigenSystem

SIX NODES are to be retained
Please enter (separated by commas) the nodes (floor levels)
that you want to retain. These are the so-called master
nodes. The remaining nodes (floors) will automatically
go into the slave set. Thus, the fundamental frequencies
will be calculated using the reduced system

The Mass and Stiffness Matrices for the Reduced System
are Determined Using the Irons-Guyan Reduction

? 1,15,14,4,8,11

xxxxx   xxxxx
xxxxx Eigenvalue Computation in Progress xxxxx
xxxxx   Please Wait   xxxxx
xxxxx   Jacobi Method at Work   xxxxx

*** from eigenvalue routine ***

These are the eigenvalues

11.07698
100.389
292.3308
635.4823
1256.914
1739.06

              Press any key to continue...

```
Number of Modes to be Used for the Analysis
Enter Integer Between 1 and MIN( N,6)  ? 3


The number of modes included accounts for
the following fraction of the total mass
of the building  .9636854


Do you want to include more modes in the analysis?


  1  proceed with the current number of modes

  2  include more modes
Enter your selection  ? 1
Results of Dynamic Method
*************************


K1  1


K2  .8


TO  .3


Number of Modes Used 3


Press any key to continue...
Mode Number  1


Participation Factor   1.291503
Period (seconds)       1.887856
Freq.   rad./sec.      3.328209
Freq.   Cps            .5297015
Modal Weight           4885275
```

Contributions of Mode 1

STORY	F HORI.	SHEAR	MOMENT	TORSION	DRIFT	DELTA
15	1.161E+04	1.161E+04	0.000E+00	-1.6981E+04	4.15E-04	2.83E-02
14	1.844E+04	3.005E+04	6.732E+04	-4.2642E+04	1.15E-03	2.79E-02
13	1.182E+04	4.187E+04	1.845E+05	-5.8173E+04	1.09E-03	2.68E-02
12	1.134E+04	5.320E+04	3.394E+05	-7.2434E+04	1.03E-03	2.57E-02
11	1.088E+04	6.409E+04	5.363E+05	-8.5452E+04	1.85E-03	2.47E-02
10	1.007E+04	7.415E+04	7.670E+05	-9.6851E+04	1.85E-03	2.28E-02
9	9.251E+03	8.341E+04	1.034E+06	-1.0666E+05	1.85E-03	2.10E-02
8	8.436E+03	9.184E+04	1.343E+06	-1.1487E+05	2.42E-03	1.91E-02
7	7.368E+03	9.921E+04	1.682E+06	-1.2130E+05	2.42E-03	1.67E-02
6	6.300E+03	1.055E+05	2.049E+06	-1.2605E+05	2.26E-03	1.43E-02
5	5.304E+03	1.108E+05	2.429E+06	-1.2936E+05	2.26E-03	1.20E-02
4	4.337E+03	1.152E+05	2.839E+06	-1.3119E+05	2.62E-03	9.76E-03
3	3.173E+03	1.183E+05	3.254E+06	-1.3158E+05	2.36E-03	7.14E-03
2	2.154E+03	1.205E+05	3.692E+06	-1.3060E+05	2.36E-03	4.78E-03
1	6.749E+02	1.212E+05	4.125E+06	-1.2802E+05	2.42E-03	2.42E-03
0			4.646E+06			

```
Mode Number  2


Participation Factor   .4578195
Period (seconds)       .6270996
Freq.   rad./sec.      10.01943
Freq.   Cps            1.594643
Modal Weight           590102.9
```

Contributions of Mode 2

STORY	F HORI.	SHEAR	MOMENT	TORSION	DRIFT	DELTA
15	-1.033E+04	-1.033E+04	0.000E+00	1.5119E+04	-3.69E-04	-2.78E-03
14	-1.445E+04	-2.479E+04	-5.994E+04	3.5177E+04	-8.31E-04	-2.41E-03
13	-6.336E+03	-3.113E+04	-1.566E+05	4.3247E+04	-7.87E-04	-1.58E-03
12	-3.188E+03	-3.431E+04	-2.718E+05	4.6714E+04	-7.48E-04	-7.97E-04
11	-1.969E+02	-3.451E+04	-3.987E+05	4.6014E+04	-7.66E-04	-4.92E-05
10	2.865E+03	-3.164E+04	-5.230E+05	4.1330E+04	-7.66E-04	7.16E-04
9	5.928E+03	-2.572E+04	-6.369E+05	3.2886E+04	-7.66E-04	1.48E-03
8	8.990E+03	-1.673E+04	-7.320E+05	2.0920E+04	-6.87E-05	2.25E-03
7	9.265E+03	-7.461E+03	-7.939E+05	9.1224E+03	-6.87E-05	2.32E-03
6	9.540E+03	2.079E+03	-8.215E+05	-2.4840E+03	-6.41E-05	2.38E-03
5	9.797E+03	1.188E+04	-8.140E+05	-1.3864E+04	-6.41E-05	2.45E-03
4	1.012E+04	2.200E+04	-7.701E+05	-2.5064E+04	6.35E-04	2.51E-03
3	7.565E+03	2.956E+04	-6.909E+05	-3.2877E+04	5.72E-04	1.88E-03
2	5.336E+03	3.490E+04	-5.815E+05	-3.7832E+04	5.72E-04	1.31E-03
1	1.856E+03	3.676E+04	-4.559E+05	-3.8840E+04	7.35E-04	7.35E-04
0			-2.978E+05			

```
Mode Number  3


Participation Factor   .29381
Period (seconds)       .3674872
Freq.   rad./sec.      17.09768
Freq.   Cps            2.721183
Modal Weight           203257.8
```

Contributions of Mode 3

STORY	F HORI.	SHEAR	MOMENT	TORSION	DRIFT	DELTA
15	7.367E+03	7.367E+03	0.000E+00	-1.0779E+04	2.63E-04	6.82E-04
14	7.293E+03	1.466E+04	4.273E+04	-2.0804E+04	3.82E-04	4.18E-04
13	4.273E+02	1.509E+04	9.991E+04	-2.0964E+04	3.62E-04	3.67E-05
12	-3.785E+03	1.130E+04	1.557E+05	-1.5388E+04	3.44E-04	-3.25E-04
11	-7.786E+03	3.517E+03	1.976E+05	-4.6891E+03	-7.69E-05	-6.68E-04
10	-6.891E+03	-3.374E+03	2.102E+05	4.4069E+03	-7.69E-05	-5.92E-04
9	-5.995E+03	-9.369E+03	1.981E+05	1.1982E+04	-7.69E-05	-5.15E-04
8	-5.100E+03	-1.447E+04	1.634E+05	1.8097E+04	-3.13E-04	-4.38E-04
7	-1.455E+03	-1.592E+04	1.099E+05	1.9470E+04	-3.13E-04	-1.25E-04
6	2.189E+03	-1.373E+04	5.095E+04	1.6408E+04	-2.92E-04	1.88E-04
5	5.591E+03	-8.144E+03	1.503E+03	9.5066E+03	-2.92E-04	4.80E-04
4	9.055E+03	9.115E+02	-2.863E+04	-1.0385E+03	1.62E-04	7.72E-04
3	7.156E+03	8.067E+03	-2.535E+04	-8.9712E+03	1.46E-04	6.10E-04
2	5.522E+03	1.359E+04	4.503E+03	-1.4731E+04	1.46E-04	4.64E-04
1	2.342E+03	1.593E+04	5.342E+04	-1.6835E+04	3.19E-04	3.19E-04
0			1.219E+05			

```
The total base shear using the number of modes you
selected is less than  0.06 * K1 * K2 * Weight
Therefore, all the values computed will be scaled
such that  base shear becomes = .06*K1*K2*Weight


Base Shear (without scaling)   150721.6
0.06 * K1 * K2 * Weight        282846
scaling factor                 1.876612


Results, combining the contributions of the modes
in this case the number of modes used is     3
```

STORY	F HORI.	SHEAR	MOMENT	TORSION	DRIFT	DELTA
15	4.364E+04	4.364E+04	0.000E+00	6.3843E+04	1.56E-03	5.66E-02
14	6.074E+04	1.043E+05	2.531E+05	1.4796E+05	3.59E-03	5.52E-02
13	3.002E+04	1.336E+05	6.595E+05	1.8564E+05	3.40E-03	5.18E-02
12	2.879E+04	1.531E+05	1.153E+06	2.0839E+05	3.23E-03	4.93E-02
11	3.026E+04	1.642E+05	1.717E+06	2.1893E+05	4.40E-03	4.69E-02
10	3.036E+04	1.782E+05	2.301E+06	2.3268E+05	4.40E-03	4.40E-02
9	3.161E+04	1.935E+05	2.908E+06	2.4751E+05	4.40E-03	4.13E-02
8	3.365E+04	2.041E+05	3.543E+06	2.5526E+05	4.92E-03	3.85E-02
7	2.816E+04	2.096E+05	4.175E+06	2.5624E+05	4.92E-03	3.38E-02
6	2.784E+04	2.137E+05	4.814E+06	2.5529E+05	4.59E-03	2.94E-02
5	3.111E+04	2.276E+05	5.449E+06	2.6571E+05	4.59E-03	2.55E-02
4	3.544E+04	2.395E+05	6.174E+06	2.7292E+05	5.74E-03	2.17E-02
3	2.700E+04	2.610E+05	6.846E+06	2.9026E+05	5.17E-03	1.60E-02
2	1.969E+04	2.769E+05	7.520E+06	3.0018E+05	5.17E-03	1.08E-02
1	7.447E+03	2.828E+05	8.243E+06	2.9888E+05	5.66E-03	5.66E-03
0			9.124E+06			

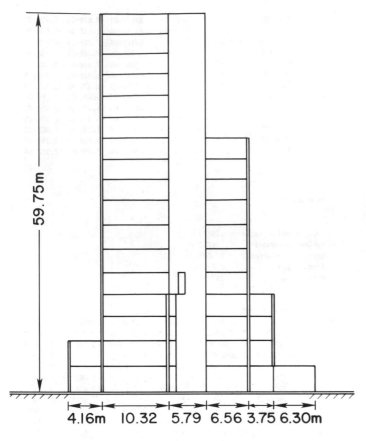

$$59.75\text{m}$$

4.16m 10.32 5.79 6.56 3.75 6.30m

Fig. 10.4. Seventeen-story concrete building for Example 10.3

10.9.3 Concrete Building (Example 10.3)

This example deals with a 17-story reinforced concrete building as shown in Fig. 10.4. This is an approximate (and simplified) version of building 1 described in the study performed by the Catholic University of Chile (Hidalgo, Arias, and Cruz 1990). The story weights shown in the computer listing account for the dead load plus 25% of the live load. The eccentricities at the story levels are also shown in the listing. It is assumed in this case that $K_1 = K_2 = 1.0$. The value for T_0 is 0.3 sec.

The value for base shear in the dynamic analysis was governed by ($0.06 \times K_1 \times K_2 \times$ weight), which coincides with the value computed with the static method because $K_1 = K_2 = 1.0$. The distribution of lateral forces in this case is roughly the same for both methods because the dynamic response is governed by the first mode. From the computer output for this example given in the following pages, the base shear force is $Q_0 = 788,531$ kg. Again, for comparison, the corresponding values for base shear computed using the UBC formulas are given for UBC-85:

$$Q_0 = ZIKCSW$$

$$= 1.0 \times 1.0 \times 0.8 \times 0.06 \times 1.2 \times 131.3 \times 10^5$$
$$= 756,288 \text{ kg};$$

and for UBC-88:

$$Q_0 = (ZICW)/R_w$$

$$= (0.4 \times 1.0 \times 1.32 \times 131.3 \times 10^5)/9$$
$$= 770,293 \text{ kg}.$$

If this building had been located in Spain (a country with modest seismic activity compared to Chile) the computed value for base shear—assuming a comparable foundation and for the highest risk zone in Spain—would have been approximately 400,000 kg, that is, roughly half the value obtained with the Chilean code. In all three examples presented in this chapter, the values for base shear computed with the static method were more conservative than those obtained using the dynamic method. Incidentally, this was the case for the five buildings described in Table 8.1 of the previously cited report (Hidalgo, Arias, and Cruz 1990).

10.9.4 Selected Computer Results for Example 10.3

```
Results of Static Method
************************

K1    1
K2    1
TO    .3
T1 in seconds   1.2

    T1 is the estimate of the fundamental period used
    in the determination of C

C   .06

Minimum Value for Base Shear
(0.06 * Weight)              788531.3


QO, base shear    788531.3
```

STORY	F HORI.	SHEAR	MOMENT	TORSION	DRIFT	DELTA
17	1.177E+05	1.177E+05	0.000E+00	2.1782E+05	1.47E-03	4.04E-02
16	4.993E+04	1.677E+05	4.015E+05	3.0699E+05	2.10E-03	3.89E-02
15	4.643E+04	2.141E+05	9.619E+05	3.8793E+05	2.14E-03	3.68E-02
14	4.893E+04	2.630E+05	1.664E+06	4.7714E+05	1.64E-03	3.47E-02
13	4.311E+04	3.061E+05	2.509E+06	5.5330E+05	1.75E-03	3.31E-02
12	3.897E+04	3.451E+05	3.474E+06	6.1993E+05	1.97E-03	3.13E-02
11	4.301E+04	3.881E+05	4.540E+06	7.3224E+05	2.22E-03	2.93E-02
10	4.003E+04	4.281E+05	5.714E+06	8.3305E+05	2.45E-03	2.71E-02
9	3.760E+04	4.657E+05	6.983E+06	9.2419E+05	2.45E-03	2.47E-02
8	3.556E+04	5.013E+05	8.334E+06	1.0070E+06	2.64E-03	2.22E-02
7	3.382E+04	5.351E+05	9.756E+06	1.0825E+06	2.82E-03	1.96E-02
6	3.232E+04	5.674E+05	1.124E+07	1.1514E+06	2.84E-03	1.68E-02
5	3.100E+04	5.984E+05	1.278E+07	1.2145E+06	2.99E-03	1.39E-02
4	2.982E+04	6.283E+05	1.436E+07	1.2722E+06	3.14E-03	1.09E-02
3	2.878E+04	6.570E+05	1.598E+07	1.3251E+06	3.29E-03	7.80E-03
2	6.545E+04	7.225E+05	1.764E+07	4.5065E+06	2.26E-03	4.51E-03
1	6.605E+04	7.885E+05	1.970E+07	4.2134E+06	2.25E-03	2.25E-03
0			2.190E+07			

```
        Results of Dynamic Method
        *************************



        Mode Number   1


        Participation Factor    1.394911
        Period (seconds)        1.199857
        Freq.   rad./sec.       5.236608
        Freq.   Cps              .8334328
        Modal Weight            9375911



        Mode Number   2


        Participation Factor     .7569086
        Period (seconds)         .4264345
        Freq.   rad./sec.       14.73422
        Freq.   Cps             2.345027
        Modal Weight            1442561



        Mode Number   3


        Participation Factor     .6688468
        Period (seconds)         .269138
        Freq.   rad./sec.       23.34557
        Freq.   Cps             3.715567
        Modal Weight            906341.1
```

```
Results, combining the contributions of the modes
in this case the number of modes used  is         3
STORY   F HORI.   SHEAR     MOMENT    TORSION    DRIFT     DELTA
  17   9.032E+04  9.032E+04 0.000E+00 1.6710E+05 1.95E-03  3.57E-02
  16   7.334E+04  1.634E+05 3.116E+05 2.9915E+05 1.95E-03  3.40E-02
  15   6.761E+04  2.295E+05 8.750E+05 4.1586E+05 1.56E-03  3.23E-02
  14   7.481E+04  2.918E+05 1.665E+06 5.3032E+05 2.08E-03  3.11E-02
  13   6.943E+04  3.386E+05 2.665E+06 6.1318E+05 1.90E-03  2.94E-02
  12   6.631E+04  3.722E+05 3.821E+06 6.6950E+05 1.90E-03  2.79E-02
  11   8.650E+04  4.214E+05 5.086E+06 8.2529E+05 2.66E-03  2.69E-02
  10   8.063E+04  4.727E+05 6.425E+06 9.6079E+05 2.66E-03  2.50E-02
   9   7.678E+04  5.081E+05 7.832E+06 1.0501E+06 2.45E-03  2.31E-02
   8   7.533E+04  5.295E+05 9.293E+06 1.1501E+06 2.87E-03  2.13E-02
   7   6.900E+04  5.407E+05 1.080E+07 1.2450E+06 2.87E-03  1.86E-02
   6   7.121E+04  5.734E+05 1.237E+07 1.3309E+06 2.73E-03  1.61E-02
   5   7.492E+04  6.041E+05 1.418E+07 1.3940E+06 2.73E-03  1.37E-02
   4   8.005E+04  6.437E+05 1.594E+07 1.4566E+06 3.43E-03  1.13E-02
   3   5.817E+04  6.903E+05 1.764E+07 1.5529E+06 3.43E-03  7.82E-03
   2   7.392E+04  7.499E+05 1.930E+07 4.8307E+06 2.14E-03  4.39E-03
   1   4.855E+04  7.885E+05 2.147E+07 4.3518E+06 2.25E-03  2.25E-03
   0                        2.377E+07
```

10.10 ADDITIONAL REGULATIONS

The Chilean seismic code also provides some general guidelines regarding building repairs, building instrumentation (strong motion accelerographs are mandatory for buildings of 20 stories or more), analysis of nonstructural elements, and separation between buildings and parts of buildings. Also, the code states that the story drift or relative displacement between two consecutive floors should not exceed 0.2% of the story height (distance between the two floors).

10.11 MODIFICATIONS AND ENHANCEMENTS TO THE CHILEAN SEISMIC CODE

As of November 1992, a committee of Chilean engineers headed by Arturo Arias, Elias Arze, Rodrigo Flores, and Pedro Hidalgo has undertaken a review of the current code (NCh.433.Of72). A tentative draft of the new code—a project still in progress—known as *Diseño Sísmico de Edificios* or simply NCh433.cR89 (INN 1989) can be obtained from INN (see footnote 1). Additional information can be found in a recent paper (Hidalgo and Arias 1990). Finalization of the revised code is likely to be a lengthy process, as controversy surrounds some proposed revisions. However, it is safe to say that this new code will retain the essential flavor of the current code in the sense that it will be based on two approaches: the static method and the dynamic method.

The major potential modifications are: (a) division of the country into three zones on the basis of seismic risk, (b) a new expression to compute the base shear in the static method (the expression to compute the lateral loads remains the same), (c) new expressions for the acceleration spectrum to be used in the dynamic method, (d) specific guidelines as to when the static or dynamic method should be used, (e) a requirement to include three degrees of freedom per floor (two translations plus one rotation) when modeling the building, and (f) a requirement that the number of modes included in the dynamic method should account for at least 90% of the total mass.

REFERENCES

ARIAS, A., and HUSID, R. (1962) *Fórmula Empírica para el Cálculo del Período Propio de Vibración de Edificios de Hormigón Armado con Muros de Rigidez.* Revista IDIEM, Volume 1. Santiago, Chile.

BATHE, K. J. (1982) *Finite Element Procedures in Engineering Analysis.* Prentice-Hall, Englewood Cliffs, NJ.

BELES, A., IFRIM, M., and GARCÍA, A. (1975) *Elementos de Ingeniería Sísmica.* Ediciones Omega, Barcelona, Spain.

CHOPRA, A. K. (1980) *Dynamics of Structures, A Primer.* Earthquake Engineering Research Institute, Berkeley, CA.

CRAIG, R. R. (1981) *Structural Dynamics.* John Wiley, New York, NY.

GUYAN, R. J. (1965) "Reduction of Stiffness and Mass Matrices." *AIAA Journal* Vol. 3: 380.

HIDALGO, P., ARIAS, A., and CRUZ, E. (1990) *Evaluación de las Disposiciones del Anteproyecto de la Nueva Norma NCh 433 Cálculo Antisísmico de Edificios.* Publication DIE 90-8, Pontificia Universidad Católica de Chile, Departamento de Ingeniería Estructural, Santiago, Chile.

HIDALGO, P., and ARIAS, A. (1990) "New Chilean Code for Earthquake-Resistant Design of Buildings." Proceedings of the 4th U.S. National Conference on Earthquake Engineering. Vol. 2, pp. 927–936. Palm Springs, CA.

Housner, G. W., and Jennings, P. C. (1982) *Earthquake Design Criteria*. Earthquake Engineering Research Institute, Berkeley, CA.

Hughes, T. J. R. (1987) *The Finite Element Method*. Prentice-Hall, Englewood Cliffs, NJ.

INN, *Cálculo Antisísmico de Edificios* (1985) Publication NCh.433.Of72, 2d ed. Santiago, Chile.

———, *Diseño Sísmico de Edificios* (1989) (Proyecto de Norma en Consulta Pública). Publication NCh.433.c89, Santiago, Chile.

International Conference of Building Officials (1985) *Uniform Building Code (UBC)*. Whittier, CA.

——— (1988) *Uniform Building Code (UBC)*. Whittier, CA.

Irons, B. M. (1963) "Eigenvalue Economisers in Vibration Problems." *Journal of the Royal Aeronautical Society* 67: 526.

——— (1965) "Structural Eigenvalue Problems: Elimination of Unwanted Variables." *AIAA Journal* 3: 961.

The London Folio Society (1977) *A Narrative of the Voyage of H.M.S. Beagle*. W. & F. Mackay Limited, Chathman, U.K.

Moorehead, A. (1969) *Darwin and the Beagle*. Harper & Row, New York, NY.

Newmark, N. M., and Rosenblueth, E. (1971) *Fundamentals of Earthquake Engineering*. Prentice-Hall, Englewood Cliffs, NJ.

Paz, M. (1991) *Structural Dynamics: Theory and Computation*. 3d ed. Van Nostrand Reinhold, New York, NY.

Sarrazin, M. (1979) *Regulaciones Sismorresistentes: La Experiencia Chilena*. Publication SES I-2/79 (143). Universidad de Chile, Facultad de Ciencias Físicas y Matemáticas, Departamento de Obras Civiles, Sección Ingeniería Estructural, Santiago, Chile.

11

China

Ye Yaoxian

Coauthors:
Niu Zezhen and Huang Jiguang

11.1 INTRODUCTION

Before 1964, there was no seismic-resistant design code for buildings or other structures in China. Earthquake-resistant designs were not considered for most buildings. However, designers of some important structures and buildings located in areas of known high seismic activity were required by the authorities to consider earthquake effects. A draft for a seismic-resistant design code in China was prepared in 1964; but unfortunately, this draft was never officially approved or enforced.

The first official seismic code for China was issued for trial implementation in 1974 with the title, *Seismic Design Code for Industrial and Civil Buildings* (1974) (TJ 11-74). In 1975 and 1976, China suffered two strong earthquakes, the 1975 earthquake in Haicheng with a magnitude of 7.3 and the 1976 Tangshan, Hebei earthquakes of magnitudes 7.8. These earthquakes were considered the most catastrophic disasters in China in recent decades; a total of 242,829 people died and 4,332,700 rooms[1] collapsed. Work was initiated to revise the 1974 code soon after the 1976 Tangshan earthquake. This revision produced a new code for

---

[1]The floor area of a room in China is about 15 to 20 square meters.

seismic design in 1978 (TJ 11-78) that was implemented in 1979; a further revision (*Seismic Design Code for Buildings and Structures*, GBJ 11-89) was published in 1989 and put into effect in January 1991. It was partially revised in 1993 and put into effect on July 4, 1993. The partially revised 1989 Code consists of 11 chapters and seven appendices. The first four chapters decribe general requirements for earthquake-resistant design of all types of buildings and structures. The remaining seven chapters specify the seismic countermeasures that are required for specific types of buildings and structures.

This chapter presents the main provisions for seismic design requirements of the partially revised 1989 code, calculations of seismic actions for a five-story building with a computer program to implement the code specification, and two examples to apply the program to multistory buildings.

11.2 GENERAL PROVISIONS

11.2.1 Classification of Buildings and Structures

The Code classifies buildings and structures into the following four types:

Type A. Buildings and structures that should not fail beyond repair during an earthquake because of their functional importance and the severe consequences of their failure. The design of these structures must be approved by competent authorities.

Type B. Buildings and structures of lifeline systems in the main cities of the country.

Type C. Buildings and structures not included under Types A, B, and D.

Type D. Buildings and structures of less importance, where damage is not likely to cause deaths or injury to people and/or considerable economic losses.

11.2.2 Site Categories

The following four categories are based on the characteristics of the soil at the site of the structure:

Site Category I. A soil profile with either (a) a rock-like material characterized by a shear wave velocity greater than 1,840 feet per second or by other suitable means of classification, where the soil depth is zero, or (b) a stiff or dense soil condition characterized by a shear wave velocity greater than 820 feet per second and less than 1,600 feet per second, where the soil depth is less than 30 feet, or (c) other soil conditions where the soil depth is less than 10 feet.

Site Category II. A soil profile with either (a) dense or stiff soil conditions, where the soil depth exceeds 30 feet, or (b) medium-stiff clay characterized by a shear wave velocity greater than 460 feet per second and less than 820 feet per second where the soil depth is 10 to 282 feet, or (c) a soil profile containing 10 to 30 feet of soft clay characterized by a shear wave velocity less than 460 feet per second.

Site Category III. A soil profile with either (a) medium-stiff clay where the soil depth is more than 282 feet, or (b) soft clay where the soil depth is 30 to 282 feet.

Site Category IV. A soil profile containing more than 282 feet of soft clay characterized by a shear wave velocity less than 460 feet per second.

11.2.3 Principles of Seismic Action

The principles of seismic action are as follows:

- For ordinary buildings and structures, separate computations of horizontal seismic actions should be conducted along the two main orthogonal directions of the building; the horizontal seismic action should be resisted totally in each direction by the corresponding lateral force-resisting elements.

- For structures with braced lateral force-resisting elements, the horizontal seismic action in the direction of the lateral force-resisting element should be considered separately.

- For structures with obvious asymmetry and nonuniformity of mass and stiffness distribution, the torsion effects caused by horizontal seismic action should be considered.

- For large-span structures, long cantilevered structures, chimneys, and similar tall structures, in regions with an intensity of VIII or IX[2], vertical seismic action should be considered.

- For high-rise buildings in regions with an intensity of IX, vertical seismic action should be considered.

Intensity ratings are obtained in China from the Earthquake Intensity Zoning Map issued by the State Seismological Bureau (SSB) that has jurisdiction in the region where the structure will be located.

11.2.4 Seismic Computation Methods

The following computation methods are to be used:

- For structures less than 40 m high, with deformations predominantly due to shear and generally uniform distribution of mass and stiffness in elevation, or for structures modeled as a single-mass system, a simplified method such as the equivalent lateral force procedure can be used.

---

[2]Seismic intensity is determined according to the seismic intensity scale of China, which is similar to the Modified Mercalli Intensity (MMI) scale.

Table 11.1. Combination Coefficients for Live Loads

Type of Live Load	Combination Coefficient
Snow load	0.5
Dust load on the roof	0.5
Live load on the roof	(Need not be considered)
Actual live load on floor	1.0
Equivalent uniform live load on floor of stack room, or file storage room	0.8
Equivalent uniform floor live load for other civil buildings	0.5
Crane load (with rigid hook)	0.3
Crane load (with flexible hook)	(Need not be considered)

- For freestanding chimneys less than 100 m high, an approximate procedure is allowed.
- In general, the modal analysis procedure is acceptable.
- For the following buildings and structures, the time-history analysis procedure should be used as an additional safeguard:
  - Buildings with extremely irregular configurations
  - Type A buildings
  - High-rise buildings more than 80 m high in a region with an intensity of VII or VIII and a Site Category I or II
  - High-rise buildings more than 60 m high in a region with an intensity of VIII and with site categories of III and IV, or an intensity of IX and any site category.
- The number of acceleration records or synthesized acceleration histories for time-history analysis should be selected according to intensity, epicenter distance (near or far earthquake), and category at the site of the building or structure. However, the base shear obtained from time-history analysis should be no less than 80% of that calculated by the equivalent lateral force procedure or by the modal analysis procedure.

11.2.5 Seismic Loads

When calculating seismic actions, the dead loads of the structure and permanent equipment should be combined with a fraction of the live loads to obtain the values of seismic weights W_i at each level of the building. The coefficients for variable live loads to be used in obtaining load combinations are shown in Table 11.1.

11.3 EQUIVALENT LATERAL FORCE METHOD

11.3.1 Base Shear Force

In applying the equivalent lateral force method, each level of a building can be considered as possessing only one lateral degree of freedom in the direction considered for the seismic action. Fig. 11.1 shows

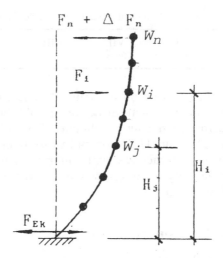

Fig. 11.1. Sketch indicating general lateral displacements

a sketch indicating typical movements at the various levels of a multistory building. The total horizontal seismic action F_{Ek} (i.e., base shear force) is given by

$$F_{Ek} = \alpha W_{eq} \qquad (11.1)$$

$$W_{eq} = \sum_{i=1}^{n} W_i$$

where W_{eq} is the total equivalent seismic weight of a building or a structure and α is the seismic coefficient which can be determined by eq.(11.2), based on fundamental period T_1.

$$\alpha = (5.5T_1 + 0.45)\alpha_{max} \qquad \text{for } T_1 \leqslant 0.1 \text{ sec}$$
$$\alpha = \alpha_{max} \qquad \text{for } 0.1 < T_1 \leqslant T_g \qquad (11.2)$$
$$\alpha = (T_g/T_1)^{0.9}\alpha_{max} \qquad \text{for } T_g < T_1 < 3.0 \text{ sec}$$

where T_g is the characteristic period of vibration of the soil given in Table 11.2 for various site categories and epicenter distance (near-earthquake or far-earthquake) and α_{max} is given in Table 11.3 as a function of intensity.

The value of total seismic weight should be used when the structure is modeled as a single-degree-of-freedom system. When the structure is modeled as a multidegree-of-freedom system, 85% of the value of total seismic weight is acceptable.

Table 11.2. Characteristic Period of Vibration T_g (sec)

Epicenter Distance	Site Category			
	I	II	III	IV
Near-surface earthquake	0.20	0.30	0.40	0.65
Remote earthquake	0.25	0.40	0.55	0.85

Table 11.3. Maximum Values of the Seismic Coefficient α_{max}

Intensity	VI	VII	VIII	IX
α_{max}	0.04	0.08	0.16	0.32

Table 11.4. Additional Seismic Action Coefficient (δ_n)

	Reinforced Multistory Concrete Building		Multistory Inner-framed Brick Building	Other Building
T_g (sec)	$T_1 > 1.4T_g$	$T_1 \leq 1.4T_g$		
≤ 0.25	$0.08T_1 + 0.07$	No need to consider	0.2	No need to consider
0.3–0.4	$0.08T_1 + 0.01$			
≥ 0.55	$0.08T_1 - 0.02$			

Alternatively, based on epicenter distance, intensity, and fundamental natural period of vibration, the seismic coefficient α may be obtained from Fig. 11.2 and Tables 11.2 and 11.3. For multistory brick buildings, multistory structures with interior frames, or framed single-story buildings, $\alpha = \alpha_{max}$; where α_{max} is given in Table 11.3.

11.3.2 Horizontal Seismic Forces

The horizontal seismic forces F_i applied at any level i of the building are given by

$$F_i = \frac{W_i H_i}{\displaystyle\sum_{j=1}^{n} W_j H_j} F_{Ek}(1 - \delta_n) \tag{11.3}$$

with the additional seismic force ΔF_n applied to the top level of the building:

$$\Delta F_n = \delta_n F_{Ek} \tag{11.4}$$

where

W_i, W_j = Values of seismic weights concentrated at the levels of the building i and j respectively, taken as the combination of the dead load of structural components and a fraction of the live load, as indicated in Table 11.1.

H_i, H_j = Height of levels i, j from the base of the building.

δ_n = Additional seismic action coefficient for the top level of the building, calculated as indicated in Table 11.4.

11.3.3 Modal Analysis Procedure Ignoring Torsional Effects

The horizontal force F_{ji} at level i, corresponding to the jth mode, can be determined by the following formula:

$$F_{ji} = \alpha_j \gamma_j X_{ji} W_i \quad (i = 1, 2, \ldots, n; j = 1, 2, \ldots, m) \tag{11.5}$$

where the mode participation factor γ_j is given by

$$\gamma_j = \frac{\displaystyle\sum_{i=1}^{n} X_{ji} W_i}{\displaystyle\sum_{i=1}^{n} X_{ji}^2 W_i} \tag{11.6}$$

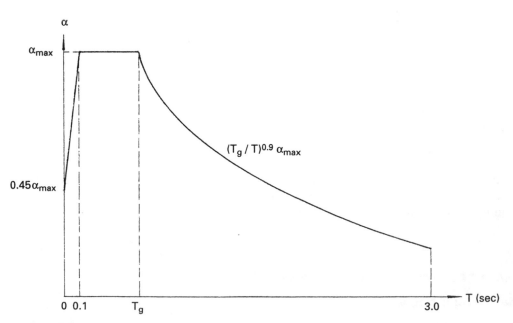

Fig. 11.2. Seismic coefficient (α)
  T = Natural period of vibration of a structure
  T_g = Soil characteristic period (Table 11.2)
  α_{max} = Maximum value of the seismic coefficient (Table 11.3)

and

α_j = the seismic coefficient corresponding to the natural period T_j of the jth mode of the structure [determined by eq.(11.2) or by Fig. 11.2 and Tables 11.2 and 11.3]

X_{ji} = the horizontal modal displacement of the mass at level i, corresponding to the jth mode of the natural vibration of the structure.

The total effect of the horizontal seismic action S (bending moment, shear, axial force, or deformation) is calculated by combining the modal contributions using the SRSS procedure (Square Root of Sum of Squares) as

$$S = \sqrt{\sum_{j=1}^{N} S_j^2} \qquad (11.7)$$

In eq.(11.7) S_j is the modal effect caused by the seismic forces of the jth mode of natural vibration of the structure and N is the number of modes considered. Usually, only the first two or three modes are included in obtaining the total effect. However, when the fundamental period of natural vibration $T_1 > 1.5$ sec, or the ratio of height to width of the building exceeds five, a larger number of modes should be included in the computations using eq.(11.7).

11.3.4 Seismic Action of Projecting Components on the Roof

The effect of horizontal seismic action on a penthouse, parapet, and/or stack projecting above the roof should be multiplied by a factor of 3.0; the increase obtained should be assigned to the roof, not the lower part of the structure. The projecting part can be considered as a mass point when the modal analysis procedure is used.

11.3.5 Distribution of the Horizontal Seismic Shear Force

The horizontal seismic shear force at each floor level of the structure should be distributed to the lateral force-resisting members (such as walls, columns, and shear walls) according to the following principles:

- For buildings with rigid floors and roofs (such as cast-in-place and/or prefabricated reinforced concrete floors and roof), the distribution may be done in proportion to the equivalent stiffness of the lateral force-resisting members.
- For buildings with flexible (such as wooden) floors and roof, the distribution may be carried out in proportion

to the values of seismic weights acting on the floor area, braced by the lateral force-resisting members.
- For buildings with ordinary prefabricated reinforced concrete floors and roof, the average values of the results obtained from the two methods of distribution given above should be used.

11.3.6 Torsional Effects

When considering torsional effects, the building can be modeled with three degrees of freedom, including two orthogonal horizontal displacements, and one angular rotation for each level of the building. The horizontal seismic forces and the response can be calculated by the modal analysis procedure. The simplified method for the determination of seismic effects is acceptable provided that the structure fulfills the conditions for regularity and height limitation.

11.3.7 Horizontal Seismic Action

The design values for the horizontal seismic force applied to the ith floor corresponding to the jth mode of natural vibration of the structure can be determined by the following equations:

$$\begin{aligned} F_{xji} &= \alpha_j \gamma_{tj} X_{ji} W_i \\ F_{yji} &= \alpha_j \gamma_{tj} Y_{ji} W_i \\ F_{tji} &= \alpha_j \gamma_{tj} r_i^2 \phi_{ji} W_i \end{aligned} \qquad (11.8)$$

$$(i = 1, 2, \ldots, n; j = 1, 2, \ldots, m)$$

where

$F_{xji}, F_{yji}, F_{tji}$ = the design values of horizontal seismic force applied to the ith floor, corresponding to the jth mode of natural vibration in the directions of x and y, and angular rotation, respectively

X_{ji}, Y_{ji} = the horizontal displacements of the center of mass of the ith floor, corresponding to the jth mode of natural vibration, in the x and y directions, respectively

ϕ_{ji} = the angular rotation of the ith floor corresponding to the jth mode of natural vibration

r_i = the mass radius of gyration for the ith floor, which is the square root (positive value) of the quotient obtained by dividing the rotational moment of inertia of the mass in the ith level by the total mass of that level

α_j = the seismic coefficient corresponding to mode j [eq.(11.2)]

γ_{tj} = the mode participation factor of the jth mode of natural vibration of the structure, considering torsion effects, which can be determined by the following formulas

When only the seismic action in the x direction is considered:

$$\gamma_{tj} = \frac{\sum_{i=1}^{n} X_{ji} W_i}{\sum_{i=1}^{n} (X_{ji}^2 + Y_{ji}^2 + \phi_{ji}^2 r_i^2) W_i} \qquad (11.9a)$$

When only the seismic action in the y direction is considered:

$$\gamma_{tj} = \frac{\sum_{i=1}^{n} Y_{ji} W_i}{\sum_{i=1}^{n} (X_{ji}^2 + Y_{ji}^2 + \phi_{ji}^2 r_i^2) W_i} \qquad (11.9b)$$

When torsional effects are considered, the Complete Quadratic Combination (CQC), described in Chapter 4, Section 4.8, can be used to obtain the response S (force, moment displacement, etc.) given by the following formula:

$$S = \sqrt{\sum_{j=1}^{n} \sum_{k=1}^{n} \rho_{jk} S_j S_k} \qquad (11.10)$$

where ρ_{jk} is the coupling coefficient of both the jth mode and the kth mode given by

$$\rho_{ik} = \frac{0.02(1 + \lambda_T)(\lambda_T)^{1.5}}{(1 - \lambda_T^2)^2 + 0.01(1 + \lambda_T)^2 \lambda_T} \qquad (11.11)$$

and

S_j, S_k = effects caused by seismic forces of the jth and the kth modes, respectively, taking the first nine to 15 modes as sufficient

λ_T = the ratio of the periods of the kth and the jth modes of natural vibration of the structure.

11.3.8 Interaction of Subsoil and Structure

In general, the interaction of subsoil and structure can be ignored. The horizontal seismic action for reinforced concrete of high-rise buildings with deep foundation or a rigid raft foundation of sites in Categories III or IV can be decreased by 10–20%, based on the assumption of rigid subsoil conditions; the story drifts can be calculated on the basis of the reduced shear force.

11.4 COMPUTATION OF VERTICAL SEISMIC ACTION

11.4.1 High-Rise Buildings and Structures

The design values of vertical seismic action should be determined using eqs.(11.12) and (11.13) for chim-

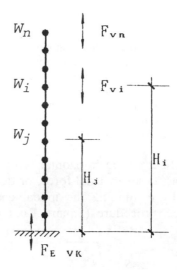

Fig. 11.3. Sketch for computation of vertical seismic action

neys and similar structures, as well as for high-rise buildings. The effects of vertical seismic action should be distributed in proportion to the values of seismic weight supported by each structural element. (See Fig. 11.3.)

$$F_{Evk} = \alpha_{v\max} W_{eq} \qquad (11.12)$$

$$F_{vi} = \frac{W_i H_i}{\sum_{j=1}^{n} W_j H_j} F_{Evk} \qquad (11.13)$$

where

F_{Evk} = the total design force of vertical seismic action

F_{vi} = the design value of vertical seismic force at level i

$\alpha_{v\max}$ = the maximum value of the coefficient of vertical seismic action effect, which may be taken as 65% of the maximum value of the horizontal seismic coefficient $\alpha_{\max}$ (Table 11.3).

W_{eq} = the total equivalent design weight of the structure, which may be taken as 75% of the total seismic weight of the structure [eq.(11.1)].

11.4.2 Flat Network Roof Structure and Large Span Trusses

For a flat network roof[3] structure, and for trusses with a span larger than 24 m, the design value of vertical seismic action can be taken as the product of the total seismic weight and the coefficient of vertical seismic action (α_v). Values for the coefficient of vertical seismic action (α_v) are shown in Table 11.5.

---

[3]Flat network roof is a roof that uses a steel joist system with composite paper-gravel cover.

Table 11.5. Coefficients of Vertical Seismic Action (α_v)

Type of Roof Structure	Intensity	Site Category		
		I	II	III, IV
Flat network structure	VIII	No need to consider	0.08	0.10
	IX	0.15	0.15	0.20
Reinforced concrete roof structure	VIII	0.10	0.13	0.13
	IX	0.20	0.25	0.25

11.4.3 Long Cantilever and Other Large-Span Structures

In regions of seismic intensity of VIII and IX, for structures or structural elements with a long cantilever or large spans, the design value of the vertical seismic force can be taken as 10% or 20%, respectively, of the seismic weight of the structure.

11.5 SEISMIC CHECKING IN CROSS-SECTION

11.5.1 Combination of Effects of Seismic Actions and Loads

The combined effect of seismic action and other loads on structural elements should be calculated by the following formula:

$$S = \gamma_G C_G W_E + \gamma_{Eh} C_{Eh} E_{hk}$$
$$+ \gamma_{Ev} C_{Ev} E_{vk} + \psi_w \gamma_w C_w W_k \qquad (11.14)$$

where

S = combined loads for design of structural elements

γ_G = modification coefficient for weight. 1.2 should be used in ordinary conditions; when the effect of weight is beneficial to the load-bearing capacity of structural elements, 1.0 can be used

γ_{Eh}, γ_{Ev} = modification coefficients for horizontal and vertical seismic action, respectively (numerical values for these coefficients are given in Table 11.6)

γ_w = modification coefficient for wind load, for which 1.4 should be used

W_E = seismic weight (dead load plus reduced live load, as per Table 11.1); when a crane is available, the reduced value of the hanging load should be included

E_{hk} = design value of horizontal seismic action

E_{vk} = design value of vertical seismic action

W_k = design code value of wind load

ψ_w = combination coefficient for wind load, which need not be considered for ordinary structures; for chimneys, tall water towers, and high-rise buildings $\psi_w = 0.2$ can be used

Table 11.6. Coefficients of Seismic Actions

Seismic Actions	γ_{Eh}	γ_{Ev}
Considering horizontal seismic action only	1.3	Need not consider
Considering vertical seismic action only	Need not consider	1.3
Considering both horizontal and vertical actions	1.3	0.5

$C_G, C_{Eh},$ = the effect coefficients for gravity load, horizontal C_{Ev}, C_w seismic action, vertical seismic action, and wind load, respectively (in ordinary cases, they should be determined by the national standard, *Uniform Standard for Structural Design of Buildings* (1984) GBJ 68-84).

11.5.2 Design Strength

When checking the element cross-section for seismic resistance, the following expression for design should be used:

$$S \le R/\gamma_{RE} \qquad (11.15)$$

where

S = load to be supported (eq.11.14)

R = the design value of load-bearing capacity of the structural element, calculated by the pertinent code provisions

γ_{RE} = seismic adjustment coefficients for load-bearing capacity (given in Table 11.7)

When only vertical seismic action is considered, the seismic adjustment coefficient should be taken as 1.0 for all of the structural elements.

11.6 CHECKING SEISMIC DEFORMATION

11.6.1 Frequent and Rare Earthquakes

The code establishes that there should be "no damage during a frequent moderate earthquake and no collapse during a rare strong earthquake." A probability plot of earthquake frequency, within a period of 50 years, versus intensity is shown in Fig. 11.4. In this plot, earthquakes of intensity less than IV were omitted. The design intensity for a "rare (strong) earthquake" I_s is defined as the intensity value that is exceeded by only 2–3% of all the earthquakes and the design intensity for a "frequent (moderate) earthquake" I_0 as the intensity with a probability of exceedance of 10–13% as indicated in Fig. 11.4. This figure also indicates the mode intensity I_m defining the "minor (weak) earthquake."

Table 11.7. Seismic Adjustment Coefficients for Load-Bearing Capacity (γ_{RE})

Material	Type of Structural Element	Stress Type	γ_{RE}
Steel	Column	Eccentric compression	0.7
	Column braces in factory buildings with steel structural systems		0.8
	Column braces in factory buildings with reinforced concrete structural system		0.9
	Welded-joint elements		1.0
Masonry	Seismic shear walls with integral columns at both ends	Shear	0.9
	Other seismic shear walls	Shear	1.0
Reinforced Concrete	Beam	Bending	0.75
	Column with axial compression ratio <0.15	Eccentric compression	0.75
	Column with axial compression ratio >0.15	Eccentric compression	0.80
	Seismic shear wall	Eccentric compression	0.85
	All types of elements	Shear, eccentric tension	0.85

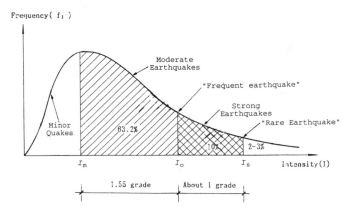

Fig. 11.4. Frequency distribution of earthquakes as function of their intensity

Table 11.8. Elastic Drift Limitation

Type of Structure	Conditions	$[\theta_e]$
Frame	Considering lateral earthquake-resistant action of brick infill walls	1/550
	Others	1/450
	Public buildings with ornamental decorations	1/800
Frame-shear wall	Others	1/650

should be taken from Table 11.3, all of the reduction coefficients of various seismic actions shall be taken as 1.0, and the elastic modulus may be used for reinforced concrete elements

$[\theta_e]$ = elastic drift limitation, to be taken in accordance with Table 11.8

H = story height

11.6.2 Elastic Drift Caused by Frequent Earthquakes

For frames (including frames with infill walls) and frame-shear wall structures (including frames with soft stories), the seismic deformation of a structure must be checked for the frequent-earthquake effect for the region (intensity I_0 in Fig. 11.4). The story elastic relative displacement should not exceed the limit given by the following formula:

$$\Delta U_e \leqslant [\theta_e]H \qquad (11.16)$$

where

ΔU_e = elastic story relative displacement caused by the design value of frequent earthquake action. The maximum value of the horizontal seismic coefficient

11.6.3 Elastoplastic Deformation Caused by Rare Large Earthquakes

The following structures should be checked for seismic deformation at their weak locations after the occurrence of a rare large earthquake:

- Single-story factory buildings with large reinforced concrete columns, and long beam spans situated in areas with intensity of VIII and a Site Category of III and IV, or with intensity of IX

- Frame structures and brick wall buildings with a frame in the first story, situated in areas with intensity of VII to IX when the yield strength coefficient is less than 0.5.[4]

- Reinforced concrete structures of building type A.

---

[4] The yield strength coefficient is the ratio between the story load-bearing capacity, calculated by the actual steel bar arrangement and the design strength of the building material, and the story shear force.

The weak locations of a structure can be identified as follows:

- For structures with a uniform distribution of story yield strength coefficients in elevation, the weak location is the first story of the building.
- For structures with nonuniform distribution of story yield strength coefficients in elevation, the weak locations can be identified as the stories with minimum, or smaller, story yield strength coefficients (in general, no more than two or three locations need be considered).
- For single-story factory buildings, the weak location is at the top of the columns.
  [The elastoplastic deformation at the weak location of a structure caused by a rare earthquake can be calculated by eq.(11.17a) or by eq.(11.17b).]
- For frame and frame-with-infill wall structures, and for single-story factory buildings with reinforced concrete columns that are no more than 12 stories high, without abrupt change in story rigidity, the elastoplastic story relative displacement can be calculated by the following formula:

$$\Delta U_p = \eta_p \Delta U_e \qquad (11.17a)$$

or

$$\Delta U_p = \mu \Delta U_y = \eta_p \Delta U_y / \xi_y \qquad (11.17b)$$

where

ΔU_p = elastoplastic story relative displacement
ΔU_y = story yield displacement
μ = story ductility coefficient
ΔU_e = story displacement calculated by elastic analysis caused by rare earthquake action
η_p = amplification coefficient of elastoplastic displacement (from Table 11.9). When the yield strength coefficient of the weak location is no less than 80% of the average value of the coefficients of neighboring locations, but when the yield strength coefficient is less than 50% of the above-mentioned average value, the figures in the Table shall be multiplied by 1.5 (in other conditions, the coefficient can be determined by interpolation)
ξ_y = story yield strength coefficient, which can be determined as follows:

$$\xi_y = \frac{F_y}{Q_e}$$

where F_y is the story yield strength, which for a frame building is given by

$$F_y = 2Q_{y1} + (m - 2)Q_{y2}$$

in which

Q_{y1}, Q_{y2} = the average story yield strength of the exterior and interior columns, respectively

Table 11.9. Amplification Coefficient of Elastoplastic Displacement (η_p)

Type of Structure	Number of Stories or Location	ξ_y 0.5	0.4	0.3	0.2
Multistory structure with uniform elevation	2–4	1.30	1.40	1.60	2.10
	5–7	1.50	1.65	1.80	2.40
	8–12	1.80	2.00	2.20	2.80
Single-story factory building	Upper part of column	1.30	1.60	2.00	2.60

Table 11.10. Elastoplastic Drift Limitation

Type of Structure	$[\theta_p]$
Single-story reinforced concrete frame bents	1/30
Frame and frame-with-infill wall	1/50
Frame in first story of a brick wall building	1/70

m = the total number of columns in the story considered
Q_e = the elastic story shear, which can be determined by elastic analysis or obtained from Seismic Design Code for Industrial and Civil Buildings (1978).

The elastoplastic-story relative displacement at the weak locations of a structure should not exceed the limit given by the following formula:

$$U_p \leq [\theta_p]H \qquad (11.18)$$

where

H = Story height at weak story or location, or the height of the upper part of the columns of single-story industrial buildings
$[\theta_p]$ = Elastoplastic drift limitation obtained from Table 11.10.

For frame structures, when the axial compression ratio is less than 0.40, the values in Table 11.10 can be increased by 10%. When maximum spiral or hoop rebars are used for the entire height of the columns, the values in Table 11.10 can be increased by 20%, but by no more than 25%, as a combination of both these allowances.

- For buildings more than 12 stories high and buildings of Type A, the time-history analysis procedure should be used to calculate displacements.

The maximum value of the horizontal seismic coefficient $\alpha_{\max}$ must be determined by using Table 11.11.

Table 11.11.   Maximum Value of Horizontal Seismic Coefficient Under Rare Earthquake Action

Intensity	VII	VIII	IX
α_{max}	0.50	0.90	1.40

Example 11.1

To illustrate the application of the seismic code, a simple example is given. Fig. 11.5 shows a schematic elevation of a reinforced concrete plane frame which serves as a model for a five-story building. Design weights of 400 kN have been assigned to each floor of the building, except the roof where the load is 250 kN. The plane frame width is 6 m from center to center of the exterior columns. The frame rises 21.5 m above the ground level. All story heights are 4 m with the exception of the first story, which is 5.5 m. The dimensions of the rectangular columns are 0.45 m × 0.75 m for the first story and the second story and 0.40 m × 0.60 m for the rest of the stories. The cross-section of all girders is 0.30 m × 0.70 m. The modulus of elasticity for concrete is 3.0 10⁴ MPa. The building is located in an area with seismic intensity of VIII and the site soil category is II; a far earthquake is considered.

Determine according to the *Seismic Design Code for Buildings and Structures* (GBJ 11-89) the following:

1. Fundamental period
2. Horizontal seismic forces at each level of the building
3. Effective shear force at each story
4. Overturning moments at each level of the building
5. Torsional moments for an assumed eccentricity of 1 m at each level
6. Elastic story drifts.

Solution

1. Fundamental period
Empirical formula for reinforced concrete frame building:

$$T = 1.7\beta\sqrt{\Delta_T} \qquad (11.19)$$

where Δ_T denotes displacement of the top level of the building produced by the action of lateral forces equal to seismic weights at each level of the building, and β is a coefficient representing the effect of fill walls in the reinforced concrete frame; in this case, $\beta = 0.9$.

Δ_T can be calculated by the *D*-value method, as follows:

$$\Delta_T = \sum_{i+1}^{N} \delta_i = \sum_{i+1}^{N} \frac{Q_i}{2D_i} \qquad (11.20)$$

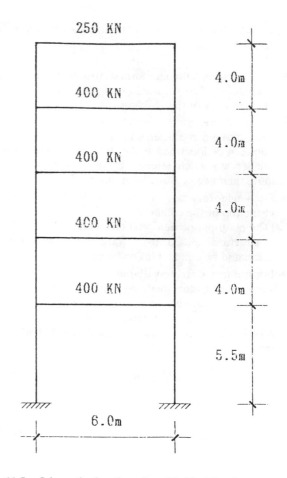

Fig. 11.5.   Schematic elevation of modeled building for Example 11.1

where

Δ_T = displacement at top level
δ_i = drift for story i
Q_i = shear force for story i

and

$$D_i = \left(\alpha_z K_z \frac{12}{h^2}\right)_i \qquad (11.21)$$

where

(a) For first story:

$$\alpha_z = \frac{0.5 + K}{2 + K}$$

$$K = \frac{K_{L1} + K_{L2}}{K_c}$$

(b) For other stories:

$$\alpha_z = \frac{K}{2 + K}$$

$$K = \frac{K_{L1} + K_{L2}K_{L3} + K_{L4}}{2K_c}$$

in which

$$K_L = \frac{1.5E_c J}{L} \text{ (beam stiffness factor)}$$

$K_C = \dfrac{E_c J}{h}$ (column stiffness factor)

$E_c = 3 \times 10^4$ MPa $= 3 \times 10^7$ kN/m^2
(modulus of elasticity of concrete)

$J_1 = \dfrac{0.3 \times 0.7^3}{12} = 8.575 \times 10^{-3}$ m^4
          (cross-sectional moment of inertia)

$J_{c1} = \dfrac{0.4 \times 0.6^3}{12} = 7.200 \times 10^{-3}$ m^4

$J_{c2} = \dfrac{0.45 \times 0.7^3}{12} = 12.8625 \times 10^{-3}$ m^4

L = beam length

h = column length

K = stiffness of beams framing to the ends of a column

Then

$K_L = \dfrac{1.5 \times 3 \times 10^7 \times 8.575 \times 10^{-3}}{6} = 64,312.5$ kN-m

$K_{c1} = \dfrac{3 \times 10^7 \times 7.200 \times 10^{-3}}{5.5} = 70,159.1$ kN-m

$K_{c2} = \dfrac{3 \times 10^7 \times 12.8625 \times 10^{-3}}{4} = 96,468.75$ kN-m

$K_{c3} = K_{c4} = K_{c5} = \dfrac{3 \times 10^7 \times 7.2 \times 10^{-3}}{4} = 54,000$ kN-m

Calculated values are given in Table 11.12.

Table 11.12. Calculated Story Drifts δ_i

Story	K_z	h	K	α_z	D_i(kN/m)	Q_i(kN)	δ_i(m)
5	54,000.0	4.0	1.19097	0.37323	15,115.82	250	0.008269
4	54,000.0	4.0	1.19097	0.37323	15,115.82	650	0.021500
3	54,000.0	4.0	1.19097	0.37323	15,115.82	1,050	0.034730
2	96,468.8	4.0	0.66667	0.25000	18,087.90	1,450	0.040080
1	70,159.1	5.5	0.91667	0.48571	13,518.10	1,850	0.068430

$\Delta_T = \Sigma\delta_i = 0.173009$

$T_1 = 1.5 \times 0.9 \times \sqrt{0.173009} = 0.636$ sec   [eq.(11.19)]

2. Seismic lateral forces

$T_g = 0.4$ sec (from Table 11.2 for Site Category II, far earthquake)
$\alpha_{max} = 0.16$ (from Table 11.3 for Intensity VIII)

$\alpha = \left(\dfrac{T_g}{T}\right)^{0.9} \alpha_{max} = \left(\dfrac{0.4}{0.636}\right)^{0.9} 0.16 = 0.1054$

                              [eq.(11.2)]

Table 11.13. Calculation of Seismic Forces F_i

Level	H_i	W_i	W_iH_i	$W_iH_i/\Sigma W_iH_i$	$F_{Ek}(1-\delta_n)$	F_i(kN)
5	21.5	250	5,375	0.2261	155.646	45.282*
4	17.5	400	7,000	0.2944	155.646	45.822
3	13.5	400	5,400	0.2271	155.646	35.347
2	9.5	400	3,800	0.1598	155.646	24.872
1	5.5	400	2,200	0.0925	155.646	14.397

$\Sigma W_iH_i = 23,775$
*Includes the top force ΔF_n

$F_i = \dfrac{W_iH_i}{\displaystyle\sum_{j=1}^{n} W_jH_j} F_{Ek}(1-\delta_n)$ [eq.(11.3)]

$F_{Ek} = \alpha W_{eq}$ [eq.(11.1)]

$W_{eq} = 0.85\Sigma W_i = 0.85(250 + 4 \times 400) = 1,572.5$ kN

$F_{Ek} = \alpha W_{eq} = 0.1054 \times 1572.5 = 165.74$ kN

$\Delta F_n = \delta_n F_{ek}$ [eq.(11.4)]

$\delta_n = 0.08T + 0.01 = 0.08 \times 0.636 + 0.01$
     $= 0.0609$ (Table 11.4)

$\Delta F_n = 0.0609 \times 165.74 = 10.09$ kN
     (additional force at top)

Calculation of the seismic lateral forces is shown in Table 11.13.

3. Story shear forces

$$Q_i = \sum_{j=1}^{n} F_j$$

Results in Table 11.14.

4. Overturning moments

$$M_i = \sum_{j=i+1}^{n} F_j(H_j - H_i)$$

Results in Table 11.14.

Table 11.14. Story Shear Forces, Overturning Moments, and Torsional Moments

Level (i)	Q_i (kN)	M_i (kN-m)	M_{Ti} (kN-m)
5	45.282	—	45.282
4	91.104	181.128	91.104
3	126.451	545.544	126.451
2	151.323	1,051.348	151.323
1	165.720	1,056.640	165.720
0	—	2,568.100	—

5. Torsional moments

$$M_{Ti} = Q_i e_i$$

where $e_i = 1.0$ m (eccentricity at level i).
Results in Table 11.14.

6. Elastic displacements and story drifts

$$\delta_i = \frac{Q_i}{\sum D_i} \times 1,000 \text{ (mm)}$$

$\sum D_i$ = total column stiffness (from Table 11.12)

Calculations and results are shown in Table 11.15.

Table 11.15. Calculation of Elastic Displacements

Level (i)	Q_i (kN)	*ΣD_i (kN/m)	δ_i (mm)	δ_i/h_i	θ_e
5	45.282	30,231.64	1.4978	1/2,670.6	
4	91.104	30,231.64	3.0135	1/1,327.4	
3	126.451	30,231.64	4.1827	1/956.3	1/450
2	151.323	36,175.80	4.1830	1/956.2	
1	165.720	27,036.20	6.1296	1/897.3	

*Two columns per story.

The maximum permissible elastic story drift for frames is given in Table 11.8 as $\theta_e = 1/450$. Values for story-drifts δ_i/h_i shown in Table 11.15 are less than this maximum permissible value.

Example 11.2

Example 11.1 is solved by using the computer program that was prepared for the implementation of the Chinese Seismic Design Code (GBJ 11-89). The first three modes are included in the solution.

```
        INPUT DATA AND OUTPUT RESULTS FOR EXAMPLE 11.2

INPUT DATA:

    NUMBER OF STORIES:  N = 5

    NUMBER OF COLUMNS IN EACH STORY:  2

    SPAN OF BUILDING FRAME (M):  6

MEMBER AREAS (M):

    TYPE        B           D

     1        0.45        0.70
     2        0.40        0.60
     3        0.30        0.70

STORY DATA:

TYPE   STORIES CONCRETE STRENGTH    COLUMN TYPE    GIRDER TYPE

 1      1-2        30 X E4              1              3
 2      3-5        30 X E4              2              3

LEVEL      HEIGHT (M)     WEIGHT (KN)    ECCENTRICITY (M)

  5          4.0             250             1.0
  4          4.0             400             1.0
  3          4.0             400             1.0
  2          4.0             400             1.0
  1          4.0             400             1.0
```

```
SEISMIC ACTION:

INTENSITY:  I = 8

SITE CATEGORY:  TS = 2

FAR OR NEAR EARTHQUAKE:  DN = FAR

COEFFICIENT OF PERIOD FOR INFILL-WALL:  CP = 0.9

OUTPUT RESULTS:

NATURAL PERIODS:

    MODE        PERIOD (SEC)

     1            0.654
     2            0.913
     3            0.120

SHEAR FORCE (KN):

LEVEL     MODE 1      MODE 2      MODE 3      SRSS

  5       32.981     -16.879      8.078      37.919
  4       81.965     -28.580      3.428      86.873
  3      122.784     -17.608     -7.959     124.295
  2      151.395       6.823     -4.382     151.612
  1      167.465      26.550      5.822     169.656

STORY DRIFT (MM):

LEVEL     MODE 1      MODE 2      MODE 3      SRSS

  5        1.24       -0.53       0.19       1.36
  4        2.67       -0.79       0.07       2.79
  3        4.00       -0.47      -0.16       4.03
  2        4.11        0.16      -0.07       4.12
  1        5.27        0.68       0.11       5.32

OVERTURNING MOMENTS (KN-M):

LEVEL     MODE 1      MODE 2      MODE 3      SRSS

  4      131.922     -67.515     32.313     151.677
  3      459.782    -181.837     46.024     496.570
  2      950.918    -252.270     14.187     983.913
  1     1556.499    -224.977     -3.340    1572.678
  0     2477.555     -78.955     28.682    2478.978

TORSIONAL MOMENTS (KN-M)

LEVEL     MODE 1      MODE 2      MODE 3      SRSS

  5       32.981     -16.879      8.078      37.919
  4       81.965     -28.580      3.428      86.873
  3      122.784     -17.608     -7.959     124.295
  2      151.395       6.823     -4.382     151.612
  1      167.465      26.550      5.822     169.656
```

Example 11.3

A 15-story reinforced concrete frame with plan dimensions of 18 m by 36 m, shown in Fig. 11.6, will be used to model an office building for an analysis of earthquake resistance. The building is located in a zone with intensity VII, and is designed for a far earthquake in Site Category I. Seismic weights of 1,380 kN are distributed to each floor of the building, except to the roof and the first and second story, where the loads are 1,200 kN, 1,680 kN, and 1,530 kN, respectively. The building's plane has sidespans of 6.5 m and a mid-span of 5.0 m and rises 54.20 m above ground level. The story height of all stories is 3.40 m, except for the first and second stories which are 5.50 m and 4.50 m high, respectively. The eccentricity of load at each floor is 1.0 m.

---

[5]SRSS: Effective values obtained by Square Root of Sum of Squares procedure.

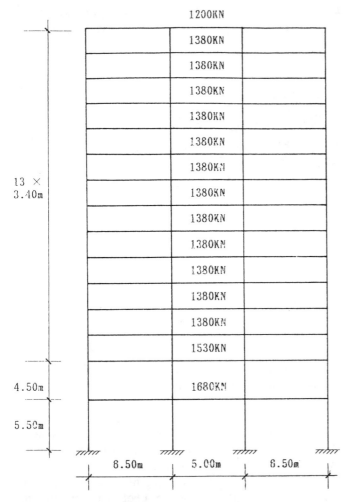

Fig. 11.6. Modeled building for Example 11.3

According to the *Seismic Design Code for Buildings and Structures* (GBJ 11-89), determine the following:

1. First three natural periods of the building
2. Design value of horizontal seismic forces
3. Effective shear force at each story of the building
4. Overturning moment at each level
5. Torsional moments for an assumed eccentricity of 1 m at each level
6. Lateral displacements
7. Elastic story drifts at each level floor.

INPUT DATA AND OUTPUT RESULTS FOR EXAMPLE 11.3

NUMBER OF STORIES: N = 15

NUMBER OF COLUMNS IN EACH STORY: 4

SPAN OF BUILDING FRAME (M): 6.5

MEMBER AREAS (M):

TYPE	B	D
1	0.50	0.80
2	0.45	0.70
3	0.40	0.60
4	0.40	0.50
5	0.30	0.75
6	0.30	0.70
7	0.30	0.60

STORY DATA:

TYPE	STORIES	CONCRETE STRENGTH
1	1-2	30 E4
2	3-5	30 E4
3	6-10	25 E4
4	11-15	20 E4

LEVEL	HEIGHT (M)	WEIGHT (KN)	ECCENTRICITY (M)
15	3.4	1200	1.0
14	3.4	1380	1.0
13	3.4	1380	1.0
12	3.4	1380	1.0
11	3.4	1380	1.0
10	3.4	1380	1.0
9	3.4	1380	1.0
8	3.4	1380	1.0
7	3.4	1380	1.0
6	3.4	1380	1.0
5	3.4	1380	1.0
4	3.4	1380	1.0
3	3.4	1380	1.0
2	4.5	1530	1.0
1	5.5	1680	1.0

SEISMIC ACTION:

INTENSITY: I = 7

SITE CATEGORY: TS = 1

FAR OR NEAR EARTHQUAKE: DN = FAR

COEFFICIENT OF PERIOD FOR INFILL-WALL: CP = 0.8

OUTPUT RESULTS:

NATURAL PERIODS:

MODE	PERIOD (SEC)
1	1.718
2	0.635
3	0.379

RESPONSE (EFFECTIVE VALUES):

LEVEL	SHEAR FORCE (KN)	STORY DRIFT (MM)	TORSIONAL MOMENT (KN-M)	OVERTURNING MOMENT (KN-M)
15	36.324	0.52	36.324	123.500
14	73.297	0.99	73.297	372.549
13	102.576	1.39	102.576	719.948
12	124.117	1.68	124.117	1136.322
11	140.512	1.85	140.512	1598.911
10	154.904	1.55	154.904	2096.082
9	169.493	1.67	169.493	2628.218
8	183.347	1.80	183.347	3196.518
7	195.431	1.90	195.431	3799.872
6	205.751	1.80	205.751	4434.851
5	215.529	1.27	215.529	5098.385
4	226.302	1.31	226.302	5790.850
3	237.659	1.42	237.659	6513.680
2	249.584	2.42	249.584	7517.706
1	257.256	2.48	257.256	8802.949

REFERENCES

Seismic Design Code for Buildings and Structures (1989) (GBJ 11-89), Chinese Academy of Building Research, Beijing, China.

Seismic Design Code for Industrial and Civil Buildings (1974) (TJ 11-74), Chinese Academy of Building Research, Beijing, China.

Seismic Design Code for Industrial and Civil Buildings (1978) (TJ 11-78), Chinese Academy of Building Research, Beijing, China.

Uniform Standard for Structural Design of Buildings (1984) (GBJ 68-84), Chinese Academy of Building Research, Beijing, China.

12

Colombia

Luis E. García

12.1 INTRODUCTION

Since it was founded in 1974, one of the main objectives of the Asociación Colombiana de Ingeniería Sísmica, AIS, (Colombian Association for Earthquake Engineering) has been the development of a meaningful set of earthquake-resistant design requirements that eventually could be adopted by the Colombian government as part of a general building code. The first step was to translate into Spanish the 1974 SEAOC Requirements (Seismology Committee 1974). This Spanish version was widely distributed throughout the country.

On November 23, 1979, the central part of Colombia was affected by a $6.4M_s$[1] magnitude earthquake, followed by a $7.8M_s$ event that occurred on the Pacific coast on December 12, 1979. (Appendix A12 contains more information on these events.) The effects of these earthquakes showed the deficiencies in building practice, both in design and construction (García 1984b). This prompted the AIS, which had translated the ATC-3-06 document (Applied Technology Council 1978), to develop a set of regulations for earthquake-resistant design and construction, *Norma AIS 100-81, Requisitos Sísmicos para Edificios* (AIS 1981).

After the Popayán earthquake of March 31, 1983, the AIS revised the AIS 100-81 document. The Colombian government enacted the Norma AIS 100-83, which mandated earthquake-resistant design requirements, and allowed one year for its development (García 1984b). The official name of the Code is *Código Colombiano de Construcciones Sismo Resistentes*, CCCSR-84, (Colombian Code for Seismic-Resistant Construction).

The Code has the following main sections:

Title A – General Requirements for Seismic-Resistant Construction
Title B – Loads
Title C – Reinforced Concrete
Title D – Structural Masonry
Title E – One- and Two-Story Dwellings

---

[1]M_s is a scale for measuring the magnitude of earthquakes. Specifically, the M_s (surface wave) magnitude is proportional to the source spectral amplitude of the 20-second-period wave (see Appendix of this book).

Title F – Metal Structures
Title G – Penalties

The AIS document was used as the basis for Title A. This section defines the scope of the Code, outlines acceptable design procedures, gives requirements for the approval of materials and design procedures not covered by the scope of the Code, defines mandatory supervision during construction, specifies the design spectra and the seismic design parameters, sets the type of structural systems and methods of analysis permitted, and defines the allowable drift limits and sets the specific requirements of the Code that must be followed in each of the seismic risk zones.

Title B, developed from previous recommendations of the Colombian Society of Engineers, covers all loads with differing seismic effects, such as dead, live, and wind loading.

Title C includes provisions for reinforced concrete design given in ACI 318-83 (ACI 1983), but varies from ACI 318-83 by considering different Colombian construction practices.

Title D covers all types of structural masonry. Unreinforced masonry is prohibited except in specific regions of the low-seismic-risk zones. It gives requirements for different types of masonry and imposes mandatory supervision on all structural masonry construction operations. It is based on three documents suitably adapted for local materials and quality of construction: Chapter 12 of ATC-3-06 (ATC 1978), Standard ACI 531 (ACI 1981), and the Mexican Building Code (UNAM 1977). Previous experimental research on Colombian masonry materials has also been incorporated into this section.

Title E gives requirements for the construction of one- and two-story dwellings built on masonry bearing walls. The intent of this section was to lessen the socio-economic impact of the Title D ban on unreinforced masonry by giving a set of empirical rules to guide a person, not necessarily an architect or engineer, on how to build an earthquake-resistant house.

Title F is based on the specifications of the American Institute of Steel Construction (AISC 1978). Metal structures are not used widely in Colombia, with the possible exception of long-span roof trusses and other light structures.

Title G states the legal penalties for non-compliance with the Code.

The Code was enacted by Decree 1400 of June 7, 1984 by the Ministerio de Obras Públicas y Transporte, CCCSR-84 (1984). The code became fully mandatory by the end of 1984. Three recent earthquakes have had a great influence on the further development of seismic-resistant regulations. These quakes are described in the Appendix (A12).

12.2 DESIGN PROCEDURE

The Code establishes nine steps that must be followed in the design of a building:

Step 1. Location of the building site on the seismic zone map in Fig. 12.1.

Step 2. Determination of the seismic risk level (Low, Intermediate, or High) as a function of A_a and A_v which are parameters obtained from the maps shown in Figs. 12.2 and 12.3.

Step 3. Definition of the design spectra for the site using A_a, A_v, the importance of the building for post-earthquake function, and the soil profile at the site.

Step 4. Definition of the analysis procedure to be used depending on the regularity or irregularity of the building layout both in plan and elevation. The analysis method must be either the equivalent lateral load method or the modal analysis method.

Step 5. Determination of the lateral loads to be used as a function of design spectra for the site, and the response modification factor R, which depends on the lateral load-resisting system and the design requirements for the structural material used. There are restrictions on the design requirements that can be used depending on the level of seismic risk of the building site allowed for reinforced concrete, structural masonry, and structural steel.

Step 6. Analysis of the structure for the lateral loads obtained using a linear elastic mathematical model and the method of analysis defined in Step 4.

Step 7. Evaluation of the total elastic and inelastic lateral deflections of the building obtained by multiplying the elastic lateral displacements from the analysis by a displacement amplification factor C_d, which depends on the response modification factor R used to obtain the lateral loads.

Step 8. Verification that the structure complies with the drift limits for the total elastic and inelastic story drifts. The structure must be stiffened if the story drift limits are exceeded.

Step 9. Design of the structural elements following the structural material requirements compatible with the response modification factor R used to obtain the lateral loads.

12.3 SEISMIC ZONING AND DESIGN SPECTRA

The seismic zoning and risk assessment are based on a three-year study (García et al. 1984a) that was com-

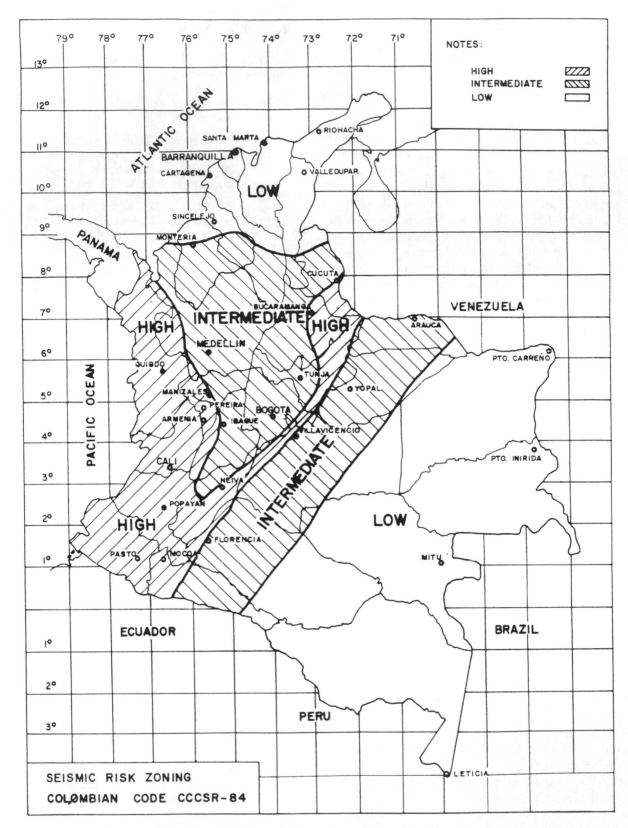

Fig. 12.1.   Seismic risk zone Map (Colombian Code CCCSR-84)

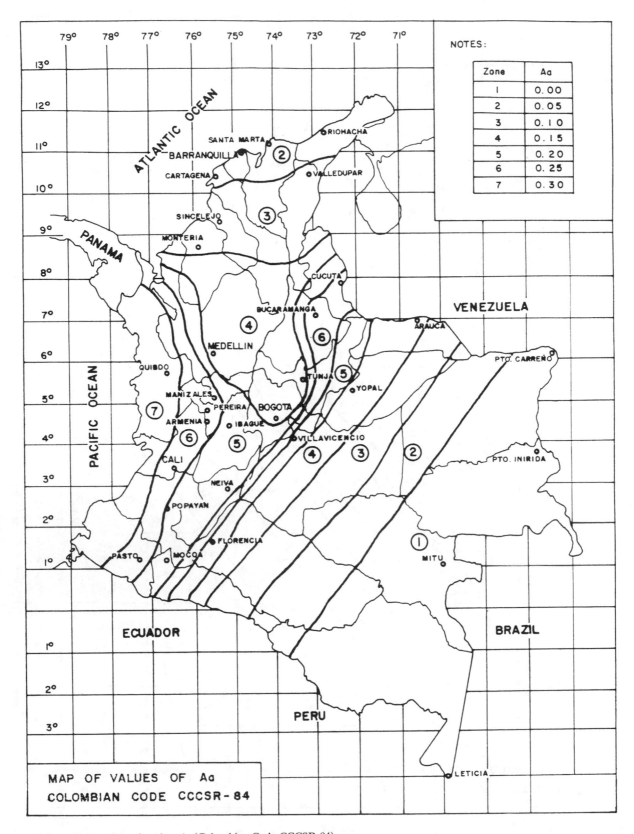

Fig. 12.2.   Map of spectral acceleration A_a (Colombian Code CCCSR-84)

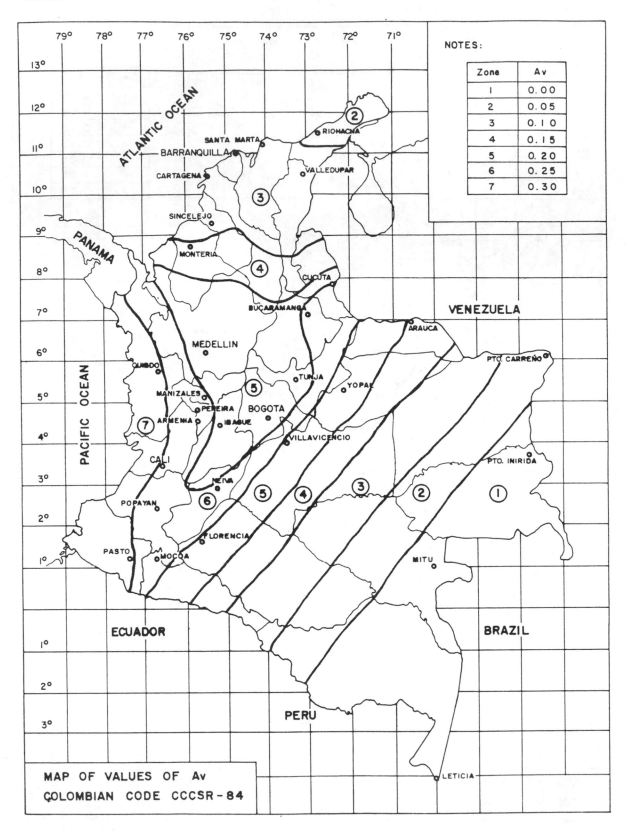

Fig. 12.3. Map of spectral velocity A_v (Colombian Code CCCSR-84)

pleted in time for the enactment of the Code in 1984. The country is divided into three zones: High risk, Intermediate risk, and Low risk. These zones are shown in Fig. 12.1. The Code gives different construction requirements for each zone, with the strictest mandatory requirements given for the high-risk zone. Two additional maps, one for values of effective peak acceleration, EPA, (A_a), Fig. 12.2, and one for values of effective peak velocity, EPV, (A_v), Fig. 12.3, both in terms of acceleration, are given by the Code. These two maps allow the construction of a smooth design spectra for any site in the country. The values of A_a and A_v defining the design earthquake have a 10% probability of being exceeded in a period of 50 years.

Low-seismic-risk zones cover those sites where both A_a and A_v are less than or equal to 0.10. Intermediate-seismic-risk zones are those where both A_a and A_v are greater than 0.10 and neither is greater than 0.20. High-seismic-risk zones are those where either A_a or A_v is greater than 0.20.

The soil site effects depend on the soil profile at the site of the building. The Code defines the following three soil profile types:

Soil Profile Type S_1
(a) Rock, either shale-like or crystalline, with a shear wave velocity greater than 750 meters per second.
(b) Stiff soil conditions with a depth of 60 meters or less of stable deposits of sand, gravel, or stiff clays.

Soil Profile Type S_2. Rock overlain by more than 60 meters of stable deposits of stiff clay or cohesionless soils.

Soil Profile Type S_3. Rock overlain by more than 10 meters of soft to medium-stiff clays, with or without intervening layers of sand or other cohesionless soils.

The site coefficient **S** depends on the soil profile type. The values of **S** are given in Table 12.1.

All buildings must be classified in one of the following risk exposure use groups:

Use Group III. Buildings that must be operational after an earthquake, such as hospitals, fire stations, police stations, telecommunications centers, and power and water supply buildings.

Use Group II. Buildings where more than 200 people could gather in the same room at one time, outdoor grandstands for more than 2,000 persons, schools, universities, stores with more than 500 square meters per story, and all buildings where the occupants could be restricted in their exit if a general panic occurred.

Table 12.1. Site Coefficient **S**

Soil Profile Type:	S_1	S_2	S_3
Site Coefficient (**S**):	1.0	1.2	1.5

Table 12.2. Importance Coefficient *I*

Use Group:	I	II	III
Importance Coefficient (*I*):	1.0	1.1	1.2

Use Group I. All buildings not covered in Groups II and III.

The importance coefficient *I* modifies the design spectra for each of the use groups. Values for *I* are given in Table 12.2.

The shape of the elastic design acceleration spectra for 5% of critical damping is given by the following equation:

$$S_a = \frac{1.2A_v SI}{T^{2/3}} \leqslant 2.5A_a I \qquad (12.1)$$

where

S_a = maximum horizontal acceleration, as a fraction of the acceleration of gravity, that a single-degree-of-freedom system with a vibration period T will sustain during the design earthquake
A_a = effective peak ground acceleration given in Fig. 12.2, as a fraction of the acceleration of gravity
A_v = effective peak ground velocity, in terms of acceleration, given in Fig. 12.3, as a fraction of the acceleration of gravity
S = site coefficient, dimensionless, given in Table 12.1
I = importance coefficient, dimensionless, given in Table 12.2
T = fundamental period of the system in seconds

For Soil Profile S_3 with A_a greater than or equal to 0.30, the value of S_a need not exceed $2.0A_a I$. Fig. 12.4 shows the elastic design acceleration spectra in rock for four major cities in Colombia.

12.4 LATERAL LOAD-RESISTING STRUCTURAL SYSTEMS

The Code recognizes three lateral load-resisting structural systems. For each system, the Code defines a value of the response modification factor R and of the lateral displacement amplification factor C_d, depending on the seismic risk level of the location of the building and the structural material requirements. The lateral load-resisting structural systems recognized by the Code are:

Frames. Unbraced moment-resisting space frames that resist both lateral and vertical loads.
Structural walls. The vertical loads are resisted by structural walls and/or unbraced frames. The lateral loads are resisted by structural walls and/or braced frames.

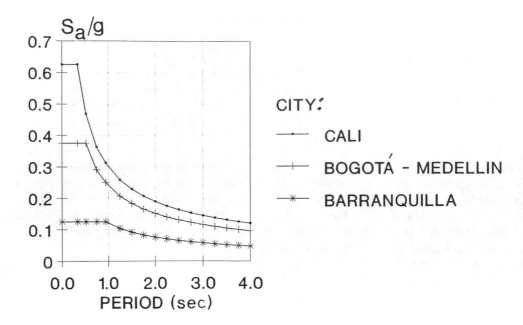

Fig. 12.4. Design acceleration spectra S_a in rock for four major Colombian cities

Dual system. Combination of moment-resisting unbraced space frame and structural walls or braced frames. This system must comply with the following requirements:

(a) The moment-resisting unbraced space frame and the structural walls or braced frames must be able to support lateral seismic loads proportional to their lateral stiffness, taking into account the interaction between frame and walls.

(b) The structural walls or braced frames, acting independently of the unbraced moment-resisting space frame, must be able to resist all of the lateral seismic loads.

(c) The moment-resisting unbraced space frame, acting independently of the structural walls or braced frames, must be able to support 25% of the seismic lateral loads.

Any building must be classified in one of these three structural systems. When structural systems are combined in elevation, the lowest value of R and the highest value of C_d corresponding to the combined structural systems must be used. Table 12.3 gives the values of R and C_d for the different structural systems and material design requirements.

12.5 BUILDING CONFIGURATION

The building must be classified either as regular or irregular for seismic-resistant design purposes. The Code recognizes two types of irregularities: plan irregularity and elevation irregularity. The method of analysis is determined from this classification of the building. For irregular buildings, the modal analysis

method is recommended, although the equivalent lateral load method is permitted. The selection is made, based on engineering criteria, by the designer. For regular structures either of the two methods is acceptable.

12.6 EQUIVALENT LATERAL LOAD METHOD

The fundamental vibration period of the building must be obtained from the dynamic properties of the building using a linear elastic model, assuming that the building is fixed at the base. When such a procedure is not used, an approximate value for the fundamental period that subsequently could be used to obtain the spectral acceleration may be calculated. The Code provides the following equations to calculate an approximate value for the fundamental period:

For reinforced concrete and structural steel unbraced moment-resisting frames:

$$T_a = 0.08 h_n^{3/4} \tag{12.2}$$

where

h_n = height of the building measured from the base, in meters
T_a = approximate fundamental period of the building in seconds.

For all other structural systems the approximate fundamental period T_a can be obtained using:

$$T_a = \frac{0.09 h_n}{L^{1/2}} \tag{12.3}$$

Table 12.3. Structural Systems Permitted and Values of R and C_d

Structural System	Lateral Load-Resisting System	Note	Material Design Requirements (Chapters)	Seismic Risk Zone					
				Low		Intermediate		High	
				R	C_d	R	C_d	R	C_d
Frame	Reinforced concrete unbraced frames								
	– High ductility	(1)	C.21	6.0	5.0	6.0	5.0	6.0	5.0
	– Moderate ductility	(2)	C.20	4.0	3.5	4.0	3.5	N.P.	N.P.
	– Low ductility	(3)	C.1 to C.19	2.0	2.0	N.P.	N.P.	N.P.	N.P.
	Structural steel frames	(4)	Title F	4.5	4.0	4.5	4.0	4.5	4.0
Structural Walls	Reinforced concrete walls								
	– High ductility	(1)	C.21	5.0	4.5	5.0	4.5	5.0	4.5
	– Low ductility	(3)	C.1 to C.19	4.0	3.5	4.0	3.5	N.P.	N.P.
	Reinforced concrete braced frames								
	– High ductility	(1)	C.21	4.5	4.0	4.5	4.0	4.5	4.0
	– Moderate ductility	(2)	C.20	3.5	3.0	3.5	3.0	N.P.	N.P.
	– Low ductility	(3)	C.1 to C.19	2.0	2.0	N.P.	N.P.	N.P.	N.P.
	Unreinforced masonry	(5)	D.4	1.0	1.0	N.P.	N.P.	N.P.	N.P.
	Partially reinforced masonry	(6)	D.5	2.0	2.0	2.0	2.0	N.P.	N.P.
	Reinforced masonry	(6)	D.6	3.5	3.0	3.5	3.0	3.5	3.0
	Confined masonry								
	– With internal reinforcement	(7)	D.7	2.5	2.5	2.5	2.5	2.5	2.5
	– Without internal reinforcement	(7)	D.7	1.5	1.5	1.5	1.5	1.5	1.5
Dual System	Reinforced concrete walls								
	– High ductility	(1)	C.21	7.0	6.0	7.0	6.0	7.0	6.0
	– Low ductility	(3)	C.1 to C.19	5.0	4.5	5.0	4.5	N.P.	N.P.
	Reinforced concrete braced frames								
	– High ductility	(1)	C.21	5.0	4.5	5.0	4.5	5.0	4.5
	– Moderate ductility	(2)	C.20	4.0	3.5	4.0	3.5	N.P.	N.P.
	– Low ductility	(3)	C.1 to C.19	2.0	2.0	N.P.	N.P.	N.P.	N.P.
	Partially reinforced masonry walls	(6)	D.5	3.0	3.0	N.P.	N.P.	N.P.	N.P.
	Reinforced masonry walls	(6)	D.6	4.5	4.0	4.5	4.0	5.5	5.0

Notes:

N.P. = Not Permitted

(1) Chapter C.21 is based on Appendix A of ACI 318-83 and is equivalent to Chapter 21 of ACI 318-89. These structures must also comply with the reinforced concrete requirements of Chapters C.1 to C.19.

(2) Chapter C.20 is based on Section A.9 of Appendix A of ACI 318-83 and is equivalent to Section 21.9 of ACI 318-89. These structures must also comply with the reinforced concrete requirements of Chapters C.1 to C.19.

(3) Chapters C.1 to C.19 are based on Chapters 1 to 19 of ACI 318-83.

(4) Title F of the Code is based on the AISC Specifications.

(5) Unreinforced masonry is permitted only in those parts of zones of Low seismic risk where A_a and A_v are both less than 0.05.

(6) Chapters D.5 and D.6 cover reinforced masonry made out of hollow concrete block and hollow clay brick, in which the reinforcement is grouted within the vertical cells. The only difference between Chapter D.5 and Chapter D.6 is that the latter requires higher reinforcement ratios.

(7) Confined Masonry consists of a masonry wall that carries the vertical loads and is surrounded by a light reinforced concrete frame that confines the masonry and gives strength for lateral loads. The light concrete frame is cast after the wall is built. The concrete frame elements have longitudinal and transverse reinforcement. The masonry wall can be built with or without internal reinforcement.

where

L = length, in meters, of the structural system at the base in the direction under consideration.

In order to obtain the value of L to be used in eq.(12.3) the Code Commentary suggests the following formula:

$$L = L_{s\max} \sum_{s=1}^{N} \left\{ \frac{L_s}{L_{s\max}} \right\}^2 \qquad (12.4)$$

where

L = length, in meters, of the structural system at the base in the direction under consideration

L_s = length, in meters, of a segment of wall or braced frame, measured in the direction under consideration

$L_{s\max}$ = length, in meters, of the larger segment of wall or braced frame in the direction under consideration

N = number of segments of walls or braced frames in the direction under consideration.

The seismic design coefficient C_s is calculated as

$$C_s = \frac{S_a}{R} \qquad (12.5)$$

where

S_a = value of the acceleration obtained from the design spectra corresponding to the fundamental period of the building

R = response modification factor for the structural system and material detail requirements as given in Table 12.3.

The building must be designed to resist a horizontal design base shear given by

$$V = C_s W \qquad (12.6)$$

where

V = horizontal design base shear

W = total vertical weight of the building. W must be equal to the total weight of the structure plus the weight of all other elements such as partitions, permanent equipment, tanks, and their contents, etc. In warehouses, 25% of the live load on the floor must be included.

The design base shear V must be distributed along the height of the building, obtaining at level x a horizontal design force F_x given by the following equation:

$$F_x = C_{vx} V \qquad (12.7)$$

where

$$C_{vx} = \frac{w_x h_x^k}{\sum_{i=1}^{n} W_i h_i^k} \qquad (12.8)$$

and

k = exponent related to the fundamental period of the structure as follows:
for T less than or equal to 0.5 seconds: $k = 1.0$
for T greater than 2.5 seconds: $k = 2.0$
for T between 0.5 and 2.5 seconds: $k = 0.75 + 0.5T$,

and where

w_i, w_x = part of W located at level i or x, respectively

h_i, h_x = height, measured from the base, of level i or x, respectively.

The lateral forces corresponding to each level must be applied to the structure as a whole using a linear elastic mathematical model in order to obtain the forces in the structural elements. The internal forces obtained from this analysis must be combined with the other forces prescribed by the Code in order to obtain the full design forces. The seismic forces are prescribed according to stress level; using an ultimate strength design method, the load factor for the seismic forces is 1.0. When using a working stress design method, the seismic forces must be divided by a factor equal to 1.4, because the allowable stresses have been set at lower levels. The lateral displacements of the structure obtained from linear elastic analysis are used to determine the elastic story drifts.

12.7 MODAL ANALYSIS METHOD

The symbols used in the modal analysis method have the same meaning as in the equivalent lateral load method; the subscript "m" is added to indicate that they refer to mode m.

Regular buildings can be modeled as a system with masses concentrated at each level, where each mass has a horizontal degree of freedom. Buildings with plan irregularities may be modeled with concentrated masses at each level having three degrees of freedom, two orthogonal horizontal translations and a rotation about a vertical axis. The modes and corresponding periods must be calculated using the principles of structural dynamics for a fixed base, linear elastic model. The analysis must include in each of the principal orthogonal directions at least three modes, or all modes corresponding to vibration periods greater than 0.4 second.

For each mode, a modal seismic coefficient C_{sm} must be evaluated for the corresponding vibration period, using the following formula:

$$C_{sm} = \frac{S_{am}}{R} \quad (12.9)$$

where

S_{am} = maximum horizontal acceleration obtained from the design spectra for the period T_m of mode m
R = response modification factor for the structural system and material detail requirements as given in Table 12.3.

For soil profile class S_3, the value of C_{sm} for modes different from the fundamental mode and for vibration periods less than 0.3 second may be multiplied by $(0.4 + 2.0T_m)$. Also, for structures where T_m exceeds 4.0 seconds, the value of C_{sm} for that mode must be multiplied by $(2.5/[T_m]^{2/3})$.

The portion of the base shear V_m contributed by mode m is determined using

$$V_m = C_{sm}\bar{W}_m \quad (12.10)$$

where $\bar{W}$ = effective modal weight for mode m given by the following equation:

$$\bar{W}_m = \frac{\left[\sum_{i=1}^{n} w_i\phi_{im}\right]^2}{\sum_{i=1}^{n} w_i\phi_{im}^2} \quad (12.11)$$

where

ϕ_{im} = displacement amplitude at level i of the building when vibrating in mode m
w_i = seismic weight at level i of the building.

The modal force F_{xm} at level x is given by the following equation:

$$F_{xm} = C_{vxm}V_m \quad (12.12)$$

in which

$$C_{vxm} = \frac{w_x\phi_{xm}}{\sum_{i=1}^{n} w_i\phi_{im}} \quad (12.13)$$

The lateral elastic displacement δ_{xem} at level x is evaluated using:

$$\delta_{xem} = \frac{gT_m^2 F_{xm}}{4\pi^2 w_x} \quad (12.14)$$

where g = acceleration of gravity.

The design values of the base shear force V_t, the story shear force V_x, and story elastic lateral displacements δ_{xe}, must be obtained by combining the corresponding modal values. These combinations may be obtained using the Square Root of the Sum of the Squares (SRSS) method of each of the modal values.

The value of the design base shear force V_t must be compared with the base shear force V_0, calculated using the equivalent lateral load method, but using a value of T equal to $1.4T_a$. If V_t is less than V_0, then the story shears and the story displacements must be scaled up by a factor equal to V_0/V_t. The value of the design base shear force need not exceed the one obtained using the equivalent lateral load method.

12.8 LIMITS ON STORY DRIFT

The story drifts must be calculated from the elastic lateral displacements at each level of the building using the same mathematical model adopted to obtain the element forces. The total elastic and inelastic lateral displacement at level x is obtained from:

$$\delta_x = C_d\delta_{xe} \quad (12.15)$$

where

δ_x = actual lateral displacement, in meters, of the structure at level x, taking into account the inelastic effects
δ_{xe} = elastic lateral displacement, in meters, at level x
C_d = lateral displacement amplification factor (Table 12.3).

When cracked sections are used to evaluate the moment of inertia of the elements in reinforced concrete structures, the value of C_d can be multiplied by 0.7. The value of C_d thus obtained cannot be less than 2.

The design story drift is obtained as the algebraic difference between the actual inelastic lateral displacements of two consecutive stories:

$$\Delta_x = \delta_x - \delta_{x-1} \quad (12.16)$$

where

Δ_x = actual design story drift, in meters, of the structure at level x, taking into account the inelastic effects
δ_x = actual inelastic lateral displacement, in meters, at level x
δ_{x-1} = actual inelastic lateral displacement, in meters, at level $x-1$.

The actual design story drift cannot exceed 1.5% (0.015) of the story height. If the calculated drift is greater than this limit, the structure must be stiffened until the drift complies with the limit.

12.9 EXAMPLE

Example 12.1

The seismic lateral loads must be evaluated for a regular six-story apartment building as shown in Fig. 12.5. The structural system is a moment-resisting reinforced concrete space frame. The slabs consist of one-way joists. The building is located in the city of Bogotá.

Solution

We follow the steps prescribed by the Code for the design of the building as follows:

Steps 1   Location of the building site on the seismic zone
and 2.   maps and determination of the seismic risk level (Low, Intermediate, or High) as a function of A_a and A_v:

The city of Bogotá is located in Zone 4 (Fig. 12.2) with a value of $A_a = 0.15$ and in Zone 5 (Fig. 12.3) with $A_v = 0.20$. For these values of A_a and A_v, the seismic risk level is Intermediate as shown in Fig. 12.1.

Step 3.   Definition of the design spectra for the site using A_a, A_v, the importance of the building for post-earthquake function, and the soil profile at the site:

An apartment building is in Use Group I, and the corresponding importance coefficient $I = 1.0$ (Table 12.2).

The soil profile at the site is shown in Fig. 12.5(c). It has 30 m of soft clay between rock and surface; therefore the soil profile is classified as Type S_3 and the site coefficient $\mathbf{S} = 1.5$ (Table 12.1).

The design spectrum is given by

$$S_a = \frac{1.2 A_v S I}{T^{2/3}} \leqslant 2.5 A_a I \qquad \text{[eq.(12.1)]}$$

Substituting numerical values of $A_a = 0.15$, $A_v = 0.20$, $\mathbf{S} = 1.5$, and $I = 1.0$ yields

$$S_a = \frac{1.2 \times 0.20 \times 1.5 \times 1.0}{T^{2/3}} \leqslant 2.5 \times 0.15 \times 1.0$$

$$= \frac{0.36}{T^{2/3}} \leqslant 0.375$$

The value of S_a is expressed as a fraction of the acceleration of gravity.

Step 4.   Definition of the analysis procedure to be used. The building is regular both in plan and in elevation and therefore, it is possible to use the equivalent lateral load method.

Step 5.   Determination of the seismic lateral loads to be used, as a function of design spectra for the site, and a response modification factor R which depends on the lateral load-resisting system and the design requirements for the structural material used:

The structural system of the building for lateral and vertical load resistance is a reinforced concrete moment-resisting frame. The building is located in an intermediate seismic risk zone; thus, it is permitted to use the moderate ductility material design requirements (Table 12.3). Consequently, the value of the response modification factor is $R = 4.0$ and the elements of the structure must be designed following the requirements of Chapter C.20 of the Code for reinforced concrete structures with moderate ductility.

In order to evaluate the lateral design forces, the equivalent lateral load method is followed. The Code permits the use of an approximate fundamental period T_a, obtained from the following equation:

$$T_a = 0.08 h_n^{3/4} \qquad \text{[eq.(12.2)]}$$

From Fig. 12.5(a), $h_n = 4.0 + 5 \times 3.0 = 19.0$ m, hence:

$$T_a = 0.08(19.0)^{3/4} = 0.728 \text{ sec}$$

The seismic coefficient C_s is obtained from $C_s = S_a/R$. For $T_a = 0.728$ sec, $S_a = 0.36/(0.728)^{2/3} = 0.445 > 0.375$; thus $S_a = 0.375$ and

$$C_s = S_a/R = 0.375/4.0 = 0.09375 \qquad \text{[eq.(12.5)]}$$

The weight W of the building must be evaluated. W includes the weight of the structure plus all other permanent loads. These include the following:

- Selfweight of the slab, obtained as the volume of concrete times the unit weight of reinforced concrete (2.4 tonne/m³) = $[(18.4 \text{ m} \times 15.3 \text{ m} \times 0.45 \text{ m}) - 90$ void units $\times 0.844 \text{ m} \times 2.75 \text{ m} \times 0.40 \text{ m})] \times 2.4$ tonne/m³ = $[126.68 - 83.56] \times 2.4 = 103.51$ tonnes.

- Selfweight of the columns per floor. Assuming a column section of 0.40 m × 0.35 m for all columns in the building = 0.40 m × 0.35 m × 2.55 m × 2.4 tonne/m³ × 16 units = 13.71 tonnes per level (neglecting the greater height, 4 m, of the lowest columns).

- Partitions. The Code (Sec. B.3.4) requires that the selfweight of the partitions be evaluated using the architectural plans and sets a minimum value of 0.15 tonne/m² for tile partitions and 0.20 tonne/m² for solid brick partitions. In this case a value of 0.25 tonne/m², typical for apartment buildings, is used. Therefore the weight of partitions per floor = 18.4 m × 15.3 m × 0.25 tonne/m² = 70.38 tonnes.

EXAMPLE I

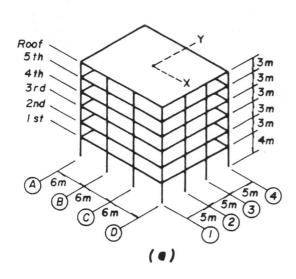

Roof
5th
4th
3rd
2nd
I st

3m
3m
3m
3m
3m
4m

A 6m
  B 6m
    C 6m
      D

5m 4
5m 3
5m 2
   I

(a)

ALL DIMENSIONS IN METERS

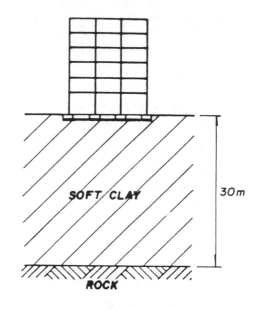

SOFT CLAY

ROCK

30m

SOIL PROFILE AT THE SITE
(c)

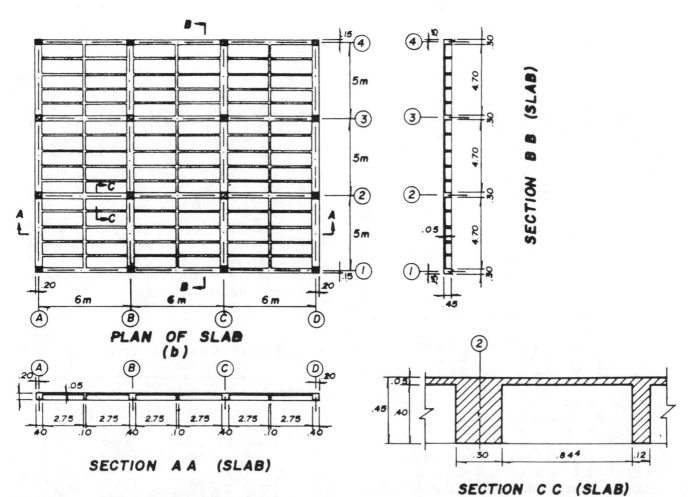

PLAN OF SLAB
(b)

SECTION A A (SLAB)

SECTION B B (SLAB)

SECTION C C (SLAB)

Fig. 12.5.   Six-story regular framed building for Example 12.1

• Architectural elements, including ceiling paneling, soffit, floor tile, etc. The Code (Sec. B.3.5) requires a minimum of 0.10 tonne/m². A value of 0.15 tonne/m² typical for apartment buildings is used here. The weight per floor is $= 18.4\,\text{m} \times 15.3\,\text{m} \times 0.15\,\text{tonne/m}^2 = 42.23$ tonnes.

The following weights per floor thus are obtained:

Roof:

Slab	103.51
Columns × 0.5	6.86
Partitions	0.00
Architectural elements	42.23
TOTAL	**152.60 tonnes**

Typical floor:

Slab	103.51
Columns	13.71
Partitions	70.38
Architectural elements	42.23
TOTAL	**229.83 tonnes**

Accordingly, the total weight of the building is:

$W = 152.60$ tonnes $+ 229.83$ tonnes $\times 5 = 1{,}301.75$ tonnes.

The design base shear is obtained from

$$V = C_s W = 0.09375 \times 1{,}301.75 = 122.04 \text{ tonnes.} \qquad [\text{eq.}(12.6)]$$

The base shear force, $V = 122.04$, must be distributed in the height of the building, obtaining at level x a horizontal design force F_x given by the following equation:

$$F_x = C_{vx} V \qquad [\text{eq.}(12.7)]$$

and

$$C_{vx} = \frac{w_x h_x^k}{\displaystyle\sum_{i=1}^{n} w_i h_i^k} \qquad [\text{eq.}(12.8)]$$

where

k = exponent related to the fundamental period of the structure:

    for T less than or equal to 0.5 second: $k = 1.0$
    for T greater than 2.5 seconds: $k = 2.0$
    for T between 0.5 and 2.5 seconds: $k = 0.75 + 0.5T$

w_i, w_x = part of W located at levels i and x, respectively.

h_i, h_x = height, measured from the base, of levels i and x, respectively.

Because $T_a = 0.728$ sec, the value of $k = 0.75 + 0.5 \times 0.728 = 1.114$.

Calculated values for the vertical distribution of the lateral loads are shown in Table 12.4.

Table 12.4. Calculation of Lateral Seismic Forces

Level	h_x (m)	w_x (tonnes)	$w_x(h_x)^k$	C_{vx}	F_x (tonnes)	V_x (tonnes)
Roof	19.0	152.60	4,055.889	0.212	25.81	25.81
5	16.0	229.83	5,044.247	0.263	32.10	57.91
4	13.0	229.83	4,002.576	0.209	25.48	83.39
3	10.0	229.83	2,988.180	0.156	19.02	102.41
2	7.0	229.83	2,008.380	0.105	12.78	115.19
1	4.0	229.83	1,076.717	0.056	6.85	122.04
		$\Sigma = 1{,}301.75$	19,175.989	1.000	122.04	

Step 6. Analysis of the structure for the lateral loads obtained, using a linear elastic mathematical model and the method of analysis defined in Step 4. Any mathematical model that complies with the principles of structural mechanics is allowed. In this case, the analysis was performed using the frame program developed for this chapter.

The lateral elastic displacements and story drifts (lateral displacement between contiguous floors), using eqs.(12.14) and (12.16), are calculated for the two principal plan directions of the building as shown in Table 12.5.

Step 7. Evaluation of the total elastic and inelastic lateral displacements of the building obtained by multiplying the elastic lateral deflections by the displacement amplification coefficient C_d, which is given in Table 12.3 and is related to the response modification factor R used to obtain the lateral loads. Hence

$$\delta_x = \delta_{xe} C_d \quad \text{and} \quad \Delta_x = \Delta_{xe} C_d$$

From Table 12.3, for reinforced concrete frames of moderate ductility, the corresponding value for C_d is 3.5. The total displacements and drifts are shown in Table 12.6. The relative drift is also shown as a percentage of the story height.

Table 12.5. Calculation of Elastic Lateral Displacements and Story Drifts

Level	δ_{xe} (m)	Δx_e (m)	δ_{ye} (m)	Δ_{ye} (m)
Roof	0.0463	0.0027	0.0409	0.0022
5	0.0436	0.0049	0.0387	0.0041
4	0.0387	0.0071	0.0346	0.0059
3	0.0316	0.0087	0.0287	0.0072
2	0.0229	0.0100	0.0215	0.0082
1	0.0129	0.0129	0.0133	0.0133

Table 12.6. Calculation of Total Lateral Displacements and Story Drifts

Level	δ_x (m)	Δ_x (m)	Δ_x/h_i (%)	δ_y (m)	Δ_y (m)	$\Delta y/h_i$ (%)
Roof	0.1621	0.0095	0.32	0.1432	0.0077	0.26
5	0.1526	0.0172	0.57	0.1355	0.0144	0.48
4	0.1355	0.0249	0.83	0.1211	0.0207	0.69
3	0.1106	0.0305	1.02	0.1005	0.0252	0.84
2	0.0802	0.0350	1.17	0.0753	0.0287	0.96
1	0.0452	0.0452	1.13	0.0466	0.0466	1.16

Step 8.   Verification of the compliance of the drift requirements for the inelastic story drifts. The structure must be stiffened if the story drift limits are not met.

The story drift limit set by the Code is 1.5% of h_i. In this case, the largest story drift in the X direction is 1.17% in the second story, and in the Y direction, 1.16% in the first story. Both are well within the limit set by the Code; thus, the structure has adequate stiffness.

Step 9.   The structural elements are designed following the material requirements corresponding to the response modification factor R used in obtaining the lateral loads.

In this case, the requirements of Chapters C.1 to C.20 of the Code relating to reinforced concrete must be followed, with special attention given to Chapter C.20. Chapter C.20 of the Code contains the requirements for reinforced concrete structures of moderate ductility.

Orthogonal or directional effects must be taken into account by combining the effects on the elements caused by the lateral load acting in each of the principal plan directions plus 30% of the effect caused by the lateral load that acts in the orthogonal direction.

Seismic loads are defined by the Code in terms of strength design level. For ultimate strength design, the seismic load factor is unity (1.0). The load cases that must be used in the design of the elements are the following:

$$U = 1.60D$$
$$U = 1.40D + 1.70L$$
$$U = 1.05D + 1.28L + 1.0E_x + 0.3E_y$$
$$U = 1.05D + 1.28L - 1.0E_x - 0.3E_y$$
$$U = 1.05D + 1.28L + 0.3E_x + 1.0E_y$$
$$U = 1.05D + 1.28L - 0.3E_x - 1.0E_y$$
$$U = 0.90D + 1.0E_x + 0.3E_y$$
$$U = 0.90D - 1.0E_x - 0.3E_y$$
$$U = 0.90D + 0.3E_x + 1.0E_y$$
$$U = 0.90D - 0.3E_x - 1.0E_y$$

where

D = effect of the dead load
L = effect of the live load
E_x = effect of the seismic lateral load acting in the X direction
E_y = effect of the seismic lateral load acting in the Y direction.

12.10 DESCRIPTION OF THE COMPUTER PROGRAM

A computer program has been developed to perform the seismic analysis of frame or wall buildings (the latter are treated as single-column frames) following the requirements of the Colombian Seismic Code (CCCSR-84). The program assumes rigid horizontal diaphragms at the levels of the floors with three degrees of freedom per story: two orthogonal lateral translations and a rotation about a vertical axis. These movements are referenced to the center of mass of the story diaphragm. The frames and resisting walls are planar elements that are linked at each story to the diaphragm. The program takes into account axial deformations for the columns and shear deformations for both columns and beams. It also takes into account, depending on user preferences, the rigidity of connections between beams and columns. The program determines the lateral stiffness matrix for each of the frames or walls, and based on these matrices, assembles the lateral stiffness matrix of the whole structure. The program calculates the lateral seismic loads as required by the Code when the equivalent lateral load method is selected by the user. If a dynamic analysis is selected, the program calculates the mode shapes and frequencies for the structure using Jacobi's method, and then by applying the procedure prescribed by the Code, calculates a set of modal equivalent lateral loads. The program calculates the lateral displacements for the structure as a whole using a Gaussian frontal algorithm. Then, the lateral displacements for each of the frames or walls are evaluated, verifying compliance with the drift requirements. The output forces on the elements are suppressed if the drift requirements are not met. The program then calculates the internal forces for the frames or walls. Forces are evaluated at the face of the joint when rigid connections are selected by the user in the input data.

The program requires the following input data:

(a) General information for seismic analysis. Spectral values (A_a, A_v), Soil Profile Type (S_1, S_2, or S_3), Use Group (I, II, or III), response modification factor (R), and displacement amplification coefficient (C_d).

(b) General information on building. The total number of stories of the building, the number of different frames, and the total number of frames.

(c) Information on stories. The following information must be given for each of the diaphragms of the building from top to bottom: number or name of the story; interstory height in meters (measured from the top of the slab of the floor immediately below to the top of the slab of the story); weight of the building at that level, *w*, in metric tonnes; coordinates of the mass center of the diaphragm in meters; area of the diaphragm in square meters; and polar moment of inertia J_0 in m^4 of the diaphragm at the center of mass. The translational mass is obtained as $M_t = w/g$ where *g* is the acceleration of gravity. The polar moment of inertia, $J_0 = I_{xx} + I_{yy}$ is used to obtain the rotational mass moment of inertia of the diaphragm, $M_r = J_0$ (M_t/Area).

(d) Information on frames. Frame types are defined sequentially starting from the first. When two or more frames are identical, the information defining the frame must be given only once. The frames must be planar. The first story is the first diaphragm of the building and the first column line is the one located to the left. The frame can have fewer stories than the building, but some of the frames must reach the top diaphragm in order to have stiffness at that level. The information to be given is the following:

• **General information on frame.** Number of stories in this frame; number of column lines (a wall is a one-column-line frame); number of column section properties, number of beam section properties; and the selection (or not) of rigid connections at the joints of the frame (dimensions of the rigid zones are calculated by the program from the dimensions of the elements that are connected at the joint).

• **Distance between column lines.** The distances in meters between column lines from left to right correspond to the spans of the beams at each bay.

• **Column section properties.** Column sections are numbered sequentially from one. For each section type, the following information must be given: height (*H*) of the section in meters measured in the plane of the frame; width (*B*) of the section in meters measured in a direction perpendicular to the plane of the frame; Young's modulus (*E*) in tonne/m^2; effective inertia coefficient I_{eff}, a constant used by the program to obtain the moment of inertia of the section to be used as $I_{eff}H^3B/12$); effective area coefficient A_{eff}, a constant used to obtain the section area to be used as $A_{eff}HB$.

• **Beam section properties.** Beam sections are numbered sequentially from one. For each section type, the following information must be given: height (*H*) of the section in meters measured in the plane of

the frame; width (*B*) of the section in meters measured in a direction perpendicular to the plane of the frame; Young's modulus (*E*) in tonne/m^2; and effective inertia coefficient I_{eff}, a constant used by the program to obtain the moment of inertia of the section to be used as $I_{eff}H^3B/12$.

• **Definition of column sections.** For each column line, from top story to bottom and from left to right, the number of the column section type to be used must be given. In the case where no column exists at that story, a column section type equal to zero must be given.

• **Definition of beam sections.** For each bay, from top story to bottom and from left to right, the number of the beam section type to be used must be given. In the case where no beam exists at that story, a beam section type equal to zero must be given.

(e) Information on location of frames. For each of the total number of frames given at the beginning, the following information must be provided sequentially starting from one: name of the frame; frame type number; coordinates of the first and last column lines (X_a, Y_a) and (X_b, Y_b); and definition of output of the forces in the elements for this frame. When the frame is a resisting wall, the coordinates of the edges must be given as (X_a, Y_a) and (X_b, Y_b).

12.11 EXAMPLES USING THE COMPUTER PROGRAM

Example 12.2

A dynamic analysis is performed for the building used in Example 12.1. The following results are obtained (Selected output):

```
* * * * * * * * * * * * * * * * * * * * * * * * * * * * * * * * *
*                                                               *
*  E A R T H Q U A K E      R E S I S T A N T    D E S I G N    *
*                                                               *
*     PROGRAM BASED ON THE COLOMBIAN SEISMIC CODE CCCSR-84      *
*                                                               *
*    PROGRAM FOR THE THREE DIMENSIONAL ANALYSIS OF BUILDINGS    *
*           COMPOSED OF PLANAR FRAMES AND WALLS.                *
*                                                               *
*  PROGRAMMED BY: LUIS E. GARCIA            JANUARY 1992         *
* * * * * * * * * * * * * * * * * * * * * * * * * * * * * * * * *

DATE ---->01-08-1992
TIME ---->14:32:13

PATH OF FILES ------>A:\DATA\EXAMPLE-2
NAME OF DATA ------->
=================================================================

INFORMATION FOR SEISMIC ANALYSIS
********************************

    VALUE OF Aa                           =    0.15
    VALUE OF Av                           =    0.20
    SOIL PROFILE TYPE                     =    S3
      Soil Profile Type S1 (S=1.0)
      Soil Profile Type S2 (S=1.2)
      Soil Profile Type S3 (S=1.5)
    USE GROUP                             =    1
      Use Group III (I=1.2)
      Use Group II (I=1.1)
      Use Group  I (I=1.0)
    RESPONSE MODIFICATION FACTOR - R      =    4.00
    DISPLACEMENT AMPLIFICATION FACTOR - Cd =   3.50
    EQUATION FOR CALCULATION OF Ta        =    1
      Ta=0.08hn^.75  (for frames) = 1
      Ta=0.09hn/L^.5 (for walls)  = 2
    TYPE OF ANALYSIS                      =    2
      Equivalent Lateral Load = 1
      Modal Analysis          = 2
=================================================================
```

```
GENERAL INFORMATION OF THE BUILDING
***********************************

   TOTAL NUMBER OF STORIES OF THE BUILDING  =    6
   NUMBER OF TYPES OF FRAMES                =    2
   TOTAL NUMBER OF FRAMES                   =    8


BUILDING STORY INFORMATION:
***************************

   STORY      HEIGHT    WEIGHT     Xo       Yo      AREA       Jo
               [m]      [ton]     [m]      [m]     [m^2]      [m^4]


   6 ROOF      3.000    152.60    0.000    0.000   281.52   13434.37
   5 5th       3.000    229.83    0.000    0.000   281.52   13434.37
   4 4th       3.000    229.83    0.000    0.000   281.52   13434.37
   3 3rd       3.000    229.83    0.000    0.000   281.52   13434.37
   2 2nd       3.000    229.83    0.000    0.000   281.52   13434.37
   1 1st       4.000    229.83    0.000    0.000   281.52   13434.37
===========================================================================
INFORMATION OF THE DIFFERENT FRAMES:
************************************

FRAME TYPE No. ---------> 1
***************************

   No. of Stories                =    6
   No. of Column Lines           =    4
   No. of Column Properties      =    1
   No. of Beam Properties        =    1
   Use of Rigid Zones            =    1
       Use Rigid Zones -----------> 1
       Do Not Use Rigid Zones ----> 2

DISTANCE BETWEEN COLUMN LINES
   BAY  From Col   To Col    Span [m]
    1      1         2        6.000
    2      2         3        6.000
    3      3         4        6.000

COLUMN SECTION PROPERTIES
  Col Type    H        B         E          Ieff    Aeff
             [m]      [m]    [ton/m^2]
     1       0.400    0.350  2200000.00    1.000   1.000

BEAM SECTION PROPERTIES
  Beam Type   H        B         E          Ieff
             [m]      [m]    [ton/m^2]
     1        0.450    0.300  2200000.00    1.000

DEFINITION OF THE COLUMN SECTIONS
          COLUMN LINE
  STO   1   2   3   4

   6    1   1   1   1
   5    1   1   1   1
   4    1   1   1   1
   3    1   1   1   1
   2    1   1   1   1
   1    1   1   1   1

DEFINITION OF THE BEAM SECTIONS
          BAY LINE
  STO   1   2   3

   6    1   1   1
   5    1   1   1
   4    1   1   1
   3    1   1   1
   2    1   1   1
   1    1   1   1
```

```
DESIGN STORY SHEAR AND LATERAL LOAD
***********************************
(Square root of the sum of the squares)
   ------------------ CASE 1 - (Direction X) ----------------
     STORY       SHEAR      DEFLECTION     DRIFT     LAT.LOAD
                 [ton]         [m]          [m]       [ton]

       6         20.64       0.0414        0.0022     20.64
       5         48.55       0.0393        0.0042     27.91
       4         72.39       0.0352        0.0062     23.84
       3         91.87       0.0291        0.0078     19.48
       2        106.64       0.0213        0.0091     14.77
       1        115.68       0.0122        0.0122      9.03


   ------------------ CASE 2 - (Direction Y) ----------------
     STORY       SHEAR      DEFLECTION     DRIFT     LAT.LOAD
                 [ton]         [m]          [m]       [ton]

       6         19.71       0.0363        0.0017     19.71
       5         46.95       0.0346        0.0034     27.25
       4         70.63       0.0313        0.0050     23.67
       3         90.29       0.0264        0.0063     19.67
       2        105.62       0.0201        0.0075     15.32
       1        115.68       0.0125        0.0125     10.06

   ------ CASE 3 - (Rotational) ------
     STORY     ACCUMULATED      MOMENT
               MOMENT [ton-m]   [ton-m]

       6          0.00          0.00
       5          0.00          0.00
       4          0.00          0.00
       3          0.00          0.00
       2          0.00          0.00
       1          0.00          0.00

===========================================================================


LATERAL DISPLACEMENTS
*********************

   STORY   DIREC.    CASE 1      CASE 2      CASE 3
                      [m]         [m]         [m]

   ROOF      X      0.041822    0.000000    0.000000
             Y      0.000000    0.036555    0.000000
            ROT     0.000000    0.000000    0.000000

   5th       X      0.039638    0.000000    0.000000
             Y      0.000000    0.034849    0.000000
            ROT     0.000000    0.000000    0.000000

   4th       X      0.035412    0.000000    0.000000
             Y      0.000000    0.031441    0.000000
            ROT     0.000000    0.000000    0.000000

   3rd       X      0.029210    0.000000    0.000000
             Y      0.000000    0.026431    0.000000
            ROT     0.000000    0.000000    0.000000

   2nd       X      0.021374    0.000000    0.000000
             Y      0.000000    0.020093    0.000000
            ROT     0.000000    0.000000    0.000000

   1st       X      0.012208    0.000000    0.000000
             Y      0.000000    0.012534    0.000000
            ROT     0.000000    0.000000    0.000000
===========================================================================
```

```
-----------------------------------------------------------
FRAME TYPE No. ---------> 2
***************************
   No. of Stories                =    6
   No. of Column Lines           =    4
   No. of Column Properties      =    1
   No. of Beam Properties        =    1
   Use of Rigid Zones            =    1
       Use Rigid Zones -----------> 1
       Do Not Use Rigid Zones ----> 2
-----------------------------------------------------------
MODES AND FREQUENCIES*********************
MODE   PERIOD       FREQUENCY
        [sec]       [rad/sec]

   1    1.151         5.461
   2    1.094         5.746
   3    0.881         7.132
   4    0.365        17.197
   5    0.351        17.909
   6    0.282        22.299
   7    0.202        31.037
   8    0.198        31.694
   9    0.158        39.744
  10    0.136        46.317
  11    0.134        47.034
  12    0.107        58.925
  13    0.104        60.548
  14    0.098        64.299
  15    0.088        71.717
  16    0.080        78.502
  17    0.079        79.432
  18    0.066        94.763


DESIGN BASE SHEAR (SRSS) DIRECTION X    =     95.51
APPROXIMATE PERIOD (Ta x [sec])         =      0.7280
BASE SHEAR FOR PERIOD = 1.4 Ta x        =    115.68
BASE SHEAR FOR EQUIVENT LATERAL LOAD    =    122.04
CORRECTION FACTOR X DIRECTION           =      1.2111

DESIGN BASE SHEAR (SRSS) DIRECTION Y    =    100.77
APPROXIMATE PERIOD (Ta y [sec])         =      0.7280
BASE SHEAR FOR PERIOD = 1.4 Ta y        =    115.68
BASE SHEAR FOR EQUIVALENT LATERAL LOAD  =    122.04
CORRECTION FACTOR Y DIRECTION           =      1.1480
```

```
FORCES IN THE ELEMENTS OF THE FRAMES SELECTED FOR OUTPUT
*******************************************************
-----------------------------------------------------------
FRAME No.   5  LINE-A         (FRAME TYPE 2)
************

STORY    1    1st
*********
   ------------------ COLUMN FORCES -------------------
   COL CASE MOM BOT    MOM TOP      AXIAL      SHEAR
            [ton-m]    [ton-m]      [ton]      [ton]
   --------------------------------------------------------
    1   1    0.000      0.000       0.00        0.00
    1   2  -13.297     -9.529      21.31        6.43
    1   3    0.000      0.000       0.00        0.00
   --------------------------------------------------------
    2   1    0.000      0.000       0.00        0.00
    2   2  -15.364    -13.142      -2.29        8.03
    2   3    0.000      0.000       0.00        0.00
   --------------------------------------------------------
    3   1    0.000      0.000       0.00        0.00
    3   2  -15.364    -13.142       2.29        8.03
    3   3    0.000      0.000       0.00        0.00
   --------------------------------------------------------
    4   1    0.000      0.000       0.00        0.00
    4   2  -13.297     -9.529     -21.31        6.43
    4   3    0.000      0.000       0.00        0.00
   --------------------------------------------------------

   -------------------- BEAM FORCES -------------------
   BAY CASE LEFT MOM   LEFT SHEAR  RIGHT MOM  RIGHT SHEAR
            [ton-m]     [ton]       [ton-m]     [ton]
   --------------------------------------------------------
    1   1    0.000      0.00        0.000       0.00
    1   2   16.424     -6.52       13.916       6.52
    1   3    0.000      0.00        0.000       0.00
   --------------------------------------------------------
    2   1    0.000      0.00        0.000       0.00
    2   2   11.727     -5.04       11.727       5.04
    2   3    0.000      0.00        0.000       0.00
   --------------------------------------------------------
    3   1    0.000      0.00        0.000       0.00
    3   2   13.916     -6.52       16.424       6.52
    3   3    0.000      0.00        0.000       0.00
   --------------------------------------------------------
```

Example 12.3

A dynamic analysis is performed on an irregular, nine-story building, shown in Fig. 12.6. The structural system is a moment-resisting reinforced concrete space frame. The slabs consist of one-way joists. The building is located in the city of Medellín. The results are as follows (Selected output):

```
* * * * * * * * * * * * * * * * * * * * * * * * * * * * *
*                                                       *
*   E A R T H Q U A K E    R E S I S T A N T    D E S I G N  *
*                                                       *
*     PROGRAM BASED ON THE COLOMBIAN SEISMIC CODE CCCSR-84  *
*                                                       *
*     PROGRAM FOR THE THREE DIMENSIONAL ANALYSIS OF BUILDINGS *
*         COMPOSED OF PLANAR FRAMES AND WALLS.          *
*                                                       *
*   PROGRAMMED BY: LUIS E. GARCIA           JANUARY 1992  *
* * * * * * * * * * * * * * * * * * * * * * * * * * * * *

DATE ---->01-08-1992
TIME ---->14:33:24
PATH OF FILES ------>A:\DATA\EXAMPLE-3
NAME OF DATA ------->
=======================================================================

INFORMATION FOR SEISMIC ANALYSIS
********************************

   VALUE OF Aa                           =     0.15
   VALUE OF Av                           =     0.20
   SOIL PROFILE TYPE                     =     S2
     Soil Profile Type S1 (S=1.0)
     Soil Profile Type S2 (S=1.2)
     Soil Profile Type S3 (S=1.5)
   USE GROUP                             =     1
     Use Group III (I=1.2)
     Use Group  II (I=1.1)
     Use Group   I (I=1.0)
   RESPONSE MODIFICATION FACTOR - R      =     4.00
   DEFLECTION AMPLIFICATION FACTOR - Cd  =     3.50
   EQUATION FOR CALCULATION OF Ta        =     1
   Ta=0.08hn^.75  (for frames) = 1
   Ta=0.09hn/L^.5 (for walls)  = 2
   TYPE OF ANALYSIS                      =     2
     Equivalent Lateral Load = 1
     Modal Analysis          = 2
=======================================================================

GENERAL INFORMATION OF THE BUILDING
***********************************

   TOTAL NUMBER OF STORIES OF THE BUILDING =   9
   NUMBER OF TYPES OF FRAMES               =   4
   TOTAL NUMBER OF FRAMES                  =   7


BUILDING STORY INFORMATION

   STORY    HEIGHT   WEIGHT    Xo      Yo     AREA       Jo
            [m]      [ton]     [m]     [m]    [m^2]     [m^4]
   9 ROOF   3.000    100.80   6.000   7.000  168.00    4760.00
   8 8th    3.000    134.40   6.000   7.000  168.00    4760.00
   7 7th    3.000    134.40   6.000   7.000  168.00    4760.00
   6 6th    3.000    134.40   6.000   7.000  168.00    4760.00
   5 5th    3.000    134.40   6.000   7.000  168.00    4760.00
   4 4th    3.000    201.60   9.000   7.000  252.00   10920.00
   3 3rd    3.000    201.60   9.000   7.000  252.00   10920.00
   2 2nd    3.000    201.60   9.000   7.000  252.00   10920.00
   1 1st    4.000    201.60   9.000   7.000  252.00   10920.00
=======================================================================

PERIODS AND FREQUENCIES (MODES NOT PRINTED)
***********************************

MODE  PERIOD   FREQUENCY
      [sec]    [rad/sec]

  1   1.286      4.887
  2   0.977      6.428
  3   0.726      8.653
  4   0.476     13.187
  5   0.383     16.419
  6   0.316     19.857
  7   0.247     25.439
  8   0.210     29.936
  9   0.172     36.554
 10   0.158     39.787
 11   0.137     45.935
 12   0.124     50.817
 13   0.105     60.074
 14   0.102     61.312
 15   0.089     70.314
 16   0.079     79.897
 17   0.077     81.367
 18   0.072     86.895
 19   0.059    106.649
 20   0.058    108.288
 21   0.057    111.098
 22   0.048    130.152
 23   0.045    139.436
 24   0.044    144.352
 25   0.035    177.231
 26   0.034    185.401
 27   0.027    234.205
```

```
DESIGN BASE SHEAR (SRSS) DIRECTION X   =     81.22
APPROXIMATE PERIOD (Ta x [sec])        =     0.9738
BASE SHEAR FOR PERIOD = 1.4 Ta x       =     84.61
BASE SHEAR FOR EQUIVALENT LATERAL LOAD =    105.89
CORRECTION FACTOR X DIRECTION          =     1.0417

DESIGN BASE SHEAR (SRSS) DIRECTION Y   =     68.82
APPROXIMATE PERIOD (Ta y [sec])        =     0.9738
BASE SHEAR FOR PERIOD = 1.4 Ta y       =     84.61
BASE SHEAR FOR EQUIVALENT LATERAL LOAD =    105.89
CORRECTION FACTOR Y DIRECTION          =     1.2293

DESIGN STORY SHEAR AND LATERAL LOAD
***********************************
(Square root of the sum of the squares)

------------- CASE 1 - (Direction X) ----------------
  STORY     SHEAR     DISPLACEMENT    DRIFT     LAT.LOAD
            [ton]         [m]          [m]       [ton]

   9        13.87       0.0264       0.0017      13.87
   8        28.73       0.0249       0.0030      14.86
   7        40.61       0.0221       0.0033      11.88
   6        50.28       0.0191       0.0038       9.68
   5        57.96       0.0154       0.0038       7.68
   4        67.41       0.0117       0.0030       9.45
   3        75.45       0.0087       0.0031       8.04
   2        81.39       0.0056       0.0030       5.94
   1        84.61       0.0026       0.0026       3.22


--------------- CASE 2 - (Direction Y) --------------
  STORY     SHEAR     DISPLACEMENT    DRIFT     LAT.LOAD
            [ton]         [m]          [m]       [ton]

   9        13.58       0.0434       0.0023      13.58
   8        28.20       0.0414       0.0040      14.62
   7        40.01       0.0378       0.0048      11.81
   6        49.76       0.0333       0.0057       9.75
   5        57.58       0.0278       0.0071       7.81
   4        67.03       0.0209       0.0054       9.45
   3        75.16       0.0156       0.0056       8.13
   2        81.27       0.0100       0.0054       6.11
   1        84.61       0.0046       0.0046       3.34

------ CASE 3 - (Rotational) ------
  STORY    ACCUMULATED        MOMENT
           MOMENT [ton-m]     [ton-m]

   9        20.99             20.99
   8        40.07             19.08
   7        54.12             14.05
   6        66.49             12.37
   5        76.24              9.75
   4        91.84             15.60
   3       103.72             11.88
   2       113.12              9.40
   1       119.18              6.06
=======================================================================
```

APPENDIX A12.   RECENT EARTHQUAKES IN COLOMBIA

November 23, 1979 Earthquake

This $6.4M_s$ magnitude earthquake (García and Sarria 1980) had coordinates of lat. 4°30′N and long. 75°54′W and a depth of 80 km. It affected the western-central part of the country, specifically the coffee-growing region. It caused 37 deaths and 493 injuries. There was extensive structural damage including the collapse of several four-to-six-story buildings. The cities of Manizales (population 280,000), Pereira (population 370,000), and Armenia (population 280,000) particularly were affected. Well-designed and -constructed buildings behaved very well; they suffered minor or no damage.

December 12, 1979 Earthquake

This event (García and Sarria 1980) occurred in the Pacific Ocean (lat. 2°30′N and long. 79°30′W) close to the Colombian city of Tumaco (population 35,000). The focal depth was 40 km; the magnitude M_s was 7.8. In spite of the fact that the Pacific coast of Colombia is sparsely inhabited, the combination of tsunami and earthquake caused 453 deaths and 1,047 injuries. Because of the low degree of development of the area, few engineered structures were affected, although

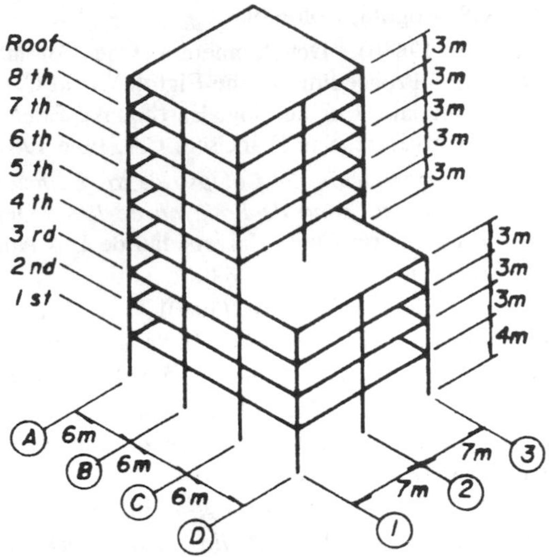

(a)

COLUMN SIZE

	LINES 1,2,3 X LINES A,B,C		LINES 1,2,3 X LINE D	
	H(m)	B(m)	H(m)	B(m)
Roof	0.40 x	0.40		
8 th	0.40 x	0.40		
7 th	0.60 x	0.40		
6 th	0.60 x	0.40		
5 th	0.60 x	0.40		
4 th	1.00 x	0.40	0.40 x	0.40
3 rd	1.00 x	0.40	0.40 x	0.40
2nd	1.00 x	0.40	0.40 x	0.40
1 st	1.00 x	0.40	0.40 x	0.40

BEAM SIZE

LINES 1, 2, 3

H(m) = 0.50   B(m) = 0.40

LINES A, B, C, D

H(m) = 0.50   B(m) = 0.30

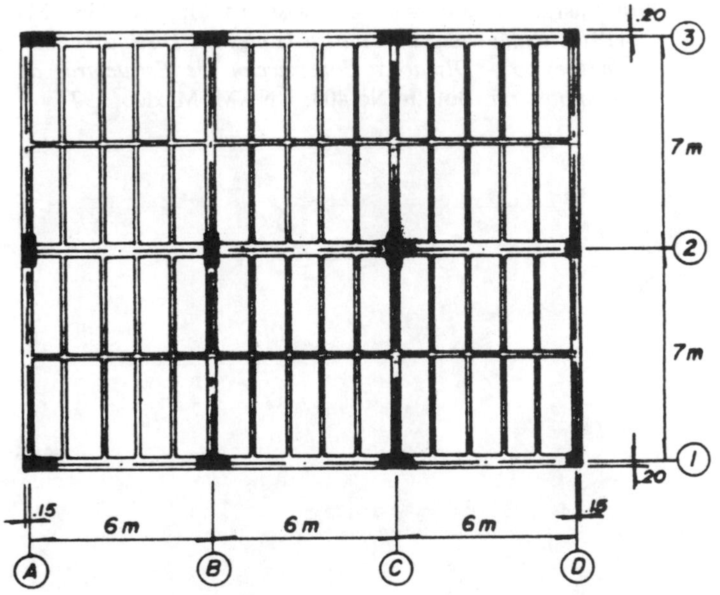

STORY WEIGHT

Roof	100.8 ton.
8 th	134.4 ton.
7 th	134.4 ton.
6 th	134.4 ton.
5 th	134.4 ton.
4 th	201.6 ton.
3 rd	201.6 ton.
2 nd	201.6 ton.
1 st	201.6 ton.

TYPICAL FLOOR PLAN

Fig. 12.6. Nine-story irregular framed building for Example 12.3

some damage was reported in the city of Cali (population 1,200,000), located 280 km from the epicenter.

March 31, 1983 Popayán Earthquake

This event (García and Sarria 1983) destroyed one of the most beautiful cities in Colombia, which contained some of the best examples of colonial architecture in the country. The earthquake had a $5.5M_s$ magnitude and the epicenter was located 10 km from the city, at a focal depth of 5 km. It caused 250 deaths and 1,508 injuries in a city with a population of 220,000, and affected all types of construction (García and Sarria 1983). The old colonial part of the city was damaged beyond repair. The earthquake caused extensive damage and collapse of several three-to-five-story buildings that were engineered (García 1984b, 1986).

REFERENCES

AMERICAN CONCRETE INSTITUTE (ACI) (1981), *Building Code Requirements for Concrete Masonry Structures (ACI 531-81)*. ACI, Detroit, MI.

—— (1983) *Building Code Requirements for Reinforced Concrete (ACI 318-83)*. ACI, Detroit, MI.

AMERICAN INSTITUTE OF STEEL CONSTRUCTION (AISC) (1978) *Specifications for the Design, Fabrication and Erection of Structural Steel for Buildings*. AISC, Chicago, IL.

APPLIED TECHNOLOGY COUNCIL (1978) *Tentative Provisions for the Development of Seismic Regulations for Buildings (ATC-3-06)* (Spanish Translation by AIS, Bogotá), ATC, Palo Alto, CA.

ASOCIACIÓN COLOMBIANA DE INGENIERÍA SÍSMICA (AIS) (1981) *Requisitos Sísmicos para Edificios (NORMA AIS 100-81)*. AIS, Bogotá, Colombia.

GARCÍA, L. E. (1984b) "Development of the Colombian Seismic Code." Proceedings of the Eighth World Conference on Earthquake Engineering, Earthquake Engineering Research Institute, San Francisco, CA, June 1984.

—— (1986) *Interpretación del Colapso de los Edificios de Pubenza Durante el Sismo de Popayan de 1983*. Departamento de Ingeniería Civil, Universidad de Los Andes, Bogotá, Colombia, Agosto de 1986.

GARCÍA, L. E., and SARRIA, A. (1980) "Los Temblores Colombianos de 1979." *Revista Análes de Ingeniería*, Sociedad Colombiana de Ingenieros, Bogotá, Colombia.

—— (1983) "The March 31, 1983 Popayán Earthquake – Preliminary Report." *Earthquake Engineering Research Institute Newsletter*, Palo Alto, CA.

GARCÍA, L. E.; SARRIA, A.; ESPINOSA, A.; BERNAL, C. E.; and PUCCINI, M. (1984a) *Estudio General del Riesgo Sismico de Colombia*. Asociación Colombiana de Ingeniería Sísmica, Bogotá, Colombia.

MINISTERIO DE OBRAS PÚBLICAS Y TRANSPORTE (1984) *(CCCSR-84), Código Colombiano de Construcciones Sismo Resistentes, Decreto 1400 de Junio 7 de 1984*. Bogotá, Colombia, 1984.

SEISMOLOGY COMMITTEE (1974) *Recommended Lateral Force Requirements and Commentary*. (Spanish translation by AIS, Bogotá.) Structural Engineers Association of California, San Francisco, CA.

UNIVERSIDAD NACIONAL AUTÓNOMA DE MÉXICO (UNAM) (1977) *Reglamento de Construcciones del Distrito Federal de Mexico – Diseño y Construccion de Estructuras de Mamposteria*. Boletín No.403, UNAM, México, 1977.

13

Costa Rica

José L. Barzuna

13.1 INTRODUCTION

Costa Rica is in Central America on the Caribbean tectonic plate. The western coast of the country is a region of subduction where the interaction of three tectonic plates (the Caribbean, the Cocos, and the Nazca Plates) creates a zone of intense seismicity. Also, the country is crossed by a series of local faults that contribute further to the seismicity.

The mitigation of hazards caused by construction inadequate to withstand strong earthquakes has been a major concern since very early in the history of Costa Rica. As early as 1841, then-President Braulio Carrillo decreed the first building code of Costa Rica, which adopted an empirical approach to ensuring the quality of construction. President Cleto Gonzalez Viquez (1906–1910) took a personal interest in the matter by having all available information on earthquakes in Costa Rica compiled into a single volume. This work was one of the first seismic reports in the world, and preserved valuable information that otherwise would have been lost. President Ricardo Jiménez (1910–1914) approved a new building code for the zone of Cartago following the major earthquake in that city in 1910. The code was used widely at the time by practicing engineers and contractors throughout the country. This code prohibited the use of adobe and regulated the use of *bahareque* (walls of wooden sticks interwoven with cane and covered with mud) and unreinforced masonry because of the poor performance of buildings made of these materials during earthquakes. One of the major reasons why Costa Rica has experienced a relatively low death rate in recent major earthquakes is that there are fewer of these types of buildings in comparison with other Latin American countries. The engineering community realized the need for a modern seismic code as a result of the 1972 earthquake that originated in Managua, Nicaragua, which is north of Costa Rica.

The first edition of the modern Costa Rican Seismic Code (*Código Sísmico de Costa Rica*) was written by Jorge Gutiérrez G. and published by the Colegio Federado de Ingenieros y Arquitectos (1974). This Code was revised and approved by the Comisión Permanente de Estudio y Revisión del Código Sísmico de Costa Rica. This special commission was created by the national engineering regulatory agency to provide the country with a seismic code that they would keep current. In 1986, a new edition of the Code was adopted; it incorporated 12 years of experience in the use of the first edition of the Code and the worldwide technological advances applied to the customary methods of construction in Costa Rica.

Since 1982, Costa Rica has been immersed in a rather intensive seismic cycle that has included four major earthquakes and many small to medium quakes. The structures designed and constructed to the 1974 Code standards have performed very well. However, it is necessary to revise the 1986 seismic risk maps because a $7.4M_s$[1] scale event occurred in a region previously assumed to be a low-seismicity zone.

The latest version of the seismic code of Costa Rica, *Código Sísmico de Costa Rica, 1986*, is written in a simple, self-explanatory format, in contrast to other Codes that are more legalistic. This Code attempts to guide the engineer through the design phase, encouraging the development of structures that will perform well under seismic excitation. Authors of the Code have recognized that in zones of high seismicity, factors such as symmetry and regularity are more important than other aspects of design.

The Code is organized into three major parts. The first part considers those matters that pertain to all types of structures: Code philosophy, common definitions, handling of special situations not covered by the general body of the Code, general lateral load application criteria, site selection criteria, minimum strength criteria for the construction phase, seismic zoning and isoacceleration maps for different risk periods, and selection criteria for the risk period.

The second part of the Code is divided into five main sections and 17 subsections. Subsections 1 and 2: general aspects of the analysis, design and construction of buildings, and local soil characteristics as they affect seismic response; Subsections 3 through 10: analysis, including such aspects as structure classification, seismic coefficient, loading and load participation factors, static method of analysis, dynamic method of analysis, displacement and deformation computation procedure and limits, general foundation requirements, and lateral load analysis criteria and proce-

dures for systems and components; Subsections 11 through 15: design requirements for specific types of material, including concrete, masonry, steel, wood, and prefabricated structures and components; Subsection 16: remodeling, repairing, and instrumentation; Subsection 17: documentation of the structural design in the blueprints for the building.

The third major part provides a simplified design approach for one- and two-story residential dwellings; it also includes some general guidelines for the detailing and the constructing of such buildings.

13.2 CODE PHILOSOPHY

The Seismic Code of Costa Rica seeks to fulfill two main objectives: (1) protecting life; and (2) mitigating economic and social costs associated with the occurrence of a medium-to-strong earthquake. Structures designed under the provisions of this Code are expected to suffer little or no damage when subjected to minor or moderate earthquakes that might occur during the life of the structure, and not collapse during a major earthquake; significant but repairable structural damage is expected as a result of a major quake. Structures and facilities that must remain operable throughout a major earthquake are protected by special provisions relating to risk period, importance factor, and lateral displacement limitation. The critical aspects of the behavior of structures under earthquake loading are the resultant displacements and deformations. The Code anticipates the occurrence of deformations beyond the elastic response range, and allows permanent deformation as long as the proper provisions are taken to ensure that the structure will not suffer a significant degradation in strength due to the formation of hinges. The Code provides only minimum requirements for analysis and design of structures under normal circumstances. These minimum requirements must be increased if the designer foresees the occurrence of conditions more severe than those anticipated by the Code.

13.3 LATERAL FORCE SYSTEM

For the purpose of analysis and design, the Code establishes that the mass of the building may be considered to be lumped at discrete points, each located at the center of gravity of each floor level of multistory buildings. The mass for each story is calculated as the mass of the elements contained

---

[1]M_s is a scale for measuring the magnitude of earthquakes. Specifically, the M_s (surface wave) magnitude is proportional to the source spectral amplitude of the 20-seconds period wave (see Appendix in this book).

Table 13.1. Live Load Seismic Participation Factor

Structure Type	Factor
Warehouses	0.25
General buildings	0.15
Roof structures	0.00

between the mid-height of the story below and the mid-height of the story above the floor under consideration, which are attached in such a way as to participate with the mass of that floor. For the purpose of mass computation, it is necessary to include the masses of the building elements and any permanent architectural, mechanical, or electrical equipment and systems anchored to the structure to resist seismic forces, and the mass of contained liquids assuming that they are full. Likewise, it is necessary to include a fraction of the live load, which varies according to the use of the structure as shown in Table 13.1.

If the floors of the structure are flexible, the Code requires that each member of the lateral-load resisting system be loaded by its tributary mass in the two main orthogonal directions for the building. On the other hand, if the floors are rigid, then the loads are to be applied horizontally in each of the two principal inertial directions of the building, or in the least favorable orthogonal directions. The analysis for each direction is to be performed independently; then the effects on the structural elements of the vector sum of the full force in one direction, plus 30% of the force in the other direction, must be considered. This requirement must be met unless all the resisting elements in the building are oriented in one or two well-defined orthogonal directions. In the latter case, the analysis may be done independently in each direction.

13.4 VERTICAL SEISMIC LOADING

Vertical seismic excitation may be disregarded except when the structure has significant flexibility under vertical loading, such as might be the case in flat plates, long-span beams, long cantilever beams, and columns sustaining heavy axial stresses. The Code stipulates that the vertical excitation component may be taken as two-thirds of the horizontal component. The resultant vertical forces are to be combined with the horizontal forces in the manner previously described: the vector sum of forces in one direction, plus 30% of the forces in the other direction, plus 30% of the vertical forces. If, in addition, vertical displacement is determined to be the primary mode of response, then it is necessary to consider the full vertical forces combined with 30% of each of the two horizontal force components.

13.5 LOAD COMBINATION FACTORS

In the calculation of internal forces and structural displacements, the applied loads must be combined and multiplied (or reduced), as indicated in eqs.(13.1) through (13.6), to determine the appropriate forces that the structure must be designed to resist. Equations (13.1) through (13.3) correspond to the ultimate load design method, while eqs.(13.4) through (13.6) refer to the working stress method of design.

$$UL = 1.4DL + 1.7LL \tag{13.1}$$

$$UL = 0.75(1.4DL + 1.7LL) \pm SL \tag{13.2}$$

$$UL = 0.95DL \pm SL \tag{13.3}$$

$$WL = DL + LL \tag{13.4}$$

$$WL = 1.1(DL + LL) \pm SL \tag{13.5}$$

$$WL = 0.95DL \pm SL \tag{13.6}$$

where

UL = ultimate load
WL = working load
DL = permanent or dead load
LL = temporary or live load
SL = seismic load

13.6 SEISMIC ZONES

The determination of the maximum acceleration for the calculation of the seismic coefficient requires the evaluation of the design risk period PR given by the following equation:

$$PR = \frac{1}{1 - (1 - PE)^{1/N}} \tag{13.7}$$

where

PE = probability of exceedance
N = required structure economic life (years)

Four isoacceleration maps are provided in the Code for risk periods of 50, 100, 500 and 1,000 years (Figs. 13.1 through 13.4). The maximum design acceleration can be obtained from the appropriate map or by interpolation between maps (using logarithmic values) for seismic risk periods other than those on which the maps were based.

The determination of the probability of exceedance is entirely up to the design engineer, but general guidelines are provided in the Code as to the appropriate values to use under normal circumstances. The selection criteria are functions of the importance of the structure and the structural type. Acceptable

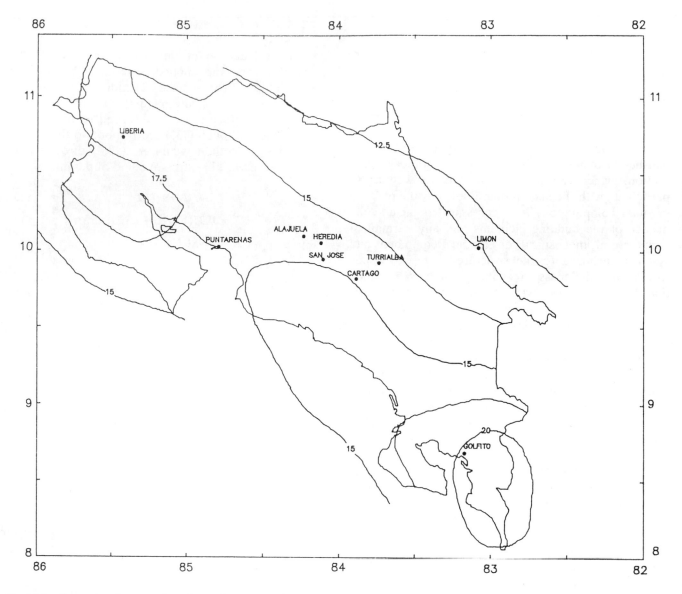

Fig. 13.1. Isoacceleration map given the maximum ground acceleration ($a_{max}\%g$) for 50-year risk period. (Adapted from *Código Sísmico de Costa Rica*, 1986)

values of probability of exceedance range from 0.05, for a very important nonductile structure, to 0.75 for a temporary structure of high ductility, with normal values between 0.20 and 0.45.

13.7 CLASSIFICATION OF STRUCTURES

The Code requires that structures be classified under three different independent criteria: importance of structure, structural type, and regularity of structure.

13.7.1 Importance of Structure

Three levels of importance are established by the Code, as shown in Table 13.2.

The importance of a structure is assessed in consideration of the post-earthquake function it performs, the value of the building contents, the risk of creating a hazard for much of the population of the country, and the social and economic impact of the failure of the structure. The importance factor affects the probability of exceedance used in determining the applied lateral forces, thereby affecting the seismic risk

Table 13.2. Classification by Importance

Importance Level	Classification
High importance	A
Normal importance	B
Temporary or rural uninhabited constructions	C

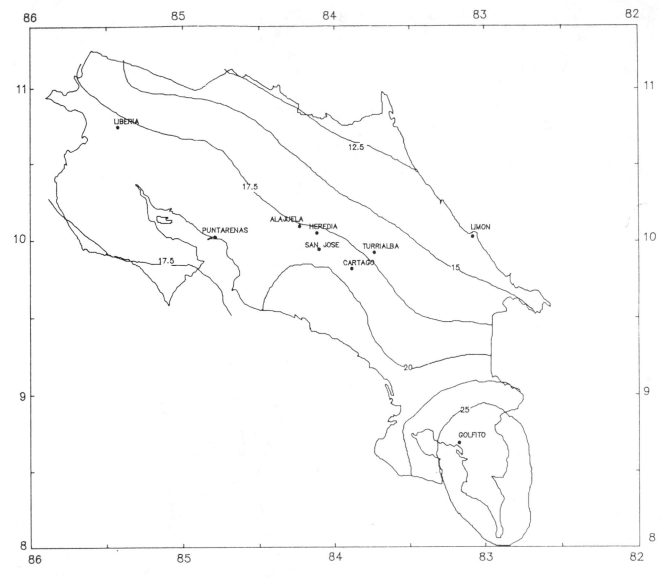

Fig. 13.2.   Isoacceleration map given the maximum ground acceleration ($a_{max}\%g$) for 100-year risk period. (Adapted from *Código Sísmico de Costa Rica*, 1986)

period. The greater the importance of the structure, the longer is the seismic risk period and the lower is the allowable probability of exceedance.

13.7.2 Structural Type

Structures are classified according to their ductility and rigidity into five different categories. The identification of the structural type affects the level of forces that the structure is required to resist elastically at the expense of ductility. Using a flexible structural system will require the structure to accommodate large deformations, which will in turn require special detailing. Using stiffer structures will reduce the ductility requirements, but will require the structure to resist a

higher level of forces (such resistance usually is easy to achieve). The Code intent is that designers select stiffer systems in which deformations will not exceed the limits stipulated by the Code. Table 13.3 shows a compilation of the requirements and characteristics of the five different types of structures identified by the Code.

13.7.3 Regularity

The Code requires that structures meet minimum criteria of regularity in plan and height because of the importance of stiffness distribution and mass distribution to the dynamic behavior of structures. Full

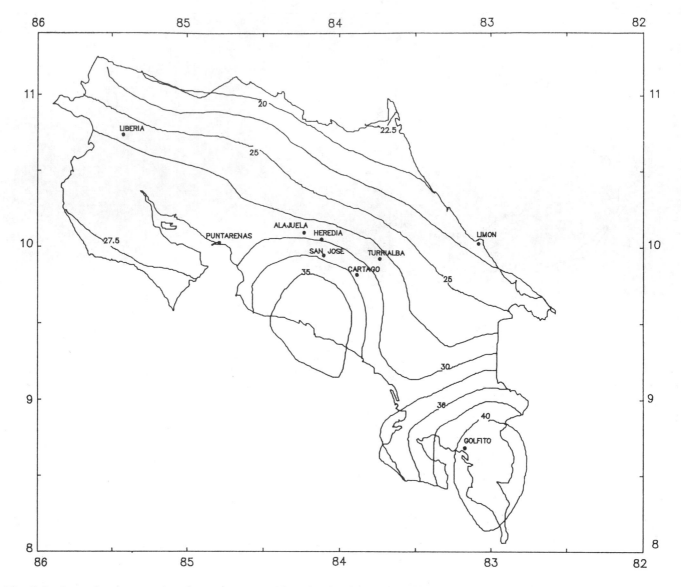

Fig. 13.3. Isoacceleration map given the maximum ground acceleration ($a_{max}\%g$) for 500-year risk period. (Adapted from *Código Sísmico de Costa Rica*, 1986)

compliance with the plan and height regularity requirements will improve the behavior of the structure and reduce the risk of damage.

13.7.3.1 Height regularity.
Six different criteria must be examined to establish height regularity. The intent of these requirements is to provide a nearly equal distribution of mass and stiffness throughout the height of the structure. The requirements are as follows:

1. All structural-resisting systems must be continuous from the foundation to the highest level of the building.

2. The weight of two adjacent stories must not differ by more than 15%, except for the roof level and those stories located in the first 20% of the height of tall buildings.

3. The stiffness of a story must not be less than 50% of the stiffness of the story below. In the computation of the stiffness of a story, the vertical elements may be assumed to be fixed at their upper and lower ends. Appropriate consideration must be given to interrupted floors such as mezzanines in the computation of the stiffness of the elements attached to these levels.

4. The interstory height must not differ from the heights of adjacent stories by more than 20%, except for the first story.

5. The horizontal projection of the centers of mass of all stories must be within the limits of a rectangle defined by 10% of the maximum plan dimensions in each direction.

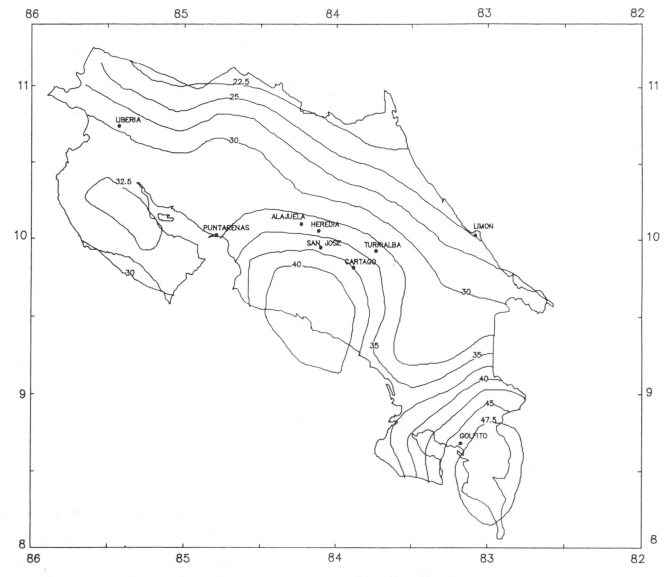

Fig. 13.4. Isoacceleration map given the maximum ground acceleration ($a_{max}\%g$) for 1,000-year risk period. (Adapted from *Código Sísmico de Costa Rica*, 1986)

6. The horizontal projection of the centers of stiffness of all stories must be within the limits of a rectangle defined by 10% of the maximum plan dimensions in each direction.

13.7.3.2 Plan regularity. The intent of these regulations is to avoid torsion in horizontal planes and to provide appropriate torsional stiffness to the structure. The requirements for plan regularity are as follows:

1. At each level of the building, the total eccentricity in each of the principal orthogonal directions e_{xi} and e_{yi} must not exceed 5% of the plan dimension in that direction; that is,

$$\frac{e_{xi}}{D_{xi}} \leq 0.05 \qquad (13.8)$$

$$\frac{e_{yi}}{D_{yi}} \leq 0.05 \qquad (13.9)$$

where

$$e_{xi} = \frac{\sum_j k_{yi,j} x_j}{\sum_j k_{yi,j}} \qquad (13.10)$$

$$e_{yi} = \frac{\sum_j k_{xi,j} y_j}{\sum_j k_{xi,j}} \qquad (13.11)$$

Table 13.3. Characteristics and Requirements of Structures According to Structural Type

Characteristic	Type					
	1	2	3	4	5a	5b,c
Steel	A	A	A	A	A	
Concrete	A	A	A	A	A	
Reinforced masonry	P	P	A	A	A	
Confined masonry	P	P	A	A	A	
Wood	P	P	A	A	A	
Plan regularity	R	R	NR	NR	NR	
Height regularity	R	R	NR	NR	NR	
Moment resisting frames	R	C	C	NA	C	
Braced frames	NA	C	C	C	C	
Concrete shear walls	NA	C	C	C	C	
Masonry shear walls	NA	P	C	C	C	
Description of structural system	DMRF	DMRF and DSW/BR	MRF and SW/BR	BOX	ANY	
Rigid diaphragms	R	R	A	A	A	
Flexible diaphragms	P	P	A	A	A	
Ductility required	6	4	2	1.2	1	
Inelastic disp. factor	6	4	2	1.2	1	
Damping	0.05	0.05	0.07	0.10	0.05	
Max. drift ratio (Level A)	0.010	0.01	0.01	0.008	0.01	0.008
Max. drift ratio (Levels B, C)	0.016	0.014	0.011	0.008	0.016	0.008

NA = Non-Applicable   DMRF = Ductile Moment-Resisting Frame
 A = Acceptable   MRF = Moment-Resisting Frame
 R = Required   DSW = Ductile Shear Wall
NR = Not Required   SW = Shear Wall
 P = Prohibited   BR = Braced Frame
 C = Combinations possible   BOX = Box System
ANY = Any system not complying with the requirements of Types 1 through 4

D_{xi}, D_{yi} = plan dimension in x and y directions

$k_{xi,j}, k_{yi,j}$ = lateral stiffness in x, y directions of vertical elements j at level i

x_j, y_j = x, y distance of the center of mass at level i from the element j.

2. The following inequalities must be satisfied at each level i of the building:

$$\frac{K_{\Theta i}}{K_{xi} r_{ci}^2} \geq 2 \qquad \frac{K_{\Theta i}}{K_{yi} r_{ci}^2} \geq 2 \qquad (13.12)$$

$$K_{xi} = \sum_j k_{xi,j} \qquad K_{yi} = \sum_j k_{yi,j} \qquad (13.13)$$

$$r_{ci} = \sqrt{\frac{I_{ci}}{M_i}} \qquad (13.14)$$

in which

r_{ci} = radius of gyration of level i with respect to its center of mass
M_i = mass at level i
I_{ci} = polar moment of inertia of the mass at level i with respect to its center of mass

$K_{\Theta i}$ = rotational stiffness of level i with respect to the center of mass

$$= \sum_j [k_{\Theta i,j} + (k_{xi,j} y_j^2 + k_{yi,j} x_j^2)] \qquad (13.15)$$

$k_{\Theta i,j}$ = torsional stiffness of vertical element j at level i.

The maximum allowed plan eccentricity for irregular structures is limited to 30% of the plan dimension in any of the principal directions; structures with higher eccentricities are not permitted.

13.8 SOIL EFFECTS

Soil response effects are included in the response spectrum curves [Dynamic Amplification Factor (FAD)] provided in the Code. Three types of soil conditions are considered:

1. Rock (seismic propagation velocity >750 m/sec)
2. Firm soil (dense sands, gravels, stiff clays)
3. Soft soil (layers of sand and clay of medium to loose density or soft to moderate stiffness, more than 10 m deep).

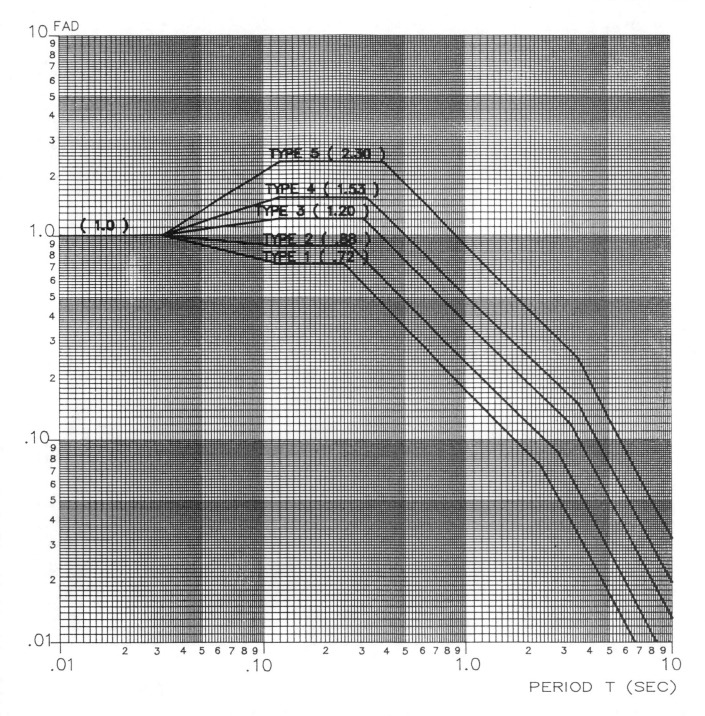

Fig. 13.5. Dynamic amplification factor (FAD) for rock profile. (Adapted from *Código Sísmico de Costa Rica*, 1986)

Three graphs are provided in the code (Figs. 13.5 through 13.7), one for each soil condition. Each graph shows a curve for each type of structure in the Code. The soft soil spectrum is to be used when any doubt exists regarding soil conditions. Other seismic effects that should be considered are liquefaction potential and slope instability, as they might affect foundation stability.

13.9 FUNDAMENTAL PERIOD

The exact fundamental period of vibration of the structure will not be known until the final design is completed. However, the Code recommends that the fundamental period be estimated initially using the expressions shown in Table 13.4 for the purpose of computing the seismic coefficient. Then, when the first

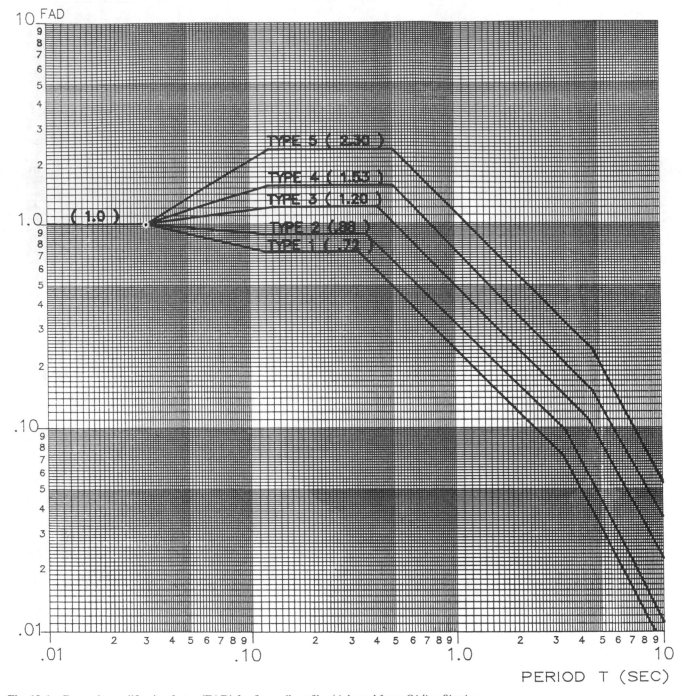

Fig. 13.6. Dynamic amplification factor (FAD) for firm soil profile. (Adapted from *Código Sísmico de Costa Rica*, 1986)

Table 13.4. Preliminary Fundamental Period *T*

Period *T* (sec)	Structural System
$T = 0.12N$*	Steel rigid frames
$T = 0.10N$	Concrete rigid frames
$T = 0.08N$	Combined frames and shear walls or braced frames or masonry walls
$T = 0.05N$	Structures composed exclusively of reinforced concrete walls

*N = Total number of stories

cycle of analysis is completed, and the internal forces and displacements are calculated, the Code requires that the fundamental period be recalculated by the Rayleigh formula:

$$T = 2\pi \sqrt{\frac{\sum\limits_{i=1}^{N} W_i \delta_{ei}^2}{g \sum\limits_{i=1}^{N} F_i \delta_{ei}}} \qquad (13.16)$$

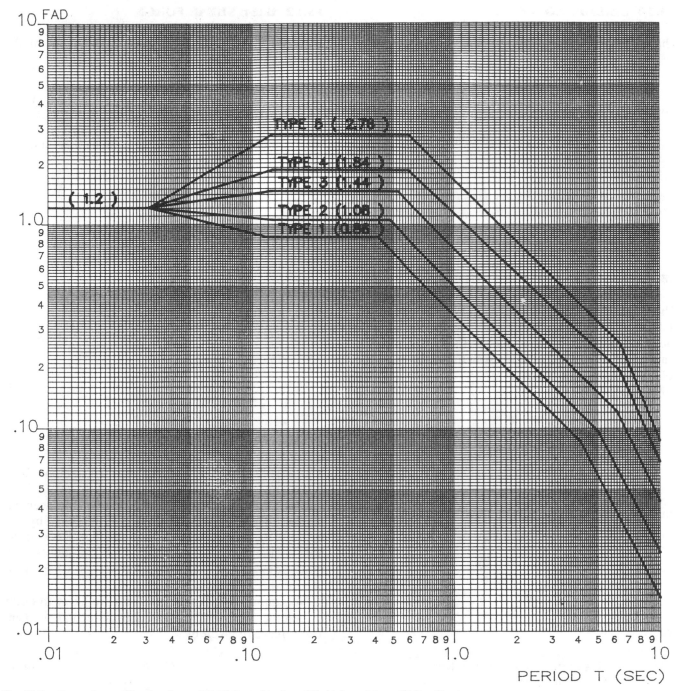

Fig. 13.7. Dynamic amplification factor (FAD) for soft soil profile. (Adapted from *Código Sísmico de Costa Rica*, 1986)

where

δ_{ei} = elastic displacement at level i
F_i = seismic force at level i [eq.(13.18)]

When the new calculated period causes a change in seismic coefficient, the effects must be reduced or increased in the same proportion as the change in this seismic coefficient. If a reduction in effects is appropriate, the Code stipulates that values obtained on the basis of the initially estimated fundamental period cannot be reduced by more than 20%.

To account for the softening effect in reinforced concrete structures due to cracking during strong earthquakes, the Code stipulates that an equivalent moment of inertia I_e be used in the determination of the internal forces and elastic displacements as follows:

$I_e = 1.00 I_{\text{gross}}$ for elements sustaining combined flexural and compressive axial loads
$I_e = 0.50 I_{\text{gross}}$ for flexural elements including slabs.

13.10 SEISMIC COEFFICIENT

The seismic coefficient C is given by

$$C = Ra_{max}\,FAD \qquad (13.17)$$

where

R = spatial acceleration distribution factor, taken as 0.80

a_{max} = maximum ground acceleration (expressed as a fraction of the acceleration of gravity) as obtained from the isoacceleration maps (Figs. 13.1–13.4)

FAD = dynamic amplification factor (Figs. 13.5–13.7).

13.11 LATERAL SEISMIC FORCES

The Code provides two methods for calculating the induced inertia forces at each level: (1) the static method and (2) the dynamic method. The simpler static method may be used for structures that meet the following minimum criteria:

- Height regularity
- Plan regularity
- No more than seven stories or 30 meters high above street level

Structures that do not meet these minimum criteria must be analyzed by the dynamic method.

For the static method, the story lateral load at each level is calculated assuming a distribution of laterally applied seismic forces F_i given by

$$F_i = C\eta h_i W_i \qquad (13.18)$$

where

F_i = seismic force acting on level i

C = seismic coefficient

η = constant factor defined as:

$$\eta = \frac{\sum_{k=1}^{N} W_k h_k}{\sum_{k=1}^{N} W_k h_k^2} \qquad (13.19)$$

where

W_i = design weight at level i

h_i = height of level i with respect to base level

N = total number of stories.

13.12 BASE SHEAR FORCE

The base shear force V is given by the sum of the lateral seismic forces as

$$V = \sum_{i=1}^{N} F_i \qquad (13.20)$$

which, after using eq.(13.18), may be expressed as

$$V = C\,\frac{\left(\sum_{k=1}^{N} W_k h_k\right)^2}{\sum_{k=1}^{N} W_k h_k^2} \qquad (13.21)$$

13.13 STORY SHEAR FORCE

The story shear force V_i for story i is equal to the sum of the lateral seismic forces at level i and above that level; that is,

$$V_i = \sum_{k=i}^{N} F_i \qquad (13.22)$$

13.14 DYNAMIC METHOD

The objective in applying the dynamic method is to determine a set of static forces at each level of the building, for each of the significant displacement modes. These forces are then combined by an appropriate procedure such as the Square Root of the Sum of the Squares (SRSS) or the Complete Quadratic Combination (CQC) described in Chapter 4, Section 4.8.

No significant torsional moments are likely to occur by seismic loading if the building is regular in plan. The forces are given by

$$F_{ij} = C_j \eta_j \phi_{ij} W_i \qquad (13.23)$$

where

F_{ij} = seismic force acting on level i due to mode j

C_j = seismic coefficient corresponding to the period T_j for mode j

ϕ_{ij} = mode shape j at level i (sign included)

W_i = seismic weight at level i

η_j = constant for each mode defined as:

$$\eta_j = \frac{\sum_{k=1}^{N} (W_k \phi_{kj})}{\sum_{k=1}^{N} W_k \phi_{kj}^2} \qquad (13.24)$$

with

W_k = seismic weight at level k

ϕ_{kj} = mode shape j at level k (sign included).

The analysis must include at least a number of modes equal to one-fourth (plus any fraction thereof) of the number of stories for the first eight stories and one additional mode for every five additional stories (or fraction thereof). If the structure does not meet the plan regularity requirements, then it must be analyzed by the dynamic method as a three-dimensional system. This analysis should consider at least three times the number of modes used in the case of regular buildings, with three degrees of freedom at each level (two orthogonal horizontal translational modes and a plan rotational mode).

13.15 MODAL COMBINATION

The internal forces, displacements, and base reactions are calculated independently for each mode. These modal components are then combined using the SRSS method, provided that the modes are well separated. If two or more modes are closely spaced (less than 10% difference), then the modes must be combined by a procedure that considers the closeness of modes, such as the CQC technique as described by Wilson et al. (1981).

13.16 MAXIMUM ALLOWED DRIFT RATIO

The inelastic displacement δ_{ij} at level i corresponding to mode j is calculated as the product of inelastic displacement coefficient **K** (Table 13.5) and the elastic displacement δ_{eij} at that level; that is,

$$\delta_{ij} = \mathbf{K}\delta_{eij} \tag{13.25}$$

where the elastic displacement may be obtained by

$$\delta_{eij} = \frac{gT_j^2 F_{ij}}{4\pi^2 W_i} \tag{13.26}$$

where

T_j = period for mode j

F_{ij} = seismic force at level i for mode j

g = acceleration due to gravity

W_i = seismic weight at level i.

The elastic story drift Δ_{eij} for story i corresponding to mode j is then given by

$$\Delta_{eij} = \delta_{eij} - \delta_{e(i-1)j} \tag{13.27}$$

with $\delta_{e0j} = 0$

Table 13.5. Maximum Allowed Interstory Drift Ratios

Structure Type	Inelastic Displacement Factor **K**	Allowed Drift Ratio*	
		Importance† Level A	Importance Level B, C
1	6	0.010	0.016
2	4	0.010	0.014
3	2	0.010	0.011
4	1.2	0.008	0.008
5a	1	0.010	0.016
5b,c	1	0.008	0.008

*Drift ratio = story drift/interstory height

†Importance Level (Table 13.2)

Correspondingly, the inelastic story drift Δ_{ij} is

$$\Delta_{ij} = \mathbf{K}\Delta_{eij} \tag{13.28}$$

In recognition of the importance of lateral displacements of buildings subjected to seismic forces, the code limits the inelastic interstory displacements (or story drift). Table 13.5 shows the maximum values permitted by the code for the interstory relative displacements or drift ratios [Δ_{ij}/H_i (H_i = interstory height)] for different structure types and importance levels.

13.17 NONSTRUCTURAL ELEMENTS

The effects of past earthquakes have shown the importance of properly designing nonstructural elements such as architectural, electrical, and mechanical elements that otherwise may become major hazards during earthquakes. The Code requires that both the element and the anchorage of the element to the main structure be designed to resist seismic forces.

Two different formulas are provided in the Code for calculation of the design seismic load, one for architectural elements [eq.(13.29)], and one for electrical and mechanical elements [eq.(13.30)]. The load is assumed to act at the center of mass of the element.

(a) For architectural elements

$$F_p = \left[\frac{2h_i}{H+h_1}\frac{V_t}{W_t} + a_{\max}\right]X_p W_p \tag{13.29}$$

where

F_p = element seismic load

h_i = height above base level

h_1 = height of first story

H = total height of the building

V_t = total base shear force

W_t = total seismic weight of the structure

Table 13.6.   Seismic Coefficient for Architectural Elements (X_p)

Element	X_p
Attachments	
Nonbearing exterior walls	1.2
Elements anchored to walls or roofs	2.0
Tiles	1.5
Roofs	1.2
Containers and other related elements	1.0
Partitions and bearing walls	
Stairs and elevators	1.3
Vertical ducts	1.2
Catwalks	1.2
Corridors	1.0
Full height partitions	1.0
Other components	1.2

Table 13.7.   Seismic Coefficient for Mechanical or Electrical Elements (X_c)

Element	X_c
Emergency lighting systems	2.0
Fire and smoke detectors	2.0
Fire-fighting equipment	2.0
Human safety equipment	2.0
Boilers, ovens, incinerators	2.0
water heaters, other types of equipment that use combustible materials, or heat sources	
Communications systems	1.5
Primary wiring systems	2.0
Motor control panels, motor control mechanisms, transformers	1.5
Rotary or reciprocating equipment	1.5
Pressurized vessels	2.0
Manufacturing equipment	1.2
Duct work	1.2
Monitor screens and electric panels	1.5
Conveyor belts	1.2
Lamps	1.5

a_{max} = maximum site acceleration
X_p = seismic coefficient from Table 13.6
W_p = weight of architectural element.

(b) For electrical and mechanical elements

$$F_c = \left[\frac{2h_i}{H + h_1} \frac{V_t}{W_t} + a_{max} \right] \beta X_c W_c \qquad (13.30)$$

where

F_c = element seismic load
X_c = seismic coefficient from Table 13.7
W_c = weight of mechanical or electrical element
β = dynamic amplification factor
    a. Rigid anchorage above base level
      $\beta = 1.0$

b. Flexible anchorage above base level
    $\beta = 1.0$ for $T_c/T_1 \leqslant 0.60$
    $\beta = 1.0$ for $T_c/T_1 > 1.40$
    $\beta = 2.0$ $0.60 < T_c/T_1 \leqslant 1.40$
c. On ground level
    $\beta = 2$
T_1 = fundamental period of the structure
T_c = fundamental period of the component element

$$T_c = 2\pi \sqrt{\frac{W_c}{gK}} \qquad (13.31)$$

where K = Stiffness constant of element support

Example 13.1

A four-story office building located in downtown Heredia is to be designed. Soil studies have shown that soil conditions at the site fit the description of the Code for a soft soil profile. The structure is to be of reinforced concrete, composed of four frames in the x direction and three frames in the y direction. In addition, the center frame in the y direction contains an infilled reinforced concrete panel in the center bay. Columns are 50 cm by 50 cm square, and the wall panel is 30 cm thick. The estimated permanent floor loading of each story is 750 kg/m^2, except for the roof level, which is to be designed for a loading of 450 kg/m^2. The estimated live load on each story is 300 kg/m^2, except for the roof, which carries 40 kg/m^2. (See Table 13.8.) All occupation levels are to be used as office space, except for the second level, which will be used as a warehouse. Fig. 13.8 shows the general layout of the structure. The design life is 50 years, and the allowed probability of exceedance of design seismic event is 0.40.

Solution

It may be concluded from Tables 13.9 and 13.10 that the structure is regular in height, but that the eccentricity does not meet the plan regularity requirements [e_x/D_x = 9.43% >5%, eq.(13.8)] and does not meet the torsional stiffness requirements [$K_\Theta/K_y r_c^2 > 2$,

Table 13.8.   Design Floor Loads for Example 13.1

Level	Computation	Loading (kg/m^2)
4 (Roof)	450 + (0) (40) =	450
3	750 + (0.15) (300) =	795
2	750 + (0.25) (300) =	825
1	750 + (0.15)* (300) =	795

*Live load seismic factors from Table 13.1

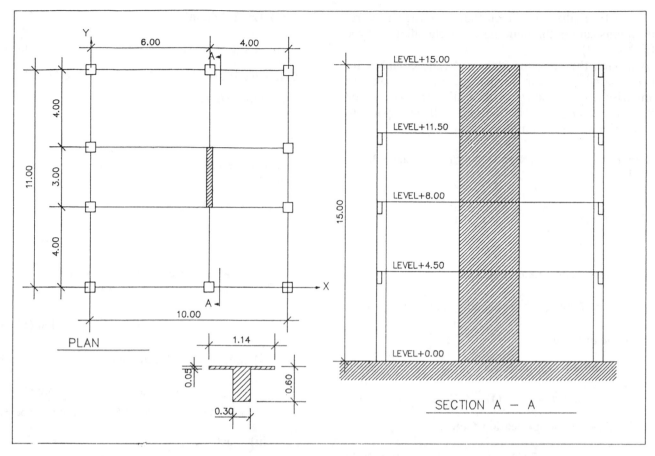

Fig. 13.8. Plan and elevation for a four-story building for Example 13.1 (all dimensions are in meters)

Table 13.9. Mass Properties [eq.(13.14)]

Level	Load/Area (kg/m²)	Mass/Area (kgm/m²)	Center X (m)	Center Y (m)	Dim. x (m)	Dim. y (m)	Total Mass (kgm)	Mass Difference (%)	I_{ci} (kgm.sec²)	r_{ci} (m)
4 (Roof)	450	45.89	5.00	5.50	10.00	11.00	5,047	−43.40	92,956	4.29
3	795	81.06	5.00	5.50	10.00	11.00	8,917	−3.63	164,223	4.29
2	825	84.12	5.00	5.50	10.00	11.00	9,253	+3.77	170,420	4.29
1	795	81.06	5.00	5.50	10.00	11.00	8,917		164,223	4.29

Table 13.10. Stiffness Properties [eq.(13.12)]

Level	h_i (m)	K_x (kg/m)	K_y (kg/m)	K_Θ (kg-m/rad)	e_x (m)	e_x/D_x (%)	e_y (m)	e_y/D_y (%)	$\dfrac{K_\Theta}{K_x r_{ci}^2}$	$\dfrac{K_\Theta}{K_y r_{ci}^2}$
4 (Roof)	3.50	3.62E + 07	4.47E + 08	1.70E + 09	0.94	9.43	0.00	0.00	2.55	0.21
3	3.50	7.23E + 07	8.94E + 08	3.41E + 09	0.94	9.43	0.00	0.00	2.55	0.21
2	3.50	7.23E + 07	8.94E + 08	3.41E + 09	0.94	9.43	0.00	0.00	2.55	0.21
1	4.50	5.32E + 07	6.57E + 08	2.51E + 09	0.94	9.43	0.00	0.00	2.56	0.21

eq.(13.12)]. Failure to meet the plan regularity requirements causes the building to be classified as Type 3 (Table 13.3).

Although the structure does not fully comply with all the conditions for regularity, for the sake of comparison, both analyses (the static method and the two-dimensional dynamic method) will be applied to the solution of this example.

Importance Level: B (Normal importance, Table 13.2)
Risk Period:

$$PR = \frac{1}{1-(1-PE)^{1/N}} = \frac{1}{1-(1-0.40)^{1/50}} \quad [\text{eq.}(13.7)]$$

$$= 98 \text{ years}$$

Maximum ground acceleration:

$$a_{max} = 17.50\%g \qquad (\text{isoacceleration map, Fig. 13.2})$$

1. Static analysis

(a) Fundamental period T

$$T = 0.08N = (0.08)(4) = 0.32 \text{ sec.} \quad (\text{Table 13.4})$$

(b) Dynamic amplification factor

$$\text{FAD} = 1.44 \qquad (\text{Fig. 13.7})$$

(c) Seismic coefficient

$$C = Ra_{max}\text{FAD} \qquad [(\text{eq.}13.17)]$$

$$= (0.80)(0.175)(1.44)$$

$$= 0.2016$$

(d) Constant factor η

$$\eta = \frac{\sum\limits_{k=1}^{N}(W_k)(h_k)}{\sum\limits_{k=1}^{N}(W_k)(h_k^2)} = \frac{2,867,700}{30,281,633} = 0.0947$$

$$[\text{eq.}(13.19)]$$

(e) Lateral seismic forces

$$F_i = C\eta h_i W_i \qquad [\text{eq.}(13.18)]$$

Calculated seismic forces are given in Table 13.11.

(f) Story shear forces

$$V_i = \sum_{j=i}^{N} F_j \qquad [\text{eq.}(13.22)]$$

Calculated values of story shear forces are shown in Table 13.11.

(g) Base shear force

$$V = \sum_{k=1}^{N} F_k = 54,749 \text{ kg} \qquad [\text{eq.}(13.20)]$$

(h) Lateral displacements and drift ratios
(i) elastic displacements

$$\delta_{eij} = \frac{gT_j^2 F_{ij}}{4\pi^2 W_i} \qquad [\text{eq.}(13.26)]$$

(ii) inelastic displacements

$$\delta_i = K\delta_{ei} \qquad [\text{eq.}(13.25)]$$

$$\mathbf{K} = 2.0 \qquad (\text{Table 13.5 for structure Type 3})$$

(iii) elastic drifts

$$\Delta_{ei} = \delta_{ei} - \delta_{e,(i-1)} \qquad [\text{eq.}(13.27)]$$

(iv) inelastic drifts

$$\Delta_i = K\Delta_{ei} \qquad [\text{eq.}(13.28)]$$

Values for lateral displacements and drift ratios, Δ/H, (H = interstory height) are shown in Table 13.12.

To verify the fundamental period:

$$T = 2\pi \sqrt{\frac{\sum\limits_{i=1}^{N} W_i \delta_{ei}^2}{g \sum\limits_{i=1}^{N} F_i \delta_{ei}}} \qquad [\text{eq.}(13.16)]$$

$$T = 0.237 \text{ sec}$$

Table 13.11. Story Seismic Loads

Level	W_i (kg)	H_i (m)	h_i (m)	$W_i h_i$ (kg-m)	$W_i h_i^2$ (kg-m²)	F_i (kg)	A_c (%g)	V_i (kg)
4 (Roof)	49,500	3.50	15.00	742,500	11,137,510	14,175	28.64	14,175
3	87,450	3.50	11.50	1,005,675	11,565,262	19,200	21.96	33,375
2	90,749	3.50	8.00	725,999	5,807,997	13,860	15.27	47,236
1	87,450	4.50	4.50	393,525	1,770,862	7,513	8.59	54,749
Σ =	315,150			2,867,700	30,281,631	54,748		

Table 13.12. Elastic and Inelastic Displacements

	Elastic		Inelastic	
Level	Displ. (m)	Drift Ratio	Displ. (m)	Drift Ratio
4 (Roof)	0.0041	0.00030	0.0082	0.00060
3	0.0032	0.00038	0.0064	0.00076
2	0.0020	0.00042	0.0040	0.00084
1	0.0009	0.00032	0.0018	0.00064

Table 13.13. Mode Shapes (3 Modes)

Level	Mode 1	Mode 2	Mode 3
Roof	+0.00853	+0.00806	+0.00592
3	+0.00629	−0.00089	−0.00548
2	+0.00383	−0.00622	−0.00140
1	+0.00156	−0.00467	−0.00720
Period (sec):	0.224	0.057	0.027
Factor η:	169.22	−74.88	−103.93
FAD	1.44	1.30	1.20

Table 13.14. Story Lateral Loads (kg)

Level	Mode 1	Mode 2	Mode 3	Effective*
Roof	14,405	−5,437	−5,116	16,225
3	18,765	+1,060	+8,367	20,573
2	11,857	+7,693	+2,218	14,307
1	4,654	+5,565	10,994	13,172

*Modal combination using the SRSS method.

Table 13.15. Story Shears (kg)

Level	Mode 1	Mode 2	Mode 3	Effective*
Roof	14,405	−5,437	−5,116	16,225
3	33,170	−4,377	+3,251	33,615
2	45,027	+3,316	+5,469	45,479
1	49,681	+8,881	16,463	53,086
Base				

*Modal combination using the SRSS method.

The new period does not change the value of the dynamic amplification factor, FAD; therefore, the seismic forces remain unchanged.

2. Two-dimensional dynamic analysis. The dynamic analysis of this four-story building with three degrees of freedom per story was performed using a computer program that implements the provisions of the Seismic Code of Costa Rica. Tables 13.13 through 13.17 show the results of the dynamic analysis by modal superposition based on the first three modes of vibration.

Table 13.16. Overturning Moments (kg-m)

Level	Static	Mode 1	Mode 2	Mode 3	Effective*
3	49,612	50,417	−19,029	−17,906	56,785
2	166,425	166,512	−34,349	−6,527	170,143
1	331,747	324,107	−22,743	+12,614	325,148
Base	578,114	547,671	+17,221	+86,697	554,758

*Modal combination using the SRSS method.

Table 13.17. Elastic and Inelastic Displacements and Drift Ratios

	Elastic		Inelastic	
Level	Displ. (m)	Drift Ratio	Displ. (m)	Drift Ratio
4 Roof	0.0036	0.00027	0.0072	0.00054
3	0.0027	0.00030	0.0054	0.00060
2	0.0016	0.00028	0.0032	0.00056
1	0.0007	0.00015	0.0014	0.00030

13.18 COMPUTER PROGRAM

A computer program has been written to implement the provisions of the Code. It handles most of the analysis requirements of the Code as they pertain to earthquake forces. The program has been designed to permit the user to create a basic library of vertical elements section properties and mass configurations, which will then be placed at the appropriate locations in each story, allowing the user to modify parts of the input to obtain a new analysis, or to modify an old analysis for a new project.

The input file that the program creates for later runs of the same structure is fully labeled, so that the user might easily modify its contents with the help of a word processor. The program will always create a copy of the input file, and a successful execution of the program will create a file with the same name as the input file with a file extension "ECC" in which the story eccentricities are stored. The program also creates a file with a file extension "OUT", with all of the information that would normally be sent to the printer.

13.18.1 Mass Configurations

Usually, the various stories of a building are similar in mass distribution. For this reason the program has been prepared to ask only for the information related to those few different types of mass configurations in the building. The definition of the mass configurations assumes that the mass of each story can be approximated as an arrangement of a number of rectangular segments. The information required to compose the mass of a story is then the mass per unit area of floor space of each segment, the X and Y coordinates of the center of the segment, and the x and y dimensions of the segment.

13.18.2 Section Properties of Vertical Elements

Most columns, walls, and bracings in a building can be defined from a very small set of different section property types. The program will calculate the required properties for a rectangular, circular, and H section oriented in the X and the Y directions. If the section property is not covered by this basic set, its properties might be entered directly. It is possible to change the default material properties of a section at this level.

13.18.3 Column Coordinates

The column coordinates define the location of columns at any level of the building. Even if a column exists at only one level, it must be defined for all levels; however, the story data will have a section property associated only in that floor.

13.18.4 Story Data

In this section each story is defined by assigning to it a typical mass configuration of the set previously defined, and a section property to the column lines that connect to the slab of that level from above and below. If a connecting column is not connected to any of the adjacent stories, then the height of the column

must be increased to account for this. A value of zero in the story height will indicate the program to use the default values.

13.18.5 Structural Type

The Code defines the different structural types possible given certain criteria as to the framing type and the capacity of the structure to dissipate energy through deformations (ductility).

13.18.5.1 Fundamental period structural type. The fundamental of the structure might be approximated, using specific Code equations for framing type and structural materials. The program does not have a built-in eigenvalue solver to determine the periods of vibration and corresponding modal shapes of the structure. However, the Code program implements the approximation formulated for the first period of vibration and a commonly used approximation for the second and third modes. The modal shapes assumed for these approximations are shown in Fig. 13.9.

Example 13.2

A 20-story office building located in downtown San Jose is to be analyzed. The structure will house among its tenants a bank and a major archeological museum, which requires that the structure remain fully operational after a major earthquake with minimum damage. Soil studies have shown that soil conditions fit the description of the Code's firm soil profile. The structure is a concrete shear wall/ductile moment-resisting steel frame structure in both principal directions composed of three frames in the x direction and three frames in the y direction. Columns are H sections built

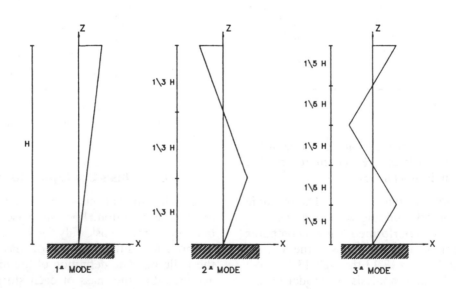

Fig. 13.9. Program's assumption of modal shapes for lateral displacements

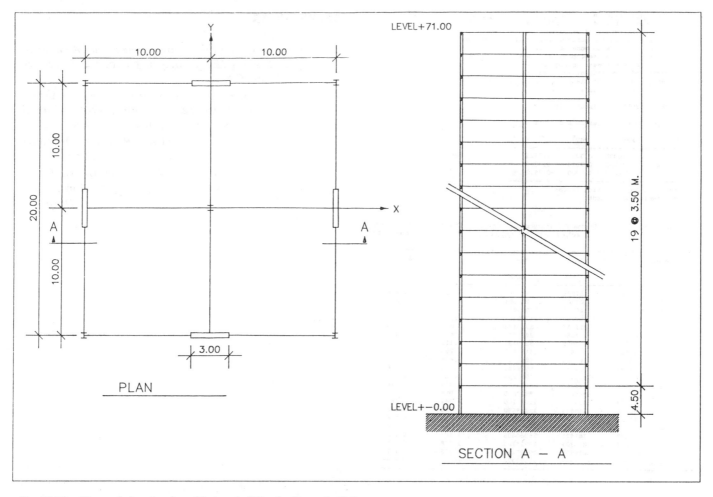

Fig. 13.10. Plan and elevation for a 20-story building for Example 13.2

out of 4-cm-thick steel plates through the height of the structure with a depth of 55 cm and a flange width of 55 cm, the wall panels are 40 cm thick by 3 m long. The estimated permanent unit weight of each story is 750 kg/m^2; the estimated temporary unit weight of each story is 300 kg/m^2. All living levels are to be used as office space. The specified concrete compressive strength is 210 kg/cm^2, the steel is A36. Fig. 13.10 shows the general layout of the structure.

Solution

The implementation of the computer program for seismic analysis of a building by the dynamic method requires input data consisting of (1) general data such as Importance Level, probability of exceedance, risk period, soil profile, etc., and (2) detailed information on the geometry of the building, the mass distribution at the various levels, and the properties and location of the lateral resisting elements at each story of the building. The computer program prepared for this chapter contains instructions to prepare and input the necessary data to execute the program. However, to

save space, the input/output information presented for this problem provides only selected portions of the pertinent information. The computer output also includes detailed calculations of the torsional eccentricities at each story of the building. These calculations are not reproduced here.

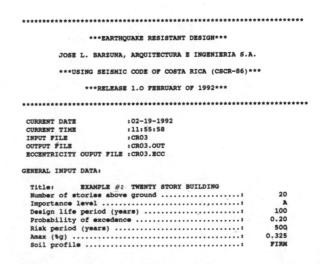

```
***************************************************************
          ***EARTHQUAKE RESISTANT DESIGN***

     JOSE L. BARZUNA, ARQUITECTURA E INGENIERIA S.A.

     ***USING SEISMIC CODE OF COSTA RICA (CSCR-86)***

          ***RELEASE 1.0 FEBRUARY OF 1992***

***************************************************************

    CURRENT DATE          :02-19-1992
    CURRENT TIME          :11:55:58
    INPUT FILE            :CR03
    OUTPUT FILE           :CR03.OUT
    ECCENTRICITY OUPUT FILE :CR03.ECC

    GENERAL INPUT DATA:

      Title:     EXAMPLE #2 TWENTY STORY BUILDING
      Number of stories above ground ...................:    20
      Importance level .................................:    A
      Design life period (years) .......................:    100
      Probability of excedence .........................:    0.20
      Risk period (years) ..............................:    500
      Amax (%g) ........................................:    0.325
      Soil profile .....................................:    FIRM
```

```
Structural type in X direction ....................:        2
Fundamental Period structural type in X direction :        3
Structural type in Y direction ....................:        2
Fundamental Period structural type in Y direction :        3
Number of modes to consider ......................:        3
Number different sec. prop. for vertical elements :        3
Number of story mass configurat ..................:        1
Number of vertical column lines ..................:        9
Default Modulus of Elasticity ....................: 2.040E+10
Default Shear Modulus ............................: 7.860E+09

OUTPUT RESULTS (X direction):

   Dynamic properties in X direction:

Mode    Period    FAD*      Cx**
-----------------------------------------
  1     1.6000    0.3000    1.4621
  2     0.5333    0.9000    0.3880
  3     0.3200    1.2000   -0.3512
-----------------------------------------
*Dynamic Amplification Factor
**Seismic coefficient
```

Equivalent lateral forces F_i, Story shear forces V_i, Overturning moments M_i, Lateral displacements δ_i, and Story drifts Δ_i in X direction

Level i	F_i kg	V_i kg	M_i kg-m	δ_i cm	Δ_i cm
20	57998	57998	------	7.289	0.376
19	49842	107707	202994	6.923	0.376
18	42367	149525	579829	6.557	0.376
17	36001	183990	1102498	6.193	0.376
16	31426	211834	1744458	5.831	0.376
15	29487	234056	2481030	5.470	0.376
14	30688	252008	3289932	5.112	0.376
13	34704	267498	4151953	4.756	0.376
12	40711	282835	5051815	4.404	0.376
11	47966	300789	5979241	4.056	0.376
10	55987	324357	6930233	3.714	0.376
9	64488	356344	7908518	3.379	0.376
8	73039	398941	8927035	3.054	0.376
7	82329	453492	10009200	2.742	0.364
6	90214	520458	11189590	2.424	0.376
5	97114	599007	12517890	2.095	0.376
4	104470	688198	14054490	1.786	0.376
3	112193	787421	15860408	1.506	0.376
2	120212	896305	17994250	1.278	0.376
1	128472	1014635	20509240	1.132	1.132
0	------	------	24323920	------	------

REFERENCES

ASHRAF HABIBULLAH (1991) *Three Dimensional Analysis of Buildings Systems (ETABS)*. Computers and Structures Inc., Berkeley, CA.

COLEGIO FEDERADO DE INGENIEROS Y ARQUITECTOS (1974) *Código Sísmico de Costa Rica, 1974*. Editorial Tecnológica de Costa Rica, San Jose, Costa Rica.

—— (1987) *Código Sísmico de Costa Rica, 1986*. Editorial Tecnológica de Costa Rica, San Jose, Costa Rica.

WILSON, E. L.; KIUREGHIAN, A.D.; and BAYO, E. P. (1981) "A Replacement for the SRSS Method in Seismic Analysis." *Int. J. Earthquake Eng. and Struct. Dynamics* 9: 187–194.

14

Egypt

Fouad H. Fouad

14.1 INTRODUCTION

The *Regulations for Earthquake-Resistant Design of Buildings in Egypt* were published by the Egyptian

Society for Earthquake Engineering in 1988. A basic consideration in earthquake-resistant design is to predict the intensity of the future seismic events for the site of a proposed structure. The Republic of Egypt is divided into four zones according to the level of seismic activity, as shown in Fig. 14.1. Regions of high seismic activity in Zone 3 include counties that are adjacent to the Red Sea shore and southern Sinai Peninsula, and Aswan County. These regions are assigned a value of VIII on the Modified Mercalli Intensity (MMI) scale (this intensity corresponds to moderate damage to buildings and other structures). Zone 2 is assigned an MMI value of VII, Zone 1 an MMI value of VI, and Zone 0 an MMI value $\leqslant$V.

14.2 METHODS OF ANALYSIS

The 1988 Egyptian Regulations provide two methods for the analysis and design of earthquake-resistant buildings: (1) The equivalent static force method, and (2) the spectral modal analysis method.

A detailed presentation of the equivalent static force method is given in *The Egyptian Code for the Design and Construction of Reinforced Concrete Buildings*, published by the Egyptian Society of Engineering (1989); the method is widely used because of its simplicity. The spectral modal analysis method is based on dynamic analysis; this method must be used for irregular buildings or for structures of particular importance to the community that require a more exact analysis. In this chapter, the presentation will be limited to the equivalent static force method.

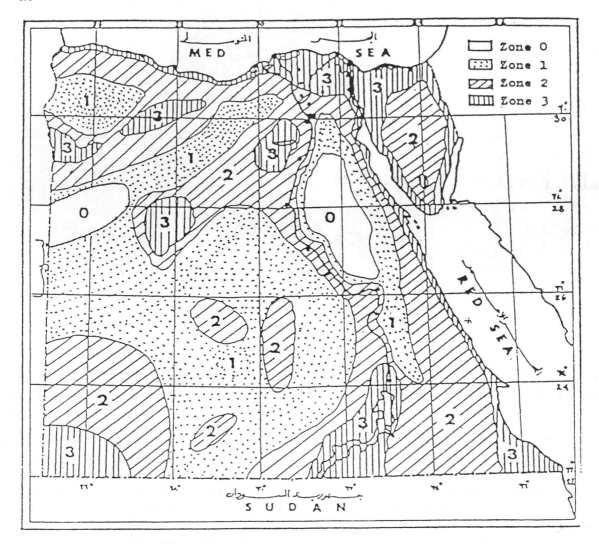

Fig. 14.1.   Seismic zoning map for Egypt. (From Regulations for Earthquake Resistant Design of Buildings in Egypt, 1988)

14.3 EQUIVALENT STATIC FORCE METHOD OF ANALYSIS

14.3.1 General

The 1988 Egyptian Regulations require that buildings subjected to seismic action be analyzed under the action of an equivalent system of static forces applied horizontally at each floor or roof level. For buildings that have at least one axis of symmetry and have resisting elements along two perpendicular directions, equivalent seismic forces are applied independently along each of these two horizontal directions. For unsymmetrical buildings, the equivalent seismic forces must be applied in the direction that will produce the most unfavorable effect on any structural element.

14.3.2 Total Horizontal Seismic Force (Base Shear)

The total horizontal seismic force or base shear force V in the direction under consideration is given by the following formula:

$$V = C_s W_t \tag{14.1}$$

where C_s is the seismic design coefficient and W_t is the total seismic weight. A procedure for evaluating each of these quantities is given in the sections that follow.

14.3.2.1 Seismic design coefficient C_s.   The seismic design coefficient C_s shall be determined from the following formula:

$$C_s = ZISMRQ \tag{14.2}$$

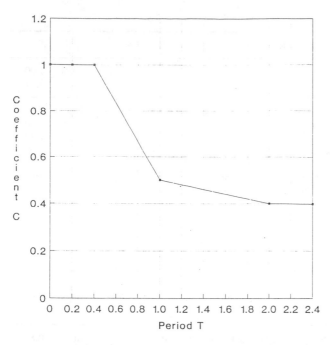

Fig. 14.2. Standardized response spectrum coefficient for average damping of 5%

Table 14.1. Horizontal Acceleration Ratio A

Zone	Intensity (MMI)	$A(g)$
0	≤V	0
1	VI	0.02
2	VII	0.04
3	VIII	0.08

Table 14.2. Soil Foundation Factor F

Soil Type	Description	F
1	Rock; dense and very dense coarse-grained soils; very stiff and hard fine-grained soils; compact coarse-grained soils; and firm and stiff fine-grained soils with a depth of less than 15 m.	1.0
2	Compact coarse-grained soils; firm and stiff fine-grained soils with a depth greater than 15 m. Very loose and loose coarse-grained soils; and very soft and soft fine-grained soils from zero to 15 m. deep.	1.3
3	Very loose and loose coarse-grained soils; and very soft and soft fine-grained soils with depth greater than 15 m.	1.5

where

Z = seismic zoning factor
I = importance factor
S = structural system type factor
M = material factor
R = risk factor
Q = construction quality factor.

C_s shall not be taken as greater than $4.8 \times ZIMQ$, nor less than 0.02 for sites with seismic zone factors greater than zero.

14.3.2.2 Seismic zoning factor Z. The seismic zoning factor Z is given by

$$Z = ACF \qquad (14.3)$$

where

A = standard value for horizontal acceleration ratio (expressed as fraction of the acceleration of gravity g) to be determined from Table 14.1 and the seismic zoning map shown in Fig. 14.1
C = coefficient of the standardized response spectrum for average damping of 5%, to be determined from Fig. 14.2, using the estimated fundamental period T (see Sect. 14.3.3)
F = foundation soil factor to be determined from Table 14.2.

14.3.2.3 Importance Factor I. Values for the importance factor I are given in Table 14.3, as functions of building use.

Table 14.3. Importance Factor I

Class	Description	I
I	(a) Structures and buildings to be used during or immediately after an earthquake (such as hospitals, fire stations, broadcasting buildings, power stations, terminals). (b) Buildings housing valuable and important items (such as museums and banks).	1.50
II	Buildings and structures with high occupancy (such as schools, stadiums, theaters, cinemas, workshop buildings).	1.30
III	Buildings and structures with low occupancy (such as residential buildings, hotels, office buildings, restaurants).	1.00

14.3.2.4 Structural system-type factor S. Values for the structural system-type factor S must be determined separately for each direction of the building. Unless more precise values are determined, the recommended values for this factor that are given in Table 14.4 may be used.

Table 14.4. Structural System Type Factor *S*

Structural System	Type	Description	S
Frames	1	Moment-resisting frame.	1.00
	2	Ductile moment-resisting frame.	0.67
Shear walls	3	Two or more parallel and approximately symmetrically arranged ductile cantilever shear walls.	1.00
	4	Ductile coupled shear walls.	0.80
	5	Single ductile cantilever shear walls.	1.20
	6	Shear walls not designed for ductile flexural yielding but having the ability to dissipate a significant amount of seismic energy. (This includes walls other than those listed above.)	1.60
Diagonal bracing	7	Buildings with diagonal bracing capable of plastic deformation in tension only: (a) Single-story (b) Two or three stories (c) More than three stories	2.0 2.5 *
	8	Buildings with diagonal bracing capable of plastic deformation in both tension and compression.	*
Box system	9	Both horizontal and gravity forces resisted by wall system.	1.33 in each direction
Shared force-resisting system	10	Buildings with a dual bracing system consisting of a ductile moment-resisting frame plus shear walls conforming with the following conditions: (a) The frames and shear walls shall resist the total lateral force in accordance with their relative rigidities considering the interaction between the shear walls and frames; (b) The shear walls acting independently shall resist the total lateral force; (c) The ductile moment-resisting frame shall have the capacity to resist at least 25% of the total lateral force.	0.80
	11	Shear walls plus a ductile moment resisting frame that do not satisfy the conditions of Type 10.	1.33
	12	Buildings with bracing of Type 7 plus a frame of Type 2.	*
Miscellaneous	13	Small tanks on the ground.	2.00
	14	Freestanding elevated tanks with full contents, on four or more cross-braced legs and not supported by a building.	2.50
	15	Freestanding structure of an inverted pendulum type, including elevated tanks on three or less cross-braced legs, or on unbraced legs.	3.20
Other structural concepts	16	Structural systems other than those of Types* 1 to 15 above may be accepted where evidence shows that equivalent ductility and energy absorption are available sufficient to relate behavior of such alternative structural systems to any of Types 1 to 15.	

*Values to be determined by analysis.

14.3.2.5 Material factor M. Table 14.5 provides recommended values for the material factor *M*.

Table 14.5. Material Factor *M*

Item	Material	M
1	Reinforced concrete	1.0
2	Structural steel	0.8
3	Prestressed concrete	1.2
4	Reinforced masonry	1.2

14.3.2.6 Risk factor R. Risk factors are given in Table 14.6. The Code requires that the higher values shall be used for cases in which two different values of *R* are applicable.

14.3.2.7 Construction quality factor Q. The construction quality factor *Q* shall be taken as 1.0 if the contract provides for inspection and material testing at

Table 14.6. Risk Factor *R*

Item	Description	R
1	Buildings other than those given in this table and presenting no unusual risk.	1.0
2	Distribution facilities for natural gas, coal gas, or petroleum products.	2.0
3	Structures and installations for the direct containment of toxic liquids or gases, spirits, acids, alkalis, molten metal, or poisonous substances, including substances that could form dangerous gases if released.	3.0

important stages of the construction. If such provisions do not exist, Q shall be taken as 1.2.

14.3.2.8 Total seismic weight W_t. The total seismic weight W_t shall be evaluated as follows:

$$W_t = \sum_{i=1}^{n} W_i \qquad (14.4)$$

and

$$W_i = D_i + pL_i \qquad (14.5)$$

where

D_i = total dead load for the ith level of the building
L_i = total design live load for the ith level of the building
p = incidence factor for live load as given in Table 14.7.

Table 14.7. Incidence Factor for Live Loads p

Type of Structure	p
Residential buildings, hotels, offices, hospitals, public buildings, etc.	0.25
Storage areas and warehouses.*	0.50

*Tanks, reservoirs, silos, and the like shall be considered with their full contents.

14.3.3 Fundamental Period T

Two procedures are given in the code for calculating the fundamental period T of a building. In determining the total horizontal seismic force V, the Code gives two procedures for estimating the fundamental period of the building, the simplified procedure and the detailed procedure.

14.3.3.1 Simplified procedure. The fundamental period T of a building may be estimated by the following empirical formula:

$$T = \frac{0.09H}{\sqrt{d}} \text{ (sec)} \qquad (14.6)$$

where

H = the total height of the building in meters
d = the dimension of the building in a direction parallel to the applied seismic forces, in meters.

For buildings in which the total horizontal force is resisted by a moment-resisting space frame, the value of the fundamental period T may be obtained from

$$T = 0.1n \qquad (14.7)$$

where n is the total number of stories in the building, above the base.

14.3.3.2 Detailed procedure. The value of the fundamental period T may be determined from a detailed analysis that considers the structural properties of the building and the deformational characteristics of its resisting elements. One such method of analysis is Rayleigh's formula:

$$T = 2\pi \sqrt{\frac{\sum_{i=1}^{n} W_i \delta_i^2}{g\left(\sum_{i=1}^{n-1} F_i \delta_i + (F_t + F_n)\delta_n\right)}} \qquad (14.8)$$

where

δ_i = the elastic horizontal displacement of the ith level of the building due to the forces F_i
g = the acceleration due to gravity
F_i = the part of the total horizontal force V that is assigned to the ith level (Section 14.3.4)
F_t = additional horizontal force applied to the top level of the building (Section 14.3.4).

14.3.4 Distribution of Horizontal Seismic Forces

The total horizontal seismic force V is distributed over the height of the building according to the following formula:

$$F_i = \frac{W_i h_i}{\sum_{i=1}^{n} W_i h_i} V \qquad (14.9)$$

where

F_i = the part of the total horizontal force V that is assigned to the ith floor
h_i = the height over the base to the ith level
W_i = the total seismic load on the ith level of the building [eq.(14.5)].

The following special requirements must be considered in the distribution of the base shear force.

(a) For buildings with a height-to-width ratio H/d equal to or greater than 3, an additional force $F_t = 0.1V$ shall be applied as a concentrated force at the top level of the building, and the remaining $0.9V$ shall be distributed according to eq.(14.9).
(b) For chimneys and smokestacks resting on the ground, a force of $0.2V$ shall be considered as concentrated at the top of the structure, and the remaining $0.8V$ shall be distributed according to eq.(14.9).

The force F_i at each level shall be applied over the building area according to the mass distribution at that level. The total shear in any story shall be distributed to the various resisting elements in proportion to their stiffnesses. To reduce the torsional effects developed by these forces, the main resisting elements should be located symmetrically about the center of mass of the building.

The method of horizontal seismic force distribution given by eq.(14.9) is not applicable to buildings with highly irregular shapes, large differences in lateral resistance or stiffness between stories, or other unusual structural features. For such cases, the force distribution should be determined using dynamic analysis (see Section 14.4).

14.3.5 Horizontal Torsional Moments

According to the Egyptian Seismic Regulations, there are three methods for considering torsional effects:

1. Static method of analysis
2. Two-dimensional modal analysis
3. Three-dimensional modal analysis.

The particular method to be used depends on the type of structure and the eccentricity between the center of mass and the center of rigidity of the building. Any of the methods may be used for structures that are not more than five stories high. These methods may also be used for regular structures that have more than five stories. However, if the eccentricity exceeds one-fifth the plan dimension perpendicular to the direction of the applied seismic forces, the three-dimensional modal analysis method is recommended; also, this recommendation applies to irregular structures of more than five stories. A presentation of the three-dimensional modal analysis, which includes dynamic torsional effects, is beyond the scope of this chapter, but may be found in other references (Paz 1991). The static method for torsional moment is presented in the next section.

14.3.5.1 Static method for torsional moment calculation.
In the static method of analysis, the horizontal forces at any level are assumed to act eccentrically from the center of rigidity at that level. The eccentricity e_a, at any level of the building measured in the direction perpendicular to the direction of the seismic forces, shall be computed from eqs.(14.10) and (14.11). The most critical value given by these equations should be used in the analysis.

$$e_a = 1.5e + 0.1b \qquad (14.10)$$

$$e_a = e - 0.1b \qquad (14.11)$$

where

e = the distance from the center of rigidity to the center of mass at the level considered, measured perpendicularly to the direction of the seismic forces
b = the maximum horizontal dimension of the building at the level considered, measured perpendicularly to the direction of the seismic forces.

14.3.6 Overturning Moments

The overturning moment M_i at any level of a building is determined from statics as the moment produced at that level by the seismic forces:

$$M_i = \sum_{k=i+1}^{n} F_k(h_k - h_i) + F_t(h_n - h_i) \qquad (14.12)$$

where $i = 0, 1, 2, 3, \ldots, n-1$
and

F_k = horizontal seismic force at level k
F_t = additional force at the top of the building
h_k = height to level k
h_i = height to level i
h_n = height to top level of the building.

14.3.7 Setbacks

The effect of setbacks in buildings may be ignored if the plan dimension of the tower portion is not less than 75% of the plan dimension of the lower part of the building, in each principal direction. For larger setbacks, dynamic analysis should be used.

14.3.8 Parts or Portions of Buildings

Any part or portion of a building shall be designed for a seismic force F_p applied at its center of mass, for each direction considered. The seismic force F_p on a part of a building is given by

$$F_p = C_p W_p \qquad (14.13)$$

where

C_p = seismic design coefficient for the part or portion of the building
W_p = weight of that part or portion of the building, calculated in a similar manner to that of Section 14.3.2.8.

14.3.8.1 Seismic design coefficient C_p.
The seismic design coefficient for a part or portion of a building C_p is given by

$$C_p = C_s P_p R_p \qquad (14.14)$$

Table 14.8. Risk Factor for Part or Portion of a Building R_p

Item	Description	R_p
1	Walls and partitions:	
	(a) Adjacent to an exit-way, street, or public place, or required to have a fire resistance rating	1.1
	(b) Cantilevered walls and parapets	1.5
	(c) All walls not specified in items 1(a), 1(b), or 2.	1.0
2	Stairs and their enclosing shaft walls, and shaft walls for lifts.	2.0
3	Horizontally cantilevered floors, beams, etc. (the force acts vertically upward or downward).	2.0

where

C_s = seismic design coefficient, determined according to Section 14.3.2.1

P_p = position factor = $1.0 + h_p/H$ where h_p is the height at which the part or portion of the building is located, and H is the total height of the building

R_p = risk factor for the part or portion of the building (recommended values for the risk factor are given in Table 14.8).

14.4 SPECTRAL MODAL ANALYSIS

The spectral modal analysis is a more detailed method based on dynamic analysis. It should be used in cases where the equivalent static force method is not applicable, such as for highly irregular buildings or buildings of particular importance. A synopsis of this method is given in Part I of this book: in Chapter 2, Seismic Response and Design Spectra; and Chapter 4, Structures Modeled as Multidegree-of-Freedom Systems.

14.5 DESIGN METHOD

The Egyptian Code uses the working stress method for the design of both steel and reinforced concrete structures. Allowable material stresses are specified by the relevant material code. When earthquake forces are considered in addition to other loads, the code permits an increase in the allowable stresses by one-third of the value used for design under gravitational loads. However, these increases in allowable stresses are subjected to the following limitations:

(a) The maximum steel stress should not exceed the yield stress for steel with a definite yield point, and shall not exceed the smaller of the two values of the proof-stress or 80% of the ultimate strength for steel without a definite yield point.

(b) Bond stresses in reinforced concrete shall not be increased.

(c) The allowable concrete tensile stress for prestressed concrete members shall not exceed two-thirds of the modulus of rupture of the concrete.

(d) The allowable soil bearing pressure shall not be increased for soils of loose sands or soft clays, and no increase in stresses is permitted for reinforced concrete foundations bearing on these soils.

14.6 DESIGN LOADS

The 1988 Egyptian Regulations specify various load combinations; the design load is required to be the load combination determined by eqs.(14.15) that is applicable and produces the greatest effect. In designing for seismic forces, the following two combinations should be considered:

$$A = D + L_1 + E$$

$$A = 0.85D + E \tag{14.15a}$$

The other load combinations are as follows:

For gravity loads: $\quad A = D + L_1 \tag{14.15b}$

For wind loads: $\quad A = D + L_1 + W \tag{14.15c}$

$$A = 0.85D + W$$

For earth pressure loads: $A = D + L_1 + Q \tag{14.15d}$

$$A = D + Q$$

For liquid pressure: $\quad A = D + L_1 + F \tag{14.15e}$

$$A = D + F$$

where

A = design load using the working stress method
D = dead load
E = earthquake load
F = liquid pressure load
L_1 = reduced live load including the effect of impact
Q = earth pressure load
W = wind load.

14.7 EXAMPLE 14.1

A three-story rigid steel-frame office building is to be constructed in Alexandria, Egypt. The weight of the building is assumed to be lumped at each level of the building, as follows: $W_1 = W_2 = 320,000$ kg for the first two levels, and $W_3 = 250,000$ kg for the top level, as shown in Fig. 14.3. The seismic forces are assumed to be applied in the X direction. The foundation consists of isolated column footings in loose, sandy soil.

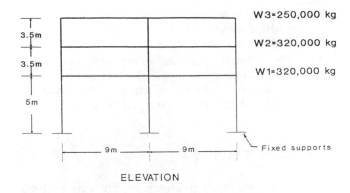

W3=250,000 kg

W2=320,000 kg

W1=320,000 kg

ELEVATION

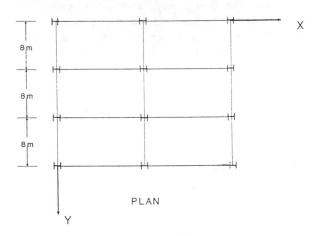

PLAN

Fig. 14.3. Modeled building for Example 14.1

According to the Egyptian Regulations, determine the following:

Fundamental period T
Base shear force V
Distribution of horizontal seismic forces F_i
Story shear forces V_i
Horizontal torsional moments M_{ti}
Overturning moment at the various levels of the building M_i
Lateral displacements δ_i
Story drifts Δ_i.

(a) Fundamental period T

$$T = \frac{0.09H}{\sqrt{d}} \qquad \text{[eq.(14.6)]}$$

$$T = \frac{0.09 \times 12}{\sqrt{18}} = 0.255 \text{ sec.}$$

The city of Alexandria is located in Zone 3, with intensity VIII (Fig. 14.1), horizontal acceleration ratio $A = 0.08$ (Table 14.1), and a standardized response spectrum coefficient, $C = 1.0$ (Fig. 14.2).

$F = 1.5$ for loose sandy soil   (Table 14.2)
$Z = ACF = 0.08 \times 1.0 \times 1.5 = 0.12$ [(eq.(14.3)]
$I = 1.0$ for office building   (Table 14.3)
$S = 1.0$ for moment-resisting frame   (Table 14.4)
$M = 0.8$ for structural steel   (Table 14.5)
$R = 1.0$ for no unusual risk   (Table 14.6)
$Q = 1.0$ for specified inspection during construction

Thus,

$$C_s = ZISMRQ \qquad \text{[eq.(14.2)]}$$
$$= 0.12 \times 1.0 \times 1.0 \times 0.8 \times 1.0 \times 1.0$$
$$= 0.096 > 0.02$$

Check:

$$C_s \le 4.8ZIMQ = 4.8 \times 0.12 \times 1.0 \times 0.8 \times 1.0 = 0.46$$

Total seismic weight W_t:

$$W_t = \sum_{i=1}^{n} W_i$$
$$= W_1 + W_2 + W_3$$
$$= 320{,}000 + 320{,}000 + 250{,}000$$
$$= 890{,}000 \text{ kg}$$

(b) Base shear V

$$V = C_s W_t$$
$$= 0.096 \times 890{,}000$$
$$= 85{,}440 \text{ kg}$$

(c) Distribution of horizontal seismic forces F_i

$$F_i = \frac{W_i h_i}{\sum\limits_{i=1}^{n} W_i h_i} V$$

$$F_t = 0 \quad (H/d = 12/18 = 0.67 < 3.0) \quad \text{[eq.(14.9)]}$$

Calculated values for the horizontal seismic forces are given in Table 14.9.

(d) Story shear forces V_i

$$V_i = \sum_{j=i}^{n} F_j$$

Because all bents are identical and the floor diaphragms are quite rigid, the forces are shared equally by the bents; that is, each of the four frames will receive one-fourth of the total seismic force in the X direction. Likewise, each of the three frames will receive one-third of the total seismic force in the Y direction.

Table 14.9. Calculated Values for Lateral Forces F_i, Story Shear Forces V_i, Torsional Moments M_{ti}, Overturning Moments M_i, Lateral Displacements δ_i, and Story Drifts Δ_i for Example 14.1

Level	F_i (kg)	V_i (kg-m)	M_{ti} (kg-m)	M_i (kg-m)	δ_i (mm)	Δ_i (mm)
3	35,016	35,016	84,038	——	2.26	0.66
2	31,748	66,764	160,233	122,556	1.60	0.66
1	18,676	85,440	205,056	356,230	0.94	0.94
0	——	——	——	783,430	——	——

Calculated values of the story shear forces are shown in Table 14.9.

(e) Horizontal torsional moments M_{ti}. Assuming that the center of rigidity and the center of mass coincide, a minimum eccentricity for the Y direction shall be computed as

$$e_a = 0.1b$$
$$= 0.1 \times 24 = 2.4m \qquad \text{[eq.(14.10)]}$$
$$M_{ti} = V_i e_a$$

Values calculated for the torsional moments for the Y direction are shown in Table 14.9. Figure 14.4 shows the torsional forces for both the X and the Y directions.

(f) Overturning moments M_i

$$M_i = \sum_{k=i+1}^{n} F_k(h_k - h_i) \qquad \text{[eq.(14.12)]}$$
$$i = 0, 1, 2, \ldots, n-1.$$

Calculated values for overturning moments are given in Table 14.9.

(g) Lateral displacements δ_i. The lateral displacements can be calculated approximately from eq.(4.78) of Chapter 4:

$$\delta_i = \frac{g}{4\pi^2} \frac{T^2 F_i}{W_i}$$
$$= \frac{9.8}{4\pi^2} (0.255)^2 \frac{F_i}{W_i}$$
$$= 0.0161 \frac{F_i}{W_i}$$

Calculated story displacements are given in Table 14.9.

(h) Story drifts Δ_i. Story drifts Δ_i are given by the difference between the displacements of the upper and lower levels of the story. Hence

$$\Delta_i = \delta_i - \delta_{i-1}$$

with $\delta_0 = 0$.

Calculated values of story drifts are shown in Table 14.9.

Although the Egyptian Regulations do not provide a limit on drift, the engineer must set a practical limit to ensure acceptable performance. For this example, the limit on drift is set at 0.005 times the story height, according to the *Standard Building Code* (1991).

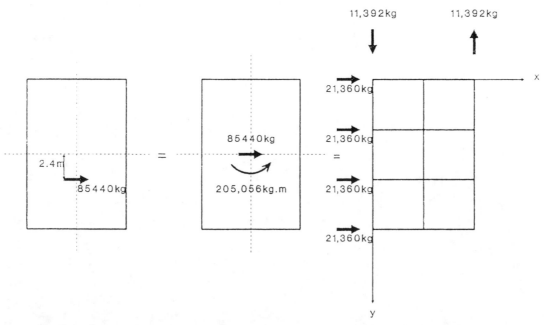

Fig. 14.4. Horizontal torsional forces for level 1

REFERENCES

EGYPTIAN SOCIETY FOR EARTHQUAKE ENGINEERING (1988) *Regulations for Earthquake-Resistant Design of Buildings in Egypt.* Cairo, Egypt.

EGYPTIAN SOCIETY OF ENGINEERING (1989) *The Egyptian Code for the Design and Construction of Reinforced Concrete Buildings, 1989.* Cairo, Egypt.

PAZ, MARIO (1991) *Structural Dynamics: Theory and Computation.* 3d ed. Van Nostrand Reinhold, New York, NY.

SOUTHERN BUILDING CODE CONGRESS INTERNATIONAL (1991) *Standard Building Code.* Birmingham, AL.

15

El Salvador

Celso S. Alfaro

15.1 INTRODUCTION

El Salvador is located in the western part of the Caribbean Plate which contains several tectonic regions (see Fig. 15.1): a very narrow chain of active volcanoes (Carr et al. 1982) along the Pacific coast of Central America; the Cocos Plate which subducts

under the Caribbean Plate southwest of El Salvador forming the Middle America Trench (White and Cifuentes 1988); the North American-Caribbean Plate boundary in the northwest which crosses Guatemala (it contains the Chixoy-Polochic and Motagua faults); and a series of grabens in the northeast that form the Honduras Depression (Kiremidjian et al. 1979).

A narrow band of volcanoes (the Volcano Chain) along the Pacific coast of Central America is in a region of similar tectonic and seismic patterns. The seismic events within this band show strike-slip displacements with one plane parallel to the Volcano Chain (White and Harlow 1988). Left-lateral fault motions have occurred in faults perpendicular to the path of the chain, and right-lateral slip motions have probably occurred along faults parallel to the chain.

Epicentral locations of earthquakes fall within a very narrow band, focal depths of very well-constrained events are within 5.0 and 15.0 km (White and Harlow 1988), and maximum magnitudes of earthquakes range from magnitude M_s (surface wave magnitude) of 5.4 to 6.8. Earthquakes with these characteristics caused severe damage to San Salvador, the capital of El Salvador (1965 with magnitude M_L of 6.0 and 1986 with M_L of 5.8).[1]

At the Middle America Trench, the Cocos Plate subducts under the Caribbean Plate at moderate angles and defines a Benioff zone. This zone appears to be formed by two parts separated at an average depth of 60 km. The shallower part is inclined at about 15° to 25° and the deeper part is inclined at approx-

---

[1] M_L is the magnitude of an earthquake (original Richter magnitude). (See Appendix of this Handbook.)

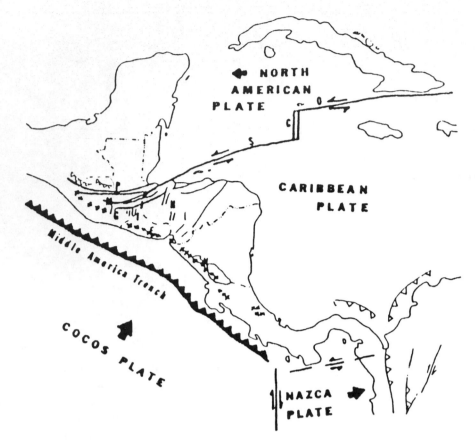

Fig. 15.1. Major structures and boundaries of the western Caribbean Plate

imately 40° to 55°. The subduction zone is suggested by an interpretation of the hypocentral cross-sections of the Middle America Trench (Burbach et al. 1984). White and Cifuentes (1988) have estimated that the subduction zone has the potential to develop earthquakes of magnitude $M_s = 8.1$.[2] The maximum registered earthquake magnitude (M_s of 7.2) occurred in 1982 off the coast of El Salvador; it caused moderate damage to cities in El Salvador.

The North American-Caribbean Plate boundary contains two parallel, left-lateral, strike-slip faults, the Chixoy-Polochic and Motagua faults. However, another subparallel fault has been considered as part of this boundary for purposes of seismic hazard and seismic code development, the Jocotan-Chamelecon fault (Kiremidjian et al. 1979; Alfaro et al 1990). These faults are characterized by left-lateral, strike-slip displacements. Their seismic activity is shallow. The two main faults have ruptured at least once; an

earthquake of magnitude M_w (moment magnitude) of 7.3 to 7.5 (White 1988) ruptured the eastern portion of the Chixoy-Polochic fault, and another earthquake of magnitude M_w of 7.5 to 7.7 ruptured its western portion (White 1985). In 1976, an earthquake of magnitude M_w of 7.5 ruptured the Motagua fault (White 1988); however, there was limited damage to cities in El Salvador.

The Honduras Depression is a group of disconnected grabens or basins that extend from the Caribbean to the Pacific. Its seismicity is very low, magnitudes range from $M_s = 3.0$ to 6.3. Earthquakes from this source have not caused any damage to cities in El Salvador.

Occurrences of earthquakes bring changes in engineering practices and seismic code regulations in most countries of the world, and El Salvador is not an exception. A brief history of the development of seismic code regulations is given in the next section.

---

[2]M_s is the surface wave magnitude of an earthquake (source spectral amplitude of the 20-second-period wave). (See Appendix of this Handbook.)

15.2 SEISMIC CODE DEVELOPMENT

Only earthquakes occurring in the Volcanic Chain and the Benioff zone have caused significant damage to cities in El Salvador. Two earthquakes in the Volcanic Chain have prompted significant changes in the seismic code regulations of El Salvador: the 1965 earthquake (magnitude M_L of 6.0), which resulted in the 1965 Seismic Code (White and Harlow 1988); and the 1986 earthquake (magnitude M_L of 5.8), which destroyed San Salvador and caused significant changes to be made to the 1965 version of the code. At the present time, a slightly modified version of the 1986 seismic code is in use (*Reglamento de Diseño Sísmico de El Salvador*, 1989). However, a major study for the development of a new version of the seismic code was undertaken by the government of El Salvador in 1992.

15.3 CLASSIFICATION OF BUILDINGS

The Salvadoran Seismic Code of 1989 classifies buildings according to their use, structural system, and structural configuration.

Table 15.1. Use Classification of Structures

Category	Function of Structure
I	Important facilities necessary for life, care, and safety after an earthquake. This category includes hospitals and other medical facilities having surgery and emergency treatment areas, communication centers, fire and police stations, and other facilities required for emergency response.
II	Facilities with high occupancy or nonpermanent high occupancy, necessary for continued operation immediately after an earthquake. This category includes government buildings, schools, universities, day-care centers, markets, shopping centers with an area greater than 3,000 m^2 (not including the parking area), department stores with either areas greater than 500 m^2/ story or heights greater than ten meters, stadiums with a capacity for more than 2,000 persons, office buildings greater than four stories in height or with a plan area greater than 1,000 m^2 per story; buildings with more than four stories, parking buildings for emergency vehicles, museums, monuments, bus terminals, medical facilities not included in Category I, and facilities with expensive equipment.
III	Structures with low occupancy. This category includes family residences, commercial and industrial structures, hotels, office buildings, and all structures not included in Categories I and II.
IV	Nonpermanent structures whose failure will not injure persons, or damage either expensive equipment or structures included in Categories I, II, and III.

15.3.1 Use of the Structure

The Code classifies the use of structures into four categories based on the importance and need of the building, before, during, and after a strong earthquake. Table 15.1 provides a general description of the characteristics that help to classify a building in one of these four categories. Buildings are assigned a numerical value, the importance factor I, corresponding to each category (see Table 15.2).

15.3.2 Structural Systems

The Code of 1989 provides a classification of buildings and other structures similar to the one given by the Applied Technology Council – ATC 3-06 (1978) containing five types of structural systems as described in Table 15.3. Table 15.4 provides numerical

Table 15.2. Importance Factor I

Category	Factor I
I	1.5
II	1.3
III	1.0
IV	0.2

Table 15.3. Structural Type Systems

Type Number	System	Description
1	Frames	Unbraced moment-resisting frames that have the capacity to resist both gravity and lateral loads.
2	Structural Walls	A structural system with unbraced space frames that resists both gravity loads and lateral loads.
3	Dual System	A structural system consisting of ductile moment-resisting space frames and either structural walls or braced frames analyzed according to the following criteria: 1. The frames and structural walls or braced frames shall resist the lateral loads in accordance with their relative rigidities considering the interaction between both frames and walls. 2. The moment-resisting space frames have the capacity to resist at least 25% of the seismic lateral forces. 3. In both cases, gravity loads are supported by the frames.
4	Bearing Wall	A structural system with bearing walls that resist the gravity loads. Seismic lateral forces are resisted by structural walls or braced frames.
5	Isolated Systems	Structures supported by a group of columns (water tanks), chimneys, and structures whose columns and walls are not properly tied at the roof or the floors by elements with sufficient rigidity.

Table 15.4.   Structural Factor R and Displacement Amplification Factor C_d

Type Number	Structural System	R	C_d
1	Reinforced concrete MRSF*	8 1/3	8
	Steel MRSF	10	9
2	Reinforced concrete walls	9 1/8	7
	Reinforced masonry walls	6	4
	Reinforced concrete or steel braced frames	7 3/4	6 1/2
3	Reinforced concrete walls	10	9
	Reinforced masonry walls	6 3/4	4 1/2
	Reinforced concrete or steel braced frames	9 1/8	7
4	Reinforced concrete walls	8 1/3	5 1/2
	Reinforced masonry walls	5 1/3	3
	Reinforced concrete or steel braced frames	7 1/8	5
5	Reinforced concrete or steel isolated structures	3 1/3	2 1/2

* Moment-resisting space frames

values for the structural factor R as well as values for the displacement amplification factor C_d, corresponding to each type of structural system defined in Table 15.3.

15.3.3 Structural Configuration

The Code of 1989 classifies buildings as regular or irregular. In this classification, both plan and vertical configurations have to be considered. Structures shall be classified as regular if they comply with the requirements listed in Table 15.5. Structures that do not fulfill these requirements shall be considered as irregular.

15.4 SEISMIC ZONES

The seismic hazard condition for different regions of El Salvador is provided by the seismic zone map shown in Fig. 15.2 with the corresponding values of the peak ground acceleration (PGA) listed in Table 15.6. The PGA values for these zones should not be

Table 15.6.   Peak Ground Acceleration (PGA) Values

Zone	PGA
1	1.0
2	0.5

Table 15.5.   System Configuration

1. Plan Regularity
   - Mass and stiffness regularity shall be considered if they are approximately symmetric at any story.
   - A structure shall be considered as regular if its larger plan dimension is less than 2.5 times its shorter plan dimension.
   - At any story, the torsional eccentricity statically calculated shall not exceed by 10% the plan dimension of the story in the direction under consideration.
   - Projections of re-entrant corners and out-of-plane offsets shall not exceed by more than 20% the plan dimension of the structure in the direction under consideration.

2. Vertical Regularity
   - Each level has either a roof structure or rigid diaphragm.
   - Dimension of openings shall not exceed by 20% the plan dimension of the structure in the direction under consideration. Additionally, these openings shall not change in location at different floor levels and their area shall not exceed by 20% the floor area.
   - Weight of a story shall be neither greater nor smaller (by 20%) than the weight of the story immediately below. An exception shall be made at the highest story.
   - At any level, columns must be properly connected at the floor system by rigid diaphragms, flat slabs, or beams.
   - Lateral stiffness at any story shall not exceed by more than 100% the stiffness of the story immediately below.
   - A structure shall be considered as regular if its height is less than 2.5 times its shorter plan dimension.

interpreted as being consistent with a given acceptable risk. For an updated seismic macrozonation for El Salvador see Alfaro, Kiremidjian, and White (1990).

15.5 SEISMIC COEFFICIENT

The Code of 1989 defines the seismic coefficient C_s by the following formula:

$$C_s = \frac{IA_g D}{R} \tag{15.1}$$

where

I = importance factor (Table 15.2)
A_g = peak ground acceleration (PGA) (Table 15.6)
D = response spectrum factor (Section 15.6)
R = structural type factor (Table 15.4).

15.6 RESPONSE SPECTRUM FACTOR

The response spectrum factor D of a region is defined as the average acceleration response spectrum. The soil-structure interaction effect is not included in this factor. Updated response spectrum charts for the metropolitan area of San Salvador are given by Consorzio (1988).

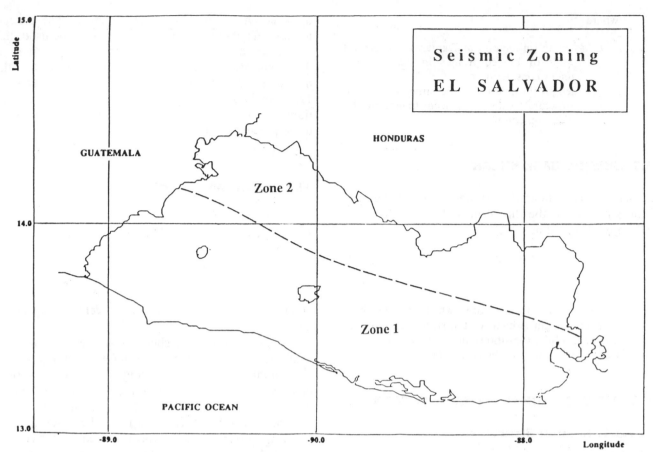

Fig. 15.2. Simplified version of the seismic map for El Salvador

The response spectrum factor D shall be determined by the following formula:

$$D = \frac{0.72}{T^{2/3}} \leqslant 1.0 \tag{15.2}$$

where T = fundamental period (sec) of the structure in the direction considered for the seismic forces.

The fundamental period T may be determined using Rayleigh's formula.

$$T = 2\pi \sqrt{\frac{\sum\limits_{i=1}^{n} W_i \delta_i^2}{g\left[F_t \delta_n + \sum\limits_{i=1}^{n} F_i \delta_i\right]}} \tag{15.3}$$

where

δ_i = the elastic lateral displacement at level i calculated by applying the lateral forces F_i
F_i = system of lateral forces given by eqs.(15.8) and (15.9)
g = acceleration of gravity
W_i = portion of the building weight allocated at level i.

Because the lateral displacements δ_i are not known, eq.(15.3) cannot be used initially to determine the fundamental period T. However, the 1989 Code provides the following empirical formulas for estimating the fundamental period:

(a) For buildings that are classified as Type 1 in Table 15.3,

$$T = C_t h_n^{3/4} \tag{15.4}$$

where

$C_t = 0.85$ for steel moment-resisting space frames
$C_t = 0.073$ for reinforced concrete moment-resisting space frames
h_n = height of the building in meters.

(b) For buildings other than those considered in eq.(15.4),

$$T = \frac{0.09 h_n}{\sqrt{L_e}} \tag{15.5}$$

where

h_n = height of the building in meters
L_e = effective dimension of the building in meters, with

$$L_e = L_{s\,max} \sum \frac{L_s^2}{L_{s\,max}^2} \tag{15.6}$$

where

L_s = length (in meters) of segments of structural walls or braced frames in the direction under consideration

$L_{s\,max}$ = length (in meters) of the larger segment of structural walls or braced frames in the direction considered.

15.7 SEISMIC BASE SHEAR

The seismic base shear V, assumed to act at the base of the structure in the direction under consideration, shall be determined as

$$V = C_s W \tag{15.7}$$

where

W = total weight of the structure, which includes both permanent weight (weight of the structure) and non-permanent weight (partitions and equipment)

C_s = seismic coefficient given by eq.(15.1).

15.8 DISTRIBUTION OF LATERAL FORCES

15.8.1 Regular Structures

The seismic base shear V calculated in eq.(15.7) shall be distributed as lateral forces F_{xi} over the height of the building according to the following formulas:

$$F_x = \frac{(V - F_t) W_x h_x}{\sum_{i=1}^{n} W_i h_i} \tag{15.8}$$

with an additional concentrated force F_t applied at the top level of the building:

$$F_t = 0.07TV > 0.25V \qquad \text{for } T > 0.7 \text{ sec} \tag{15.9}$$

$$F_t = 0 \qquad \text{for } T \leqq 0.7 \text{ sec}$$

where

W_x, W_i = weight of the structure allocated at level x, i

h_x, h_i = height above the base to level x, i

n = number of levels in the building.

At each level x the lateral forces F_x and the force at the top F_t shall be applied over the area, considering the mass distribution at that level.

15.8.2 Irregular Structures

For buildings classified as irregular, the distribution of lateral forces shall be determined considering the dynamic characteristics of the structure. The code is not specific about the dynamic method to be used; this is left to the designer's judgment. For purposes of describing the designing practices in El Salvador, the modal superposition method (Clough and Penzien 1975; Wakabayashi 1986; Paz 1991) and the regulations of ATC 3-06 (1978) will be used for the examples to be presented in this chapter.

15.9 HORIZONTAL SHEAR FORCE

At each level of the building, the horizontal shear force V_x is determined from statics as

$$V_x = F_t + \sum_{i=x}^{n} F_i \tag{15.10}$$

where F_x and F_t are given, respectively, by eqs.(15.8) and (15.9).

The total horizontal shear in any horizontal plane shall be distributed, considering the rigidity of the horizontal diaphragm or bracing system, to the various elements of the lateral force-resisting system in proportion to their rigidities. Rigid elements that have not been assumed as a part of the lateral force-resisting system may be incorporated into buildings provided their effect on the action of the resisting system is accounted for in the analysis.

15.10 HORIZONTAL TORSIONAL MOMENTS

The increase in shear resulting from a horizontal torsion due to an eccentricity between the center of mass and the center of rigidity should be considered in the analysis. Negative torsional shears shall be ignored. The shear-resisting elements shall be capable of resisting a minimum torsional moment resulting by assuming an eccentricity of 5% of the building dimension perpendicular to the direction of the forces under consideration.

15.11 OVERTURNING MOMENTS

The overturning moment M_x at any level x shall be determined by the following formula:

$$M_x = k\left[F_t(h_n - h_x) + \sum_{i=x+1}^{n} F_i(h_i - h_x) \right] \tag{15.11}$$

where

$$x = 0, 1, 2, \ldots, n - 1$$

and

$k = 1.0$ for the top 10 stories
$k = 0.8$ for all stories, except the top 19 stories
$k =$ a value between 1.0 and 0.8 determined by linear interpolation for stories between the top 10 and the top 20 stories.

In determining the overturning moment M_0 at the base of the building, k shall be taken as 0.75, except for inverted pendulum structures. At any level of the building, the increment of overturning moment in the story under consideration shall be distributed to the resisting elements in the same proportion as the distribution of horizontal shear force.

15.12 LATERAL DISPLACEMENTS AND STORY DRIFT

The elastic lateral displacement δ_x at level x shall be computed using an elastic analysis with the structure fixed at the base. The lateral forces to be applied are the forces F_x and F_t determined by eqs.(15.8) and (15.9). The total displacement δ_{tx} at level x shall be determined by the following formula:

$$\delta_{tx} = C_d \delta_x \qquad (15.12)$$

The displacement amplification factor C_d is obtained from Table 15.4 for the different types of structural systems.

The design story drift Δ_x at any level shall be calculated as the difference of the total displacements of two consecutive stories. At any level, the design story drift $C_d \Delta_x$ shall not exceed the limit of 0.015 times the height of the story under consideration. If the calculated drift exceeds this limiting value, the structure shall be stiffened to satisfy this condition.

15.13 P-DELTA EFFECTS

P-delta (P-Δ) effects on story shear, overturning moments, and story drifts shall be neglected if the stability coefficient and any level of the building is equal to or less than 0.1. The stability coefficient θ_x at level x shall be calculated by the following formula:

$$\theta_x = \frac{P_x \Delta_x}{C_d V_x h_{sx}} \qquad (15.13)$$

where

P_x = total gravity load at and above level x
Δ_x = design story drift for story x
V_x = shear force at level x
h_{sx} = story height below level x.

When the stability coefficient θ_x at any level of the building is greater than 0.1, P-delta effects shall be determined. The total story shear forces and moments shall be determined by adding shears and moments due to the increase in story drift to those determined without considering the P-delta effects.

15.14 ADDITIONAL REGULATIONS

The Salvadorian Seismic Code also provides general regulations for consideration of live and dead loads, orthogonal effects, soil mechanics and foundation design, reinforced concrete, steel, reinforced masonry, building repairs, inspection and material quality, alternate methods of analysis, and construction that will not be covered in this chapter. A copy of the Seismic Code for Buildings for El Salvador can be obtained by request to the Salvadorian Society of Civil Engineers and Architects.

15.15 NUMERICAL EXAMPLE 15.1

The four-story reinforced concrete building shown in Figs. 15.3 and 15.4 is analyzed for seismic forces. The building has the following characteristics:

- Structure Use: Office – Category II
- Structure System: Unbraced moment-resisting space frame – Type 1
- Structure Configuration: Regular – mass and stiffness symmetry
- Seismic Zoning: Zone 1
- Material: Concrete ($E_c = 2.1 \times 10^9$ kg/m^2)
- Cross-section for all columns: 0.45 m $\times$ 0.55 m

Solution

(a) *Fundamental period T*

$$T = C_t h_n^{3/4} \qquad [eq.(15.4)]$$
$$= 0.073(12)^{3/4}$$
$$= 0.47 \text{ sec}$$

(b) *Base shear V*

$$V = E_s W \qquad [eq.(15.7)]$$

where

$$C_s = \frac{I A_g D}{R} \qquad [eq.(15.1)]$$

$I = 1.3$ (from Tables 15.1 and 15.2, office)
$A_g = 1.0$ (for site in Zone 1, Table 15.6)
$R = 8 \ 1/3$ (from Table 15.4, moment-resisting space frame)
$W = 1,160$ (tonnes) (from Fig. 15.5)

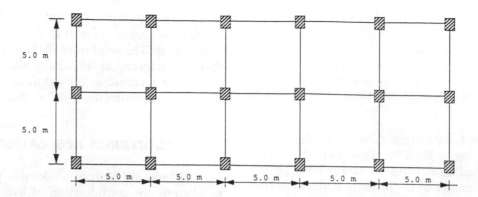

Fig. 15.3. Plan view of building for Example 15.1

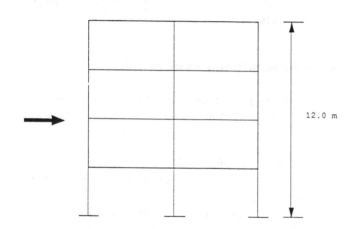

Fig. 15.4. Elevation of building for Example 15.1

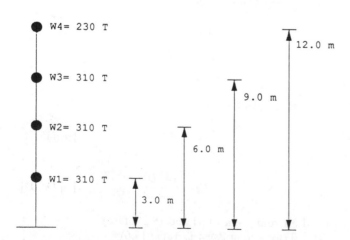

Fig. 15.5. Model of building of Example 15.1 for the seismic analysis

Table 15.7. Lateral Forces F_x and Story Shear Force V_x

Level	W_x (tonnes)	h_x (m)	$W_x h_x$ (tonne-m)	F_x (tonnes)	V_x (tonnes)
4	230	12	2,760	59.9	59.9
3	310	9	2,790	60.5	120.4
2	310	6	1,860	40.3	160.7
1	310	3	930	20.2	180.9
			$\Sigma =$ 8,340		

and

$$D = \frac{0.72}{T^{2/3}} \qquad \text{[eq.(15.2)]}$$

$$= 1.0$$

Hence,

$$C_s = \frac{1.3 \times 1.0 \times 1.01}{8^{2/3}} = 0.156$$

$$V = 0.156 \times 1,160$$

$$= 180.9 \text{ (tonnes)}$$

(c) Lateral forces F_x

$$F_x = \frac{(V - F_t) W_x h_x}{\sum_{i=1}^{n} W_i h_i} \qquad \text{[eq.(15.8)]}$$

with $F_t = 0$ since $T = 0.47 < 0.7$ sec.
Calculated lateral forces F_x are given in Table 15.7.

(d) Overturning moments M_x

$$M_x = k\left[F_t(h_n - h_x) + \sum_{i=x+1}^{n} F_i(h_i - h_x) \right] \qquad \text{[eq.(15.11)]}$$

Table 15.8. Overturning and Torsional Moments

Level	F_x (tonnes)	M_x (tonne-m)	V_x (tonnes)	M_{tx} (tonne-m)
4	59.9	—	59.9	74.9
3	60.5	179.7	120.4	150.5
2	40.3	540.9	160.7	200.9
1	20.2	1,023.0	180.9	226.1
0	—	1,565.7	—	—

Table 15.9. Calculation of Design Story Drift

Level	V_x (tonnes)	k_x (tonne/m)	Δ_x (m)	$C_d\Delta_x$ (m)
4	59.9	1.04×10^5	0.00057	0.00456
3	120.4	1.04×10^5	0.00115	0.00920
2	160.7	1.04×10^5	0.00153	0.01224
1	180.9	1.04×10^5	0.00173	0.01381

where $x = 0, 1, 2, \ldots, n-1$ and with $k = 1$ (for four-story building).

Calculated values are shown in Table 15.8.

(e) Torsional moments M_{tx}

$$M_{tx} = 0.05 L_x V_x$$

$$M_{tx} = (0.05)(25)V_x$$

where

L_x = building dimension perpendicular to the direction of the forces under consideration
V_x = story shear forces.

Calculations are shown in Table 15.8.

(f) Story drift and lateral displacements. Assuming that the building is a shear building (Paz 1991), the stiffness for a column between two consecutive stories can be calculated as:

$$k = \frac{12EI}{L^3}$$

where

$E = 2.1 \times 10^9$ kg/m^2
$L = 3$ m
I = moment of inertia for a section of $0.45\,\text{m} \times 0.55\,\text{m}$

The total stiffness for any story is

$$k_1 = k_2 = k_3 = k_4 = 1.04816 \times 10^5 \text{ tonne/m}$$

The story drift can be calculated as

$$\Delta = \frac{V_x}{k_x}$$

From Section 15.12, the design story drift $C_d\Delta_x$ shall not exceed 0.015 times the height of the story under consideration ($0.015 \times 3 = 0.045$ m). The value for $C_d = 8.0$ was taken from Table 15.4.

Results of calculations are presented in Table 15.9.

The allowable drift 0.045 m is larger than the design story drift calculated in Table 15.9. A recalculation of the fundamental period of vibration T using eq.(15.3) is required. If the recalculated value for T is not close enough to the one calculated by eq.(15.4) or by eq.(15.5), the analysis should be repeated using the new value for T.

REFERENCES

ALFARO, C. S.; KIREMIDJIAN, A. S.; and WHITE, R. A. (1990) *Seismic Zoning and Ground Motion Parameters for El Salvador*. Report No. 93, The John A. Blume Earthquake Engineering Center, Stanford University, Stanford, CA.

APPLIED TECHNOLOGY COUNCIL – ATC 3-06 (1978) *Tentative Provisions for the Development of Seismic Regulations for Buildings*. National Bureau of Standards, Special Publication 510, Washington, D.C.

A.S.I.A., Asociación Salvadoreña de Ingenieros y Arquitectos (1989) *Reglamento de Diseño Sísmico de El Salvador*. A.S.I.A., San Salvador, El Salvador.

BURBACH, G.; VAN, N.; FROHLICH, C.; PENNINGTON, W. D.; and MATUMOTO, T. (1984) "Seismicity and Tectonics of the Subducted Cocos Plate." *Journal of Geotechnical Research* 98: 7719–7735.

CARR, M. J.; ROSE, Jr., W. I.; and STOIBER, R. E. (1982) "Regional Distribution and Character of Active Andesitic Volcanism: Central America." In *Andesites, Orogenic Andesites and Related Rocks*, R. S. Thorpe, ed., 149–166. John Wiley, New York, NY.

CLOUGH, R. W., and PENZIEN, J. (1975) *Dynamics of Structures*. McGraw-Hill, New York, NY.

CONSORZIO, SALVADOR E. (1988) *San Salvador – Programa di Ricostruzione allo Sviluppo*. Rupubblica Italiana, Ministerio degli affari Esteri, Rome, Italy.

KIREMIDJIAN, A. S.; SUTCH, P.; and SHAH, H. C. (1979) *Seismic Hazard Analysis of Honduras*. Report No. 38, The John A. Blume Earthquake Engineering Center, Stanford University, Stanford, CA.

PAZ, M. (1991) *Structural Dynamics: Theory and Computation*. 3d ed. Van Nostrand Reinhold, New York, NY.

WAKABAYASHI, M. (1986) *Design of Earthquake Resistant Buildings*. McGraw-Hill, New York, NY.

WHITE, R. A. (1985) "The Guatemala Earthquake of 1816 on the Chixoy-Polochic Fault." *Bull. Seis. Soc. Am.* 75, No. 2: 455–73.

———— (1988) *Catalog of Historic Seismicity in the Vicinity of the Chixoy-Polochic and Motagua Faults, Guatemala.* Open-File Report 84–88, V.S.G.S., Menlo Park, CA.

WHITE, R. A., and CIFUENTES, Q. (1988) "Seismic History of the Middle America Trench Spanning the Guatemala Triple Junctionand an Earthquake Forecast for Western El Salvador." Submitted to B.S.S.A.

WHITE, R. A., and HARLOW, D. H. (1988) "Hypocentral Parameters of Significant Shallow Volcanic Chain Earthquakes of Central America Since 1890." Submitted to B.S.S.A.

16

France

Auguste Boissonnade*

16.1 INTRODUCTION

Although most of France has experienced little or no seismicity, records starting in A.D. 1021 indicate that tremors with intensities up to X on the International Macroseismic Intensity Scale (MSK)[1] have occurred in the southern and southeastern parts of France. (The

---

*The author wishes to acknowledge the assistance of the French Earthquake Engineering Association and more specifically of its president, Victor Davidovici.

[1]The International Macroseismic Intensity Scale (MSK) which is similar to the Modified Mercalli Intensity Scale (MMI), provides an empirical appraisal of the effects of an earthquake based on observed damage. (See the Appendix of this book.)

most recent damaging earthquake in these areas occurred in 1909 in the county of Nice.)

Seismic recommendations (SOCOTEC 1955) were developed in the mid-1950's following the 1954 catastrophic earthquake at Orleansville (later named El Asnam and more recently Ech Cheliff) in Algeria. French recommendations for seismic design of buildings and other structures were published in 1962, 1964, and 1969. The 1969 seismic design recommendations, modified in 1982 (DTU 1982), became compulsory in 1987 for all new structures built in seismic zones. The 1982 modifications did not remedy all the Code shortcomings, such as the need to include the seismic risk posed to modern facilities and major civil structures located in seismic zones (AFPS 1990), so the French authorities have undertaken a major revision of the current seismic code of 1982.[2]

Pending the new Code, structures proposed for construction in seismic zones should be designed according to the provisions of the PS 69 modified 1982 Code (DTU 1982). Structures that are not considered to be within the purview of the 1982 Code may be designed with the AFPS-90 recommendations. Whether or not a structure is in a seismic region can be determined by reference to the official decree published in 1991 (*Delegations aux Risques Majeurs*). Because of this transition period in the French seismic regulations, this chapter will first present the Seismic Code PS 69/82 and then introduce the main features of the AFPS-90 recommendations.

16.2 SEISMIC CODE PS 69, MODIFIED 1982

16.2.1 Introduction

The main topics included in PS 69/82 are the determination of seismic design forces, basic principles of seismic design, overall rules of good seismic construction, and minimum detailing requirements for concrete and masonry structures. However, this chapter is limited to Code determination of design seismic forces.

The Code PS 69/82 requires that buildings be analyzed under the action of an equivalent system of static forces applied horizontally at various levels of the structure. Dynamic methods of analysis may be used provided the input seismicity level is comparable to the seismicity level specified in the Code. However, no definite instructions are given regarding the analysis of structures by dynamic methods; thus, PS 69/82 contains requirements only for the evaluation of equivalent static seismic forces.

The Code PS 69/82 does not provide for an initial estimate of the total base shear force to be distributed at various levels of the building. The design seismic forces are applied directly to each structural element as a function of the dynamic characteristics of the structure. Three types of equivalent systems of forces are considered:

1. An equivalent system of static horizontal forces (S_H);
2. An equivalent system of static vertical forces (S_V);
3. An equivalent system of global horizontal torsional moments (S_T).

The equivalent system of static horizontal forces is composed of elementary parallel forces applied to the center of gravity of the elements of the structure. The intensity of each force, acting along a given direction x, is given by:

$$F_x = \sigma_x W \qquad (16.1)$$

in which σ_x is the horizontal seismic coefficient specified in Section 16.2.4 and W is the seismic weight defined in Section 16.2.2. The Code specifies that the horizontal forces F_x should be applied in the most unfavorable direction to obtain the seismic horizontal force ($F_H = \sigma_H W$). However, the Code permits the designer to design for the application of seismic horizontal forces independently along the two orthogonal main directions of the building, in lieu of determining the most unfavorable direction.

The equivalent system of static vertical forces is composed of elementary forces, each one acting on an individual element of the structure through its center of gravity. These forces act all in the same direction, either upward or downward. The intensity of each of these forces is given by

$$F_v = \sigma_v W \qquad (16.2)$$

in which σ_v is the vertical seismic coefficient defined in Section 16.2.4.

The design torsional moment at any story of the building shall be the moment resulting from the eccentricity between the resultant horizontal force applied at the center of mass, and the center of rigidity of that story. The PS 69/82 Code also requires that an *accidental torsional moment* be added to the actual torsional moment in either of the following cases:

- When η, the ratio between the square of the largest dimension of a planar section of the structure perpendicular to the direction of the horizontal forces, and the area of that section, is greater than 2.5

---

[2]Nuclear power plants are enforced under special seismic regulations.

- When more than two-thirds of the lateral loads are carried in the same plane by one element of the lateral load-resisting system.

The accidental eccentricity e to be added to the actual eccentricity is given by

$$e = \left(\frac{1}{20}\right)(\lambda + \mu)(L_y + d) \tag{16.3}$$

in which

L_y = the largest dimension in plan of the structure at a given level, perpendicular to the direction of the horizontal forces

d = the distance between the center of rigidity and the center of mass of the story considered

$\lambda = 0$ if $\eta \leqslant 2.5$
$\lambda = \eta - 2.5$ if $\eta > 2.5$, with a maximum value of $\lambda = 1.0$;
$\mu = 0$ for $\theta \leqslant 2/3$
$\mu = 6(\theta - 2/3)$ for $\theta > 2/3$

in which θ is the largest fraction of the lateral force carried by a single structural element of the lateral force-resisting system, at a given horizontal plane of the building.

16.2.2 Seismic Weights

The seismic weight W in eq.(16.1) includes dead load (the total weight of the structure and permanent equipment) and applicable portions of other loads as follows:

- Snow loads[3] in excess of 0.35 kN/m^2
- Live loads multiplied by a reduction factor given in Table 16.1.

16.2.3 Load Combinations

The following loads should be considered in the seismic design of structures in their most unfavorable combinations:

1. Gravitational loads
   - Dead loads
   - Live loads multiplied by a reduction factor (Table 16.1)
   - Snow loads as specified in Section 16.2.2.
2. Dynamic loads
   - Seismic loads
   - Loads caused by movement or operation of permanent equipment.
3. Loads caused by differential ground motions
4. Thermal loads.

---

[3]The Code specifies this load as 35 dN/m^2, where $1 \text{ dN} = 10$ newtons.

Table 16.1. Reduction Coefficient For Live Load

Case	Description	Reduction Coefficient
1	Dwellings and similar buildings (offices, schools, hospitals, etc.)	0.20
2	Public spaces, libraries, industrial buildings, garages, etc:	
	• machinery, mobile or not	1.0
	• other loads	0.5
3	Storage buildings and warehouses	1.0

Note that in some instances the most unfavorable combination of loads may occur when reduction factors are not applied to the live loads. For example, in the determination of the strength of reinforced concrete columns, in the analysis of the steel reinforcement under tension, a reduction factor on the live load should be used; no such reduction factor should be used for the strength verification of the concrete in compression.

16.2.4 Seismic Coefficients

The PS 69/82 Code considers seismic coefficients for the following three idealized structures: (1) regular shear buildings with uniform damping for all the modes of vibration, (2) flexible structures with low damping, and (3) structures with several degrees of freedom in a single horizontal plane.

The seismic coefficients for structures that can be idealized as regular shear building models are presented in Section 16.2.4.1. Regular shear buildings are those in which the entire mass can be considered concentrated at the floor levels, for which the floor systems and beams are rigid, whereas the columns are flexible to lateral deformation but rigid in the vertical direction. These structures should exhibit uniform damping for all degrees of freedom and, in addition, their dynamic analysis need consider only the fundamental mode of vibration.

Seismic coefficients for flexible and relatively lightly damped structures with periods exceeding 0.5 second (for example chimneys) and structures having several degrees of freedom in a single horizontal plane are given in Sections 16.2.4.2 and 16.2.4.3, respectively.

16.2.4.1 Seismic coefficients for regular shear building models

16.2.4.1.1 HORIZONTAL SEISMIC COEFFICIENT. The horizontal seismic coefficient σ_{ix} in eq.(16.1) corresponding to level i, is defined as the ratio between the seismic acceleration at that level and the acceleration of gravity. This coefficient is expressed by the product of four factors as indicated in the following formula:

$$\sigma_x = \alpha\beta\delta\gamma \qquad (16.4)$$

where

α = intensity factor, defined in Section 16.2.6
β = response factor, defined in Section 16.2.7
δ = foundation factor, defined in Section 16.2.8
γ = distribution factor, defined in Section 16.2.9.

The horizontal seismic coefficient σ_{ix} depends on the fundamental period of the structure T as the factor β is a function of T.

16.2.4.1.2 VERTICAL SEISMIC COEFFICIENT. The PS 69/82 Code establishes that the elements of the lateral load resisting systems should be considered subject to vertical forces applied to element centers of gravity, in order to evaluate the stability of the whole system. In the determination of these vertical forces, the total of the live load and dead load is considered. The vertical seismic coefficient is defined by

$$\sigma_V = \pm \frac{1}{\sqrt{\alpha}} \sigma_H \qquad (16.5)$$

in which σ_H is the maximum horizontal seismic coefficient obtained for this element in the various horizontal directions (see Section 16.2.1) and α is the intensity factor defined in Section 16.2.6. For cases where α is less than 1.0, σ_V may be taken to be equal to σ_H.

The vertical seismic coefficient defined in eq.(16.5) is not applicable to elements for which strength is not appreciably affected by horizontal stresses. This includes cantilevered elements and corbels, hanging elements with low inertia, and ties of vaults or arches. For those elements, the vertical seismic coefficient is given by

$$\sigma_V = \pm \frac{2}{\delta} \sigma_H \leq 1.0 \qquad (16.6)$$

In addition, cantilevers and corbels and other overhanging components should be designed using a seismic vertical coefficient

$$\sigma_V = \frac{0.15\alpha}{T} \qquad (16.7)$$

but not exceeding 0.4α or unity. In eq.(16.7), T is the vertical period of the particular element and α is the intensity factor.

16.2.4.2 Seismic coefficients for flexible structures with light damping. The design of flexible structures with relatively light damping must take into account the effects of higher vibration modes when the fun-

damental period is greater than 0.5 second. The higher modes that should be considered, in addition to the fundamental mode, are the second mode, when the period of the fundamental mode is between 0.5 and 0.75 second, and the second and third modes, when the same fundamental period is greater than 0.75 second. The periods and mode shapes can be determined by theoretical calculations or by experimental tests on structural models or on actual structures.

16.2.4.2.1 HORIZONTAL SEISMIC COEFFICIENT. The horizontal seismic coefficient for the ith mode of vibration is defined for each element of the structure by

$$\sigma_i = \alpha\beta_i\delta\gamma_i \qquad (16.8)$$

where

α = intensity factor (Section 16.2.6)
β_i = response factor, dependent on the mode of vibration (Section 16.2.7)
δ = foundation factor (Section 16.2.8)
γ_i = distribution factor, dependent on the mode of oscillation (Section 16.2.9).

16.2.4.2.2 VERTICAL SEISMIC COEFFICIENT. The vertical seismic coefficient is defined by

$$\sigma_V = \frac{1}{\sqrt{\alpha}} (\sigma_H)_1 \qquad (16.9)$$

in which $(\sigma_H)_1$ is the horizontal seismic coefficient in the most unfavorable direction, corresponding to the fundamental mode.

16.2.4.3 Seismic coefficients for structures with several degrees of freedom in a single horizontal plane. Structures having several degrees of freedom in one horizontal plane may be idealized as a series of linked one-degree-of-freedom elementary oscillators constrained to show the same deformations in the horizontal planes where they are connected; i.e., the movements of all oscillators must be identical.

16.2.4.3.1 HORIZONTAL SEISMIC COEFFICIENT. The horizontal seismic coefficient for the ith mode of oscillation is defined by

$$\sigma_i = \alpha\beta_i\delta\gamma_i \qquad (16.10)$$

in which the factors α and δ are defined in Sections 16.2.6 and 16.2.8, respectively. The response factor β_i is defined in Section 16.2.7 for flexible structures with light damping [see eqs.(16.19) and (16.20)]. For other structures, the response factor for the first mode β_1 is presented in Section 16.2.7, eqs.(16.16), (16.17), and (16.18). The response factors β_2 and β_3 for the second and third modes are defined in Section 16.2.7 for flexible structures with light damping [eq.(16.19) or (16.20)] and reduced to 80% or 60% depending on whether or not these structures have moderate or normal damping, respectively.

The distribution factor γ_i is determined using eqs.(16.21) or (16.22).

16.2.4.3.2 VERTICAL SEISMIC COEFFICIENT. The vertical seismic coefficient is evaluated in the same manner as for structures defined in Sections 16.2.4.1 and 16.2.4.2.

16.2.5 Fundamental Period

The PS 69/82 Code indicates that the fundamental period T of the structure may be calculated using exact theoretical procedures or determined experimentally in a structure, or in a structural model with similar characteristics.

The Code also provides empirical formulas for the estimation of the fundamental period of buildings with simple geometry and regularity among the stories:

- For buildings with masonry or cast-in-place concrete structural walls:

$$T = 0.06 \frac{H}{\sqrt{L_x}} \sqrt{\frac{H}{2L_x + H}} \qquad (16.11)$$

- For buildings with reinforced concrete shear walls or with steel or concrete bracing

$$T = 0.08 \frac{H}{\sqrt{L_x}} \sqrt{\frac{H}{L_x + H}} \qquad (16.12)$$

- For buildings with reinforced concrete moment-resisting frames

$$T = 0.09 \frac{H}{\sqrt{L_x}} \qquad (16.13)$$

- For buildings with steel moment-resisting frames

$$T = 0.10 \frac{H}{\sqrt{L_x}} \qquad (16.14)$$

For buildings with dual lateral force-resisting systems, the most unfavorable of the empirical formulas should be used if the lateral load-resisting system is composed of any of the systems described above.

In these formulas, H is the height of the structure (in meters), L_x is the dimension in plan (in meters) in the direction x of the seismic force considered, and T is the fundamental period of the structure in the same direction x, expressed in seconds.

16.2.6 Intensity Factor α

The intensity factor α is a function of the seismicity level of the region where the structure is located, and

Table 16.2. Correlation Between Intensity Levels and Nominal Intensity

Nominal Intensity (i_N)	6	7	8	9	10
Intensity Level		VII	VIII	IX	X	XI

of the importance of the structure; that is, of the associated level of risk.

France is subdivided into seismic zones on the basis of maximum probable seismic intensity level in the International Macroseismic Intensity Scale. The Code provides a correlation between the seismic intensity levels and the nominal intensity i_N, given in Table 16.2.

The intensity factor α is correlated with the nominal intensity i_N by the following formula:

$$\alpha = 2^{(i_N - 8)} \qquad (16.15)$$

Equation (16.15) establishes a minimum value of the intensity factor α for each seismic zone. Structures designed using eq.(16.15) and other provisions in the Code are assumed to be protected against specified ground motion levels characterized by intensity levels of the International Macroseismic Intensity Scale. For example, a structure designed with a value of α equal to unity is assumed to be protected against a nominal ground motion shaking intensity of 8 corresponding to level IX in the International Macroseismic Intensity Scale.

The minimum values of the intensity factor to be used in a specific locality are given by official decree on seismic zonation (Delegation aux Risques Majeurs 1991). This decree lists the classification of each city and county within the French territory.

France is divided into the following five seismic zones (1991 classification):

Zone 0.   No or negligible seismicity
Zone Ia.   Very low but nonnegligible seismicity
Zone Ib.   Low seismicity
Zone II.   Moderate seismicity
Zone III. Strong seismicity.

The minimum intensity factors to be used for the different zones are:

Zone I. $\alpha = 0.5$
Zone II. $\alpha = 1.0$
Zone III. $\alpha = 1.5$.

It is recommended that structures be classified by seismic risk levels although the PS 69/82 Code does not so specify. These risk levels define the level of desired protection, dependent upon the anticipated

Table 16.3.   Classification of Structures by Importance

Structure Class	Definition	Occupancy
D	Structures whose safety is fundamental for emergency response	Communication centers, hospital and other medical facilities, structures and equipment required for emergency response, drinking water production facilities, power-generating and power-distribution facilities
C	Structures whose primary occupancy is greater than 300 persons	Housing projects, collective use structures, offices, industrial facilities
B	Structures whose primary occupancy is less than 300 persons	Dwellings, apartment houses, offices, industrial facilities
A	Structures with no permanent occupants	—

Table 16.4.   Values of Intensity Factor α

Seismicity Zone		Structure Class		
		B	C	D
0	Negligible seismicity	—	—	—
Ia	Very low seismicity	0.50	0.50	0.75
Ib	Low seismicity	0.50	0.75	1.00
II	Moderate seismicity	1.00	1.20	1.50
III	Strong seismicity	1.50	1.70	2.00

use and importance of the structure. Suggestions for such classifications are given in the PS 69/82 Code. The classification given in Table 16.3 is based on a recent bulletin published by the French Earthquake Engineering Association (AFPS 1990).

The recommended values for the intensity factor α for different classes of structures and seismic zones are given in Table 16.4.

16.2.7 Response Factor β

The response factor β depends on the fundamental period of the structure, on its damping characteristics, and on the nature of the soils at the site. Two categories of soils are considered in the evaluation of the response factor: (1) soft soils of thickness in excess of 15 meters and (2) soft soils of thickness less than 15 meters or other types of soils of any thickness. The PS 69/82 Code provides expressions for the response factor β for the following damping conditions:

Normal damping. Structures with normal damping such as ordinary buildings used for housing and offices. For these structures the response factor is given by

$$\beta = \frac{0.065}{\sqrt[3]{T}} \qquad (16.16)$$

with a minimum value of 0.050 and a maximum value of 0.085 for soft soils and of 0.100 for other soils.

Moderate damping. Structures or parts of buildings (e.g., large halls with few subdivisions), buildings with light partitions or partitions weakly connected to the structural elements, high-rise buildings with curtain walls, etc. For these structures the response factor is given by

$$\beta = \frac{0.085}{\sqrt[3]{T}} \qquad (16.17)$$

with a minimum value of 0.065 and a maximum value of 0.110 for soft soils and of 0.130 for other soils.

Light damping. Structures or parts of buildings that cannot be classified in either of the two above classes; for example, large open halls, industrial floors on unclad structural frames, factory chimneys, water towers, antennas, etc. For these structures the response factor is given by

$$\beta = \frac{0.105}{\sqrt[4]{T^3}} \qquad (16.18)$$

with a minimum value of 0.075 and a maximum value of 0.175 for soft soils and a value of 0.200 for other soils.

Note: Application of eq.(16.18) is limited to structures for which the fundamental period is less than or equal to 0.5 second.

16.2.7.1 Flexible structures with light damping. For flexible structures with light damping and for which the period is greater than 0.5 second, the coefficient of response β_i for the ith mode of vibration and for soils other than those classified as "soft" depends on the value of the corresponding natural period T_i:

$$\begin{aligned} \beta_i &= 1.25 T_i & &\text{for } 0 < T_i \leqslant 0.2 \text{ sec} \\ \beta_i &= 0.25 & &\text{for } 0.2 < T_i \leqslant 0.26 \text{ sec} \end{aligned}$$
$$(16.19)$$

$$\beta_i = \frac{0.09}{\sqrt[4]{T_i^3}} \qquad \text{for } T_i > 0.26 \text{ sec}$$

with a minimum value of $\beta = 0.06$.

For soft soils, the response factor is given by

$$\begin{aligned} \beta_i &= 1.25 T_i & &\text{for } 0 < T_i \leqslant 0.16 \text{ sec} \\ \beta_i &= 0.20 & &\text{for } 0.16 < T_i \leqslant 0.35 \text{ sec} \end{aligned}$$
$$(16.20)$$

$$\beta_i = \frac{0.09}{\sqrt[4]{T_i^3}} \qquad \text{for } T_i > 0.35 \text{ sec}$$

with a minimum value of $\beta = 0.06$.

Table 16.5 Foundation Factor δ

Type of Foundation	Type of Soil			
	A Rock	B Medium Consistency	C Loose, fairly high water content	D Waterlogged clays and silts
Spread Footing	1.00	1.15	1.25	—
Deep Foundation	0.90	1.00	1.15	—
Mat	—	1.00	1.10	1.20
Piles:				
Bearing Type	—	1.10	1.15	1.30
Friction Type	—	1.10	1.30	—

16.2.8 Foundation Factor δ

The value of the foundation factor δ is given in Table 16.5 as a function of the nature of the soils and types of foundations. Note that the foundation factor δ is independent of the dynamic properties of the structure.

16.2.9 Distribution Factor γ

For structures modeled with discrete concentrated masses $(M(Z))$ or distributed masses $(m(z))$ at elevations Z or z above the fixed base, the distribution factor for the fundamental mode is given by

$$\gamma(h) = A(h) \frac{\Sigma M(Z) A(Z) + \int m(z) A(z) dz}{\Sigma M(Z) A^2(Z) + \int m(z) A^2(z) dz} \qquad (16.21)$$

where $A(Z)$ and $A(z)$ are the lateral displacements of the structure at elevations Z and z, respectively.

For structures where the second and higher modes need to be considered, eq.(16.21) can be used for each mode of oscillation. Hence, the distribution factor γ_i for the ith mode of oscillation is given by

$$\gamma_i(h) = A_i(h) \frac{\Sigma M(Z) A_i(z) + \int m(z) A_i(z) dz}{\Sigma M(Z) A_i^2(z) + \int m(z) A_i^2(z) dz} \qquad (16.22)$$

in which the function $A_i(z)$ represents the lateral displacement at level z of the structure in the ith mode.

16.2.9.1 Simplified calculations of the distribution factor. The Code provides a simplified formula for the value of the distribution factor corresponding to the fundamental mode based on the assumption of a linear vibration in displacement along the height of the building. Under this assumption, eq.(16.21) can be replaced by

$$\gamma(h) = h \frac{\int Z M(Z)}{\int Z^2 M(Z)} = h \frac{S}{I} \qquad (16.23)$$

in which h is the height of a given level, and S and I are respectively the static moment [eq.(16.24a)] and the moment of inertia [eq.(16.24b)] of the concentrated weights with respect to the base of the building.

$$S = \sum_{k=1}^{N} W_k h_k \qquad (16.24a)$$

and

$$I = \sum_{k=1}^{N} W_k h_k^2 \qquad (16.24b)$$

where

W_k = concentrated weight at level k
h_k = height of level as measured from the base of the building
N = number of levels above the base of the building
H = height of the building

For structures where heights and weights are equal among levels (or not too different), eq.(16.23) reduces to

$$\gamma_i = \frac{3i}{2n + 1} \qquad (16.25)$$

in which n is the total number of levels above the base and i is the level considered.

16.2.10 Evaluation of Seismic Forces

Structural elements must be designed to resist the most unfavorable combination of the equivalent lateral force S_H, the equivalent vertical force S_V, and the equivalent torsional moment S_T. It is acceptable to estimate the horizontal seismic forces applied independently in only two perpendicular horizontal directions.

For normal structures, defined in Section 16.2.4 as regular shear buildings, the equivalent forces are estimated for the fundamental mode only. For flexible structures with light damping for which the fundamental period is greater than 0.5 second (see Section 16.2.4.2), the lateral forces for each mode applied to each structural element are combined, using eq.(16.26), to obtain the effective force P.

If the most unfavorable result from the application of the horizontal seismic forces for the first mode and the vertical seismic forces is designated by P_1, and the most unfavorable result from the application of the horizontal seismic forces in the second and third modes are denoted by P_2 and P_3, respectively, then the combined effect P is calculated by

$$P = \sqrt{P_1^2 + \lambda_2 P_2^2 + \lambda_3 P_3^2} \qquad (16.26)$$

where

$$\lambda_2 = 4T_1 - 2$$

with a maximum value of 1 for $T \geqslant 0.75$ sec and a minimum of 0 for $T \leqslant 0.5$ sec, and

$$\lambda_3 = 4T_1 - 3$$

with a maximum of 1 for $T \geqslant 1$ sec and a minimum of 0 for $T \leqslant 0.75$ sec.

16.2.11 Story Drift

The story drift, that is, the relative displacement between two adjacent stories, should not exceed $\sqrt{\alpha}/1{,}000$ times the story height, in which α is the intensity factor.

16.2.12 Seismic Joints

In general, the lateral displacements at the various levels of the structures should be determined with consideration for the deterioration of the rigidity of the structural elements. However, unless special criteria are available, lateral displacements may be determined by static analysis of the structure subjected to the equivalent system of forces. In such instances, the resulting displacements should be multiplied by 2.0 for structures with shear walls or bracing, and by 4.0 for structures with lateral load-resisting frames.

The minimum distance between two adjacent structures for which the individual total lateral displacements are d_1 and d_2, respectively, is obtained by

$$d = \sqrt{d_1^2 + d_2^2} \qquad (16.27)$$

but should not be less than 4 cm.

16.2.13 Material Strengths

For limit state design, it should be verified that the limit stresses will not be exceeded under the most unfavorable load combinations. For allowable stress design, the worst load combination should not cause stresses higher than 150% of the allowable strength of ductile materials and 125% of the strength of nonductile materials (no increases are allowed for hollow masonry blocks), with the exception of reinforced concrete where the maximum concrete strength is taken as 85% of the nominal 28-day concrete uniaxial compression strength and the maximum steel strength is taken as the nominal steel-yield stress in tension.

16.2.14 Example 16.1

The simple example introduced in this section illustrates the applications of the provisions of the PS 69/82 Code. The structure considered is a five-story reinforced concrete apartment building with three

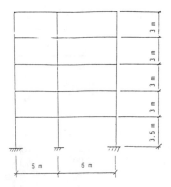

b) Elevation

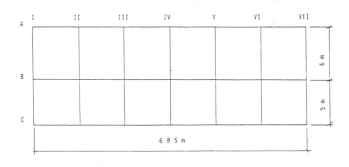

a) Plan

Fig. 16.1. Plan and elevation for a five-story building for Example 16.1

frames in the longitudinal direction (*x* direction) and seven frames in the transverse direction (*y* direction) as shown in Fig. 16.1. The system floors and beams are assumed to be rigid at each floor, and the columns are assumed to be rigid in the vertical direction. This building can be idealized as a shear building with masses lumped at each floor. The floor height is 3.00 meters except for the height of the first floor, which is 3.50 meters. The total dead load for each floor is 2,300 kN and the live load for each floor is 600 kN, with the exception of the roof where the live load is 400 kN. No snow loads in excess of 0.35 kN/m² are considered in this analysis.

The foundation system consists of continuous and spread footings on consolidated sand and gravel (medium density). The foundation system is located several tens of meters above the water table. The building site is located in a zone of strong seismicity.

For this building, a preliminary seismic analysis is performed in accordance with the provisions of the PS 69/82 Code.

Solution

1. Seismic weights. The design load for houses includes 20% of the live loads. Hence,

$$W_1 = W_2 = W_3 = W_4 = 2,300 + 0.2 \times 600 = 2,420 \text{ kN}$$

$$W_5 = 2,300 + 0.2 \times 400 = 2,380 \text{ kN}$$

2. Fundamental period

$$T = 0.09 \frac{H}{\sqrt{L}} \qquad [31.(16.13)]$$

in which $H = 15.5$ m and $L = 30$ m or 11 m, for x or y direction,

$$T_x = 0.26 \text{ sec}$$

$$T_y = 0.42 \text{ sec}$$

3. Intensity factor α. For structures located in strong seismicity zones (Zone III), the minimum value of the intensity factor is $\alpha = 1.5$ (see Section 16.2.6). In addition, the building is judged to belong to Class B (see Table 16.3). From Table 16.4, the intensity factor is equal to 1.5 for zones of strong seismicity. Therefore the intensity factor is

$$\alpha = 1.5$$

4. Response factor β. For this type of structure, a "normal" damping is judged to be consistent with the specifications given in the PS 69/82 Code. The response factors along the x and y directions are given by eq.(16.16) as

$$\beta_x = \frac{0.065}{\sqrt[3]{T_x}} = 0.103 \qquad \text{(limited to 0.100)}$$

$$\beta_y = \frac{0.065}{\sqrt[3]{T_y}} = 0.087$$

5. Foundation factor δ. For spread footings on medium-dense soils (Soil Type B), the foundation factor from Table 16.5 in Section 16.2.8 is

$$\delta = 1.15$$

6. Distribution factor γ_i. For a regular structure, the simplified formula given in eq.(16.23) of Section 16.2.9.1 can be used for the determination of the distribution factors at the ith level for the fundamental mode:

$$\gamma_i = h_i \frac{S}{I} \qquad [\text{eq.}(16.23)]$$

with $\quad S = \sum_{k=1}^{N} W_k h_k \quad$ and $\quad I = \sum_{k=1}^{N} W_k h_k^2 \quad$ [eq.(16.24)]

Table 16.6. Distribution Factors

Level	Distribution Factor eq.(16.23)	Distribution Factor eq.(16.25)
5	1.363	1.364
4	1.099	1.091
3	0.835	0.818
2	0.572	0.545
1	0.308	0.273

Table 16.7. Seismic Coefficients

Level	Seismic Coefficients x direction	y direction	Vertical
5	0.235	0.204	0.192
4	0.190	0.164	0.155
3	0.144	0.125	0.118
2	0.099	0.086	0.081
1	0.053	0.046	0.043

in which W_k are the concentrated weights at the kth level and h_k is the elevation of this level above the base. The distribution factors at each level of the building are listed in Table 16.6. The distribution factors obtained with the simplified formula in eq.(16.25) are also listed in this table for comparison.

7. Seismic coefficients σ_{ix}, σ_{iy}. The horizontal seismic coefficients at the ith level and for the x and y directions are calculated using the formula in eq.(16.4):

$$\sigma_{ix} = \alpha \beta_x \delta \gamma_i$$

$$\sigma_{iy} = \alpha \beta_y \delta \gamma_i$$

The vertical seismic coefficient at the ith level is calculated using the formula given in eq.(16.5)

$$\sigma_{iV} = \pm \frac{1}{\sqrt{\alpha}} \alpha_{iH} = 0.816 \sigma_{iH}$$

where σ_{iH} is the greater of σ_{ix} or σ_{iy} at the ith level.

Table 16.7 shows the calculated horizontal and vertical seismic coefficients corresponding to the different levels of the building.

8. Equivalent lateral seismic forces. The equivalent horizontal system of forces at the ith level is estimated using the formula of eq.(16.1):

$$F_{ix} = \sigma_x W_i$$

$$F_{iy} = \sigma_y W_i$$

Table 16.8. Equivalent Seismic Lateral Forces

Level	Lateral Forces (kN)	
	x direction	y direction
5	559.3	385.5
4	459.8	317.0
3	348.5	239.1
2	239.6	164.0
1	128.3	88.9

Table 16.9. Story Shear Forces

Level	Shear Forces (kN)	
	x direction	y direction
5	559.3	385.5
4	1,019.1	702.5
3	1,367.6	941.6
2	1,607.2	1,105.6
1	1,735.5	1,194.5

in which W_i is the weight of the ith level. The values of the horizontal seismic forces at the various levels of the building are listed in Table 16.8.

9. Story shear forces. The story shear forces are given by

$$V_{ix} = \sum_{j=i}^{N} F_{jx}$$

$$V_{iy} = \sum_{j=i}^{N} F_{jy} \qquad (16.28)$$

for the x and y directions with N equal to the total number of levels. These values are listed in Table 16.9.

10. Torsional moments. It is assumed that the actual eccentricity (distance between center of mass and center of rigidity) is equal to 0.5 meter in the longitudinal direction and negligible in the transverse direction. However, a check must be made to determine whether an accidental eccentricity should be introduced in either of these two directions.

For the longitudinal x direction, η is found to be equal to 0.37 (11 meters divided by 30 meters). Because η is less than 2.5, no accidental eccentricity is considered in the longitudinal direction. However, η in the transverse direction is greater than 2.5 (30/11 = 2.73). Consequently, an accidental eccentricity in the transverse direction is calculated using eq.(16.3):

$$e = \left(\frac{1}{20}\right)(\lambda + \mu)(L_y + d)$$

Table 16.10. Torsional Moments

Level	Torsional Moment (kN-m)	
	x direction	y direction
5	279.7	131.1
4	509.6	238.8
3	683.8	320.2
2	803.6	376.0
1	867.8	406.2

With $L_y = 30$ m, $d = 0$, $\lambda = \eta - 2.5 = 0.23$, and $\mu = 0$; $e = 0.34$ m.

Torsional moments at each level due to the natural and accidental longitudinal and transverse eccentricities are calculated as

$$M_{tix} = \sum_{j=i}^{N} F_{jx} e_{jx}$$

$$M_{tiy} = \sum_{j=i}^{N} F_{jy} e_{jy} \qquad (16.29)$$

with $e_{ix} = 0.5$ m and $e_{iy} = 0.34$ m at each level. The torsional moments caused by the horizontal seismic forces are given in Table 16.10.

11. Torsional shear forces. Torsional shear forces due to the torsional moments should be estimated and added to the story shear forces. Hence, the total shear applied to a frame of the building is the sum of the story shear and the shear resulting from torsion:

$$V_L = \left(\frac{K_L}{\Sigma K_L}\right) V_s + \left(\frac{K_L d_L}{\Sigma K_L d_L^2}\right) M_t \qquad (16.30)$$

where

V_L = total shear force on the Lth frame
V_s = story shear force
K_L = lateral stiffness of the Lth frame
d_L = distance between the Lth frame and the center of rigidity
M_t = torsional moment.

For this example, the lateral stiffnesses of the frames at each level are assumed to be identical. Thus, the total maximum shears applied at each level i to frames A and I are

$$V_{ixA} = \frac{1}{3} V_{ix} + \frac{d_{yA}}{\sum_{l=1}^{3} d_{yl}^2} M_{tix}$$

$$V_{iyI} = \frac{1}{7} V_{iy} + \frac{d_{xI}}{\sum_{l=1}^{7} d_{xl}^2} M_{tiy}$$

Table 16.11. Total Shear Forces

| Level | Shear Forces (kN) | |
	Frame A	Frame I
5	213.7	58.0
4	389.3	105.6
3	522.3	141.5
2	614.0	166.2
1	663.0	179.6

with $d_{yA} = 6$, $d_{yB} = 0$, $d_{yC} = -5$, $d_{xI} = -15.34$, $d_{xII} = -10.34$, $d_{xIII} = -5.34$, $d_{xIV} = 0.34$, $d_{xV} = 4.66$, $d_{xVI} = 9.66$, $d_{xVI} = 14.66$, (all in meters) and

$$V_{ixA} = 0.333V_{ix} + 0.098M_{tix}$$

$$V_{iyI} = 0.143V_{iy} + 0.022M_{tiy}$$

The total shear forces applied to frames A and I are listed in Table 16.11.

12. Determination of overturning moments. The overturning moment at level i on a frame is calculated by

$$M_{Oi} = \sum_{j=i+1}^{N} F_j(h_j - h_i) \qquad (16.31)$$

where $i = 0, 1, 2, \ldots, n - 1$, and in which F_j is the lateral force, including torsional effects, on the frame. Table 16.12 shows the calculated overturning moments on frame A.

13. Story drifts and lateral displacements. Assuming columns fixed at both ends, stiffnesses are

$$K = \frac{12EI}{H^3} \qquad (16.32)$$

where H is the story height. For 40 cm by 40 cm columns at levels 2 through 5 and 45 cm by 45 cm columns at the first level, and $E = 3.0 \times 10^7$ kN/m^2, the column stiffnesses at levels 2 through 5 are $K = 28,444$ kN/m, and at the first level

Table 16.12. Overturning Moments on Frame A

Level	Total Lateral Force (kN)	Story Height (meters)	Overturning Moment (kN-m)
5	213.7	3.0	—
4	389.3	3.0	641.1
3	522.3	3.0	1,809.0
2	614.0	3.0	3,375.9
1	663.0	3.5	5,217.9
Base	—	—	7,538.4

Table 16.13. Story Drifts and Lateral Displacements

| Level | Frame A | | Frame I | |
	Story Drift (cm)	Displacement (cm)	Story Drift (cm)	Displacement (cm)
5	0.107	1.206	0.068	0.761
4	0.196	1.099	0.124	0.693
3	0.262	0.903	0.165	0.569
2	0.308	0.641	0.195	0.404
1	0.333	0.333	0.209	0.209

$K = 28,692$ kN/m. The stiffnesses of frames A and I are equal to the sums of the column stiffnesses; that is,

$$K_A = 7K$$

$$K_I = 3K$$

At each level, the story drifts are estimated as $\Delta_i = V_i/K$ where V_i is the shear acting upon the frames. The story drifts and lateral displacements are shown in Table 16.13.

As indicated in Section 16.2.11, the allowable story drift is $(\sqrt{\alpha}/1,000)H$, that is, 0.367 cm and 0.429 cm for story heights of 3.00 m and 3.50 m, respectively.

The total lateral displacements at the top of building calculated using the static equivalent systems of forces are 1.206 cm and 0.761 cm in the longitudinal and transverse directions, respectively. These displacements should be multiplied by 4.0 for structures with lateral load-resisting frames (Section 16.2.12). Therefore, the total design lateral displacements are 4.82 cm and 3.04 cm in the longitudinal and transverse directions, respectively.

If the building in question is part of an ensemble of similar buildings, the distance between these structures should not be less than 4 cm or the square root of the sum of the squares of their total lateral displacements. For example, suppose that the total lateral displacement of each of two adjacent blocks is $d_1 = d_2 = 4.82$ cm; then the gap between these structures should be at least

$$d = \sqrt{d_1^2 + d_2^2} = 6.8 \text{ cm}. \qquad \text{[eq.(16.27)]}$$

16.2.15 Computer Program

A computer program has been developed to implement the provisions of the French Seismic Code PS 69/82 using the equivalent static lateral force method. The program contains provisions for the calculation of the fundamental period using the empirical formulas of the code. All the factors necessary to determine the equivalent seismic forces are calculated and printed by the program.

16.2.16 Example 16.2

The program is used to estimate the seismic forces on frame A of the five-story moment-resisting frame shown in Fig. 16.1.

INPUT DATA AND OUTPUT RESULTS

SEISMIC INPUT DATA

```
Seismic Zone Number    : NZ = 3 (III)
Soil Type              : S = 2 (B)
```

BUILDING INPUT DATA

```
Structure Class        : I = 1 (B)
Foundation Type        : F = 1 (Spread)
Damping Type           : G = 1 (Normal)
Building Dimension,
Force Direction (m)    : 30
Normal to Force (m)    : 11
Number of Stories      : 5
```

Level No.	Height (m)	Story Weight (KN)	Stiffness (KN/m)	eccentricity (m)	Stiffness Factor Translation	Rotation
5	3.00	793.33	199108	.500	.333	.098
4	3.00	806.67	200844	.500	.333	.098
3	3.00	806.67	200844	.500	.333	.098
2	3.00	806.67	200844	.500	.333	.098
1	3.50	806.67	200844	.500	.333	.098

ESTIMATE NATURAL PERIOD

```
Use code formulas
1. T = 0.06 * (H/SQRT(L)) * SQRT(H/(2*L+H))  (masonry/cast-in-place walls)
2. T = 0.08 * (H/SQRT(L)) * SQRT(H/(L+H))   (reinforced concrete walls)
3. T = 0.09 * (H/SQRT(L))                    (reinforced concrete frames)
4. T = 0.10 * (H/SQRT(L))                    (steel frames)
5. Use estimated value for period

      Select number?   3
```

OUTPUT RESULTS

```
Intensity Factor    : alpha  = 1.5
Fundamental Period  : T (sec) = .25
Response Factor     : beta   = .100
Foundation Factor   : delta  = 1.15
```

DISTRIBUTION FACTOR, gamma

Level No.	gamma
5	1.363
4	1.099
3	.835
2	.572
1	.308

SEISMIC COEFFICIENT, sigma

Level No.	sigma
5	.235
4	.190
3	.144
2	.099
1	.053

DISTRIBUTION SEISMIC LATERAL FORCES WITHOUT ECCENTRICITY

Level No.	Lateral Force (KN)	Shear Force (KN)
5	186.4	186.4
4	153.3	339.7
3	116.2	455.9
2	79.9	535.7
1	42.8	578.5

TORSIONAL AND OVERTURNING MOMENTS

Level No.	Eccentricity (m)	Torsional Moment (KNm)	Total Shear (KN)	Overturning Moment (KNm)
5	.5	279.7	213.7	
4	.5	509.6	389.3	641.1
3	.5	683.8	522.3	1809.0
2	.5	803.6	614.0	3375.9
1	.5	867.8	663.0	5217.9
0				7538.4

LATERAL DISPLACEMENTS

Level No.	Drift (cm)	Displacement (cm)	Max. Drift permitted (cm)
5	.11	1.21	.37
4	.20	1.10	.37
3	.26	.90	.37
2	.31	.64	.37
1	.33	.33	.43

```
Total design lateral displacement = 4.8 cm
Period by Rayleigh's formula = .27 sec
```

16.3 AFPS-90 RECOMMENDATIONS

16.3.1 Introduction

The PS 69/82 Seismic Code is greatly expanded by the tentative seismic provisions in the AFPS-90 Recommendations. In theory, all structures (with the exception of tunnels, dams, levees, and other structures, including offshore, that cannot safely be designed with somewhat simplified solutions) are included in the provisions of AFPS-90.

As in the 69/82 Code, simplified methods for regular structures are proposed in the AFPS-90 Recommendations. However, for cases where the simplified rules do not apply, the AFPS-90 Recommendations stipulate guidelines for using spectral, time-history, or random vibration analyses. Other main differences between the AFPS-90 Recommendations and the PS 69/82 Code occur in the definitions of the seismic forces. AFPS-90 includes a new structural factor that is dependent on the type of structure and on its ductility. In part, this factor replaces the PS 69/82 response factor β (see Section 16.2.7), which is independent of the type of structure, material, and other mechanical properties. The PS 69/82 response factor also is combined with the design spectral shape. This second component of the PS 69/82 response factor is defined explicitly in the AFPS-90 Recommendations as a spectral shape dependent on nominal ground acceleration values.

The nominal ground acceleration values are specified in the AFPS-90 Recommendations as peak ground accelerations. These nominal values will be stipulated in the forthcoming code as functions of the site seismicity. By specifying a spectral shape dependent on nominal peak ground accelerations, response spectrum guideline analyses are defined explicitly in the rules. The AFPS-90 Recommendations also specify target spectra to be used in conjunction with time-history analyses.

Different damping values are specified in the determination of the design spectral shapes, which are frequency dependent on the soil at the site. In the PS 69/82 Code, the spectral shapes are dependent only on a soil factor, γ. The number of soil categories has been greatly expanded from the number specified in the PS 69/82 Code. In addition, a topographic factor has been introduced, which accounts for the potential amplification effects of certain superficial or subterranean geologic features.

As in the PS 69/82 Code, the AFPS-90 Recommendations provide means to estimate the seismic forces on the components of the lateral load-resisting system as functions of the dynamic characteristics of the structure. The following sections summarize the provisions of AFPS-90 Recommendations.

16.3.2 Seismic Weights

The seismic weights include dead loads of the structure and applicable portions of other loads as follows:

1. 20% of normal live loads
2. 50% of live loads for public-use floors and for other buildings with a large number of occupants
3. 100% of semipermanent or permanent live loads
4. Minimum of 30% of snow loads
5. Other loads as specified for particular structure uses.

16.3.3 Load Combinations

The following loads are considered in the seismic design of structures:

1. Permanent loads
2. Dynamic loads
3. Prestress and earth loads, if any
4. Fractions of live loads, as defined in each material-specific code (reinforced concrete, steel, etc.)
5. Climatic effect loads.

16.3.4 Equivalent Static Force Method

The equivalent static force method may be used for the following:

1. Structures regular in plan and elevation
2. Structures for which the vertical loads can be carried down to the foundations by continuous vertical elements
3. Structures for which each level can be modeled by discrete masses
4. Structures with rigid lateral load-resisting horizontal elements.

The AFPS-90 Recommendations stipulate that the equivalent static forces to be applied at the rth level of the structure along the seismic direction (horizontal or vertical) are to be of the form:

$$f_r = \frac{1}{q}\, m_r u_r \frac{M}{\Sigma m_s u_s}\, R(T) \qquad (16.33)$$

in which

M = total mass
m_r, m_s = concentrated masses at the rth and sth levels, respectively
u_r, u_s = horizontal or vertical displacements of the rth and sth levels, respectively
T = fundamental period
$R(T)$ = design spectral acceleration at the fundamental period
q = behavior factor.

The total displacements at the rth levels are given by the formula

$$d_r = \frac{T^2}{4\pi^2}\, u_r \frac{M}{\Sigma m_s u_s}\, R(T) \qquad (16.34)$$

16.3.4.1 Fundamental period. The fundamental period T may be calculated using either empirical formulas derived from experiential data or from methods of dynamics (Rayleigh, Stodola, etc.).

16.3.4.2 Displacements. Empirical formulas can be used to estimate the displacements at each level of the structure. These formulas specify displacements (u) at any elevation (z) as a function of elevation above the base, defined as the zero-displacement plane:

$$u = z^k \qquad (16.35)$$

in which the exponent k can take values between 1 and 2.

16.3.4.3 Design spectral acceleration. The design spectral acceleration at the period T is given by

$$R(T) = a_N \tau \rho R_D(T) \qquad (16.36)$$

in which

T = period of the structure
a_N = nominal acceleration
τ = topographic factor
ρ = damping factor
$R_D(T)$ = normalized design spectral value at T.

16.3.4.4 Nominal acceleration. France is subdivided into four seismic zones similar to those described in Section 16.2.6. Structures are also classified into four classes, as defined in Table 16.14.

Table 16.14.   Classification of Structures

Structure Class	Definition
O	Structures for which collapse is not a hazard to population
A	Standard occupancy structures
B	Structures with large public occupancy or of great economic importance
C	Essential structures

Table 16.15.   Nominal Acceleration a_N (m/s²)

Seismicity Zone	Values of a_N for Structure Class			
	O	A	B	C
0	—	—	—	—
Ia	—	1.0	1.5	2.0
Ib	—	1.5	2.0	2.5
II	—	2.5	3.0	3.5
III	—	3.5	4.0	4.5

With the above structure classification, AFPS-90 recommends nominal acceleration values a_N, which are given in Table 16.15.

16.3.4.5 Topographic factor.

Observations from past earthquakes have indicated that ground motions are amplified for certain topographic features, including, but not limited to, hills, bluffs, ridges, and peaks. Although no theory is available to assess this amplification in a quantitative manner, the AFPS-90 Recommendations use an empirical approach to quantify this aspect.

The topographic factor τ is defined for the situation shown in Fig. 16.2 where the elevation H is greater than or equal to 10 meters, and the slope i is less than

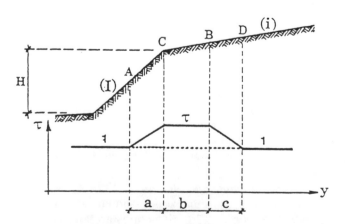

Fig. 16.2.   Topographic factor

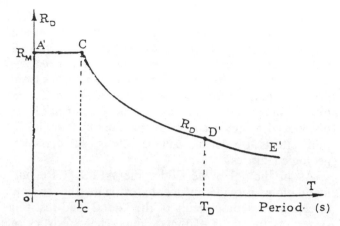

Fig. 16.3.   Normalized design spectral shape

or equal to 1/3 of the slope I. Depending on the change in slope $\Delta_i = |I - i|$, τ takes on values equal to:

$$\tau = 1 \qquad \text{for} \qquad \Delta i \leq 0.4$$
$$\tau = 1 + 0.8(\Delta i - 0.4) \quad \text{for} \quad 0.4 \leq \Delta i \leq 0.9 \qquad (16.37)$$
$$\tau = 1.4 \qquad \text{for} \qquad \Delta i \geq 0.9$$

with

b = the lesser of $20I$ or $(H + 10)/4$
$a = H/3$
$c = H/4$.

16.3.4.6 Normalized design spectra.

The normalized design spectral shape is given in Fig. 16.3 and is defined by:

Branch A'C: $R_D(T) = R_M$

Branch CD': $R_D(T) = R_M \left[\dfrac{T_C}{T} \right]^{2/3}$

Branch D'E': $R_D(T) = R_M \left[\dfrac{T_C}{T_D} \right]^{2/3} \left[\dfrac{T_D}{T} \right]^{5/3}$

The AFPS-90 Recommendations stipulate normalized design response spectra for different soil types:

- S_0: A soil profile with either:
  - (a) A rock-like material characterized by a shear-wave velocity greater than 800 m/sec (or by other suitable means of classification), or
  - (b) Stiff or dense soil conditions characterized by a shear-wave velocity greater than 400 m/sec and with a soil depth of less than 15 m.
- S_1: A soil profile with either:
  - (a) Stiff or dense soil conditions with shear-wave velocity greater than 400 m/sec and where the soil depth exceeds 15 m, or
  - (b) Medium-stiff clay or medium-dense granular soil with a depth of less than 15 m.

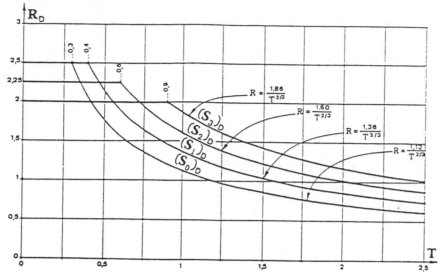

Fig. 16.4. Normalized horizontal design spectra for different soil types

- S₂: A soil profile with either:
    (a) Medium-stiff clay or medium-dense granular soil more than 15 m deep but not more than 50 m deep, or
    (b) Soft clay or loose sandy soil with a depth of less than 15 m.

- S₃: A soil profile with either:
    (a) Medium-stiff clay or medium-dense granular soil with a depth in excess of 50 m, or
    (b) Soft clay or loose sandy soil with a depth of more than 15 m but not exceeding 100 m.

For soft clays or loose sandy soils with depths in excess of 100 m, AFPS-90 recommends site-specific spectral analysis.

For each soil type defined above, the AFPS-90 Recommendations give a normalized 5% damping elastic response spectrum shape based on a peak ground acceleration of unity. The general shapes of the horizontal spectra are given in Fig. 16.4 and their characteristics are listed in Table 16.16.

The vertical normalized design response spectra are similar to the horizontal design spectra for Soil Types S₀ or S₁ (see Figure 16.5). For the other soil types, the slopes of the spectra equal the horizontal spectral slope for Soil Type S₁.

For spectra with damping values different from 5%,

Table 16.16. 5% Damped Horizontal Normalized Design Response Spectra

Soil Type	T_C (sec)	T_D (sec)	R_M
S₀	0.30	2.67	2.5
S₁	0.40	3.20	2.5
S₂	0.60	3.85	2.25
S₃	0.90	4.44	2.0

Table 16.17. Recommended Damping Values

Construction System	Damping Value (%)
Welded steel	2
Bolted steel	4
Reinforced masonry	5–6
Nonreinforced concrete	3
Reinforced concrete	4
Prestressed concrete	5
Wood	4–5

AFPS-90 introduces a correction factor, not to exceed 30%, on the spectral ordinates, defined as:

$$\rho = \left(\frac{5}{\xi}\right)^{0.4} \tag{16.38}$$

where ξ is the damping ratio.

16.3.4.7 Damping values. The recommended values for damping expressed as percentage of critical damping are listed in Table 16.17.

16.3.4.8 Behavior factor. The behavior factor is a measure of the capacity of the structural system to absorb energy in the inelastic range through ductility and redundancy. It is taken to be unity for elastic analysis and is dependent on the structure type and materials used. This factor will be specified in the forthcoming seismic code.

16.3.5 Dynamic Analysis

The AFPS-90 Recommendations stipulate that when a dynamic analysis method is used it shall conform to criteria established in the Recommenda-

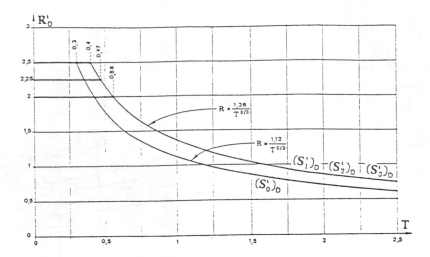

Fig. 16.5. Normalized vertical design spectra for different soil types

tions. This stipulation is implemented generally by the application of the modal superposition method described in Part I, Chapter 4 of this book. In addition, spectral analysis can be performed only for structures without strong irregularities in geometry or discontinuities in the lateral load-resisting system. The criteria established by the recommendations include the provisions defined below.

16.3.5.1 Ground motion representation

16.3.5.1.1 RESPONSE SPECTRUM ANALYSIS. In this type of analysis, the provisions of AFPS-90 recommend that the structure be modeled as an elastic system, as described in Section 16.3.4.3. The ground motion is represented by either an elastic spectrum or by a design response spectrum (see eq.16.36), depending on whether the structure should or should not remain in the elastic domain. The elastic spectrum is defined by

$$R(T) = a_N \tau \rho R_E(T) \qquad (16.39)$$

in which $R_E(T)$ = the elastic spectral value at period T, and the other variables are as defined in eq.(16.35).

The normalized elastic design spectral shape is given in Fig. 16.6 and is defined by:

Branch AB: $R_E(T) = R_A + (R_M - R_A)\left[\dfrac{T}{R_B}\right]$

Branch BC: $R_D(T) = R_M$

Branch CD: $R_D(T) = R_M\left[\dfrac{T_C}{T}\right]$

Branch DE: $R_D(T) = R_M\left[\dfrac{T_C}{T_D}\right]\left[\dfrac{T_D}{T}\right]$

For each soil type defined in Section 16.3.4.6, the characteristics of the spectra are given in Table 16.18.

The vertical elastic spectrum and correction for damping are selected in the same fashion as for the horizontal design spectrum.

16.3.5.1.2 TIME-HISTORY ANALYSIS. Ground motion time histories can be developed for the specific site using either artificial time histories or time histories recorded on sites similar to the specific site. The ordinates of the 5% damped response spectrum developed using artificial earthquake records or using

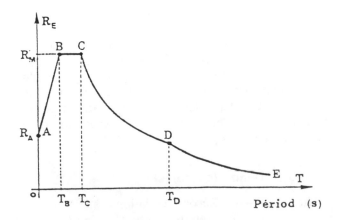

Fig. 16.6. Normalized elastic design spectral shape

Table 16.18. 5% Damping Horizontal Normalized Elastic Response Spectra

Soil Type	T_B (sec)	T_C (sec)	T_D (sec)	R_A	R_B
S_0	0.15	0.30	2.67	1.0	2.5
S_1	0.20	0.40	3.20	1.0	2.5
S_2	0.30	0.60	3.85	0.9	2.25
S_3	0.45	0.90	4.44	0.8	2.0

actual earthquake records, should not be less, respectively, than 5% or 20% of the corresponding ordinates of the elastic spectrum in the rules. Also, the AFPS-90 Recommendations specify limitations on the spectral energies of these time histories in relation to the elastic spectra defined in the same rules. In addition, artificial time histories must have minimum "durations," defined as the duration over which 90% of the total energy (root-mean-square) of the shaking is recorded, specified as:

- Between 10 and 20 seconds for Zones I_a and I_b
- Between 15 and 30 seconds for Zone II
- Between 20 and 40 seconds for Zone III.

Finally, the time steps should be less than or equal to 0.02 sec and the correlation coefficient between the two horizontal components should be between 0 and 0.3.

16.3.5.2 Model representation. The adopted mathematical model should represent the spatial distribution of the masses and stiffnesses of the structure.

16.3.5.3 Modal combination. If the dynamic analysis uses an elastic response spectrum method, all the significant modes should be combined in a statistical manner, such as the Complete Quadratic Combination (CQC) method (see Chapter 4) to obtain an approximate total structural response. The rules specify that a minimum of three modes are to be considered and that the total effective modal mass should be equal to at least 90% of the total modal mass. In cases where this condition is not achieved, a residual mode should be considered in the analysis.

If a time-history analysis is used, a time-history response analysis of the structure should be obtained at each increment in time.

16.3.6 Seismic Joint Limitations

The minimum distance between two adjacent structures should not be less than 4 cm in Zone I and 6 cm in Zones II and III.

16.3.7 P-Δ Effect

The P-Δ effect refers to the additional moment produced by the vertical loads acting at the lateral displacements of columns and other elements of the structure lateral force-resisting system (see Chapter 4, Section 4.9). The AFPS-90 Recommendations specify that the resulting forces, displacements, and story drifts induced by this effect should be considered if the ratio θ, defined below, exceeds 0.05.

$$\theta = \frac{\Delta_r P_r}{h_r F_r} \qquad (16.40)$$

where

P_r = total seismic weight at level r and above
Δ_r = drift of the rth level
F_r = shear force at the rth level
h_r = height of the rth story

16.4 BUILDING CODES

Groupe de Coordination des Textes Techniques (1987), "Regles neige et vent (NV 65)," Editions Eyrolles, Paris, France.

Groupe de Coordination des Textes Techniques (1987), "Regles de calcul des constructions en acier et additifs 1980(CM 66)," CTICM, Paris, France.

Groupe de Coordination des Textes Techniques (1983), "Regles techniques de conception et de calcul des ouvrages et constructions en béton armé suivant la méthode des états limites (BAEL 83)," Numéro spécial 83-35 bis du Bulletin Officiel du ministere de l'Urbanisme et du Logement, du ministere des Transports et du ministere de l'Environnement, Paris, France.

Groupe de Coordination des Textes Techniques (1983), "Regles techniques de conception et de calcul des ouvrages et constructions en beton precontraint suivant la methodes des etats limites (BPEL 83)," Numero special 83-35 bis du Bulletin Officiel du ministere de l'Urbanisme et du Logement, du ministere des Transports et du ministere de l'Environnement, Paris, France.

Groupe de Coordination des Textes Techniques (1983), "Regles de calcul des charpentes en bois et modificatifs 1975 (CB 71)," Editions Eyrolles, Paris, France.

REFERENCES

ASSOCIATION FRANCAISE DU GENIE PARASISMIQUE (AFPS) (1990) *Recommendations AFPS-90 pour la redaction des regles relatives aux ouvrages et installations a realiser dans les regions sujettes aux seismes*. Presse des Ponts et Chaussées, Paris, France.

DELEGATIONS AUX RISQUES MAJEURS (1991) *Nouveau zonage sismique de la France*. Journal officiel, Paris, France.

GROUPE DE COORDINATION DES TEXTES TECHNIQUES (DTU) (1982) *Regles parasismiques 1969, annexes et addenda 1982. (Regles PS 69, modifiée 82)* Editions Eyrolles, Paris, France.

SOCOTEC (1955) *Regles Antisismiques (AS 55)*. Société de Contole Technique, Paris, France.

17

Greece

George C. Manos

17.1 INTRODUCTION

Greece is located over one of the most active seismic belts on the earth; seismologists have estimated that 50% of the energy released annually by earthquake activity on the European continent is located in Greece. The first code for seismic-resistant design for Greece was published in 1959 (Royal Decrees 1959). Since the publication of that first code, the Greek governments have continued promoting methods to mitigate seismic risk, particularly after the 1978 Thessaloniki earthquake, which caused widespread damage to the city of Thessaloniki, with one million inhabitants, and the 1981 Alkionides earthquake, which also caused serious damage to the metropolitan area of Athens, with four million inhabitants. A new version of the seismic code was published after the latter quake [Seismic Code for Building Structures (1984), a decree of the Minister of the Environment]. The relevant provisions of this Code, which is still currently valid, are included in Sections 17.2 through 17.8.

New studies to revise further the seismic code of

1984 intensified after the very damaging 1986 earthquake that occurred in the city of Kalamata in the Peloponnese [Proposed Revisions to the New Greek Seismic Code, Edition 1988 (1989)]. After the publication of these proposed revisions, a review period commenced with the participation of various professional bodies, research institutions and individual academics, and designers. In 1990, a special committee was appointed by the Earthquake Planning and Protection Organization of Greece (OASP) to study the contributed comments and then rewrite the final provisions and the explanatory report which was to be sent to the Ministry of the Environment. This effort was completed during the latter part of 1992 and received Ministerial approval, thus becoming law of the land as the New Seismic Code of Greece. The provisions of this new Code are included in Appendix A of this chapter. These 1992 provisions were published in the Government's Gazette of October 1992; for a transition period, provisionally set for two years, either the 1984 or the 1992 Seismic Code Provisions may be employed in design. It is expected that when this transition period expires, the 1992 Seismic Code Provisions will become mandatory for the design of earthquake-resistant structures in Greece.

Similarly, the Reinforced Concrete Code has been recently revised (1991) and is expected to become the sole mandatory code in 1994; it includes certain provisions for the seismic-resistant design of concrete structures. In the meantime, the designer may use either the old or the recently revised provisions of the Reinforced Concrete Code.

Two computer programs have been developed: GREECE1 is used to implement the provisions of the 1984 Greek Seismic Code; and GREECE2 is used for seismic design of buildings using the provisions of the New 1992 Greek Seismic Code (the programs only use the equivalent static method). These computer programs serve only to determine the equivalent lateral forces and the story shear forces. A 10-story building is presented as an application of the program GREECE2 using the new 1992 Seismic Code (Example A17.2).

17.2 SEISMIC ZONES

The Greek Seismic Code (1984) divides the territory of the country into three seismic zones: I, II, and III. The first (Zone I) corresponds to a seismic zone of least intensive earthquake activity, the second (Zone II) corresponds to moderate activity, and the third (Zone III) designates regions with the most intense earthquake activity. The Greek Seismic Code is neither based on a specific seismic zone map nor on contour lines of maximum expected ground acceleration levels:

Seismic Zone Numbers I, II, or III have been assigned to such places as the main cities and towns of Greece. Table 17.1 gives a list, with the corresponding seismic zone, of these main cities and towns and other areas included in the Greek Seismic Code. Any place in Greece not included in Table 17.1 is assumed to be in a seismic zone equivalent to the seismic zone of the nearest place listed.

17.3 SEISMIC COEFFICIENT

The value of the seismic coefficient ε is determined by considering three main parameters: (1) the seismic zone; (2) the soil category, and (3) the importance factor. The Code provides for four soil categories (A,B,C,D) listed in Table 17.2, with the corresponding values for the soil factor S shown in Table 17.3. In Table 17.4 the structure is rated according to its importance. The value of the seismic coefficient ε is found by multiplying the importance factor I from Table 17.4 by the soil factor S, taken from Table 17.3. The value of the soil factor depends on the seismic zone and soil category, as indicated in Table 17.3.

17.4 LOAD COMBINATIONS AND SEISMIC WEIGHT

The Greek Seismic Code (1984) requires that the seismic horizontal loads be considered as static forces acting simultaneously with the other static permanent or temporary loads. Specifically, the Code requires consideration of the following load combinations at each level i of the building:

$$G_i + Q_i$$
$$G_i + Q_i + F_i \qquad (17.1)$$
$$G_i + Q_i - F_i$$

where G_i is the dead load at level i due to the weight of the structure and the permanent loads, Q_i is the live load, and F_i is the lateral force generated by the action of the earthquake.

The structural analysis for determining the structural member forces is based on elastic theory. The Code provides for the use of allowable stresses for the materials when the member forces are determined from load combinations that exclude the seismic forces. However, when the member forces are found from load combinations that include forces produced by an earthquake, the allowable stresses can be increased by 20%, except for the allowable shear stress. The effect of the vertical ground acceleration is considered only in the design of slabs.

Table 17.1. Seismic Zones of Greek Cities and Towns (Codes of 1984 and 1992)*

Name	Seismic Zone (1984)	Seismic Zone (1992)	Name	Seismic Zone (1984)	Seismic Zone (1992)
Agia (Larissa)	I	III	Agios Kirikos (Ikaria)	I	II
Agios Nikolaos (Kriti)	I	III	Agrinio	II	II
ATHINA	I	II	Alexandroupoli	I	I
Almiros	II	III	Amfilohia	II	II
Amfissa	II	III	Amorgos	II	II
Andritsena	II	II	Andros	I	I
Argostoli	III	IV	Areopoli	I	III
Aridea	I	I	Arta	II	II
Arnea	II	III	Didimotiho	I	I
Atalanti	III	III	Drama	II	II
Domokos	II	III	Egina	I	I
Edessa	I	I	Elassona	I	II
Egio	II	III	Filiatra (Kiparissia)	III	III
Erinoupolis	I	I	Florina	I	I
Farsala	III	III	Githio	II	II
Filiates	II	III	Halkis	II	III
Giannitsa	I	I	Hios	III	III
Goumenissa	I	I	Iraklio (Crete)	III	III
Grevena	I	I	Ithaki	II	IV
Hania	II	III	Kalamata	II	III
Igoumenitsa	II	III	Kalavrita	I	III
Ioannina	II	II	Karpenissi	II	II
Istiea	II	III	Kastoria	I	I
Karditsa	II	II	Katerini	I	I
Karistos	I	I	Kerkira	II	III
Kastelorizo	III	III	Kimi	I	I
Kavala	I	II	Kithira	II	III
Kilkis	I	III	Komotini	I	II
Kiparissia	II	II	Korinthos	III	III
Konitsa	I	II	Lamia	II	III
Kos	III	III	Lavrio	I	I
Kozani	I	I	Livadia	II	III
Lagadas	II	III	Megalopolis	II	I
Larissa	II	III	Mesologi	I	III
Lefkada	III	IV	Milos	I	I
Limin (Thasos)	I	I	Nafpaktos	II	III
Megara	II	III	Nea Orestiada	I	I
Metsovon	I	II	Nea Zihni	I	II
Mitilini	III	III	Patra	II	III
Nafplio	I	II	Pirgos	II	III
Naoussa	II	I	Preveza	II	III
Nigrita	I	II	Rodos	III	III
Paros	II	I	Sami	III	IV
Pilos	II	II	Samos	III	III
Poligiros	II	II	Skiathos	II	II
Rethimno	II	III	Sparti	II	II
Salamina	II	II	Trikala	II	II
Serres	I	I	Tripoli	II	II
Sitia	III	III	Veria	II	II
Skiros	II	II	Xanthi	II	II
THESSALONIKI	II	II			
Tirnavos	I	II			
Volos	III	III			
Vonitsa	II	III			
Zakinthos	III	IV			

*Note: To save space, only those localities listed in both Greek Seismic Codes (1984 and 1992) are reproduced in this table. Also, note that the 1992 Greek Seismic Code (see Appendix A17) has four Seismic Zones (I–IV).

Table 17.2. Soil Categories

Soil Category	Description
A Low seismic hazard	Extensive horizontal layer of compacted soil more than 15 m deep
B Moderate seismic hazard	Not very well compacted, horizontal, or slightly inclined soil layers
C High seismic hazard	Very recent alluvial deposits near the sea or at the bottom of drained lakes
D Exceptionally high seismic hazard	Nonhomogeneous and/or loose foundation material located on slopes or above cavities

Table 17.3. Soil Factor (S)

Seismic Zone	Values of S for Soil Category		
	A	B	C
I	0.04	0.06	0.08
II	0.06	0.08	0.12
III	0.08	0.12	0.16

Table 17.4. Importance Factor (I)

Seismic Zone	Importance of Structure	
	Large*	Usual
I and II	1.5	1.0
III	1.2	1.0

*The selection for the importance of the structure (Large or Usual) is based on the following factors: (a) the social, financial, and functional consequences of failure, (b) the initial cost, (c) the need to have the structure functioning after the earthquake, (d) the expected life of the structure, and (e) the number of times that the same design will be used.

At each level of the building, the seismic weight W_i includes the dead load plus the live load.

17.5 SEISMIC LATERAL FORCES

The total static seismic shear force V at the base of the structure is determined by the formula:

$$V = \varepsilon W \qquad (17.2a)$$

where $W = \Sigma W_i$ is the total seismic weight of the building and $\varepsilon = IS$ is the seismic coefficient (Section 17.3).

Note: The base shear force V in the 1991 Concrete Code is determined as

$$V = 1.70\varepsilon(G + \psi_2 Q) \qquad (17.2b)$$

where G and Q are, respectively, the dead and the live loads of the building, and ψ_2 is a reduction coefficient with the following values:

$\psi_2 = 0.3$ for apartment buildings
$\psi_2 = 0.4$ for office buildings or shopping centers
$\psi_2 = 0.5$ for public buildings (schools, theaters, stadiums, etc.)
$\psi_2 = 0.6$ for long-term storage buildings (libraries, warehouses, repositories, and parking garages).

When the base shear is determined according to eq.(17.2b), the use of limit state design is allowed.

The base shear force V is distributed as lateral forces F_i applied at the various levels of the building according to:

$$F_i = V \frac{W_i h_i}{\sum\limits_{j=1}^{n} W_j h_j} \qquad (17.3)$$

where

n = number of levels in the structure
W_i, W_j = seismic weights at levels i, j
h_i, h_j = heights from the base of the building for levels i, j.

17.6 STORY SHEAR FORCE

The shear force V_i that must be resisted by structural elements at story i is given by the sum of the lateral seismic forces above that story as shown in Fig. 17.1; that is,

$$V_i = \sum_{j=i}^{n} F_j \qquad (17.4)$$

The lateral seismic forces F_i, for structures in which masses are concentrated mostly at the story levels, act at the centers of mass of the stories. The Greek Seismic Code requires the structure to be designed to resist, in combination with other forces, the seismic forces applied along each of the two main perpendicular directions of the building independently.

17.7 LATERAL DRIFT LIMITATIONS

The Greek Seismic Code limits the maximum interstory drift to prevent excessive damage to masonry

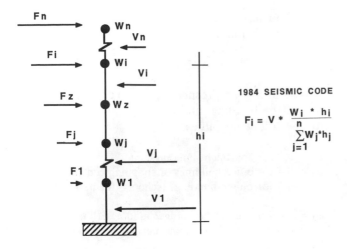

Fig. 17.1. Modeled building showing lateral seismic forces F_i, base shear force V, and story shear forces V_i

infill panels, which may be used as partition walls in multistory buildings. This maximum drift limit need not be satisfied for buildings with flexible partitioning panels. The story drift, e.g. the relative displacement between the two adjacent levels of the building, is given by

$$\Delta_i = X_i - X_{(i-1)} \qquad (17.5)$$

where Δ_i is the drift of the ith story, and X_i and X_{i-1} are the lateral displacements (calculated assuming elastic behavior) at levels i and $i - 1$, respectively. The seismic code imposes the following limitation on story drift:

$$\Delta_i \leq 0.002 L_i \qquad (17.6)$$

where L_i is the story height.

The draft revisions to the 1984 Code (later incorporated in the Seismic Code of 1992) provide the following simplified formula for use in estimating the story drift:

$$\Delta_i = \frac{1.5 V_i}{(\Sigma D_c + \Sigma D_{sw})_i} \qquad (17.7)$$

where

Δ_i = story drift
V_i = story shear force
D_c = column lateral stiffness (which can be assumed with both ends fixed at story levels)
D_{sw} = lateral shear wall stiffness (which can be assumed to be simply supported at one end and fixed at the other).

17.8 SECOND-ORDER P-Δ EFFECT

The Greek Seismic Code (1984) includes an additional provision in relation to second-order P-delta (P-Δ)

effects. These second-order effects must be considered when the value of the parameter θ_i, defined by eq.(17.8) for any level of the building, is greater than 0.10. Furthermore, the value of θ_i for any story should not exceed 0.20 ($\theta_i \leq 0.20$).

$$\theta_i = \frac{W_i \Delta_i}{V_i L_i} \qquad (17.8)$$

where

W_i = seismic weight at level i
V_i = shear force for story i
L_i = height of story i
Δ_i = story drift for story i.

In structural systems with shear walls that have the same cross-section through the height of the building, there is no need to consider second-order effects when the parameter α, defined by eq.(17.9), satisfies the following limitations:

$$\alpha < 0.2 + 0.1n \qquad \text{for } n \leq 3$$
$$\alpha < 0.6 \qquad \text{for } n > 3$$

$$\alpha = h_n \sqrt{\frac{W}{\Sigma(EI)}} \qquad (17.9)$$

in which

n = the number of stories
h_n = the height of the building
W = the total seismic weight of the building
$\Sigma(EI)$ = the sum of the flexural stiffnesses for all the shear wall sections (assuming no cracks in the walls) with respect to the direction being considered.

17.9 EXAMPLE 17.1

The calculation of seismic loads for a six-story reinforced concrete building is presented. The building has a configuration typical of the majority of concrete multistory buildings constructed in Greece. Figs 17.2 and 17.3 show, respectively, the plan view of a typical story and the elevation along the transverse (y) direction. The structure is symmetric in the transverse (y) direction and asymmetric in the longitudinal (x) direction. The seismic analysis is performed for the transverse (y) direction. The foundation, which is assumed to be flexible, consists of separate footings linked by beams and a perimeter shear wall; this perimetric shear wall has a height of 3.00 m, which extends from the foundation level up to zero grade, and a length in the y direction of 18.0 m. The lateral force resisting system consists of frames and a shear wall coupled with the frames. It is assumed that the

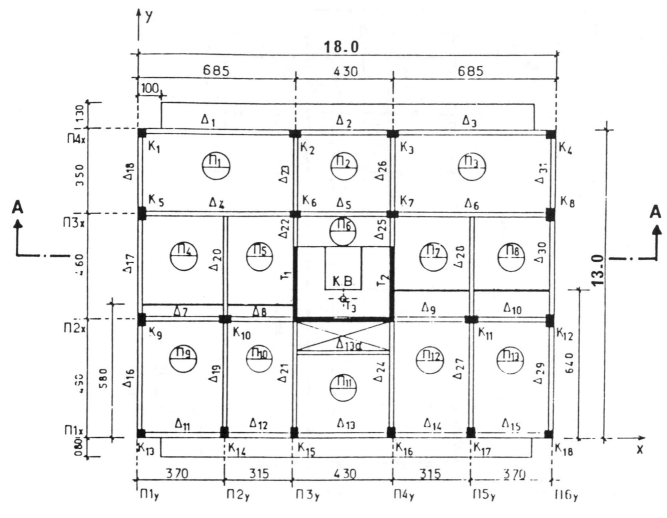

Fig. 17.2. Plan view for a typical story of the building for Example 17.1 (all dimensions are centimeters)

building is regular, that the shear wall resists less than 50% of the total seismic force, and that no additional resistance arises from masonry infills. The building, located in Iraklio, Crete, is in Seismic Zone III (Table 17.1). The importance factor is 1.0 which corresponds to "Usual" as given in Table 17.4. The weight (dead and live load) at a typical level of the building, is calculated as 401.2 tonnes, except for the first level which is 180.4 tonnes. The soil at the site is soft marine sediment.

Solution

Seismic Zone: Z = III (for Iraklio from Table 17.1)

Soil Category: C (from Table 17.2)

Importance factor: I = 1.0 (from Table 17.4)

Seismic weight: W = 5 × 401.2 + 180.4 = 2,187 tonnes

Soil factor: S = 0.16 (from Table 17.3)

Seismic coefficient

$$\varepsilon = IS \text{ (Section 17.3)}$$
$$= 1.0 \times 0.16 = 0.16$$

Base shear force

$$V = \varepsilon W \qquad \text{[eq.(17.2a)]}$$
$$= 0.16 \times 2,187 = 350 \text{ tonnes}$$

Seismic lateral forces

$$F_i = V \frac{W_i h_i}{\sum_{j=1}^{n} W_j h_j} \qquad \text{[eq.(17.3)]}$$

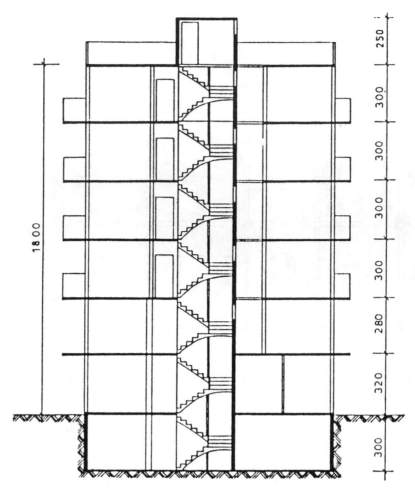

Fig. 17.3. Elevation of the six-story building for Example 17.1 (all dimensions are in centimeters)

Calculation results for the lateral seismic forces are shown in Table 17.5.

Story shear forces V_i

$$V_i = \sum_{j=i}^{n} F_j \qquad \text{[eq.(17.4)]}$$

Results for story shear forces are shown in Table 17.5.

Table 17.5. Calculation of Lateral Seismic Forces for Example 17.1

Level i	Height h_i (m)	Weight W_i (tonnes)	Lat. Force F_i (tonnes)	Story Shear V_i (tonnes)	Overturning Moment (tonne-m)
6	18.0	401.2	102.5	102.5	—
5	15.0	401.2	85.4	187.9	307.5
4	12.0	401.2	68.3	256.2	871.2
3	9.0	401.2	51.2	307.5	1,639.9
2	6.0	401.2	34.2	341.7	2,562.4
1	3.2	180.4	8.2	349.8	3,519.0
0	0.0	—	—	—	4,638.5

Preliminary check of story drifts

$$\Delta_i = \frac{1.5 V_i}{(\Sigma D_c + \Sigma D_{sw})_i} \qquad \text{[eq.(17.7)]}$$

$(\Sigma D_c + \Sigma D_{sw})_i = 0.55 \times 10^6 \, \text{t/m}$ (estimated lateral stiffness)

$$\Delta_1 = \frac{1.5 \times 350}{0.55 \times 10^6} = 0.95 \, \text{mm}$$

$$\Delta_1 < 0.002 \times 3.2 = 6.4 \, \text{mm} \qquad \text{[eq.(17.6)]}$$

Second-order P-Δ effect

$$\alpha = h_n \sqrt{\frac{W}{\Sigma(EI)}} \qquad \text{[eq.(17.9)]}$$

$\Sigma(EI) = 4.66 \times 10^6 \, (\text{t-m}^2)$ (calculated flexural stiffness)

$$\alpha = 18 \sqrt{\frac{2,187}{4.66 \times 10^6}} = 0.39$$

$\alpha = 0.39 < 0.6$ [no need to consider P-Δ effect (Section 17.8)].

Appendix A17

Provisions of the New Greek Seismic Code (1992)

A17.1 INTRODUCTION

The provisions of the 1992 Greek Seismic Code apply to the analysis and design of buildings subjected to earthquake excitation. These provisions are intended to ensure satisfactory performance of the structure based on the following criteria: (1) to resist a minor or moderate earthquake with no structural damage and acceptable damage to nonstructural components of the building, and (2) to resist a major earthquake with no collapse of the structure, although it may suffer structural damage in addition to general damage to nonstructural elements. To prevent collapse, the design must be based on the highest anticipated seismic excitation of the base rock for a specified probability of not being exceeded for a certain period of years. Definite numerical values for this probability will depend on the type and function of the structure. The application of more accurate design and analysis methods for a structure may become acceptable after endorsement by the appropriate public authority. An alternative design method must be based on scientifically established concepts and its applicability must attain at least the same level of safety as the provisions of the 1992 Seismic Code. The provisions of the seismic code are valid only with accurate application of the design provisions recommended by the appropriate code for the material that is used in the structure (e.g., Reinforced Concrete, Masonry, Steel, Wood). Alterations of structural or nonstructural members are not allowed, nor is alteration of the function of the structure without the prior study of the consequences of these alterations.

In order to meet sufficiently the seismic hazard, the design, construction, and use of a structure are all considered. The intent is to ensure that there will be limited and repairable damage to the structural members in the case of a "design earthquake," and that the damage will be further minimized for earthquakes that are less intense than the "design earthquake." These less intense earthquakes have a higher probability of occurrence. With an acceptable level of exceedance from the design seismic actions to be defined in the code, the following requirements of seismic behavior should be fulfilled:

- The probability of structural collapse must be sufficiently small, as defined by the specific criteria included in the Seismic Code (1992) and the specific Construction Material Code, which must be combined to ensure a sufficient level of remaining strength and integrity of the structure after an earthquake.
- The structural damage for the design earthquake must be of a limited degree and of a repairable nature, whereas the damage for less intense earthquakes (with a higher probability of occurrence) must be minimal. There must be a minimum level of functionality that is ensured to withstand the effects of the design earthquake. This must be determined in terms of the type of use and the importance of the structure.
- The seismic design actions are either global actions to the structure as a whole or local actions to certain structural or nonstructural parts.
- There must be sufficient control over the design and construction stages as well as over the stage of final use of the structure.

The requirements for seismic design are considered to be satisfied when the following criteria are met:

239

- The actions upon each member of the structure are reliably transferred to the soil without large remaining deformations.
- It has been ensured that sufficient strength is available to all structural members, taking into account second-order effects as necessary.
- The development of plastic mechanisms has been checked and is satisfactory for the design earthquake. This is done by proper designing and detailing to avoid brittle-type failures, concentrating the use of plastic hinges to a few structural members (e.g., in the case of a soft story), and providing at the locations of the plastic hinges levels of ductility that are higher than the ones required.
- Sufficient ductility is provided to any critical area where there is the possibility that a plastic hinge may develop (such areas are the toe and top of all the columns).
- The actual response of the structure is sufficiently in accordance with the one resulting from the numerical modeling; all necessary care must be taken to minimize the uncertainties of the computational assumptions.
- Protective measures are considered to safeguard the structure from pounding by any adjacent structures during the earthquake excitation.
- The interstory drift values do not exceed the limits that are considered to be sufficient to ensure the minimization of damage to nonstructural elements such as masonry partitions.
- The support points of the various secondary items, which are necessary for the proper functioning of the structure during its final use, are properly designed to withstand the design earthquake.

A17.2 MODELING CONSIDERATIONS

A17.2.1 Degrees of Freedom

The number and type of the degrees of freedom to be considered must be chosen so that a good approximation is achieved of all the important deformation modes and the corresponding inertial forces. For buildings with a slab in each story that can be assumed to act as a rigid horizontal diaphragm, it is sufficient to consider two normal horizontal translational and one rotational degrees of freedom per story.

In general, the ground support of a structure can be considered as rigid. However, additional degrees of freedom at the support points are allowed to be introduced, in certain cases.

A17.2.2 Modeling Distributed Mass and Definition of Seismic Weight

In the modeling of the distributed masses for building structures it is sufficient to replace the distributed mass at each story by a concentrated mass at the center of mass, plus a mass moment of inertia with respect to a vertical axis through the center of mass.

The seismic weight W_i at any level i of the building is calculated as the dead load G_i attributed to that level, plus a fraction of the design live load Q_i; that is,

$$W_i = G_i + \psi_2 Q_i \qquad (A17.1)$$

where ψ_2 is a reduction coefficient with the following values:

- $\psi_2 = 0.3$ for apartment buildings, office buildings, and shopping-centers
- $\psi_2 = 0.5$ for buildings of public use (schools, theaters, stadiums, etc.)
- $\psi_2 = 0.6$ for parking-garages
- $\psi_2 = 0.8$ for structures used in long-term storage (e.g., libraries, warehouses, tanks, silos).

A17.2.3 Modeling the Stiffnesses of Structural Elements

In modeling the structure, all elements with significant stiffness shall be considered. When the structural members are made from reinforced concrete, their stiffnesses must be calculated on the basis of cracked sections. If no detailed calculations are made for the cracked sections, the bending stiffnesses of the various elements are allowed to be calculated on the basis of the corresponding stiffnesses of the uncracked sections. The contribution of the reinforcement is omitted as indicated below:

- Column bending stiffness = 100% bending stiffness of uncracked section with no rebars
- Beams bending stiffness = 50% bending stiffness of uncracked section with no rebars
- Shear walls bending stiffness = 67% bending stiffness of uncracked section with no rebars
- Torsional stiffness of all members = 10% corresponding stiffness of uncracked section with no rebars.

A17.3 LOAD COMBINATION

The seismic forces are considered to be exceptional loads: A special loading combination is formed from the seismic forces, the forces resulting from gravitational loads, and the prestress operations. Other forces, such as those due to wind action, snow, etc., are not included in the special loading combination when seismic forces are considered. In determining the resultant forces in the structural elements of the building, the effects of the seismic forces are combined with the effects due to other loads on the structures, such as dead weight, the appropriate fraction of the design live load and prestress forces.

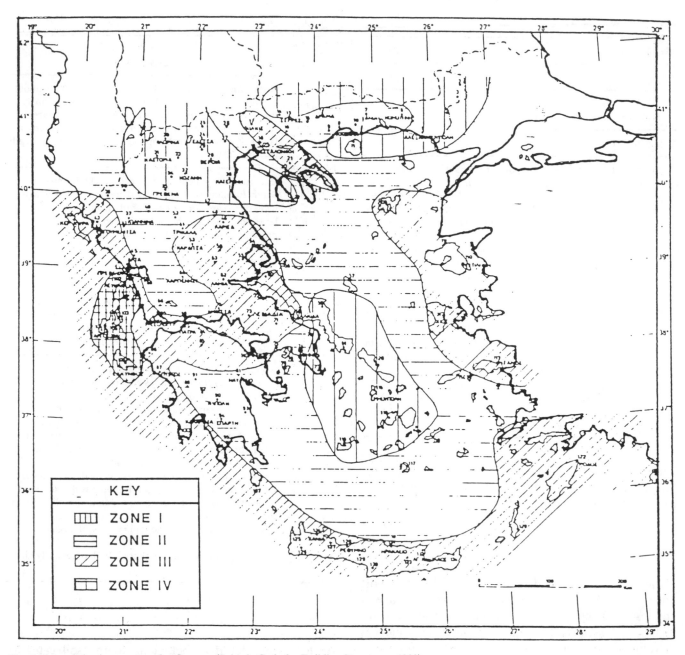

Fig. A17.1.   Seismic zone map for Greece. (Seismic Code for Building Structures, 1992)

The 1992 Greek Seismic Code states that the effects of the seismic forces can be considered independently for the two main perpendicular directions of the building. However, it specifies that the seismic force effect in one direction must be combined with 30% of the effect, on the same members, from the seismic force in the perpendicular direction.

A17.4 SEISMIC ZONES AND GROUND ACCELERATION COEFFICIENT (α)

The 1992 Greek Seismic Code provides for four seismic zones: I, II, III, IV. Zone I and Zone II

correspond to the same classifications as used in the 1984 Code; that is, Zone I represents low intensity and Zone II represents moderate intensity of earthquake activity. Zone III corresponds to a zone of strong earthquake activity. Finally, Zone IV refers to very intense earthquake activity. Figure A17.1 depicts the seismic zone map showing these seismic zones [*Seismic Hazards in Greece* (1989)]. Table 17.1 provides a list of the main cities and other places included in the 1992 Greek Seismic Code, with their corresponding seismic zone. Any place in Greece not included in Table 17.1 is assumed to be in a seismic zone of the nearest place in the table. A specific value of the earthquake ground acceleration coefficient α is assigned for each seismic

Table A17.1. Seismic Coefficient (α)

Seismic Zone	I	II	III	IV
α	0.12	0.16	0.24	0.36

Table A17.2. Importance Factor (I)

Category	I	Description
"a"	0.85	Buildings not important for public safety (e.g. farming sheds, warehouses).
"b"	1.0	Industrial buildings and ordinary buildings housing apartments, offices, hotels, etc.
"c"	1.25	Schools, educational establishments, buildings for public use, airport terminals, and buildings where large numbers of people congregate at all times. Also buildings housing facilities of great financial importance (e.g., computer centers, special industries).
"d"	1.30	Buildings housing services very essential during the aftermath of an earthquake (e.g., hospitals, fire stations, telecommunications centers, police stations, civil service headquarters). Also structures housing works of great artistic value (e.g., museums), and buildings used to store hazardous materials.

Table A17.3. Soil Categories

Soil Category	Description
A	Rock formations of considerable area and depth without extensive fragmentation. Extensive layers of extremely dense sandy formations with a very small percentage of silty clay and with a layer depth less than 7 m. Extensive layers of very hard precompressed clay with a layer depth less than 70 m.
B	Extensively fragmented and eroded rocky formations. Sandy formations of relatively moderate density with a layer depth greater than 5 m or of high density with a layer depth greater than 70 m. Moderately hard precompressed clay with a layer depth greater than 70 m.
C	Loose sandy formations of relatively low density with a layer depth greater than 5 m or of medium density with a layer depth greater than 7 m. Silty clay soils of small bearing capacity with a layer depth greater than 5 m.
D	Soil with relatively soft clays, high plasticity index ($1p > 60$), and total layer depth greater than 5 m.
X	– Soil formations with slope stability problems. – Rocky or earth formations with cavities or with potential to develop cavities. – Loose silty sands under the water level with high potential for liquefaction (unless a special study can testify that such a danger does not exist or measures are taken for improving the soil mechanical characteristics). – Earth formations near active faults. – Formations with high slope covered with loose material. – Loose sandy or silty clay soil formations that are known to be dangerous with respect to dynamic consolidation or loss of strength. – Recently deposited loose refillments with organic material. – All the formations of Category C with excessively high slope.

zone, as shown in Table A17.1. According to seismological data, these values have a 10% probability of exceedance in a period of 50 years.

A17.5 IMPORTANCE FACTOR (I)

Structures are classified in four categories of importance related to the eventuality of their damage or collapse and the disruption of their function, the risk of human loss, and the social/economic consequences. A value of the importance factor I is assigned for each category, as shown in Table A17.2.

A17.6 CLASSIFICATION OF SOILS

Five categories are used to classify the soils where the foundation of a structure is to be built. The classification is based on the hazard potential that each soil category presents. A brief description of these five soil categories is given in Table A17.3.

In order to confront efficiently the nature of the existing hazards, the foundation of a structure in a soil of Category X is permitted only after special detailed investigations and studies, and only after the introduction of certain techniques for soil improvement. When the thickness of a soil layer is less than 5 m, this layer

can be considered as belonging to the soil category immediately above the soil category listed in Table A17.3. Of course, this consideration is not allowed for soil layers belonging to Category X.

A17.7 FUNDAMENTAL PERIOD

The fundamental period of the structure in the direction of the seismic action is allowed to be estimated following any recognized approximate method of engineering science. For buildings of orthogonal plan the following formula may be applied for estimating the value of the fundamental period:

$$T = 0.09 \frac{h_n}{\sqrt{B}} \sqrt{\frac{h_n}{h_n + \rho B}} \quad \text{(sec)} \qquad \text{(A17.2)}$$

Table A17.4. Values of β_0 and T_2

Soil Category	A	B	C	D
β_0	2.5	2.5	2.5	2.5
T_2 (sec)	0.4	0.6	0.8	1.2

Table A17.5. Maximum Values of Damping Ratio Coefficient ($\zeta\%$)

Type of Structure		ζ^*
Steel structures	Welded	2.0%
	Bolted	4.0%
Reinforced Concrete		5.0%
Prestressed Concrete		3.0%
Masonry	Reinforced with	6.0%
	horizontal R.C. joint beams	6.0%
Wooden	Trusses	9.0%
	Frames	7.0%

*Listed values are for structures with no masonry partitions responding in the elastic range; in the inelastic range behavior, the value of the damping corrective factor (n) in eq.(A17.4) is allowed to be taken equal to one irrespective of the actual damping ratio value, since the damping influence is included in the response modification coefficient.

where

n = number of stories in the building
h_n = total height of the building
B = building dimension in the direction of the seismic forces
ρ = ratio of the area of reinforced concrete shear walls to the total area of shear walls and reinforced concrete columns, in a typical story of the building.

A17.8 ELASTIC SPECTRAL AMPLIFICATION COEFFICIENT $\beta(T)$

The value of the spectral amplification coefficient $\beta(T)$ is obtained from Fig. A17.2 for each of the four soil categories (Section A17.6) by the following relationships:

$$\beta(T) = \beta_0 \qquad \text{for } T \leq T_2$$
$$\beta(T) = \beta_0(T_2/T)^{2/3} \qquad \text{for } T > T_2 \qquad \text{(A17.3)}$$

where

β_0 = initial amplification coefficient with a value equal to 2.5 (see Table A17.4 and Fig. A17.2)
T = the fundamental period of the structure in seconds (Section A17.7)
T_2 = a particular period value characterizing the initiation of the descending part of the spectral curve corres-

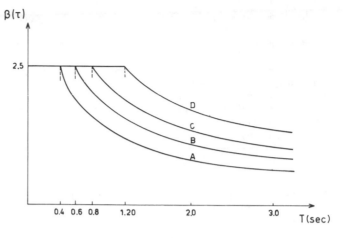

Fig. A17.2. Elastic spectral amplification coefficient $\beta(T)$

ponding to each particular soil category (see Fig. A17.2 and Table A17.4).

A17.9 DAMPING CORRECTIVE FACTOR (n)

The spectral acceleration amplification curves in Fig. A17.2 correspond to a damping ratio coefficient with a value equal to 5% of the critical damping. When the structure under consideration is assumed to have a damping ratio different than 5%, the spectral acceleration amplification coefficient $\beta(T)$ must be multiplied by the damping corrective factor n, as included in eq.(A17.11). The damping corrective factor is given by the following relationship:

$$n = \sqrt{\frac{0.05}{\zeta}} \qquad \text{(A17.4)}$$

in which ζ is the maximum value of the damping ratio coefficient. Table A17.5 shows maximum values of ζ for different types of structures.

A17.10 RESPONSE MODIFICATION COEFFICIENT (q)

The response modification coefficient introduces the reduction of the seismic forces due to the inelastic behavior that develops in the actual structure during realistic strong earthquake motions, when these forces are compared with the forces that would have developed in an ideally elastic system. Values of the response modification coefficient q, which depend on both the material used and on the type of structural system, are given in Tables A17.6a, A17.6b, and A17.6c, respectively, for reinforced concrete structures, masonry structures, and steel structures.

Table A17.6a.   Response Modification Coefficient (q):* Reinforced Concrete Structures

Structural System	q
(a) Structures with frames or frames and shear walls	3.5
(b) Structures that consist solely of shear wall systems	3.0
(c) Structures of the type of inverted pendulum or structures where the lateral force-resisting system acts as isolated cantilever (water tanks, chimney, elevated silos, etc.)	2.0

*The values for q are for monolithic types of construction.

Table A17.6b.   Response Modification Coefficient (q): Masonry Structures

Structural Systems	q
(a) With continuous horizontal R.C. joint beams	1.5
(b) With continuous horizontal R.C. joint beams as well as vertical R.C. members	2.0
(c) With horizontal and vertical reinforcement in all structural components	2.5

Table A17.6c.   Response Modification Coefficient (q): Steel Structures

Structural Systems	q
(a) Frames with no bracing	4.0
(b) Diagonally braced frames	3.0
(c) Frames with V-type concentric bracing	1.5
(d) Frames with K-type bracing	1.0
(e) Eccentrically braced frames	4.0

A17.11 FOUNDATION COEFFICIENT (θ)

The value of the foundation coefficient θ depends on the depth and stiffness of the foundation. The value of this coefficient must be taken equal to 1 for foundations on soil layers belonging to Categories A and B. The value of the foundation coefficient on soil layers belonging to Categories C and D is shown in Table A17.7 for various conditions. However, the product of the amplification factor $\beta(T)$ and the foundation coefficient θ must have a value equal to or larger than the value that would have been obtained if the soil category were assigned as B.

A17.12 TORSIONAL CONSIDERATIONS

The inertial force induced by the earthquake at a specific story acts through the center of mass at that story, whereas the resultant resisting force is at the center of stiffness of the story. The center of stiffness at a story and the center of mass above that story may not coincide if the structure is not dynamically sym-

Table A17.7.   Foundation Coefficient θ

Conditions	θ
1a. The building has one basement level.	0.90
1b. The building has a relatively stiff mat foundation.	
1c. The foundation consists of piles with pile caps linked with beams.	
2a. The building has at least two basement levels.	0.80
2b. The building has at least one basement level and a relatively stiff foundation mat.	
2c. The piles of the foundation of the building are connected with one pile cap (not necessarily of the same thickness).	

metric (this would induce a torsional moment). The Code specifies that an *accidental eccentricity* should be considered in addition to the actual eccentricity. The seismic action at each story level i shall be considered acting not only at the center of mass but also at a distance e_{ai} on either side of the center of the mass.

A17.12.1 Torsional Flexibility Coefficient

The evaluation of the accidental eccentricity e_{ai} for story i of the building depends on the value of the story stiffness coefficient, ξ_i, which is given by

$$\xi_i = \left(\frac{\delta_{max}}{1.2\delta_{avr}}\right)_i \leqslant 3.0 \qquad (A17.5)$$

with

$$\delta_{avr} = \frac{\delta_{max} + \delta_{min}}{2}$$

where

δ_{max} = maximum lateral displacement of story i
δ_{min} = minimum lateral displacement of story i.

The accidental eccentricity e_{ai} for a story i is calculated by

$$\begin{aligned} e_{ai} &= 0.05L & \text{when } \xi_i \leqslant 1.0 \\ e_{ai} &= 0.05LA_i & \text{when } 1.0 < \xi_i \leqslant 3.0 \end{aligned} \qquad (A17.6)$$

where

L = the dimension of the story in the direction perpendicular to the seismic action considered
A_i = the accidental eccentricity amplification factor given by

$A_i = 1.5\xi_i$ when the Equivalent Force Method is used

$A_i = \xi_i$ when the dynamic Spectral Method is used

(A17.7)

The 1992 Code states that the *accidental* eccentricity amplification factor A_i need not exceed 3.0, as indicated in eq.(A17.5). Alternatively to the values of e_{ai} specified in eq.(A17.6), the accidental eccentricity can be taken equal to $e_{ai} = 0.10L$. However, in this case the story torsional flexibility ξ_i must be smaller than or equal to 2.5 ($\xi_i \leqslant 2.5$). The amplification of the accidental eccentricity by the factor A_i is intended to represent the increased eccentricity caused by perimeter elements that yield; factor A_i provides a simple yet effective control of excessive torsional yielding in a given story.

A17.12.2 Torsional Moments

Because of the possibility of an eccentricity between the center of mass and the center of stiffness for any story, it is necessary to consider torsional moments at each story of the building. The torsional moment M_{ti} for any story i is given by

$$M_{ti} = \sum_{j=i}^{n} F_j e_{dj} \qquad (A17.16)$$

where

F_j = the lateral seismic force at story j
e_{dj} = the total design eccentricity

The torsional effects must be superimposed to the effects resulting from the combination of other loads (section A17.3) irrespective of the method used to determine the horizontal seismic action (equivalent force method or dynamical spectral method).

A17.13 EARTHQUAKE-RESISTANT DESIGN METHODS

The 1992 Greek Seismic Code provides two methods for earthquake-resistant design: (1) the equivalent force method, and (2) the dynamic spectral method. The equivalent force method can be applied to regular buildings up to 10 stories high. It can also be applied to irregular buildings up to six stories high, except in the following cases:

- Irregular buildings more than three stories high in Importance Category "c" (Table A17.2) located in Seismic Zones III and IV (Table A17.1)
- Irregular buildings more than two stories high classified in Importance Category "d" (Table A17.2), located in any seismic zone.

The dynamic spectral method may be used for any structure, but *must* be used for all structures for which the equivalent force method is not applicable.

A building is considered regular when all stories are torsionally stiff ($\xi_i \leqslant 1$, see Section A17.12), and when the increase or decrease in the stiffness ($\Delta K_i = K_{i+1} - K_i$) for the two main horizontal directions or in the mass ($\Delta m_i = m_{i+1} - m_i$) of a story i, when compared with the story above $i+1$, is within the following limits:

$\Delta K_i < 0.35 K_i$, $\Delta m_i < 0.35 m_i$ (in the case of an increa

$$(A \quad 9)$$

$\Delta K_i < 0.50 K_i$, $\Delta m_i < 0.50 m_i$ (in the case of a decreas

A17.14 EQUIVALENT FORCE METHOD

In the equivalent force method the response of the structural system is estimated by applying *equivalent seismic forces*, which are calculated according to eq.(A17.13) in Section A17.14.2. These equivalent seismic forces are applied as static forces at the corresponding locations of the centers of mass of the various levels of the building.

A17.14.1 Base Shear Force

The base shear force V is determined by the following formula:

$$V = \varepsilon_h W \qquad (A17.10)$$

where

$$W = \sum_{i=1}^{n} W_i \qquad (A17.11)$$

in which W_i is given by eq.(A17.1) and $\varepsilon_h(T)$ is the seismic design coefficient given by

$$\varepsilon_h(T) = \frac{\alpha I \beta(T)}{q} n\theta \qquad (A17.12)$$

in which

α = the seismic coefficient corresponding to a seismic zone (Table 17.1), section A17.4
I = the importance factor given in Table A17.2
$\beta(T)$ = the spectral amplification factor obtained from the design spectral curves (Fig. A17.2), Section A17.8 and Table A17.6 as a function of the fundamental period (Section A17.7) and the soil categories A, B, C, D
θ = the foundation coefficient reflecting the influence of the type of foundation given in Table A17.7
q = the response modification coefficient given in Tables A17.6a, A17.6b, and A17.6c (Section A17.10)
n = the damping corrective factor (Section A17.9).

The 1992 Code imposes the following *design* limitation to the seismic coefficient $\varepsilon_h(T)$ [eq.(A17.12)]:

$$\frac{\varepsilon_h(T)}{\alpha I \theta} \geqslant 0.20 \qquad (A17.13)$$

The 1992 Code also allows the value of $\beta(T)$, the elastic response design spectral amplification coefficient, to be taken as 2.5, irrespective of the soil category and the fundamental period value.

A17.14.2 Equivalent Seismic Forces

The total base shear force V given by eq.(A17.10) is distributed among the various levels of the building according to the following formula:

$$F_i = (V - V_h) \frac{W_i \phi_i}{\sum\limits_{j=i}^{N} W_j \phi_j} \qquad (A17.14)$$

where

F_i = lateral seismic force at level i
W_i = seismic weight attributed to level i
ϕ_i = modal component at level i for the first mode estimated by an acceptable approximate method
N = number of stories
V = base shear force [eq.(A17.10)]
V_h = additional force applied at the top of the building equal to 0.07 TV but no greater than 0.25 V in which T is the fundamental period

For regular buildings the distribution of the equivalent seismic forces is given by the following formula:

$$F_i = (V - V_h) \frac{W_i h_i}{\sum\limits_{j=i}^{N} W_j h_j} \qquad (A17.15)$$

where

h_i = height of level i measured from the foundation of the building

The determination of the equivalent seismic forces according to eq.(A17.5) is also allowed for irregular buildings in the following cases:

(a) Irregular buildings of importance category a, b, or c up to two stories height for all seismic zones
(b) Irregular buildings of importance categories a or b, up to three stories height in seismic zones I, II and III
(c) Irregular buildings of importance category a or b, up to four stories height in seismic zones I and II.

A17.14.3 Overturning Moments

The first-order overturning moment M_{Ii} at level i due to the seismic forces is equal to the sum of the moments of seismic forces F_i above that level, namely

$$M_{Ii} = \sum_{j=i+1}^{n} F_j(h_j - h_i) \qquad (A17.16)$$

where $i = 0, 1, 2, \ldots, n-1$

A17.15 DYNAMIC SPECTRAL METHOD

The Greek Seismic Code (1992) introduces the dynamic method for seismic design. As presented in Chapter 4 of this handbook, the dynamic spectral method requires the previous solution of an eigenproblem to determine the natural frequencies (or natural periods) and the corresponding modal shapes. Then, the response for each significant mode is obtained from design spectral charts such as those in Fig. A17.2 (Greek Seismic Code 1992). Finally, the modal maximum responses are combined by a suitable method, the Square Root of the Sum of Squares (SRSS) or the Complete Quadratic Combination (CQC) [Chapter 4, Section 4.8]. The 1992 Code indicates that for ordinary buildings, the SRSS method may be used to determine the base shear force, the seismic force distribution, and other response quantities of interest. However, the Code requires that the base shear force obtained from the dynamic method should not be less than 90% of the corresponding value for the base shear force derived from the application of the static method. The Code requires that responses obtained by the dynamic method be adjusted linearly in the same proportion as the base shear force obtain by the dynamic method which must be scaled in order to be equal to 90% of the base shear calculated by the static method.

A17.16 SECOND-ORDER P-Δ EFFECT

The P-Δ effect is considered using the following parameter θ_i, evaluated at each level i of the building:

$$\theta_i = \frac{N_i q \Delta_i}{V_i L_i} \qquad (A17.17)$$

where

Δ_i = story drift for the combined action of loads (section A17.3)
q = response modification coefficient (Tables A17.6a–A17.6c)
N_i = total axial load at story i for the combined action of loads
V_i = total shear force at story i for the combined action of loads
L_i = story height.

The value of this parameter θ_i is not allowed to be larger than 0.3 in any story ($\theta_i < 0.3$). If this limit is exceeded in at least one story, a modification in the stiffness of the structural system is required so that the values of θ_i become smaller than 0.3. The check for the second-order effects is the following:

for $\theta_i \leqslant 0.1$ second-order effects can be ignored for all stories for $0.1 < \theta_i \leqslant 0.2$ second-order effects can be accounted approximately by amplifying any calculated value for response (force, moment, displacement, etc.) by the factor $1/(1 - \theta_i)$

for $0.2 < \theta_i \leqslant 0.3$ second-order effects must be accounted by a more accurate evaluation than the application of the factor $1/(1 - \theta_i)$.

A17.17 STORY DRIFT LIMITATIONS

To avoid excessive damage, the shear strain of the infill panels in any story under the combined action of the loads defined in Section A17.3 must not exceed the limit specified by eq.(A17.18). The shear strain γ of the infill panels for a story height of L_i is given by the following relationship:

$$\gamma = \frac{q\Delta_i}{2.5L_i} < \gamma_{\text{limit}} \qquad (A17.18)$$

where

Δ_i = story drift for the combined action of loads (Sect. A17.3)

q = modification response coefficient (Table A17.6a–A17.6c)

L_i = story height.

For conventional infill panels in the value of γ_{limit} can be taken equal to 0.005. For infills less sensitive to shear strains the value of γ_{limit} can be taken equal to 0.007.

A17.18 VERTICAL SEISMIC ACTION

It is generally allowed by the 1992 Seismic Code to ignore the vertical seismic action except in cases of structural prestressed concrete members or girders acting as supports of overlying columns. In these cases, it is allowed to examine the vertical seismic action on stsructural members independently from the rest of the structure by considering that the vertical seismic action is applied at the support of the structural member.

The fundamental period in the vertical direction (T_v) for the member can be obtained by applying the Rayleigh formula:

$$T = 2\pi \sqrt{\frac{\sum\limits_{i=1}^{N} m_i y_i^2}{\sum\limits_{i=1}^{N} m_i y_i}} \qquad (17.19)$$

where

N = number of concentrated masses m_i at various locations of the member

y_i = the vertical displacement of the concentrated mass m_i under the action of the gravity load $m_i g$.

The vertical seismic force F_{vi} on mass m_i is given by

$$F_{vi} = \frac{W\varepsilon_v m_i y_i}{\sum\limits_{i=1}^{N} m_i y_i} \qquad (A17.20)$$

where

W = total vibrating weight

ε_v = seismic design coefficient for vertical action given by

$$\varepsilon_v = 0.7 \frac{\alpha I \beta(T_v)}{q_v} n \qquad (A17.21)$$

in which

α = seismic zone coefficient (Table A17.1)

I = importance factor (Table A17.2)

$\beta(T_v)$ = spectral amplification factor (Fig. A17.2)

T_v = fundamental period in the vertical direction [eq.(A17.19)]

q_v = vertical response modification coefficient which can be taken equal to half the value of the response modification coefficient specified in Tables A17.6a–A17.6c, that is

$$q_v = 0.5q > 1 \qquad (A17.22)$$

This simplified method of taking into account the vertical seismic action may be applied irrespective of the method employed for the horizontal seismic action (the equivalent force method or dynamic spectral method). Also, it is allowed to use simple superposition procedures to combine the effects of the horizontal and the vertical seismic actions.

A17.19 DESIGN CRITERIA FOR NO-COLLAPSE

During the response of a structure for the design earthquake, the formation of an elastoplastic mechanism with a predictable safe behavior is required. Such a behavior is believed to be ensured by the following criteria.

A17.19.1 Strength Design

The present Code requires that the minimum level of strength (R_d) in all structural elements be equal or greater than the demand S_d resulting from the action of the combined loads (Section A17.3). Hence,

$$S_d \leqslant R_d \qquad (A17.23)$$

The Code also requires that the resulting demand includes, when necessary, the influence from the second-order effects (Section A17.16). The design strength R_d is calculated according to the provisions of the corresponding "Material Codes" using prescribed safety factors for combined loads. When the demand consists of more than one type of action (e.g. bending and axial force), then the extreme values (maximum or minimum) of the effects of the actions should satisfy eq.(A17.23) using the appropriate interaction diagram.

A17.20 COMPUTER PROGRAMS

Two computer programs have been developed: GREECE1 is used to implement the provisions of the 1984 Greek Seismic Code, and GREECE2 is used for seismic design of buildings using the 1992 Greek Seismic Code. These computer programs serve to determine the equivalent lateral forces and the story shear forces.

A17.20.1 Example 17.2

A 10-story reinforced concrete building with the plan configuration shown in Fig. 17.2 for Example 17.1 is to be analyzed according to the provisions of the Greek Seismic Code (1992). The building has the same general properties that were described for the six-story building in Example 17.1, except for the addition of four 3-m stories.

```
INPUT DATA AND OUTPUT RESULTS FOR EXAMPLE 17.2

INPUT DATA:

A) GENERAL DATA:

   SEISMIC ZONE:  3 (Iraklio, Table 17.1)

   SOIL FACTOR:  C (from Table A17.3)

   IMPORTANCE CATEGORY:  I = 1 (from Table A17.2)

   NUMBER OF STORIES:  10

   BUILDING LENGTH (m):  13

   BUILDING HEIGHT (M):  30.0

   COLUMN AREA  (M**2):  3.06

   SHEAR WALL AREA (m**2):  0.86

   BUILDING LENGTH (M):   13.0

   MATERIAL:  REINFORCED CONCRETE

   REGULARITY (Y/N):  Y

   STRUCTURAL SYSTEM:  q = 3.5  (Table A17.5.1)

   FOUNDATION:     THETA = 0.9 (One basement level, Table A17.6)

   B)  STORY DATA:
```

LEVEL (i)	HEIGHT H(m)	DEAD WEIGHT G(Tonne)	LIVE WEIGHT Q(Tonne)	ECCENTRICITY (m)
10	30.0	334.6	66.6	0.9
9	27.0	334.6	66.6	0.9
8	24.0	334.6	66.6	0.9
7	21.0	334.6	66.6	0.9
6	18.0	334.6	66.6	0.9
5	15.0	334.6	66.6	0.9
4	12.0	334.6	66.6	0.9
3	9.0	334.6	66.6	0.9
2	6.0	334.6	66.6	0.9
1	3.2	334.6	66.6	0.9

```
OUTPUT RESULTS:

   GROUND ACCELERATION:  ALPHA = 0.24

   FOUNDATION FACTOR:  THETA = 0.9

   FUNDAMENTAL PERIOD (SEC):   T = 0.716

   SPECTRAL AMPLIFICATION:  BETA = 2.5

   DAMPING CORRECTIVE FACTOR:   N = 1

   RESPONSE MOD. FACTOR:   Q = 3.5

   IMPORTANCE FACTOR:  I = 1

   SEISMIC COEFFICIENT:  EPSILON = 0.1543

   BASE SHEAR FORCE (TONNE):  V = 547.1
```

LEVEL (i)	LATERAL FORCE F_i (tonne)	STORY SHEAR V_i (tonne)	OVERTURNING MOMENT M_i(tonne-m)	TORSIONAL MOMENT*** M_{ti}(tonne-m)	STORY* DRIFT (mm)	PERMITTED** DRIFT (mm)
10	104.1	104.1	----	93.7	7.1	12.8
9	92.7	196.8	312.3	83.4	7.0	12.8
8	81.3	278.1	902.7	73.2	6.9	12.8
7	70.1	348.2	1737.0	63.1	6.1	12.8
6	59.1	407.4	2781.6	53.2	6.7	12.8
5	48.3	455.7	4003.8	43.2	6.8	12.8
4	37.7	493.4	5370.9	33.9	6.1	12.8
3	27.4	520.8	6851.1	24.7	6.1	12.8
2	17.5	538.3	8413.5	15.8	5.4	12.8
1	8.7	547.1	9920.7	7.8	5.4	13.6
0	----	----	11671.5	----	-----	----

```
*    Calculated using eqs.(4.72) and (4.73) of Chapter 4.
**   Calculated from eq.(A17.18) [β = 0.67 and θ_g = 1%, Section A17.14.6]
***  Assuming ζ_i < 1.0 for all stories (see Section A17.12)
```

REFERENCES

ANAGNOSTOPOULOS, S. A., and LEKIDIS, B. A. (1986) *Seismic Design According to the Provisions of New Seismic Code, Application Example Rules and Parametric Study.* Special Edition of the Institute of Engineering Seismology and Earthquake Engineering and the Technical Chamber of Greece, Thessaloniki, Greece (in Greek).

"Decree of the Minister of the Environment on the Revision of the 1959 Seismic Code for Building Structures" (1984) *Government's Gazette*, Issue B, No. 239, April 16, 1984, Greece (in Greek).

"Proposed Revisions to the New Greek Seismic Code, Edition 1988" (1989) *Weekly Publication of the Technical Chamber of Greece*, Issue No. 1575, July 17 (in Greek).

"Royal Decree Revising Article 6 of the Seismic Code for Building Structures" (1959) *Government's Gazette*, Issue A, No. 190, September 14, 1959, Greece (in Greek).

"Royal Decree on the Seismic Code for Building Structures" (1959) *Government's Gazette*, Issue A, No. 36, February 19, 1959, Greece (in Greek).

Seismic Hazards in Greece (1989) Final Report submitted to the Earthquake Protection and Planning organizations of Greece by a working group that included the Geophysical Laboratory of Aristotle University of Thessaloniki, the Geophysical Laboratory of Kapodistrian University of Athens, the Seismological Laboratory of the National Observatory of Athens, and the Institute of Engineering Seismology and Earthquake Engineering of Thessaloniki, Greece (in Greek).

TASIOS, T. P., and GAZETAS, G. (1979) "A Proposal for a Possible Revision to the Greek Seismic Code." *Laboratory of Reinforced Concrete Structures Report*, Technical University of Athens, Athens, Greece.

18
Hungary

György Vértes

18.1 INTRODUCTION

The first Hungarian provisions for earthquake-resistant design and construction of buildings published as *Technical Guiding Principles* (1978) remained in use until 1993. These provisions, which are not mandatory requirements, contain general guidance for earthquake-resistant design and for executing legal contracts for the construction of buildings.[1] In this chapter, the main provisions of the *Technical Guiding Principles* are presented with specific modifications that will be included in the new seismic code. An example is provided to show how the code provisions will be applied.

18.2 GENERAL REGULATIONS

The provisions of the *Technical Guiding Principles* are applicable to buildings 50 or more meters high in regions of the country where earthquakes of intensities $N = 6$ to $N = 9$ of the MSK scale[2] are likely. In the absence of earthquake records, the MSK intensity N may be calculated by the following formula:

$$N \leq \log \frac{3H(H+10)}{10B\gamma(H+5)} + 1 \qquad (18.1)$$

where

H = height of the building in meters (m)
B = the smaller dimension of the building in plan (m)
γ = specific weight of the building (kN/m^3)

The provisions of 1978 do not apply to structures of Importance Category I (as defined in Section 18.4), to tower-like structures, or to structures with limited deformation capabilities.

The provisions of 1978 require consideration of the following:

1. The seismic weight should include the dead load and a fraction of the design value used for live load.
2. Any secondary system not considered as part of the earthquake resistant structure should be designed to retain its structural integrity and residual strength after the seismic action has ceased.
3. The entire building, including structural and nonstructural elements, must be designed to resist adequately earthquakes of moderate intensity without structural damage, and strong earthquakes without collapse or loss of lives.

---

[1]A new seismic code containing mandatory requirements for earthquake-resistant design of buildings was in preparation as this book went to press.

[2]The MSK is the International Macroseismic Intensity scale similar to the Mercalli Modified Intensity scale (MMI). (See Appendix of this Handbook.)

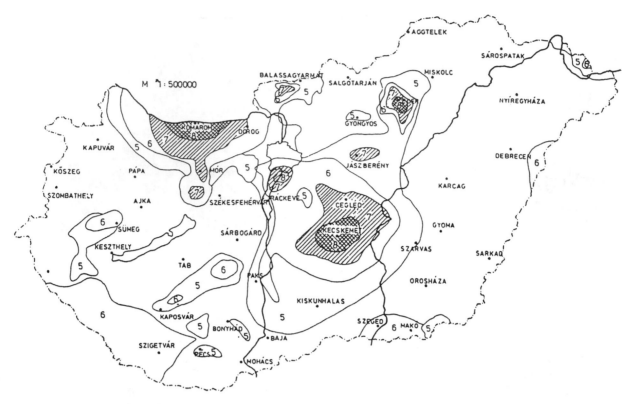

Fig. 18.1. Hungarian seismic intensity map in MSK scale

4. Special consideration should be given to the use of materials with ductility that will adequately dissipate energy during strong earthquakes.

5. The building's fundamental frequency and second natural frequency values should not be close to the dominant frequency of past earthquakes in the region.

6. The building's structural configuration and mass distribution, and the location of the lateral force-resisting elements, should be such as to avoid torsional loading and deformation in the structure.

7. The effects due to torsion should be considered in the design of buildings when the distance between the center of mass and the center of rigidity (or stiffness center) for any story is greater than 10% of the dimension of the building normal to the direction considered for the seismic forces.

8. The load-bearing structural system should be configured to respond as a three-dimensional structure.

9. The building, both in plan and elevation, should be regular in the distribution of stiffness and mass.

18.3 SEISMIC INTENSITY ZONES

Hungary has been divided into seismic intensity zones as shown in the seismic intensity map of Fig. 18.1. This map shows contour lines of seismic intensity on the MSK scale. Linear interpolation between these contour lines is permitted for intermediate locations on the map.

18.4 IMPORTANCE CATEGORIES

The provisions of 1978 classify structures according to their occupancy and use.

Importance Category I. This category is used for structures of extraordinary importance; serious damage to these structures would be catastrophic to the community. The following types of structures are included:

• Nuclear reactors and atomic power plants
• Unusual industrial buildings in which damage may cause dangerous blasts or fires (oil refineries, industrial chemical plants, etc.)
• Warehouses or factories in which dangerous or poisonous materials are stored or processed
• Establishments that must remain operational after an earthquake (telecommunication centers, electric-energy-distribution centers, dams, etc.).

Importance Category II. This category includes buildings and structures in which damage may cause severe loss in terms of human lives and materials.

• Structures of industrial/commercial importance (industrial complexes, railway stations, airports, etc.)
• Structures with high occupancy (governmental offices, theaters, movie houses, etc.)
• Hospitals, army or police buildings, fire stations, etc.

- Food stores and other commodity outlets or warehouses.

Importance Category III. This category includes buildings in which some damage may be sustained and that does not need to be repaired immediately after the earthquake.

- Universities, schools, and daycare/nursery buildings
- Industrial factory buildings
- Ordinary buildings such as dwellings, hotels, and student dormitories
- Barns, silos, and other structures for agriculture or animal husbandry
- Office buildings not included in Importance Category II.

Importance Category IV. This category includes buildings of moderate importance and structures of little economical value such as workshops and detached family houses.

Importance Category V. This group includes buildings of minimal importance, whose collapse would not involve loss of life or great financial loss.

- Small industrial or agricultural buildings
- Temporary buildings.

18.5 SEISMIC FORCES AND INTERNAL STRESSES

The following issues have been considered in the provisions of the new Hungarian Seismic Code:

- Element stresses must be calculated using the provisions of Hungarian Code MSZ 15021-15028.
- The modulus of elasticity of the material used in the structure may be increased by a factor of 1.1 to 1.2.
- The design of the structural system supporting gravitational loads must consider the formation of plastic hinges in critical sections due to seismic action. This requirement implies that the formation of anticipated plastic hinges should not affect the stability of the building and should not produce a progressive failure of the structure.
- The seismic design of the building must take into consideration the fact that the earthquake motion may occur in any direction.

18.6 SEISMIC WEIGHT

The seismic weight includes the dead weight of the building, the weight of permanent equipment, and, for warehouses, the total weight of material stored. Seismic analysis generally is conducted only for horizontal actions. The analysis disregards the vertical components of seismic force produced by earthquakes, except for structures that are sensitive to vertical vibratory motion, such as cantilevers, shells, etc. When it must be considered, the vertical action may be assumed to be two-thirds of the value determined for the horizontal effects.

18.7 METHODS OF SEISMIC ANALYSIS

The proposed Code provides the following three methods for seismic design of buildings: (1) equivalent static method, (2) modal analysis method, and (3) time-history analysis method. The equivalent static method is applicable to buildings classified as "regular." In this method, the dynamic loads induced in the structure in response to severe ground shaking are represented by equivalent static loads applied horizontally at the various levels of the building. These latter forces are obtained from a design response spectrum that is a suitably modified elastic response spectrum. The distribution of horizontal seismic actions is assumed to follow that of the first mode of vibration.

The modal analysis method must be used for more irregular structures or for tall buildings where the equivalent static analysis method does not apply. The dynamic analysis method assumes elastic behavior in order to determine the inertial forces acting at each floor level of the building. The modal responses are determined using the same design response spectrum as used in the application of the equivalent static method. In the modal analysis method, the building is modeled as a structural system with discrete masses lumped at each floor level. The inertial forces at each level of the building are obtained separately for each mode of vibration. Usually, the first three normal modes are included in the analysis.

For important or unusual structures, a time-history linear or nonlinear dynamic analysis may be justified. In this type of analysis, the response of the structure to a particular accelerogram record from a severe earthquake is computed numerically at small-time steps of the earthquake record.

18.8 EQUIVALENT LATERAL FORCES

The equivalent lateral forces applied at each floor level of the building are given by

$$F_i = C_d \eta_i W_i \qquad (18.2)$$

where

C_d = the design seismic coefficient defined in eq.(18.4)
η_{i1} = the distribution factor at level i of the building corresponding to the first mode of vibration
W_i = the seismic design weight attributed to the ith level.

The distribution factor η_{ir} at level i for mode r is given by

$$\eta_{ir} = h_i \frac{\sum\limits_{i=1}^{n} W_i Y_{ir}}{\sum\limits_{i=1}^{n} W_i Y_{ir}^2} \qquad (18.3)$$

in which

h_i = the height of the ith level measured from the base

Y_{ir} = the normal mode displacement at level i in the mode r.

The design seismic coefficient C_d in eq.(18.2) is given as the product of several factors:

$$C_d = k_g k_s k_t \alpha \beta_r \psi \qquad (18.4)$$

where

α = the amplification factor

β_r = the spectral response factor, which depends on the shape of the design response spectrum and on the fundamental period of the structure:

$$\beta_r = 1 \quad \text{for} \quad T_0 \leq T_z$$

$$\beta_r = \left(\frac{T_z}{T_0}\right)^\beta \quad \text{for} \quad T_0 \geq T_z \qquad (18.5)$$

in which T_0 = the fundamental period of the structure, and T_z = the characteristic period of the soil at the site.

ψ = the reduction factor, which takes into account the fact that the seismic forces were obtained from an elastic analysis. The reduction factor considers the effect of the potential nonlinear behavior of the structure, ductility, internal force redistribution, and energy dissipation through damping. Numerical values for ψ are given in Table 18.4.

k_s = the seismic zone factor (given in Table 18.1)

k_g = the importance factor (given in Table 18.2)

k_t = the soil profile factor given in Table 18.3 for soil profile types S_1, S_2, and S_3.

The characteristics of the soil profile types are as follows:

Soil Profile S_1. Rock, of any type, or stiff soil where the soil depth is less than 60 m and the soil types have overlying rocks that are stable deposits of sand, gravel, or clay.

Soil Profile S_2. Deep cohesionless soil or stiff clay soil, including sites where the soil depth exceeds 60 m and soil types that have overlying rock and stable deposits of sand, gravel, or stiff clays.

Soil Profile S_3. Soft to medium-stiff clays and sand, characterized by 10 m or more of soft to medium-stiff clay with or without intervening layers of sand or other cohesionless soil.

Table 18.1. Seismic Zone Factor (k_s)

MKS Intensity	k_s
6	0.15
7	0.22
8	0.26
9	0.32

Table 18.2. Importance Factor (k_g)

Importance Category	k_g
I	1.6
II	1.4
III	1.0
IV	0.7
V	0.5

Table 18.3. Soil Profile Factor (k_t)

Soil Profile Type	k_t
S_1	1.0
S_2	1.2
S_3	1.5

Table 18.4. Reduction Factor (ψ)

Type of Building	ψ
Multistory steel-framed building	0.20
Reinforced concrete multistory buildings without frame–shear wall interaction	0.25
Reinforced concrete multistory frame buildings with structural walls	0.30
Reinforced concrete buildings with walls, columns, and slab floors but without beams	0.35

When specific site-related information[3] for α, β, and T_z is not available, the following values may be assigned to these factors:

$$\alpha = 2.5 \qquad \beta = 2/3 \qquad T_z = 0.4 \, \text{sec}$$

Example 18.1

A seven-story office building with two bays is modeled by the plane frame shown in Fig. 18.2. The site of the building is in a region of intensity 7 on the

---

[3]Recommended values for the factors α and β are given in *Model Code for Seismic Design of Concrete Structures* (1985).

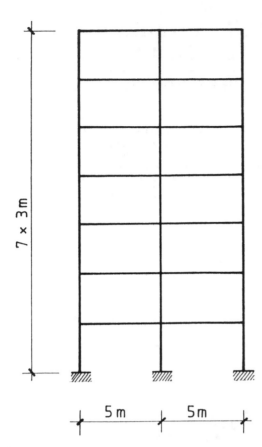

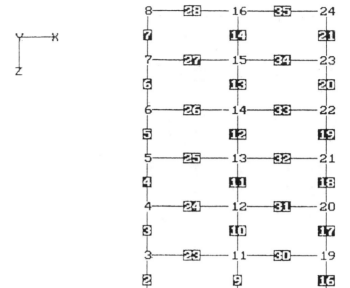

Fig. 18.3. Elements and nodes for the frame used in Example 18.1

Fig. 18.2. Six-story plane frame used to model building of Example 18.1

MSK scale. The soil at the site (stiff clay) has been classified as soil profile S_1, with a characteristic period $T_z = 0.25$ sec. The cross-sectional area and corresponding moment of inertia of all beam elements in the frame are respectively, 0.12 m^2 and $5.4 \times 10^{-3} \text{ m}^4$. For the columns, the cross-sectional area is 0.12 m^2 and the moment of inertia is $1.6 \times 10^{-3} \text{ m}^4$. The total weight of the structure, 20 tonnes, is concentrated at the nodes at each level of the building. Fig. 18.3 shows the elements and nodes.

Solution

The computer program developed for this chapter is used to perform the seismic analysis of the building.

Seismic zone factor: $k_s = 0.22$ (Table 18.1 for Intensity 7)
Importance factor: $k_g = 1.00$ (Table 18.2 for Category III)
Soil profile factor: $k_t = 1.00$ (Table 18.3 for Soil Profile S_1); $\alpha = 2.50$ (standard value); $\beta = 0.66$ (standard value)
Characteristic period: $T_z = 0.25$ sec

The program results for Example 18.1 are shown in Tables 18.5 through 18.11.

Table 18.5 Joint Coordinates and Joint Concentrated Weights

J	X [m]	Z [m]	Weight (tonnes)	J	X [m]	Z [m]	Weight (tonnes)
1	0.000	0.000	0	2	0.000	−3.000	5.00
3	0.000	−6.000	5.00	4	0.000	−9.000	5.00
5	0.000	−12.000	5.00	6	0.000	−15.000	5.00
7	0.000	−18.000	5.00	8	0.000	−21.000	5.00
9	5.000	−0.000	0	10	5.000	−3.000	10.00
11	5.000	−6.000	10.00	12	5.000	−9.000	10.00
13	5.000	−12.000	10.00	14	5.000	−15.000	10.00
15	5.000	−18.000	10.00	16	5.000	−21.000	10.00
17	10.000	−0.000	0	18	10.000	−3.000	5.00
19	10.000	−6.000	5.00	20	10.000	−9.000	5.00
21	10.000	−12.000	5.00	22	10.000	−15.000	5.00
23	10.000	−18.000	5.00	24	10.000	−21.000	5.00

Table 18.6 Properties of Elements

Element		A [cm^2]	I$_\mu$ [cm^4]	I$_v$ [cm^4]	I$_w$ [cm^4]	E (kN/m^2)	G (kN/m^2)	Gamma kN/m^3	Alpha
column	-	1200.00	1.00	160000.00	1.00	30E6	12.SE6	0.0	12E-6
beam		1800.00	1.00	540000.00	1.00	30E6	12.SE6	0.0	12E-6

Table 18.7 Element Connectivity

Elm.	J1	J2	Type	V-Y* axes	Beta** (deg)	Elm.	J1	J2	Type	V-Y*	Beta (deg)
1	1	2	1	Y	0	2	2	3	1	Y	0
3	3	4	1	Y	0	4	4	5	1	Y	0
5	5	6	1	Y	0	6	6	7	1	Y	0
7	7	8	1	Y	0	8	9	10	1	Y	0
9	10	11	1	Y	0	10	11	12	1	Y	0
11	12	13	1	Y	0	12	13	14	1	Y	0
13	14	15	1	Y	0	14	15	16	1	Y	0
15	17	18	1	Y	0	16	18	19	1	Y	0
17	19	20	1	Y	0	18	20	21	1	Y	0
19	21	22	1	Y	0	20	22	23	1	Y	0
21	23	24	1	Y	0	22	2	10	2	Y	0
23	3	11	2	Y	0	24	4	12	2	Y	0
25	5	13	2	Y	0	26	6	14	2	Y	0
27	7	15	2	Y	0	28	8	16	2	Y	0
29	10	18	2	Y	0	30	11	19	2	Y	0
31	12	20	2	Y	0	32	13	21	2	Y	0
33	14	22	2	Y	0	34	15	23	2	Y	0
35	16	24	2	Y	0						

* Local V-axis parallel to GLOBAL Y-axis.

** Rolling angle between V and Y axes.

OUTPUT RESULTS:

Table 18.8 Natural Periods

Mode -	Radian fr. [Rad/sec]	Frequency [Hz]	Period [Sec]
1	8.83035	1.40539	0.71154
2	26.79346	4.26431	0.23450
3	45.92305	7.30888	0.13682

Table 18.9 Displacements

Joint -	Linear [m]	[m]	[m]	Angular [rad]	[rad]	[rad]
1	0.00000	0.00000	0.00000	0.00000	0.00000	0.00000
2	0.00239	0.00000	0.00012	0.00000	0.00056	0.00000
3	0.00541	0.00000	0.00021	0.00000	0.00055	0.00000
4	0.00824	0.00000	0.00028	0.00000	0.00051	0.00000
5	0.01074	0.00000	0.00032	0.00000	0.00045	0.00000
6	0.01279	0.00000	0.00035	0.00000	0.00037	0.00000
7	0.01431	0.00000	0.00036	0.00000	0.00027	0.00000
8	0.01522	0.00000	0.00037	0.00000	0.00015	0.00000
9	0.00000	0.00000	0.00000	0.00000	0.00000	0.00000
10	0.00239	0.00000	0.00000	0.00000	0.00026	0.00000
11	0.00541	0.00000	0.00000	0.00000	0.00030	0.00000
12	0.00824	0.00000	0.00000	0.00000	0.00028	0.00000
13	0.01074	0.00000	0.00000	0.00000	0.00026	0.00000
14	0.01279	0.00000	0.00000	0.00000	0.00022	0.00000
15	0.01431	0.00000	0.00000	0.00000	0.00017	0.00000
16	0.01522	0.00000	0.00000	0.00000	0.00011	0.00000
17	0.00000	0.00000	0.00000	0.00000	0.00000	0.00000
18	0.00239	0.00000	0.00012	0.00000	0.00056	0.00000
19	0.00541	0.00000	0.00021	0.00000	0.00055	0.00000
20	0.00824	0.00000	0.00028	0.00000	0.00051	0.00000
21	0.01074	0.00000	0.00032	0.00000	0.00045	0.00000
22	0.01279	0.00000	0.00035	0.00000	0.00037	0.00000
23	0.01431	0.00000	0.00036	0.00000	0.00027	0.00000
24	0.01522	0.00000	0.00037	0.00000	0.00015	0.00000

Table 18.10 Reactions

Joint -	F$_x$ [kN]	F$_y$ [kN]	F$_z$ [kN]	M$_x$ [kNm]	M$_y$ [kNm]	M$_z$ [kNm]
1	33.24	0.00	140.38	0.00	58.75	0.00
9	42.77	0.00	0.00	0.00	68.31	0.00
17	33.24	0.00	140.38	0.00	58.75	0.00

Table 18.11 Element Joint Forces

Elm.	F_u [kN]	F_v [kN]	F_w [kN]	M_u [kNm]	M_v [kNm]	M_w [kNm]	F_u [kN]	F_v [kN]	F_w [kN]	M_u [kNm]	M_v [kNm]	M_w [kNm]
1	140.38	0.00	33.24	0.00	58.75	0.00	140.38	0.00	33.24	0.00	40.98	0.00
2	111.33	0.00	29.04	0.00	43.50	0.00	111.33	0.00	29.04	0.00	43.64	0.00
3	82.35	0.00	26.64	0.00	39.33	0.00	0.00	0.00	26.64	0.00	40.61	0.00
4	56.16	0.00	23.17	0.00	33.79	0.00	0.00	0.00	23.17	0.00	35.74	0.00
5	33.73	0.00	18.98	0.00	27.20	0.00	0.00	0.00	18.98	0.00	29.76	0.00
6	16.13	0.00	13.79	0.00	19.03	0.00	0.00	0.00	13.79	0.00	22.35	0.00
7	4.63	0.00	7.22	0.00	9.05	0.00	0.00	0.00	7.22	0.00	12.61	0.00
8	0.00	0.00	42.77	0.00	68.31	0.00	0.00	0.00	42.77	0.00	60.00	0.00
9	0.00	0.00	46.64	0.00	70.54	0.00	0.00	0.00	46.64	0.00	69.40	0.00
10	0.00	0.00	42.24	0.00	63.12	0.00	0.00	0.00	42.24	0.00	63.60	0.00
11	0.00	0.00	36.82	0.00	54.82	0.00	0.00	0.00	36.82	0.00	55.64	0.00
12	0.00	0.00	30.11	0.00	44.57	0.00	0.00	0.00	30.11	0.00	45.76	0.00
13	0.00	0.00	21.76	0.00	31.87	0.00	0.00	0.00	21.76	0.00	33.42	0.00
14	0.00	0.00	11.88	0.00	16.80	0.00	0.00	0.00	11.88	0.00	18.84	0.00
15	140.38	0.00	33.24	0.00	58.75	0.00	0.00	0.00	33.24	0.00	40.98	0.00
16	111.33	0.00	29.04	0.00	43.50	0.00	0.00	0.00	29.04	0.00	43.64	0.00
17	82.35	0.00	26.64	0.00	39.33	0.00	0.00	0.00	26.64	0.00	40.61	0.00
18	56.16	0.00	23.17	0.00	33.79	0.00	0.00	0.00	23.17	0.00	35.74	0.00
19	33.73	0.00	18.98	0.00	27.20	0.00	0.00	0.00	18.98	0.00	29.76	0.00
20	16.13	0.00	13.79	0.00	19.03	0.00	0.00	0.00	13.79	0.00	22.35	0.00
21	4.63	0.00	7.22	0.00	9.05	0.00	0.00	0.00	7.22	0.00	12.61	0.00
22	3.07	0.00	29.93	0.00	84.42	0.00	0.00	0.00	29.93	0.00	65.24	0.00
23	0.33	0.00	29.78	0.00	82.76	0.00	0.00	0.00	29.78	0.00	66.15	0.00
24	0.56	0.00	26.62	0.00	74.06	0.00	0.00	0.00	26.62	0.00	59.05	0.00
25	0.54	0.00	22.52	0.00	62.64	0.00	0.00	0.00	22.52	0.00	49.95	0.00
26	0.56	0.00	17.47	0.00	48.62	0.00	0.00	0.00	17.47	0.00	38.73	0.00
27	0.82	0.00	11.29	0.00	31.35	0.00	0.00	0.00	11.29	0.00	25.09	0.00
28	0.64	0.00	4.40	0.00	12.61	0.00	0.00	0.00	4.40	0.00	9.42	0.00
29	3.07	0.00	29.93	0.00	65.24	0.00	0.00	0.00	29.93	0.00	84.42	0.00
30	0.33	0.00	29.78	0.00	66.15	0.00	0.00	0.00	29.78	0.00	82.76	0.00
31	0.56	0.00	26.62	0.00	59.05	0.00	0.00	0.00	26.62	0.00	74.06	0.00
32	0.54	0.00	22.52	0.00	49.95	0.00	0.00	0.00	22.52	0.00	62.64	0.00
33	0.56	0.00	17.47	0.00	38.73	0.00	0.00	0.00	17.47	0.00	48.62	0.00
34	0.82	0.00	11.29	0.00	25.09	0.00	0.00	0.00	11.29	0.00	31.35	0.00
35	0.64	0.00	4.40	0.00	9.42	0.00	0.00	0.00	4.40	0.00	12.61	0.00

REFERENCES

Design and Construction of Seismic Resistant Reinforced Concrete Buildings (1984) UNDP/UNIDO Project RER/79/015, Unido Center, Vienna, Austria.

Hungarian Code (MSZ 15021-15028) (1989) Magyar Szabványugyi Hivatal H-1450, Budapest, Hungary.

Model Code for Seismic Design of Concrete Structures (1985) CEB, No. 165, Comité Euro-International du Béton, Geneva, Switzerland.

Technical Guiding Principles (1978) MI-04 133-78, Magyar Szabványugyi Hivatal H-1450, Budapest, Hungary.

19

India

Sudhir K. Jain, Brijesh Chandra, and D. K. Paul*

19.1 INTRODUCTION

India has been subjected to some of the most severe earthquakes in the world. The strongest earthquakes that occurred in the country in the last one hundred years were the Assam earthquake of 1897 ($M = 8.7$),[1] the Kangra earthquake of 1905 ($M = 8.6$), the Bihar-Nepal earthquake of 1934 ($M = 8.4$), and the Assam-Tibet earthquake of 1950 ($M = 8.7$).

The Military Engineer Service (MES) made the first attempt at earthquake-resistant construction in India after the 1935 Quetta earthquake (Quetta is now in Pakistan). The MES required strengthening of brick or stone masonry buildings by providing reinforced concrete bands at plinth, lintel, and roof levels.[2]

The first seismic design code in India was published in 1962 [IS:1893 (1962)]; the code has since been revised in 1966, 1970, 1975, and 1984. A code that specifies the design and the required detailing for seismic construction of buildings was published in 1967 [IS:4326 (1967)]; that code was revised in 1976. As of 1992, current design seismic forces for buildings, elevated liquid storage tanks, stacks, concrete and masonry dams, embankments, bridges, and retaining walls are specified by IS:1893 (1984) (hereinafter referred to as "the Code"), while detailing and other construction aspects for seismic resistance are covered in IS:4326 (1976). The Code IS:4326 of 1984 covers seismic design and detailing requirements for concrete, steel, masonry, and timber buildings. In addition, the Bureau of Indian Standards has published an explanatory handbook on the two codes [SP: 22

---

*The authors wish to acknowledge the assistance provided by Mr. Vijay K. Saraf of the Indian Institute of Technology Kanpur in preparation of the computer program.

[1]M is the magnitude of an earthquake as measured on the Richter scale (see Appendix on Magnitude and Intensity of Earthquakes).

[2]Many of the major structures built in the seismic regions of the country immediately after independence in 1947 were designed for an empirically fixed horizontal force of 10% of the weight of the structure (Krishna 1985).

(1982)]; besides offering explanations of the two codes, the handbook also provides some examples with their solutions (Rajasankar and Jain 1988).

In this chapter, the seismic design regulations for buildings provided by the Code are presented. To provide an overall view of the seismic safety provisions in the Code, the safety factors and load factors specified in other Indian codes are also presented. Neither the design and detailing criteria for structures nor the seismic design provisions for structures other than buildings are presented in this chapter.

19.2 PERMISSIBLE STRESSES, SAFETY FACTORS, AND LOAD FACTORS

Most multistory building construction in India is done in reinforced concrete. Steel is usually used only for industrial structures because of the high cost. Construction that involves the use of reinforced concrete, prestressed concrete, and steel is governed by IS:456 (1978), IS:1343 (1980), and IS:800 (1984), respectively. IS:875 (1987) contains design load specifications (except seismic loads) for buildings.

The limit state design method is commonly used for design of buildings, although IS:456 (1978) allows the use of either the limit state or the working stress design methods. IS:1343 (1980) prescribes the limit state design procedure for prestressed concrete structures; in addition, a check on stresses caused by service loads is required. In the limit state design method, for both the reinforced concrete and prestressed concrete, the material strength partial safety factor is prescribed at 1.5 on concrete strength and at 1.15 on the yield stress of steel. The partial safety load factors for limit design are

(a) $1.5(DL + LL)$
(b) $1.2(DL + LL + EQ/WL)$
(c) $1.5(DL + EQ/WL)$
(d) $0.9DL + 1.5EQ/WL$

where DL, LL, EQ, and WL stand for dead, live, earthquake, and wind loads, respectively.

In working stress design, factors of safety assigned to concrete in direct compression and bending compression are 4.0 and 3.0, respectively, on 150-mm cube-crushing strength; the factor of safety is 1.80 on yield stress for reinforcement bars in tension.

In the design of steel structures, IS:800 (1984) allows the use of the working stress or the plastic methods of design; however, in practice the working stress method is usually followed. The factor of safety for the working stress method is 1.50 for direct stress and 1.67 for bending stress. Load factors for plastic design are

(a) $1.7(DL)$
(b) $1.7(DL + LL)$
(c) $1.7(DL + EQ/WL)$
(d) $1.3(DL + LL + EQ/WL)$.

In combinations with seismic loads, the Code allows an increase of 33.33 percent for allowable stresses in the elastic method of design, subject to the conditions that (i) for steel with a definite yield point, the stress is to be limited to its yield stress, (ii) for steel without a definite yield point, the stress is to be limited to the 0.2% proof stress or 80% of the ultimate strength, whichever is less, and (iii) for prestressed concrete members, the tensile stress in the concrete must not exceed two-thirds of the modulus of rupture of concrete.

The Code also allows an increase in the allowable bearing pressure for design of foundations when considering the seismic loads (Table 19.1). This increase depends on the type of foundation and the soil conditions. Because the allowable bearing pressure could be governed by either settlement limitations or shear failure considerations, the Code allows a lower increase (or no increase) in allowable bearing pressure for those footings and soil types that are more vulnerable to differential settlement.

19.3 OVERVIEW OF THE CODE

The Code provides both static (seismic coefficient method) and dynamic (response spectrum method) procedures for the determination of seismic design forces for buildings. Depending upon the height of the building, the Code recommends the use of the seismic coefficient method, the response spectrum method, or even a time-history analysis (Table 19.2). The Code requires that modal analysis be used for buildings that have unusual configurations, or irregular shapes and/

Table 19.1. Permissible Increase in Allowable Bearing Pressure or Resistance of Soils (%)

Serial No.	Type of Foundation	Rock/ Hard soils	Medium soils	Soft soils
1	Piles passing through any soil but resting on rock or hard soil	50	50	50
2	Piles not covered above	–	25	25
3	Raft foundations	50	50	50
4	Combined or isolated R.C. footings with tie beams	50	25	25
5	Isolated R.C. footings without tie beams or unreinforced strip foundations	50	25	–
6	Caisson foundations	50	25	25

Table 19.2. Recommended Method for Seismic Design of Buildings

Serial No.	Building Height	Seismic Zone*	Recommended Method
1	Greater than 40 m	III, IV, and V	Detailed dynamic analysis (either modal analysis or time-history analysis based on expected ground motion for which special studies are required). For preliminary design, modal analysis using response spectrum method may be employed.
2	Greater than 90 m	I and II	Modal analysis using response spectrum method.
3	Greater than 40 m and up to 90 m	All zones	Modal analysis using response spectrum method. Use of seismic coefficient method permitted for Zones I, II, and III.
4	Less than 40 m	All zones	Modal analysis using response spectrum method. Use of seismic coefficient method permitted in all zones.

*See Section 19.4.

or irregular distribution of mass or stiffness, as well as for industrial buildings and frame structures with large spans or height. For the seismic coefficient method to be used, the story heights should be approximately uniform, ranging between 2.7 m and 3.6 m, except that one or two stories may be up to 5 m high.

The Code generally requires that the design for horizontal seismic forces be considered only in any one direction at a time. However, if the stability of the building is a criterion for design, vertical seismic forces must be considered simultaneously with horizontal forces in any one direction. The Code also states that the design earthquake forces are assumed not to occur simultaneously with maximum flood, wind, or wave loads.

Wherever the floors of the building are capable of providing rigid diaphragm action, the lateral load is to be distributed to the various lateral-load-resisting elements, assuming the floors to be absolutely rigid in their horizontal planes. Otherwise, the Code states that frames are to be designed to behave independently, with the seismic force on each frame assessed in accordance with tributary mass. The latter requirement seems to have been an oversight because numerous situations occur where the floor is not completely rigid in its own plane but still has considerable in-plane stiffness, and for such situations, assumption of zero in-plane stiffness of floors is not appropriate (e.g., Jain and Mandal, 1992). The Code also prescribes that when a combination of shear walls and

moment-resisting frames is used for lateral load resistance, the frames are to be designed for at least 25% of the seismic design force.

In both the seismic coefficient and the response spectrum methods, due consideration is given to the seismic zone where the structure is located, importance of the structure, soil-foundation system, ductility of construction, flexibility of the structure, and weight of the building.

19.4   ZONING MAP AND BASIC COEFFICIENT

India may be divided into three subregions based on geological considerations: (1) the Alpine Himalayan belt, (2) the southern peninsula, and (3) the intervening Indo-Gangetic plains. While peninsular India is an ancient stable area, the Alpine Himalayan belt is one of the most earthquake-prone regions in the world. Crustal instability in this belt is ascribed to the movement of the Indian Plate towards the Eurasian Plate, which occurs at a rate of about 50 mm per year.

The seismic zone map for the country was developed based on the epicentral distribution of significant past earthquakes and on the isoseismal configurations of such events. The original map demarcated areas that had potential for ground shaking of intensities of less than V, V, VI, VII, VIII, IX, X, and more than X in the Modified Mercalli Intensity (MMI) scale. The map was revised in the 1966 and 1970 editions of the Code based on the geological and geophysical data obtained from tectonic mapping and aeromagnetic and gravity surveys. [Krishna (1992) provides a brief historical view of zoning in India.] The current Indian zone map (Fig. 19.1) divides the country into five seismic zones (I to V) with the associated MMI of V (or less), VI, VII, VIII, and IX (and above), respectively. This zoning map is based on expected maximum seismic intensity in a region and does not consider frequency of occurrence; in this sense, the map does not divide the country into areas of equal seismic risk. But, this particular limitation of the zoning map can be an advantage in that it provides for a direct comparison between the maximum seismic intensity of an earthquake and the intensity assigned to the region.

The basic seismic coefficient (α_0) is equal to 0.01, 0.02, 0.04, 0.05, and 0.08, respectively, for the five zones (Table 19.3). The seismic zone factor F_0 used in the response spectrum method is simply five times the factor α_0. While observations on building performance in severe shaking during past earthquakes formed the basis of assigning $\alpha_0 = 0.08$ to Zone V, the value of α_0 for other zones was fixed more or less arbitrarily. Although much observational, experimental, and analytical information is available on the required

Fig. 19.1. Seismic zone map of India

seismic forces for zones of severe shaking, such data for areas of low or medium shaking is still lacking.

For underground structures and foundations at depths of 30 m or greater, the basic seismic coefficient may be taken as one-half of that in Table 19.3. Linear interpolation is allowed for depths less than 30 m. This specification of the Code recognizes the fact that seismic waves are amplified as they are reflected from a free boundary, i.e., ground surface. For situations where consideration of vertical acceleration is required, the vertical coefficient may be taken as one-half of that given in Table 19.3.

Table 19.3. Values of Basic Seismic Coefficient and Seismic Zone Factor

Serial No.	Zone No.	Basic Horizontal Seismic Coefficient* α_0	Seismic Zone Factor† F_0
1	V	0.08	0.40
2	IV	0.05	0.25
3	III	0.04	0.20
4	II	0.02	0.10
5	I	0.01	0.05

*For seismic coefficient method.
†For response spectrum method.

19.5 SOIL-FOUNDATION SYSTEM

The soil-foundation system has several important effects on the seismic behavior of a structure. First, the expected ground motion varies for different soil profiles. This is explicitly accommodated for in many codes by specifying a somewhat different design spectrum for different soil profiles (e.g., NEHRP 1991). Second, the flexibility due to soil and foundation deformation leads to a higher natural period and increased damping, and thus, in most cases a reduced seismic force. This is accommodated for in some codes by considering soil-structure interaction effects (e.g., Appendix to Chapter 6 of NEHRP 1991). The Indian Code IS:1893 does not account for these two effects, although it is also well recognized that those buildings for which the foundation system behaves as an entity, with minimal differential settlement, behave better in earthquakes. The Indian Code emphasizes the need to have a foundation system that will show minimal differential settlement by prescribing a factor β (Table 19.4) with a higher value for those soil and foundation systems that are liable to show more differential settlement. Therefore, a building on soft soil with isolated-untied footings is to be designed for 50% higher seismic loads than if the same building is supported on a raft foundation.

19.6 IMPORTANCE FACTOR

The Code prescribes an importance factor I of 1.0 for ordinary buildings and 1.5 for important service and community buildings. Table 19.5 gives the values of the importance factor for different structures. The Code indicates that these values are meant only for guidance and that the designer can choose a suitable value depending upon the importance of the structure

Table 19.4. Values of β for Different Soil-Foundation Systems

Serial No.	Type of Foundation	Rock/ Hard Soils	Medium Soils	Soft Soils
1	Piles passing through any soil but resting on rock or hard soil	1.0	1.0	1.0
2	Piles not covered above	–	1.0	1.2
3	Raft foundations	1.0	1.0	1.0
4	Combined or isolated R.C. footings with tie beams	1.0	1.0	1.2
5	Isolated R.C. footings without tie beams or unreinforced strip foundations	1.0	1.2	1.5
6	Caisson foundation	1.0	1.2	1.5

Table 19.5. Values of Importance Factor I

Serial No.	Structure	I
1	Dams (all types)	3.0
2	Containers of inflammable or poisonous gases or liquids	2.0
3	Important service and community structures, such as hospitals, water towers and tanks; schools, important bridges, important power houses, monumental structures; emergency buildings like telephone exchanges and fire bridges; large assembly structures like cinemas, assembly halls, and subway stations	1.5
4	All others	1.0

based on economy, design strategy, and other considerations.

19.7 PERFORMANCE FACTOR

Prior to 1984, there was no explicit Code consideration given to ductility in the determination of design forces. The only requirement was for ductile detailing as per IS:4326 whenever the factor $(\beta I \alpha_0)$ exceeded 0.05, which always happened in Seismic Zones IV and V. Since then, the Code explicitly recognizes the advantages of ductile construction and specifies a performance factor K that depends on the ductility of the structure. Values of this factor for different types of building construction are given in Table 19.6. Performance factor values for systems corresponding to Serial Nos. 1b, 2a, and 2b in Table 19.6 are applicable only if (a) the steel bracing members and the infill panels are considered in stiffness, as well as lateral strength calculations, and (b) the frame acting alone will be able to resist at least 25% of the design seismic forces. However, even in the 1984 Code the performance factor is not included for structures other than buildings.

Table 19.6. Values of Performance Factor K

Serial No.		Structure	K
1	a.	Moment-resistant frame with appropriate ductility details as given in IS:4326-1976 in reinforced concrete or steel	1.0
	b.	Frame as above with R.C. shear walls or steel bracing members designed for ductility	1.0
2	a.	Frame as in 1a with either steel bracing members or plain or nominally reinforced concrete infill panels	1.3
	b.	Frame as in 1a in combination with masonry infills	1.6
3		R.C. framed buildings not covered by 1 or 2 above	1.6

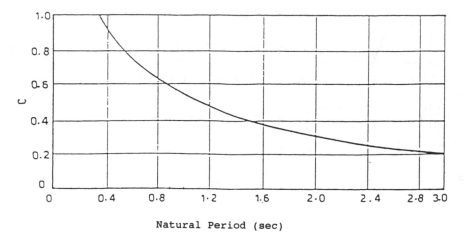

Natural Period (sec)

Fig. 19.2. Seismic coefficient C versus T

19.8 STRUCTURAL FLEXIBILITY

19.8.1 Seismic Coefficient Method

In the seismic coefficient method, the Code specifies a seismic factor C, which through its fundamental period T depends on the flexibility of the structure (see Fig. 19.2). Except in the low period range, the shape of the curve for C is the same as the average response spectrum curve for 5% damping proposed by Housner using four earthquake time histories (e.g., Housner and Jennings 1982).

The Code allows for the estimation of the fundamental period T by (a) experimental observations on similar buildings (this almost never happens in practice), or (b) any rational method of analysis (dynamic analysis). Alternately, for regular buildings, the Code suggests the following empirical relationships for the estimation of the fundamental period:

(i)   For moment-resisting frames without shear walls or bracings

$$T = 0.1n \qquad (19.1)$$

(ii) For other buildings

$$T = \frac{0.09H}{\sqrt{d}} \qquad (19.2)$$

where n = number of stories including basements, H = total building height in meters, and d = maximum base dimension of building in meters in the direction parallel to the applied seismic force.

Although dynamic analysis is the preferred method for the estimation of the fundamental period, no provision is built into the Code to prevent the design for a natural period that, although obtained by dynamic analysis, may have a value unrealistically large. A designer may perform a dynamic analysis ignoring the filler walls (which usually are of unreinforced brick masonry) and obtain a rather large value for the natural period, and thus consequently low values for the seismic design forces. Future editions of the Code should incorporate a minimum design force based on empirical estimation of the fundamental period, as it is found in the codes of several other countries (e.g., UBC 1991).

19.8.2 Response Spectrum Method

In this method, a set of design spectrum curves (the Code terms them average acceleration spectra, denoted by S_a/g) are provided that account for flexibility of the structure (Fig. 19.3). These spectra are based on the average spectrum curves obtained by Housner using four earthquake time histories. A comparison of $\alpha_0 C$ and $F_0 S_a/g$ (for 5% damping) curves (Figs. 19.2 and 19.3) shows that the two match rather well except in the very low period range of 0–0.1 sec.; this range rarely governs design of a building. Thus, if the same fundamental period is used, both methods will give about the same overall design seismic force, provided 5% damping is considered.

The Code does not provide an explicit specification of damping for buildings. However, Appendix F of the Code recommends the use of the following percentages of critical damping for different types of structures:

(a)   Steel structures   2 to 5%
(b)   Concrete structures   5 to 10%
(c)   Brick structures in cement mortar   5 to 10%
(d)   Timber structures   2 to 5%
(e)   Earthen structures   10 to 30%.

Also, the Code provides definite values of damping

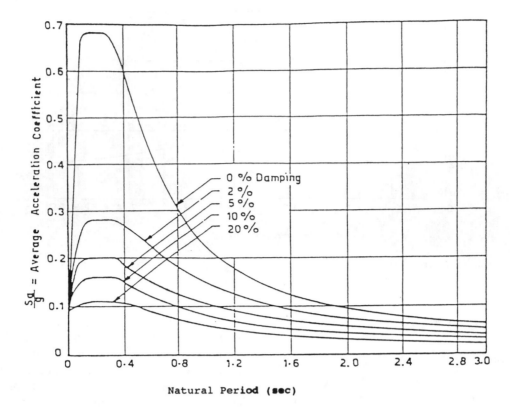

Fig. 19.3. Average acceleration spectra S_a/g versus T

for some special structures such as elevated tanks, gravity dams, and embankments.

19.9 WEIGHT OF THE STRUCTURE

The Code considers the fact that the live loads in the building do not contribute fully to the development of seismic forces, because generally such loads are not caused by masses firmly attached to the structure. Also, at a time of severe shaking, a building may not be loaded by its full design live load. Consequently, the Code specifies that only a fraction of the design live load need be considered in earthquake-resistant design (Table 19.7). The fractions of the live load indicated in Table 19.7 are to be used for the calculation of (a) lumped weights for determining the seismic forces, and (b) stresses caused by combined effects of gravitational and seismic forces. The Code also specifies that no live load for the roof of the

Table 19.7. Percentage of Design Live Load to be Considered for Seismic Load Calculation

Live Load Class (kg/m²)	Design Live Load Percentage
200, 250, and 300	25
400, 500, 750, and 1,000	50

building be included in the calculations. The Code states that when the live load is reduced for earthquake effects, no further reduction in live load can be made, such as specified by IS:875-1987 (the usual reduction in live load allowed for the design of foundation and columns of lower stories).

19.10 BASE SHEAR FORCE AND ITS DISTRIBUTION WITH HEIGHT

19.10.1 Seismic Coefficient Method

In the seismic coefficient method, the design base shear V is obtained from the following formula:

$$V = KC\beta I\alpha_0 W \qquad (19.3)$$

where W is the total dead load plus the appropriate live load (Section 19.9). The Code specifies a parabolic distribution of seismic force with respect to height, given by

$$F_i = V \frac{W_i h_i^2}{\sum_{j=1}^{n} w_j h_j^2} \qquad (19.4)$$

where

F_i = lateral force at the ith floor (or roof)
W_i = gravity load (dead load plus appropriate amount of live load) at the ith floor
h_i = height measured from the base of the building to the ith floor
n = number of stories.

The gravity load W_i at any floor is to be obtained by equally distributing the weight of walls and columns, in any story, to the floor above and the floor below.

The Code provides that when the basement walls are not connected with the ground floor deck or the basement walls are not fitted between building columns, the number of stories (n) in eq.(19.4) is to include the basement stories. Otherwise n excludes the basement stories. This amounts to assuming that the building is to be fixed at the ground floor, if the basement walls are connected to the ground floor or between the columns.

19.10.2 Response Spectrum Procedure

In this method, natural frequencies and mode shapes are to be obtained by a free vibration analysis. For each significant natural mode, the average acceleration coefficient S_a/g is obtained from Fig. 19.3. The seismic design lateral load F_{ir} applied at the ith floor level corresponding to the rth mode of vibration is given by the following equation:

$$F_{ir} = K\beta I F_0 \phi_{ir} C_r \frac{S_{ar}}{g} W_i \qquad (19.5)$$

where

ϕ_{ir} = mode shape coefficient at ith floor in rth mode of vibration
C_r = modal participation factor for the rth mode given by

$$C_r = \frac{\displaystyle\sum_{j=1}^{n} W_j \phi_{jr}}{\displaystyle\sum_{j=1}^{n} W_j [\phi_{jr}]^2} \qquad (19.6)$$

If the absolute values of the forces F_{ir} were combined for the different modes, the resultant force would be too conservative because in such a combination the sign of opposing forces in the higher modes is lost. Instead, the Code specified that modal story *shears* be combined first, and that the final lateral forces be obtained from this combination. The Code provides for combination of different modes by using the Square Root of Sum of Squares (SRSS) for buildings taller than 90 m, and a modified version of the SRSS for buildings under 90 m in height. The

Table 19.8. Values of the Coefficient γ

Height, H (m)	γ
Up to 20	0.40
40	0.60
60	0.80
90	1.00

shear force V_i for the ith story is to be obtained by superposition of the first three modes as

$$V_i = (1 - \gamma) \sum_{r=1}^{3} |V_{ir}| + \gamma \sqrt{\sum_{r=1}^{3} [V_{ir}]^2} \qquad (19.7)$$

where V_{ir} = maximum shear at the ith story corresponding to the rth mode; and where the value for the coefficient γ is given in Table 19.8. For buildings of intermediate height, values of γ may be obtained by linear interpolation.

The total lateral loads F_n acting at roof level n and F_i acting at the ith floor level, are back-calculated from the story shear using the following equations:

$$F_n = V_n \qquad (19.8a)$$

$$F_i = V_i - V_{i+1} \qquad (19.8b)$$

19.11 TORSION

The Code does not provide for a minimum design eccentricity due to accidental torsion. In case of an eccentricity between the center of stiffness at a story and the center of the above mass, the Code stipulates that torsional moments shall be calculated with a design eccentricity equal to 1.5 times the actual eccentricity. However, reduction in seismic shear in a frame due to torsion is to be ignored. The requirement of torsion analysis is particularly emphasized for buildings more than 40 m high.

19.12 STORY DRIFT

The Code requires that the maximum relative displacement (story drift) between two successive floors due to design seismic forces must not exceed 0.004 times the story height. The Code emphasizes that this check is particularly necessary for buildings more than 40 m high.

19.13 CANTILEVERS AND PROJECTIONS

The vertical and horizontal projections of a building, e.g., towers, parapets, stacks, and balconies, are

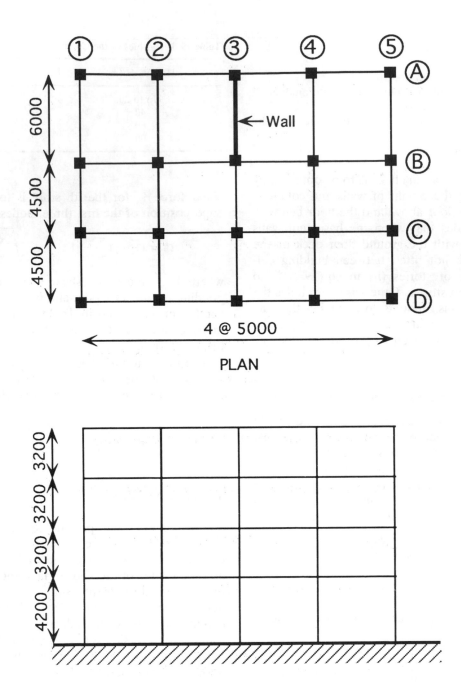

PLAN

ELEVATION

Fig. 19.4(a).   Configuration of the building for Example 19.1

particularly vulnerable to damage during an earthquake. Therefore, the Code requires that the vertical projections and their connections with the structure (not the supporting building) be designed for five times the basic seismic coefficient (α_0). Similarly, horizontal projections and their connections are to be designed for five times the vertical seismic coefficient (e.g., $5 \times 0.5\alpha_0$).

19.14 EXAMPLE 19.1

Consider a four-story reinforced concrete office building as shown in Fig. 19.4(a). The building is located in Shillong (Seismic Zone V). The soils are medium stiff and the entire building is supported on a raft foundation. The column size is 40 cm × 40 cm, all outer beams are 25 cm × 60 cm, and all inner beams are

30 cm × 45 cm. The shear wall in the Y direction is 20 cm thick. All construction is in M20 concrete (cube-crushing strength = 20 MPa). The lumped weight due to dead loads is 12 kN/m² on floors and 10 kN/m² on the roof, and the live loads are 4 kN/m² on floors and 1.5 kN/m² on the roof. The floors are of cast-in-place reinforced concrete and provide rigid diaphragm action.

Determine the design seismic forces on the structure by (a) the seismic coefficient method (and distribute these forces to different frames and the wall), and (b) the response spectrum method. Also, compare the results obtained with these two methods.

Solution

Design parameters. For Seismic Zone V, the basic horizontal seismic coefficient α_0 for the seismic coefficient method is 0.08, and the seismic zone factor F_0 for the response spectrum method is 0.40 (Table 19.3). Because the building is supported on a raft foundation, factor β for the soil-foundation system is 1.0 (Table 19.4). For an office building, the importance factor I is 1.0 (Table 19.5). It is assumed that the building will be designed for ductility according to the provisions of IS:4326, hence the performance factor K is 1.0 (Table 19.6).

Lumped weights. The floor area is $15 \times 20 = 300$ m². The live load class is [3]400 kg/m², thus only 50% of the live load is lumped at the floors (Table 19.7). At the roof, no live load is to be lumped. Hence, the total lumped weight on the floors and the roof is:

Floors: $W_1 = W_2 = W_3 = 300 \times (12 + 0.5 \times 4)$
$= 4,200$ kN
Roof: $W_4 = 300 \times (10) = 3,000$ kN
Total weight of the structure:
$W = \Sigma W_i = 3 \times 4,200 + 3,000 = 15,600$ kN.

(a) Seismic coefficient method

- **Fundamental period.** For earthquake motion in the X direction, lateral load resistance is provided by moment-resisting frames without bracing or shear walls, hence eq.(19.1) is applicable. In the Y direction, the building has a shear wall, hence eq.(19.2) has to be applied. Thus,

In X direction: $T = 0.1n = 0.1 \times 4 = 0.40$ sec

In Y direction: $T = \dfrac{0.09H}{\sqrt{d}} = \dfrac{0.09 \times 13.8}{\sqrt{15}} = 0.32$ sec

- **Design base shear.** From Figure 19.2:

In X direction: $C = 0.9$ (for $T = 0.40$ sec)
In Y direction: $C = 1.0$ (for $T = 0.32$ sec)

Also, $\alpha_0 = 0.08$, $I = 1.0$, $K = 1.0$, and $\beta = 1.0$. Hence, from eq.(19.3), the design base shear is
In X direction:
$V = 1.0 \times 0.9 \times 1.0 \times 1.0 \times 0.08 \times 15,600$
$= 1,123$ kN
In Y direction:
$V = 1.0 \times 1.0 \times 1.0 \times 1.0 \times 0.08 \times 15,600$
$= 1,248$ kN

- **Force distribution with height.** The design base shear is to be distributed with height as per eq.(19.4). Table 19.9 gives the calculations and the resulting lateral forces in the X and Y directions.

- **Force distribution in frames (Y direction).** In the Y direction, the building is symmetrical, therefore no torsion takes place. The monolithic reinforced concrete slab provides a rigid diaphragm action. Therefore, the lateral forces calculated in Table 19.9 should be distributed to different frames and the wall such that the lateral displacements in the wall and in the frames, at each level, will be the same. The mode of deformation in the wall and the in frames differs; the wall tends to deform like a flexural cantilever beam, while the frames have a tendency to deform like a

Table 19.9. Lateral Load Distribution with Height by the Seismic Coefficient Method

Story Level i	W_i (kN)	h_i (m)	$W_i h_i^2$ (×10³)	$\dfrac{W_i h_i^2}{\Sigma W_i h_i^2}$	Lateral Force for direction (kN)	
					X	Y
4	3,000	13.8	571.3	0.424	476	529
3	4,200	10.6	471.9	0.350	393	437
2	4,200	7.4	230.0	0.171	192	213
1	4,200	4.2	74.1	0.055	62	69
Σ			1,347.3	1.000	1,123	1,248

[3]One kilogram weight = 9.8 Newtons. However, in engineering practice, the conversion factor is rounded to 10.

shear beam. An accurate computer analysis ensures force distribution among the frames/walls such that displacement compatibility is enforced at all levels. However, when using an approximate manual calculation, it is common practice to ensure only equal displacement at the roof level. This procedure then consists of (Macleod 1971) (i) lumping the walls into one equivalent wall and the frames into one equivalent frame, (ii) applying the vertically distributed seismic forces to the equivalent wall, which is assumed to interact with the equivalent frame only at the roof level, (iii) calculating the interaction force at roof level by ensuring equal displacement at the roof, and (iv) distributing the resulting forces on the equivalent wall and the equivalent frame to the different real walls and frames, respectively, in proportion to their lateral stiffness. Such a calculation may underestimate the shear force for the middle stories of the frames by as much as 30%. For this reason it is the usual practice to increase the calculated force in the frame by 30%.

The lateral stiffness of frames can be determined either by using a computer program analysis or by using approximate manual calculations. Herein the stiffness has been calculated by an approximate procedure (Macleod 1971). The modeling assumptions include (i) columns rigidly fixed at the base, (ii) gross-section moment of inertia (rectangular section for beams), (iii) finite flexibility of beams, and (iv) finite size of beam-column joints, which provide rigid zones at either end of the beams and the columns. The wall stiffness has been obtained considering (i) flexural as well as shear deformations, (ii) gross area of cross-section of the wall including the two columns at both ends of the wall, and (iii) the wall rigidly fixed at the base because of the raft foundation. The modulus of elasticity of concrete has been assumed equal to 25,500 MPa. Under these conditions, the following values of lateral displacement at roof level that are due to lateral forces, and the resulting lateral stiffnesses in individual resisting elements, are obtained [see Fig. 19.4(b)]:

1,000 kN at roof	Deflection (m)	Stiffness (kN/m)
Frames 1 and 5	0.0822	12,200
Frames 2 and 4	0.1320	7,580
Frame 3 excluding wall	0.220	4,540
Wall	0.00704	
Wall with 1,248 kN distributed (Fig.19.4b)	0.00645	

Net deflection in wall (Δ_w) and in the equivalent frame (Δ_f) are [(Fig. 19.4(b)]:

$$\Delta_w = 0.00645 - \frac{P \times 0.00704}{1,000}$$

$$\Delta_f = \frac{P}{44,100}$$

$$\Delta_w = \Delta_f \quad \text{gives} \quad P = 217 \text{ kN}$$

As mentioned earlier, the shear in frames is increased by 30% to account for approximations. Hence, the frames would be designed for a seismic shear of 282 kN, which is 22.6% of the total base shear. Because the Code requires that frames be designed for a minimum of 25% of the total base shear, the design force for frames will be 312 kN (= 0.25 × 1,248 kN). This force is distributed further to the five frames in proportion to the lateral stiffness. Thus, design forces for different frames and the wall are [(Fig. 19.4(c)]:

Frames 1 and 5:

$$\frac{312 \times 12,200}{44,100} = 86.3 \text{ kN (force applied at roof)}$$

Frames 2 and 4:

$$\frac{312 \times 7,580}{44,100} = 53.6 \text{ kN (force applied at roof)}$$

Frame 3 (excluding wall):

$$\frac{312 \times 4,540}{44,100} = 32.1 \text{ kN (force applied at roof)}$$

Wall: At roof = (529 − 217) = 312 kN
At third floor = 437 kN
At second floor = 213 kN
At first floor = 69 kN

- **Force distribution in frames (*X* direction).** The earthquake force in the *X* direction is resisted by four moment resisting frames that are not symmetrically placed; this placement causes torsion. The lateral stiffness of these frames calculated by the approximate method of Macleod is shown in Fig. 19.4(d).

Total stiffness in *X* direction = 2 × 15,850 + 2 × 9,940
= 51,580 kN/m

The distance *d* of the center of stiffness from frame *D* is

$$d = \frac{15,850 \times 15 + 9,940 \times 9 + 9,940 \times 4.5}{51,580} = 7.21 \text{ m}$$

Calculated eccentricity between the center of mass and the center of stiffness is $e = 7.5 - 7.21 = 0.29$ m. The building will be designed for an eccentricity of $1.5e$ (= $1.5 \times 0.29 = 0.435$ m). Thus, the lateral force V at the center of mass C_m can be represented as a lateral force V at the center of stiffness C_s, and with a torsional moment of magnitude $M_t = 1.5eV$ [Fig. 19.4(e)].

The lateral load at the center of stiffness is to be distributed in proportion to the frame stiffness. Hence,

Force in frames A and D $= \dfrac{15,850}{51,580} V = 0.307V$

Force in frames B and C $= \dfrac{9,940}{51,580} V = 0.193V$

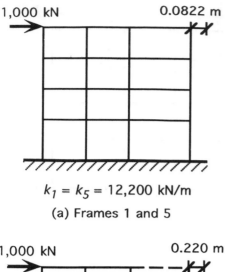

$k_1 = k_5 = 12,200$ kN/m

(a) Frames 1 and 5

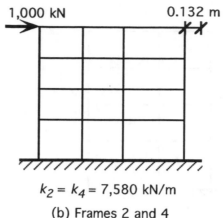

$k_2 = k_4 = 7,580$ kN/m

(b) Frames 2 and 4

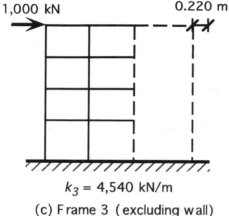

$k_3 = 4,540$ kN/m

(c) Frame 3 (excluding wall)

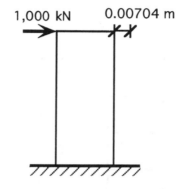

(d) Wall with point load at top

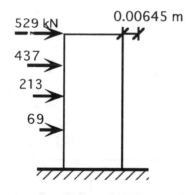

(e) Wall with distributed load

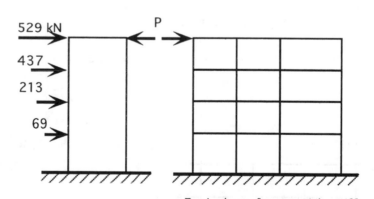

Equivalent frame with stiffness
$k = \Sigma k_i = 44,100$ kN/m

Fig. 19.4(b). Lateral stiffness of frames and wall in Y direction

The torsional moment applied at the center of stiffness will be resisted by frames A, B, C, D, 1, 2, 4, and 5. Frame 3 and the shear wall are at zero distance from the center of stiffness and do not provide resistance to torsion. Force in the ith frame is given by $M_t(k_i r_i)/(\Sigma k_i r_i^2)$; where k_i = stiffness of the ith frame; r_i = distance of the ith frame from the center of stiffness; and M_t = torsional moment ($= 1.5eV$). Table 19.10 shows the calculations necessary to determine

the forces developed by torsion on the various resisting elements of the building. The negative sign for force in frames C and D indicates that the force due to torsion is in a direction opposite to that due to direct lateral force. Such a reduction in frame force due to torsion is to be ignored as stipulated by the code. The forces in frames 1, 2, 4, and 5 caused by torsion as a result of earthquake motion in the X direction are small as compared to the forces in these frames resulting from

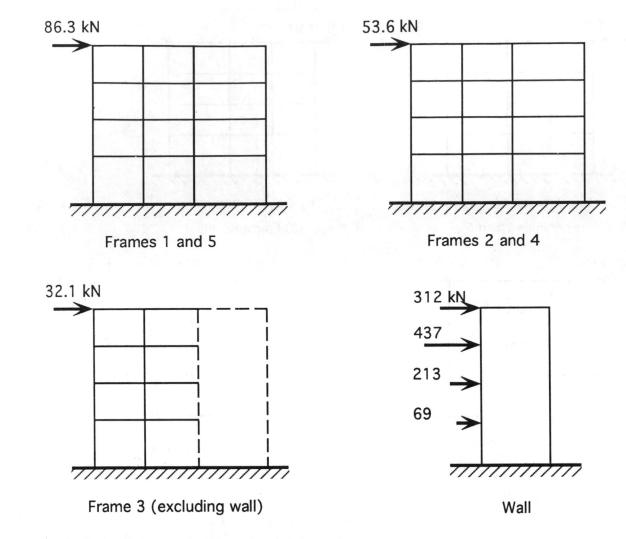

86.3 kN

Frames 1 and 5

53.6 kN

Frames 2 and 4

32.1 kN

Frame 3 (excluding wall)

312 kN

437

213

69

Wall

Fig. 19.4(c).   Design force in frames and wall for earthquake in Y direction

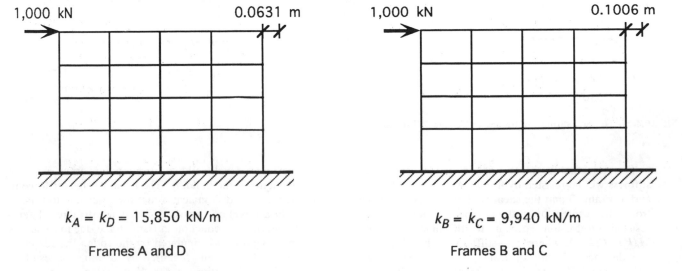

1,000 kN   0.0631 m

$k_A = k_D = 15,850$ kN/m

Frames A and D

1,000 kN   0.1006 m

$k_B = k_C = 9,940$ kN/m

Frames B and C

Fig. 19.4(d).   Lateral stiffness of frames in X direction

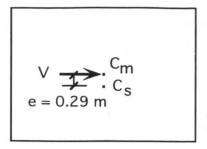

Fig. 19.4(e). Calculated eccentricity and design torsional moment

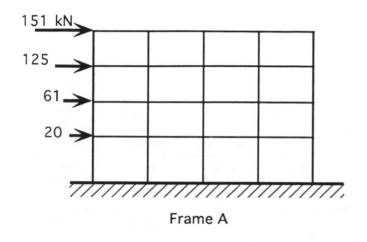

Frame A

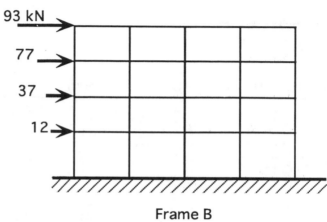

Frame B

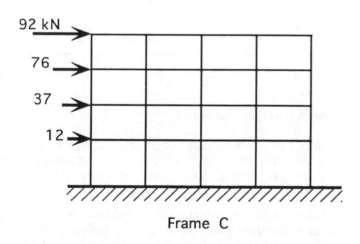

Frame C

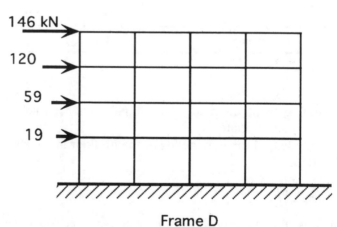

Frame D

Fig. 19.4(f). Design forces in frames for earthquake in X direction

earthquake motion in the Y direction. Therefore, forces caused by torsion do not affect the design of these frames. Fig.19.4(f) shows the design forces in frames and in the wall for earthquake motion in the X direction.

(b) Response spectrum method. This method requires a free-vibration analysis to determine the natural periods and corresponding modal shapes of the building. The Code requires three modes to be

superposed; therefore, the free-vibration analysis must yield at least three modes in each of the two main directions of the building. A free-vibration analysis of the building was performed using a computer program; the building properties and the modeling assumptions were the same as those used in the solution of part (a). The natural periods, the mode shapes, and the modal participation factor thus obtained are shown in Table 19.11. The Code does not specify the value of

Table 19.10. Calculation of Force in Different Frames due to Torsional Moment

Frame	Stiffness ($\times 10^3$ kN/m)	r_i (m)	$k_i r_i$ ($\times 10^3$ kN)	$k_i r_i^2$ ($\times 10^3$)	$\dfrac{k_i r_i}{\Sigma k_i r_i^2}$	Force in Frame due to Torsion	Force in Frame due to Direct Force	Total Design Force
A	15.85	7.79	123.5	961.8	0.0206	0.0114V	0.307V	0.318V
B	9.94	1.79	17.8	31.8	0.0038	0.0017V	0.193V	0.195V
C	9.94	−2.71	−26.9	73.0	−0.0057	−0.0025V	0.193V	0.193V
D	15.85	−7.21	−114.3	823.9	−0.0241	−0.0105V	0.307V	0.307V
1	12.20	−10.0	−122.0	1,220.0	−0.0258	0.0112V	—	—
2	7.58	−5.0	−37.9	189.5	−0.0080	0.0035V	—	—
4	7.58	5.0	37.9	189.5	0.0080	0.0035V	—	—
5	12.20	10.0	122.0	1,220.0	0.0258	0.0112V	—	—
Σ				4,709.6				

Table 19.11. Free-Vibration Properties of the Building of Example 19.1

	X direction			Y direction		
	Mode 1	Mode 2	Mode 3	Mode 1	Mode 2	Mode 3
Natural Period (sec)	0.860	0.265	0.145	0.303	0.057	0.021
	Mode Shape					
Roof	1.000	1.000	1.000	1.000	1.000	1.000
3rd Floor	0.904	0.216	−0.831	0.690	−0.327	−1.407
2nd Floor	0.716	−0.701	−0.574	0.393	−0.986	0.040
1st Floor	0.441	−0.921	1.016	0.147	−0.694	1.636
Modal Participation Factor	1.240	−0.329	0.118	1.423	−0.568	0.183

damping to be adopted; 5% damping has been considered appropriate.

The lateral force F_{ir} acting at the ith floor in the rth mode is

$$F_{ir} = K\beta I F_0 \phi_{ir} C_r \frac{S_{ar}}{g} W_i$$

$$= 1.0 \times 1.0 \times 1.0 \times 0.40 \times C_r \frac{S_{ar}}{g} W_i \phi_{ir} \quad \text{[eq.(19.5)]}$$

$$= 0.40 C_r \frac{S_{ar}}{g} W_i \phi_{ir}$$

The modal participation factor C_r is given by eq.(19.6) while the acceleration coefficient (S_a/g) for different modes is obtained from Fig. 19.3. Calculated lateral forces at different levels in each mode are as follows:

Earthquake in X direction.

Mode 1 ($T = 0.860$): $S_a/g = 0.12$; $F_{i1} = 0.05952 W_i \phi_{i1}$
Mode 2 ($T = 0.265$): $S_a/g = 0.20$; $F_{i2} = 0.02632 W_i \phi_{i2}$
Mode 3 ($T = 0.145$): $S_a/g = 0.20$; $F_{i3} = 0.00944 W_i \phi_{i3}$

Earthquake in Y direction

Mode 1 ($T = 0.303$): $S_a/g = 0.20$; $F_{i1} = 0.11384 W_i \phi_{i1}$
Mode 2 ($T = 0.057$): $S_a/g = 0.18$; $F_{i2} = 0.04090 W_i \phi_{i2}$
Mode 3 ($T = 0.021$): $S_a/g = 0.12$; $F_{i3} = 0.00878 W_i \phi_{i3}$

Table 19.12 summarizes the calculation of lateral forces at different floors in each mode for earthquake motion in the X direction and the resulting story shear for corresponding modes. The contributions of different modes are combined by eq.(19.7), with building height = 13.8 m the value of α is 0.4. Thus, for earthquakes in the X direction,

$$V_1 = (1 - 0.4)(693.8 + 76.4 + 12.9)$$
$$+ 0.4[693.8^2 + 76.4^2 + 12.9^2]^{1/2} = 749.1 \text{ kN}$$

$$V_2 = 615.7 \text{ kN}$$

$$V_3 = 474.2 \text{ kN}$$

$$V_4 = 250.5 \text{ kN}$$

Table 19.12. Lateral Force Calculation by Response Spectrum Method (Earthquake in X Direction)

Floor	Weight W_i (kN)	Mode 1			Mode 2			Mode 3		
		ϕ_{i1}	F_{i1} (kN)	V_{i1} (kN)	ϕ_{i2}	F_{i2} (kN)	V_{i2} (kN)	ϕ_{i3}	F_{i3} (kN)	V_{i3} (kN)
4	3,000	1.000	178.6	178.6	1.000	79.0	79.0	1.000	28.3	28.3
3	4,200	0.904	226.0	404.5	0.216	23.9	102.9	−0.831	−32.9	4.6
2	4,200	0.716	179.0	583.5	−0.701	−77.5	25.4	−0.574	−22.8	27.4
1	4,200	0.441	110.2	693.8	−0.921	−101.8	76.4	1.016	40.3	12.9

Table 19.13. Lateral Force Calculation by Response Spectrum Method (Earthquake in Y Direction)

Floor	Weight W_i (kN)	Mode 1			Mode 2			Mode 3		
		ϕ_{i1}	F_{i1} (kN)	V_{i1} (kN)	ϕ_{i2}	F_{i2} (kN)	V_{i2} (kN)	ϕ_{i3}	F_{i3} (kN)	V_{i3} (kN)
4	3,000	1.000	341.5	341.5	1.000	122.7	122.7	1.000	26.4	26.4
3	4,200	0.690	329.9	671.4	−0.327	−56.2	66.5	−1.407	−51.9	25.5
2	4,200	0.393	187.9	859.3	−0.986	−169.4	102.8	−0.040	1.5	24.0
1	4,200	0.147	70.3	929.6	−0.694	−119.2	222.0	1.636	60.4	36.4

Table 19.14. Comparison of Design Seismic Forces

Floor	Seismic Coefficient Method		Response Spectrum Method	
	X direction (kN)	Y direction (kN)	X direction (kN)	Y direction (kN)
4	476	529	251	440
3	393	437	224	288
2	192	213	142	210
1	62	69	133	157
Base Shear (kN)	1,123	1,248	750	1,095
Base Moment (kN-m)	12,416	13,798	7,448	11,338

The externally applied design forces are obtained from eq.(19.8) as

$$F_4 = V_4 = 250.5 \text{ kN}$$

$$F_3 = V_3 - V_4 = 474.2 - 250.5 = 223.7 \text{ kN}$$

$$F_2 = V_2 - V_3 = 615.7 - 474.2 = 141.5 \text{ kN}$$

$$F_1 = V_1 - V_2 = 749.1 - 615.7 = 133.4 \text{ kN}$$

In a similar manner, the lateral force and story shear forces are calculated in the Y direction for different modes (Table 19.13). The combined story shears are

$$V_1 = 1,095 \text{ kN}, V_2 = 938 \text{ kN}, V_3 = 728.1 \text{ kN},$$
$$V_4 = 440.0 \text{ kN}$$

The externally applied design loads are obtained from eq.(19.8) as

$$F_4 = 440.0 \text{ kN}, F_3 = 288.1 \text{ kN}, F_2 = 210.0 \text{ kN},$$
$$F_1 = 157.0 \text{ kN}$$

Table 19.14 compares the design seismic forces obtained by the two procedures: the seismic coefficient method and the response spectrum method. It should be apparent that the response spectrum method underestimates the design force by 33% and 12%, respectively, for the X and the Y directions, as compared to the design values obtained by the seismic coefficient method. The discrepancy is large in the X direction because the free-vibration analysis on a bare frame (ignoring effects of filler walls) gives a high value of the fundamental period (0.86 sec) as compared to 0.40 sec used in the seismic coefficient method. In the Y direction, the reinforced concrete wall provides significant stiffness, hence the bareframe fundamental period by free-vibration analysis (0.303 sec) is close to the value of 0.32 sec used in the static method. As a result, in the Y direction, the difference in design shear using the two procedures is not so large.

19.15 COMPUTER PROGRAM AND EXAMPLES

A computer program has been developed for the evaluation of the design seismic force on a multistory building according to the provisions of IS:1893 (1984). The program can be used in an interactive mode or through an input data file. In the seismic coefficient method, the program allows users to calculate the fundamental period by either of the expressions prescribed in the Code, or by the use of a value that the user obtains through dynamic analysis or experimentation, which are options provided by the Code. In the response spectrum method, the natural frequencies and mode shapes of the building are input as data. The program calculates the design base shear, design base moment, and design seismic forces at the different levels of the building. The user may opt for an approximate distribution (Macleod 1971) of design seismic forces to different frames and/or walls, based on the rigid floor diaphragm assumption. In this case, the user must provide properties such as dimensions and concrete grade for each frame and wall. The procedure is valid for uniform frames and walls; if there is some nonuniformity in span, story height, or member sizes, average properties should be given.

Example 19.2

Solve the problem of Example 19.1 using the computer program.

Solution

The results obtained by the program differ slightly from those obtained by manual calculations because the values of C (in the seismic coefficient method) and S_a/g (in the response spectrum method) interpolated by the program differ slightly from those interpolated from the plot for S_a/g. The following is the output file for this example, which also echoes the input data.

```
(a) Seismic Coefficient Method

    Input file name  : EX2A.IN
    Output file name : EX2A.OUT
    Title of problem : EXAMPLE 19.2A
```

```
Method selected : Seismic Coefficient Method

Performance factor                          K =  1
Coefficient for soil foundation system      ß =  1
Importance factor                           I =  1
Seismic zone (1,2,3,4,5)                     Z =  5
Number of stories                           N =  4
Maximum base length in X-direction          Dx = 20 m
Maximum base length in Y direction          Dy = 15 m

   Level        Story height (m)        Weight (kN)
     4               3.2                  3000.0
     3               3.2                  4200.0
     2               3.2                  4200.0
     1               4.2                  4200.0


              Fundamental period (sec)      Value of C
  X-direction        0.400                    0.920
  Y-direction        0.321                    1.000

                    Base shear (kN)        Base moment (kNm)
  X-direction         1148.16                 12697.22
  Y-direction         1248.00                 13801.33

   Level                    Design seismic force (kN)
                           X-direction        Y-direction
     4                        487                529
     3                        402                437
     2                        196                213
     1                         63                 69

Design Force Distribution to Different Frames and Walls
(with Rigid Floor Diaphragms)
(MacLeod, 1971)
------------------------------------------------------------
FRAMES IN X-DIRECTION:
---------------------
    Number of frames = 4  :  Types of frame = 2

Frames Type 1

    Total  width of frame in X-direction = 20.000
    Distance between column center lines =  5.000
    ΣI of columns at top of frame         = 0.1067E-01
    ΣI of columns at bottom of frame      = 0.1067E-01
    ΣI of beams at top of frame           = 0.1800E-01
    ΣI of beams at bottom of frame        = 0.1800E-01
    Area of ext. cols at top of frame     = 0.160
    Area of ext. cols at bottom of frame  = 0.160
    Width of column                       = 0.400
    Depth of beam                         = 0.600
    Grade of concrete used in columns     = M20
    Grade of concrete used in  beams      = M20

Frames Type 2 :

    Total  width of frame in X-direction = 20.000
    Distance between column center lines =  5.000
    ΣI of columns at top of frame         = 0.1067E-01
    ΣI of columns at bottom of frame      = 0.1067E-01
    ΣI of beams at top of frame           = 0.9112E-02
    ΣI of beams at bottom of frame        = 0.9112E-02
    Area of ext. cols at top of frame     = 0.160
    Area of ext. cols at bottom of frame  = 0.160
    Width of column                       = 0.400
    Depth of beam                         = 0.450
    Grade of concrete used in columns     = M20
    Grade of concrete used in  beams      = M20

  Frame    Type    Y location
    A        1       15.000
    B        2        9.000
    C        2        4.500
    D        1        0.000

FRAMES IN Y-DIRECTION:
---------------------
    Number of frames = 5  :  Types of frame = 3

Frames Type 1 :

    Total  width of frame in Y-direction = 15.000
    Distance between column center lines =  5.000
    ΣI of columns at top of frame         = 0.8533E-02
    ΣI of columns at bottom of frame      = 0.8533E-02
    ΣI of beams at top of frame           = 0.1350E-01
    ΣI of beams at bottom of frame        = 0.1350E-01
    Area of ext. cols at top of frame     = 0.160
    Area of ext. cols at bottom of frame  = 0.160
    Width of column                       = 0.400
    Depth of beam                         = 0.600
    Grade of concrete used in columns     = M20
    Grade of concrete used in beams       = M20
```

Frames Type 2 :

      Total width of frame in Y-direction = 15.000
      Distance between column center lines = 5.000
      ΣI of columns at top of frame   = 0.8533E-02
      ΣI of columns at bottom of frame = 0.8533E-02
      ΣI of beams at top of frame   = 0.6834E-02
      ΣI of beams at bottom of frame   = 0.6834E-02
      Area of ext. cols at top of frame = 0.160
      Area of ext. cols at bottom of frame = 0.160
      Width of column   = 0.400
      Depth of beam   = 0.450
      Grade of concrete used in columns = M20
      Grade of concrete used in beams   = M20

Frames Type 3 :

      Total width of frame in Y-direction = 4.500
      Distance between column center lines = 4.500
      ΣI of columns at top of frame   = 0.4267E-02
      ΣI of columns at bottom of frame = 0.4267E-02
      ΣI of beams at top of frame   = 0.4556E-02
      ΣI of beams at bottom of frame   = 0.4556E-02
      Area of ext. cols at top of frame = 0.160
      Area of ext. cols at bottom of frame = 0.160
      Width of column   = 0.400
      Depth of beam   = 0.450
      Grade of concrete used in columns = M20
      Grade of concrete used in beams   = M20

Frame	Type	X location
1	1	0.000
2	2	5.000
3	3	10.000
4	2	15.000
5	1	20.000

WALLS IN Y-DIRECTION:

      Number of walls = 1 : Types of wall = 1

Walls Type 1 :

      Moment of inertia =   5.811
      Shear area   =   1.200

Wall	Type	X location
1	1	10.000

ECCENTRICITY CALCULATION:

Center of mass   : (10.000, 7.500) : given
Center of stiffness   : (10.000, 7.211) : calculated
Eccentricity (e)   : (0.000, 0.289)
Design eccentricity (1.5e): (0.000, 0.434)

FRAME AND WALL STIFFNESSES:

Frames in X-direction

Frame	Stiffness (Point load (kN/m) at top)
A	15843.44
B	9933.17
C	9933.17
D	15843.44

Frames in Y-direction

Frame	Stiffness (Point load (kN/m) at top)
1	12154.84
2	7578.51
3	4540.91
4	7578.51
5	12154.84

Walls in Y-direction

Wall	Stiffness (Point load (kN/m) at top)	Stiffness (Actual seismic load (kN/m) distribution)
1	142010.00	193288.80

FORCE DISTRIBUTION AMONG FRAMES IN X-DIRECTION

	Frame A	Frame B	Frame C	Frame D
4	155.17	94.61	93.81	149.63
3	128.17	78.15	77.49	123.59
2	62.46	38.09	37.76	60.23
1	20.12	12.27	12.17	19.40

FORCE DISTRIBUTION AMONG FRAMES IN Y-DIRECTION

Level	Frame 1	Frame 2	Frame 3	Frame 4	Frame 5
4	59.91	37.36	22.38	37.36	59.91
3	0.00	0.00	0.00	0.00	0.00
2	0.00	0.00	0.00	0.00	0.00
1	0.00	0.00	0.00	0.00	0.00

For wall-frame system, the frame force thus obtained are usually increased by 30%. These are:

Level	Frame 1	Frame 2	Frame 3	Frame 4	Frame 5
4	77.89	48.56	29.10	48.56	77.89
3	0.00	0.00	0.00	0.00	0.00
2	0.00	0.00	0.00	0.00	0.00
1	0.00	0.00	0.00	0.00	0.00

The seismic force in frames as above is less than 25% of the total design seismic force. Hence, as per code requirement, the design force for frames is to be increased to:

Level	Frame 1	Frame 2	Frame 3	Frame 4	Frame 5
4	86.17	53.73	32.19	53.73	86.17
3	0.00	0.00	0.00	0.00	0.00
2	0.00	0.00	0.00	0.00	0.00
1	0.00	0.00	0.00	0.00	0.00

FORCE DISTRIBUTION AMONG SHEAR WALLS IN Y-DIRECTION

Level	Wall 1
4	312.29
3	437.13
2	213.04
1	68.63

(b) Response Spectrum Method

Input file name : EX2B.IN
Output file name : EX2B.OUT
Title of problem : EXAMPLE 19.2B

            Method selected : Response Spectrum Method

Performance factor	K = 1
Coefficient for soil foundation system	ß = 1
Importance factor	I = 1
Seismic zone (1,2,3,4,5)	Z = 5
Number of stories	N = 4
Percentage of damping (0,2,5,10,20)	= 5

Story level	Story height (m)	Weight (kN)
4	3.2	3000.0
3	3.2	4200.0
2	3.2	4200.0
1	4.2	4200.0

Mode Shapes in X-direction (given)

Level	MODE 1	MODE 2	MODE 3
4	1.000	1.000	1.000
3	0.904	0.216	-0.831
2	0.716	-0.701	-0.574
1	0.441	-0.921	1.016

Mode Shapes in Y-direction (given)

Level	MODE 1	MODE 2	MODE 3
4	1.000	1.000	1.000
3	0.690	-0.327	-1.407
2	0.393	-0.986	0.040
1	0.147	-0.694	1.636

	Time period (sec)	Value of Sa/g	Participation factor
X-direction:			
Mode 1	0.860	0.121	1.240
Mode 2	0.265	0.200	-0.329
Mode 3	0.145	0.200	0.118
Y-direction			
Mode 1	0.303	0.200	1.423
Mode 2	0.057	0.182	-0.568
Mode 3	0.021	0.132	0.183

LATERAL LOAD CALCULATION (Fi = Force, Vi = Story Shear)

(X-direction)

Level	Mode 1		Mode 2		Mode 3		SRSS value	
	Fi(1)	Vi(1)	Fi(2)	Vi(2)	Fi(3)	Vi(3)	Vi	Fi
4	180.3	180.3	-79.0	79.0	28.1	28.1	252.0	252.0
3	228.2	408.5	-23.9	102.9	-32.7	4.6	478.1	226.1
2	180.7	589.2	77.6	25.4	-22.6	27.2	621.3	143.1
1	111.3	700.6	101.9	76.5	40.0	12.8	755.9	134.6

LATERAL LOAD CALCULATION (Fi = Force, Vi = Story Shear)

(Y-direction)

Level	Mode 1		Mode 2		Mode 3		SRSS value	
	Fi(1)	Vi(1)	Fi(2)	Vi(2)	Fi(3)	Vi(3)	Fi	Vi
4	341.5	341.5	-124.2	124.2	29.0	29.0	442.6	442.6
3	329.9	671.4	56.8	67.3	-57.1	28.1	730.2	287.6
2	187.9	859.3	171.4	104.1	1.6	26.5	940.3	210.1
1	70.3	929.5	120.6	224.7	66.3	39.9	1099.3	159.1

	Base shear (kN)	Base moment (kNm)
X-direction	755.88	7499.28
Y-direction	1099.33	11378.92

Level	Design seismic force (kN)	
	X-direction	Y-direction
4	252	443
3	226	288
2	143	210
1	135	159

Design Force Distribution to Different Frames and Walls
(with Rigid Floor Diaphragms)
MacLeod (1971)

FRAMES IN X-DIRECTION:

       Number of frames = 4 : Types of frame = 2

Type 1 :

        Total width of frame in X-direction = 20.000
        Distance between column center lines = 5.000
        7S1I of columns at top of frame   = 0.1067E-01
        7S1I of columns at bottom of frame   = 0.1067E-01
        7S1I of beams at top of frame   = 0.1800E-01
        7S1I of beams at bottom of frame   = 0.1800E-01
        Area of ext. cols at top of frame   = 0.160
        Area of ext. cols at bottom of frame = 0.160
        Width of column   = 0.400
        Depth of beam   = 0.600
        Grade of concrete used in columns   = M20
        Grade of concrete used in beams   = M20

Type 2 :

        Total width of frame in X-direction = 20.000
        Distance between column center lines = 5.000
        7S1I of columns at top of frame   = 0.1067E-01
        7S1I of columns at bottom of frame   = 0.1067E-01
        7S1I of beams at top of frame   = 0.9112E-02
        7S1I of beams at bottom of frame   = 0.9112E-02
        Area of ext. cols at top of frame   = 0.160
        Area of ext. cols at bottom of frame = 0.160
        Width of column   = 0.400
        Depth of beam   = 0.450
        Grade of concrete used in columns   = M20
        Grade of concrete used in beams   = M20

Frame	Type	Y location
A	1	0.000
B	2	4.500
C	2	9.000
D	1	15.000

FRAMES IN Y-DIRECTION:

       Number of frames = 5 : Types of frame = 3

Type 1 :

        Total width of frame in Y-direction = 15.000
        Distance between column center lines = 5.000
        7S1I of columns at top of frame   = 0.8533E-02
        7S1I of columns at bottom of frame   = 0.8533E-02

        7S1I of beams at top of frame   = 0.1350E-01
        7S1I of beams at bottom of frame   = 0.1350E-01
        Area of ext. cols at top of frame   = 0.160
        Area of ext. cols at bottom of frame = 0.160
        Width of column   = 0.400
        Depth of beam   = 0.600
        Grade of concrete used in columns   = M20
        Grade of concrete used in beams   = M20

Type 2 :

        Total width of frame in Y-direction   = 15.000
        Distance between column center lines = 5.000
        7S1I of columns at top of frame   = 0.8533E-02
        7S1I of columns at bottom of frame   = 0.8533E-02
        7S1I of beams at top of frame   = 0.6834E-02
        7S1I of beams at bottom of frame   = 0.6834E-02
        Area of ext. cols at top of frame   = 0.160
        Area of ext. cols at bottom of frame = 0.160
        Width of column   = 0.400
        Depth of beam   = 0.450
        Grade of concrete used in columns   = M20
        Grade of concrete used in beams   = M20

Type 3 :

        Total width of frame in Y-direction   = 4.500
        Distance between column center lines = 4.500
        7S1I of columns at top of frame   = 0.4267E-02
        7S1I of beams at top of frame   = 0.4556E-02
        7S1I of beams at bottom of frame   = 0.4556E-02
        Area of ext. cols at top of frame   = 0.160
        Area of ext. cols at bottom of frame = 0.160
        Width of column   = 0.400
        Depth of beam   = 0.450
        Grade of concrete used in columns   = M20
        Grade of concrete used in beams   = M20

Frame	Type	X location
1	1	0.000
2	2	5.000
3	3	10.000
4	2	15.000
5	1	20.000

WALLS IN Y-DIRECTION:

       Number of walls = 1 : Types of wall = 1

Type 1 :

        Moment of inertia =   5.811
        Shear area   =   1.200

Wall	Type	X location
1	1	10.000

ECCENTRICITY CALCULATION:

Center of mass   : (10.000, 7.500) : given,
Center of stiffness   : (10.000, 7.211) : calculated
Eccentricity (e)   : (0.000, 0.289)
Design eccentricity (1.5e): (0.000, 0.434)

FRAME AND WALL STIFFNESSES:

Frames in X-direction

Frame	Stiffness (Point load at top) (kN/m)
A	15843.44
B	9933.17
C	9933.17
D	15843.44

Frames in Y-direction

Frame	Stiffness (Point load at top) (kN/m)
1	12154.84
2	7578.51
3	4540.91
4	7578.51
5	12154.84

Walls in Y-direction

Wall	Stiffness (Point load at top) (kN/m)	Stiffness (Actual seismic load distribution) (kN/m)
1	142010.00	210408.30

FORCE DISTRIBUTION AMONG FRAMES IN X-DIRECTION

Level	Frame A	Frame B	Frame C	Frame D
4	77.46	48.56	48.98	80.33
3	69.48	43.56	43.93	72.06
2	43.99	27.58	27.81	45.61
1	41.37	25.94	26.16	42.91

FORCE DISTRIBUTION AMONG FRAMES IN Y-DIRECTION

Level	Frame 1	Frame 2	Frame 3	Frame 4	Frame 5
4	48.48	30.23	18.11	30.23	48.48
3	0.00	0.00	0.00	0.00	0.00
2	0.00	0.00	0.00	0.00	0.00
1	0.00	0.00	0.00	0.00	0.00

For wall-frame system, the frame force thus obtained are usually increased by 30%. These are

Level	Frame 1	Frame 2	Frame 3	Frame 4	Frame 5
4	63.03	39.30	23.55	39.30	63.03
3	0.00	0.00	0.00	0.00	0.00
2	0.00	0.00	0.00	0.00	0.00
1	0.00	0.00	0.00	0.00	0.00

The seismic force in frames as above is less than 25% of the total design seismic force. Hence, as per code requirement, the design force for frames is to be increased to

Level	Frame 1	Frame 2	Frame 3	Frame 4	Frame 5
4	75.91	47.33	28.36	47.33	75.91
3	0.00	0.00	0.00	0.00	0.00
2	0.00	0.00	0.00	0.00	0.00
1	0.00	0.00	0.00	0.00	0.00

FORCE DISTRIBUTION AMONG SHEAR WALLS IN Y-DIRECTION

Story	Wall 1
4	267.05
3	287.62
2	210.06
1	159.07

Seismic zone (1,2,3,4,5)	= 4
Number of stories (N)	= 16
Maximum base length in X-direction (Dx)	= 30 m
Maximum base length in Y direction (Dy)	= 15 m

Story level	Story height (m)	Weight (kN)
16	3.2	3600.0
15	3.2	5400.0
14	3.2	5400.0
13	3.2	5400.0
12	3.2	5400.0
11	3.2	5400.0
10	3.2	5400.0
9	3.2	5400.0
8	3.2	5400.0
7	3.2	5400.0
6	3.2	5400.0
5	3.2	5400.0
4	3.2	5400.0
3	3.2	5400.0
2	3.2	5400.0
1	4.4	5400.0

	Fundamental time period (sec)	Value of C
X-direction	1.600	0.380
Y-direction	1.600	0.380

	Base shear (kN)	Base moment (kNm)
X-direction	1607.40	63913.84
Y-direction	1607.40	63913.84

Story level	Design seismic force (kN) X-direction	Y-direction
16	190	190
15	251	251
14	220	220
13	190	190
12	163	163
11	138	138
10	115	115
9	94	94
8	75	75
7	58	58
6	43	43
5	31	31
4	20	20
3	12	12
2	6	6
1	2	2

Example 19.3

A 16-story office building of reinforced concrete is located in New Delhi (Seismic Zone IV). The building is founded on a raft foundation. The lumped weight is 5,400 kN on the floors and 3,600 kN on the roof. The story height is 4.4 m for the first story and 3.2 m for the remaining stories. Calculate the design seismic force on the building by the seismic coefficient method according to IS:1893 (1984). Assume $I = 1.0$ and $K = 1.0$.

Solution

```
Input file name  : ex3.in

Output file name : EX3.OUT

Title of problem : EXAMPLE 19.3

Method selected  : Seismic Coefficient Method

Performance factor (K)                      = 1
Coefficient for soil foundation system (β)  = 1
Importance factor (I)                       = 1
```

19.16 EVALUATION

The Code does not yet have any regulation to control the use of the value for the fundamental period obtained by dynamic analysis, which may be too large. This enables the designer to perform a bare-frame analysis by excluding the stiffness of nonstructural members, which will result in an unrealistically large value for the fundamental period, in turn resulting in low design forces. The seismic codes of many countries now avoid such a situation by (1) specifying the use of a fundamental period obtained from an empirical formula, (2) establishing a lower limit on design seismic force based on empirical formulas for the fundamental period, or (3) establishing an upper limit on the fundamental period based on empirical formulas.

The revision of the Code IS:1893 has begun, but it

[4]To save space, detailed computer output has not been reproduced for this example.

will take some time to be finalized. It is expected that the next revision will incorporate one of the aforementioned provisions. Also, it appears that the Code may be revised to specify a higher elastic force, to be reduced by the response factor, to obtain the design force, as is now prevalent in many other codes.

REFERENCES

HOUSNER, G. W., and JENNINGS, P. C. (1982) *Earthquake Design Criteria*. Earthquake Engineering Research Institute, CA.

IS:1893 (1962, 1966, 1970) *Recommendations for Earthquake Resistant Design of Structures*. Bureau of Indian Standards, New Delhi, India.

IS:1893 (1975, 1984) *Indian Standard Criteria for Earthquake Resistant Design of Structures*. Bureau of Indian Standards, New Delhi, India.

IS:4326 (1967, 1976) *Indian Standard Code of Practice for Earthquake Resistant Design and Construction of Buildings*. Bureau of Indian Standards, New Delhi, India.

IS:456 (1978) *Indian Standard Code of Practice for Plain and Reinforced Concrete*. Bureau of Indian Standards, New Delhi, India.

IS:1343 (1980) *Indian Standard Code of Practice for Prestressed Concrete*. Bureau of Indian Standards, New Delhi, India.

IS:800 (1984) *Indian Standard Code of Practice for General Construction in Steel*. Bureau of Indian Standards, New Delhi, India.

IS:875 (1987) *Indian Standard Code of Practice for Design Loads (Other Than Earthquake) for Buildings and Structures*. Bureau of Indian Standards, New Delhi, India.

JAIN, S. K., and MANDAL, U. K. (1992) "Dynamics of Buildings with V-Shaped Plan." *Journal of Engineering Mechanics, ASCE* Vol. 118, No. 6: 1093–1112.

KRISHNA, J. (1985) "Historical Developments in Growth of Earthquake Engineering Research in India." Keynote address, Workshop on Earthquake Engineering: Past, Present and Future, Roorkee, India.

——— (1992) "Seismic Zoning Maps of India." *Current Science*, Current Science Association, Indian Academy of Science, Vol. 62, Nos. 1 & 2: 17–23, Bangalore, India.

MACLEOD, I. A. (1971) *Shear Wall-Frame Interaction, A Design Aid*. Portland Cement Association, Skokie, IL.

NEHRP (1991) *Recommended Provisions for the Development of Seismic Regulations for New Buildings*. Building Seismic Safety Council, Washington, D.C.

RAJASANKAR, J., and JAIN, S. K. (1988) "A Review of Part I of Explanatory Handbook on Codes for Earthquake Engineering." *Bulletin of the Indian Society of Earthquake Technology* Vol. 25, No. 2: 91–114.

SP:22 (1982) *Explanatory Handbook on Codes for Earthquake Engineering (IS:1893-1975 and IS:4326-1976)*. Bureau of Indian Standards, New Delhi, India.

UBC (1991) *Uniform Building Code*. International Conference of Building Officials, Whittier, CA.

20

Indonesia

Suradjin Sutjipto

20.1 INTRODUCTION

The first Indonesian provisions for seismic building design were incorporated in the 1970 Indonesian Loading Code (NI-18). Unlike other seismic design code, it introduced a method based on a base force coefficient rather than on a base shear coefficient. This base force coefficient is not dependent directly on the period of the building. The code also introduced the first official seismic zoning map of Indonesia.

Twenty years ago, a sequence of severe earthquakes caused heavy loss of life and material damage in some apparently low-risk seismic regions in Indonesia. Indonesian structural engineers and government officials began to give serious attention to earthquake-resistant design. In 1976, an earthquake engineering study was carried out as part of a bilateral program of the governments of Indonesia and New Zealand. The objectives of the study were to develop a new seismic design code and seismic zoning (see Fig. 20.1) for Indonesia. The Indonesian Seismic Code for Building Design (*Peraturan Perencanaan Tahan Gempa Indonesia Untuk Gedung*) published in 1983 was based on this study. With minor changes, this code was issued in 1987 as one of the Indonesian Construction Standards entitled, *Pedoman Perencanaan Ketahanan Gempa Untuk Rumah Dan Gedung – SKBI 1.3.53.1987 UDC: 699.841*. It was ratified in 1989 as one of the Indonesian National Standards and registered as SNI 1726-1989-F. The latest version of these standards has been retained as the current seismic provisions for building design in Indonesia.

The 1983 Indonesian Seismic Code for Building Design introduced some important concepts that had not appeared in the seismic provisions of the 1970 Indonesian Loading Code. The Indonesian Seismic Code now requires the lateral-resisting system of the building to be designed and detailed to behave in a ductile manner; strict requirements for design and detailing are mandatory. Two methods of analysis for earthquake-resistant design are provided: (1) equivalent static load analysis and (2) dynamic analysis. The equivalent static load analysis is applicable only for regular structures under 40 m in height. Dynamic analysis may be used for any structure but *must* be used for structures more than 40 m in height. It must also be used for structures with highly irregular shapes, large differences in lateral stiffness between adjacent stories, large setbacks, or other unusual features. However, as stated in the Code commentary, if a building has wings (L-, T-, U-shaped or other irregular planform) of significant size, they should be designed separately from the rest of the building, regardless of the method of analysis.

Another significant improvement relates to the

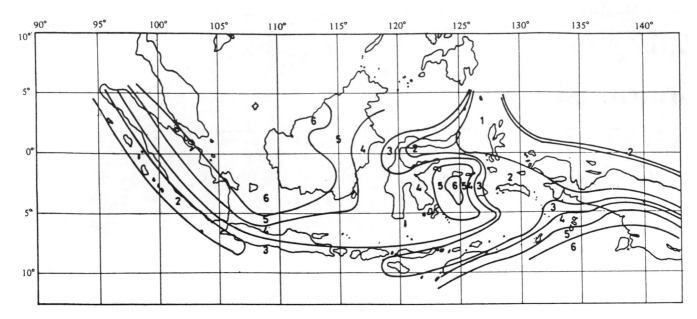

Fig. 20.1. Indonesian seismic zones. (Reproduced from the *Pedoman Perencanaan Ketahanan Gempa Untuk Rumah Dan Gedung SNI 1726-1989-F*, with permission of Departement Pekerjaan Umum.)

design of orthogonal effects of the seismic forces. The Code specifies that the horizontal seismic loads determined from the analysis shall be applied at each floor and roof level in both principal directions of the building simultaneously. The code requires that primary elements, especially columns or other vertical elements of the lateral-resisting system, should be designed for 100% of the effects of seismic design loads in one principal direction in combination with 30% of the effects of seismic design loads in the orthogonal direction.

This chapter is not intended as a comprehensive description of all the provisions in the Indonesian Seismic Code. Therefore, only the major provisions for determining seismic forces and associated responses for buildings are presented.

20.2 EQUIVALENT STATIC LOAD ANALYSIS

The Indonesian Seismic Code for Building Design establishes that every structure should be designed and constructed to resist a total lateral seismic base shear force given by the following formula:

$$V = CIKW_t \tag{20.1}$$

The term C is the *basic seismic coefficient*, obtained from Fig. 20.2, that corresponds to the appropriate seismic zone (Fig. 20.1) for the fundamental period of the corresponding structure. The Code specifies two

types of subsoil to be considered in the selection of coefficient C, soft and firm. A structure shall be considered on a soft subsoil if it is on soil deposits deeper than the following limits:

> 6 m: For a cohesive soil with an average undrained shear strength not exceeding 0.5 kg/cm².
> 9 m: For any site where the soils are either cohesive with an average undrained shear strength of 0.5 kg/cm² to 1 kg/cm², or very dense granular materials.
> 12 m: For a cohesive soil with an average undrained shear strength of 1 kg/cm² to 2 kg/cm².
> 20 m: For a very dense cemented granular soil.

For soil deposits less deep than these limits, the subsoil shall be considered firm. These provisions apply regardless of whether or not foundation piles extend to a deeper hard layer. Depths shall be measured from the level where the ground provides effective lateral restraint to the building.

The term I is the *importance factor*, related to the anticipated use of the structures as classified in Table 20.1. A factor higher than 1.0 is applied to monumental buildings, essential facilities, and buildings whose loss would pose an abnormal risk for the community.

The term K is the *structural-type factor* which reflects the varying degrees of ductility of different types of structural systems associated with the construction material of the seismic-energy-dissipating elements. The Code establishes values of K for different types of construction as shown in Table 20.2.

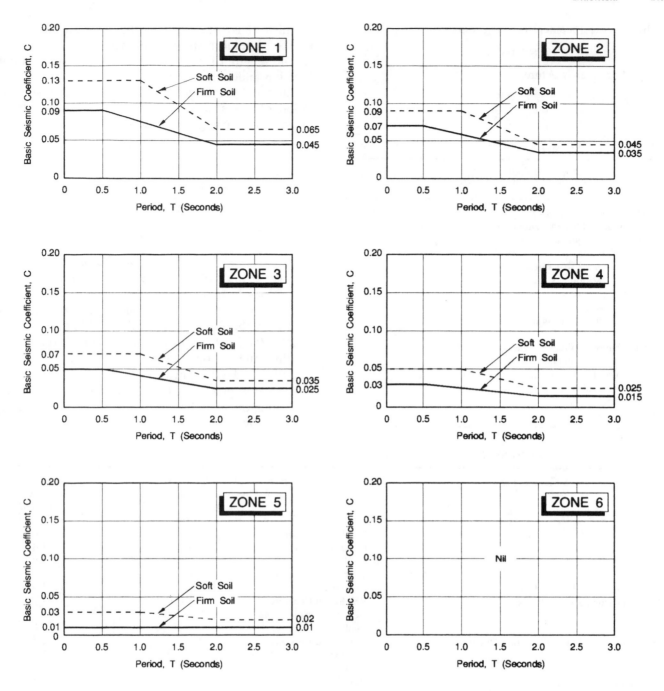

Fig. 20.2.   Basic seismic coefficient

When more than one structural system is used in the building, the K value should be selected on the basis of the element providing the greatest part of the seismic load resistance.

The term W_t is the combination of the total vertical dead load and the reduced vertical live load of the structure. Table 20.3 shows the coefficient of reduction for live loads associated with the types of buildings as specified in the Indonesian Loading Code for Housing and Buildings, *SNI 1727-1989-F*.

20.3 FUNDAMENTAL PERIOD

For the purposes of determining the basic seismic coefficient C initially, the code provides empirical formulas for estimating the fundamental period T of the structure in the direction under consideration, as follows:

(a)   For concrete moment-resisting frames:

$$T = 0.060H^{3/4} \qquad (20.2)$$

(b) For steel moment-resisting frames:

$$T = 0.085H^{3/4} \qquad (20.3)$$

(c) For other structures:

$$T = \frac{0.090H}{\sqrt{B}} \qquad (20.4)$$

Table 20.1. Importance Factor I

Type of Building	Importance Factor I
(a) Monumental Buildings	1.5
(b) Essential facilities that should remain functional after an earthquake, such as	1.5
– Hospitals	
– Food storage structures	
– Emergency relief stores	
– Power stations	
– Water works	
– Radio and television facilities	
– Places of assembly	
(c) Distribution facilities for gas or petroleum products in urban areas	2.0
(d) Structures for the support or containment of dangerous substances (such as acids, toxic substances, etc.)	2.0
(e) Other structures	1.0

in which H is the height of the structure above the level of lateral restraint (in meters) and B is the overall length of the building at the base, in the direction under consideration (in meters).

However, the Code requires that the fundamental period of the structure as finally detailed be determined from Rayleigh's formula:

$$T = 6.3 \sqrt{\frac{\sum_{i=1}^{N} W_i d_i^2}{g \sum_{i=1}^{N} F_i d_i}} \qquad (20.5)$$

In this formula, N is the total number of stories above the base of the building, W_i are the seismic weights at the various levels of the building, F_i are the lateral seismic loads in the direction under consideration applied at the floor levels of the building, d_i are the actual lateral displacements at the floor levels of the building due to the lateral forces F_i, and g is the acceleration due to gravity.

The Code precludes the possible use of an excessively long calculated period to justify an unreasonably low base shear; it requires that the value of C obtained

Table 20.2. Structural Type Factor K

Structural Type	Construction Material of the Seismic-Energy-Dissipating Elements	Structural Type Factor K
Ductile Frames	– Reinforced Concrete	1.0
	– Prestressed Concrete[3]	1.4
	– Steel	1.0
	– Timber	1.7
Coupled Shear Walls with Ductile Walls[1]	– Reinforced Concrete	1.0
Cantilever Ductile Shear Walls[1]	– Reinforced Concrete	1.2
	– Reinforced Hollow Masonry	2.5
	– Timber[2]	2.0
Cantilever Shear Walls of Limited Ductility[1]	– Reinforced Concrete	1.5
	– Reinforced Hollow Masonry	3.0
	– Timber[2]	2.5
Diagonally Braced Frames	– Reinforced Concrete	2.5
	– Steel	2.5
	– Timber	3.0
Cantilevered Single-Story Frames[5]	– Reinforced Concrete	2.5
	– Steel	2.5
Chimneys, Small Tanks	– Reinforced Concrete	3.0
	– Steel	3.0
Other Structures		See Note (4)

Notes:
1. Ductile shear walls are those in which the height-to-width ratio of individual wall exceeds 2. Walls of limited ductility are those that do not comply with this ratio.
2. Timber shear walls are those in which a timber frame is clad with timber sheathing in such a way as to form a diaphragm capable of resisting horizontal loads.
3. For structures in which members employ both prestressing and nonprestressing reinforcement to resist the seismic load a K value shall be determined by interpolating between that for reinforced concrete and prestressed concrete giving consideration to the proportion of the seismic load resisted by the two materials.
4. Structures not falling within the categories listed in this table may be assigned a K factor by interpolation between the tabulated K values by a comparison of the structure's ductility and energy dissipating capacity. A value of $K = 4.0$ should be used for structures not designed to resist earthquakes.
5. This value of K is applied where the horizontal restraint is provided by only one or two columns or there is no roof diaphragm capable of transferring the seismic load between the frames. If there are three or more columns and there is a roof diaphragm the frames may fall within the category of ductile frames.

Table 20.3. Coefficient of Reduction for Live Load

Type of Building	Coefficient of Reduction for Live Load
(a) Residential, hotels, hospitals	0.30
(b) Schools	0.50
(c) Churches, mosques, theaters, restaurants	0.50
(d) Office buildings	0.30
(e) Retail stores	0.80
(f) Warehouses, libraries	0.80
(g) Industrial buildings	0.90
(h) Garages	0.50
(i) Corridors, stairs:	
– Residential	0.30
– Schools, offices	0.50
– Community buildings, stores, warehouse, libraries, industrial buildings, garages	0.50

by using T given by eq.(20.5) shall not be less than 80% of the value obtained by using T from eqs.(20.2), (20.3), or (20.4).

20.4 DISTRIBUTION OF LATERAL SEISMIC FORCES

The total lateral seismic base shear V calculated from eq.(20.1) is distributed over the height of the structure as a set of forces, with one F_i at each level, in accordance with the following formula:[1]

$$F_i = \frac{W_i h_i}{\sum_{j=1}^{N} W_j h_j} (V - F_t) \qquad (20.6)$$

where

$F_t = 0$	for $H/B < 3$
$F_t = 0.1V$	for $H/B \geqslant 3$
$F_t = 0.2V$	for chimneys and smokestacks
$F_t = V$	for elevated tanks

in which

N = total number of stories above the base of the building
H = total height of the lateral-resisting system
B = width of the lateral-resisting system
F_i, F_j = lateral force applied at level i or j
F_t = additional portion of the base shear V acting at the top of the structure
W_i, W_j = seismic weight of ith or jth level
h_i, h_j = height of level i or j above the base.

The application of the additional force F_t at the top of the structure is done to compensate flexible struc-

---

[1]For clarity, this formula has been modified from the original given in the Code.

tures, where the influence of modes of vibration higher than the fundamental mode may be so significant that a nonlinear distribution of the total seismic base shear will occur.

The shear force V_i at any story i is given by the sum of the lateral seismic forces at that story and above; that is,

$$V_i = F_t + \sum_{j=i}^{N} F_j \qquad (20.7)$$

20.5 OVERTURNING MOMENTS

The overturning moment at any level of the building is determined from statics as the moment produced at that level by the lateral forces applied above that level. Therefore, the overturning moment M_i at any level i of the building is given by

$$M_i = F_t(h_N - h_i) + \sum_{j=i+1}^{N} F_j(h_j - h_i) \qquad (20.8)$$

for $i = 0, 1, 2, \ldots, N-1$.

The overturning moment at the base of the structure is M_0.

20.6 HORIZONTAL TORSIONAL MOMENTS

When the building floors are considered as rigid diaphragms in their own planes, the Indonesian Seismic Code requires horizontal torsion at the various levels of the structure to be taken into account in the design. The torsional moment at any level under consideration is produced by the story seismic shear V_i applied at a design eccentricity e_d at that level. Expressed mathematically,

$$M_{ti} = V_i e_{di} \qquad (20.9)$$

The design eccentricity e_d is determined based on the magnitude of the computed eccentricity e_c. The Code defines e_c as a theoretical distance between the center of mass and the center of rigidity of the story, measured perpendicular to the direction of loading. It specifies the eccentricity e_d as follows:

(a) If e_c is less than $0.1b$ and the building is four stories or less in height, e_d may be taken as zero.

(b) If e_c is less than $0.3b$ and clause (a) does not apply, then

$$e_d = 1.5e_c + 0.05b \qquad (20.10)$$

or

$$e_d = e_c - 0.05b \qquad (20.11)$$

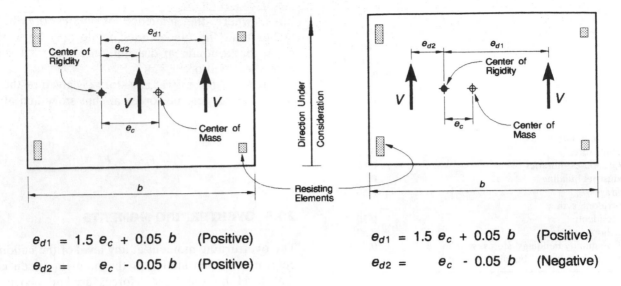

$$e_{d1} = 1.5\ e_c + 0.05\ b \quad \text{(Positive)}$$

$$e_{d2} = \quad e_c - 0.05\ b \quad \text{(Positive)}$$

$$e_{d1} = 1.5\ e_c + 0.05\ b \quad \text{(Positive)}$$

$$e_{d2} = \quad e_c - 0.05\ b \quad \text{(Negative)}$$

Fig. 20.3. Design eccentricity

whichever is the more severe for the element under consideration (see Fig. 20.3). The torsional effects caused by earthquake loading need be considered in only one direction at a time.

(c) If e_c is greater than $0.3b$ the structure shall be analyzed using a three-dimensional dynamic modal analysis.

In the expressions above, the term b is the maximum building dimension at the particular level, measured perpendicular to the loading direction under consideration.

The 1.5 coefficient of the e_c term in eq.(20.10) represents the amplification effects due to interaction between torsional and translational modes. The term $0.05b$ in eqs.(20.10) and (20.11) represents an additional eccentricity corresponding to accidental torsional moment. This accidental torsional moment is proposed to compensate for inaccuracies in determining the center of rigidity, uncertain live load distribution, uncertain dead load distribution (due to variations in workmanship, materials, etc.), inelastic behavior of lateral-resisting elements, and effects of torsional ground motion.

It should be noted that the Code defines the center of rigidity of a floor as the centroid of the resultant of seismic shear forces in all vertical elements of the lateral-resisting system present at that level, and the center of mass as the centroid of the resultant of the sum of all gravity loads above the particular level (cumulative). The procedure for calculating the center of mass is described more fully and shown in Example 20.1.

20.7 DYNAMIC ANALYSIS

The Code requires dynamic analysis for structures

with the following features: over 40 m in height, highly irregular shape, large differences in lateral stiffness between adjacent stories, large setbacks or other unusual features. The code indicates that dynamic analysis should be used if the ratio of the floor mass to the stiffness of the particular story of a building differs by more than 25% of the average of this ratio for the structure. However, a difference of more than 50% is not allowed. Basically, the main objective of the dynamic analysis is to obtain a more reliable distribution of the story seismic shear force. The Code indicates two methods of dynamic analysis: the spectral modal analysis and the time-history response analysis. The first method is commonly used in design practice in Indonesia. The second method is used only in the design of unusual or special structures.

The code stipulates the curve of basic seismic coefficient C (Fig. 20.2) to be used as the design spectrum in determining the contribution of each mode to the total response. The number of translational modes to be considered need not be more than 3 for regular structures with the computed eccentricity e_c less than $0.1b$ or 5 for slightly imbalanced buildings. However, for irregular structures or particularly tall buildings, the number of modes must equal the number of stories in the building; alternatively, the number of modes should be sufficient to contain at least 90% of the seismic energy dissipated in the building. As indicated in Section 4.7.2 of Chapter 4, this last requirement can be satisfied by simply adding a sufficient number of modal contributions [eq.(4.65)] until the total effective weight is 90%, or more, of the seismic design weight of the building.

The Code suggests use of the Square Root of the Sum of Squares (SRSS) technique to obtain the combined modal responses. This technique for combining modal responses is simple to apply and widely accepted. However, it may produce unconservative results if two modes have closely spaced natural periods. In such a case, the technique of taking absolute summation should be used for combining modal responses. Alternatively, it is recommended that the Complete Quadratic Combination (CQC) method [eq.(4.78), Chapter 4] be used.

In any case, the Code requires that the total seismic base shear determined by the spectral modal analysis be not less than $0.9CIKW_t$, where C is determined for the value of the first mode period of the structural model. This 10% reduction of the total seismic base shear is allowed because the use of modal analysis ensures a more accurate seismic shear distribution.

In the application of time-history response analysis, the Code requires at least four suitable time-history records to be used in determining the lateral seismic forces. The forces obtained from each record shall be scaled so that the total seismic base shear is not less than $0.9CIKW_t$, where C is determined for the fundamental period of the structure. The resulting shear forces are then averaged to determine the design seismic shear forces. The Code stipulates that if the distribution for any one record is significantly different from the other distributions, that record should be disregarded and replaced by a distribution based on spectral modal analysis.

To include horizontal torsional effects in the dynamic analysis, the Code indicates that, for reasonably regular structures in which the distance between the centers of mass and rigidity is less than $0.3b$, the torsionally induced shears may be determined from the static analysis procedure given in Section 20.6. However, for irregular structures with computed eccentricity e_c exceeding $0.3b$, the torsional effects of at least 3 modes of vibration shall be considered in a three-dimensional modal analysis. It should be noted, however, that in very irregular buildings even a three-dimensional modal analysis may underestimate earthquake effects. In such situations, where high torsions are present, the energy dissipation mechanism may be concentrated in a few elements, thereby placing excessive demands on their ductility. Accordingly, the accidental torsion eccentricity of $\pm 0.1b$ should be used in the analysis in addition to the computed eccentricity e_c.

20.8 LATERAL DEFORMATION DUE TO EARTHQUAKE LOADS

The difference between the design load level set by the coefficient C and the load level associated with a severe earthquake having a return period of about 300

Table 20.4. Coefficient of Reduction of Reinforced Concrete Structure Stiffness

Type of R.C. Structure	Coefficient of Reduction for Gross Uncracked Moment of Inertia
(a) Moment resisting frame	0.75
(b) Cantilever shear wall	0.60
(c) Coupled shear wall:	
– wall under axial tension	0.50
– wall under compression	0.80
– diagonally reinforced coupling beams	0.40
– conventionally reinforced coupling beams	0.20

to 400 years is virtually constant. During a severe earthquake, this additional load probably will be accommodated by the ductility of the structure. If a ductility of 4 is required to compensate for this additional loading, a structural deformation of 4 times the deformation at yield is implied. If the ductility is only 3, a K factor of 4/3 is included to bring the product to the same value of 4. However, the deformations caused by the severe earthquake are still 4 times those calculated for the load level determined without consideration of K. Therefore, the factor K must be removed in calculation of deformations, or unnecessarily high values will be obtained for the deformations resulting from loads amplified by a constant value. Similarly, the scaling factor of 0.9 included in dynamic analysis must be removed before deflections are calculated. Therefore, the Code requires that the computed structure lateral displacements be multiplied by factor $1/K$ for the equivalent static load analysis and $1/(0.9K)$ for a dynamic analysis.

Also, for computing the lateral deformation for reinforced concrete structures, the Code requires consideration of the reduction of the lateral resisting-element stiffness caused by cracking. Table 20.4 lists the coefficients of reduction for the gross uncracked moment of inertia associated with various types of reinforced concrete structures suggested by the Code. Furthermore, the stiffness of composite structures with a steel frame may be calculated using a reduction to 75% in the moment of inertia and in the cross-sectional area.

The Code defines the interstory displacement (story drift) as a horizontal displacement of any point on a floor relative to the corresponding point on the floor below. The interstory displacement is restricted not to exceed 0.5% of the story height, or 2 cm. This limit is intended to reduce discomfort to the building occupants and also to reduce secondary moments due to

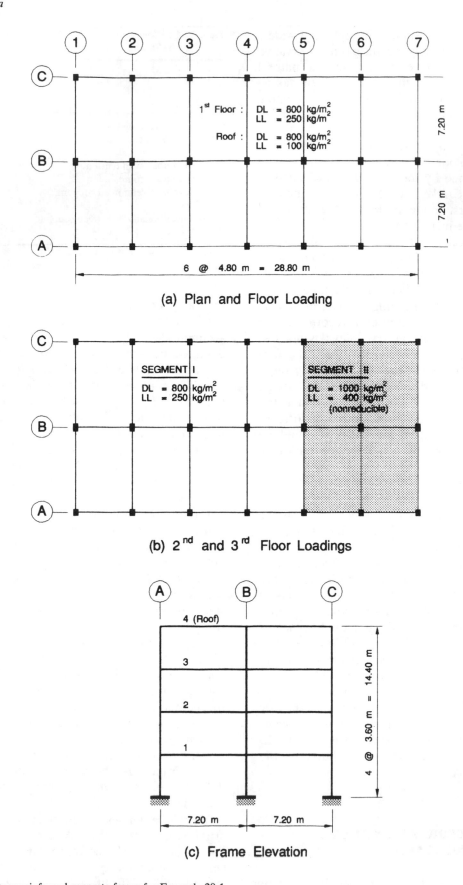

(a) Plan and Floor Loading

(b) 2nd and 3rd Floor Loadings

(c) Frame Elevation

Fig. 20.4.   Four-story reinforced concrete frame for Example 20.1

P-delta (P-Δ) effects. Secondary moments are especially important for a slender tall building with heavy gravity loads. Although the interstory displacements are limited, an explicit check of P-Δ effects is recommended in cases where this effect might be critical.

In order to avoid significant damage and alteration of the assumed response caused by pounding of adjacent buildings, the Code requires each building above the ground level to have either a separation from the boundary of 4 times the computed displacements or 0.002 times its height, whichever is larger, but not less than 3.75 cm. Buildings on the same site that are not designed to act as an integral unit or parts of buildings shall be separated from each other by a distance of at least 4 times the sum of the computed displacements or 0.004 times their height, whichever is larger, but not less than 7.5 cm.

20.9 IMPLEMENTATION OF INDONESIAN SEISMIC CODE

Example 20.1

A four-story reinforced-concrete-framed building has the plan dimensions shown in Fig. 20.4. The sizes of the exterior columns (seven each on lines A and C) are 40 cm × 50 cm; the sizes of the interior columns (seven on line B) are 40 cm × 60 cm for the bottom two stories, and, respectively, 40 cm × 40 cm and 40 cm × 50 cm for the top and second-highest stories. The height between floors is 3.60 meters. The dead load (consists of the weight of floor slab, beam, half of columns above and below the floor, partition walls, etc.) and the live load are indicated in the figure. The soil below the foundation is assumed to be soft soil. The building site is located in Seismic Zone 4. The building is intended to be used as an office. Perform the seismic analysis for this structure (in the short direction of the building) in accordance with the Indonesian Seismic Code.

Solution

1. Effective weight at various floors. The seismic design load should include parts of the reduced live load. A coefficient of reduction of 0.30 is used for an office building (Table 20.3). Therefore, the effective weights at various floors are:

$$W_1 = ((800 + 0.30 \times 250) \times 28.8 \times 14.4)/1{,}000$$
$$= 362.88 \text{ ton}$$

$$W_2 = W_3 = ((800 + 0.30 \times 250) \times 19.2 \times 14.4)/1{,}000 +$$
$$((1{,}000 + 400) \times 9.6 \times 14.4)/1{,}000 = 435.46 \text{ ton}$$

$$W_4 = ((800 + 0.30 \times 100) \times 28.8 \times 14.4)/1{,}000 = 344.22 \text{ ton.}$$

The total effective weight of the building is then

$$W_t = 362.88 + 2 \times 435.46 + 344.22 = 1{,}578.02 \text{ ton}$$

2. Fundamental period. For a reinforced concrete moment-resisting frame,

$$T = 0.060H^{3/4} \qquad \text{[eq.(20.2)]}$$

for

$$H = 4 \times 3.60 = 14.40 \text{ m (total height of the building)}$$

thus

$$T = 0.06 \times 14.40^{3/4} = 0.44 \text{ sec.}$$

3. Base shear force.

$$V = CIKW_t \qquad \text{[eq.(20.1)]}$$

$C = 0.05$ for site in Zone 4, soft soil, and T = 0.44 sec. (Figs. 20.1 and 20.2)
$I = 1.0$ for office (Table 20.1)
$K = 1.0$ for ductile reinforced concrete frame (Table 20.2)

thus

$$V = 0.05 \times 1.0 \times 1.0 \times 1{,}578.02 \text{ ton} = 78.90 \text{ ton}$$

4. Lateral seismic forces

$$F_i = \frac{W_i h_i}{\displaystyle\sum_{j=1}^{N} W_j h_j} (V - F_t) \qquad \text{[eq.(20.6)]}$$

where $F_t = 0$ ton for $H/B = 14.40/14.40 = 1.0 < 3.0$.

The values of lateral seismic forces are shown in Table 20.5.

5. Shear forces

$$V_i = F_t + \sum_{j=1}^{N} F_j \qquad \text{[eq.(20.7)]}$$

Calculated values of shear forces are shown in the last column of Table 20.5.

Table 20.5. Lateral Forces

Level	W_i (ton)	h_i (m)	$W_i h_i$ (tm)	F_i (ton)	V_i (ton)
4	344.22	14.40	4,956.77	27.73	27.73
3	435.46	10.80	4,702.97	26.31	54.04
2	435.46	7.20	3,135.31	17.54	71.58
1	362.88	3.60	1,306.37	7.31	78.89
			$\Sigma = 14{,}101.42$		

Table 20.6. Overturning Moments

Level	F_i (ton)	h_i (m)	M_i (ton-m)
4	27.73	14.40	—
3	26.31	10.80	99.83
2	17.54	7.20	294.37
1	7.31	3.60	552.06
Base	—	—	836.06

6. Overturning moments

$$M_i = F_t(h_N - h_i) + \sum_{j=i+1}^{N} F_j(h_j - h_i) \qquad [\text{eq.}(20.8)]$$

for $i = 0, 1, 2, \ldots, N - 1$.

Calculated overturning moments are shown in Table 20.6.

7. Torsional moments

$$M_{ti} = V_i e_{di} \qquad [\text{eq.}(20.9)]$$

where

for $e_c < 0.1b$ and $N \leqslant 4$: $e_d = 0$
for $0.1b \leqslant e_c < 0.3b$:
$e_{d1} = 1.5e_c + 0.05b$ [eq.(20.10)]

or

$e_{d2} = e_c - 0.05b$ [eq.(20.11)]

The lateral distance to the center of mass of a floor containing various segments of loading is given by

$$X_{CMi}^0 = \frac{\sum_{j=1}^{NS} W_{Sj} X_{Sj}}{\sum_{j=1}^{NS} W_{Sj}} \qquad (20.12)$$

where

NS = number of segments in the floor under consideration
W_S = weight of segment
X_S = distance of center of segment weight from a reference line

Because loading is symmetrical on the first floor and on the roof, the centers of mass at the first floor and at the roof are located at the center of the floor plan, thus

$$X_{CM4}^0 = 14.40 \, \text{m}$$

and

$$X_{CM1}^0 = 14.40 \, \text{m}$$

The centers of mass of the second and the third floor are obtained as follows:

$$W_I = (19.2 \times 14.4 \times (800 + 0.3 \times 250))/1,000 = 241.92 \, \text{ton}$$

$$W_{II} = (9.6 \times 14.4 \times (1,000 + 400))/1,000 = 193.54 \, \text{ton}$$

$$X_{CM2}^0 = X_{CM3}^0 = \frac{241.92 \times 9.6 + 193.54 \times 24.0}{241.92 + 193.54} = 16.0 \, \text{m}$$

The center of mass X_{CMi} of the ith level of the building, as defined by the Code, is the centroid of the resultant of the sum of all gravity loads above that level. Therefore,

$$X_{CMi} = \frac{\sum_{j=i}^{N} W_j X_{CMj}^0}{\sum_{j=i}^{N} W_j} \qquad (20.13)$$

in which

N = number of stories in the building
W_j = total weight of the jth floor
X_{CMj}^0 = center of mass of the jth floor calculated by eq.(20.12)

Table 20.7 shows results of the calculations for the centers of mass.

The center of rigidity of the ith story in the building may be calculated approximately by the following formula:

$$X_{CRi} = \frac{\sum_{j=1}^{NV} I_{Vxj} X_{Vj}}{\sum_{j=1}^{NV} I_{Vxj}} \qquad (20.14)$$

Table 20.7. Floor Center of above Mass

Level	W_i (ton)	$\sum_{j=i}^{N} W_j$ (ton)	X_{CMi}^0 (m)	$W_i \times X_{CMi}^0$ (ton-m)	$\sum_{j=i}^{N} W_j X_{CMj}^0$ (ton-m)	X_{CMi} (m)
4	344.22	344.22	14.40	4,956.77	4,956.77	14.40
3	435.46	779.68	16.00	6,967.36	11,924.13	15.29
2	435.46	1,215.04	16.00	6,967.36	18,891.49	15.55
1	362.88	1,578.02	14.40	5,225.47	24,116.96	15.28

where

NV = number of vertical elements in the ith story
I_{Vx} = moment of inertia of vertical element about its x-axis
X_V = distance of the center of vertical element from a reference line

The framing system of this case study building is symmetrically configured. Therefore, the center of rigidity of each floor is located at the center of the floor plan, i.e., $X_{CRi} = 14.40$ m. The computed eccentricity e_c of each floor is presented in Table 20.8, for each floor $e_c < 0.1b$. Because the building is four stories high, the design eccentricity e_d may be taken as zero for each floor; thus no torsional moments need be considered.

8. Story drift and lateral displacement. The structure has been modeled as a shear building to simplify the hand calculations for this example. In this case, the stiffness for a column between two consecutive floors is given by the formula

$$k = \frac{12EI_c}{h^3} \tag{20.15}$$

where

h = story height
E = modulus of elasticity of concrete = 2.8×10^6 ton/m^2
I_c = moment of inertia of column

Thus

$I_c = 1/12 \times 0.40 \times 0.40^3 = 2.133 \times 10^{-3}$ m^4
for 40 cm × 40 cm columns

$I_c = 1/12 \times 0.40 \times 0.50^3 = 4.167 \times 10^{-3}$ m^4
for 40 cm × 50 cm columns

$I_c = 1/12 \times 0.40 \times 0.60^3 = 7.200 \times 10^{-3}$ m^4
for 40 cm × 60 cm columns

By using a coefficient of reduction of 0.75 (see Table 20.4), the flexural stiffnesses of cracked sections are:

$EI_c = 0.75 \times 2.8 \times 10^6 \times 2.133 \times 10^{-3} = 4,480$ t-m^2
for 40 cm × 40 cm columns

$EI_c = 0.75 \times 2.8 \times 10^6 \times 4.167 \times 10^{-3} = 8,750$ t-m^2
for 40 cm × 50 cm columns

$EI_c = 0.75 \times 2.8 \times 10^6 \times 7.200 \times 10^{-3} = 15,120$ t-m^2
for 40 cm × 60 cm columns

The total flexural stiffness at

3rd and 4th stories is
$EI_{cTOTAL} = (14 \times 4,480) + (7 \times 8.750) = 123,970$ t-m^2

1st and 2nd stories is
$EI_{cTOTAL} = (14 \times 8,750) + (7 \times 15,120) = 228,340$ t-m^2

Table 20.8. Design Eccentricity

Level	X_{CRi} (m)	X_{CMi} (m)	e_{ci} (m)	$0.1b$ (m)	e_{di} (m)
4	14.40	14.40	0.00	2.88	0.00
3	14.40	15.29	0.89	2.88	0.00
2	14.40	15.55	1.15	2.88	0.00
1	14.40	15.28	0.88	2.88	0.00

The total stiffness at

3rd and 4th stories $k_3 = k_4 = \dfrac{12 \times 123,970}{3.60^3} = 31,885.29$ ton/m

1st and 2nd stories $k_1 = k_2 = \dfrac{12 \times 228,340}{3.60^3} = 58,729.42$ ton/m

The story drift at the ith story of the building is given by

$$\Delta_i = \frac{V_i}{k_i} \tag{20.16}$$

where

V_i = shear force at ith story
k_i = total story stiffness at ith story

As mentioned in Section 20.8, the Code requires that the structure lateral displacements computed by the equivalent static load analysis be divided by the structural-type factor K; thus,

$$\Delta_i = \frac{V_i}{Kk_i} \tag{20.17}$$

The results of the calculations are shown in Table 20.9. The story drift permitted is equal to 0.5% of the story height or 2 cm. Thus, for this four-story example, the permissible drift is $0.005 \times 360 = 1.80$ cm (0.018 m), which is much larger than the drifts shown in Table 20.9. The lateral displacement of the ith level is the sum of the story drifts below that level. Expressed mathematically,

$$d_i = \sum_{j=1}^{i} \Delta_j \tag{20.18}$$

The values of lateral displacement are shown in the last column of Table 20.9.

Table 20.9. Story Drift and Lateral Displacement

Level	V_i (ton)	k_i (ton/m)	Δ_i (m)	d_i (m)
4	27.73	31,885.29	0.00087	0.00513
3	54.04	31,885.29	0.00170	0.00426
2	71.58	58,729.42	0.00122	0.00256
1	78.89	58,729.42	0.00134	0.00134

Table 20.10. Calculations from Rayleigh's Formula

Level	W_i (ton)	d_i (m)	$W_i d_i^2$ (ton-m^2)	F_i (ton)	$F_i d_i$ (tm)
4	344.22	0.00513	9.024 E-3	27.73	0.142
3	435.46	0.00426	7.865 E-3	26.31	0.11
2	435.46	0.00256	2.854 E-3	17.54	0.045
1	362.88	0.00134	6.516 E-4	7.31	0.010
			$\Sigma = 2.040$ E-2		$\Sigma = 0.309$

9. Recalculation of fundamental period using Rayleigh formula. Table 20.10 shows detailed calculations for determining the fundamental period by using eq.(20.5). With $g = 9.81$ m/sec^2,

$$T = 6.3 \sqrt{\frac{2.040 \times 10^{-2}}{9.81 \times 0.309}} = 0.52 \text{ sec.}$$

This value of the fundamental period is greater than the value previously estimated for the natural period; i.e., $T = 0.44$ sec. However, it should be noted that since this new fundamental period is still less than 1.0 sec, the basic seismic coefficient C will remain equal to 0.05 (see Fig. 20.2, Zone 4 for soft soil curve). Accordingly, a recalculation of C based on the new fundamental period need not be considered.

10. Uniformity of story mass and stiffness. Table 20.11 presents the calculation for examining the uniformity of story mass and stiffness of the building. Note that $m_i = W_i/g$, where $g = 9.81$ m/sec^2. The average ratio of the story mass to the stiffness for the structure is

$$\left(\frac{m}{k}\right)_{\text{AVG}} = \frac{0.00388}{4} = 0.00097$$

As indicated in the last column of Table 20.11, the ratios of the story mass to the stiffness of the 1st level and the 3rd level differ by more than 25% of the average ratio for the structure. Therefore, dynamic analysis should be performed.

Table 20.11. Uniformity of Story Mass and Stiffness

Level	W_i (ton)	m_i (t-sec^2/m)	k_i (ton/m)	m_i/k_i	$\frac{(m_i/k_i)}{(m/k)_{\text{AVG}}}$
4	344.22	35.09	31,885.29	0.00110	1.13
3	435.46	44.39	31,885.22	0.00139	1.43 (>1.25)
2	435.46	44.39	58,729.42	0.00076	0.78
1	362.88	36.99	58,729.42	0.00063	0.65(< 0.75)
				Σ 0.00388	

Example 20.2

Consider again the four-story reinforced concrete building of Example 20.1. Model this building as a shear building and perform the spectral modal dynamic analysis according to the Indonesian Seismic Code. Use the response spectra provided by the Code (Fig. 20.2).

Solution

Problem data (from Example 20.1)

Seismic weights: $W_1 = 362.88$ ton
  $W_2 = W_3 = 435.46$ ton
  $W_4 = 344.22$ ton
Total seismic weight: $W_t = 1,578.02$ ton
Importance factor: $I = 1.0$ (office)
Structural-type factor: $K = 1.0$
  (ductile reinforced concrete frame)

1. Modeling the structure. Since this structure is modeled as a shear building, the stiffness matrix associated with the four lateral displacement coordinates of the building is given by

$$[K] = \begin{bmatrix} k_4 & -k_4 & 0 & 0 \\ -k_4 & k_3+k_4 & -k_3 & 0 \\ 0 & -k_3 & k_2+k_3 & -k_2 \\ 0 & 0 & -k_2 & k_1+k_2 \end{bmatrix} \quad (20.19)$$

$[K] =$

$$\begin{bmatrix} 31,885.29 & -31,885.29 & 0 & 0 \\ -31,885.29 & 63,770.58 & -31,885.29 & 0 \\ 0 & -31,885.29 & 90,614.71 & -58,729.42 \\ 0 & 0 & -58,729.42 & 117,458.84 \end{bmatrix} \text{(t/m)}$$

and the mass matrix (W_i/g) is given by

$$[M] = \begin{bmatrix} m_4 & 0 & 0 & 0 \\ 0 & m_3 & 0 & 0 \\ 0 & 0 & m_2 & 0 \\ 0 & 0 & 0 & m_1 \end{bmatrix} \quad (20.20)$$

thus

$$[M] = \begin{bmatrix} 35.09 & 0 & 0 & 0 \\ 0 & 44.39 & 0 & 0 \\ 0 & 0 & 44.39 & 0 \\ 0 & 0 & 0 & 36.99 \end{bmatrix} \text{(t sec}^2\text{/m)}$$

Note that $g = 9.81$ m/sec^2.

2. Natural periods and modal shapes. The natural frequencies and the normalized modal shapes are obtained by solving the eigenproblem

$$[[K] - \omega^2 [M]]\{\phi\} = \{0\} \quad (20.21)$$

The roots of the corresponding characteristic equation

$$|[K] - \omega^2[M]| = 0 \qquad (20.22)$$

are $\omega_1^2 = 148.158 \qquad \omega_3^2 = 2{,}179.062$
$\omega_2^2 = 1{,}003.955 \qquad \omega_4^2 = 4{,}230.634$

yielding the natural frequencies ($f = \omega/2\pi$)

$$f_1 = 1.937 \text{ cps} \qquad f_3 = 7.431 \text{ cps}$$
$$f_2 = 5.042 \text{ cps} \qquad f_4 = 10.352 \text{ cps}$$

and the natural periods ($T = 1/f$)

$$T_1 = 0.516 \text{ sec} \qquad T_3 = 0.135 \text{ sec}$$
$$T_2 = 0.198 \text{ sec} \qquad T_4 = 0.097 \text{ sec}$$

The corresponding mode shapes associated with each floor level, arranged in the columns of the modal matrix, are

$$[\Phi] = \begin{bmatrix} \phi_{41} & \phi_{42} & \phi_{43} & \phi_{44} \\ \phi_{31} & \phi_{32} & \phi_{33} & \phi_{34} \\ \phi_{21} & \phi_{22} & \phi_{23} & \phi_{24} \\ \phi_{11} & \phi_{12} & \phi_{13} & \phi_{14} \end{bmatrix} \qquad (20.23)$$

$$[\Phi] = \begin{bmatrix} 0.1119 & -0.0962 & 0.0818 & -0.0066 \\ 0.0936 & 0.0100 & -0.1144 & 0.0241 \\ 0.0561 & 0.1023 & 0.0365 & -0.0871 \\ 0.0294 & 0.0748 & 0.0583 & 0.1311 \end{bmatrix}$$

3. Spectral accelerations. The spectral accelerations for the above natural periods, obtained from the spectral chart in Fig. 20.2, are

$$S_{a1} = 0.05g \qquad S_{a3} = 0.05g$$
$$S_{a2} = 0.05g \qquad S_{a4} = 0.05g$$

4. Modal effective weights. The modal effective weights are given by eq.(4.65) of Chapter 4:

$$W_m = \frac{\left[\sum_{i=1}^{N} \phi_{im} W_i \right]^2}{\sum_{i=1}^{N} \phi_{im}^2 W_i} \qquad (20.24)$$

Values obtained for W_m ($m = 1, 2, 3, 4$) are shown in Table 20.12. This table also shows the modal effective weights as a percentage of the total seismic weight of the building.

5. Modal base shear. The total base shear force is defined by the following equation

$$V_m = W_m S_{am} \qquad (20.25)$$

Table 20.12. Modal Effective Weight and Modal Base Shear

Mode	Modal Effective Weight		Modal Base Shear
m	W_m (ton)	(%)	V_m (ton)
1	1,333.59	84.51	66.68
2	187.75	11.90	9.39
3	24.13	1.53	1.21
4	32.55	2.06	1.63
	$\Sigma = 1{,}578.02$		

Numerical values of V_m are also given in Table 20.12. Based on the SRSS technique, the total base shear force can be calculated by

$$V = \sqrt{\sum_{m=1}^{N} V_m^2} \qquad (20.26)$$

hence,

$$V = \sqrt{66.68^2 + 9.39^2 + 1.21^2 + 1.63^2} = 67.37 \text{ ton}$$

6. Scaling modal effective weight and modal base shear. The Code requires that the modal values for the effective weight and for the base shear in Table 20.12 are scaled up to a value of 90% of the base shear determined by the equivalent static load method based on the first period of the structural model. For $T = 0.516$ sec, the basic seismic coefficient C for sites in Zone 4 with soft subsoil (Fig. 20.2) is 0.05. Therefore, the base shear is

$$V = CIKW_t$$
$$= 0.05 \times 1.0 \times 1.0 \times 1{,}578.02 = 78.90 \text{ ton}$$

The design base shear is

$$V = 0.9 \times 78.90 = 71.01 \text{ ton}$$

The scaling ratio A' is

$$A' = \frac{71.01}{67.37} = 1.054$$

Table 20.13 shows the scaled results for effective modal weight and modal base shear.

Table 20.13. Scaled Values of Modal Effective Weight and Modal Base Shear

Mode	Modal Effective Weight		Modal Base Shear
m	W_m (ton)	(%)	V_m (ton)
1	1,405.60	84.51	70.28
2	197.89	11.9	9.899
3	25.43	1.53	1.27
4	34.31	2.06	1.72
	$\Sigma = 1{,}663.23$		

7. Modal seismic force. The modal seismic lateral force at the level under consideration i is expressed by eq.(4.67) of Chapter 4 as

$$F_{im} = C_{im} V_m \qquad (20.27)$$

where C_{im} is the modal seismic coefficient at level i given by eq.(4.68) as

$$C_{im} = \frac{\phi_{im} W_i}{\sum_{j=1}^{N} \phi_{jm} W_j} \qquad (20.28)$$

By application of the SRSS technique, the design seismic force can be calculated as

$$F_i = \sqrt{\sum_{m=1}^{N} F_{im}^2} \qquad (20.29)$$

The numerical values for the basic seismic coefficient C_{im} and for the seismic lateral force F_{im} are shown in Tables 20.14 and 20.15, respectively. The design seismic forces are shown in the last column of Table 20.15.

8. Modal shear force. The modal shear force V_{im} at level i of the building is equal to the sum of the seismic forces F_{im} above that level. Expressed mathematically,

$$V_{im} = \sum_{j=i}^{N} F_{jm} \qquad (20.30)$$

Table 20.16 shows the values of modal shear forces calculated by using the values of seismic forces F_{im} from Table 20.15. Using the SRSS technique, the design values for the story shear force can be calculated by using eq.(20.31). These values are shown in the last column of Table 20.16.

$$V_i = \sqrt{\sum_{m=1}^{N} V_{im}^2} \qquad (20.31)$$

9. Modal overturning moment. The modal overturning moment M_{im} at level i of the building, calculated as the sum of the moments of the seismic forces above that level, is given by

$$M_{im} = \sum_{j=i+1}^{N} F_{jm}(h_j - h_i) \qquad (20.32)$$

where $i = 0, 1, 2, \ldots, N-1$.

Table 20.17 shows the values for the modal overturning moment calculated by using eq.(20.32). The last column of this table gives the design values for overturning moments M_i calculated using the SRSS technique as

$$M_i = \sqrt{\sum_{m=1}^{N} M_{im}^2} \qquad (20.33)$$

Table 20.14. Modal Seismic Coefficient

Level	Mode 1	Mode 2	Mode 3	Mode 4
4	0.3368	−0.7713	1.8295	−0.1271
3	0.3564	0.1014	−3.2368	0.5873
2	0.2136	1.0376	1.0327	−2.1227
1	0.0933	0.6322	1.3746	2.6625

Table 20.15. Modal Seismic Force

Level	Mode 1	Mode 2	Mode 3	Mode 4	F_i (ton)
4	23.67	−7.63	2.32	−0.22	24.97
3	25.05	1.01	−4.11	1.01	25.43
2	15.01	10.26	1.31	−3.65	18.59
1	6.56	6.25	1.75	4.58	10.30

Table 20.16. Modal Shear Force

Level	Mode 1	Mode 2	Mode 3	Mode 4	V_i (ton)
4	23.67	−7.63	2.32	−0.22	24.97
3	48.72	−6.63	−1.79	0.79	49.21
2	63.73	3.63	−0.48	−2.86	63.90
1	70.29	9.88	1.27	1.72	71.01

Table 20.17. Modal Overturning Moment

Level	Mode 1	Mode 2	Mode 3	Mode 4	M_i (ton-m)
4	—	—	—	—	—
3	85.21	−27.47	8.35	−0.79	89.84
2	260.60	−51.34	1.91	2.05	265.62
1	490.03	−38.27	0.18	−8.24	491.59
Base	743.08	−2.70	4.75	−2.05	743.10

10. Modal torsional moment. The modal torsional moment M_{tim} at level i due to a design eccentricity e_{di} is calculated as

$$M_{tim} = V_{im} e_{di} \qquad (20.34)$$

where V_{im} is the modal shear force at level i. The design value for the torsional moment M_{ti} is calculated by

$$M_{ti} = \sqrt{\sum_{m=1}^{N} M_{tim}^2} \qquad (20.35)$$

However, as indicated previously in Example 20.1, the computed eccentricity e_c for each level is less than $0.1b$. Since the building is four stories in height, the design eccentricity e_d may be taken as zero. Therefore, no torsional moments need be calculated.

11. Modal story drift. The modal story drift Δ_{im} for the *i*th story of the building, modeled as a shear building, is given by

$$\Delta_{im} = \frac{V_{im}}{k_i} \qquad (20.36)$$

where

V_{im} = modal shear force of story *i*
k_i = total lateral stiffness of story *i*

As stated in Section 20.8, the Code requires the designer to multiply the computed structure lateral deformations determined from dynamic analysis by a factor of $1/(0.9K)$, where *K* is the structural-type factor. Therefore, the formula for modal story drift becomes

$$\Delta_{im} = \frac{V_{im}}{0.9Kk_i} \qquad (20.37)$$

Using the SRSS technique, the design story drift is given by

$$\Delta_i = \sqrt{\sum_{m=1}^{N} \Delta_{im}^2} \qquad (20.38)$$

Table 20.18 shows the values of modal story drift calculated by eq.(20.37). The values in the last column of this table are the design story drifts determined by eq.(20.38). It should be noted that the maximum story drift permitted by the Indonesian Code is $0.005h_i$ ($0.005 \times 3.60 = 0.0018$ m) or 2 cm (0.02 m). The design values Δ_i shown in Table 20.18 for this example are far less than these story drift limits.

12. Modal lateral displacement. The modal lateral displacement for the *i*th level of the building is given by

$$d_{im} = \sum_{j=1}^{i} \Delta_{jm} \qquad (20.39)$$

where Δ_{jm} is the modal story drift. Values for the modal lateral displacement d_{im} at level *i* calculated by using this equation are shown in Table 20.19. In the last column, this table also shows the values of design

Table 20.18. Modal Story Drift

Level	Mode 1	Mode 2	Mode 3	Mode 4	Δ_i (m)
4	0.00082	−0.00027	0.00008	−0.00001	0.00087
3	0.00170	−0.00023	−0.00007	0.00002	0.00171
2	0.00121	0.00007	−0.00001	−0.00006	0.00121
1	0.00133	0.00019	0.00002	0.00003	0.00134

Table 20.19. Modal Lateral Displacement

Level	Mode 1	Mode 2	Mode 3	Mode 4	d_i (m)
4	0.00506	−0.00024	0.00002	−0.00002	0.00513
3	0.00424	0.00003	−0.00006	−0.00001	0.00426
2	0.00254	0.00026	0.00001	−0.00003	0.00255
1	0.00133	0.00019	0.00002	0.00003	0.00134

lateral displacement d_i calculated using the SRSS technique from the following equation:

$$d_i = \sqrt{\sum_{m=1}^{N} d_{im}^2} \qquad (20.40)$$

20.10 COMPUTER PROGRAM AND EXAMPLES

An interactive computer program, called INSEC, has been developed to implement the provisions of the Indonesian Seismic Code for Building Design for earthquake-resistant design of buildings. The program provides two methods of analysis: (1) equivalent static load analysis and (2) dynamic analysis, in accordance with the provisions of the Indonesian Seismic Code. All the necessary data to determine the lateral seismic forces are calculated and printed by the program. Values for lateral seismic forces, story shear forces, overturning moments, torsional moments, story drifts, and lateral displacements are obtained for each level of the building. The following examples are presented to illustrate the application of the program.

Example 20.3

Perform the seismic analysis using the INSEC computer program for the four-story reinforced concrete building of Example 20.1. Use the response spectra provided by the Code (Fig. 20.2) for the spectral modal dynamic analysis.

Input Data and Output Results for Example 20.3

```
BUILDING INPUT DATA

NUMBER OF STORIES ---------------------------------------- 4

BUILDING DIMENSION IN FORCE DIRECTION (M) ---------------- 14.400

STORY INPUT DATA

STORY    HEIGHT    WEIGHT    STORY FLEXURAL    COMPUTED      DIM., NORMAL
                             STIFFNESS (EI)    ECCENTRICITY  TO FORCE DIR.
  #      (M)       (TON)     (TON-M^2)         (M)           (M)

  4      3.600     344.220   1.2397E+05        0.000         28.800
  3      3.600     435.460   1.2397E+05        0.890         28.800
  2      3.600     435.460   2.2834E+05        1.150         28.800
  1      3.600     362.880   2.2834E+05        0.880         28.800
```

SEISMIC INPUT DATA:

SEISMIC ZONE --- 4

TYPE OF SUBSOIL --- 1

CODE FOR ESTIMATED BUILDING PERIOD ---------------------- 1

CODE FOR IMPORTANCE FACTOR ------------------------------ 5

CODE FOR STRUCTURAL TYPE FACTOR ------------------------- 1

STATIC ANALYSIS RESULTS:

BUILDING TOTAL HEIGHT (M) --------------------------- H = 14.400

BUILDING TOTAL WEIGHT (TON) ----------------------- W = 1578.020

BUILDING ESTIMATED NATURAL PERIOD (SEC) ------------- T = 0.444

BASIC SEISMIC COEFFICIENT --------------------------- C = 0.0500

IMPORTANCE FACTOR ---------------------------------- I = 1.000

STRUCTURAL TYPE FACTOR ----------------------------- K = 1.000

TOTAL BASE SHEAR (TON) ----------------------------- V = 78.901

BUILDING NATURAL PERIOD BY RAYLEIGH'S FORMULA (SEC) -- T = 0.518

STATIC ANALYSIS LATERAL AND SHEAR FORCES

LEVEL	LATERAL FORCE (TON)	SHEAR FORCE (TON)
4	27.734	27.734
3	26.314	54.049
2	17.543	71.592
1	7.309	78.901

STATIC ANALYSIS OVERTURNING AND TORSIONAL MOMENTS

LEVEL	OVERTURNING MOMENT (TON-M)	DESIGN ECC. ED-1 (M)	TORSIONAL MOMENT-1 (TON-M)	DESIGN ECC. ED-2 (M)	TORSIONAL MOMENT-2 (TON-M)
4	-	0.000	0.000	0.000	0.000
3	99.844	0.000	0.000	0.000	0.000
2	294.419	0.000	0.000	0.000	0.000
1	552.149	0.000	0.000	0.000	0.000
BASE	836.192	-	-	-	-

STATIC ANALYSIS STORY DRIFTS AND DISPLACEMENTS

LEVEL	STORY DRIFT (M)	DRIFT RATIO	STORY DISPL. (M)
4	0.0009	0.00024	0.0051
3	0.0017	0.00047	0.0043
2	0.0012	0.00034	0.0026
1	0.0013	0.00037	0.0013

UNIFORMITY OF STORY MASS AND STIFFNESS

LEVEL #	STORY MASS (TON-SEC^2/M)	STORY STIFF. (TON/M)	M/K RATIO	AVERAGE M/K = 0.00097
4	35.089	3.1885E+04	0.00110	OK
3	44.389	3.1885E+04	0.00139	> 125%
2	44.389	5.8729E+04	0.00076	OK
1	36.991	5.8729E+04	0.00063	< 75%

STATIC ANALYSIS WARNING :

RATIO OF STORY MASS TO STIFFNESS DIFFERS MORE THAN 25% OF THE AVERAGE
---> RUN DYNAMIC ANALYSIS !!

DYNAMIC ANALYSIS:

STRUCTURAL TIME PERIODS, FREQUENCIES AND RELATIVE SPECTRAL ACCELERATIONS

MODE #	PERIOD (SEC)	FREQUENCY (CYCLES/SEC)	CIRCULAR-FREQ (RADIANS/SEC)	REL. SPECTRAL ACC. (g)
1	0.516	1.937	12.172	0.0500
2	0.198	5.042	31.679	0.0500
3	0.135	7.430	46.687	0.0500
4	0.097	10.352	65.043	0.0500

STRUCTURAL MODE SHAPES (PRINTED ONLY THE FIRST FIVE)

LEVEL	MODE 1	MODE 2	MODE 3	MODE 4
4	0.11188	-0.09620	0.08176	-0.00659
3	0.09364	0.01004	-0.11436	0.02410
2	0.05608	0.10225	0.03654	-0.08713
1	0.02941	0.07475	0.05827	0.13109

STATIC   BASE SHEAR (TON) ---------------------------- V = 78.901

DYNAMIC BASE SHEAR (TON) --------------------------- V = 67.369

DESIGN   BASE SHEAR (TON) --------------------------- V = 71.011

SCALING FACTOR ------------------------------------- A'= 1.054

SCALED MODAL EFFECTIVE WEIGHTS AND MODAL BASE SHEAR

MODE	MODAL EFFECTIVE WEIGHT (TON)	%	ACCUMULATED %	MODAL BASE SHEAR (TON)
1	1405.718	84.51	< 84.51>	70.286
2	197.870	11.90	< 96.41>	9.894
3	25.487	1.53	< 97.94>	1.274
4	34.245	2.06	<100.00>	1.712

MODAL LATERAL FORCES

LEVEL	MODE 1	MODE 2	MODE 3	MODE 4	MODE 5	LATERAL FORCE (TON)
4	23.664	-7.634	2.329	-0.218	0.000	24.975
3	25.056	1.008	-4.121	1.006	0.000	25.432
2	15.007	10.265	1.317	-3.639	0.000	18.589
1	6.559	6.254	1.750	4.563	0.000	10.296

MODAL SHEAR FORCES

LEVEL	MODE 1	MODE 2	MODE 3	MODE 4	MODE 5	SHEAR FORCE (TON)
4	23.664	-7.634	2.329	-0.218	0.000	24.975
3	48.720	-6.626	-1.792	0.789	0.000	49.208
2	63.727	3.640	-0.475	-2.850	0.000	63.896
1	70.286	9.894	1.274	1.712	0.000	71.011

MODAL OVERTURNING MOMENTS

LEVEL	MODE 1	MODE 2	MODE 3	MODE 4	MODE 5	OVERTURNING MOMENT (TON-M)
3	85.191	-27.483	8.384	-0.783	0.000	89.910
2	260.584	-51.335	1.933	2.056	0.000	265.607
1	490.001	-38.232	0.222	-8.205	0.000	491.559
BASE	743.031	-2.616	4.810	-2.041	0.000	743.053

MODAL TORSIONAL MOMENTS DUE TO ED-1

LEVEL	MODE 1	MODE 2	MODE 3	MODE 4	MODE 5	TORSIONAL MOMENT-1 TON-M
4	0.000	0.000	0.000	0.000	0.000	0.000
3	0.000	0.000	0.000	0.000	0.000	0.000
2	0.000	0.000	0.000	0.000	0.000	0.000
1	0.000	0.000	0.000	0.000	0.000	0.000

MODAL TORSIONAL MOMENTS DUE TO ED-2

LEVEL	MODE 1	MODE 2	MODE 3	MODE 4	MODE 5	TORSIONAL MOMENT-2 (TON-M)
4	0.000	0.000	0.000	0.000	0.000	0.000
3	0.000	0.000	0.000	0.000	0.000	0.000
2	0.000	0.000	0.000	0.000	0.000	0.000
1	0.000	0.000	0.000	0.000	0.000	0.000

MODAL STORY DRIFTS

LEVEL	MODE 1	MODE 2	MODE 3	MODE 4	MODE 5	STORY DRIFT (M)
4	0.0008	-0.0003	0.0001	-0.0000	0.0000	0.0009
3	0.0017	-0.0002	-0.0001	0.0000	0.0000	0.0017
2	0.0012	0.0001	-0.0000	-0.0001	0.0000	0.0012
1	0.0013	0.0002	0.0000	0.0000	0.0000	0.0013

DYNAMIC ANALYSIS STORY DRIFT RATIOS

LEVEL	DRIFT RATIO
4	0.00024
3	0.00048
2	0.00034
1	0.00037

MODAL STORY DISPLACEMENTS

LEVEL	MODE 1	MODE 2	MODE 3	MODE 4	MODE 5	STORY DISPL. (M)
4	0.0051	-0.0002	0.0000	-0.0000	0.0000	0.0051
3	0.0042	0.0000	-0.0000	0.0000	0.0000	0.0042
2	0.0025	0.0003	0.0000	-0.0000	0.0000	0.0025
1	0.0013	0.0002	0.0000	0.0000	0.0000	0.0013

Example 20.4

A 12-story reinforced-concrete-framed building has the plan dimensions shown in Fig. 20.5. The sizes of the exterior columns (seven each on lines A and C) are $40 \, \text{cm} \times 50 \, \text{cm}$; the sizes of the interior columns (seven on line B) are $40 \, \text{cm} \times 60 \, \text{cm}$ for the bottom six stories, and respectively, $40 \, \text{cm} \times 40 \, \text{cm}$ and $40 \, \text{cm} \times 50 \, \text{cm}$ for the top six stories. The height between floors is 3.60 meters. The dead load, which consists of the weight of floor slab, beam, half of columns above and half of columns below the floor, partition walls, etc., is estimated typically to be $800 \, \text{kg/m}^2$, and the live loads for the floor and the roof are assumed to be $250 \, \text{kg/m}^2$ and $100 \, \text{kg/m}^2$, respectively. The soil below the foundation is assumed to be firm soil. The building site is located in Seismic Zone 1. The building is intended to be used as a hospital. Perform the seismic analysis for this structure (in the short direction of the building) by using the INSEC computer program.

Input Data and Output Results for Example 20.4

BUILDING INPUT DATA

NUMBER OF STORIES --- 12

BUILDING DIMENSION IN FORCE DIRECTION (M) ---------------- 14.400

STORY INPUT DATA

STORY #	HEIGHT (M)	WEIGHT (TON)	STORY FLEXURAL STIFFNESS (EI) (TON-M^2)	COMPUTED ECCENTRICITY (M)	DIM., NORMAL TO FORCE DIR. (M)
12	3.600	344.200	1.2397E+05	0.000	28.800
11	3.600	362.900	1.2397E+05	0.000	28.800
10	3.600	362.900	1.2397E+05	0.000	28.800
9	3.600	362.900	1.2397E+05	0.000	28.800
8	3.600	362.900	1.2397E+05	0.000	28.800
7	3.600	362.900	1.2397E+05	0.000	28.800
6	3.600	362.900	2.2834E+05	0.000	28.800
5	3.600	362.900	2.2834E+05	0.000	28.800
4	3.600	362.900	2.2834E+05	0.000	28.800
3	3.600	362.900	2.2834E+05	0.000	28.800
2	3.600	362.900	2.2834E+05	0.000	28.800
1	3.600	362.900	2.2834E+05	0.000	28.800

SEISMIC INPUT DATA

SEISMIC ZONE --- 1

TYPE OF SUBSOIL --- 2

CODE FOR ESTIMATED BUILDING PERIOD ---------------------- 1

CODE FOR IMPORTANCE FACTOR ------------------------------ 2

CODE FOR STRUCTURAL TYPE FACTOR ------------------------- 1

STATIC ANALYSIS RESULTS

BUILDING TOTAL HEIGHT (M) ---------------------------- H = 43.200

BUILDING TOTAL WEIGHT (TON) ------------------------- W = 4336.100

BUILDING ESTIMATED NATURAL PERIOD (SEC) ------------- T = 1.011

BASIC SEISMIC COEFFICIENT --------------------------- C = 0.0747

IMPORTANCE FACTOR ----------------------------------- I = 1.500

STRUCTURAL TYPE FACTOR ------------------------------ K = 1.000

TOTAL BASE SHEAR (TON) ------------------------------ V = 485.659

BUILDING NATURAL PERIOD BY RAYLEIGH'S FORMULA (SEC) -- T = 1.358

STATIC ANALYSIS LATERAL AND SHEAR FORCES

LEVEL	LATERAL FORCE (TON)	SHEAR FORCE (TON)
12	112.856	112.856
11	62.134	174.989
10	56.485	231.475
9	50.837	282.312
8	45.188	327.500

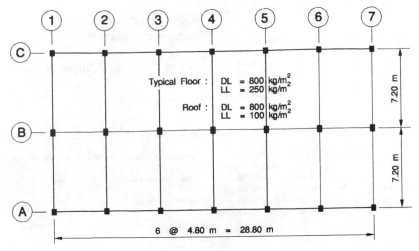

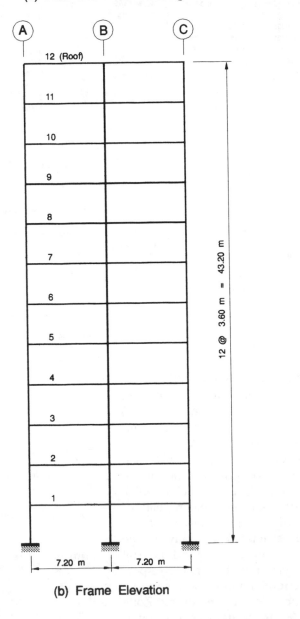

Fig. 20.5.   12-story reinforced concrete frame for Example 20.4

7	39.540	367.040
6	33.891	400.931
5	28.243	429.174
4	22.594	451.768
3	16.946	468.713
2	11.297	480.010
1	5.649	485.659

STATIC ANALYSIS OVERTURNING AND TORSIONAL MOMENTS

LEVEL	OVERTURNING MOMENT (TON-M)	DESIGN ECC. ED-1 (M)	TORSIONAL MOMENT-1 (TON-M)	DESIGN ECC. ED-2 (M)	TORSIONAL MOMENT-2 (TON-M)
12	-	1.440	162.512	-1.440	-162.512
11	406.280	1.440	251.985	-1.440	-251.985
10	1036.242	1.440	333.324	-1.440	-333.324
9	1869.552	1.440	406.529	-1.440	-406.529
8	2885.874	1.440	471.600	-1.440	-471.600
7	4064.874	1.440	528.537	-1.440	-528.537
6	5386.217	1.440	577.341	-1.440	-577.341
5	6829.568	1.440	618.010	-1.440	-618.010
4	8374.593	1.440	650.546	-1.440	-650.546
3	10000.957	1.440	674.947	-1.440	-674.947
2	11688.325	1.440	691.215	-1.440	-691.215
1	13416.363	1.440	699.349	-1.440	-699.349
BASE	15164.736	-	-	-	-

STATIC ANALYSIS STORY DRIFTS AND DISPLACEMENTS

LEVEL	STORY DRIFT (M)	DRIFT RATIO	STORY DISPL. (M)
12	0.0035	0.00098	0.0932
11	0.0055	0.00152	0.0896
10	0.0073	0.00202	0.0841
9	0.0089	0.00246	0.0769
8	0.0103	0.00285	0.0680
7	0.0115	0.00320	0.0578
6	0.0068	0.00190	0.0463
5	0.0073	0.00203	0.0394
4	0.0077	0.00214	0.0321
3	0.0080	0.00222	0.0244
2	0.0082	0.00227	0.0164
1	0.0083	0.00230	0.0083

UNIFORMITY OF STORY MASS AND STIFFNESS

LEVEL #	STORY MASS (TON-SEC^2/M)	STORY STIFF. (TON/M)	M/K RATIO	AVERAGE M/K
12	35.087	3.1885E+04	0.00110	OK
11	36.993	3.1885E+04	0.00116	> 125%
10	36.993	3.1885E+04	0.00116	> 125%
9	36.993	3.1885E+04	0.00116	> 125%
8	36.993	3.1885E+04	0.00116	> 125%
7	36.993	3.1885E+04	0.00116	> 125%
6	36.993	5.8729E+04	0.00063	< 75%
5	36.993	5.8729E+04	0.00063	< 75%
4	36.993	5.8729E+04	0.00063	< 75%
3	36.993	5.8729E+04	0.00063	< 75%
2	36.993	5.8729E+04	0.00063	< 75%
1	36.993	5.8729E+04	0.00063	< 75%

STATIC ANALYSIS WARNING :

*> TOTAL BUILDING HEIGHT IS GREATER THAN 40 M ---> RUN DYNAMIC ANALYSIS !!

*> RATIO OF STORY MASS TO STIFFNESS DIFFERS MORE THAN 25% OF THE AVERAGE
--> RUN DYNAMIC ANALYSIS !!

REFERENCES

BATHE, K.J., and WILSON, E.L. (1976) *Numerical Methods in Finite Element Analysis*. Prentice-Hall, Englewood Cliffs, NJ.

Beca Carter Hollings & Ferner Ltd. and The Indonesian Counterpart Team (1978) *Indonesian Earthquake Study*, Vols. 1–7, Bandung, Indonesia.

BERG, GLEN V. (1989) *Elements of Structural Dynamics*. Prentice-Hall, Englewood Cliffs, NJ.

CHOPRA, A. K. (1980) *Dynamics of Structures – A Primer*. Earthquake Engineering Research Institute, Berkeley, CA.

CLOUGH, R. W., and PENZIEN, J. (1975) *Dynamics of Structures*. McGraw-Hill, New York, NY.

PAZ, MARIO (1986) *Microcomputer-Aided Engineering: Structural Dynamics*. Van Nostrand Reinhold Company, New York, NY.

—— (1991) *Structural Dynamics*. Van Nostrand Reinhold Company, New York, NY.

Pedoman Perencanaan Ketahanan Gempa Untuk Rumah Dan Gedung SKBI-1.3.53.1987 UDC : 699.841 / SNI 1726-1989-F. Departement Pekerjaan Umum, Bandung, Indonesia.

Pedoman Perencanaan Pembebanan Untuk Rumah Dan Gedung SKBI-1.3.53.1987 UDC : 624.042 / SNI 1727-1989-F. Departemen Pekerjaan Umum, Bandung, Indonesia.

Peraturan Muatan Indonesia (1970) *NI-18*. Direktorat Penyelidikan Masalah Bangunan, Bandung, Indonesia.

Peraturan Perencanaan Tahan Gempa Indonesia Untuk Gedung (1983) Direktorat Penyelidikan Masalah Bangunan, Bandung, Indonesia.

SUTJIPTO, SURADJIN (1986a) "An Overview of Microcomputer Softwares for Structural Analysis." Computer Seminar – Bandung Institute of Technology, Bandung, Indonesia.

—— (1986b) "Structural Analysis on Microcomputers." Computer Symposium – Trisakti University.

—— (1987) "Computer Application Engineering on Structural Analysis." Computer Seminar – Parahyangan University.

Editor's Note: In order to save space, the computer output for dynamic analysis of this example is not reprinted here.

21

Iran

J. P. Mohsen

21.1 INTRODUCTION

The official provisions for earthquake-resistant design of buildings in Iran are contained in the *Iranian Code for Seismic Resistant Design of Buildings* (1988). The purpose of this code is to establish the minimum criteria for the seismic design and construction of buildings so that they will withstand seismic effects to the extent that loss of life is prevented and damage to buildings is minimized. This code is intended to provide structural damage resistance for buildings against an earthquake with an intensity up to VII on the Modified Mercalli Intensity (MMI) scale[1] and resistance to collapse for more severe earthquakes, up to intensity IX.

The Iranian Seismic Code is applicable to the design and construction of reinforced concrete, steel, wood, and masonry buildings and other structures. There are, however, certain special structures (dams, bridges, piers, marinas, offshore structures, and nuclear power plants) whose design procedures deviate from those prescribed in the current code. The design of these special structures should be given specific considerations. The Code states that the design base

---

[1]See Appendix I of this handbook.

acceleration shall not be less than the values prescribed by the Code. The result of an investigation to identify the seismic characteristics of a region may be used as a basis for design provided that the design base acceleration is not less than two-thirds of the value provided by the Code. The Code further recommends that traditional building materials like sun-baked brick or clay be avoided. However, the Code recognizes that the use of this type of construction in certain regions of the country may be unavoidable because of the limited availability of approved construction materials. In such cases, the Code recommends increasing the seismic resistance of buildings by following the special guidelines for reinforcement using construction materials such as wood, steel, or concrete. Construction of buildings is not allowed in any region where there is a likelihood of ground subsidence, sliding, or liquefaction due to earthquake activity. When the construction of buildings in the vicinity of a fault is unavoidable, this hazard condition should be considered in the design.

21.2 GENERAL DESIGN CRITERIA

In the event of an earthquake, the building shall behave as one unit in which the bearing elements shall be suitably tied to one another so that they will not become disjointed. The horizontal diaphragms at the floor levels of the structure should be designed to transmit the seismic forces to the vertical resisting elements. The building should be capable of resisting horizontal seismic forces in the two main orthogonal directions of the structure.

For buildings higher than 12 meters or of more than four stories, a separation joint should be placed between adjoining buildings to prevent or minimize damage due to pounding of adjacent buildings. The minimum width of such separation joints at each story level shall be equal to 1/100 of the height of that story measured from the foundation. If it becomes necessary, these joints should be filled with low-strength materials that would easily crush in the event of an earthquake.

The Code stipulates that buildings should be symmetrical in both main directions, avoiding excessive protrusions or recessed sections; symmetry should also bc rctaincd in thc building elevation. All vertical elements should be lined up so that the vertical loads carried by such elements do not have to be transmitted through any horizontal elements. All elements resisting horizontal seismic forces should be designed to transmit such forces directly to the foundation.

In order to reduce the torsional effects developed during an earthquake, the center of mass of each story should coincide with its center of stiffness. The distance between the two centers (eccentricity) should be not more than 5% of the building plan dimension in the direction normal to the seismic forces. The length of a cantilever element should be limited to 1.5 meters and should not be heavily loaded either by other structural elements or by any external installations; these limitations also apply to slender members with long unsupported spans. In order to have the center of mass of the building as low as possible, placement of heavy loads in the upper stories should be avoided. High-strength structural materials and lightweight nonstructural materials should be used to minimize building weight. The building design should exhibit a high degree of ductility with the columns specially designed to outlast the beam elements.

21.3 STRUCTURE CLASSIFICATIONS

21.3.1 Structural Categories

Buildings are classified in three categories according to their degree of occupancy importance.

Category 1 – High-Priority Buildings. There are four subgroups in this category:

(a) Those buildings whose collapse would result in heavy loss of life. Examples in this category include, schools, mosques, sport stadiums, movie theaters, department stores, airports, railroad stations, bus terminals; in general, any confined space accommodating in excess of 300 people.

(b) Buildings that serve an important function in the rescue efforts subsequent to the occurrence of an earthquake, such as hospitals, fire stations, power stations, power transmission installations, communication centers, radio and television stations, water treatment plants, police headquarters, and facilities used as rescue stations.

(c) Buildings that house national archives or important works of art such as museums and libraries.

(d) Buildings of an industrial nature such as refineries, oil or fuel storage tanks, and gas distribution centers, which could cause widespread air pollution or fire, if damaged.

Category 2 – Medium-Priority Buildings. This category includes all buildings whose collapse would result in considerable damage and heavy loss of life: residential properties, commercial buildings, hotels, warehouses, multilevel parking garages, and industrial buildings not included in Category 1.

Category 3 – Low-Priority Buildings. Two groups of buildings are included in this category:

(a) Buildings whose collapse would result in relatively minor damage and minimal loss of life, such as grain silos and other agricultural structures.

(b)   Temporary structures intended for less than two years of service.

21.3.2 Classification Based on Geometric Shape

Buildings are classified into two categories, based on regular or irregular geometric shape. To be considered regular a building should satisfy the following criteria for plan and elevation.

21.3.2.1 Regularity in Plan

(a)   Buildings should be symmetrical or nearly symmetrical about the main axes of the structure. Generally, the primary members resisting seismic forces are placed in the direction of these axes. Any existing recess shall not exceed 25% of the dimension of the building in that same direction.

(b)   The eccentricity between the center of mass and the center of stiffness in any story shall not be greater than 20% of the building dimension in any of the two main directions.

21.3.2.2 Regularity in Elevation

(a)   The mass distributed along the elevation of the building should be generally uniform, such that any variation of mass between adjacent floors does not exceed 50%, except for the attic of the building.

(b)   Lateral stiffness of any story shall not be less than 30% of that of the story directly below; furthermore, it shall not be less than 50% of the lateral stiffness of the three stories immediately below.

21.4 DESIGN PRINCIPLES

Buildings should be designed to resist seismic forces and wind forces independently, with the prevailing forces controlling the design. Generally, only the horizontal component of the earthquake forces is considered in the design. The vertical component of such forces is taken into account in the design of balconies constructed as cantilevered elements, as discussed in Section 21.7.8. Calculations of the lateral forces shall be carried out for two orthogonal directions. For regular buildings, calculations are made in each of these two directions regardless of the applied force in the other directions. Lateral seismic forces shall be resisted by shear walls, bracings, moment-resisting frames, or a combination thereof. For buildings taller than 50 meters or those with 15 stories or more, lateral seismic forces should not be expected to be fully resisted by shear walls or bracing members. In such buildings, the moment-resisting frames should be designed to carry at least 25% of these forces.

Table 21.1.   Percentage of Design Live Load Added to Dead Load

Live Load Location	Live Load Percentage
Roofs with a slope of 20% or more	0
Roofs with snow accumulation possibilities having a slope of less than 20%	20
Hotels, residential, and office buildings	20
Hospitals, schools, store buildings, and assembly halls	40
Warehouses and libraries	60
Reservoirs	100

21.4.1 Live Loads

The portion of the live load to be taken into account in evaluating the lateral seismic forces is summarized in Table 21.1 as a percentage of the design vertical live load.

21.5 DESIGN METHODOLOGIES

The Code provides the following three methods for the seismic analysis and design of buildings:

(a)   Equivalent static analysis method
(b)   Pseudo-dynamic analysis method incorporating modal analysis using design response spectrum
(c)   Dynamic analysis method using accelerogram data from past earthquakes.

The equivalent static analysis method is deemed appropriate for regular buildings, as long as the overall height does not exceed 80 meters. The pseudo-dynamic analysis method or the dynamic analysis method must be used for irregular buildings or for buildings that exceed 80 meters. However, the equivalent static analysis can be applied to an irregular building if its height does not exceed 18 meters and there are no more than five stories. The sections that follow present only the equivalent static analysis method. The other two methods contain provisions that are based on material described in Chapter 4.

21.6 EQUIVALENT STATIC ANALYSIS

In this method, lateral seismic forces that depend on the fundamental period of vibration of the building are evaluated using the design response spectrum.

21.6.1 Base Shear Force

The minimum base shear force V in each main direction of the building is calculated using eq.(21.1).

$$V = CW \qquad (21.1)$$

where

W = total seismic weight of the building (includes dead load plus percentage of live load), indicated in Table 21.1

C = seismic coefficient obtained by eq.(21.2).

$$C = \frac{ABI}{R} \qquad (21.2)$$

where

A = design base acceleration in terms of gravity acceleration g

B = response coefficient obtained through the design response spectrum

I = importance factor

R = behavior coefficient.

The Code specifies that the value of the seismic coefficient C should not be less than 10% of the design base acceleration A.

21.6.2 Base Acceleration

The design base acceleration is evaluated in various geographical regions of the country according to the seismic risk level identified in each region, as indicated in Table 21.2. The relative seismic risk for various cities and regions of the country is given in the Appendix (A21) of this chapter.

21.6.3 Response Coefficient

Building response coefficient B is an indication of the building response relative to the design base

Table 21.2. Design Base Acceleration (A)

Region	Description	A
1	High relative seismic risk	0.35
2	Intermediate relative seismic risk	0.25
3	Low relative seismic risk	0.20

Table 21.3. Characteristic Period for Different Type Categories

Soil Type Category	T_0
I	0.3
II	0.4
III	0.5
IV	0.7

acceleration. It is calculated using eq.(21.3).

$$B = 2.0\left(\frac{T_0}{T}\right)^{2/3} \qquad (21.3)$$

where $0.6 \leqslant B \leqslant 2.0$

and

T = the fundamental period of the building, in seconds

T_0 = the characteristic period of the site based on the type of soil (Table 21.3).

Fig. 21.1 shows a plot of the design response spectrum for each soil type category.

21.6.4 Soil Classification

The four soil type categories are defined in Table 21.4.

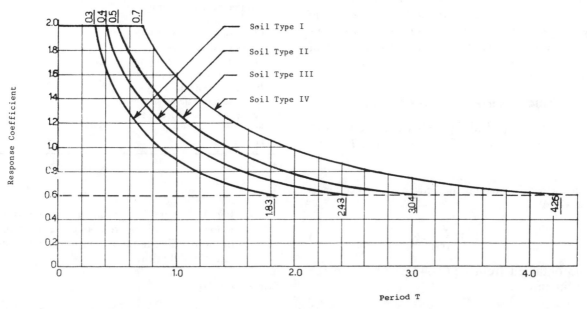

Fig. 21.1. Response spectral coefficient

Table 21.4. Soil Type Categories

Category	Soil Type
I	(a) Igneous rocks (coarse- and fine-grained texture), hard and stiff sedimentary rocks, and massive metamorphic rocks (gneiss-crystalline silicate rocks).
	(b) Conglomerate beds, compact sand and gravel and stiff clay (argillite) beds up to 60 meters from the bedrock.
II	(a) Loose igneous rocks (such as tuff), friable sedimentary rocks, foliated metamorphic rocks, and rocks that have been loosened by weathering.
	(b) Conglomerate beds, compact sand and gravel and stiff clay (argillite) beds where the soil thickness extends more than 60 meters from the bedrock.
III	(a) Rocks that have been disintegrated by weathering.
	(b) Beds of gravel and sand with weak cementation and/or uncemented unindurated clay (clay stone) where the soil thickness extends less than 10 meters from the bedrock.
IV	(a) Soft and wet deposits resulting from high level of water table.
	(b) Gravel and sand beds with weak cementation and/or uncemented unindurated clay (clay stone) where the soil thickness extends more than 10 meters from the bedrock.

21.6.5 Building Importance Factor

The building importance factor I is given in Table 21.5 for the three building categories presented in Section 21.3.1.

21.6.6 Building Behavior Coefficient

The building behavior coefficient is an indication of the energy-absorbing capacity of the building. This coefficient reflects such characteristics as the structural materials used, damping, structural type, and the ductility of the building. The building behavior coefficient is presented in Table 21.6 for various types of structures, based on the lateral resistance system.

21.6.7 Fundamental Period of Vibration

The following empirical equations may be used to estimate the fundamental period of vibration T of the structure.

(a) The general equation applied to all structures is eq.(21.4).

$$T = 0.09 \frac{H}{\sqrt{D}} \qquad (21.4)$$

The value obtained from eq.(21.4) is subject to the following limitation:

$$T \leqslant 0.06 H^{3/4}$$

Table 21.5. Building Importance Factor (I)

Building Category	Importance Factor
Category 1	1.2
Category 2	1.0
Category 3	0.8

Table 21.6. Behavior Coefficient of Buildings (R)

System of Structure	Lateral Force Resisting System	R
1. Bearing Walls Systems: The walls support all or the most parts of the vertical loads, and resistance to seismic lateral force is provided by shear walls.	(a) Reinforced concrete shear walls	5
	(b) Reinforced masonry shear walls	4
2. Simple Space Frame System: The space frame with simple connections supports all vertical loads and resistance to seismic lateral forces is provided by shear walls shear walls or bracings.	(a) Reinforced concrete shear walls	7
	(b) Bracings	7
	(c) Reinforced masonry shear walls	5
3. Moment-Resisting Space Frame System: The space frame alone supports all vertical loads and seismic lateral forces (without shear walls or bracings).	(a) Steel moment-resisting space frame	6
	(b) Reinforced concrete moment-resisting space frame	5
4. Combined System of Moment-Resisting Space Frame and Shear Walls or Bracings: The space frame supports the vertical loads and resistance to seismic lateral forces is provided jointly by the space frame and the shear walls, or by the space frame and the bracings. The seismic lateral forces are distributed between the space frame and the bracings in proportion to their rigidities; but in any case, the frame alone should be capable of supporting at least 25% of the seismic lateral forces.	(a) Moment-resisting frame + reinforced concrete shear walls	8
	(b) Moment-resisting frame + bracings	8

(b) For buildings with moment-resisting frames, the following equations are applicable if the movement of the building frame is not prevented by any other structural element.

For steel frames:

$$T = 0.08 H^{3/4} \qquad (21.5)$$

For reinforced concrete frames:

$$T = 0.07H^{3/4} \qquad (21.6)$$

In the above equations, H indicates the height of the building measured from the base level, and D represents the dimension of the building in the direction under consideration. Both D and H in these equations are expressed in meters. The fundamental period of vibration of the building can also be evaluated using a dynamic analysis method based on both the structural characteristics and the deformation characteristics of the load-bearing elements. However, the fundamental period used in the design should not exceed 1.25 times that of the value obtained using the empirical relationships expressed as eqs.(21.4) through (21.6).

21.7 LATERAL FORCE DISTRIBUTION

21.7.1 Distribution Along Building Height

The calculated base shear force V in eq.(21.1) is distributed along the height of the building, according to eq.(21.7).

$$F_i = (V - F_t)\frac{W_i h_i}{\sum_{j=1}^{n} W_j h_j} \qquad (21.7)$$

where

F_i = lateral force at level i
W_i = weight of level i including the weight of the floor and its live load, and half of the weight of the walls and columns immediately above and below that floor
h_i = height of level i measured from the base
n = number of stories
F_t = additional lateral force at the top level calculated by eq.(21.8).

$$F_t = 0.07TV \qquad (21.8)$$
$$F_t \leq 0.25V$$

It should be noted that if $T \leq 0.7$ sec, then F_t can be taken as zero.

21.7.2 Distribution Among Lateral Load-Resisting Elements

In each story, the shear force shall be distributed among the vertical elements of the lateral force-resisting system in proportion to their respective rigidities.

21.7.3 Torsional Moment Caused by Lateral Forces

The effect of torsional moment shall be evaluated using eq.(21.9).

$$T_i = \sum_{j=1}^{n} e_{ij}F_j + T_a \qquad (21.9)$$

where

T_i = torsional moment at the ith story level
e_{ij} = horizontal eccentricity between the center of rigidity in story i and the center of mass at level j
F_j = lateral seismic force at level j
T_a = accidental torsional moment.

The accidental torsional moment accounts for possible random shifting of masses at various levels of the building. The magnitude of the torsional moment shall be taken to be at least equal to the product of the story shear force times 5% of the dimension of the building in the direction perpendicular to the lateral seismic forces.

It is not necessary to consider the torsional moment for buildings of no more than five stories or no more than 18 meters high if the horizontal distance between the center of rigidity of each story and the center of mass of the stories above is less than 5% of the building dimension normal to the action of the lateral forces. If these requirements are not met, the torsional effects need to be considered. The accidental torsional moment, however, may be disregarded.

21.7.4 Story Shear Force

The shear force V_i at any story i is given by the sum of the lateral seismic forces above that story; that is,

$$V_i = \sum_{j=i}^{N} F_j \qquad (21.10)$$

21.7.5 Overturning Moments

The stability of the building against overturning must be ensured. The overturning moment M_i at level i of the building is calculated by summing the products of the lateral forces above that level by the corresponding height measured from the same level as follows:

$$M_i = F_t(h_N - h_i) + \sum_{j=i+1}^{N} F_j(h_j - h_i) \qquad (21.11)$$

where $i = 0, 1, 2, \ldots, N-1$.

A minimum factor of safety against overturning shall be taken as 1.75. In calculating the resisting moment, the equilibrium load is taken to be the same as the vertical load used to calculate the lateral forces plus the weight of the foundation and the backfill, if any.

21.7.6 Lateral Displacement and Story Drift

The story drift or lateral displacement of each story, relative to those immediately above and below, shall not exceed 0.005 of the story height. In calculating this displacement, the effect of the lateral forces is coupled with the effect of torsional moment.

21.7.7 Building Components and Building Additions

The seismic lateral force for building components and building additions shall be calculated using eq.21.12.

$$F_p = AB_pIW_p \qquad (21.12)$$

where

A = design base acceleration in terms of gravity acceleration evaluated in accordance with Section 21.6.2

I = importance factor, evaluated in accordance with Section 21.6.5

W_p = weight of the component or the building addition; in the case of warehouses and libraries, all shelf spaces should be considered fully stocked when their weight is added to the dead load

B_p = the component factor, a constant identified in Table 21.7.

21.7.8 Vertical Seismic Action

The vertical component of the seismic force shall also be considered in the case of cantilevered elements

Table 21.7. Values of Component Factor (B_p)

Building Components or Added Portions	Direction of Horizontal Force	B_p
Outer and inner walls of the building, and partition walls	Perpendicular to wall surface	0.7
Parapets and cantilever walls	Perpendicular to wall surface	2.0
Outside and inside ornamental elements or components of the building	Any direction	2.0
Reservoirs, towers, chimneys, machinery, and equipment if attached to the building, or made part of it, and suspended ceilings	Any direction	1.0
Connections of prefabricated structural elements	Any direction	1.0

of structures. The vertical component of the seismic force for these elements shall be determined by eq.(21.13).

$$F_v = \frac{2AI}{R_v} W_p \qquad (21.13)$$

where

A = design base acceleration in terms of gravity acceleration evaluated in accordance with Section 21.6.2

I = importance factor, evaluated in accordance with Section 21.6.5

W_p = dead load plus total live load

R_v = reaction coefficient which shall be taken equal to 2.4 for cantilevered steel beams, and 2.0 for reinforced concrete elements.

It is necessary to consider the vertical component of the seismic force in both directions, upward and downward.

21.8 EXAMPLE 21.1

Fig. 21.2 represents a model for a six-story braced steel building to be analyzed according to the equivalent static analysis method of the Iranian Seismic Code. Weights of 20,000 kN are attributed to each level of the building except at the roof where the load is 10,000 kN. The structure is classified as a Category 2 building in a high seismic hazard region with Type I foundation soil. The lateral force resisting system is reinforced concrete shear walls. The building, with a plan dimension of 6 m by 48 m, is regular in plan and in elevation.

Solution

Fundamental period

$T = 0.08H^{3/4}$ [eq.(21.5)] (for braced steel frame)
$T = 0.08 \times 33^{3/4} = 1.10 \sec$

Characteristic period of the soil

$T_0 = 0.3$ (Table 21.3 for soil type I)

Building importance factor

$I = 1.0$ (Table 21.5 for Category 2)

Behavior coefficient

$R = 7$ (Table 21.6 for steel braced frame)

Design base acceleration

$A = 0.35$ (Table 21.2 for high seismic risk)

Response coefficient

$$B = 2.0\left(\frac{T_0}{T}\right)^{2/3} \qquad \text{[eq.(21.3)]}$$

where $0.6 \leqslant B \leqslant 2.0$

$$B = 2.0\left(\frac{0.3}{1.10}\right)^{2/3} = 0.84$$

Seismic coefficient

$$C = \frac{ABI}{R} \qquad \text{[eq.(21.2)]}$$

$$C = \frac{0.35 \times 0.84 \times 1.0}{7.0} = 0.042 > 0.1 \times 0.35 = 0.035$$

Base shear force

$$V = CW \qquad \text{[eq.(21.1)]}$$
$$= 0.042 \times 110{,}000 = 4{,}622 \text{ kN}$$

Seismic lateral forces

$$F_t = 0.07TV \qquad \text{[eq.(21.8)]}$$
$$= 0.07 \times 1.1 \times 4{,}622 = 356 \text{ kN} < 0.25V = 1{,}155 \text{ kN}$$

$$F_i = (V - F_t)\frac{W_i h_i}{\displaystyle\sum_{j=1}^{n} W_j h_j} \qquad \text{[eq.(21.7)]}$$

$$V - F_t = 4{,}266 \text{ kN}$$

Calculated values of the seismic lateral forces are given in Table 21.8.

Story shear forces

$$V_i = \sum_{j=i}^{N} F_j \qquad \text{[eq.(21.10)]}$$

Calculated values of the story shear forces are shown in Table 21.8.

Overturning moments

$$M_i = F_t(h_N - h_i) + \sum_{j=i+1}^{N} F_j(h_j - h_i) \qquad \text{[eq.(21.11)]}$$

where $i = 0, 1, 2, \ldots, N - 1$.
   Calculated values of the overturning moments are given in Table 21.8.

Torsional moments

$$T_i = 0.05DV_i \qquad \text{(accidental torsional moment)}$$
$$= 0.05 \times 48V_i = 2.4V_i$$

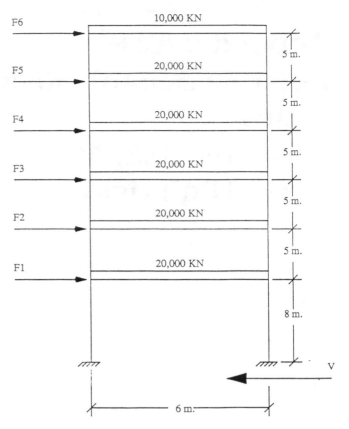

Fig. 21.2. Plane frame used to model building of Example 21.1

Table 21.8. Seismic Lateral Forces (F_i), Story Shear Forces (V_i), Overturning Moments (M_i), and Torsional Moments (T_i)

Level i	H_i (m)	W_i (kN)	F_i (kN)	V_i (kN)	M_i (kN-m)	T_i (kN-m)
6	33	10,000	1,017.1*	1,017.1	—	2,441
5	28	20,000	1,121.5	2,138.6	5,085	5,133
4	23	20,000	921.2	3,059.8	15,778	7,344
3	18	20,000	720.9	3,780.7	31,077	9,074
2	13	20,000	520.7	4,301.4	49,981	10,323
1	8	20,000	320.4	4,621.8	71,490	11,092
0	0	—	—	—	108,462	—

*Including the additional force F_t at the top of the building.

Calculated values of the torsional moments are shown in Table 21.8.

REFERENCES

CAKMAK, A. S. (1987) *Ground Motions and Engineering Seismology.* pp. 59–75. Elsevier, New York, NY.

Iranian Code for Seismic Resistant Design of Buildings (1988) Building and Housing Research Center, Publication No. 82, 4th ed. Iran.

OKAMOTO, SHUNZO (1973) *Introduction to Earthquake Engineering.* John Wiley and Sons, New York, NY.

Appendix A21:
Relative Seismic Risk in Various Iranian Cities
(* = Low risk, ** = Intermediate risk, *** = High risk)

Abadan, Khoozestan	*	Baneh, Kordestan	***
Abadeh, Fars	**	Bastam, Semnan	***
Abarkooh, Fars	**	Bazman, Seestan Baloochestan	**
Abegarm, Zanjan	***	Beejar, Kordestan	**
Abhar, Zanjan	***	Beerjand, Khorasan	***
Aghajari, Khoozestan	***	Behabad, Yazd	***
Agh Ghale, Mazandaran	***	Behbahan, Khoozestan	***
Aghleed, Fars	***	Behshahr, Mazandaran	***
Ahar, East Azarbaejan	**	Bojnoord, Khorasan	***
Ahrom, Booshehr	***	Booshehr, Booshehr	**
Ahvaz, Khoozestan	*	Borazjan, Booshehr	***
Aligoodarz, Lorestan	**	Boroojen, Charmahal Bakhtyari	***
Amol, Mazandaran	***	Boroojerd, Lorestan	***
Anar, Yazd	***	Boshravyek, Khorasan	***
Anarak, Esfahan	***	Bostan, Khoozestan	**
Andymeshk, Khoozestan	***	Bostan Abad, East Azarbaejan	***
Arak, Markazi	**	Boyeen Zahra, Zanjan	***
Ardebeel, East Azarbaejan	***	Chah Bahar, Seestan Baloochestan	***
Ardekan, Yazd	**	Chaloos, Mazandaran	***
Ardel, Charmahal Bakhtyari	***	Charak, Hormozgan	***
Ardestan, Esfahan	***	Chenaran, Khorasan	***
Asfarayn, Khorasan	***	Dahbeed, Fars	***
Astara, Gilan	***	Damavand, Tehran	***
Babol, Mazandaran	***	Damghan, Semnan	***
Babolsar, Mazandaran	***	Darab, Fars	***
Bafgh, Yazd	***	Daran, Esfahan	**
Baft, Kerman	***	Dargaz, Khorasan	**
Baj Gyran, Khorasan	***	Dehloran, Ilam	***
Bakhtaran, Bakhtaran	***	Delijan, Markazi	***
Bam, Kerman	***	Dezfool, Khoozestan	***
Bandar Abas, Hormozgan	***	Dorood, Lorestan	***
Bandar Anzaly, Geelan	***	Esfahan, Esfahan	**
Bandar Daylam, Booshehr	**	Eshtehard, Tehran	***
Bandar Gaz, Mazandaran	***	Fariman, Khorasan	**
Bandar Khameer, Hormozgan	***	Farooj, Khorasan	***
Bandar Lengeh, Hormozgan	***	Farsan, Charmahal Bakhtyari	***
Bandar Torkaman, Mazandaran	***	Fasa, Fars	***

Fasham, Tehran	***	Malaer, Hamedan	**
Ferdos, Khorasan	***	Maragheh, East Azarbaejan	**
Firooz Abad, East Azarbaejan	***	Marand, East Azarbaejan	***
Firooz Abad, Fars	***	Mard Abad, Markazi	***
Firooz Kooh, Tehran	***	Marivan, Kordestan	***
Gachsaran, Kohkilueh Boer Ahmad	***	Marvdasht, Fars	***
Garmsar, Semnan	***	Mashad, Khorasan	***
Ghaem, Shahr, Mazandran	***	Masjed Solayman, Khoozestan	***
Ghaen, Khorasan	***	Mehran, Ilam	**
Ghasreshireen, Bakhtaran	***	Meshgeen Shahr, East Azarbaejan	**
Ghazveen, Zanjan	***	Myaneh, East Azarbaejan	***
Gheshm, Hormozgan	***	Mynab, Hormozgan	***
Ghom, Tehran	***	Naghadeh, West Azarbaejan	**
Ghomsheh, Esfahan	**	Nahavand, Hamadan	***
Ghoochan, Khorasan	***	Najaf Abad, Esfahan	**
Golbaf, Kerman	***	Natanz, Esfahan	
Golpaygan, Esfahan	**	Nayn, Esfahan	***
Gon Abad, Khorasan	***	Nayreez, Fars	***
Gonaveh, Booshehr	**	Nayshaboor, Khorasan	***
Gonbad Kavoos, Mazandaran	***	Noor, Mazandaran	***
Gorgan, Mazandran	***	Noshahr, Mazandaran	***
Haftgel, Khoozestan	***	Nosrat Abad, Seestan Baloochestan	***
Haji Abad, Hormozgan	***	Orumyeh, West Azarbaejan	**
Hamedan, Hamedan	***	Pars Abad, East Azarbaejan	***
Hashtpar, Geelan	***	Peeran Shahr, West Azarbaejan	***
Hendijan, Khoozestan	*	Pol Dokhtar, Lorestan	***
Hovayzeh, Khoozestan	*	Polur, Mazandaran	***
Ilam, Ilam	***	Rafsanjan, Kerman	***
Iranshahr, Seestan Baloochestan	***	Ramsar, Mazandran	***
Islam Abad, Bakhtaran	***	Rasht, Geelan	***
Ivanak, Semnan	***	Rashtkhar, Khorasan	***
Izeh, Khoozestan	***	Ravar, Kerman	***
Jahrom, Fars	***	Ray, Tehran	***
Jasak, Hormozgan	***	Robat, Khorasan	***
Jiroft, Kerman	***	Roodhen, Tehran	***
Jolfa, East Azarbaejan	**	Roodsar, Geelan	***
Kangan, Booshehr	***	Sabzevar, Khorasan	***
Karaj, Tehran	***	Saghand, Yazd	**
Kashan, Esfahan	***	Saghez, Kordestan	**
Kashmar, Khorasan	***	Salafchegan, Markazi	***
Kazeroon, Fars	***	Salmas, West Azarbaejan	***
Keesh, Hormozgan	***	Sanandaj, Kordestan	**
Kerman, Kerman	***	Sarab, East Azarbaejan	***
Khalkhal, East Azarbaejan	***	Saravan, Seestan Baloochestan	***
Khansar, Esfahan	**	Sarcheshmeh, Kerman	***
Khark, Booshehr	*	Sarakhs, Khorasan	**
Khash, Seestan Baloochestan	***	Sary, Mazandaran	***
Khomain, Markazi	**	Saveh, Markazi	***
Khoram Abad, Lorestan	***	Sedeh, Khorasan	***
Khoram Shahr, Khoozestan	*	Seerjan, Kerman	**
Khoy, West Azarbaejan	***	Semnan, Semnan	***
Koohak, Seestan Baloochestan	***	Shadegan, Khoozestan	*
Koohpaeh, Esfahan	***	Shahdad, Kerman	***
Laheejan, Geelan	***	Shahre Kord, Charmahal Bakhtyari	***
Lar, Fars	***	Shahrood, Semnan	***
Lavan, Hormozgan	***	Shazand, Markazi	**
Mahabad, West Azarbaejan	**	Shiraz, Fars	***
Mahalat, Markazi	**	Shirvan, Khorasan	***
Mahmood Abad, Mazandaran	***	Shooshtar, Khoozestan	***
Mahshahr, Khoozestan	*	Soltanyeh, Zanjan	***
Makoo, West Azarbaejan	***	Susangerd, Khoozestan	*

Tabas, Khorasan	***	Toyserkan, Hamadon	**
Tabriz, East Azarbaejan	***	Varameen, Tehran	***
Tafresh, Central	***	Yasooj, Kohkylueh Boer Ahmad	***
Teeran, Esfahan	**	Yazd, Yazd	**
Tehran, Tehran	***	Zabol, Seestan Baloochestan	**
Tonekabon, Mazandaran	***	Zahedan, Seestan Baloochestan	**
Torbat Haydaryeh, Khorasan	***	Zanjan, Zanjan	***
Torbat Jam, Khorasan	***	Zarand, Kerman	***

22

Israel

Jacob Gluck

22.1 INTRODUCTION

Israel is located between the Mediterranean Sea on the west and an almost continuous desert belt on the south and east; it is about 280 miles long and about 70 miles wide. Israel contains all the major characteristics of neighboring countries; coastal plains, mountain ranges, plateaus, and basins that culminate in the Rift Valley (*Encyclopedia Judaica, Vol. 9* 1972).

Accounts of destructive earthquakes in Israel extend far into the past and into biblical times. The descriptions of earthquakes in the Bible, especially those given by the prophets, indicate that such cataclysms occurred from time to time and that the people were therefore familiar with their consequences (*Encyclopedia Judaica, Vol. 6* 1972). The almost scientific description of the phenomenon of earth dislocation and cracking related to the prophecy of wrath (Zech. 14:4–5) might have been based on personal experience of an earthquake. In the year 31 B.C., a disastrous earthquake in Judea claimed 10,000 to 30,000 victims according to biblical accounts.

During the last 2,000 years, earthquakes in the region of Israel have been recorded in greater detail. These records reveal that, on the average, several damaging earthquakes have occurred in each century. Seismological observations have been operated by the Geological Survey of Israel since 1955, by the Weizmann Institute of Science since 1969 and by the Seismological Division in the Institute for Petroleum Research and Geophysics Science since 1979. Recent studies indicate that most earthquake epicenters are situated in, or near, the Jordan Rift Valley, an area where the two most destructive earthquakes since the eighteenth century originated. The earthquake of January 1, 1837, whose epicenter was near Safed, claimed about 5,000 victims and ruined most of the old city; it was strongly felt from Beirut to Jerusalem. On July 11, 1927, an earthquake occurred north of Jericho violently affecting vast areas from Lebanon to the

Negev, killing 350 people in Transjordan, and ruining some 800 buildings (mainly in Shechem). The 1927 earthquake was estimated to have had a magnitude $M = 6\frac{1}{4}$ on the Richter scale. Several other strong earthquakes have since occurred in the region.

The first Israeli regulation for earthquake-resistant design of buildings was published in 1961 (*Israel Standard, IC 413* 1961). The 1961 Code introduced in Israel the constant seismic coefficient method. This code was not official, thus its use was optional. A revised version of the IC-413 Code was published in 1975 (*Israel Standard, IC 413* 1975, with amendment, 1982); the first seismic risk map for Israel (Fig. 22.1) was prepared in 1982 by the Seismological Division of the Institute for Petroleum Research and Geophysics and subsequently adopted as the official seismic map (*Israel Standard, IC-413*, 1975).

A revised version of the 1975 Code for seismic-resistant design of dwellings and public buildings in Israel has recently been released (*Israel Standard, IC 413* 1994); the main provisions contained in it will be presented in this chapter.

22.2 CONFIGURATION CHARACTERISTICS OF REGULAR STRUCTURES

The Israeli Seismic Code defines a "regular" structure as one having the following characteristics:

- A nearly symmetric layout in plan of the lateral resisting structural elements
- The distance between mass and rigidity centers is less than 15% of the structure's plan dimension in the direction normal to the seismic action considered
- No horizontal or vertical discontinuities in the lateral structural resisting systems
- The maximum calculated story drift, including the torsional effect, at one end of the story is not larger than 1.5 times the maximum drift at the opposite end
- No soft or weak stories
- For framed structures, the ratio between the story shear capacity and its design shear force does not differ from story to story by more than 20%, except for the uppermost story in all types of building structures or the uppermost two stories in structures higher than six stories
- No "heavy" floor, with a mass 50% larger than that of the story below.

Structures that violate any of the above conditions are classified as "irregular" structures.

A soft story is defined either as a story having a horizontal stiffness smaller than 70% of the stiffness of the story above it, or as a story having a total length of infill or concrete walls, in the two main directions, smaller than 50% of that in the story above it.

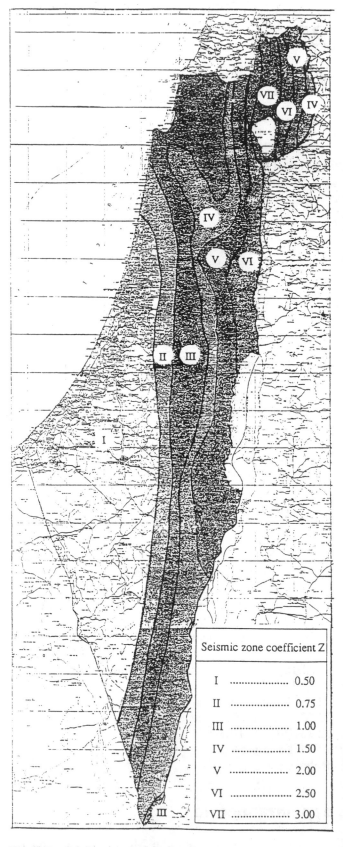

Fig. 22.1.   Seismic zone map for Israel

Seismic zone coefficient Z	
I	0.50
II	0.75
III	1.00
IV	1.50
V	2.00
VI	2.50
VII	3.00

Table 22.1. Reduction Factor for Reinforced Concrete Structures (*K*)

Resisting System	Ductility Level		
	Low	Medium	High
Moment resisting frames	4	5.5	7
Braced frames	3.5	4.5	6
Shear walls	4	5.5	7
Dual systems	3.5	5	6

Table 22.2. Reduction Factor for Steel Structures (*K*)

Resisting system	*K*
Ductile space frames	8
Ductile dual systems	7
Braced frames	4

A weak story is defined as a story having, in any direction, a shear capacity less than 80% of that of the story above it. Structures having a weak story, with a shear capacity of 65% to 80% of that of the story above it, are allowed, provided that there are no more than two stories above with a total height of not more than 9 m. This limitation in building height is not applicable to a building with a weak story where that story and the next stories are capable of resisting a lateral load equal to 0.75*K* times the calculated reduced shear force. The reduction factor *K* is given in Tables 22.1 and 22.2.

The Israeli Seismic Code establishes that structures should be designed to resist the action of seismic equivalent lateral loads applied statically at the mass centers at each level of the building. In establishing these lateral forces, the weight W_i considered at each level *i* is determined by

$$W_i = G_i + k_g(Q_i + A_i q_i) \tag{22.1}$$

where G_i is the dead load attributed to the level *i*, Q_i is any concentrated load at that level, and q_i is the design live load applied over the area A_i at level *i*. In eq.(22.1), k_g is the live load factor which takes the values of 0.2 for dwellings, 0.5 for libraries and stores, and 1.0 for grain or liquid storage structures.

22.3 FUNDAMENTAL PERIOD

The fundamental period of vibration of a structure may be determined by any accepted method in structural dynamics. The analytical model adopted should include all the relevant components influencing the stiffness and mass characteristics of the structure;

the model for a concrete structure may be based on a linear behavior of uncracked sections.

In the application of the equivalent static method, the fundamental period *T* may be determined by the following expressions:

For concrete frames:

$$T = 0.073H^{3/4} \tag{22.2}$$

For steel frames:

$$T = 0.085H^{3/4} \tag{22.3}$$

For other structures:

$$T = 0.049H^{3/4} \tag{22.4}$$

where *H* is the total height of the structure, in meters. When the natural period is calculated by any structural dynamics method, the value of the fundamental period shall not be larger than the value obtained by the following formulas:

For concrete frames:

$$T = 0.102H^{3/4} \tag{22.5}$$

For steel frames:

$$T = 0.119H^{3/4} \tag{22.6}$$

For other structures:

$$T = 0.068H^{3/4} \tag{22.7}$$

This requirement ensures that the base shear force considered in the analysis is not less than 80% of the base shear force determined using the period obtained by eqs.(22.2) through (22.4).

22.4 ANALYTICAL MODEL

The analytical model for the evaluation of the structural seismic response should include all relevant structural elements that influence the response, a linear elastic behavior being assumed. The code prescribes either a static or a dynamic lateral load procedure, depending on the type of structure.

The equivalent static lateral load procedure is applicable to the following cases:

- Regular structures of Importance Categories B and C (see Table 22.3), less than 80 m high, and having a fundamental period not longer than 2 sec
- Irregular dwelling structures of Importance Category C, with height and fundamental period limitations as above, located in a seismic zone with an acceleration factor $Z \leqslant 0.075$

Table 22.3. Importance Factor I

Category	Description	I
A	Essential buildings, required to be operative during and after an earthquake, like hospitals, power stations, fire stations, communication facilities, etc.	1.4
B	Public buildings characterized by a large concentration of persons, like schools, etc.	1.2
C	All other buildings not included in A or B	1.0

- Regular structures of Importance Category C with a soft story, containing up to five stories
- Regular or irregular structures with up to five stories, limited to 20 m height above the foundation level, provided that they have no soft or weak stories and that the eccentricity (distance between mass center and stiffness center) is less than 15% of the structural plan dimension in the direction perpendicular to the seismic action considered.

Whenever the equivalent static lateral load procedure is not applicable, a dynamic analysis is required.

22.5 STATIC LATERAL LOAD PROCEDURE

22.5.1 Base Shear Force

The total lateral seismic force V, or base shear force, acting on the structure is evaluated as

$$V = C_d \sum_{i=1}^{N} W_i \qquad (22.8)$$

where W_i is the gravity load at level i calculated by eq.(22.1) and C_d is the seismic coefficient selected as the larger value from the following two formulas:

$$C_d \geqslant \frac{R_a I Z}{K} \qquad (22.9)$$

$$C_d \geqslant \frac{S I Z}{\sqrt{3}K} \qquad (22.10)$$

where

R_a is the spectral amplification factor calculated from

$$R_a(T) = \frac{1.25S}{T^{2/3}} \qquad (22.11)$$

limited in both extremes by

$$2.5 \geqslant R_a(T) \geqslant 0.2K \qquad (22.12)$$

However, C_d shall not exceed $0.3I$, $0.2I$, and $0.1I$, for structures with low, medium, and high ductility levels, respectively; and $0.1I$ for ductile steel structures.

Table 22.4. Normalized Maximum Ground Acceleration Factor

Seismic Zone (Fig. 22.1)	Normalized Factor Z
I	0.075
II	0.075
III	0.10
IV	0.15
V	0.25
VI	0.30

Table 22.5. Site Coefficient S

Coefficient S	Description
1.0	A soil profile with either: (a) A rock-like material characterized by a shear wave velocity greater than 800 meters per second or by other suitable means of classification, or (b) Stiff or dense soil condition where the soil depth is less than 60 meters.
1.2	A soil profile with dense or stiff soil conditions, where the soil depth exceeds 60 meters.
1.5	A soil profile 20 meters or more in depth and containing more than 6 meters of soft to medium-stiff clay but not more than 12 meters of soft clay.
2.0	A soil profile containing more than 12 m of soft clay.

I is the importance factor, ranging from 1.0 to 1.4 as given in Table 22.3

Z is the normalized maximum ground acceleration factor, obtained by dividing by ten the seismic zone coefficient given in Fig. 22.1, but not less than 0.075 or directly from Table 22.4

K is the reduction factor as given in Tables 22.1 and 22.2

S is the site coefficient, which may take the values 1.0, 1.2, 1.5 and 2.0, depending on the qualitative characteristics of the soil profile, as described in Table 22.5.

A low ductility level is allowed only in (1) "simple structures," (2) buildings of Importance Category C located in zones with acceleration factor $Z \leqslant 0.20$, (3) buildings of Importance Category B located in zones with $Z \leqslant 0.10$, and (4) buildings of Importance Category A, located in zones with $Z = 0.075$. In this context a "simple structure" is any of the following:

1. One- or two-story residential buildings, with a total built area less than $400\,\text{m}^2$, located in zones with $Z = 0.075$, without open first story, and with a concrete or steel framing
2. One-story storage buildings, with spans less than 10 m, and no overhead crane
3. Temporary structures.

Structures of Importance Category A, in zones where $Z \geqslant 0.10$, will be designed assuming a medium ductility level only.

During earthquakes, well-behaving structures are allowed an increase in the reduction factor, as follows:

- For regular structural systems, the reduction factor K may be increased by 15%
- For dual systems (concrete or steel), in which walls are capable of resisting the total horizontal load, and the frames can resist at least 25% of the total horizontal force, the reduction factor K may also be increased by 15%.
- For structures satisfying simultaneously the previous two items, the reduction factor may be increased by 25%.

22.5.2 Equivalent Lateral Forces

The total lateral seismic force V [eq.(22.8)] is distributed as equivalent lateral forces at the various levels of the building. If the fundamental period T is longer than 0.7 sec, a concentrated lateral load F_t equal to

$$F_t = 0.07TV \leqslant 0.25V \qquad (22.13)$$

is applied at the uppermost level. The lateral forces F_i at the various levels of height H_i, from the base of the building, are then calculated from the remaining lateral load $(V - F_t)$ as follows:

$$F_i = \frac{(V - F_t)W_i H_i}{\displaystyle\sum_{i=1}^{N} W_i H_i} \qquad (22.14)$$

At the top level, the lateral load F_N is added to the force F_t calculated by eq.(22.13).

22.5.3 Story Shear Force

The shear force V_i at story i of the building, is equal to the sum of the lateral seismic forces above that story; that is,

$$V_i = \sum_{j=i}^{N} F_j \qquad (22.15)$$

22.6 TORSIONAL PROVISIONS

The Israeli Code requires that both the accidental torsion and the torsion resulting from the eccentricity between the location of the mass and the stiffness center be included in the analysis (see Fig. 22.2). The lateral force at each level is applied with an accidental eccentricity e given by

$$e = \pm 0.05a \qquad (22.16)$$

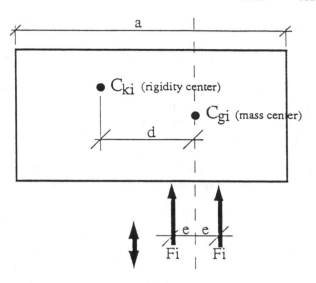

Fig. 22.2. Torsion provisions: eccentricities

where a is the horizontal plan dimension of the building perpendicular to the direction of the force considered. The influence of torsion should be increased by the following amplifying factor A_T:

$$3.0 \geqslant A_T = 2.78 \left(\frac{\delta_{max}}{\delta_{max} + \delta_{min}} \right)^2 \geqslant 1.0 \qquad (22.17)$$

in which δ_{max} is the maximum horizontal floor displacement, including the torsional influence but not amplified by A_T; and δ_{min} is the minimum horizontal displacement at the same floor.

In symmetric structures (in regard to both the mass and stiffness distributions), the influence of torsion may be accounted for in a simpler way. A multiplying factor ξ given by eq.(22.18) is applied to the resulting lateral load on each stiffness element.

$$\xi = 1.0 + 0.6 \frac{y}{a} \qquad (22.18)$$

where a is the horizontal plan dimension of the structure, perpendicular to the considered seismic action, and y is the plan distance from the stiffening element to the stiffness center, perpendicular to the seismic action.

22.7 VERTICAL SEISMIC ACTION

A vertical static analysis is required only for those structural elements (cantilevers and prestressed beams) that are sensitive to the vertical seismic load. The vertical load to be applied on cantilevered elements is given by

$$F_V = \pm \frac{2}{3} ZW \qquad (22.19)$$

and on prestressed beams by

$$F_V = W_{min} - 1.5ZISW \qquad (22.20)$$

as a load combination including gravity. In this equation W_{min} is the minimum load acting on the prestressed beam and W is the dead load.

The horizontal design loads F_i acting at each floor of the structure will be distributed to various lateral resisting elements (assuming that the floors are rigid diaphragms) by any method that takes into account their stiffness and distribution in the plan.

The bearing capacity of the vertical elements that are not part of the lateral resisting system, carrying only the gravity loads, must be checked at their deformed configuration for the effect of the computed lateral story drifts multiplied by the reduction factor K (Tables 22.1 and 22.2).

22.8 P-DELTA EFFECT

P-delta (P-Δ) analysis is required if the distortion coefficient for any story θ_i is larger than 0.1. The distortion coefficient is calculated for each floor level as

$$\theta_i = \frac{W\Delta_{el,i}K}{V_i h_i} \qquad (22.21)$$

where W is the total vertical load at and above level i, $\Delta_{el,i}$ is the computed maximum elastic story drift, K is the reduction factor, V_i is the total shear force at story i, and h_i is the height of the story below level i. A distortion coefficient greater than 0.2 at any story is not allowed.

22.9 DRIFT LIMITATIONS

The maximum interstory displacement, or story drift, is limited by $\Delta_{i,lim}$ given for a fundamental period $T \leq 0.7$ sec by

$$\Delta_{i,lim} = min\left(\frac{h_i}{40K}; \frac{h_i}{200}\right) \qquad (22.22)$$

or, for a fundamental period $T > 0.7$ sec, by

$$\Delta_{i,lim} = min\left(\frac{0.75h_i}{40K}; \frac{h_i}{250}\right) \qquad (22.23)$$

In these expressions, h_i is the height from the base to the level i of the building.

The maximum expected actual displacement $\delta_{i,max}$ in each direction is determined by

$$\delta_{i,max} = \pm K \sum_{i=0}^{i} \Delta_i \qquad (22.24)$$

in which Δ_i is the computed interstory displacement resulting from the elastic linear analysis. This maximum displacement $\delta_{i,max}$ is to be considered in determining the minimum gap between adjacent structures to avoid hammering.

22.10 DYNAMIC LATERAL LOAD PROCEDURE

22.10.1 Planar and Three-Dimensional Modeling

In structures with no significant torsional coupling, the analysis may be carried out in two independent orthogonal planar models. In that case all the vibration modes, in each direction, with vibration periods longer than 0.4 sec should be considered in the analysis. If there are less than three modes satisfying this condition, at least three vibration modes shall be considered, including those with vibration periods shorter than 0.4 sec. For structures that have significant torsional effect, a three-dimensional analysis considering at least the four lowest vibration modes should be performed (including vibration periods shorter than 0.4 sec) with at least two of the modes predominantly torsional.

22.10.2 Modal Effective Participation Factor

The Code requires that, in applying the dynamic method of analysis, all the significant modes of vibration must be included. This requirement can be satisfied by including a sufficient number of modes so that the total modal effective weight considered in the analysis is at least 90% of the total seismic weight of the structure.

The modal effective weight[1] W_m is given as

$$W_m = \frac{\left(\sum_i W_i \phi_{im}\right)^2}{\sum_i W_i \phi_{im}^2} \qquad (22.25)$$

where W_i is the seismic weight at level i and ϕ_{im} is the amplitude at that level corresponding to the mth mode of vibration.

---

[1]See eq.(4.65), in Chapter 4.

The modal effective participation factor η_m is then calculated as a fraction of the total seismic weight of the structure as follows:

$$\eta_m = \frac{W_m}{\sum\limits_{i=1}^{N} W_i} \qquad (22.26)$$

22.10.3 Total Modal Shear Force

The total modal shear force or base shear force V_m is calculated as

$$V_m = C_{dm} W_m \qquad (22.27)$$

in which the modal seismic coefficient C_{dm} is evaluated, for the corresponding natural period T_m, with eqs.(22.9) and (22.10), and W_m is the modal effective weight given by eq.(22.25).

The modal lateral force F_{im} at level i is then evaluated by

$$F_{im} = V_m \frac{W_i \phi_{im}}{\sum\limits_{i=1}^{N} W_i \phi_{im}} \qquad (22.28)$$

22.10.4 Modal Shear Force

The modal shear force V_{im} directly below level i is calculated from statics as the sum of the modal lateral forces F_{im} for levels above and including the lateral force at that level as follows:

$$V_{im} = \sum\limits_{j=i}^{N} F_{jm} \qquad (22.29)$$

22.10.5 Modal Displacements

The maximum expected modal displacement[2] $\delta_{im,max}$ in each direction is determined by

$$\delta_{im,max} = \pm K \sum\limits_{j=0}^{i} \Delta_{jm} \qquad (22.30)$$

in which Δ_{jm} is the computed interstory modal displacement resulting from the elastic modal analysis and K is the reduction factor given in Tables 22.1 and 22.2.

22.10.6 Modal Overturning Moment

The modal overturning moment M_{im} at level i of the building is calculated as the sum of the moments of the modal lateral forces F_{im} above that level; that is,

$$M_{im} = \sum\limits_{j=i+1}^{N} F_{jm}(h_j - h_i) \qquad (22.31)$$

where $i = 0, 1, 2, 3, \ldots, N-1$.

22.10.7 Modal Torsional Moment

In the case of structures with no significant torsional coupling, for which a planar model is allowed, the torsional moment M_{tim} at level i is calculated as

$$M_{tim} = (d_i \pm e_i) V_{im} \qquad (22.32)$$

where V_{im} is the modal shear force at level i, d_i is the eccentricity between the center of the above mass and the center of stiffness at that level (measured normal to the direction considered) and $\pm e_i$ is the accidental eccentricity, in the same normal direction.

22.11 TOTAL DESIGN VALUES

The design values for the total shear force, story shear, lateral deflections, story drift, overturning moment, and torsional moments are obtained by combining corresponding modal responses. Such combination is performed using the Square Root of the Sum of Squares method (SRSS),[3] namely

$$Q = \sqrt{\sum\limits_{m=1}^{N} Q_m^2} \qquad (22.33)$$

---

[2]Editor's note: Alternatively, the modal displacements δ_{im} for the level i of the building may be determined using eq.(4.72), Chapter 4, as

$$\delta_{im} = \frac{g}{4\pi^2} \cdot \frac{T_m^2 F_{im}}{W_i}$$

where T_m is the mth natural period, g the acceleration of gravity, F_{im} the modal lateral force, and W_i the seismic weight at level i. The modal drift Δ_{im} for story i, defined as the relative displacement of two consecutive levels, is then given by

$$\Delta_{im} = \delta_{im} - \delta_{(i-1)m}$$

with $\delta_{0m} = 0$.

[3]Editor's note: This method of combining modal values generally produces a satisfactory estimation of maximum response. However, the SRSS method may result in relatively large errors when some of the natural periods are closely spaced. A more refined method, the Complete Quadratic Combination (CQC) (Section 4.8, Chapter 4) is becoming the accepted method for implementation in computer programs.

where Q is the total design value and Q_m are the corresponding modal values for the response of interest.

22.12 SCALING OF RESULTS

When the total design shear force V_c, obtained by combining the total modal shear forces, is less than the product of the total shear force V obtained in the static lateral procedure multiplied by the factor μ ($\mu = 0.8$ for regular structures and $\mu = 1.0$ for irregular structures), then all the combined or design values should be multiplied by the scale factor ν given by

$$\nu = \frac{\mu V}{V_c} \qquad (22.34)$$

where the combined total shear force V_c is given by

$$V_c = \nu \sqrt{\sum_{m=1}^{N} V_m^2} \qquad (22.35)$$

22.13 P-DELTA EFFECT

The P-delta analysis is performed similarly as in the case of the equivalent static lateral force method. However, the distortion coefficient θ_i corresponding to level i is calculated from

$$\theta_i = \frac{WK\Delta_{el,i}}{h_i V_i} \qquad (22.36)$$

in which

$$\Delta_{el,i} = \sqrt{\sum_{m=1}^{N} (\Delta_{el,im})^2} \qquad (22.37)$$

$$V_i = \sqrt{\sum_{m=1}^{N} (V_{im})^2} \qquad (22.38)$$

where V_{im} is the modal shear force at level i, $\Delta_{el,im}$ is the elastic modal drift at that level, W is the total seismic weight at the same level, h_i is the height of story i, and K is the reduction factor given in Tables 22.1 and 22.2.

22.14 EXAMPLE 22.1

The dynamic modal analysis of a 21-story nonsymmetric building using a standard computer program (ETABS 1990) is presented. The resisting system consists of shear walls and frames spanning in two orthogonal directions, and the seismic weight per unit area is $10\,kN/m^2$ at all levels. Story height and plan dimensions are shown in Fig. 22.3.

The following general data are provided for this example:

- Reduction factor $K = 5.5$ (medium ductility level)
- Importance factor $I = 1.0$ (dwelling)

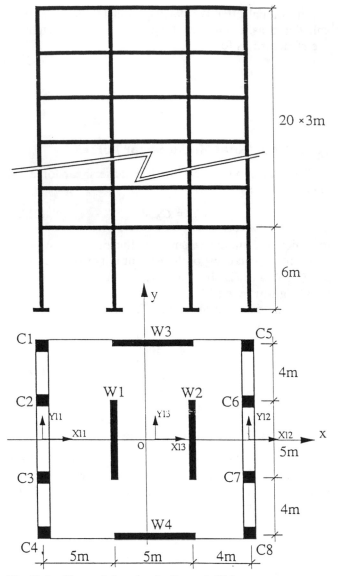

Fig. 22.3. Plan and elevation for Example 22.1

- Acceleration factor $Z = 0.1$
- Site coefficient $S = 1.2$ (noncohesive soil resting on rock bedrock).

Solution

(a) Static method
Total weight:

$$\Sigma W = 14 \times 13 \times 10 \times 21 = 38{,}220\,kN$$

Natural period:

$$T = 4.72\,sec \qquad \text{(from the dynamic method)}$$

but maximum value:

$$T = 0.102H^{3/4} \qquad \text{[eq.(22.5)]}$$
$$= 0.102 \times 66^{3/4}$$
$$= 2.362\,sec.$$

Spectral amplification factor:

$$R_a(T) = \frac{1.25S}{T^{2/3}} \qquad [\text{eq.}(22.11)]$$

$$= \frac{1.25 \times 1.2}{2.362^{2/3}} = 0.846$$

but

$$R_a(T) \geq 0.2K = 1.1$$

then

$$R_a(T) = 1.1$$

Seismic coefficient:

$$C_d \geq \frac{R_a IZ}{K} = 0.62 \qquad [\text{eq.}(22.9)]$$

$$C_d \geq \frac{SIZ}{\sqrt{3}K} = 0.02954 \qquad [\text{eq.}(22.10)]$$

Select:

$$C_d = 0.02954$$

Static base shear force:

$$V = C_d \sum_{i=1}^{N} W_i = 1{,}129 \text{ kN} \qquad [\text{eq.}(22.8)]$$

(b) Dynamic method. The analysis presented herein considers the center of mass displaced 5% to the left of its calculated position. It results in a base shear equal to 846.00 kN. The static base shear force, eq.(22.8), evaluated for the fundamental period, is 1,129 kN. The scale factor ν given by eq.(22.34) is

$$\nu = \frac{\mu V}{V_c} = \frac{1.0 \times 1{,}129}{846} = 1.33 \qquad [\text{eq.}(22.34)]$$

in which $\mu = 1.0$ was considered (irregular structure).

The analysis results given by the computer program (displacements, forces, and drifts) must then be multiplied by this factor, $\nu = 1.33$.

22.14.1 Input Data and Output Results for Example 22.1[4]

```
21-STORY THREE DIMENSIONAL BUILDING STRUCTURE - (UNITS: KN - METERS)
DYNAMIC SEISMIC RESPONSE SPECTRUM ANALYSIS - IC413.

NUMBER OF STORIES---------------------------   21
NUMBER OF DIFFERENT FRAMES-----------------    2
NUMBER OF TOTAL FRAMES---------------------    3
NUMBER OF MASS TYPES-----------------------    1
```

```
NUMBER OF LOAD CASES------------------------   1
NUMBER OF STRUCTURAL PERIODS----------------   9
NUMBER OF MATERIAL PROPERTIES---------------   2
NUMBER OF SECTION PROPERTIES FOR COLUMNS-----  2
NUMBER OF SECTION PROPERTIES FOR BEAMS-------  1
NUMBER OF SECTION PROPERTIES FOR DIAGONALS---  0
NUMBER OF SECTION PROPERTIES FOR PANELS------  1

GRAVITATIONAL ACCELERATION-----------------   .9800E+01

STORY MASS TYPE NUMBER---------------------        1
NUMBER OF MASS SEGMENTS--------------------        1
MASS SCALE FACTOR-------------------------    .100E+01
```

SEGMENT NUMBER	SEGMENT MASS	COORDINATES OF CENTER X	Y	DIMENSIONS OF SEGMENT X	Y
1	1.020400	-.70	.00	14.00	13.00

```
CALCULATED STORY MASS PROPERTIES

STORY MASS--------------------------------    185.71
MASS MOMENT OF INERTIA--------------------    5648.8
X-ORDINATE OF CENTER OF MASS--------------     -.70
Y-ORDINATE OF CENTER OF MASS--------------      .00
```

MATERIAL PROPERTIES

ID	TYPE	ELASTIC MODULUS	UNIT WEIGHT	POISSONS RATIO
1	C	.250E+08	.000E+00	.200
2	W	.250E+08	.000E+00	.200

SECTION PROPERTIES FOR COLUMNS

ID	MAT ID	SECTION TYPE	MAJOR DIM	MINOR DIM	FLANGE THICK	WEB THICK
1	1	RECT	.600	.350	.000	.000
2	1	L-SECT	.800	.800	.350	.350

ANALYSIS SECTION PROPERTIES FOR   COLUMNS

ID	AXIAL A	MAJOR AV	MINOR AV	TORSION J	MAJOR I	MINOR I
1	.210	.175	.175	.5454E-02	.6300E-02	.2144E-02
2	.438	.280	.280	.1156E-01	.2164E-01	.2164E-01

ANALYSIS SECTION PROPERTIES FOR   BEAMS

ID	AXIAL A	MAJOR AV	MINOR AV	TORSION J	MAJOR I	MINOR I
1	.175	.000	.000	.0000E+00	.3646E-02	.1786E-02

SECTION PROPERTIES FOR   PANELS

ID	MAT ID	PANEL THICK
1	2	.350

```
DYNAMIC RESPONSE SPECTRUM ANALYSIS

ISRAEL CODE 413 - SPECTRUM CURVE

NUMBER OF EXCITATION DIRECTIONS------------        1

NUMBER OF POINTS ON SPECTRUM CURVE---------       36

MODAL COMBINATION TECHNIQUE----------------      SRSS

SCALE FACTOR FOR SPECTRUM CURVE------------     9.800

DAMPING ASSOCIATED WITH SPECTRUM CURVE------     .000
```

STRUCTURAL TIME PERIODS AND FREQUENCIES

MODE NUMBER	PERIOD (TIME)	FREQUENCY (CYCLES/UNIT TIME)	CIRCULAR FREQ (RADIANS/UNIT TIME)
1	4.72127	.21181	1.33082
2	2.46256	.40608	2.55149
3	1.84847	.54099	3.39913
4	.76073	1.31453	8.25946
5	.59929	1.66865	10.48443
6	.44727	2.23579	14.04787
7	.27598	3.62343	22.76667
8	.25243	3.96153	24.89102
9	.18736	5.33731	33.53534

EFFECTIVE MASS FACTORS

MODE NUMBER	X — TRANSLATION— %-MASS	%-SUM	Y — TRANSLATION— %-MASS	%-SUM	Z-ROTATION--- %-MASS	%-SUM
1	65.74	65.7	.00	.0	.00	.0
2	.00	65.7	67.70	67.7	.21181	3.2
3	.00	65.7	3.14	70.8	67.69	70.9
4	20.09	85.8	.00	70.8	.00	70.9
5	.00	85.8	15.22	86.1	.92	71.9
6	.00	85.8	1.00	87.1	15.21	87.1
7	6.67	92.5	.00	87.1	.00	87.1
8	.00	92.5	5.46	92.5	.39	87.5
9	.00	92.5	.42	92.9	5.48	92.9

---

[4]Editor's Note: To save space only selected portions of the computer output are reproduced.

FRAME REACTION FORCES AT BASELINE (AT EACH COLUMN LINE)

VALUES ARE IN THE LOCAL COORDINATE SYSTEM OF THE FRAME

FRAME ID .... FRAME 2 (C5-C6-C7-C8)

COL ID	OUTPUT ID	FORCE ALONG-X	FORCE ALONG-Y	FORCE ALONG-Z	MOMENT ABOUT-XX	MOMENT ABOUT-YY	MOMENT ABOUT-ZZ
1	CASE 1	.95	4.61	659.70	40.40	15.64	1.42
2	CASE 1	.04	4.46	20.78	17.95	.60	.67
3	CASE 1	.04	4.46	20.78	17.95	.60	.67
4	CASE 1	.95	4.61	659.70	40.40	15.64	1.42

SUMMATION OF FRAME REACTION FORCES AT BASELINE

VALUES ARE IN THE LOCAL COORDINATE SYSTEM OF THE FRAME

FRAME ID .... FRAME 2 (C5-C6-C7-C8)

OUTPUT ID	FORCE ALONG-X	FORCE ALONG-Y	FORCE ALONG-Z
CASE 1	.00	18.13	.00

Summary of Computer Program Output for FRAME 2 (C5-C6-C7-C8)
(Units: meters - KN)

Level	Displacement X	Y	Rot	Frame-Drift X	Y	Rot	Story Shear X	Y	Rot
21	0	.04721	.00291	0	.00077	.00005	0	140.30	11.02
20	0	.04491	.00278	0	.00078	.00005	0	112.33	15.82
19	0	.04258	.00264	0	.00080	.00005	0	121.27	15.50
18	0	.04021	.00250	0	.00081	.00005	0	126.92	17.04
17	0	.03779	.00236	0	.00083	.00005	0	133.52	17.98
16	0	.03532	.00221	0	.00085	.00005	0	140.18	18.92
15	0	.03279	.00206	0	.00086	.00005	0	146.64	19.73
14	0	.03022	.00190	0	.00088	.00005	0	152.63	20.44
13	0	.02761	.00175	0	.00088	.00005	0	157.91	21.04
12	0	.02497	.00159	0	.00089	.00005	0	162.24	21.52
11	0	.02233	.00142	0	.00088	.00005	0	165.40	21.87
10	0	.01969	.00126	0	.00087	.00005	0	167.14	22.07
9	0	.01709	.00110	0	.00085	.00005	0	167.21	22.12
8	0	.01454	.00094	0	.00082	.00005	0	165.32	22.01
7	0	.01208	.00078	0	.00078	.00005	0	161.16	21.73
6	0	.00973	.00063	0	.00073	.00005	0	154.39	21.25
5	0	.00753	.00049	0	.00067	.00004	0	144.55	20.60
4	0	.00551	.00036	0	.00060	.00004	0	131.37	19.59
3	0	.00373	.00025	0	.00050	.00003	0	112.74	18.68
2	0	.00222	.00015	0	.00039	.00003	0	100.02	17.27
1	0	.00105	.00007	0	.00017	.00001	0	18.13	16.62

REFERENCES

Encyclopedia Judaica, Vol. 6 (1972) pp. 340–341, Keter Publishing House of Jerusalem Ltd., Jerusalem, Israel.

Encyclopedia Judaica, Vol. 9 (1972) pp. 112–122, Keter Publishing House of Jerusalem Ltd., Jerusalem, Israel.

ETABS – Three-Dimensional Analysis of Building Systems, User Manual (1990) Computers and Structures Inc., Berkeley, CA.

Israel Standard, IC-413 (1961, 1975) The Standard Institution of Israel, Tel Aviv, Israel.

Israel Standard, IC-413 (1994) (Draft) The Standard Institution of Israel, Tel Aviv, Israel.

23

Italy

Gianmario Benzoni and Carmelo Gentile

23.1 INTRODUCTION

Italy has had a long history of strong seismic activity. Regulations for seismic-resistant design were promulgated in 1786 as a result of a devastating earthquake in Calabria in 1783. An interesting set of rules was an ordinance issued for the City of Norcia on April 28, 1860; this set of practical rules for safer construction included the use of steel ties and contained a provision for the "corporal punishment" of workmen who assisted in building structures that violated the provisions of the ordinance.

After an earthquake that destroyed the cities of Messina and Reggio (Sicily) on December 23, 1908, with an estimated loss of 80,000 lives, a Royal Decree (n. 193) was formulated on April 18, 1909. This decree contained provisions for the strengthening and reconstruction of existing buildings and the design of new buildings in the area affected by the earthquake. Two possible seismic actions on structures were indicated. The first was static actions due to dead and live loads, increased by a percentage to take into account the effects of seismic vertical vibrations. The second was dynamic actions applied as horizontal accelerations on the masses of the structure. These provisions were further refined in 1916 (updated in 1924) after the Avezzano earthquake of 1915 which caused the loss of about 30,000 lives. The provisions prescribed horizontal forces equal to 1/6 of the weights considered to be at the floor levels for a building higher than 12 meters; also prescribed were the minimum dimensions for structural elements and a test under vertical load.

It is difficult to understand the development of seismic regulations without knowing the main steps in the process of seismic zoning in Italy. The first formal classification of seismic areas is associated with the Royal Decree of 1909. This classification reflected an advanced approach to seismic zoning: the area covered was not limited to the damaged cities but also included a large adjacent region with evident historical seismicity. Unfortunately, the development of seismic zoning after 1909 became a means of identifying areas for reconstruction rather than an effective method to prevent damage. Not until 1962 when the zoning criteria of 1909 were reinstated did the Code begin to emphasize prevention of destruction in seismic-prone regions.

The classification of communities into two seismic categories based on damage intensity principles started in 1927 (Royal Decree of March 13, 1927, n.431). The design rules for the first seismic zone were the same as those given in the previously cited Decree of 1924; for the second seismic area, the design horizontal seismic forces were reduced to 1/10 of the weights for buildings lower than 15 meters and to 1/8 of the weights for buildings taller than 15 meters.

The framework of the current code can be found in the Code of 1975, which incorporated general design criteria and provisions for masonry buildings, framed structures, and buildings with structural walls. This code allowed two types of seismic analysis: (1) a static analysis with horizontal and vertical forces, and (2) a dynamic analysis that uses design spectra of the same shape as those included in the current Code. The 1975 Code also allowed more detailed analysis using dynamic theory, provided that it was based on a well-documented "design earthquake."

The Minister of Public Works issued a series of decrees (1980 through 1984) that fixed for Italy the boundaries of the seismic zones and their classifications. For the first time, general criteria were adopted for the whole national territory; the number of listed communities was raised to 2,960, corresponding to 36.6% of the total number of cities, and involving 39.8% of the population living within 45.2% of the territory of the country.

Since 1976, the classification system has been modified to account for the probability of exceeding assigned thresholds of ground acceleration for a given time interval. The seismic areas are classified in three seismic zones with progressively less stringent seismic design provisions from the first zone to the third zone and by seismicity degrees (S) of 12, 9, and 6, respectively.

The current seismic code for Italy was updated on January 24, 1986. Its main provisions are described in this chapter. The 1986 Code encompasses a complete set of provisions for repairing and strengthening existing buildings. Considerable attention is given to masonry buildings, which in Italy represent a major portion of urban construction. A striking peculiarity of the Code is the freedom given to the designer to choose the design approach and technical solutions that may be more appropriate for a specific case. This aspect of the Code is particularly evident in the rules provided for strengthening masonry buildings; the design of these buildings must be at the ultimate-limit state.

It is reasonable to believe that the Italian Seismic Code will be modified further in order to agree more closely with other European codes (EC8 for structures in seismic regions). The Eurocodes project was intended to provide a set of rules for the member states and to serve as a guide for the development of national rules. Some parts of the Eurocode have been published, others are at the first draft level or have been distributed for comments. Provisions very similar to those included in the Eurocode can be found in a code edited by C.N.R.-G.N.D.T. (1984) (Consiglio Nazionale delle Ricerche – Gruppo Nazionale di Difesa dai Terremoti) but never officially issued ["Norme Tecniche per le Costruzioni in Zone Sismiche," 1985]. The general description that follows is intended to present the major rules of the current Italian Seismic Code of 1986.

23.2 GENERAL PROVISIONS OF THE ITALIAN SEISMIC CODE OF 1986

Three levels of seismicity are considered on Italian territory. The seismicity index S characterizes this subdivision, with S equal to 6, 9, or 12, respectively, for low, medium, and high degrees of seismicity. The stresses caused by horizontal or vertical seismic actions must be evaluated conventionally by both static and dynamic analysis. Alternatively, a more detailed analysis could be based on a careful choice of a "design earthquake." The earthquake input must be considered at the base of the structure along each of two main horizontal orthogonal directions of the building, but not at the same time in both directions. The seismic motion affects the masses related to the weight of the structure, permanent loads, and a fraction of the live loads. When the live loads are not prescribed, their magnitude must be evaluated by statistical considerations. For tanks, vessels or similar structures the weight of the contents must be taken into account.

The Code provides for the following four structural types of buildings:

1. Framed structures (reinforced concrete, prestressed concrete, steel, etc.)
2. Masonry buildings
3. Buildings with structural walls (reinforced masonry, tilt-up buildings, etc.)
4. Timber buildings

Table 23.1 indicates, for each type of building, the

Table 23.1. Maximum Height for New Buildings (m)

Structural Type	m for Seismicity Intensity		
	$S = 6$	$S = 9$	$S = 12$
Framed		No limitation	
Masonry	16.0	11.0	7.5
Structural walls	32.0	25.0	15.0
Timber	10.0	7.0	7.0

maximum height permitted for new buildings in different seismic zones. The maximum height can be raised by no more than 4 meters for buildings with basements and by 1.5 meters for structures built on a slope, provided that the average height among all the sides is consistent with the provisions in Table 23.1. For timber structures, a footing no higher than 4 meters is allowed. Other limitations in height are included in the Code.

The minimum distance $d(h)$ allowed between adjacent structures is given by

$$d(h) = \frac{h}{100} \tag{23.1}$$

where h is the height of the building measured from the ground level.

The paragraphs that follow include the basic parameters defined in the Code.

23.2.1 Response Factor (R)

The response factor R is defined as a function of the fundamental period T_0 of the structure:

for $T_0 > 0.8$ sec: $R = 0.862/T_0^{2/3}$, and
for $T_0 \leqslant 0.8$ sec: $R = 1.0$.

If an estimate of the fundamental period T_0 is not available, R must be assumed to be equal to 1.0. For structures with $T_0 > 1.4$ sec, a dynamic analysis is required. Generally, the fundamental period is evaluated by modal analysis of the overall system.

The Code provides an empirical formula to estimate the fundamental period for framed structures:

$$T_0 = 0.1 \frac{H}{\sqrt{B}} \tag{23.2}$$

where H and B are, respectively, the building height and the minimum plan dimension, in meters, and where T_0 is expressed in seconds.

23.2.2 Building Importance Factor (I)

$I = 1.4$ for structures of primary importance for the
   purpose of civil defense.

$I = 1.2$ for important structures whose function is at risk.
$I = 1.0$ for all other structures.

The importance factor must be applied to forces applied in both the horizontal and vertical directions.

23.2.3 Seismic Intensity Factor (C)

$$C = \frac{(S - 2)}{100} \tag{23.3}$$

where S is the seismicity index defined in Section 23.2.

23.2.4 Foundation Factor (ε)

Usually, the foundation factor is assumed to be equal to 1.0. It must be increased to the maximum value of 1.3 for compressible soils.

23.2.5 Structural Factor (β)

The structural factor is usually equal to 1.0. When the resistance to horizontal actions is provided totally by frames or stiffening vertical elements, this value must be increased to 1.2.

23.3 FRAMED STRUCTURES

23.3.1 Static Analysis

The static analysis can be used for structures that do not contain long spans or elements that exert a horizontal thrust. This analysis is characterized by a simplified static equivalent force system. The Code stipulates that seismic actions can be represented by static forces proportional to the story weight multiplied by a seismic coefficient (K), defined in Sections 23.3.1.1, 23.3.1.2 and 23.3.1.3.

23.3.1.1 Equivalent lateral forces. The equivalent lateral forces F_i are applied at the center of mass of each floor level i, in the direction parallel to the assumed earthquake direction. These forces are obtained from the following equations:

$$F_i = K_{hi} W_i \tag{23.4}$$

with

$$K_{hi} = CR\varepsilon\beta\gamma_i I \tag{23.5}$$

and

$$W_i = G_i + sQ_i \tag{23.6}$$

Table 23.2. Reduction Factor for Live Loads

Building Description	Factor s
Dwellings, offices, roofs, balconies	0.33
Public buildings with some limitation to the number of people (hospitals, schools, stores, restaurants, etc.)	0.5
Public buildings in which limitation to the number of people may be exceeded (churches, theaters, stairs, etc.)	1.0

where C, R, ε, β, and I are as defined in Section 23.2, and

G_i = weight at the ith level including any superimposed dead loads

Q_i = maximum live load for ith level

s = reduction factor for live loads as specified in Table 23.2

γ_i = distribution factor for the ith level.

The distribution factor γ_i corresponding to level i is a function of the weight W_i at that level and of the height h_i measured from the foundation:

$$\gamma_i = h_i \frac{\displaystyle\sum_{j=1}^{N} W_j}{\displaystyle\sum_{j=1}^{N} W_j h_j} \tag{23.7}$$

Torsional moments must be considered if there is an eccentricity at any level of the building (between the stiffness center and mass center). For symmetric buildings, torsional moments must be considered if $D/B > 2.5$, where D is the long dimension and B is the short dimension of the building. In this case, a minimum torsional moment $M_{ti,\min}$ due to the horizontal forces F_i is applied at overlying levels of the building. The torsional moment shall not be smaller than

$$M_{ti,\min} = \lambda D \sum_{j=i}^{N} F_j \tag{23.8}$$

with λ assuming the following values:

for $2.5 < D/B \leqslant 3.5$
$$\lambda = 0.03 + 0.02(D/B - 2.5), \text{ and} \tag{23.9}$$
for $3.5 < D/B$
$$\lambda = 0.05.$$

23.3.1.2 Vertical forces. The Code requires the consideration of vertical seismic forces for (1) buildings with floor spans larger than 20 meters, (2) elements that exert a horizontal thrust, and (3) cantilevered elements. The vertical seismic coefficient K_v is equal to ± 0.2 for the first two cases and equal to ± 0.4 for cantilevered elements. The resulting vertical force F_{vi} is obtained as

$$F_{vi} = K_v I W_i \tag{23.10}$$

in which I is the importance factor (Section 23.2.2) and W_i is the seismic weight at level i [eq.(23.6)].

23.3.1.3 Combination of lateral and vertical effects. The design rules used to determine the forces and displacements that are the result of combining horizontal and vertical components are

$$\alpha = \sqrt{\alpha_h^2 + \alpha_v^2}, \quad \eta = \sqrt{\eta_h^2 + \eta_v^2} \tag{23.11}$$

where

α = single force component (bending moment, axial force, shear force or torsional moment)

η = single displacement component

h = index specifying the horizontal component

v = index specifying the vertical component.

23.3.2 Dynamic Analysis

The dynamic analysis method combines modal analysis with a design spectrum. In this analysis, the structure must be considered to deform in the linear elastic range. The horizontal spectral acceleration a is given by

$$\frac{a}{g} = CR\varepsilon\beta I \tag{23.12}$$

where g is the acceleration of gravity; the other parameters in eq.(23.12) assume the previously defined values (Section 23.2). The Code requires that at least the three first structural modes be included in the analysis. The computation of seismic vertical actions usually does not require dynamic analysis.

23.3.2.1 Combination of modal effects. Modal contributions in terms of forces α_i or displacements η_i are combined by the SRSS (Square Root of the Sum of Squares) method, that is

$$\alpha = \sqrt{\sum \alpha_i^2}, \quad \eta = \sqrt{\sum \eta_i^2} \tag{23.13}$$

23.3.3 Superposition of Forces and Displacements

Forces due to seismic excitation and to gravitational loads (permanent, quasi-permanent and transient, except for wind) must be superposed in order to verify their conformance to admissible values.

If α and α_p are the forces due to seismic actions and to other loads, respectively, the combined total force α_{tot} will be

$$\alpha_{\text{tot}} = \alpha \pm \alpha_p \tag{23.14}$$

where the calculation of α_p must account for both of the following possible combinations:

1. Weight of the structure, permanent loads, and live loads (α_{p1}), and
2. Weight of the structure, permanent loads, and a fraction of live loads (α_{p2}).

Each single structural section must resist the most unfavorable combination of stresses, thus:

$$\alpha_{\text{tot}} = \frac{\alpha \pm \alpha_{p1}}{\alpha \pm \alpha_{p2}} \text{ , whichever is greater} \quad (23.15)$$

If the building design is performed assuming elastic behavior, the actual displacements η_r may be estimated as

$$\eta_r = \eta_p \pm \phi\eta \quad (23.16)$$

where

η = elastic displacement due to seismic excitation
η_p = elastic displacements due to other actions (except wind)
ϕ = 6 if displacements are obtained from static analysis
ϕ = 4 if displacements are obtained from dynamic analysis.

Displacements so computed must ensure that contiguous elements remain connected and separate elements do not "pound" against each other during an earthquake.

23.3.4 Foundations

The seismic code requires that foundation structures be connected by a grid of beams. Each connecting beam must resist an axial force (tension or compression) equal to 1/10 of the more severe vertical load applied to its ends. These connections can be avoided if the maximum drift between two isolated foundation elements is lower than either 1/1,000 of their distance, or two centimeters.

Pile foundations must be designed to resist the design horizontal seismic forces developed by considering relative pile stiffness. Stability analysis of the soil and structure must consider the total stresses $\alpha_{\text{tot}} = \alpha \pm \alpha_p$ transferred from the building to the foundation system.

23.3.5 Nonstructural Walls

Internal masonry walls higher than 4 meters and with an area larger than 20 m^2 must be connected to structural elements by reinforcing columns spaced no wider than every 3 meters. The same type of connection must be provided for external walls higher than 3.5 m with an area larger than 15 m^2. In the higher intensity seismic zone ($S > 9$), openings must be bounded by a frame connected to the main structure.

23.4 MASONRY BUILDINGS

The 1986 Code includes provisions that govern the dimensions of the elements in new masonry buildings, although the type of analysis required is not mentioned in the rules for seismic design.

The Code section relative to strengthening existing buildings states that retrofitting may be avoided if the structure conforms to the rules for seismic design of new buildings. If not, the building design must be verified with reference to ultimate seismic resistance. Seismic design actions shall be identical to those specified for framed buildings, assuming a structural factor $\beta = \beta_1 \beta_2$, where

$\beta_1 = 2$, a factor that accounts for ductility
$\beta_2 = 2$, a factor consistent with design for the ultimate state.

The values for the importance factor I are those that were previously indicated. If the strengthening renovation is intended only to recover the original structural functionality, a reduction of 30% is allowed for I. It could be assumed that the material follows an elastoplastic behavior. Stiffness of walls that are orthogonal to the seismic action need not be considered; however, the seismic actions perpendicular to the wall must be considered as

(a) A distributed horizontal load, equal to βC times the weight of the wall
(b) A set of concentrated forces equal to βC times any floor load that is not securely connected to the transverse walls

The flexural effect of the seismic action orthogonal to the wall may be evaluated assuming linear behavior and an effective cross-section equal to the gross cross-section. For foundation structures, the seismic actions must be computed assuming the factor β_2 equal to 1.

The previous rules are based on base shear coefficients defined as the ratio of the total design horizontal seismic force to the vertical weight, equal to

$0.10 \times 4 = 0.4$ for Seismic Zone I
$0.07 \times 4 = 0.28$ for Seismic Zone II
$0.04 \times 4 = 0.16$ for Seismic Zone III

However, the engineer is allowed to develop any personal procedure to verify the resistance of the structure at the ultimate state, but the associated horizontal resistant forces must be larger than those imposed by the Code. The computer program written for this chapter includes a procedure (developed after the 1976 Friuli earthquake) that provides an estimate of the ultimate building resistance.

23.5 BUILDINGS WITH STRUCTURAL WALLS

The Code requires a seismic analysis for buildings with structural walls that is similar to the analysis intended for framed structures. The structural factor β must be taken equal to 1.4 and the response factor R must be equal to 1. No detailed provisions are included for the design of timber structures.

23.6 FRAMED BUILDING EXAMPLE 23.1

A four-story building with four reinforced concrete main frames in both the OX and OY directions is shown in Fig. 23.1. The mass of the building is lumped at each floor as follows:

$$M_1 = M_2 = M_3 = M_4 = 168 \text{ kN sec}^2/\text{m}$$

The story height is constant for all levels and equal to 3.50 m. All the columns have square sections 40 cm by 40 cm so that the story stiffness at each level, for all frames, is given by

$$k_x = k_y = 5,869 \text{ kN/m}$$

The seismic forces are assumed to be applied along the OY direction. According to the Italian Seismic Code (D.M. 24.01.1986), determine (a) the fundamental period, (b) the position of the center of stiffness, (c) distribution factors, (d) horizontal seismic coefficients, (e) lateral seismic forces and torsional moments for each level, (f) story shear and overturning moments, and (g) distribution of the equivalent lateral forces for each frame and story.

Solution

(a) Fundamental period. The Code suggests that the fundamental period of a reinforced concrete building be evaluated by the following empirical formula:

$$T_0 = \frac{0.1H}{\sqrt{B}} \qquad \text{[eq.(23.2)]}$$

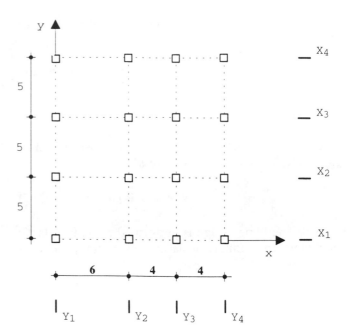

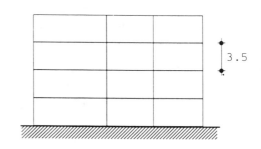

Fig. 23.1. Modeled building for Examples 23.1 and 23.2

where

H = 14.00 m (height of the building)
B = 15.00 m (dimension of the building in OY direction), and

$$T_0 = 0.1 \frac{14}{\sqrt{15}} = 0.361 \text{ sec}$$

(b) Position of the center of stiffness. The coordinates $X_{cs,i}$, $Y_{cs,i}$ of the center of stiffness at the ith level are computed using the following expressions:

$$X_{CS,i} = \frac{\sum_{j=1}^{ny} k_{y,ij} x_{ij}}{\sum_{j=1}^{ny} k_{y,ij}} \; ; \quad Y_{CS,i} = \frac{\sum_{j=1}^{nx} k_{x,ij} y_{ij}}{\sum_{j=1}^{nx} k_{x,ij}}$$

Table 23.3. Coordinates of the Center of Mass and Center of Stiffness

Story	X Center Mass (m)	Y Center Mass (m)	X Center Stiffness (m)	Y Center Stiffness (m)
1	7.000	7.500	7.500	7.500
2	7.000	7.500	7.500	7.500
3	7.000	7.500	7.500	7.500
4	7.000	7.500	7.500	7.500

Table 23.4. Distribution Factors (γ_i)

Story i	γ_i
1	0.4
2	0.8
3	1.2
4	1.6

where

$k_{x,ij}$ = story stiffness of jth frame in OX direction at level i

$k_{y,ij}$ = story stiffness of jth frame in OY direction at level i

x_{ij} = x-coordinate of jth OY frame at level i
y_{ij} = y-coordinate of jth OX frame at level i
nx = number of OX frames; and
ny = number of OY frames.

Table 23.3 shows the coordinates of mass center and stiffness center at each story of the building in terms of an XY reference frame.

(c) Distribution factors. The distribution factors γ_i are computed using eq.(23.7). The results are summarized in Table 23.4.

(d) Horizontal seismic coefficients. The seismic coefficients are given by eq.(23.5) as

$$K_{hi} = CR\varepsilon\beta\gamma_i I$$

where the following numerical values have been assumed:

$C = 0.1$ (intensity factor)
$I = 1.0$ (importance factor)
$\varepsilon = 1.0$ (foundation factor)
$\beta = 1.0$ (structural factor)
$R(T_0) = 1.0$ (response factor) because $T_0 < 0.8$ sec (Section 23.2).

Thus,

$$K_{hi} = 0.1\gamma_i \qquad \text{[eq.(23.5)]}$$

Table 23.5. Equivalent Lateral Forces F_i and Torsional Moments, M_{ti}

Level i	F_i (kN)	e_{xi} (m)	M_{ti} (kN-m)
1	65.923	−0.500	−32.9616
2	131.846	−0.500	−65.9232
3	197.770	−0.500	−98.8848
4	263.963	−0.500	−131.8464

Table 23.6. Story Shear Forces and Overturning Moments

Level i	V_i (kN)	M_i (kN-m)
1	659.232	6,921.937
2	593.309	4,614.625
3	461.462	2,538.043
4	263.693	922.925

(e) Lateral seismic forces F_i and torsional moments M_{ti} at ith level

$$F_i = K_{hi}W_i \; ; \; M_{ti} = F_i e_{xi}$$

where

W_i = effective weight at level i
e_{xi} = distance between mass center and stiffness center at level i.

Table 23.5 shows the values calculated for the equivalent lateral forces and torsional moments at each level of the building.

(f) Story shear force and overturning moments. Once lateral forces are estimated, story shear forces V_i and overturning moments M_i may be evaluated by statics as

$$V_i = \sum_{j=i}^{n} F_j \qquad (23.17)$$

and

$$M_i = \sum_{j=i+1}^{n} F_j(h_j - h_i) \qquad (23.18)$$

where $i = 0, 1, 2, \ldots, n-1$.

The computed values for story shear forces and overturning moments are shown in Table 23.6.

(g) Distribution of lateral forces among the frames. For the seismic forces applied along the OY direction, the lateral forces for story i are distributed among the frames using the following equations:

$$F_{y,ij} = F_i \left(\frac{k_{y,ij}}{\sum_{j=1}^{ny} K_{y,ij}} + \frac{e_{xi}k_{y,ij}d_{x,ij}}{K_{ti}} \right) \qquad (23.19)$$

Table 23.7. Story Forces for OX Frames

Level i	Frame X1 F_x (kN)	Frame X2 F_x (kN)	Frame X3 F_x (kN)	Frame X4 F_x (kN)
1	1.0656	0.3552	0.3552	1.0656
2	2.1311	0.7104	0.7104	2.1311
3	3.1967	1.0656	1.0656	3.1967
4	4.2623	1.4208	1.4208	4.2623

Table 23.8. Story Forces for OY Frames

Level i	Frame Y1 F_y (kN)	Frame Y2 F_y (kN)	Frame Y3 F_y (kN)	Frame Y4 F_y (kN)
1	17.5464	16.6939	16.1256	15.5573
2	35.0927	33.3878	32.2512	31.1146
3	52.6391	50.0817	48.3768	46.6719
4	70.1855	66.7757	64.5024	62.2292

and

$$F_{x,ij} = F_i \frac{e_{xi} k_{x,ij} d_{y,ij}}{K_{ti}} \qquad (23.20)$$

where

$$K_{ti} = \sum_{j=i}^{nx} k_{x,ij} d_{y,ij}^2 + \sum_{j=1}^{ny} k_{y,ij} d_{x,ij}^2 \qquad (23.21)$$

$d_{y,ij} = x$ distance of jth frame along OY direction from the center of stiffness at level i

$d_{y,ij} = y$ distance of jth frame along OX direction from the center of stiffness at level i.

The other symbols have the meanings previously specified. Tables 23.7 and 23.8 summarize the computed values, respectively, for frames in the OX and OY directions.

23.7 COMPUTER PROGRAM

A computer program has been developed to implement the provisions of the current Italian seismic Code. The program assumes that the building is modeled as an ensemble of rectangular frames in the principal directions OX and OY. It is further assumed that the horizontal diaphragms at the floor levels are rigid in their planes, and that the horizontal lateral forces are applied at the various levels of the building.

The program allows the user to input and correct the data interactively or alternatively, to use data previously prepared and stored in a file. The program can implement either the static or the dynamic method of analysis. In the static analysis, all the necessary factors to determine the equivalent lateral forces are

calculated and printed by the program. Values for seismic lateral forces, torsional moments, story shear forces, and overturning moments are evaluated and printed by the program. The distribution of equivalent lateral forces among the frames at each level is also performed. It is possible to specify the desired number of modes to be included in the dynamic analysis. The center of mass and the story weight for each level, as well as the story stiffness of each frame, must be calculated and supplied by the user. Modal forces and modal displacements are computed at each level of the building for all the significant modes, as well as for their effective values; such forces and displacements are applied at the stiffness center of each story. The effective forces are then distributed among the various frames.

The following examples are presented to illustrate the implementation and execution of the program.

Example 23.2

In Example 23.2, the computer program is used to solve the problem of Example 23.1.

```
GENERAL DATA

  INTENSITY FACTOR              (0.1, 0.07, or .05)   = 0.1
  IMPORTANCE FACTOR             (1.0, 1.2, or 1.4)    = 1.0
  FOUNDATION FACTOR             (1.0, or 1.3)         = 1.0
  STRUCTURAL FACTOR             (1.0, 1.2, or 1.4)    = 1.0
  BASE DIMENSION (X-DIRECTION)           (m)          = 14

  BASE DIMENSION (Y-DIRECTION)           (m)          = 15
  NUMBER OF STORIES                                   = 4
  NUMBER OF FRAMES (X-DIRECTION)                      = 4
  NUMBER OF FRAMES (Y-DIRECTION)                      = 4


DATA FOR EACH STORY

  STORY N. 1 - 2 - 3 - 4
  HEIGHT                        (m)       =   3.5
  TRANSLATIONAL MASS            (kN sec²/m) =  168
  ROTATIONAL MASS               (kN sec²/m) =  5894
  X OF CENTER OF MASSES         (m)       =   7.0
  Y OF CENTER OF MASSES         (m)       =   7.5
```

STORY #	STORY HEIGHT (m)	STORY WEIGHT (kN)
1	3.500	1648.080
2	3.500	1648.080
3	3.500	1648.080
4	3.500	1648.080

DATA FOR FRAMES (X-DIRECTION)

FRAME #	Y-COORDINATE (m)	STIFFNESS (KN/m)
1	0	59720
2	5	59720
3	10	59720
4	15	59720

DATA FOR FRAMES Y-DIRECTION

FRAME #	X-COORDINATE (m)	STIFFNESS (KN/m)
1	0	59720
2	6	59720
3	10	59720
4	14	59720

STORY #	X C.M.	Y C.M.	X C.S.	Y C.S.
	(m)	(m)	(m)	(m)
1	7.000	7.500	7.500	7.500
2	7.000	7.500	7.500	7.500
3	7.000	7.500	7.500	7.500
4	7.000	7.500	7.500	7.500

****************** EQUIVALENT STATIC ANALYSIS ******************

ESTIMATED VALUE OF T_x (sec) = 0.3741657
BASE SHEAR COEFFICIENT (g) = 0.1

ESTIMATED VALUE OF T_y (sec) = 0.3614784
BASE SHEAR COEFFICIENT (g) = 0.1

STORY #	Distribution Factor
1	0.40
2	0.80
3	1.20
4	1.60

EQUIVALENT LATERAL FORCES (Y DIRECTION)[1]

STORY #	F_y(kN)	E_x(m)	M_t(kN m)
1	65.923	-0.5000	-32.9616
2	131.846	-0.5000	-65.9232
3	197.770	-0.5000	-98.8848
4	263.693	-0.5000	-131.8464

STORY SHEAR & OVERTURNING MOMENT

STORY #	V(kN)	M(kN*m)
1	659.232	6921.937
2	593.309	4614.625
3	461.462	2538.043
4	263.693	922.925

EQUIVALENT LATERAL FORCES (Y DIRECTION)
FORCE ON FRAMES (KN):

STORY	FRAMES X-DIRECTION				FRAMES Y-DIRECTION			
	1	2	3	4	1	2	3	4
1	1.07	0.36	0.36	1.07	17.55	16.69	16.13	15.56
2	2.13	0.71	0.71	2.13	35.09	33.39	32.25	31.11
3	3.20	1.07	1.07	3.20	52.64	50.08	48.38	46.67
4	4.26	1.42	1.42	4.26	70.19	66.78	64.50	62.23

Example 23.3

A 10-story reinforced concrete building is shown in Fig. 23.2. The effective weight on each level is 784.80 kN. The plan dimensions of the building are 10 m by 10 m. All the columns have square cross-sections 60 cm by 60 cm at stories 1–4, 50 cm by 50 cm at stories 5–7, and 40 cm by 40 cm at stories 8–10. The building site is in a zone classified as an area of the

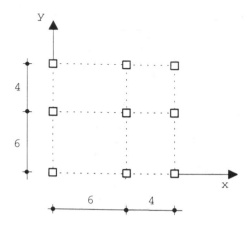

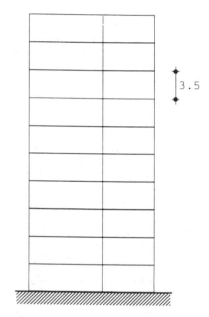

Fig. 23.2. Modeled building for Example 23.3

first seismic category ($S = 12$). The structure is supported by compact soil.

Perform the seismic analysis for the building along the OX direction using a modal analysis that accounts for the effects of the first 10 modes with the response spectrum provided by the 1986 Code.

GENERAL DATA

INTENSITY FACTOR	(0.1, 0.07, or 0.05)	= 0.1
IMPORTANCE FACTOR	(1.0, 1.2, or 1.4)	= 1.0
FOUNDATION FACTOR	(1.0, or 1.3)	= 1.0
STRUCTURAL FACTOR	(1.0, 1.2, or 1.4)	= 1.0
BASE DIMENSION (X-DIRECTION)	(m)	= 10
BASE DIMENSION (Y-DIRECTION)	(m)	= 10
NUMBER OF STORIES		= 10
NUMBER OF FRAMES (X-DIRECTION)		= 3
NUMBER OF FRAMES (Y-DIRECTION)		= 3

---

[1]To save space the computer output for equivalent forces in the X direction is not reproduced here.

DATA FOR EACH STORY

STORY N° 1-10

HEIGHT	(m) =	3.5
TRANSLATIONAL MASS	(kN sec²/m) =	80
ROTATIONAL MASS	(kN sec²/m) =	1333
X OF CENTER OF MASSES	(m) =	5.0
Y OF CENTER OF MASSES	(m) =	5.0

DATA FOR FRAMES (X-DIRECTION)   DATA FOR FRAMES (Y-DIRECTION)

STORY #	Y-COORDINATE (m)	STORY #	X-COORDINATE
1	0	1	0
2	6	2	6
3	10	3	10

STORY #	STIFFNESS (KN/m)	STORY #	STIFFNESS (KN/m)
1	226707	1	226707
2	226707	2	226707
3	226707	3	226707
4	226707	4	226707
5	109329	5	109329
6	109329	6	109329
7	109329	7	109329
8	44781	8	44781
9	44781	9	44781
10	44781	10	44781

STORY #	STORY HEIGHT (m)	STORY WEIGHT (kN)
1-10	3.500	784.800

STORY #	X C.M. (m)	Y C.M. (m)	X C.S. (m)	Y C.S. (m)
1-10	5.000	5.000	5.333	5.333

******************************MODAL ANALYSIS******************************

MODE N.	ω (rad/sec)	f (Hz)	T (sec)	$\dot{E}_x$	$\dot{E}_y$
1	10.878	1.7313	0.5776	16.6810	-16.6814
2	10.948	1.7424	0.5739	16.7815	16.7811
3	15.685	2.4963	0.4006	-1.8298	1.8298
4	25.931	4.1270	0.2423	8.2419	-8.2419
5	26.097	4.1534	0.2408	8.2913	8.2914
6	37.389	5.9506	0.1680	0.9041	-0.9041
7	43.587	6.9371	0.1442	-4.9459	4.9461
8	43.866	6.9815	0.1432	-4.9758	-4.9755
9	55.805	8.8817	0.1126	2.8771	-2.8770
10	56.162	8.9385	0.1119	2.8943	2.8943

*********** RESPONSE SPECTRUM ANALYSIS (X DIRECTION) ***********

EFFECTIVE DISPLACEMENTS

STORY #	Ux (m)	Uy (m)	θ (rad)
1	0.000590	0.000590	0.000018
2	0.001169	0.001169	0.000035
3	0.001726	0.001726	0.000052
4	0.002253	0.002253	0.000068
5	0.003270	0.003270	0.000098
6	0.004186	0.004186	0.000126
7	0.004977	0.004977	0.000149
8	0.006570	0.006570	0.000197
9	0.007721	0.007721	0.000232
10	0.008330	0.008330	0.000250

EFFECTIVE FORCES

STORY #	F_x (kN)	F_y (kN)	Mt (kNm)
1	12.8749	12.8745	12.4615
2	23.8218	23.8223	23.1873
3	31.6416	31.6416	31.0868
4	36.3135	36.3135	36.1091
5	43.2145	43.2141	43.6805
6	50.4797	50.4797	50.9392
7	55.8775	55.8771	56.1485
8	67.4445	67.4449	67.5555
9	77.4383	77.4384	78.4004
10	89.6717	89.6717	90.4231

EFFECTIVE LATERAL FORCES   (kN)

STORY #	X-FRAMES 1	2	3	Y-FRAMES 1	2	3
1	4.7509	4.2377	3.9355	4.7509	4.2377	3.9355
2	8.7913	7.8413	7.2832	8.7914	7.8413	7.2832
3	11.6787	10.4152	9.6763	11.6787	10.4152	9.6763
4	13.4055	11.9532	11.1088	13.4054	11.9532	11.1088
5	15.9570	14.2249	13.2261	15.9569	14.2249	13.2261
6	18.6393	16.6165	15.4491	18.6392	16.6165	15.4491
7	20.6309	18.3931	17.0988	20.6308	18.3931	17.0988
8	24.9004	22.2008	20.6368	24.9006	22.2008	20.6368
9	28.5950	25.4906	23.7019	28.5950	25.4906	23.7019
10	33.1102	29.5174	27.4431	33.1102	29.5174	27.4431

23.8 COMPUTER PROGRAM FOR MASONRY BUILDINGS DESIGN

A computer program has been prepared to implement a method that estimates the seismic resistance of masonry buildings at three different limit states: elastic, first cracking, and ultimate. This method is based on a distribution of lateral forces among the walls at each story, proportional to the respective stiffness. Each single masonry panel is assumed to follow an elastically perfect plastic behavior with a ductility factor specified by the user. Conventionally, it is assumed that the ultimate state is reached when the ductile resources are exhausted.

The program calculates, at discrete time steps, the displacement of the center of stiffness at each story, checking for changes in the stiffness of the resisting walls and redistributing the seismic force among the various walls. When a particular wall reaches the ultimate displacement, it is considered no longer functioning; the program then prints the situation at this step and proceeds with the calculations. In this way, it is possible to assume the total resistant force, either as the force related to the first cracking of a wall, or as a force that accounts for the residual resistance of the overall building. The first solution is more conservative, but consistent with Code philosophy; the second solution, although less conservative, compensates for the low values of ductility assumed for the resisting elements.

The ultimate horizontal force provided by the structure is computed as the sum of the shear resistances offered by all the panels. This is defined by the following formula:

$$T_u = A\tau_k \sqrt{\frac{1 + \sigma_0 + \sigma_x + \sigma_y}{1.5\tau_k} + \frac{(\sigma_0 + \sigma_y)\sigma_x}{2.25\tau_k^2}} \qquad (23.22)$$

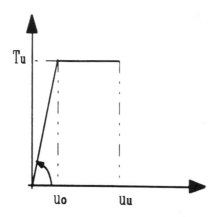

Fig. 23.3. Force versus displacement curve for masonry panels

where

τ_k = characteristic shear resistance

σ_0 = normal stress, at the center of the panel, for vertical loads

σ_y = normal stress, at the center of the panel, for eventual vertical tie rods

σ_x = normal stress, at the center of the panel, for eventual horizontal tie rods.

When the ultimate force P_u is reached, the relative displacement between the base and the top of a wall panel is defined as $u_0 = P_u/K_t$ (see Fig. 23.3). K_t is the total stiffness of the element, accounting for both the flexural and shear behavior:

$$K_t = \frac{GA}{1.2h} \cdot \frac{1}{1 + \frac{1}{1.2} \frac{G}{E} \left(\frac{h}{b} \right)^2} \qquad (23.23)$$

where

A = panel section area

t = panel thickness

b = panel width

h = panel height

G = shear modulus

E = Young's modulus.

The relation between E and G, assumed in the program $(E/G = 6)$ is suggested by the results of laboratory tests on masonry panels by Turnsek and Cacovic (1970).

The program acknowledges the principal hypotheses that the building deforms by a shear mechanism and that diaphragms at the level of each floor are considered rigid in their planes. When an eccentricity exists between a center of mass and center of stiffness, the program takes into account a torsional moment determined by the hypothesis of rigid rotation of the floor around the center of stiffness. The program detects a "torsional failure" when the ratio between

the actual displacement of the story center of mass and that of the previous iteration step exceeds the conventional value of 1.5.

In order to introduce an implicit safety factor, the design ductility factors may be taken lower than the real factor values. A reasonable value could be 1.5 for unreinforced masonry, and 2.0 for reinforced masonry. This suggestion also permits limiting the horizontal displacements of the overall building as a means of reducing the possibility of an out-of-plane failure of the wall.

23.8.1 Program Data

All the data are in free format, separated by commas.

Line 1:

Title (maximum length: 70 characters)

Line 2 (General values):

Total number of walls (max. 100)

V = height of wall (meters)

G = shear modulus (tons/meter^2)/1,000

τ_k = characteristic shear stress (tons/meter^2)

σ_y = steel yielding stress (tons/meter^2)/1,000

PA = reinforcement ratio (1/1,000)

G_{tot} = weight over the story + weight of the story (tons)

Note: The program calculates the weight of G_{tot} by using the normal stresses and the geometry of each wall. If $G_{tot} = 0$ is input as data, the calculated weight for G_{tot} will be used by the program.

Line 3 (One line for each wall):

DX = length in x direction (meters)

DY = length in y direction (meters)

X = X-coordinate of the wall center of mass (meters)

Y = Y-coordinate of the wall center of mass (meters)

So = normal stress on the sectional area of the wall (tons/meter^2)

Sor = horizontal stress due to eventual precompression (tons/meter^2)

V = height of wall (meters)

G = shear modulus (tons/meter^2)/1,000

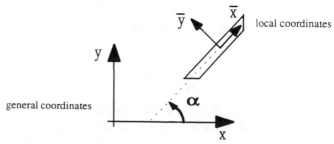

Fig. 23.4. Definition of α angle

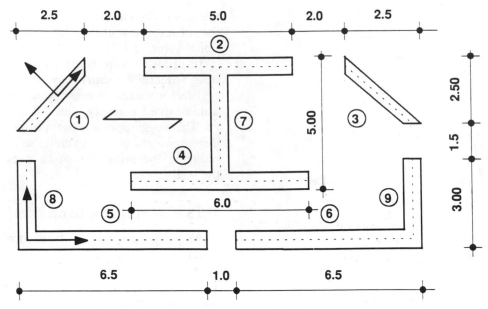

Fig. 23.5.   Plan view of building for Example 23.4

τ_k = characteristic shear stress (tons/meter^2)

ALFA = angle between local and general x axes (degree) (see Fig. 23.4)

PA = reinforcement ratio (1/1,000)

DUT = ductility ratio (1.5 for an unreinforced wall or 2.0 for a reinforced wall)

23.8.2 Example 23.4

The building in Fig. 23.5 is analyzed to illustrate the use of the program. The structure consists of three stories with an interstory height of 2.8 meters. Each wall is identified in Fig. 23.5 by a number. The thickness of the stone masonry is assumed equal to 60 cm at the first and second story and is reduced to 40 cm for the third story. The type and condition of the masonry are reported in Table 23.9.

The use of vertical and/or horizontal pretensioned tie rods is assumed to impose on the sectional areas an additional compression stress of 10 t/m^2. The values adopted in the example for the other parameters are the following:

Unit weight of stone unreinforced masonry = 1.4 t/m^3
Unit weight of brick masonry = 1.8 t/m^3
Floor loads = 500 kg/m^2
Roof loads = 400 kg/m^2

The weights for the three stories are

$$W_1 = 189\,t$$
$$W_2 = 124\,t$$
$$W_3 = 69\,t$$

In the Italian Seismic Code, the horizontal forces applied at each floor are computed as

$$F_1 = K_{hi}W_i \qquad [\text{eq.}(23.4)]$$

where

$$K_{hi} = CR\varepsilon\beta\gamma_i I \qquad [\text{eq.}(23.5)]$$

and

$$\gamma_i = h_i \frac{\displaystyle\sum_{j=1}^{N} W_j}{\displaystyle\sum_{j=1}^{N} W_j h_j} \qquad [\text{eq.}(23.7)]$$

In the example

$C = 0.1$ (Seismic Zone I)
$R = 1.0$
$\varepsilon = 1.0$
$\beta = 2 \times 2$
$I = 1.0$

Table 23.9.   Masonry Type and Conditions for Example 23.4

Wall	Original Condition	Strengthening Renovation	τ_k (t/m^2)
1	Stone, severe damage	Substitution with brick masonry	12
2	Two layers masonry, good condition	None	3.5
3	Stone, damaged	Concrete grouting	11
4	Two layers masonry, good condition	None	3.5
5	Stone, good condition	Vertical and horizontal tie rods	7
6	Stone, good condition	Vertical and horizontal tie rods	7
7	Two layers masonry, good condition	Horizontal tie rods	3.5
8	Two layers masonry, good condition	Horizontal tie rods	3.5
9	Two layers masonry, good condition	Horizontal tie rods	3.5

Table 23.10. Comparison Between Computed and Code Forces in *X* Direction

Level	Ultimate Resistant Force	Horizontal Force by Code	Walls Failed
First	262.8	151.45	2, 4
Second	235.8	107.0	2, 4
Third	126.4	48.68	2, 4

Table 23.11. Comparison Between Computed and Code Forces in *Y* Direction

Level	Ultimate Resistant Force	Horizontal Force by Code	Walls Failed
First	217.2	151.45	1, 3
Second	197.4	107.0	1, 3
Third	102.0	48.68	1, 3

thus,

$$F_i = 0.4 \, \gamma_i \, W_i$$

From the previous relations, it turns out that

$$\gamma_1 = 0.21 \times 2.8 = 0.588$$
$$\gamma_2 = 0.21 \times 5.6 = 1.176$$
$$\gamma_3 = 0.21 \times 8.4 = 1.764$$
$$F_1 = 0.4 \times 0.588 \times 189 = 44.45 \text{ t}$$
$$F_2 = 0.4 \times 1.176 \times 124 = 58.32 \text{ t}$$
$$F_3 = 0.4 \times 1.764 \times 69 = 48.68 \text{ t}.$$

The resistant force at the ultimate state, computed by the program, is compared with the total horizontal force at each story level. If the force estimated by the computer program is equal to or larger than the total force given by the Code at the same level, the building may be considered to be in compliance with the Code provisions.

In the example, the total horizontal forces at any level F_{ult} calculated by the program at the ultimate state, must be equal to or larger than the shear force at the same level, determined according to the Code:

For the first level	$F_{ult} \geqslant F_1 + F_2 + F_3 = 151.45$ t
For the second level	$F_{ult} \geqslant F_2 + F_3 = 107.0$ t
For the third level	$F_{ult} \geqslant F_3 = 48.68$ t

The results obtained using the program for the three levels of the building (one run for each level) are summarized in Tables 23.10 and 23.11.

The walls that experienced the maximum displacement consistent with the ductility factor can be detected easily by inspecting the *FI* values. These values represent the ratio between the present displacement and the displacement at the elastic limit of the wall. The ductility factor assumed in the example is 1.5. The user can accept that the ultimate state has been reached for those walls showing *FI* values close to 1.5. The ultimate displacement, in fact, is supposed to occur in the next incremental step.

The total ultimate force is computed at the last step with all the walls participating. The situation at the elastic limit and at the first cracking are printed in order to follow the evolution of the resistant element actions. The detailed results relative to the first story are presented below.

```
THREE STORY BUILDING; FIRST STORY

    DX    = length in X direction
    DY    = length in Y direction
    X,Y   = coord. wall center of masses
    SO    = normal stress
    SOR   = horizontal stress due to pre-compression
    V     = wall height
    G     = shear modulus
    TAUK  = characteristic shear stress
    SY    = steel yielding stress
    PA    = reinforcement ratio
    Tu    = maximum force
    Uo    = displacement at elastic limit
    K     = stiffness
    Uu    = ultimate displacement
    T     = actual force
    U     = actual displacement
    FI    = U/Uo
    VK    = seismic coefficient
```

WALL DATA:

Wall #	DX (m)	DY (m)	X (m)	Y (m)	SO (t/m²)	SOR* (t/m²)
1	3.535	0.450	.250	5.750	15.50	0.000
2	5.000	0.600	7.000	7.000	3.350	0.000
3	3.535	0.600	12.75	5.750	15.50	0.000
4	6.000	0.600	7.000	2.000	13.35	0.000
5	6.800	0.600	3.100	0.000	23.35	10.00
6	6.800	0.600	10.90	0.000	23.35	10.00
7	0.600	5.600	7.000	4.500	20.64	10.00
8	0.600	3.300	0.000	1.350	16.44	10.00
9	0.600	3.300	14.00	1.350	16.44	10.00

* For walls # 1, 2, 3 and 4, SOR = 0 because these walls are not provided with tie-rods; for walls # 5, 6, 7, and 8, SOR = 10 t/m², assumed additional compressive stress imposed by the tie-rods.

WALL DATA:

#	V (m)	G (t/m)/1000	TAUK (t/m²)	ANGLE (DEGREE)	SY (t/m²)/1000	PA	Uu/Uo
1	2.800	13.200	12.00	45.00	0.0000	0.000	1.500
2	2.800	3.8500	3.500	0.000	0.0000	0.000	1.500
3	2.800	12.100	11.00	-45.0	0.0000	0.000	1.500
4	2.800	3.8500	3.500	0.000	0.0000	0.000	1.500
5	2.800	7.7000	7.000	0.000	0.0000	0.000	2.000
6	2.800	7.7000	7.000	0.000	0.0000	0.000	2.000
7	2.800	3.8500	3.500	0.000	0.0000	0.000	2.000
8	2.800	3.8500	3.500	0.000	0.0000	0.000	2.000
9	2.800	3.8500	3.500	0.000	0.0000	0.000	2.000

```
WEIGHT FROM DATA = 0
WEIGHT COMPUTED  = 470.631

    X         Y
7.1`04    2.3527   center of masses
7.2843    2.3083   center of stiff
```

CHARACTERISTICS OF WALLS IN X DIRECTION:

#	Tu (t)	Uo (m)	K (t/m)
1	2,604E+01	7.741E-03	3.364E+03
2	1,976E+01	6.000E-03	3.294E+03
3	3.249E+01	7.282E-03	4.462E+03
4	2.372E+01	5.923E-03	4.004E+03

5	7.165E+01	7.844E-03	9.135E+03
6	7.165E+01	7.844E-03	9.135E+03
7	4.451E+01	4.653E-02	9.565E+02
8	2.401E+01	4.259E-02	5.637E+02
9	2.401E+01	4.259E-02	5.637E+02

CHARACTERISTICS OF STORY IN X DIRECTION

VALUES AT ELASTIC LIMIT FOR EACH STORY

```
RESISTANT FORCE : 2.103E+02 t
DISPLACEMENT    : 5.927E-03 m
STIFFNESS       : 3.548E+04 t/m
VK              : 4.468E-01
```

Editor'note: To save space the detailed computer output for the values at elastic limit and at the first cracking for each wall is not reported here.

VALUES AT ULTIMATE STATE FOR THE STORY

```
RESISTANT FORCE : 2.628E+02 t
DISPLACEMENT    : 8.358E-03 m
STIFFNESS       : 3.152E+04 t/m
VK              : 5.584E-01
```

VALUES AT ULTIMATE STATE FOR EACH WALL

	T (t)	U (m)	K (t/m)	FI
1	2.604E+01	8.635E-03	3.016E+03	0.112E+00
2	1.976E+01	8.737E-03	2.262E+03	0.146E+00
3	3.249E+01	8.635E-03	3.763E+03	0.119E+00
4	2.372E+01	8.330E-03	2.847E+03	0.141E+01
5	7.165E+01	8.167E-03	8.774E+03	0.104E+00
6	7.165E+01	8.167E-03	8.774E.03	0.104E+00
7	8.162E+00	8.533E-03	9.565E.02	0.183E+00
8	4.665E+00	8.277E-03	5.637E+02	0.194E+00
9	4.665E+00	8.277E-03	5.637E+02	0.194E+00

CHARACTERISTICS OF WALLS IN Y DIRECTION:

	T_u (t)	U_o (m)	K (t/m)
1	2.604E+01	7.741E-03	3.364E+03
2	1.976E+01	2.314E-02	8.541E+02
3	3.249E+01	7.282E-03	4.462E+03
4	2.372E+01	2.314E-02	1.025E+03
5	7.165E+01	3.084E-02	2.323E+03
6	7.165E+01	3.084E-02	2.323E+03
7	4.451E.01	1.196E-02	3.721E+03
8	2.401E+01	1.164E-02	2.063E+03
9	2.401E+01	1.164E-02	2.063E+03

CHARACTERISTICS OF STORY IN Y DIRECTION:

VALUES AT ELASTIC LIMIT FOR EACH STORY

```
RESISTANT FORCE : 1.663E+02 t
DISPLACEMENT    : 7.499E-03 m
STIFFNESS       : 2.220E+04 t/m
VK              : 3.533E-01
```

Editor's note: To save space the detailed computer output for the values at elastic limit and at the first cracking for each wall is not reported here.

VALUES FOR ULTIMATE STATE OF WALL. n. 1

```
RESISTANT FORCE : 2.177E+02 t
DISPLACEMENT    : 1.107E-02 m
STIFFNESS       : 1.967E+04 t/m
VK              : 4.626E-01
```

VALUES AT ULTIMATE STATE FOR EACH WALL

	T (t)	U (m)	K (t/m)	FI
1	2.604E+01	1.132E-02	2.300E+03	0.146E+01
2	9.461E+00	1.108E-02	8.541E+02	0.479E+00
3	3.249E+01	1.083E-02	2.999E+03	0.149E+01
4	1.135E+01	1.108E-02	1.025E+03	0.479E+00
5	2.612E+01	1.124E-02	2.323E+03	0.364E+00
6	2.535E+01	1.091E-02	2.323E+03	0.354E+00
7	4.122E+01	1.108E-02	3.721E+03	0.926E+00
8	2.346E+01	1.137E-03	2.063E+03	0.977E+00
9	2.224E+01	1.078E-02	2.063E+03	0.926E+00

VALUES FOR INCREMENTAL DISPLACEMENT

ULTIMATE STATE DETECTED AT WALLS:

```
    1
    3
RESISTANT FORCE : 2.356E+02 t
DISPLACEMENT    : 2.193E-02 m
STIFFNESS       : 1.075E+04 t/m
VK              : 5.005E-01
```

VALUES, AT LAST INCREMENTAL STEP, FOR EACH WALL

	T (t)	U (m)	K (t/m)	FI
1	0.000E+00	2.126E-02	0.000E+00	0.275E+01
2	1.872E+01	2.192E-02	8.541E+02	0.947E+00
3	0.000E+00	2.258E-02	0.000E+00	0.310E+01
4	2.247E+01	2.192E-02	1.025E+03	0.947E+00
5	4.988E+01	2.147E-02	2.323E+03	0.696E+00
6	5.196E+01	2.237E-02	2.323E+03	0.725E+00
7	4.451E+01	2.192E-02	2.030E+03	0.183E+00
8	2.401E+01	2.112E-03	1.0137+03	0.181E+00
9	2.401E+01	2.272E-02	1.056E+03	0.195E+00

REFERENCES

CHIANTINI, B., and CIPOLLINI, A. (1988) "Normativa Sismica." Edizioni Dei Roma, Tipografia del Genio Civile, Rome, Italy.

CIPOLLINI, A. (1986) "Normativa Sismica." *Ingegneria Sismica*, No. 2, Bologna, Italy.

Consiglio Superiore dei Lavori Pubblici, Servizio Sismico (1986) *Atlante della Classificazione Sismica Nazionale.* Istituto Poligrafico e Zecca dello Stato, Rome, Italy.

FREEMAN, J. R. (1932) *Earthquake Damage and Earthquake Insurance.* McGraw-Hill, New York, NY.

GAVARINI, C. (1980) *Aggiornamento della Normativa Antisismica.* Convegno Annuale C.N.R. Progetto Finalizzato Geodinamica, Rome, Italy.

Ministero dei Lavori Pubblici G. U. 12-5-1986 Decreto 24 Gennaio (1986) *Norme tecniche relative alle costruzioni sismiche.* Rome, Italy.

"Norme Tecniche per le Costruzioni in Zone Sismiche." (1985) *Ingegneria Sismica* No. 1, Bologna, Italy. Gruppo Nazionale per la Difesa dai Terremoti (National Group for Earthquake Prevention).

PETRINI, V. (1991) "Il Rischio Sismico in Italia." Private communication, Politecnico Di Milano, Milano, Italy.

PINTO, P. (1991) "Trends and Developments in the European Seismic Code for The New Reinforced Concrete Constructions." Proceedings, International Meeting on Earthquake Protection of Buildings, Ancona, Italy.

Regione Autonoma Friuli-Venezia Giulia (1976) *DT2 Raccomandazioni per la Riparazione Strutturale degli Edifici in Muratura.* Trieste, Italy.

Regione dell'Umbria (1981) *Direttive Tecniche ed Esemplificazioni delle metodologie di Interventi per la Riparazione ed il Consolidamento degli Edifici danneggiati da Eventi Sismici.* Dipartimento per l'Assetto del Territorio, Perugia, Italy.

TURNSEK, V., and CACOVIC, F. (1970) "Some Experimental Results on the Strength of Brick Masonry Walls." Proceedings, 2nd International Brick Masonry Conference, Stoke on Trent, England.

24

Japan

Yoshikazu Kitagawa and Fumio Takino

24.1 INTRODUCTION

Japan is located in the center of several seismic regions. There are both a high and a medium seismic region on the Pacific Ocean side, a medium seismic region on the Sea of Japan side, and a low seismic region in the inland areas of the country.

Since the catastrophic Kanto earthquake of 1923, Japan has experienced a number of severe earthquakes that have caused considerable damage to buildings and civil structures. It was after the Kanto earthquake that seismic forces were first considered in the design of building structures; subsequently, they were also considered in the design of other structures. The seismic coefficient method was developed and introduced for the practical design of structures in the year 1926. Seismic regulations were reviewed and amended several times as a result of severe damage during consecutive strong motion earthquakes. Since 1981, structural seismic resistant research has changed its orientation from static analysis to dynamic analysis. Dynamic analysis, aided by rapid progress in both theory and computational advances, is providing a more precise design method for structures subjected to strong motion earthquakes. However, despite advances in the theoretical modeling of structures, it is not yet possible to predict accurately the response of structures when subjected to future earthquakes whose characteristics are unknown.

This chapter presents the main provisions for seismic resistant design of buildings stipulated in both the *Building Standard Law Enforcement Order (BSLEO)*, published by the Ministry of Construction (1981), and in the *Standards for Seismic Civil Engineering Construction in Japan* (1980).

24.2 HORIZONTAL SEISMIC SHEAR

The Japanese Code for earthquake-resistant design of buildings requires that buildings be designed to withstand lateral seismic shear forces Q_i calculated for each of the main axes of the building by

$$Q_i = C_i \bar{W}_i \qquad (24.1)$$

In eq.(24.1), C_i is the horizontal seismic shear coefficient determined in accordance with eq.(24.2), and $\bar{W}_i$ is the portion of the total seismic weight of the building attributed to level i and those levels above i. The seismic weight of the building is equal to the sum of the dead load and the applicable portion of the design live load. In heavy snow regions, it also includes the effect of the snow load.

The horizontal seismic shear coefficient C_i for the ith story is determined by the following relationship:

$$C_i = Z R_t A_i C_0 \qquad (24.2)$$

The factor Z in eq.(24.2) is the seismic coefficient whose value is provided by the Minister of Construction for a specific region of the country. The seismic coefficient Z takes into account the extent of earthquake damage, seismic activity, and other seismic characteristics based on the records of past earthquakes in the specific region. Values of the seismic coefficient are obtained from the seismic map in Fig. 24.1. These values range from 0.7 to 1.0.

The factor R_t in eq.(24.2) is the design spectral coefficient, which represents the vibrational characteristics of the building. It depends on both the fundamental period T of the building and on the soil profile of the surrounding ground at the site. Fig. 24.2 shows a plot of the design spectral coefficient R_t versus the fundamental period T for three different soil profile types that are described in Table 24.1. In this table T_c is the characteristic period of the soil.

In eq.(24.2), the factor A_i is the horizontal shear distribution factor calculated by

$$A_i = 1 + \left(\frac{1}{\sqrt{\alpha_i}} - \alpha_i \right) \frac{2T}{1 + 3T} \qquad (24.3)$$

in which T is the fundamental period of the building and α_i is given by:

$$\alpha_i = \frac{\bar{W}_i}{W_0} \qquad (24.4)$$

where W_i is the weight above level i and W_0 is the weight above the ground level.

Figure 24.3 shows a parametric representation of the lateral shear distribution factor A_i as a function of α_i and the fundamental period T.

The parameter C_0 in eq.(24.2) is the standard shear coefficient, which has a value between 0.2 and 1.0; 0.2 for moderate earthquake motions and 1.0 for severe earthquake motions.

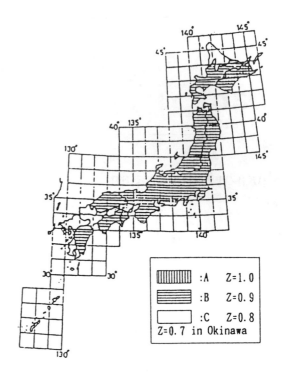

Fig. 24.1. Seismic coefficient Z

Table 24.1. Classification of Soils

Soil Type	Ground Characteristics	T_c
Type 1 (Hard Soil)	Ground consisting of rock, hard sandy gravel, etc., that is classified as tertiary or older. Ground whose period estimated by calculation or other investigation is equivalent to that of hard soil.	0.4
Type 2 (Medium Soil)	Other than Type 1 or 3.	0.6
Type 3 (Soft Soil)	Alluvium consisting of soft delta deposits, topsoil, mud, or the like (including fills, if any), whose depth is 30 m or more. Land obtained by reclamation of a marsh, muddy sea bottom, etc., where the depth of the reclaimed ground is 3 m or more, and where 30 years have not yet elapsed since the time of reclamation. Ground whose period estimated by calculation or by other investigation is equivalent to that of soft soil.	0.8

In the determination of the design spectral coefficient R_t and the horizontal shear distribution factor A_i, it is necessary to know the value of the fundamental period of the building. The Code provides the following empirical formula to estimate the fundamental period T:

$$T = h(0.02 + 0.01\gamma) \text{ sec} \qquad (24.5)$$

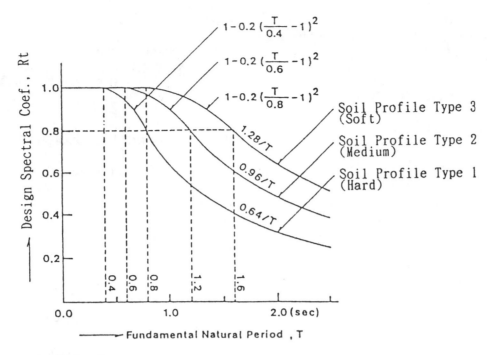

Fig. 24.2. Design spectral coefficient R_t

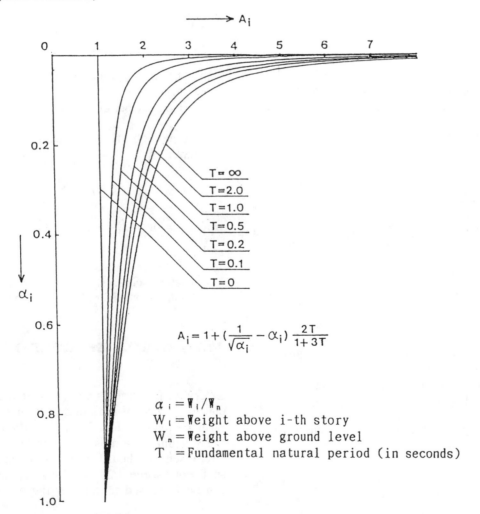

$$A_i = 1 + \left(\frac{1}{\sqrt{\alpha_i}} - \alpha_i\right)\frac{2T}{1+3T}$$

$\alpha_i = W_i/W_n$
$W_i = $ Weight above i-th story
$W_n = $ Weight above ground level
$T\ = $ Fundamental natural period (in seconds)

Fig. 24.3. Horizontal shear distribution factor A_i

where γ is the ratio of the total height of stories of steel construction to the total height of the building, and h is the height of the building.[1]

24.3 SEISMIC LATERAL FORCES

Seismic lateral forces F_i at the various levels of the building are calculated by

$$F_i = Q_i - Q_{i+1} \qquad (24.6)$$

where the horizontal seismic shear forces Q_i and Q_{i+1} are obtained from eq.(24.1).

24.4 HORIZONTAL SEISMIC SHEAR OF APPENDAGES

The horizontal seismic shear force q for penthouses, chimneys, towers, cisterns, parapets, and other appendages on buildings must be determined in accordance with the following relationship:

$$q = KW \qquad (24.7)$$

where

   K = seismic design coefficient of appendages ($K = 1.0$ but it can be reduced to 0.5 in cases where there is no risk to human lives)
   W = weight of the appendage.

24.5 HORIZONTAL SEISMIC SHEAR OF THE BASEMENT

The horizontal seismic shear Q_B at the basement of a building must be determined using the following relationship:

$$Q_B = Q_p + KW_B \qquad (24.8)$$

where

   Q_p = portion of the lateral seismic shear of the first story that extends to the basement
   K = seismic design coefficient of the basement
   W_B = weight of the basement.

The seismic design coefficient K of the basement is given by

$$K \geqslant 0.1\left(1 - \frac{H}{40}\right)Z \qquad (24.9)$$

where

   H = depth of the basement in meters (the value of H is fixed to a maximum of 20 meters)
   Z = seismic coefficient as defined in Section 24.3 and shown in Fig. 24.1.

24.6 OVERTURNING MOMENTS

The seismic lateral forces F_i produce overturning moments in the building that result in additional axial forces in the columns and other resisting elements, especially in the exterior columns of the building. While the building is subjected to earthquake motions, these overturning moments increase the gravity load effect on the exterior columns on one side at a time. The overturning moment M_i at a level i of the building is determined by statics as the moment produced at that level by the above seismic lateral forces F_i; that is,

$$M_i = \sum_{j=i+1}^{n} F_j(h_j - h_i) \qquad (24.10)$$

where $i = 0, 1, 2, \ldots, n-1$
and

   F_j = seismic lateral force at level j
   h_j = height of level j
   n = total number of stories.

Alternatively, the overturning moments M_i may be calculated by

$$M_i = \sum_{j=i+1}^{n} Q_j H_j \qquad (24.11)$$

where $i = 0, 1, 2, \ldots, n-1$
and

   Q_j = lateral seismic shear force at story j
   H_j = interstory height of story j of the building.

24.7 LATERAL DISPLACEMENTS AND STORY DRIFT

Lateral displacements and story drift are calculated by elastic analysis of the building subjected to the lateral seismic forces F_i (Fig. 24.4). The 1980 Code states that the maximum permissible story drift caused by the lateral seismic shear for moderate earthquake motions shall not exceed 1/200 of the story height. This value can be increased to 1/120 if the non-structural mem-

---

[1]Alternatively, the fundamental period may be calculated using the Rayleigh formula [see eq.(3.14) of Chapter 3].

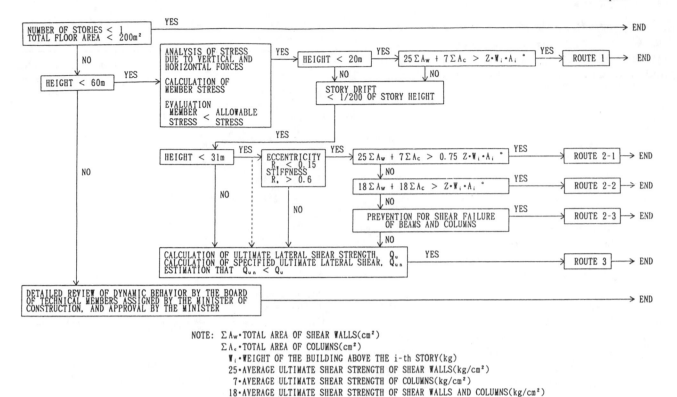

NOTE: ΣA_w·TOTAL AREA OF SHEAR WALLS(cm²)

ΣA_c·TOTAL AREA OF COLUMNS(cm²)

W_i·WEIGHT OF THE BUILDING ABOVE THE i-th STORY(kg)

25·AVERAGE ULTIMATE SHEAR STRENGTH OF SHEAR WALLS(kg/cm²)

7·AVERAGE ULTIMATE SHEAR STRENGTH OF COLUMNS(kg/cm²)

18·AVERAGE ULTIMATE SHEAR STRENGTH OF SHEAR WALLS AND COLUMNS(kg/cm²)

NOTE: 1 kg·9.8N

1 kg/cm²·98kpa

Fig. 24.4. Flow chart for design of reinforced concrete buildings

bers will incur no severe damage at the increased story drift limitation.

24.8 ECCENTRICITY RATIO AND STIFFNESS RATIO

The inertial force that results from the earthquake motion acts at each story of the building at the center of the mass above that story, while the resultant resisting force is at the center of stiffness of the story. If the structure is not dynamically symmetric, the center of stiffness at a story and the center of the mass above that story may not coincide.

The torsional stiffness J_{xy} for a story of the building is calculated as the sum of the stiffnesses K_{xi} and K_{yi} for the resisting elements of that story, multiplied by the square of the normal distance, respectively y_i and x_i, from the center of mass to the resisting element; that is,

$$J_{xy} = \sum_{i=1}^{N_x} K_{xi} y_i^2 + \sum_{i=1}^{N_y} K_{yi} x_i^2 \qquad (24.12)$$

where N_x and N_y are, respectively, the number of resisting elements of the story in the x and y directions. By analogy with the area moment of inertia, the torsional stiffness radii of gyration r_{ex} and r_{ey}, for the main directions x and y, are calculated as

$$r_{ex} = \sqrt{\frac{J_{xy}}{\sum_{i=1}^{N_x} K_{xi}}}$$

$$\qquad (24.13)$$

$$r_{ey} = \sqrt{\frac{J_{xy}}{\sum_{i=1}^{N_y} K_{yi}}}$$

The eccentricity ratios R_{ex} and R_{ey} at a story for the main directions x and y are then defined as

$$R_{ex} = \frac{e_x}{r_{ex}} \le 0.15$$

$$\qquad (24.14)$$

$$R_{ey} = \frac{e_y}{r_{ey}} \le 0.15$$

where e_x and e_y are, respectively, the eccentricities in the x and the y directions; that is, the distance between the center of mass and the center of stiffness of a story.

The 1980 Code limits the maximum eccentricity

ratios R_{ex} and R_{ey} to 0.15 as indicated in eqs.(24.14). The Code also limits the stiffness ratio R_s to a maximum value of 0.6. The stiffness ratio R_s corresponding to the x and y directions is defined as

$$R_{sx} = \frac{r_{sx}}{\overline{r}_{sx}} \geq 0.6$$

$$\text{(24.15)}$$

$$R_{sy} = \frac{r_{sy}}{\overline{r}_{sy}} \geq 0.6$$

where

r_{sx}, r_{sy} = ratio between the story height and the story drift in the x, y direction

$\overline{r}_{sx}, \overline{r}_{sy}$ = mean value of r_{sx}, r_{sy} calculated for all stories of the building above ground level.

The building code requires an ultimate strength analysis when the eccentricity ratio R_{ex} or R_{ey} exceeds the limit 0.15 indicated in eqs.(24.14) or when the relative stiffness R_{sx} or R_{sy} is less than the limit 0.6 indicated in eqs.(24.15). Such analysis uses complex empirical formulas beyond the scope of this chapter.

24.9 DESIGN CONSIDERATIONS

There are four requirements that buildings of steel construction must satisfy:

1. For steel structures, the lateral seismic shear due to moderate earthquake motions must be measured according to the following relationships:

$$Q_{bi} = (1 + 0.7\beta_i)Q_i$$

$$Q_{bi} \leq 1.5Q_i$$

$$\text{(24.16)}$$

where

Q_{bi} = increased lateral seismic shear

β_i = ratio of lateral shear in the bracing elements to the total seismic shear of the story

Q_i = horizontal seismic shear due to moderate earthquake motions (eq.24.1).

2. Each bracing element must satisfy the following relationship:

$$J_u^P \geq 1.2M_y^P \qquad \text{(24.17)}$$

where

J_u^P = strength of the bracing element at the joint

M_y^P = yield strength of the bracing element.

3. The width-to-thickness ratio of plate elements of the columns must satisfy the requirements stated in Table 24.2.

Table 24.2. Width-to-Thickness Ratio of Steel Columns and Beams

Members	Section	Portion	Steel	Width-to-Thickness Ratio	
				Standard	Maximum
Columns	⊢	Flange	SS41*	9.5	12
			SM50†	8	10
		Web	SS41	43	45
			SM50	37	39
	□		SS41	33	37
			SM50	27	32
	○		SS41	50	70
			SM50	36	50
Beams		Flange	SS41	9	11
			SM50	7.5	9.5
	I	Web	SS41	60	65
			SM50	51	55

*Steel conforming to SS41, SMA41, STK41, and STKR41 of JIS (standard strength $F = 2.4\,\text{t/cm}^2$).

†Steel conforming to SM50, SMA50, SM50Y, STK50, and STKR50 of JIS (standard strength $F = 3.3\,\text{t/cm}^2$).

4. Beam-to-column connections that are subjected to a bending moment must satisfy the following relationship:

$$M_u > \alpha M_p \qquad \text{(24.18)}$$

where

M_u = maximum bending strength of the connection

M_p = full plastic moment of the column or beam

α = safety factor (a value between 1.2 and 1.3).

24.10 ULTIMATE LATERAL SHEAR STRENGTH

The specific ultimate lateral shear strength Q_{ui} is determined by the following formula:

$$Q_{ui} = D_s F_{es} Q_i \qquad \text{(24.19)}$$

where Q_i is the lateral seismic shear force for severe earthquake motions prescribed in eq.(24.1).

The factor D_s in eq.(24.19) is a structural coefficient that represents the structural characteristics of the story. The factor D_s depends on the damping and ductile properties of the story. Tables 24.3a and 24.3b provide values of D_s for different types of construction.

The factor F_{es} in eq.(24.19) represents the stiffness and eccentricity characteristics of the story. Its value is determined by

$$F_{es} = F_e \cdot F_s \qquad \text{(24.20)}$$

The factor F_e in eq.(24.20) is given in Table 24.4 as a function of the stiffness and eccentricity of the story,

Table 24.3a. Structural Coefficient D_s for Buildings of Steel Construction

Ductility Behavior	Frame Type		
	(1) Ductile moment frame	(2) Frame other than listed in (1) and (3)	(3) Frame with compressive braces
A. Members having excellent ductility	0.25	0.30	0.35
B. Members having good ductility	0.30	0.35	0.40
C. Members having fair ductility	0.35	0.40	0.45
D. Members having poor ductility	0.40	0.45	0.50

Table 24.3b. Structural Coefficient D_s for Buildings of Reinforced Concrete or Steel-Encased Reinforced Concrete Construction*

Ductility Behavior	Frame Type		
	(1) Ductile moment frame	(2) Frame other than listed in (1) and (3)	(3) Frame with shear walls or braces
A. Members having excellent ductility	0.30	0.35	0.40
B. Members having good ductility	0.35	0.40	0.45
C. Members having fair ductility	0.40	0.45	0.50
D. Members having poor ductility	0.45	0.50	0.55

*For steel-encased reinforced concrete construction, the values shown above should be reduced by 0.05.

Table 24.4. Values of the Factor F_e as Function of Eccentricity Ratio R_e

R_e	F_e
Less than 0.15	1.0
$0.15 \leqslant R_e \leqslant 0.3$	Linear interpolation
Greater than 0.3	1.5

Table 24.5. Values of the Factor F_s as Function of Stiffness Ratio R_s

R_s	F_s
Greater than 0.6	1.0
$0.3 \leqslant R_s \leqslant 0.6$	Linear interpolation
Less than 0.3	1.5

and F_s is given in Table 24.5 as a function of the variation of the stiffness ratio R_s defined in eqs.(24.15).

24.11 STRUCTURAL REQUIREMENTS

Buildings shall be analyzed for the combined action of vertical load due to gravity and lateral forces due to seismic load. Calculated member stresses resulting from this combined load must be less than the material allowable stress.

Building structures must meet the relevant structural requirements specified in the BSLEO and in the Specifications of the Architectural Institute of Japan. Specific requirements depend on the structural type, floor area, height, and materials, with special provisions for steel and concrete structures.

24.11.1 One- or Two-Story Buildings

The following types of buildings are only required to meet the relevant structural requirements specified in the *Building Standard Law Enforcement Order (BSLEO)* (1981):

1. One- or two-story wooden buildings not exceeding 500 m² in total floor space
2. One-story buildings other than wooden ones, not exceeding 200 m² in total floor space
3. Special buildings used for schools, hospitals, etc., not exceeding 100 m² in total floor area.

24.11.2 Reinforced Concrete Buildings

24.11.2.1 Buildings not exceeding 20 m in height. For reinforced concrete buildings that do not exceed 20 m in height, it is sufficient to satisfy the following relationship at each story of the building:

$$25\Sigma A_w + 7\Sigma A_c \geqslant Z\overline{W}_i A_i \qquad (24.21)$$

where

ΣA_w = sum of the horizontal cross-sectional areas (cm²) of the reinforced concrete walls
ΣA_c = sum of the horizontal cross-sectional areas (cm²) of the reinforced concrete columns and reinforced concrete walls, except shear walls

Z = seismic coefficient (Fig. 24.1)

A_i = horizontal shear distribution factor (Fig. 24.3)

$\bar{W}_i$ = seismic weight (kg) of the building above the story considered.

24.11.2.2 Buildings not exceeding 31 m in height.
Reinforced concrete buildings not exceeding 31 m in height must satisfy the following requirements:

1. Story drift for any story in the building must not exceed 1/200 of story height.
2. Eccentricity ratio R_e must satisfy eq.(24.14): $R_e \leqslant 0.15$.
3. Stiffness ratio R_s must satisfy eq.(24.15): $R_s \geqslant 0.6$.
4. In addition, these buildings must satisfy one of the following requirements:
   (a) At each story of the building, the following relationship must be satisfied in both the longitudinal and the transverse directions:

$$25\Sigma A_w + 7\Sigma A_c + 10\Sigma A_c' \geqslant 0.75 Z A_i \bar{W}_i \quad (24.22)$$

where

ΣA_w = sum of the horizontal cross-sectional areas (cm^2) of the reinforced concrete shear walls

ΣA_c = sum of the horizontal cross-sectional areas (cm^2) of the reinforced concrete columns and reinforced concrete walls (except shear walls)

$\Sigma A_c'$ = sum of the horizontal cross-sectional areas (cm^2) of steel-encased reinforced concrete columns

Z = seismic coefficient (Fig. 24.1)

A_i = horizontal shear distribution factor (Fig. 24.3)

$\bar{W}_i$ = weight (kg) of the building above the story considered.

   (b) At each story of the building, the following relationship must be satisfied in both the longitudinal and the transverse directions:

$$18\Sigma A_w + 18\Sigma A_c \geqslant Z A_i \bar{W}_i \quad (24.23)$$

   (c) The retained shear strength of each reinforced concrete member must be greater than the retained flexural strength of the member.

24.11.2.3 Buildings not exceeding 60 m in height.
Reinforced concrete buildings not exceeding 60 m in height must satisfy the limitation that the story drift not exceed 1/200 of the story height. In addition, the specified ultimate lateral shear Q_{ui} given by eq.(24.19) must not exceed the calculated ultimate lateral shear strength Q_u:

$$Q_{ui} \leqslant Q_u$$

with

$$Q_{ui} = D_s F_{es} Q_i$$

where D_s, F_{es}, and Q_i are parameters defined in Section 24.10.

24.11.2.4 Buildings exceeding 60 m in height.
Buildings exceeding 60 m in height require special permission from the Minister of Construction. Such permission is granted after a detailed review of the dynamic analysis by a board of technical members.

Figure 24.4 is a flow chart for the design of reinforced concrete buildings according to the specifications in the *Building Standard Law Enforcement Order (BSLEO)* (1981) for seismic civil engineering construction in Japan.

24.11.3 Steel Buildings

24.11.3.1 Buildings not exceeding three stories in height.
Steel buildings that do not exceed three stories in height and that also conform to the following limitations:

Height $\leqslant 13$ m

Maximum span of beams $\leqslant 6$ m

Total floor area $\leqslant 500$ m^2

Buildings in this category should be designed for a 50% increase in seismic shear, with $C_0 = 0.3$ [eq.(24.2)].

Alternatively, buildings in this category may be designed to satisfy the following requirements for buildings not exceeding 31 m in height.

24.11.3.2 Buildings not exceeding 31 m in height.
Steel buildings not exceeding 31 m in height must satisfy the following requirements:

1. Story drift must not exceed 1/200 of the story height.
2. Eccentricity ratio R_e must not exceed the limit specified by eq.(24.14): $R_e \leqslant 0.15$.
3. Stiffness ratio R_s must satisfy eq.(24.15): $R_s \geqslant 0.6$.
4. The design seismic shear Q_i [eq.(24.1)] must be increased by the factor indicated in eq.(24.16):

$$Q_{bi} = (1 + 0.7\beta_i) Q_i$$
$$Q_{bi} \leqslant 1.5 Q_i$$

where

Q_{bi} = increased lateral seismic shear

β_i = ratio of lateral shear in the bracing elements to the total seismic shear of the story.

5. Design must satisfy the requirements to prevent joint failure of braces [eq.(24.17)].

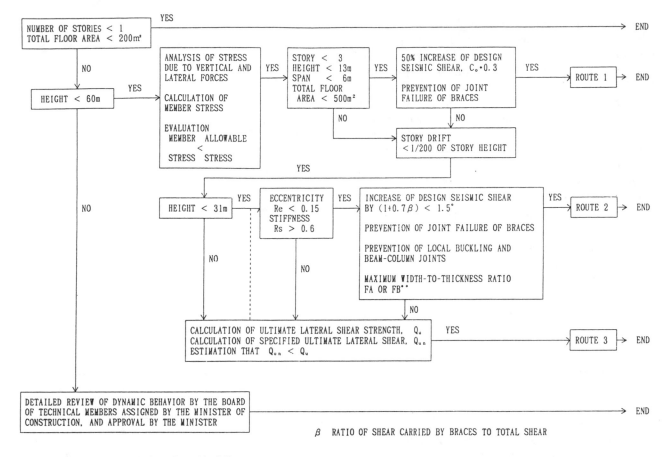

Fig. 24.5. Flow chart for design of steel buildings

6. Design must satisfy the requirement to prevent local buckling and failure of beam-column joints [eq.(24.18)].

7. Maximum width-to-thickness ratio, Table 24.2.

Alternatively, steel buildings in this category may be designed to satisfy the following requirements for buildings not exceeding 60 m in height.

24.11.3.3 Buildings not exceeding 60 m in height. Steel buildings not exceeding 60 m in height must satisfy the requirement that the specified ultimate lateral shear Q_{ui} given by eq.(24.19) does not exceed the calculated ultimate lateral shear strength Q_u:

$$Q_{ui} \leqq Q_u$$

24.11.3.4 Buildings exceeding 60 m in height. Buildings exceeding 60 m in height require special permission from the Minister of Construction. Such permission is granted after a detailed review of the dynamic analysis by a board of technical members.

Fig. 24.5 is a flow chart for the design of steel concrete buildings according to the specifications in the BSLEO.

24.12 SEISMIC LOADING FOR A HIGH-RISE BUILDING

The BSLEO requires that buildings higher than 60 m must be licensed by the Minister of Construction. Dynamic analysis is required in the design of high-rise buildings. The earthquake record for this analysis may be chosen by considering the following:

1. A record of an actual strong earthquake motion (e.g., El Centro 1940 N–S, Taft 1952 E–W, Hachinohe 1986 N–S, E–W)

2. A record of an actual strong earthquake motion with the amplitude and characteristic period modified

3. An artificial earthquake motion (e.g., simulated earthquake motions).

The seismic intensity (peak ground acceleration) should be determined statistically using expected values that consider the recurrence term that corresponds to the local conditions.

Currently, the seismic intensity may be estimated as follows:

For elastic response: 0.15–0.25g'

For elastoplastic response: 0.30–0.50g'.

The fundamental period of a high-rise building is relatively long, around three sec for a 30-story building and around five sec for a 50-story building. Thus, the fundamental mode of vibration of a high-rise building is virtually not affected by the interaction between the building and the ground. The damping for a high-rise building is about 2% of the critical damping for the fundamental mode; it is around 3% for the second mode and somewhat greater for the higher modes of vibrations.

The vertical component of the earthquake motion is not considered in the design of buildings. However, the vertical natural period in a high-rise building of 30 to 50 stories is around 0.3 to 0.6 sec. This range for the natural vertical period of a high-rise building falls in the range of greater magnification response. Consequently, for a high-rise building, vertical magnification of the response is larger than that of the horizontal motions.

24.13 APPLICATION OF THE JAPANESE SEISMIC BUILDING CODE

In Japan, generally, buildings under five or six stories are reinforced concrete constructions; higher buildings of concrete construction are steel-encased; buildings of steel construction range from low- to high-rise buildings.

Example 24.1

A three-story reinforced concrete building has the dimensions shown in Fig. 24.6. The size of the columns for the first story is $60\,cm \times 60\,cm$, for the second story $55\,cm \times 55\,cm$, and for the third story $45\,cm \times 45\,cm$. The height of the stories is $3.8\,m$. There is also a shear wall in the x direction placed eccentrically. The dead load consists of the floor slab, beams, half of the weight of the columns above and below the floor, partition walls, and a portion of the live load. The dead load is estimated to be $W_1 = 419.4$ tonne, $W_2 = 408.0$ tonne, and $W_3 = 362.5$ tonne for the three stories of the building, respectively.

The site of the building located in Tokyo is assumed to be alluvial soil (volcanic cohesive soil). The seismic analysis of this structure must be performed in accordance with the BSLEO.

Solution

In the modeling of this structure, the shear wall is replaced by an equivalent braced system. Considering the elastic characteristics of the soil under the foundations, the model building is assumed to be supported by springs. The stiffness of each spring under the columns of the building is estimated to be 200

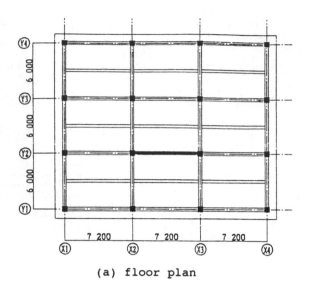

(a) floor plan

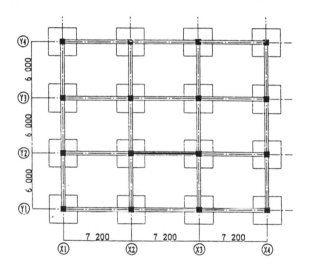

(b) footing plan

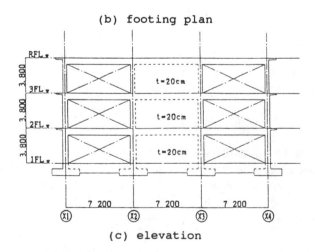

(c) elevation

Fig. 24.6. Building for Example 24.1

tonne/cm. The effective horizontal stiffness of the shear walls is reduced because of the presence of these springs, which prevents the seismic story shear force from being concentrated on the shear wall of the building. Consequently, a significant amount of this force acts upon the columns.

(a) Lateral seismic forces

$$Q_i = C_i \bar{W}_i \qquad [\text{eq.}(24.1)]$$

where

$$C_i = Z R_t A_i C_0$$

$$\bar{W}_i = \sum_{j=i}^{3} W_j$$

$Z = 1.0$ for Tokyo (Fig. 24.1)
$C_0 = 0.2$ (standard value for moderate earthquake motion)
$T = h(0.02 + 0.01\gamma)$ [eq.(24.5)]
  $= 11.4(0.02 + 0.01 \times 0)$ ($\gamma = 0$, no steel construction))
  $= 0.228$ sec
$R_t = 1.0$ (Fig. 24.2)

$$A_i = 1 + \left(\frac{1}{\sqrt{\alpha_i}} - \alpha_i \right) \frac{2T}{1 + 3T} \qquad [\text{eq.}(24.3)]$$

where

$$\alpha_i = \frac{\bar{W}_i}{W_0} \qquad [\text{eq.}(24.4)]$$

Calculations are shown in Table 24.6.
Lateral seismic forces:

$$F_i = Q_i - Q_{i+1} \qquad [\text{eq.}(24.6)]$$

Values calculated for the lateral seismic forces are shown in the last column of Table 24.6.

(b) Lateral displacements. Lateral displacements were obtained using a program for structural analysis which considers a plane frame that is acted upon by the horizontal forces F_i, shown in the last column of Table 24.6. Table 24.7 shows the resulting lateral displacements δ_i at the various levels of the building, as well as the values for the relative story drifts Δ_i/H.

Table 24.6. Calculation of Seismic Shear Force (tonne)

Level	W_i	ΣW_i	α_i	A_i	C_i	Q_i	F_i
3	362.5	362.5	0.305	1.408	0.282	102.1	102.1
2	408.0	770.5	0.648	1.161	0.232	178.9	76.8
1	419.4	1,189.9	1.000	1.000	0.200	238.0	59.1

Table 24.7. Lateral Displacements δ_i and Relative Story Drifts Δ_i/H

Level	X Direction		Y Direction	
(i)	δ_i (cm)	Δ_i/H	δ_i (cm)	Δ_i/H
3	1.133	1/1,013	1.913	1/581
2	0.758	1/987	1.259	1/573
1	0.373	1/1,019	0.596	1/638

Check the required horizontal shear strength:

$$25\Sigma A_w + 7\Sigma A_c \geq 0.75 Z A_i \bar{W}_i \qquad [\text{eq.}(24.22)]$$

$$18\Sigma A_w + 18\Sigma A_c \geq Z A_i \bar{W}_i \qquad [\text{eq.}(24.23)]$$

Table 24.8 shows the calculations necessary to check eqs.(24.22) and (24.23). Results shown in this table satisfy, for the x direction, the condition indicated by eq.(24.23); however, for the y direction, neither eq.(24.22) nor eq.(24.23) is satisfied.

REFERENCES

Building Standard Law Enforcement Order (BSLEO) (1981) Ministry of Construction, Building Center of Japan, Tokyo, Japan.

Earthquake Resistant Regulations – A World List (1988) Standards for Seismic Civil Engineering Construction pp. 579–645, *Earthquake Resistant Regulations for Building Structures* pp. 646–662. International Association for Earthquake Engineering, Gakujutsu Bunken Fukyu-kai, Tokyo, Japan.

Standards for Seismic Civil Engineering Construction in Japan (1980) Earthquake Resistant Regulations for Building Structures in Japan, Tokyo, Japan.

Standards for Structural Calculation of Reinforced Concrete Structures (1985) Architectural Institute of Japan (AIJ), Tokyo, Japan.

Table 24.8. Calculations to Check Eqs.(24.22) and (24.23)

Level (i)	X Direction				Y Direction			
	ΣA_w cm^2	ΣA_c cm^2	$\dfrac{25\Sigma A_w + 7EA_c}{ZA_iW_i}$	$\dfrac{18\Sigma A_w + 18EA_c}{ZA_iW_i}$	ΣA_w cm^2	ΣA_c cm^2	$\dfrac{25\Sigma A_w + 7EA_c}{ZA_iW_i}$	$\dfrac{18\Sigma A_w + 78\Sigma A_c}{ZA_iW_i}$
3	13,500	32,400	1.105	1.618	0	32,400	0.444	1.142
2	13,500	48,400	0.750	1.241	0	48,400	0.379	0.974
1	13,200	57,600	0.616	1.071	0	57,600	0.339	0.871

25

Mexico

Roberto Villaverde

25.1 INTRODUCTION

There is no official national code in Mexico with norms for seismic design and construction of buildings to resist the effects of earthquakes for all the regions of the country. However, what can be considered a national code is contained in the Manual for Design of Civil Works (*Manual de Diseño de Obras Civiles*, 1981) published by a federal commission. Although this manual has no official character, it is generally used in those areas of the country that lack an official seismic code.

Some regions of the country with large concentrations of population such as Mexico City, located in the Federal District, have a seismic code for design and construction of buildings. These regional building codes are published by governmental departments of the various states and of the Federal District. The municipalities of the different localities authorize and supervise planned construction and also enforce compliance with the provisions of the code by issuing construction and occupational permits. In most cases, however, the municipality delegates the responsibility for complying with code provisions to the registered engineer and architect to whom the construction license is awarded. The calculations and drawings of only some of the most important or special structures are checked or revised by city engineers (Rosenblueth 1979).

25.2 SEISMIC CODE FOR MEXICO CITY

The first building code to include seismic-resistant design provisions was issued for Mexico City in 1942. The devastating effects of the earthquake of July 28, 1957, prompted its first revision which was incorporated in the code as emergency regulations in 1957. Subsequently, the code has been modified in 1966, 1976, 1985 (emergency regulations), 1987 and 1993; the latest major modification was undertaken in the aftermath of the catastrophic earthquake of September 19, 1985. Some Mexican states have completed building codes that are primarily adaptions of the Mexico City code. The seismic code for Mexico City is divided into two parts and published in two separate documents. The first part of the seismic code is included in the general building code for the city, titled "Reglamento de Construcciones para el Distrito Federal" [Construction Regulations for the Federal District (1993)]. This part contains the general seismic design requirements and recommendations for construction of buildings. The second part contains technical details and methods of analysis pertinent to seismic design and is titled "Normas Técnicas Complementarias para Diseño por Sismo" [Auxiliary Technical Notes for Seismic Design (1987)]. A format similar to that of the seismic code for Mexico City has been adopted in other Mexican codes, such as the seismic code published by the state of Guerrero in 1988 and 1989 [Reglamento de Construcciones para los Municipios del Estado de Guerrero (1988); Normas Técnicas Complementarias para Diseño por Sismo (1989)].

This chapter focuses on the provisions of the latest version of the seismic code for Mexico City, "Reglamento de Construcciones para el Distrito Federal" (1993) and "Normas Técnicas Complementarias para Diseño por Sismo (1989)." However, seismic coefficients and spectral accelerations given in the Manual for Design of Civil Works for different seismic areas of the country will be presented because they can be applied to different regions of the country outside Mexico City. In addition to the microzonation for Mexico City, the chapter includes the classification of site soil conditions given by the recent code for the state of Guerrero (1988). This code provides general guidelines for distinguishing firm and stiff soils in the various seismic zones of the country.

The provisions in the seismic code are intended to be minimum requirements for the design of buildings to achieve an acceptable degree of safety in the event of a strong earthquake motion. The seismic design provisions of the code are based on a 50-year return period for a strong earthquake.

25.3 GENERAL REQUIREMENTS

25.3.1 Classification of Buildings According to Use or Occupancy

The code classifies structures into two groups, designated as Group A and Group B, based on function or type of occupancy.

Group A includes: (1) structures whose failure or collapse might cause a large loss of lives, an extraordinary economic loss, or a loss of irreplaceable structures or objects of cultural importance; (2) structures such as hospitals, schools, stadiums, temples, auditoriums, and hotels with meeting facilities for more than 200 people, and other public buildings that are essential during emergencies; and (3) structures used as gas stations, facilities that store flammable, hazardous, or toxic materials, transportation centers, fire stations, electric power substations, communication centers, important governmental data centers, museums, and monuments.

Group B includes those ordinary buildings and facilities that are intended for residential, office, commercial, industrial, and lodging use and do not belong to Group A.

25.3.2 Regular Structures

A building is considered to have geometric and structural regularity if it satisfies the following requirements:

1. In plan, the structure has a fairly symmetrical distribution of its masses and resisting elements with respect to two orthogonal axes.
2. The ratio of its height to least base dimension does not exceed 2.5.
3. The ratio of its base length to its base width does not exceed 2.5.
4. In plan, the structure does not have offsets and setbacks that exceed 20% of the base dimension along the direction of the offset or setback.
5. Each story has either a rigid floor or similar roof system.
6. The floors and roofs of the structure: (a) do not have openings whose dimensions exceed 20% of the corresponding dimensions of the floor or roof where the opening is located, (b) the areas of the openings do not introduce asymmetries or significantly differ from floor to floor, and (c) the area of the openings in each floor does not exceed 20% of the floor area.
7. The weight of each floor, with the specified live load for earthquake effects included, does not exceed the weight of the floor below, nor, with the exception of the top level, is it less than 70% of this weight.
8. With the exception of the top level, the area of each floor, as defined by the exterior faces of its resisting

elements, does not exceed or is less than 70% of the area of the floor below.

9. All the columns of all the floors are constrained in two orthogonal directions by a horizontal diaphragm.

10. The lateral stiffness of each story does not exceed by more than 100% that of the story below.

11. The static eccentricity in each floor does not exceed by more than 10% the dimension of the floor parallel to the direction along which the eccentricity is measured.

25.3.3 Requirements for Structural and Nonstructural Elements

To increase the safety of structural systems with a low degree of redundancy, the code requires that if in a given story one of the resisting elements contributes more than 35% of the total capacity of that story, that element should be designed with a 20% reduction in the resistance factor normally specified for otherwise similar elements.

In regard to load-bearing walls, partitions and other nonstructural elements, the code establishes the following requirements:

1. The analysis should account for the stiffness of all structural and nonstructural elements when that stiffness is of significance.

2. Load-bearing walls should be connected adequately to a structural frame or to confining beams and columns along the wall perimeter, and if confining beams and columns are used, then these elements themselves should be connected adequately to a structural frame.

3. Nonbearing walls and partitions should be connected to the structure to allow free deformation of the structure in the plane of the wall or partition.

The code requires that the seismic analysis of a structure should include calculation of all the seismic forces, deformations, lateral displacements, and torsional rotations that might affect the structure, taking into account the flexural effects in its elements, and whenever significant, the effects of axial forces, shear forces, torsional moments, and P-delta (P-Δ) effects.

25.4 SEISMIC LOADS

All structures should be designed to resist the combined effects of dead loads, live loads, and forces corresponding to two orthogonal components of horizontal ground motion. However, for live loads only instantaneous values need to be considered. Such instantaneous values are defined as the maximum values that are likely to occur in the time interval

Table 25.1. Instantaneous Live Loads in Buildings

Building Use or Occupancy	Live Load (kg/m^2)
Residential buildings such as houses, apartments, dwellings, dormitories, hotels, jails, and hospitals	90
Offices and laboratories	180
Corridors, stairways, ramps, and lobbies	150
Stadiums and public places without seats	350
Assembly buildings such as temples, theaters, gymnasiums, dancing halls, restaurants, libraries, classrooms, and game parlors	250
Commercial establishments, factories, and warehouses	Not less than 350
Roofs and covers with a slope of no more than 5%	70
Roofs and covers with a slope of more than 5%	20
Balconies and other overhanging structures	70
Garages and parking structures	100 plus a concentrated load of 1,500 kg

during which an earthquake is also likely to occur. The specified dead and live loads are defined as follows:

(a) Dead load. The weights of all structural elements, finishings, and other items or objects that are permanently fixed to the structure should be considered as dead loads. In the calculation of these dead loads, the weight per unit area of reinforced concrete cast-in-place slabs should be increased by 20 kg/m^2. Similarly, if a cast-in-place or precast slab supports a concrete fill, the weight of this fill per unit area also should be increased by 20 kg/m^2.

(b) *Live load.* The magnitude of the design live loads depends on the use or occupancy of the building. For combination with seismic loads, the code specifies the instantaneous values given in Table 25.1. These live loads are assumed to be distributed uniformly over the pertinent tributary areas.

25.4.1 Load Combinations

The code requires calculation of the ultimate internal forces (UIF) by combining loads caused by earthquake ground motion in two orthogonal axes of the building, with gravitational loads, using the most unfavorable of the following two expressions:

$$UIF = 1.1(DL \pm LL \pm ELL \pm 0.3ELT) \quad (25.1)$$

$$UIF = 1.1(DL \pm LL \pm 0.3ELL \pm ELT) \quad (25.2)$$

In these expressions, *DL* represents the internal forces induced by the dead load, *LL* those forces induced by the instantaneous live load, and *ELL* and *ELT* those loads caused by earthquake motions

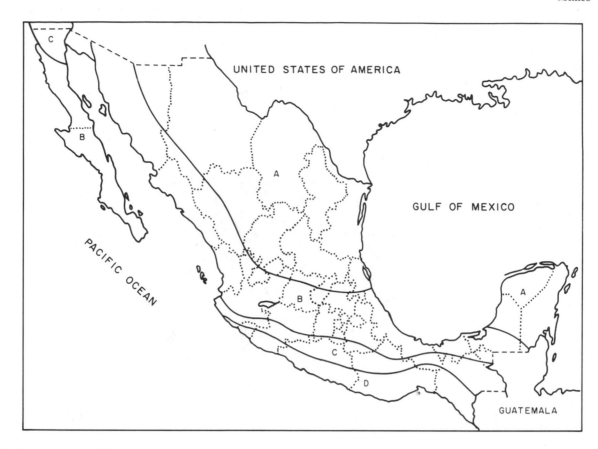

Fig. 25.1. Seismic zones of Mexico

acting, respectively, in the longitudinal and transverse directions of the structure. In each case, the signs used should be those that lead to the most critical condition.

25.5 SEISMIC ZONES

The territory of Mexico has been divided into the four seismic macrozones shown in Fig. 25.1. Zone A designates the regions of the country with the lowest seismic activity, whereas zones B, C, and D correspond to regions of progressively more intense seismic activity. The corresponding values for the seismic coefficients specified by the code for each seismic zone are given in Section 25.7.

25.6 SOIL TYPE

The seismic code identifies three different types of soils. For Mexico City, these three types of soil correspond to the soils of the three microzones shown on the map of the city in Fig. 25.2. The specific characteristics of the soil in each of these zones are described in Table 25.2. For other cities in the

country, the Manual for Design of Civil Works (1981) provides a classification of soils according to their stiffness and density as shown in Table 25.3. This classification is similar to the classification contained in the recent code for the state of Guerrero (1989). Both publications, the "Normas Técnicas Complementarias" for the state of Guerrero (1989) and the Manual for Design of Civil Works (1981), contain recom-

Table 25.2. Microzonation of Mexico City According to Soil Type*

Zone	Generic Name	Soil Type
I	Hill Zone	Rock or stiff soils deposited outside the old lakebed area, with possible lenses of loose sands and relatively soft clays
II	Transition Zone	Sands and silty sands with depths of less than 20 m and interwoven with layers of lacustrine clays with depths that vary from a few centimeters to a few meters
III	Lakebed Zone	Highly compressible lacustrine clays sandwiched between layers of silty and clayey sand; clay deposits can be over 50 m deep and thickness of sand layers can vary between a few centimeters and several meters

*From Reglamento de Construcciones para el Distrito Federal (1993).

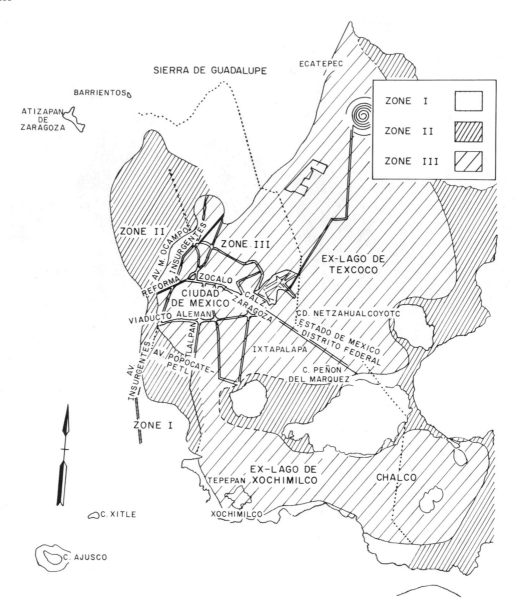

Fig. 25.2. Subzones of Zone III in Mexico City

mendations for determining the soil type for a given site on the basis of a soil exploration according to the following guidelines:

1. The soil is classified as Type I if it lies above bedrock or stiff soil, the latter defined as soil for which the shear modulus is greater than 75,000 ton/m², and requires more than 50 blows every 30 cm in the Standard Penetration Test.
2. The soil is classified as Type II if it cannot be classified as Type I according to the above criteria, and if it satisfies the following condition:

$$\bar{\beta} < 700 - 550 T_0 \qquad (25.3)$$

Table 25.3. Classification of Soils According to Rigidity*

Type	Soil
I	Bedrock and stiff soils such as compacted volcanic ashes, moderately cemented sandstones, and highly compacted clays
II	Soils of low stiffness and medium to high density, such as uncemented sands or silts, clays of medium density, and deposits of compacted alluvial soils; all with a depth of less than 40 m
III	Highly compressible clays, loose deposits of alluvial soils, and sand deposits in delta areas; all with a depth greater than 40 m

*From Manual for Design of Civil Works (1981).

where

$$\bar{\beta} = \frac{\sum_i H_i}{\sum_i \dfrac{H_i}{\beta_i}} \qquad (25.4)$$

$$T_0 = 4 \sum_i \left(\frac{H_i}{\beta_i} \right) \qquad (25.5)$$

in which

H_i = thickness of ith layer in meters
$\beta_i = \sqrt{g G_i / \gamma_i}$
G_i = shear modulus in ton/m^2
γ_i = unit weight in ton/m^3
g = acceleration of gravity in m/sec^2

3. The soil is classified as Type III if it does not fulfill the conditions to be classified as Type I or Type II.

When there is not enough information available to define all the parameters in eqs.(25.3) through (25.5), the following values may be used: $\gamma_i = 1.5$ ton/m^2 and $G_i = 0.35 E_i$, where E_i denotes the initial slope in the stress-strain curve obtained from an unconfined compression test for the ith soil layer. For small and medium-sized structures, it is also possible to define the soil type on the basis of local microzonation, if such maps are available. Mexico City and the port of Acapulco in the state of Guerrero are represented by the microzonation maps shown in Figs. 25.2 and 25.3, respectively. In the use of microzonation maps, the code requires that sites located less than 200 m from the boundary between any two zones be assumed to be located in the zone with the most unfavorable conditions.

25.7 SEISMIC COEFFICIENT

The code defines the seismic coefficient c as the ratio between the horizontal shear force which accounts for the effects of the earthquake at the base of the building, and the total weight of the structure above its base. For the purpose of this definition, the base of the structure is considered to be the lowest level of the structure for which lateral displacements, with respect to the surrounding ground, are of some significance. Similarly, the total weight of the structure includes the dead and live loads specified for seismic analysis. For Group B structures (Section 25.3.1) in Mexico City,

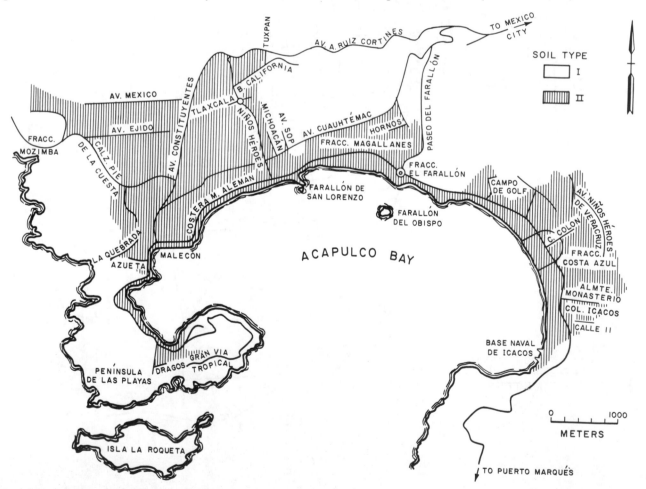

Fig. 25.3. Microzonation map of port of Acapulco

the code specifies, for each zone, the seismic coefficients given in Table 25.4. For structures in Group A, the code specifies those coefficients for Group B structures increased by 50%.

An exception is made for structures analyzed by the simplified method described in Section 25.10.1 and for those structures located in Zones II and III and within the shaded area of Fig. 25.4. In these exceptional

Table 25.4. Seismic Coefficients for Group B Structures in Mexico City

Zone	Seismic Coefficient (c)
I	0.16
II	0.32
III	0.40

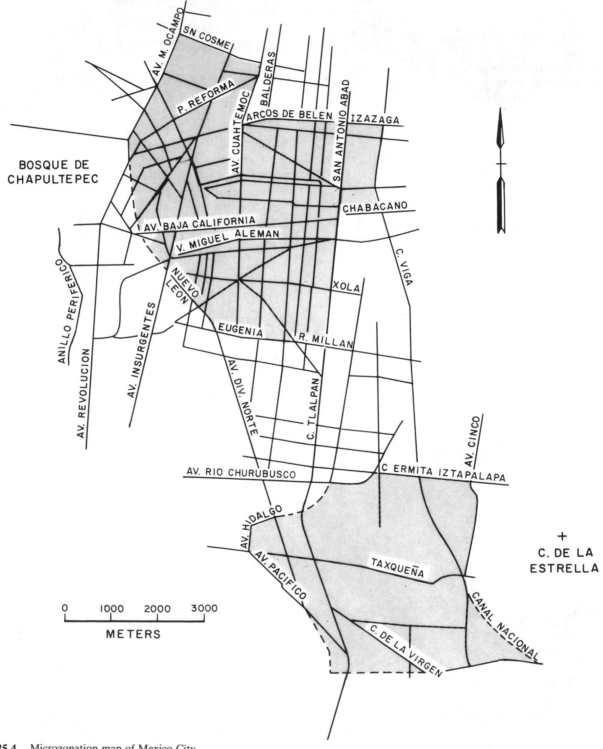

Fig. 25.4. Microzonation map of Mexico City

Table 25.5. Seismic Coefficients for Group B Structures in the Four Seismic Zones

Seismic Zone	Soil Type	Seismic Coefficient (c)
A	I	0.08
	II	0.12
	III	0.16
B	I	0.16
	II	0.20
	III	0.24
C	I	0.24
	II	0.30
	III	0.36
D	I	0.48
	II	0.56
	III	0.64

Table 25.6. Values of T_a, T_b, and r for Design Spectra in Mexico City

Zone	T_a (sec)	T_b (sec)	r
I	0.2	0.6	1/2
II*	0.3	1.5	2/3
III†	0.6	3.9	1

*Except in shaded area of Fig. 25.4.
†Including shaded area of Fig. 25.4.

Table 25.7. Values of a_0, T_a, T_b, and r for Design Spectra in the Four Seismic Zones of the Country

Seismic Zone	Soil Type	a_0	T_a (sec)	T_b (sec)	r
A	I	0.03	0.30	0.80	1/2
	II	0.045	0.55	2.00	2/3
	III	0.06	0.75	3.30	1
B	I	0.03	0.30	0.80	1/2
	II	0.045	0.50	2.00	2/3
	III	0.06	0.80	3.30	1
C	I	0.05	0.25	0.67	1/2
	II	0.08	0.45	1.60	2/3
	III	0.10	0.60	2.90	1
D	I	0.09	0.15	0.55	1/2
	II	0.14	0.30	1.4	2/3
	III	0.18	0.45	2.7	1

situations, a seismic coefficient of 0.4 should be used. For areas outside Mexico City, the Manual for Design of Civil Works recommends the seismic coefficients listed in Table 25.5. Specific seismic coefficients for 116 cities throughout the country are given by Trigos (1988).

25.8 DESIGN SPECTRA

The code for Mexico City specifies the use of the acceleration response spectrum depicted in Fig. 25.5 for the application of the static method or the dynamic method described in Section 25.10. This response spectrum is defined by the following ordinates:

$$a = (1 + 3T/T_a)c/4, \quad \text{for} \quad T \leq T_a \qquad (25.6)$$

$$a = c, \quad \text{for} \quad T_a < T \leq T_b \qquad (25.7)$$

$$a = (T_b/T)^r c, \quad \text{for} \quad T > T_b \qquad (25.8)$$

For areas outside Mexico City, the Manual for Design of Civil Works recommends the same spectral ordinates, except that instead of eq.(25.6) it specifies the use of eq.(25.9):

$$a = a_0 + (c - a_0)T/T_a, \quad \text{for} \quad T \leq T_a \qquad (25.9)$$

In the equations above, T represents one of the natural periods of the structure in seconds; c one of the seismic coefficients specified in Section 25.7; T_a, T_b, and r values that are selected from Table 25.6 for structures in Mexico City (and from Table 25.7 for structures in other areas of the country); and a_0 the spectral acceleration for a zero period, equal to $c/4$ for Mexico City (and as indicated in Table 25.7 for all the other seismic zones). Such spectral ordinates are intended to correspond to a damping ratio of 5%. Specific design spectra also are given by Trigos (1988) for the aforementioned 116 cities throughout the country.

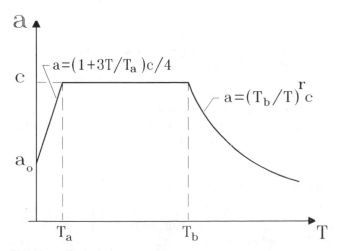

Fig. 25.5. Code design spectrum

25.9 REDUCTION FACTORS

For design purposes, the earthquake forces determined by the static method or the modal forces obtained with the dynamic method as described in

Section 25.10, may be reduced by dividing by a factor Q', which is determined as follows:

For structures classified as regular according to the requirements of Section 25.3:

$$Q' = Q \qquad \text{for } T \geq T_a \text{ (or for } T \text{ unknown)} \quad (25.10)$$

$$Q' = 1 + (T/T_a)(Q-1) \text{ for } T < T_a \qquad (25.11)$$

For structures that do not satisfy the conditions of regularity, Q' is given by the values calculated from eqs.(25.10) or (25.11) multiplied by a factor of 0.8.

In eqs.(25.10) and (25.11), T represents the fundamental natural period of the structure when the static method is used, and the natural period of the mode under consideration when the dynamic method is used; T_a is defined in Tables 25.6 and 25.7 for the different seismic zones, and Q is a factor that depends on the type and characteristics of the structure. This factor Q may be different along each of the two orthogonal main directions of the building, and may have different values for the various stories of the building.

The value of the factor Q is selected as follows:

1. Use $Q = 4$ if:
   (a) For all stories the lateral forces are resisted with unbraced moment-resisting frames of steel or reinforced concrete, or braced frames or frames with coupled reinforced concrete shear walls in which the frames are capable of resisting without the braces or the shear walls at least 50% of such lateral forces
   (b) The structure has coupled load-bearing masonry walls, and the lateral-load-resisting system is capable of resisting the lateral forces without the masonry walls; and
   (c) For any story, the ratio of the lateral capacity of the story to the corresponding shear force used in its design does not differ by more than 35% from the average of such ratios for all the stories of the building.

2. Use $Q = 3$ when the requirement in (a) above is satisfied, the requirements in (b) and (c) are not satisfied, and the lateral capacity of all the stories of the structure is provided by any of the following structural systems:
   (a) Flat plates with steel or reinforced concrete columns
   (b) Steel or reinforced concrete moment-resisting frames
   (c) Reinforced concrete shear-walls
   (d) Moment-resisting frames with coupled shear-walls
   (e) Plywood shear-walls.

3. Use $Q = 2$ when the lateral-load-resisting system consists of:
   (a) Flat plates with steel or reinforced concrete columns, braced or unbraced steel or reinforced concrete frames, or reinforced concrete walls and columns that do not satisfy some of the requirements in 1 or 2 above
   (b) Solid masonry walls with confining beams and columns of steel or reinforced concrete
   (c) Vertical and horizontal sheathed wood diaphragms combined with solid wood diagonals; or
   (d) Precast or prestressed elements.

4. Use $Q = 1.5$ when for all stories the lateral-load-resisting structure consists of:
   (a) Confined or internally reinforced hollow masonry walls
   (b) A combination of hollow masonry walls and any of the structural systems described in 2 and 3; or
   (c) Wood frames and trusses.

5. Use $Q = 1$ when the lateral-load-resisting system is composed, totally or partially, of elements or materials that are not included in any of the foregoing groups.

The final value of the Q factor for the entire structure should be equal to the lowest of the Q factors determined for the various stories of the building.

25.10 METHODS OF ANALYSIS

The code allows the use of three different methods of seismic analysis of buildings: (1) the simplified method, (2) the static method, and (3) the dynamic method. Any structure may be analyzed by means of the dynamic method. Similarly, any structure may be analyzed by the static method, provided that its height does not exceed 60 meters. The simplified method can be used only for structures that satisfy all the following requirements:

1. At least 75% of the vertical loads are supported by load-bearing walls tied together by a rigid monolithic slab or floor system, and these vertical loads are fairly symmetrically distributed along the two orthogonal directions of the structure. The walls can be of reinforced concrete, masonry, or timber.

2. The length-to-width ratio of the structure plan area does not exceed 2.0, or if, for the purpose of seismic analysis, its plan area can be divided into units with ratios that do not exceed 2.0.

3. The height of the structure does not exceed 13 meters and the ratio of height to the least dimension of the base is not greater than 1.5.

25.10.1 Simplified Method

In this method it is assumed that along each of two orthogonal main directions of the building a horizontal force is applied at each of the various levels of the structure. These horizontal forces are equal to those obtained with eqs.(25.12) and (25.13) with $Q' = 1$ and using the reduced seismic coefficient c given in Table 25.8. When the simplified method is used, it is assumed that the lateral displacements, torsional moments, and overturning moments induced by these forces are very small and, hence, that there is no need for the determination of these effects in the structure. In the application of this method, the analysis of a structure thus is reduced to computing lateral forces using eqs.(25.12) and (25.13), and verifying that at each story the resulting shear force along each of the two orthogonal main directions of the building does not exceed the total lateral strengths of the structural resisting elements in that story. The seismic coefficients shown in Table 25.8 apply to Group B structures. The seismic coefficients for Group A structures are equal to the coefficients in this table multiplied by a factor of 1.5.

25.10.2 STATIC METHOD

25.10.2.1 Lateral forces.
In the static method, it is assumed that the design lateral seismic forces act at the various levels of the building where the masses of the structure are supposed to be concentrated, and that these forces are given by

$$F_i = \frac{W_i h_i}{\sum\limits_{j=1}^{N} W_j h_j} V_0 \qquad (25.12)$$

where

$$V_0 = \frac{c}{Q'} W_0 \qquad (25.13)$$

and in which W_i and h_i represent, respectively, the weight of the ith level and its height above the base of the structure; W_0 denotes the total weight of the structure, which should include the live load specified for combination with earthquake loads; N is the total number of levels in the building; and c and Q' are, respectively, the seismic coefficient given in Tables

25.4 or 25.5 and the reduction factor specified in Section 25.9. Equation (25.12) is obtained by considering that the lateral forces are proportional to the height above the base of their points of application, and that the weight of the building can be concentrated at its floor levels. The base is the lowest level at which deflections are significant.

25.10.2.2 Reduced lateral forces.
The code allows a reduction in the lateral forces determined with eqs.(25.12) and (25.13) provided that the fundamental period of the structure is calculated and considered in the analysis. For such a purpose, it is appropriate to use the Rayleigh formula to determine such a fundamental period; that is,

$$T = 6.3 \left(\frac{\sum\limits_{i=1}^{N} W_i x_i^2}{g \sum\limits_{i=1}^{N} F_i x_i} \right)^{1/2} \qquad (25.14)$$

where F_i represents the force acting at the ith level of the structure, calculated with eqs.(25.12) and (25.13); x_i is the corresponding lateral displacement at this level; g is the acceleration of gravity; and W_i and N, respectively, are the weight for level i and the total number of levels in the building.

The reduction allowed in the lateral forces is based on the use of an acceleration design spectrum as opposed to the use of a constant seismic coefficient and as such depends on the value of the fundamental natural period T. If T is less than or equal to T_b, where T_b is one of the delimiting periods that define the response spectrum specified in Section 25.8, the reduced lateral forces may be determined as

$$F_i = \frac{W_i h_i}{\sum\limits_{j=1}^{N} W_j h_j} \frac{a}{Q'} W_0 \qquad (25.15)$$

where, as before, W_0 denotes the total weight of the structure, a the spectral acceleration given by

Table 25.8. Seismic Coefficients for Group B Structures to Be Used in Conjunction with the Simplified Method of Analysis

Zone	Masonry Walls with Solid Units or Plywood Shear-walls			Masonry Walls with Hollow Units or Sheathed Wood Diaphragms		
	Structure Height			Structure Height		
	Less than 4 m	Between 4 and 7 m	Between 7 and 13 m	Less than 4 m	Between 4 and 7 m	Between 7 and 13 m
I	0.07	0.08	0.08	0.10	0.11	0.11
II and III	0.13	0.16	0.19	0.15	0.19	0.23

eqs.(25.6), (25.7), and (25.9), and Q' the reduction factor defined by eqs.(25.10) and (25.11). If, on the other hand, T is greater than T_b, the reduced forces are calculated with the following formulas:

$$F_i = W_i(k_1 h_i + k_2 h_i^2)\frac{a}{Q'} \tag{25.16}$$

where

$$k_1 = \left(\frac{T_b}{T}\right)^r \left\{ 1 - r\left[1 - \left(\frac{T_b}{T}\right)^r\right]\right\} \frac{W_0}{\sum\limits_{j=1}^{N} W_j h_j} \tag{25.17}$$

and

$$k_2 = 1.5r\left(\frac{T_b}{T}\right)^r \left\{ 1 - \left(\frac{T_b}{T}\right)^r\right\} \frac{W_0}{\sum\limits_{j=1}^{N} W_j h_j^2} \tag{25.18}$$

in which all symbols are as previously defined.

Equation (25.16) applies to structures with long fundamental natural periods. It is intended to take into account the contributions of their higher modes and the fact that, for such structures, the fundamental mode shape approaches that of a flexural beam, which departs from the straight mode shape usually assumed for short buildings. In the application of eq.(25.16), the code requires that the value of the spectral acceleration a be always greater than $c/4$, where c is given in Tables 25.4 or 25.5.

25.10.2.3 Torsional effects.
When the static method is used, the torsional moment at a given story of a structure is assumed to be equal to the product of the shear force for that story and a design eccentricity whose value is determined by the most unfavorable of the following two formulas:

$$e = 1.5e_s + 0.1b \tag{25.19}$$
$$e = e_s - 0.1b \tag{25.20}$$

In these two equations, e_s denotes, for a given story, a "static" eccentricity defined as the distance between the point of application of the shear force for the given story and its center of twist, and b is the plan dimension of the story parallel to the direction along which e_s is being measured. The term $0.1b$ represents the so-called accidental eccentricity, while the factor 1.5 in eq.(25.19) is introduced to take into account (albeit in a crude way) the dynamic magnification of the torsional motion. The center of twist of a story is defined as the point through which the line of action of the interstory shear must pass so that the relative motion between the upper and lower ends of the story is purely a translational motion. At times, this point is referred to as the "center of rigidity." Its coordinates may be determined by means of the following equations:

$$\bar{X}_{Ti} = \frac{\sum\limits_{j=1}^{n} R_{ijy} x_{ij}}{\sum\limits_{j=1}^{n} R_{ijy}} \tag{25.21}$$

$$\bar{Y}_{Ti} = \frac{\sum\limits_{j=1}^{n} R_{ijx} y_{ij}}{\sum\limits_{j=1}^{n} R_{ijx}} \tag{25.22}$$

where R_{ijx} and R_{ijy} represent the lateral stiffnesses of the jth resisting element in the ith story of the structure along the x and y directions, respectively, and x_{ij} and y_{ij} are the centroidal coordinates of the element.

To include the effect of a torsional moment in the calculation of the internal forces in a resisting element of a story, the code requires that the most critical of the eccentricities given by eqs. (25.19) and (25.20) be considered in each case. In addition, the code specifies that this eccentricity should not be less than half the value of the largest of the eccentricities for all the stories below, and that the corresponding torsional moment should not be less than half the value of the largest of the torsional moments for all the stories above. The intention of this provision is to account for the torsional moments induced in one floor by the eccentricities at other floors. In summary, the torsional moment at each story of the building is calculated as the product of the story shear force times the story eccentricity given by eq.(25.19) or eq.(25.20), whichever is most unfavorable.

In an attempt to avoid excessively large torsional moments after the structure moves into inelastic behavior, the code also defines a story strength eccentricity e_r, considered to be the distance between the point of application of the shear force for the story in question and the centroid of the strengths of its resisting elements. For a given story, the location of this centroid coincides with the point of application of the resultant of the forces acting on the resisting elements of the story, when these resisting elements all reach their ultimate capacities. The code requires that for structures with $Q \geqslant 3$ this centroid be located on the same side from the point of application of the shear force as the center of twist, and that

$$e_r \geqslant e_s - 0.2b \quad \text{if} \quad Q = 3 \tag{25.23}$$

and

$$e_r \geqslant e_s - 0.1b \quad \text{if} \quad Q > 3 \qquad (25.24)$$

where Q is the factor defined in Section 25.9 and b is the aforementioned plan dimension.

25.10.2.4 Overturning moments. The equivalent lateral forces applied at the various levels of the building produce overturning moments that cause additional axial forces in the columns or in other vertical structural elements. The overturning moment, at any level of the building, is calculated by statics as the sum of moments exerted at that level by the lateral forces above that story. Hence, the overturning moment M_i at level i of the building is given by

$$M_i = \sum_{j=i+1}^{N} F_j(h_j - h_i) \qquad (25.25)$$

where $i = 0, 1, 2, \ldots, N-1$ and F_j is given by eq.(25.12), (25.15), or (25.16), and h_i or h_j is the height of the ith or jth level, measured from the base of the building.

The code allows a reduction in the value of the overturning moments calculated by eq.(25.25). The reduced overturning moments are obtained by multiplying the moment M_i resulting from the use of eq.(25.25) by the following reduction factor:

$$R_m = 0.8 + 0.2z \qquad (25.26)$$

where z is the ratio of the height above ground of the story under consideration to the total height of the structure. The reduced overturning moment at a given story should not be less than the product of the shear force at that story and the distance to the center of gravity of the part of the structure above that story.

25.10.2.5 Second-order effects. The code requires an explicit consideration of the additional shearing forces and bending moments induced in the elements of the structure by the so-called P-Δ effect. This effect is the action of vertical loads on a laterally displaced structure, whenever the drift in one of its stories exceeds $0.08V/W$, where V represents the calculated shear force in the story under consideration and W is the total weight of the part of the structure that is above that story.

25.10.2.6 Appendages. For an appendage or any other element whose structural system significantly differs from that of the main structure, the code specifies a lateral force calculated on the basis of the acceleration to which the appendage would be subjected if it rested directly on the ground, multiplied by a magnification factor

$$\alpha = 1 + 4c'/c \qquad (25.27)$$

where c is the seismic coefficient defined in Section 25.7, and c' denotes the coefficient by which the weight of the floor or level of the structure that supports the appendage is multiplied to determine the lateral force that acts on that floor or level. That is, $c' = F_i/W_i$, where F_i is given by eq.(25.12), (25.15), or (25.16), and W_i is the weight of such a floor or level. For the purpose of this requirement, the code considers as appendages tanks, parapets, partitions, nonstructural walls, signboards, windows, ornaments, and any other element whose stability depends directly on its own acceleration.

25.10.3 Dynamic Analysis

25.10.3.1 Methods of dynamic analysis. The code accepts the use of a modal analysis or the use of a step-by-step method to perform the dynamic analysis of a structure. However, if in the use of any of these methods the calculated shear force at the base of the structure V_0 does not satisfy the condition that

$$V_0 \geqslant 0.8aW_0/Q' \qquad (25.28)$$

the code requires that the calculated lateral forces be proportionally incremented to satisfy this relationship. In this expression, W_0, a, and Q' represent, respectively, the total weight of the structure, the spectral acceleration given by eqs.(25.6) through (25.9), and the reduction factor introduced in Section 25.9.

25.10.3.2 Modal analysis. In the use of a modal analysis, the code requires that:

1. The lateral forces in each of the modes of vibration of the structure be assumed proportional to the spectral accelerations as defined by eqs.(25.6) through (25.9), and adjusted by the reduction factors Q' given in eqs.(25.10) and (25.11)
2. All the modes of vibration with a natural period equal to or higher than 0.4 second be included in the analysis
3. The first three translational modes along the direction of analysis always be considered
4. The torsional effects in the structure be included by using, at least, the eccentricities specified in the static method
5. The total response of the structure S, where S may be a lateral force, lateral displacement, overturning moment, etc., be determined by combining the modal responses S_i in accordance with the following expression:

$$S = \sqrt{\sum S_i^2} \qquad (25.29)$$

except when the difference between the natural periods of any two modes under consideration is less than 10%. In that case, a relationship, such as the

Complete Quadratic Combination (CQC) method presented in Section 4.8 of Chapter 4, that takes into account the coupling between the different modal responses, should be used

6. For the purpose of evaluating second-order effects and complying with serviceability requirements, the calculated lateral displacements in each of the considered modes of vibration should be multiplied by the corresponding reduction factor Q'.

25.10.3.3 Step-by-step methods.

When step-by-step methods are employed, the code allows an analysis with historic or synthetic accelerograms or a combination of these accelerograms, provided that: (a) at least four of these accelerograms are used, (b) the accelerograms are independent and representative of the ground motions in the area of interest, (c) their intensities are compatible with the intensities considered when any of the other methods are used, (d) the analysis takes into account the nonlinear behavior of the structure and the uncertainties in quantifying its parameters.

25.11 DEFORMATIONS AND LATERAL DISPLACEMENTS

For the purpose of complying with the story drift and building limitations described below, and when the static method or the dynamic method introduced in Section 25.10 is used, the code requires the lateral displacements of a structure to be determined by multiplying the lateral displacements calculated with the lateral forces obtained from the application of any of those two methods by the reduction factor Q' defined in Section 25.9.

25.11.1 Story Drift Limitation

The story drift, the relative displacement between two consecutive levels of the structure, should not exceed 0.006 times the corresponding story height. However, if the brittle nonstructural elements in a story, such as masonry partitions, are adequately separated from the frame of the structure, this limit can be taken as 0.012.

25.11.2 Building Separation

All buildings must be separated from neighboring structures by a distance of not less than 5 centimeters, nor less than, at any floor level, the horizontal displacements induced by the design seismic loads. These displacements may be obtained with the reduced forces described in Section 25.10, but must be multiplied by the considered reduction factors Q'.

Additionally, these displacements must be augmented by an amount equal to 0.001 times the height above ground of the floor in question if the structure is located in Zone I, 0.003 times the height if in Zone II, and 0.006 times the height if in Zone III. Zones I, II, and III are defined in Table 25.2 for Mexico City and in Table 25.3 for sites outside that city. Similarly, the code requires that the different units or wings of a building be separated by a distance at least equal to the sum of the displacements of the adjacent units, each obtained as described for the calculation of the separation between buildings. If a building is analyzed using the simplified method, the above coefficients should be replaced by 0.007, 0.009, and 0.012, respectively.

Application of the seismic-resistant design code for Mexico is illustrated in the following examples.

Example 25.1

The building in Fig. 25.6(a) represents a five-story office building located in the soft-soil area (Zone III) of Mexico City. Its lateral force-resisting system is provided by reinforced concrete moment-resisting frames in the X direction and coupled reinforced concrete shear walls and moment-resisting frames in the Y direction (Bazan and Meli 1983). The lateral stiffnesses of the frames and shear walls are indicated in Fig. 25.6(b) through 25.6(d). The combination of the dead loads in the building and the live loads specified for use with seismic loads gives the total floor weights indicated in Fig. 25.6(a).

Determine for this structure, according to the 1993 version of the seismic building code for Mexico City, the following: (a) lateral forces, (b) interstory shear forces, (c) overturning moments, (d) story drifts, (e) lateral displacements, and (f) interstory torsional moments.

Solution

1. Classification according to use or occupancy. The building is an ordinary office structure, is located within Zone III, and is more than 15 meters in height. Therefore, the building may be considered in Group B.

2. Geometric regularity. Because of the setback of its fifth story and the uneven distribution of its resisting elements, the building violates some of the conditions specified by the code for a geometrically regular structure. The building therefore does not possess geometric regularity.

3. Method of analysis. As the building is less than 60 m in height, the static method may and will be used to estimate its lateral forces, moments, and displacements.

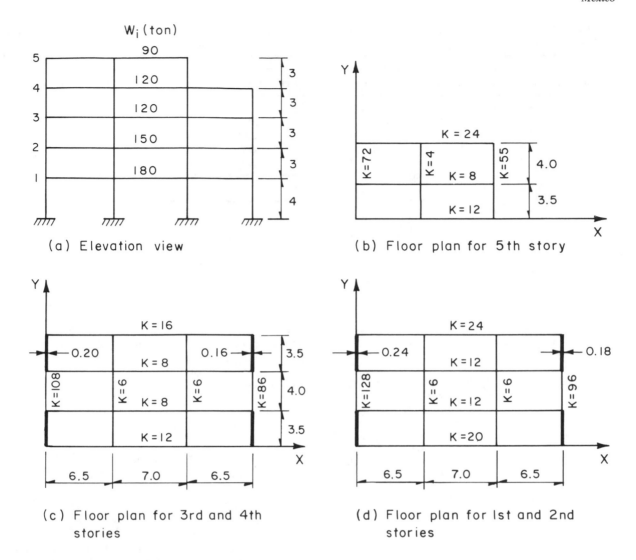

Fig. 25.6. Configuration and properties of building in Example 25.1 (lengths in meters; lateral stiffnesses in ton/cm)

4. Seismic coefficient. According to Table 25.4, the specified seismic coefficient for structures of Group B within Zone III of Mexico City is

$$c = 0.4$$

5. Reduction factor Q'. According to requirements in Section 25.9, $Q = 4$ for the moment-resisting frames along the X direction of the building, and $Q = 3$ for the coupled shear walls along the Y direction. However, these factors must be multiplied by 0.8 because the structure of the building is geometrically irregular. Accordingly, because the natural periods of the structure are unknown at this stage,

$$Q' = 0.8(4) = 3.2 \text{ along } X \text{ direction}$$

$$Q' = 0.8(3) = 2.4 \text{ along } Y \text{ direction}$$

6. Base shear. The total weight of the building is 660 tons, thus eq.(25.13) yields

$$V_0 = 0.4(660)/3.2 = 82.5 \text{ tons along the } X \text{ direction}$$

$$V_0 = 0.4(660)/2.4 = 110.0 \text{ tons along the } Y \text{ direction}$$

7. Lateral forces, interstory shears, story drifts, and floor displacements. Using the weights W_i and lateral stiffnesses K_{ix} and K_{iy} given in Fig. 25.6, eq.(25.12), and basic principles of statics, the lateral forces F_{ix} and F_{iy}, interstory shears V_{ix} and V_{iy}, story drifts Δ_{ix} and Δ_{iy}, and floor displacements x_i and y_i along the X and Y directions were calculated and then listed in Tables 25.9 and 25.10.

8. Fundamental natural period. Once the floor displacements x_i and y_i are known, eq.(25.14) can be

Table 25.9. Lateral Forces, Interstory Shears, Story Drifts, and Floor Displacements along X Direction

Level	Story	W_i (ton)	h_i (m)	$W_i h_i$ (ton-m)	F_{ix} (ton)	V_{ix} (ton)	K_{ix} (ton/cm)	Δ_{ix} (cm)	x_i (cm)
5		90	16	1,440	19.9				4.99
	5					19.9	44.0	0.45	
4		120	13	1,560	21.6				4.54
	4					41.5	44.0	0.94	
3		120	10	1,200	16.6				3.60
	3					58.1	44.0	1.32	
2		150	7	1,050	14.5				2.28
	2					72.6	68.0	1.07	
1		180	4	720	9.9				1.21
	1					82.5	68.0	1.21	
0			0						0
Sum:		660		5,970	82.5				

Table 25.10. Lateral Forces, Interstory Shears, Story Drifts, and Floor Displacements Along Y Direction

Level	Story	W_i (ton)	h_i (m)	$W_i h_i$ (ton-m)	F_{iy} (ton)	V_{iy} (ton)	K_{iy} (ton/cm)	Δ_{iy} (cm)	y_i (cm)
5		90	16	1,440	26.5				1.73
	5					26.5	131.0	0.20	
4		120	13	1,560	28.7				1.53
	4					55.2	206.0	0.27	
3		120	10	1,200	22.1				1.26
	3					77.3	206.0	0.38	
2		150	7	1,050	19.4				0.88
	2					96.7	236.0	0.41	
1		180	4	720	13.3				0.47
	1					110.0	236.0	0.47	
0			0						0
Sum:		660		5,970	110.0				

Table 25.11. Parameters for Calculation of Natural Period Along X Direction

Level	W_i (ton)	x_i (cm)	$W_i x_i^2$ (ton-cm^2)	F_{ix} (ton)	$F_{ix} x_i$ (ton-cm)
5	90	4.99	2,241.0	19.9	99.3
4	120	4.54	2,473.4	21.6	98.1
3	120	3.60	1,555.2	16.6	59.8
2	150	2.28	779.8	14.5	33.1
1	180	1.21	263.5	9.9	12.0
Sum:			7,312.9		302.3

Table 25.12. Parameters for Calculation of Natural Period Along Y Direction

Level	W_i (ton)	y_i (cm)	$W_i y_i^2$ (ton-cm^2)	F_{iy} (ton)	$F_{iy} y_i$ (ton-cm)
5	90	1.73	269.4	26.5	45.9
4	120	1.53	280.9	28.7	43.9
3	120	1.26	190.5	22.1	27.9
2	150	0.88	116.2	19.4	17.1
1	180	0.47	39.8	13.3	6.3
Sum:			896.8		141.1

used to estimate the natural periods of the building along the X and Y directions. The necessary calculations are shown in Tables 25.11 and 25.12. Hence, the following natural periods are obtained:

$$T_x = 6.3 \sqrt{\frac{7,312.9}{981(302.3)}} = 0.99 \text{ sec}$$

$$T_y = 6.3 \sqrt{\frac{896.8}{981(141.1)}} = 0.51 \text{ sec}$$

9. Reduced lateral forces. According to the code, if the natural periods of the building are known, the lateral forces obtained above may be reduced when the spectral accelerations determined by eqs.(25.6) through (25.9) are smaller than the seismic coefficient c, which in this case equals 0.4. The following procedure is used to implement this option.

- **For the X direction.** According to Table 25.6, for Zone III $T_a = 0.6$ sec and $T_b = 3.9$ sec. The natural period of the building along the X direction, $T_x = 0.99$ sec. This value lies between the two delimiting periods and, as a result, from eq.(25.7), the corresponding spectral acceleration is taken equal to the seismic coefficient c. Hence, no reduction is possible for the lateral forces along the X direction.

- **For the Y direction.** Along the Y direction, $T_y = 0.51$, which is less than T_a; therefore

$$a = [1 + 3(0.510/0.6)](0.4/4) = 0.36$$

$$V_{0y} = 0.36(660)/2.4 = 99.0$$

Reduction coefficient $= 99.0/110.0 = 0.90 > 0.80$

The reduced lateral forces, and the corresponding interstory shears, story drifts, and lateral displacements along the Y direction are listed in Table 25.13.

10. Overturning moments. The reduction factor for the overturning moments is given by eq.(25.26), which yields

$$R_{mi} = 0.8 + 0.2z = 0.8 + 0.2(h_i/16)$$

The reduced overturning moments M'_{0ix} and M'_{0iy}, and the overturning moments not reduced M_{0ix} and

Table 25.13. Reduced Lateral Forces, Interstory Shears, Story Drifts, and Floor Displacements Along Y Direction

Level	Story	W_i (ton)	h_i (m)	$W_i h_i$ (ton-m)	F_{iy} (ton)	V_{iy} (ton)	K_{iy} (ton/cm)	Δ_{iy} (cm)	y_i (cm)
5		90	16	1,440	23.8				1.55
	5					23.8	131.0	0.18	
4		120	13	1,560	25.8				1.37
	4					49.6	206.0	0.24	
3		120	10	1,200	19.9				1.13
	3					69.5	206.0	0.34	
2		150	7	1,050	17.5				0.79
	2					87.0	236.0	0.37	
1		180	4	720	12.0				0.42
	1					99.0	236.0	0.42	
0			0						0
Sum:		660		5,970	99.0				

Table 25.14. Overturning Moments Along X Direction

Level	h_i (m)	R_{mi}	F_{ix} (ton)	M_{0ix} (ton-m)	M'_{0ix} (ton-m)
4	16	1.00	19.9	59.7	59.7
3	13	0.96	21.6	184.2	176.8
2	10	0.93	16.6	358.5	333.4
1	7	0.89	14.5	576.3	512.9
0	4	0.83	9.9	906.3	752.2

Table 25.15. Overturning Moments Along Y Direction

Level	h_i (m)	R_{mi}	F_{iy} (ton)	M_{0iy} (ton-m)	M'_{0iy} (ton-m)
4	16	1.0	23.8	71.4	71.4
3	13	0.96	25.8	220.2	211.4
2	10	0.93	19.9	428.7	398.7
1	7	0.89	17.5	689.7	613.8
0	4	0.83	12.0	1,085.7	922.8

M_{0iy}, corresponding to the X and Y directions, are shown in Tables 25.14 and 25.15.

11. Static eccentricities. The code states that the static eccentricity of a story along a given direction is defined as the difference between the coordinate along the given direction of the center of application of the shear force for the story and that of its center of twist. Thus, for the Y and X directions, these eccentricities can be expressed, respectively, as

$$e_{iy} = \bar{Y}_{Ti} - \bar{Y}_{Vi}$$
$$e_{ix} = \bar{X}_{Ti} - \bar{X}_{Vi}$$

where $\bar{X}_{Ti}$ and $\bar{Y}_{Ti}$ and $\bar{X}_{Vi}$ and $\bar{Y}_{Vi}$ represent, respectively, the coordinates of the center of twist and

of the point of application of the interstory shear V_i for the ith story.

In order to obtain the static eccentricities in each of the stories of the building, it is necessary to find, first, the centers of mass of each of its floors; second, the points of application of the interstory shears; and third, the centers of twist or rigidity of each story. For each floor, the position of the center of mass along the X and Y directions is obtained using principles of statics and by taking into consideration the mass of the floor, half the mass of the shear walls above, and half the mass of the shear walls below. Similarly, the positions of the centers of rigidity are determined by means of eqs.(25.21) and (25.22). The point of application of a given interstory shear is found by considering the static equilibrium of all the forces above the story under consideration, applied at their respective centers of mass. That is, if $\bar{Y}_{mj}$ and $\bar{X}_{mj}$ represent the Y and X coordinates of the center of mass of the jth floor, the X and Y coordinates of the point of application of the interstory shear for the ith story are given by

$$\bar{Y}_{vi} = \frac{\sum\limits_{j=i}^{N} F_{jx} \bar{Y}_{mj}}{V_{ix}}$$

$$\bar{X}_{vi} = \frac{\sum\limits_{j=1}^{N} F_{jy} \bar{X}_{mj}}{V_{iy}}$$

The coordinates of the centers of twist, points of application of the interstory shears, and corresponding static eccentricities obtained for each of the stories of the building are presented in Tables 25.16 and 25.17.

12. Torsional moments. For each story, the design torsional moments are determined by multiplying the design eccentricities specified by the code and the

Table 25.16. Centers of Twist, Centers of Application of Interstory Shears, and Static Eccentricities Along X Direction

Story	F_{ix} (ton)	V_{ix} (ton)	$\bar{Y}_{mi}$ (m)	$F_{ix}\bar{Y}_{mi}$ (ton-m)	$\Sigma F_{ix}\bar{Y}_{mi}$ (ton-m)	$\bar{Y}_{Vi}$ (m)	$\bar{Y}_{Ti}$ (m)	e_{iy} (m)
5	19.9		3.75	74.6				
		19.9			74.6	3.75	4.73	0.98
4	21.6		5.50	118.8				
		41.5			193.4	4.66	6.00	1.34
3	16.6		5.50	91.3				
		58.1			284.7	4.90	6.00	1.10
2	14.5		5.50	79.8				
		72.6			364.5	5.02	5.82	0.80
1	9.9		5.50	54.5				
		82.5			419.0	5.08	5.82	0.74

Table 25.17. Centers of Twist, Centers of Application of Interstory Shears, and Static Eccentricities Along *Y* Direction

Story	F_{iy} (ton)	V_{iy} (ton)	$\bar{X}_{mi}$ (m)	$F_{iy}\bar{X}_{mi}$ (ton-m)	$\Sigma F_{iy}\bar{X}_{mi}$ (ton-m)	$\bar{X}_{Vi}$ (m)	$\bar{X}_{Ti}$ (m)	e_{ix} (m)
	23.8		6.75	160.6				
5		23.8			160.6	6.75	5.87	−0.88
	25.8		9.89	255.2				
4		49.6			415.8	8.38	8.93	0.55
	19.9		9.79	194.8				
3		69.5			610.6	8.79	8.93	0.14
	17.5		9.75	170.6				
2		87.0			781.3	8.98	8.64	−0.34
	12.0		9.66	115.9				
1		99.0			897.2	9.06	8.64	−0.42

Table 25.18. Design Eccentricities and Torsional Moments Along *X* Direction

Story	V_{ix} (ton)	e_{iy} (m)	e'_{iy} (m)	M'_{Tix} (ton-m)	e''_{iy} (m)	M''_{Tix} (ton-m)
5	19.9	0.98	2.22	44.2	0.23	4.6
4	41.5	1.34	3.11	129.1	0.24	10.0
3	58.1	1.10	2.75	159.8	0.00	0.0
2	72.6	0.80	2.30	167.0	−0.30	−21.8
1	82.5	0.74	2.21	182.3	−0.36	−29.7

interstory shears, where such design eccentricities are given by eqs. (25.19) and (25.20). For the building under consideration, these design eccentricities and the corresponding torsional moments turn out to be as shown in Table 25.18 for the *X* direction and Table 25.19 for the *Y* direction. In the tables, e'_{iy}, e'_{ix}, and M'_{Tix}, M'_{Tiy} represent, respectively, the values of these eccentricities and torsional moments when eq.(25.19) is used, and e''_{iy}, e''_{ix} and M''_{Tix}, M''_{Tiy} those values when eq.(25.20) is used.

25.12 COMPUTER PROGRAM

A computer program was developed to implement the provisions of the seismic building code of Mexico for 1993. The program is written in BASIC for its use in IBM-compatible microcomputers. It interactively prompts for the following building data: (a) number of stories, (b) floor weights, (c) floor elevations from ground level, (d) story stiffnesses along the building's longitudinal and transverse directions, (e) floor lengths and widths, (f) coordinates of the centers of mass and twist for each of the stories, (g) reduction factors along the building's longitudinal and transverse direction, and (h) the regularity and occupancy classification. It also prompts for the geographic location of the building and the microzone or type of soil where it will be built. All input data is automatically saved by the

Table 25.19. Design Eccentricities and Torsional Moments Along *Y* Direction

Story	V_{iy} (ton)	e_{ix} (m)	e'_{ix} (m)	M'_{Tiy} (ton-m)	e''_{ix} (m)	M''_{Tiy} (ton-m)
5	23.8	−0.88	−2.67	−63.5	0.47	11.2
4	49.6	0.55	2.83	140.4	−1.45	−71.9
3	69.5	0.14	2.21	153.6	−1.86	−129.3
2	87.0	−0.34	−2.51	218.4	1.66	144.4
1	99.0	−0.42	−2.63	−260.4	1.58	156.4

program using the name prompted by the program and selected by the user. The program outputs: (a) an estimate of the fundamental natural periods of the building along the building's longitudinal and transverse directions, (b) lateral seismic forces along each of these directions; (c) the corresponding interstory shears, floor displacements, and interstory drifts; (c) overturning moments; and (d) maximum and minimum torsional moments for each of its floors. Options are offered to use and modify existing input data files.

The following examples illustrate the execution of the program.

Example 25.2

Solve Example 25.1 using the computer program for implementation of the 1993 seismic building code.

```
-------------------------------------------------------------------------------
              ** EARTHQUAKE RESISTANT DESIGN **
   * PROGRAM BY R. VILLAVERDE AND H. EMADI, UNIVERSITY OF CALIFORNIA, IRVINE *
              ** BUILDING CODE OF MEXICO **          June 1994
*************************** DATA FILE INFORMATION ****************************
         1. PREPARE A NEW DATA FILE
         2. MODIFY EXISTING DATA FILE
         3. ANALYZE EXISTING DATA FILE
      Enter your choice: 1,2 or 3 ? 3
      Name of the file ? 5sb
********************************* INPUT DATA *********************************

                         Floor   Floor   Floor         Center of      Center of
         Weight  Height  Length  Width   Stiffness     Mass           Twist
                                         (ton/m)       (m)            (m)
  Level  (ton)   (m)     (m)     (m)     x      y      x      y       x      y
    1    180     4       20.00   11.00   6800   23600   9.66   5.50   8.64   5.82
    2    150     7       20.00   11.00   6800   23600   9.75   5.50   8.64   5.82
    3    120     10      20.00   11.00   4400   20600   9.79   5.50   8.93   6.00
    4    120     13      20.00   11.00   4400   20600   9.89   5.50   8.93   6.00
    5    90      16      13.50   7.50    4400   13100   6.75   3.75   5.87   4.73

********************************* INPUT DATA *********************************

*** Building location: mexico city

*** Building Zone: III

*** Reduction factor along x-direction: 4      along y-direction: 3

*** Geometric regularity: i

*** Building occupancy: b


X-Direction:

Level       Lateral      Interstory     Floor        Story
            Forces       Shears         Displ.       Drifts
            (ton )       (ton )         (m)          (m)

  1          9.95        82.50          0.012        0.012
  2         14.51        72.55          0.023        0.011
  3         16.58        58.04          0.036        0.013
  4         21.56        41.46          0.045        0.009
  5         19.90        19.90          0.050        0.005


     The fundamental natural period in sec. along x-direction is: 0.990
```

Y-Direction:

Level	Lateral Forces (ton)	Interstory Shears (ton)	Floor Displ. (m)	Story Drifts (m)
1	11.70	97.05	0.004	0.004
2	17.07	85.34	0.008	0.004
3	19.51	68.27	0.011	0.003
4	25.36	48.77	0.013	0.002
5	23.41	23.41	0.015	0.002

The fundamental natural period in sec. along y-direction is: 0.506

X-Direction:

Level	Overturning Moment (ton-m)
0	769.97
1	511.06
2	331.33
3	177.17
4	59.70

Y-Direction:

Level	Overturning Moment (ton-m)
0	905.72
1	601.17
2	389.74
3	208.41
4	70.22

X-Direction:

Level	Maximum Torsional Moment (ton-m)	Minimum Torsional Moment (ton-m)
1	182.59	29.53
2	166.87	21.77
3	159.61	0.00
4	128.93	9.95
5	44.18	4.58

Y-Direction:

Level	Maximum Torsional Moment (ton-m)	Minimum Torsional Moment (ton-m)
1	255.25	153.32
2	213.94	141.85
3	151.41	126.64
4	137.56	70.85
5	62.50	11.00

Example 25.3

Perform the seismic analysis of the 10-story office building shown in Fig. 25.7 according to the provisions of the building code of Mexico. The building will be located in Zone D on a site that lies over deposits of stiff soils. The plan dimension of the building is 12 m by 12 m. The floor weights, floor elevations, and story stiffnesses are as indicated in the figure. The lateral load-resisting systems along the building's longitudinal and transverse directions are formed by reinforced concrete moment-resisting frames.

```
            ** EARTHQUAKE RESISTANT DESIGN **

* PROGRAM BY R. VILLAVERDE AND H. EMADI, UNIVERSITY OF CALIFORNIA, IRVINE *

          ** BUILDING CODE OF MEXICO **    June 1993

************************* DATA FILE INFORMATION ***************************

   1. PREPARE A NEW DATA FILE

   2. MODIFY EXISTING DATA FILE

   3. ANALYZE EXISTING DATA FILE

Enter your choice: 1,2 or 3 ? 3

Name of the file ? 10sb
```

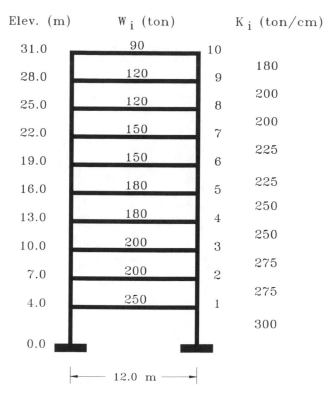

Fig. 25.7. Configuration and properties of building in Example 25.3

```
*************************** INPUT DATA ****************************
```

Level	Weight (ton)	Height (m)	Floor Length (m)	Floor Width (m)	Floor Stiffness (ton/m) x	Floor Stiffness (ton/m) y	Center of Mass (m) x	Center of Mass (m) y	Center of Twist (m) x	Center of Twist (m) y
1	250	4	12.00	12.00	30000	30000	6.00	6.00	6.00	6.00
2	200	7	12.00	12.00	27500	27500	6.00	6.00	6.00	6.00
3	200	10	12.00	12.00	27500	27500	6.00	6.00	6.00	6.00
4	180	13	12.00	12.00	25000	25000	6.00	6.00	6.00	6.00
5	180	16	12.00	12.00	25000	25000	6.00	6.00	6.00	6.00
6	150	19	12.00	12.00	22500	22500	6.00	6.00	6.00	6.00
7	150	22	12.00	12.00	22500	22500	6.00	6.00	6.00	6.00
8	120	25	12.00	12.00	20000	20000	6.00	6.00	6.00	6.00
9	120	28	12.00	12.00	20000	20000	6.00	6.00	6.00	6.00
10	90	31	12.00	12.00	18000	18000	6.00	6.00	6.00	6.00

```
*************************** INPUT DATA ****************************
```

*** Building location: Zone d

*** Soil type: I

*** Reduction factor along x-direction: 4   along y-direction: 4

*** Geometric regularity: r

*** Building occupancy: b

X-Direction:

Level	Lateral Forces (ton)	Interstory Shears (ton)	Floor Displ. (m)	Story Drifts (m)
1	4.09	118.46	0.004	0.004
2	5.90	114.37	0.008	0.004
3	8.69	108.46	0.012	0.004
4	10.46	99.77	0.016	0.004
5	13.23	89.32	0.020	0.004
6	13.46	76.08	0.023	0.003
7	16.00	62.62	0.026	0.003
8	14.92	46.63	0.028	0.002
9	17.13	31.71	0.030	0.002
10	14.58	14.58	0.031	0.001

The fundamental natural period in sec. along x-direction is: 0.970

Y-Direction:

Level	Lateral Forces (ton)	Interstory Shears (ton)	Floor Displ. (m)	Story Drifts (m)
1	4.09	118.46	0.004	0.004
2	5.90	114.37	0.008	0.004
3	8.69	108.46	0.012	0.004
4	10.46	99.77	0.016	0.004
5	13.23	89.32	0.020	0.004
6	13.46	76.08	0.023	0.003
7	16.00	62.62	0.026	0.003
8	14.92	46.63	0.028	0.002
9	17.13	31.71	0.030	0.002
10	14.58	14.58	0.031	0.001

The fundamental natural period in sec. along y-direction is: 0.970

X-Direction:

Level	Overturning Moment (ton-m)
0	1985.60
1	1631.68
2	1372.43
3	1115.56
4	869.64
5	641.07
6	439.53
7	267.96
8	136.17
9	43.73

Y-Direction:

Level	Overturning Moment (ton-m)
0	1985.60
1	1631.68
2	1372.43
3	1115.56
4	869.64
5	641.07
6	439.53
7	267.96
8	136.17
9	43.73

X-Direction:

Level	Maximum Torsional Moment (ton-m)	Minimum Torsional Moment (ton-m)
1	142.15	142.15
2	137.24	137.24
3	130.15	130.15
4	119.73	119.73
5	107.18	107.18
6	91.30	91.30
7	75.15	75.15
8	55.96	55.96
9	38.05	38.05
10	17.49	17.49

Y-Direction:

Level	Maximum Torsional Moment (ton-m)	Minimum Torsional Moment (ton-m)
1	142.15	142.15
2	137.24	137.24
3	130.15	130.15
4	119.73	119.73
5	107.18	107.18
6	91.30	91.30
7	75.15	75.15
8	55.96	55.96
9	38.05	38.05
10	17.49	17.49

REFERENCES

BAZAN, E., and MELI, R. (1983) *Manual de Diseño Sísmico de Edificios*. Mexico: Instituto de Ingeniería, UNAM, Mexico, Distrito Federal.

Manual de Diseño de Obras Civiles (1981) Mexico: Comisión Federal de Electricidad, Mexico, Distrito Federal.

"Normas Técnicas Complementarias para Diseño por Sismo" (1987) *Gaceta Oficial del Departamento del Distrito Federal*, Mexico, D.F., November 5, 1987.

"Normas Técnicas Complementarias para Diseño por Sismo" (1989) *Periódico Oficial del Estado de Guerrero*, Chilpancingo, Mexico, September 26, 1989.

"Normas Técnicas Complementarias para Diseño y Construcción de Cimentaciones" (1987) *Gaceta Ofical del Departamento del Distrito Federal*, Mexico, D.F., November 12, 1987.

"Reglamento de Construcciones para el Distrito Federal" (1993) *Diario Oficial de la Federación*, Mexico, D.F., August 2, 1993.

"Reglamento de Construcciones para los Municipios del Estado de Guerrero" (1988) *Periódico Oficial del Estado de Guerrero*, Chilpancingo, Mexico, December 27, 1988.

ROSENBLUETH, E. (1979) "Seismic Design Requirements in a Mexican Code." *Earthquake Engineering and Structural Dynamics* 7: 49–61.

TRIGOS, J. L. (1988) "Riesgo Sísmico, Construcciones y Reglamentos en México." *Memorias III Simposio Nacional de Ingeniería Sísmica*, October 30–November 1, 1988, Guadalajara, Mexico.

26

New Zealand

Thomas Paulay and Athol James Carr

26.1 INTRODUCTION

26.1.1 General Principles

Provisions for consideration of earthquake effects on buildings in New Zealand are, to a large extent, similar to those of other countries. These provisions form part of the general code prescribing requirements for the design of structures.

When this chapter was prepared, revision of the draft (DZ 4203) for a new seismic code was in its final stages. Some changes may be made in this draft before the final version of the New Zealand Standard NZS 4203 (1992) is issued. This new code will include commentary sections that outline the background considerations for each particular requirement, refer to recent research results, elaborate on applications (occasionally with the aid of diagrams), and emphasize design objectives.

26.1.2 Construction Materials

By necessity, seismic design provisions attempt to deal with the basic nature of structure response in both the elastic range and the inelastic range. The emphasis in such provisions is placed on estimation of seismic equivalent lateral forces and stipulated performance criteria to be met by buildings subjected to these equivalent forces. Specific code requirements reflect the properties of the materials employed. Thus, in addition to conforming to the provisions of the seismic code, the design also should meet the requirements of other codes for use of specific materials (steel, reinforced concrete, masonry, and timber). For example, the code for reinforced concrete [New Zealand Standard 3101 (1982)], which follows closely the nonseismic code requirements adopted in the United States [ACI 318 (1989)], sets out in each section special seismic requirements. Such seismic provisions for specific materials place great emphasis on the quality of the detailing of critical regions in the structures to ensure that the intended ductility can be reliably attained.

26.1.3 Code Developments

Mainly because of the similarity of earthquake hazards in the western United States and New Zealand, relevant U.S. developments in seismic design have strongly influenced those in New Zealand. The evolution of seismic building codes in New Zealand has also been strongly affected by the recommendations developed from numerous workshops held over the past 15 years, sponsored by the New Zealand National Society for Earthquake Engineering (NZNSEE). At these workshops, the interaction among representatives of all relevant professions has produced some innovative design concepts for earthquake resistance unique to New Zealand. Several recommendations published in the *Bulletin* of the NZNSEE have been adopted in other codes such as those of CEB (1987) and *Eurocode No. 8* (1988).

26.1.4 Limit States

The analysis of building structures considers the effects of a combination of factored loads and forces for each of two limit states: the serviceability limit state and the ultimate limit state.

(a) Serviceability limit state. The structure should have strength and stiffness adequate to resist the appropriate combination of loads and forces, including dead load, a fraction of the design live load, and the equivalent seismic forces, without excessive displacements during moderate earthquakes.

In the serviceability limit state, it is expected that the structure will remain essentially elastic. Therefore, the result obtained by the application of the combined system of loads and forces will be the same as the result obtained by superposition of effects due to gravitational loads and seismic forces applied separately. In this limit state, the lateral interstory displacement (story drift) is restricted typically to 0.33% to 0.40% of the story height.

(b) Ultimate limit state. The design of structures based on the ultimate limit state requires adequate ductility in addition to strength and stiffness. The building must be analyzed for the combination of factored loads and forces (dead load, live load, and seismic forces now at their ultimate values). Therefore, member strength should be based on the plastic state, showing that the strain capacity of materials as specified in relevant material codes is not exceeded. In this case, the strength S_u required to resist the internal actions (forces or moments) for any structural element must not exceed the dependable strength ϕS_i of the member; that is,

$$S_u \leq \phi S_i \qquad (26.1)$$

where

S_i = the nominal or ideal strength of the member based on the specified ultimate strength of constituent materials

ϕ = strength reduction factor, typically in the range 0.6 to 0.9.

In the ultimate limit design, it is expected that the dependable strength of the members essentially will be maintained during inelastic displacements corresponding to the assumed ductility demand. Because the structure may be predominantly in the plastic state, traditionally used working stress methods are inappropriate, and hence are not permitted. With few exceptions, all design of buildings in New Zealand relies on some ductility capacity. The emphasis in the ultimate limit state thus is on inelastic rather than elastic structural response, and this emphasis is reflected in the provisions of the code.

Under the earthquake effects specified for ultimate limit state design, the lateral displacements must be limited so that they do not:

(i)   Endanger life
(ii) Restrict or impair the function of structures in Categories I or II described in Table 26.1
(iii) Cause contact between parts of the structures that may endanger people or detrimentally alter the response of the structure, or reduce the strength of some structural members below the required strength
(iv) Exceed building separation from site boundaries or between neighboring buildings
(v)   Cause loss of structural integrity.

The possibility of excessive deformation of components of the foundation, or rocking or uplift of footings also must be considered when assessing displacements and energy dissipation characteristics of the structure. The code stipulates that interstory displacements (story drifts) predicted by the equivalent static or modal response spectrum methods of analyses should not exceed 2% of the corresponding story height for buildings with total height under 15 m and 1.5% for buildings with heights exceeding 30 m, with linearly interpolated values for intermediate heights. When the inelastic time-history integration method is used, computed interstory deflections should not exceed 2.5% of the corresponding story height.

26.1.5 Capacity Design

The application of *capacity design* of structures for earthquake resistance requires that specific elements of the structure be chosen and suitably designed and detailed for energy dissipation under severe imposed deformations. The critical regions of these members, usually termed *plastic hinges*, are detailed for inelastic flexural action, while shear failure is prevented by providing additional resistance against this type of failure. As a protection against brittle failure, all other structural elements are designed with strength greater than that corresponding to the strength developed in the plastic hinges. The strength of elements subjected to ductility demand is considerably less than the strength required for elastic response. Under ductility demand it is the actual rather than the nominal strength that will be developed. At maximum inelastic displacement the development of overstrength S_0 is to be expected. This is quantified by

$$S_0 = \lambda_0 S_i \qquad (26.2)$$

where $\lambda_0 > 1$ is the *overstrength factor*.

Nonductile elements, resisting actions originating from plastic hinges developed in the ductile members, must be designed for strength based on the overstrength S_0, rather than on the strength S_u used for determining required dependable strengths of hinge regions. The "capacity" design procedure ensures that the chosen means of energy dissipation can be maintained.

The following features characterize the capacity design procedure:

(a) Potential plastic hinge regions within the structure are defined clearly. These regions are designed to have dependable flexural strengths ϕS_i as close as practicable to the required strength S_u. Thus, these regions are detailed carefully to ensure that their estimated ductility demands can be accommodated reliably.

(b) Undesirable modes of inelastic deformation, such as those that may be originated from shear or anchorage failures and instability, within members containing plastic hinges are inhibited by ensuring that the strengths to resist these failure modes exceed the overstrength capacity of the plastic hinges.

(c) Potentially brittle regions, or those components not suited for stable energy dissipation, are protected by ensuring that their strengths exceed the demands originating from the overstrength of the plastic hinges. Therefore, these regions are designed to remain elastic irrespective of the intensity of the ground shaking or the magnitudes of inelastic deformations that may occur. This approach enables traditional or conventional detailing of these elements, such as that used for structures designed to resist only gravity loads and wind forces.

Capacity design is not an analysis technique, but a powerful design tool. It enables the designer to "tell the structure what to do" and to desensitize it to the characteristics of the earthquake, which are, after all, unknown. Subsequent judicious detailing of all potential plastic regions will cause the structure to fulfill the designer's intentions. The procedure is particularly relevant to the design of columns in frames in multi-story buildings when the simultaneous formation of plastic hinges at the tops and bottoms of all columns in a story must be avoided. In such frames, columns must be stronger than beams.

For example, in determining the desired reserve strength of columns with respect to adjacent beams in which two plastic hinges have developed, the designer may choose to ensure that a plastic hinge could form at only one end of a column, or that columns above the base of a building remain essentially elastic. The latter choice eliminates the need in concrete construction for the detailing of columns for ductility. The code requires that in two-way horizontal force-resisting systems, account be taken of the effects of potential concurrent yielding of beams framing into a column from all directions at a particular level. Moreover, allowance must be made for the effects of dynamic response in the allocation of resistance to moments and forces induced by the yielding members. Analysis procedures, described in Sections 26.3 and 26.4, are used to determine the required strength S_u of members in which inelastic strains are expected. For the capacity design procedures to protect members such as columns and foundation structures against yielding, their required strength must be based on the overstrength S_0 of the yielding members, as built. For the design of these members results of the initial analysis become largely irrelevant. For capacity designed elements, such as columns in multistory frames, the strength reduction factor in eq.(26.1) may be taken as 1.0.

26.1.6 Categories of Buildings

Buildings are classified according to their use in five categories as described in Table 26.1. A corresponding *risk factor R*, given in the last column of this table, is assigned to each category.

Table 26.1.   Classification of Buildings and Risk Factor *R*

Category	Description	*R*
I	Buildings dedicated to the preservation of human life or for which the loss of function would have a severe impact on society.	1.3
II	Buildings that generally hold crowds of people.	1.2
III	Publicly owned buildings whose contents are of high value to the community.	1.1
IV	Buildings not included in any other category.	1.0
V	Buildings of a secondary nature.	0.6

26.2 EARTHQUAKE PROVISIONS

26.2.1 General Requirements

(a) Resisting systems. The load path or paths, along which the inertial forces generated in an earthquake will be transferred to the supporting soil or rock, must be defined clearly.

(b) Limit states. Adequate strength, stiffness, and ductility, where appropriate, must be provided to satisfy the requirements of both the serviceability and the ultimate limit states.

(c) Ductility capacity and structural types. To satisfy the requirements of limit states, the system must be designed as a:

1. Ductile structure
2. Structure with limited ductility
3. Elastically responding structure; or
4. Combination of the above.

The structural ductility factor for the serviceability limit state is to be taken as 1.0. For the ultimate limit state, ductility factors must be assumed in accordance with the relevant material codes. Where provisions for a particular material or system are not available, the maximum ductility capacity should be deduced rationally from values given in Table 26.2, from relevant material codes, or from referenced published sources. The stringency of detailing requirements specified in material codes increases, in general, with the assumed ductility.

For mixed systems, with different forms of lateral force-resisting structures acting in parallel, the design lateral forces for each system must be determined by rational assessment, taking relative stiffness and ductility capacities into account. The overall ductility factor used must ensure that the ductility capacity of each constituent system is not exceeded. In this respect, the New Zealand code differs from the majority of similar documents because, instead of specifying global behavioral factors for various types of structures, it requires the designer to select, with guidance obtained from material codes, the most appropriate ductility capacity.

(d) Direction of design forces. For buildings with lateral-force-resisting elements located along two perpendicular directions, the specified forces may be assumed to act independently along each of these two horizontal directions. For other buildings, loading directions leading to the most unfavorable effects must be considered.

It should be noted, however, that according to capacity design principles, columns of multistory frames are required to be designed to resist simultaneous actions introduced by all beams that frame into columns of two-way frames. Moreover, these actions must consider the development of overstrength in beams, defined by eq.(26.2). This procedure is considered to provide column strength adequate to

Table 26.2.   Structural Ductility Factors μ

Structural Type	μ for Material				
	Structural Steel	Reinforced Concrete	Prestressed Concrete	Reinforced Masonry	Timber
1. Ductile structures					
(a) Braced frames (tension/compression yielding)	6	—	—	—	—
(b) Moment-resisting frames	6	6	5	4	4
(c) Walls	—	5	—	4	4
(d) Eccentrically braced frames	6	—	—	—	—
2. Structures of limited ductility					
(a) Braced frames:					
• Tension/compression yielding	3	—	—	—	3
• Tension yielding only (two stories maximum)	3	—	—	—	—
(b) Moment-resisting frames	3	3	2	2	3
(c) Walls	3	3	—	2	3
(d) Cantilevered face loaded walls (single story only)	—	2	—	2	—
3. Elastically responding structures	1.25	1.25	1.0	1.25	1.0

resist earthquake forces acting in any direction on the structure (Paulay and Priestley 1992).

(e) Seismic weights. At each level i of the building, the seismic weight W_i is to be taken as the sum of the dead load G_i plus the service live load Q_i, which is 40% of the basic live load that has been reduced to allow for tributary areas, attributed to that level. Thus, the total seismic weight W for a building is

$$W = \sum_{i=1}^{N} W_i \qquad (26.3)$$

where $W_i = G_i + Q_i$ and N is the number of levels in the building.

26.2.2 Methods of Seismic Analysis

The New Zealand code provides for three types of analysis for seismic design: (1) the equivalent static method, (2) modal response spectrum analysis, and (3) the numerical integration time-history analysis. The first method is applicable to regular structures (Section 26.3.5) for which the calculated fundamental period T is less than 2 seconds. This method may also be used if the total height of the structure does not exceed 15 m, or if the calculated fundamental period does not exceed 0.4 second. The dynamic methods (the second and third methods) may be used for any structure, but *must* be used for structures that do not conform to the requirements specified for the equivalent static method.

26.3 THE EQUIVALENT STATIC METHOD OF ANALYSIS

26.3.1 Base Shear Force

The code stipulates that the structure should be designed for a total base shear force V given by

$$V = CW \qquad (26.4)$$

where W is the total seismic weight of the structure and C is the *lateral force coefficient* given by the following equation:

$$
\begin{aligned}
C &= C_b(T, \mu)RZL > 0.025 \quad \text{for} \quad T > 0.4 \text{ sec} \\
C &= C_b(0.4, \mu)RZL > 0.025 \quad \text{for} \quad T \leq 0.4 \text{ sec}
\end{aligned} \qquad (26.5)
$$

where

> $R =$ *risk factor* for the structure given for different categories of buildings in Table 26.1
>
> $Z =$ *zone factor* accounting for the variation of seismicity in accordance with a seismic con-

tour map of New Zealand (Fig. 26.1), with a range of $0.4 \leq Z \leq 0.8$

> $L =$ *limit state factor* with values of 1.0 and 1/6, for the ultimate and serviceability limit states, respectively
>
> $C_b(T, \mu) =$ *basic seismic acceleration coefficient* obtained from the appropriate response spectral chart (Fig. 26.2) as a function of the fundamental period T and the *ductility factor* μ.

The basic response spectrum selected from Figs. 26.2(a)–26.2(c) in terms of the basic seismic acceleration coefficient C_b must be appropriate to the site Subsoil Category (1), (2), or (3) defined as follows:

Subsoil Category (1). Rock or very stiff soil sites. Bedrock with unconfined compression strength greater than 500 kPa and with a low amplitude natural period not greater than 0.25 sec. Restrictions on the thickness and strength of dense overlaying materials are also given.

Subsoil Category (2). Normal soil sites. These sites may be assumed as intermediate between categories (1) and (3), provided that the low amplitude site period does not exceed 0.6 sec.

Subsoil Category (3). Flexible or deep soil sites. For these sites, the low amplitude site period exceeds 0.6 sec. Descriptions in the commentary to the code for typical soil types with minimum depth, undrained shear strength, or standard penetration test N values provide guidance for identifying sites in this category.

Seismic risk studies have indicated that the shapes of the spectra are largely independent of the location within New Zealand, permitting the assumption of linear variation in the basic factors R, Z, and L. The smoothed inelastic response curves shown in Fig. 26.2, based on a single-degree-of-freedom oscillator with 5% equivalent viscous damping and variable yield levels, were obtained from selected earthquake records from the United States and Japan. These curves have been adjusted to allow for different soil conditions. Designers are warned that resonant soil response, likely to be encountered when recent lake or swamp deposits overlie coarser gravels or rocks, and motions in the vicinity of surface faulting may considerably exceed those movements corresponding to the code spectra.

The actual strength level of structures when the intended ductilities are developed is typically 50% in excess of the required strength. This strength margin, and the use of response spectra from Fig. 26.2, ensures that structures designed according to this code will meet the objectives of the limit state definition, i.e.:

(a) Serviceability limit state – 95% probability in any year or a return period of 20 years,

(b) Ultimate limit state – 95% probability in 50 years or a return period of 975 years.

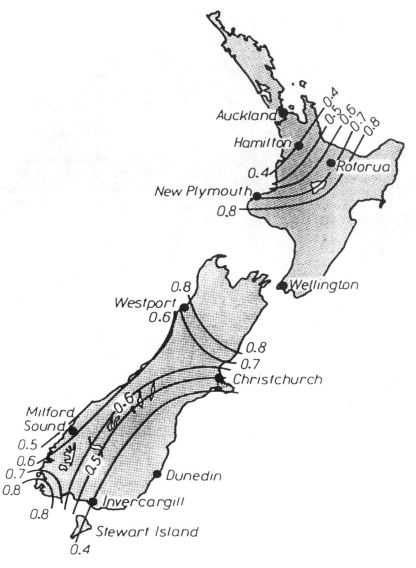

Fig. 26.1. Zoning map for New Zealand

26.3.2 Equivalent Static Forces

At each level i of the building, the equivalent static lateral force F_i is calculated from

$$F_i = 0.92V \frac{W_i h_i}{\sum_{j=1}^{N} W_j h_j} \tag{26.6}$$

where V is the base shear force given by eq.(26.4), and W_i and h_i are, respectively, the seismic weight and the height of level i of the building, with an additional force

$$F_t = 0.08V \tag{26.7}$$

to be applied at the top level of the structure.

26.3.3 Fundamental Period

The code stipulates that the fundamental period of the building must be established from substantiated data or by computation. When the equivalent static method of analysis is used, the fundamental period may be determined by Rayleigh's formula as:

$$T = 2\pi \sqrt{\frac{\sum_{i=1}^{N} W_i u_i^2}{g \sum_{i=1}^{N} W_i u_i}} \tag{26.8a}$$

where

W_i = seismic weight at level i
u_i = lateral displacement at level i due to lateral forces of magnitudes given by eqs.(26.6) and (26.7)

g = acceleration of gravity
N = number of levels.

For ductile steel structures, generally the same stiffness properties are appropriate for both the ultimate and the serviceability states. In reinforced concrete structures a smaller reduction in stiffness at the serviceability limit state, due to limited cracking, is to be considered.

For reasonably regular buildings the fundamental period may be estimated from

$$T = 0.042 \sqrt{\frac{\Delta}{C}} \qquad (26.8b)$$

where Δ = the lateral displacement in mm at the top of the main part of the building under the application

of the equivalent lateral forces defined by eqs.(26.6) and (26.7), and C = the lateral force coefficient given by eq.(26.5).

26.3.4 Lateral Displacements

The lateral displacement u_i at the various levels of the building caused by the application of the equivalent static forces are determined by elastic analysis. The displacement in each limit state should not exceed the restrictions specified in Section 26.1.4. For estimating displacements in any limit state, the stiffness properties must be those used for the determination of period and seismic actions. Because the equivalent static method of analysis simulates only the first mode response, it is considered to overestimate displacements. Thus, to comply with the requirements of Section 26.1.4, those displacements may be reduced by the following scaling factors:

(a) For buildings with a soft story: 1.0
(b) For other buildings with six or more stories: 0.85
(c) For buildings of less than six stories, linear interpolation between 1.0 (for a one-story structure) and 0.85 may be used.

When the equivalent static method or the modal response spectrum method of analysis is used, lateral displacements and corresponding interstory displacements (drifts) at the ultimate state may be estimated by multiplying the calculated elastic displacements and combined modal displacements by the structural ductility factor μ. This linear scaling of displacement generally is satisfactory for buildings of medium height (six to eight stories), or when lateral force resistance is provided primarily by structural walls. However, interstory displacements at the development of the expected maximum ductility are likely to be significantly underestimated in the lower stories of taller buildings. In such structures, no linear relationship between elastic and ultimate interstory displacements exists (Paulay and Priestley 1992). Fig. 26.3 shows computed displacement envelopes (Fenwick and Davidson 1991) for a 24-story building consisting of reinforced concrete ductile frames subjected to several earthquake records, including the artificial earthquake record referred to as ART-1. This artificial record simulates the spectra presented in Fig. 26.2. In estimating interstory displacements, consideration should be given to displacement profiles such as those obtained for the earthquake record ART-1, shown in Fig. 26.3.

When story sway mechanisms involving the simultaneous formation of plastic hinges at both ends of all columns in a story are admitted, interstory displacements in the ultimate limit state must be determined

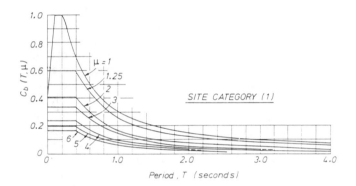

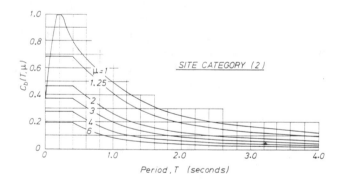

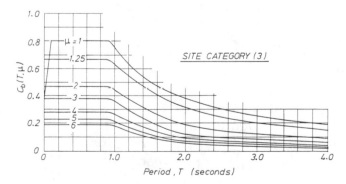

Fig. 26.2. Basic seismic acceleration coefficients

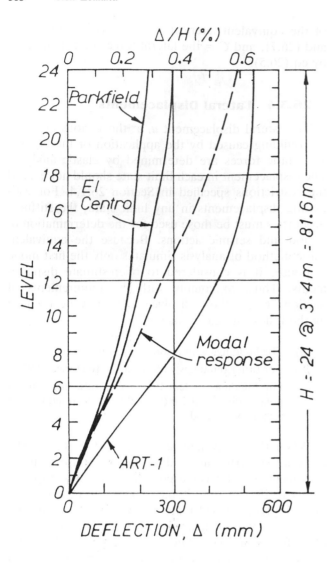

Fig. 26.3.   Determination of critical story drift

by resolving the total displacement u_t into elastic (u_e) and plastic (u_p) components. The application of this approach is shown for a three-story example structure in Fig. 26.4, where displacement profiles for three different story sway mechanisms are shown. This figure shows that the elastic displacements [Fig. 26.4(b)] scaled by the ductility factor [Fig. 26.4(c)] would lead to gross errors in the estimation of interstory displacements if story sway is allowed; compare Fig. 26.4(c) to Figs. 26.4(d)–(f). Decomposition of displacements at each level indicates the critical interstory drift. In this example, the critical interstory drift occurs for the third-story mechanism and it is equal to $(\mu u_t - u_2)/h$.

26.3.5 Structural Regularity

26.3.5.1 Horizontal regularity.   Two-dimensional analyses and the application of the equivalent lateral force methods are allowed if one of the following criteria, as illustrated in Fig. 26.5, is satisfied:

1. The horizontal distance between the center of rigidity at any level and the center of combined masses of all floors above must not exceed 0.3 times the maximum plan dimension of the structure at that level, measured perpendicular to the direction of lateral forces, nor change sign over the height of the structure.
2. Under the action of equivalent lateral static forces, calculated by eqs.(26.6) and (26.7), the ratio of horizontal displacements at the ends of the axis at any horizontal plane transverse to the direction of forces should be in the range 3/7 to 7/3.

The interpretation of these rules is illustrated in Fig. 26.5. Diaphragms should have adequate stiffness and be without major disruptions by penetrations or

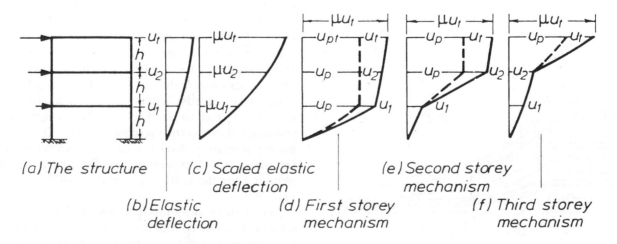

Fig. 26.4.   Displacement envelopes for a three-story frame

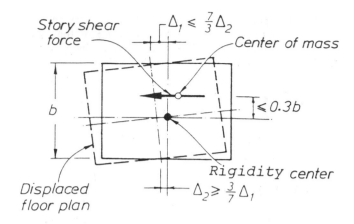

Fig. 26.5. Restriction on eccentricity and floor rotation

re-entrant corners. The purpose of these requirements is to direct the designer to use three-dimensional analyses, to ensure that contributions of torsional dynamic response are not seriously underestimated when eccentricities are significant.

26.3.5.2 Vertical regularity. The equivalent static method of analysis may be used if it can be shown that the lateral displacement at each level is reasonably proportional to the height of that level above the base.

The use of the following procedure given by the Applied Technology Council (1982) is recommended:

Step 1. Define the center of rigidity or center of stiffness for a story of the building as the point where the resultant of the resisting forces is applied.

Step 2. Evaluate the lateral displacement at each level u_i, caused by the application of equivalent static forces, using eqs.(26.6) and (26.7).

Step 3. Determine a revised set of lateral static forces F_i from the equation

$$F_i = 0.92V \frac{W_i u_i}{\displaystyle\sum_{j=1}^{N} W_j u_j} \qquad (26.9)$$

where V is the base shear force and W_i and u_i are, respectively, the seismic weight and the lateral displacement at level i, and

$$F_t = 0.08V \qquad (26.10)$$

where F_t is the additional force applied to the top level only.

Step 4. Determine the story shear forces corresponding to the revised set of lateral forces F_i from

$$V_i = F_t + \sum_{j=i}^{N} F_j \qquad (26.11)$$

If the forces derived using eqs.(26.6) and (26.7) are within +20% of those given by eqs.(26.9) and (26.10), vertical regularity may be assumed. For such a regular structure, the equivalent static force method may be used, provided that the fundamental period of vibration of the structure is 2 seconds or less. The story shear forces obtained from eq.(26.11) are considered to provide a better match to dynamic characteristics of the structure than story shear forces based on eqs.(26.6) and (26.7), and they may be used in the design for strength if desired.

26.3.6 Accidental Eccentricity

In addition to the eccentricity described in Section 26.3.5.1, an additional nominal (or accidental) eccentricity must be considered as an allowance for recognized uncertainties in the determination of stiffness, location of mass centers, and characterization of ground motions.

1. For forces applied parallel to the principal orthogonal axes of the building, this accidental eccentricity must be taken as ±0.1 times the breadth of the building b (Fig. 26.5) at right angles to the direction of forces.

2. For other directions of forces, the eccentricity should lie on the outline of an ellipse with semiaxes equal to the eccentricities specified for the orthogonal directions.

26.4 MODAL RESPONSE SPECTRUM ANALYSIS

26.4.1 General

The modal analysis, which is based on elastic structural behavior, uses the elastic response spectral charts in Fig. 26.2. These charts provide the basic seismic acceleration coefficient $C_b(T_m, \mu)$ as a function of the modal period T_m and the ductility factor μ (for the serviceability limit state, $\mu = 1.0$).

When the modal response spectrum method is used, the design spectrum is given for each mode m by

$$C_1(T_m) = S_m C_b(T_m, 1) R Z L \qquad (26.12)$$

In eq.(26.12) the response spectrum scaling factor $S_m = 1$ for the serviceability limit state. For the ultimate limit state the greater of the values S_{m1} or S_{m2} given by eq.(26.13) or eq.(26.14) is used:

$$S_{m1} = \frac{C_b(T, \mu)}{C_b(T, 1)} \quad \text{when } T > 0.4 \, \text{sec} \qquad (26.13a)$$

$$S_{m1} = \frac{C_b(0.4, \mu)}{C_b(0.4, 1)} \quad \text{when } T \leqslant 0.4 \, \text{sec} \qquad (26.13b)$$

and

$$S_{m2} = \frac{K_m C W}{V_1} \qquad (26.14)$$

where C is given by eq.(26.5), V_1 is the base shear force for the first mode, and $K_m = 0.8$ for buildings that meet the horizontal regularity provisions of Section 26.3.5.1 or when $\mu = 1.0$ (otherwise $K_m = 1.0$). In eq.(26.12), the factors R, Z, and L are the risk, zone, and limit state factors as defined in Section 26.3.1.

The modal response spectrum method usually leads to smaller base shears and overturning moments than those of the equivalent static method. Extensive experience with the equivalent static method of analysis shows that it is satisfactory when it is restricted to structures satisfying the requirements specified in Section 26.3. The scaling factor S_{m2} is used to ensure that results of the modal response spectrum analysis are not excessively smaller than those derived from the equivalent static method. The scaling should be applied to the results calculated for each mode.

When the fundamental period is greater than 1.0 sec and the structural ductility factor used in the design is $\mu = 6$, the serviceability limit state may control the minimum strength level. This level may well be less than the strength required to resist wind forces. Such cases should not divert attention from the important detailing requirements, to ensure that the intended large ductility can, in fact, be developed. For these structures, P-delta (P-Δ) effects (Section 26.6.2) may be significant in the ultimate state. A sufficient number of modes should be included in the analysis to ensure that at least 90% of the mass is participating (Clough and Penzien 1975, or Section 4.7.2 of Chapter 4).

26.4.2 Three-Dimensional Analysis

For three-dimensional analysis, except for buildings in which rigid floor diaphragms can be assumed, the location and distribution of the mass should be adjusted to account for the actual and accidental eccentricities for each direction of application of seismic forces. For rigid floor diaphragms, the effects of eccentricity in application of earthquake actions in any direction may be estimated using either of the following procedures:

(a) The general procedure described above (Section 26.4.1) may be used with the center of mass adjusted. However, the rotational inertia of the floor with respect to the center of mass need not be modified to account for the altered distribution of mass.

(b) The position and distribution of the mass need not be adjusted, but the line of action of the inertia forces must be taken eccentric to the center of mass.

26.4.3 Combination of Modal Effects for Member Design

The method for the combination of modal effects should take into account the influence of closely spaced modes, that is, modal frequencies within 15%. Design actions must be derived for each component of the structural system from the combined modal actions F_{eq} and those due to gravity for the appropriate limit state, as described in Section 26.1.4. For the ultimate limit state, when the use of capacity design procedures is required, the combination of earthquake- and gravity-induced action is used only to determine the required strength S_u of the members selected for energy dissipation, typically beams in frames. Such combined actions are irrelevant to the design of the columns except at the base of the building. Because actions obtained from modal response spectrum analyses are not in equilibrium, they cannot be used readily to quantify the desired strength hierarchy among members, the aim of which is to protect certain members, such as columns, against inelastic response. However, it may be assumed that during an earthquake, column actions at a joint, necessarily exactly balancing beam actions, bear the same relationship to beam actions as those obtained with the use of the equivalent static force analysis. Column actions so derived (and appropriately magnified to ensure that they possess adequate reserve strength with respect to those of yielding beams) can be used to proportion columns.

26.5 NUMERICAL INTEGRATION TIME-HISTORY ANALYSIS

26.5.1 General

Numerical integration time-history analysis may be used to determine the adequacy of provided component strength, or deflections, or ductility demands on components, or the forces generated in parts of the structure, or a combination of any of these. Modeling structures for this type of analysis must be cautiously appraised. Material and structural properties, including post-yield behavior where appropriate, and damping must be obtained from appropriate material codes. At least three different earthquake records should be used. P-delta effects should either be included in the analysis, or additional strength based on interstory displacements should be provided.

26.5.2 Input Earthquake Records

The chosen earthquake records should be scaled by a recognized method. Scaling must be such that over the period range of interest for the structure being analyzed, the 5% damped spectrum of the earthquake record does not differ significantly from the design spectrum appropriate for the site conditions and the limit state being considered, as specified in this code.

The input earthquake records for the ultimate limit state should contain at least 15 seconds of strong shaking, or have a strong shaking duration at least five times the fundamental period T of the structure, whichever is greater. Scaling factors also are recommended in the code if this method is intended to be used for design. When the numerical integration time-history method of analysis is used, the maximum values obtained for each required earthquake record should be considered.

26.6 SPECIAL EFFECTS

26.6.1 Torsional Effects

Where the horizontal regularity provisions are satisfied and two-dimensional modal response spectrum analysis has been used to evaluate translational effects, the static lateral method of analysis may be used to estimate torsional effects. In other cases, three-dimensional analysis methods using the provisions of Section 26.5.2 should be conducted.

When static analysis is used for torsional effects, the applied torque at each level should be based either on forces calculated by the equivalent static method, or on the combined story inertial forces obtained from the two-dimensional modal response spectrum analysis for translation. Accidental eccentricity should be assumed as required in Section 26.3.6. Torsional effects must be combined with translational effects by direct summation so as to produce the most adverse effect on each member considered.

26.6.2 P-Delta Effects

Analysis for P-delta effects at the ultimate state must be done unless at least one of the following criteria is satisfied:

(a) $T < 0.4$ sec
(b) The height of the structure does not exceed 15 m and $T < 0.8$ sec
(c) $\mu < 1.5$

(d) The ratio of computed interstory displacement to the story height does not exceed

$$\frac{u_i - u_{i-1}}{h_i - h_{i-1}} \leqslant \frac{V_i}{7.5 \sum\limits_{j=i}^{N} W_i} \qquad (26.15)$$

where u_i, u_{i-1} = horizontal displacements at levels i and $i-1$ respectively, at the development of the assumed displacement ductility μ; V_i = the design story shear force given by eq.(26.11), and all other variables have been defined previously.

Equation (26.15) emphasizes that buildings designed for relatively small lateral force resistance, as quantified by the story shear force V_i, are more vulnerable to reductions of this resistance due to P-delta moments. With increased flexibility and ductility demands and the duration of intense seismic shaking, P-delta effects become increasingly critical. When eq.(26.15) is not satisfied, rational analysis must be used to increase the shear capacity of the affected stories to compensate for P-delta moment demands.

In this analysis, realistic assumptions must be made with respect to inelastic displacement profiles over the height of the building (for example, as shown for the ART-1 earthquake record in Fig. 26.3). The code commentary suggests a technique (Fenwick and Davidson 1991) that can be used to satisfy code intentions with respect to P-delta effects.

26.7 COMPUTER APPLICATION

A computer program to illustrate the application of the New Zealand (NZ) Seismic Code to the analysis of multistory buildings was written in BASIC (IBM 1984). The program requests information on the zone factor Z, selected from the zoning map shown in Fig. 26.1, the building classification from Table 26.1, the ultimate or serviceability limit states, the site soil type described in Section 26.3.1, the type of analysis to be used, the ductility factor (Table 26.2), the elastic modulus of the material, the effective span of the girders, the width of the building normal to the earthquake forces, the length of the building in the direction of the earthquake forces, and the acceleration of gravity in terms of the selected units. Although the program uses m and kN in input prompts and output headings, any consistent set of units may be used.

Formulas for the computation of the natural periods of free vibration based on overall building dimensions (UBC) are not recommended (Paulay and Priestley 1992). The program requests information on the stiffness properties of members of the frame. An equivalent one-bay frame is analyzed using the sum of the girder *EI* (*E* = modus of elasticity and *I* = cross-sectional moment of inertia) values at each level for the stiffness of the model girders. The stiffness of the columns of the one-bay frame is based on the sum of one-half of the *EI* values for all columns in the story of the real frame. Options include the case of rigid girders or a single cantilever tower (no girders). Because the equivalent frame is symmetric and only antisymmetric forces are being considered, only one column is modeled and the girders are modeled with a horizontal roller support at mid-span. This yields a frame model with only two degrees of freedom per story. The model frame is also used to check the vertical regularity requirements given in Section 26.3.5.2.

The natural periods and mode shapes for all modes are computed by a modified Rayleigh or Vianello Stodola method (Clough and Penzien 1975) with period shifts to obtain the higher modes. Alternatively, only the first mode is obtained and the higher modes are estimated assuming that the second and third natural periods are, respectively, one-third and one-fifth of the fundamental period. As a final option, the approximate (UBC) formulas are available.

If the modal analysis method is being used with a two-dimensional frame model, there is no likelihood of close natural periods. Consequently, the modal responses may be combined by the Square Root of the Sum of the Squares (SRSS) method. Usually, for a two-dimensional frame model the response is obtained by combining up to three modes.

The next step is to select the appropriate spectral values given in Fig. 26.2. Subsequently, regularity checks are performed. The horizontal regularity check is somewhat cursory; a proper check would require some form of three-dimensional analysis. The S_{m1} and S_{m2} factors given by eqs.(26.13) and (26.14) are assessed and then the lateral forces acting on each floor and the corresponding lateral deflections are computed. The overturning moments and torsional moments, with the accidental eccentricity of 0.1 times the building width added and then subtracted from the input static eccentricities, are then computed. Finally, the interstory drifts are checked against the drift limits and, if appropriate, against the P-delta requirements.

26.8 EXAMPLES

26.8.1 Six-Story Frame Using Equivalent Static Method (Example 26.1)

The six-story reinforced concrete frame shown in Fig. 26.6 is used to illustrate the application of the equivalent static method. Except for the selection of the appropriate ductility factor, no distinction in the

derivation of earthquake actions for steel, concrete, or other structures is made in the NZ Code. The concentrated weight at each level of the building is assumed to be 500 kN, and the elastic modulus of the concrete is $E = 25 \times 10^6$ kPa. The building is assumed to be located in Wellington where a zone factor $Z = 0.8$ (Fig. 26.1) is applicable, and the building is in Category IV (Table 26.1), which gives the risk factor $R = 1.0$. The building is analyzed for the ultimate limit state; thus $L = 1.0$, relying on a ductility factor $\mu = 6.0$. The effective span of the model frame is 5.5 m. The width of the building normal to the earthquake forces is taken as 11 m.

Solution

The natural period of free vibration obtained using the modified Rayleigh method is 0.804 second. Assuming that the building is situated on a normal soil site, the basic seismic acceleration coefficient C_b obtained from equations representing the spectral curves in Fig. 26.2 is 0.107. The product of the values C_b, R, L, and Z gives the lateral force coefficient $C = 0.085$. The base shear V is computed from eq.(26.4). The weight of the building W equals 3,000 kN, resulting in a base shear of 255 kN.

The equivalent static lateral forces F_i are computed from eqs.(26.6) and (26.7). These results are shown, together with the overturning and torsional moments, in Table 26.3.

Using the equivalent lateral forces and the assumed stiffnesses, the lateral inelastic displacements at each level, and hence the inelastic interstory drifts, are computed. These results are also shown in Table 26.3. The interstory drifts have been checked in accordance with Section 26.1.4(b). In this example, the drift limit is satisfied in all stories.

26.8.2 Twelve-Story Frame Using Modal Analysis Method (Example 26.2)

(a) General description. The 12-story reinforced concrete frame shown in Fig. 26.7 is used to illustrate the application of the modal analysis method. The concentrated weight at each level is assumed to be 1,500 kN and the elastic modulus of the concrete is $E = 25 \times 10^6$ kPa. The building is assumed to be located in Wellington where a zone factor $Z = 0.8$ (Fig. 26.1) is applicable, and the building is in Category IV (Table 26.1) which gives the risk factor $R = 1.0$. The building is analyzed for the ultimate limit state; thus $L = 1.0$, relying on a ductility factor $\mu = 6.0$. The effective span of the model frame is 9.2 m. For simplicity, the width of the building normal to the earthquakes forces is taken as 9.2 m (a typical bay).

Table 26.3.   Response of Six-Story Frame

Level	Mode Shape	Lateral Forces (kN)	Story Shear (kN)	Overturning Moment (kN-m)	Torsional Moment (kN-m)	Lateral Displacement (mm)	Interstory Drift (mm)	Interstory Drift %
6	1.000	87.7		0		116		
			87.7		96		11.5	0.34
5	0.911	56.0		294		105		
			143.7		158		19.0	0.57
4	0.753	44.8		775		86		
			188.5		207		23.0	0.69
3	0.556	33.6		1,410		63		
			222.1		244		22.0	0.66
2	0.362	22.4		2,150		41		
			244.5		269		23.1	0.69
1	0.158	11.2		2,970		18		
			255.7		281		17.7	0.53
0		0.0		3,826		0		

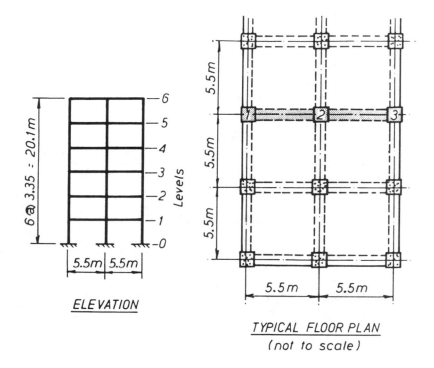

ELEVATION

TYPICAL FLOOR PLAN
(not to scale)

Properties of Members

MEMBERS	LEVEL	DIMENSIONS (mm)	INERTIA (m⁴)
BEAMS	1-3 4-6	600 x 350 550 x 350	0.005984 0.004635
EXTERIOR COLUMNS	1-3 4-6	500 x 450 450 x 450	0.003516 0.002563
INTERIOR COLUMNS	1-3 4-6	550 x 550 500 x 500	0.005719 0.003906

Fig. 26.6.   Six-story two-bay frame for Example 26.1

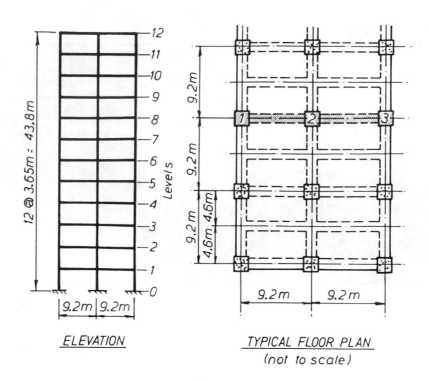

Fig. 26.7. Twelve-story two-bay frame for Example 26.2

Properties of Members

MEMBERS	LEVEL	DIMENSIONS (mm)	INERTIA (m^4)
BEAMS	1-6	900 x 400	0.02382
	7-8	850 x 400	0.02017
	9-12	800 x 400	0.01689
EXTERIOR COLUMNS	1-6	775 x 500	0.01455
	7-8	750 x 500	0.01318
	9-12	650 x 500	0.00855
INTERIOR COLUMNS	1-6	800 x 800	0.02560
	7-8	725 x 725	0.01727
	9-12	675 x 675	0.01297

(b) Input data and output results

```
********************************************************************
*                                                                  *
*           ***EARTHQUAKE RESISTANT DESIGN***                      *
*                                                                  *
*        ATHOL J. CARR. UNIVERSITY OF CANTERBURY                   *
*                                                                  *
*    ***DRAFT NEW ZEALAND LOADINGS CODE DZ 4203 (1992) ***         *
*                                                                  *
********************************************************************

   ***DATA FILE INFORMATION***

    1. PREPARE NEW DATA FILE

    2. MODIFY EXISTING DATA FILE

    3. USE EXISTING DATA FILE

        SELECT A NUMBER? 3

FILE NAME for INPUT and/or OUTPUT DATA
(Use Drive Letter and Path if required)

FILE NAME (e.g. A:\DATA\FRAME.DAT)? nz12
```

```
    ***INPUT DATA***:

NUMBER OF STORIES                              = 12
SEISMIC ZONE FACTOR          (Figure 26-1)     = .8
RISK FACTOR                  (Table 26-1)      = 1
LIMIT STATE                                    = ULTIMATE
STIFF, NORMAL or DEEP SOIL (Section 26.3.1)    = NORMAL SITE
DUCTILITY FACTOR             (Table 26-2)      = 6
ANALYSIS TYPE               (Section 26.3)     = MODAL ANALYSIS
STRUCTURE TYPE                                 = FLEXIBLE GIRDER
ELASTIC MODULUS (kN/m^2)                       = 2.5E+07
EQUIVALENT SINGLE BAY SPAN (m)                 = 9.2
WIDTH NORMAL to EARTHQUAKE DIRECTION (m)       = 9.2
LENGTH ALONG the EARTHQUAKE DIRECTION (m)      = 18.4
ACCELERATION of GRAVITY (m/sec^2)              = 9.805
HEIGHT OF BUILDING (m)                         = 43.80001
WEIGHT OF BUILDING (kN)                        = 18000
```

```
STORY  HEIGHT   WEIGHT   ECCENTRIC  I-COLUMN   I-GIRDER  COL.-ENDS BEAM-ENDS
No.     (m)      (kN)      (m)       (m^4)       (m^4)     (m)       (m)
12     3.650   1.50E+03   0.000    3.01E-02   3.38E-02   0.400    0.330
11     3.650   1.50E+03   0.000    3.01E-02   3.38E-02   0.400    0.330
10     3.650   1.50E+03   0.000    3.01E-02   3.38E-02   0.400    0.330
 9     3.650   1.50E+03   0.000    3.01E-02   3.38E-02   0.400    0.330
 8     3.650   1.50E+03   0.000    4.36E-02   4.03E-02   0.420    0.365
 7     3.650   1.50E+03   0.000    4.36E-02   4.03E-02   0.420    0.365
```

6	3.650	1.50E+03	0.000	5.47E-02	4.76E-02	0.450	0.375
5	3.650	1.50E+03	0.000	5.47E-02	4.76E-02	0.450	0.375
4	3.650	1.50E+03	0.000	5.47E-02	4.76E-02	0.450	0.375
3	3.650	1.50E+03	0.000	5.47E-02	4.76E-02	0.450	0.375
2	3.650	1.50E+03	0.000	5.47E-02	4.76E-02	0.450	0.375
1	3.650	1.50E+03	0.000	5.47E-02	4.76E-02	0.230	0.375

```
      PERIOD CALCULATION MENU

    1.- MODAL ANALYSIS for ALL 3 MODES
           Modified Rayleigh's Method using Frequency Shifts

    2.- RAYLEIGH'S METHOD for FIRST MODE
           Approximate Higher Mode Shapes and Frequencies

    3.- APPROXIMATE FIRST MODE FORMULAE
           Approximate Higher Mode Shapes and Frequencies
           Not recommended - Suitable only Preliminary Analysis

           SELECT A NUMBER ? 1
```

```
    MODAL PROPERTIES:
First  Mode Period (Seconds) =    1.779
Second Mode Period (Seconds) =    0.629
Third  Mode Period (Seconds) =    0.359
```

MODE SHAPES:

LEVEL	MODE 1	MODE 2	MODE 3
12	1.000	1.000	1.000
11	0.973	0.801	0.480
10	0.925	0.467	-0.234
9	0.854	0.052	-0.815
8	0.769	-0.347	-0.981
7	0.681	-0.638	-0.716
6	0.587	-0.824	-0.200
5	0.492	-0.885	0.331
4	0.391	-0.838	0.736
3	0.284	-0.687	0.878
2	0.174	-0.456	0.713
1	0.068	-0.188	0.329

```
    LATERAL FORCE FACTORS:
Sm1 factor . . . . . . . . . .=    0.169
Sm2 factor . . . . . . . . . .=    0.162
% (Sum Modal Weights)/Weight . .=   93.082
R*L*Z (Risk*LimitState*Zone) . .=    0.800
First  Mode Participation Factor=    1.323
Second Mode Participation Factor=   -0.483
Third  Mode Participation Factor=    0.276
```

SPECTRAL ACCELERATION COEFFICIENTS Cb:

#MODES	MODE 1	MODE 2	MODE 3
3	0.285	0.656	0.833

LATERAL FORCE COEFFICIENTS C:

#MODES	MODE 1	MODE 2	MODE 3
3	0.039	0.089	0.113

EQUIVALENT LATERAL FORCES (kN):

LEVEL	MODE 1	MODE 2	MODE 3
12	7.65E+01	-6.41E+01	4.67E+01
11	7.44E+01	-5.13E+01	2.24E+01
10	7.07E+01	-2.99E+01	-1.09E+01
9	6.53E+01	-3.33E+00	-3.80E+01
8	5.88E+01	2.23E+01	-4.58E+01
7	5.21E+01	4.09E+01	-3.34E+01
6	4.49E+01	5.28E+01	-9.34E+00
5	3.77E+01	5.68E+01	1.55E+01
4	2.99E+01	5.37E+01	3.43E+01
3	2.17E+01	4.41E+01	4.09E+01
2	1.33E+01	2.92E+01	3.33E+01
1	5.23E+00	1.21E+01	1.53E+01

STORY SHEAR FORCE (kN):

STORY	MODE 1	MODE 2	MODE 3	SRSS
12	7.65E+01	-6.41E+01	4.67E+01	1.10E+02
11	1.51E+02	-1.15E+02	6.90E+01	2.02E+02
10	2.22E+02	-1.45E+02	5.81E+01	2.71E+02
9	2.87E+02	-1.49E+02	2.01E+01	3.24E+02
8	3.46E+02	-1.26E+02	-2.57E+01	3.69E+02
7	3.98E+02	-8.56E+01	-5.90E+01	4.11E+02
6	4.43E+02	-3.28E+01	-6.84E+01	4.49E+02
5	4.80E+02	2.40E+01	-5.29E+01	4.84E+02
4	5.10E+02	7.77E+01	-1.86E+01	5.16E+02
3	5.32E+02	1.22E+02	2.23E+01	5.46E+02
2	5.45E+02	1.51E+02	5.56E+01	5.69E+02
1	5.50E+02	1.63E+02	7.09E+01	5.79E+02

OVERTURNING MOMENT (kN.m):

LEVEL	MODE 1	MODE 2	MODE 3	SRSS
12	0.00E+00	0.00E+00	0.00E+00	0.00E+00
11	2.79E+02	-2.34E+02	1.70E+02	4.02E+02
10	8.30E+02	-6.56E+02	4.22E+02	1.14E+03
9	1.64E+03	-1.19E+03	6.34E+02	2.12E+03
8	2.69E+03	-1.73E+03	7.08E+02	3.27E+03
7	3.95E+03	-2.19E+03	6.14E+02	4.56E+03
6	5.40E+03	-2.50E+03	3.99E+02	5.97E+03
5	7.02E+03	-2.62E+03	1.49E+02	7.49E+03
4	8.77E+03	-2.54E+03	-4.43E+01	9.13E+03
3	1.06E+04	-2.25E+03	-1.12E+02	1.09E+04
2	1.26E+04	-1.81E+03	-3.08E+01	1.27E+04
1	1.46E+04	-1.26E+03	1.72E+02	1.46E+04
0	1.66E+04	-6.60E+02	4.31E+02	1.66E+04

TORSIONAL MOMENT [ECC+0.1*B] (kN.m):

STORY	MODE 1	MODE 2	MODE 3	SRSS
12	7.04E+01	-5.90E+01	4.29E+01	1.01E+02
11	1.39E+02	-1.06E+02	6.35E+01	1.86E+02
10	2.04E+02	-1.34E+02	5.35E+01	2.50E+02
9	2.64E+02	-1.37E+02	1.85E+01	2.98E+02
8	3.18E+02	-1.16E+02	-2.36E+01	3.40E+02
7	3.66E+02	-7.87E+01	-5.43E+01	3.78E+02
6	4.07E+02	-3.01E+01	-6.29E+01	4.13E+02
5	4.42E+02	2.21E+01	-4.87E+01	4.45E+02
4	4.69E+02	7.15E+01	-1.71E+01	4.75E+02
3	4.89E+02	1.12E+02	2.05E+01	5.02E+02
2	5.02E+02	1.39E+02	5.12E+01	5.23E+02
1	5.06E+02	1.50E+02	6.53E+01	5.32E+02

TORSIONAL MOMENT [ECC-0.1*B] (kN.m):

STORY	MODE 1	MODE 2	MODE 3	SRSS
12	-7.04E+01	5.90E+01	-4.29E+01	1.01E+02
11	-1.39E+02	1.06E+02	-6.35E+01	1.86E+02
10	-2.04E+02	1.34E+02	-5.35E+01	2.50E+02
9	-2.64E+02	1.37E+02	-1.85E+01	2.98E+02
8	-3.18E+02	1.16E+02	2.36E+01	3.40E+02
7	-3.66E+02	7.87E+01	5.43E+01	3.78E+02
6	-4.07E+02	3.01E+01	6.29E+01	4.13E+02
5	-4.42E+02	-2.21E+01	4.87E+01	4.45E+02
4	-4.69E+02	-7.15E+01	1.71E+01	4.75E+02
3	-4.89E+02	-1.12E+02	-2.05E+01	5.02E+02
2	-5.02E+02	-1.39E+02	-5.12E+01	5.23E+02
1	-5.06E+02	-1.50E+02	-6.53E+01	5.32E+02

LATERAL DISPLACEMENTS (m):

LEVEL	MODE 1	MODE 2	MODE 3	SRSS
12	2.41E-01	-2.52E-02	5.97E-03	2.42E-01
11	2.34E-01	-2.02E-02	2.86E-03	2.35E-01
10	2.22E-01	-1.18E-02	-1.40E-03	2.23E-01
9	2.05E-01	-1.31E-03	-4.86E-03	2.06E-01
8	1.85E-01	8.75E-03	-5.86E-03	1.85E-01
7	1.64E-01	1.61E-02	-4.27E-03	1.65E-01
6	1.41E-01	2.08E-02	-1.20E-03	1.43E-01
5	1.18E-01	2.23E-02	1.98E-03	1.21E-01
4	9.41E-02	2.11E-02	4.39E-03	9.65E-02
3	6.83E-02	1.73E-02	5.24E-03	7.07E-02
2	4.18E-02	1.15E-02	4.26E-03	4.36E-02
1	1.65E-02	4.74E-03	1.96E-03	1.72E-02

INTER-STORY DRIFTS (m):

STORY	MODE 1	MODE 2	MODE 3	SRSS
12	6.41E-03	-5.02E-03	3.11E-03	8.72E-03
11	1.17E-02	-8.42E-03	4.26E-03	1.51E-02
10	1.69E-02	-1.05E-02	3.47E-03	2.02E-02
9	2.05E-02	-1.01E-02	9.91E-04	2.28E-02
8	2.12E-02	-7.32E-03	-1.58E-03	2.25E-02
7	2.27E-02	-4.69E-03	-3.08E-03	2.34E-02
6	2.27E-02	-1.56E-03	-3.17E-03	2.30E-02
5	2.44E-02	1.20E-03	-2.41E-03	2.45E-02
4	2.58E-02	3.79E-03	-8.47E-04	2.60E-02
3	2.65E-02	5.84E-03	9.81E-04	2.72E-02
2	2.54E-02	6.75E-03	2.30E-03	2.63E-02
1	1.65E-02	4.74E-03	1.96E-03	1.72E-02

```
    P-DELTA AND INTERSTORY DRIFT RATIOS:
    (% DRIFT LIMIT= 1.50  or % P-DELTA)
```

STORY	% DRIFT	% P-DELTA	OK?
12	0.24	1.37	Yes
11	0.41	1.08	Yes
10	0.55	0.96	Yes
9	0.63	0.89	Yes
8	0.62	0.83	Yes
7	0.64	0.77	Yes
6	0.63	0.73	Yes
5	0.67	0.68	Yes
4	0.71	0.64	No
3	0.74	0.59	No
2	0.72	0.55	No
1	0.47	0.51	Yes

```
TOP-FLOOR DEFLECTION/HEIGHT (%) =    0.552
```

(c) Comments on results. The last table in the computer output lists the ratio of the interstory drift to the story height expressed as a percentage, as well as the P-delta limit of eq.26.13 also expressed as a percentage. It can be seen that this structure violates the P-delta requirements in stories 2, 3, and 4, indicating that redesign will be required. One option is to increase the strength of the structure by reducing the ductility factor from $\mu = 6$ to $\mu = 4$ when all stories satisfy the P-delta drift limit. More detailed studies may show, however, that a small increase in the lateral force resistance, in the affected stories only, causing a smaller reduction in ductility demand, would suffice.

ACKNOWLEDGMENTS

The permission of the New Zealand Standards Association to collate the information from the final draft of this code, and their assistance in making the necessary information available are acknowledged.

REFERENCES

AMERICAN CONCRETE INSTITUTE (1989) *Building Code Requirements for Reinforced Concrete* (ACI 318-89) and Commentary (ACI 318R-89). 353 pages. Detroit, MI.

APPLIED TECHNOLOGY COUNCIL (1982) *Tentative Provisions for the Development of Seismic Regulations for Buildings*. ATC 3-06, NBS SP-150, NSF 78-8. National Bureau of Standards, Washington D.C.

CLOUGH, R. W., and PENZIEN, J. (1975) *Dynamics of Structures*. 652 pages. McGraw-Hill, New York, NY.

COMITÉ EURO-INTERNATIONAL DU BETON, CEB (1987) *Seismic Design of Concrete Structures*. 298 pages. Gower Technical Press, Aldershot, U.K.

COMMISSION OF THE EUROPEAN COMMUNITIES (1988) *Eurocode No. 8 – Structures in Seismic Regions-Design*, Draft, 327 pages. Directorate-General, Luxembourg.

FENWICK, R. C., and DAVIDSON, B. J. (1991) *The Seismic Response of Multi-Story Buildings*. Report No. 495. 78 pages. University of Auckland, Department of Civil Engineering, New Zealand.

IBM (1984) "General Programming Information" and "Basic Reference" in *Basic Handbook*. Personal Computer Hardware Reference Library, IBM, New York, NY.

INTERNATIONAL CONFERENCE OF BUILDING OFFICIALS (1988) *Uniform Building Code (UBC)*. Whittier, CA.

NEW ZEALAND NATIONAL SOCIETY FOR EARTHQUAKE ENGINEERING (1968) *Bulletin*, Vols. 1 to 25, 1968 to 1992, Wellington.

NEW ZEALAND STANDARD NZS 3101 (1982) *Code of Practice for the Design of Concrete Structures*. Part 1, 127 pages, Part 2: Commentary, 156 pages. Amended in 1989. Standards Association of New Zealand, Wellington, New Zealand.

NEW ZEALAND STANDARD NZS 4203 (1992) *General Structural Design and Design Loadings for Buildings*. Draft DZ 4203, Standards Association of New Zealand, Wellington, New Zealand.

PAULAY, T., and PRIESTLEY, M. J. M. (1992) *Seismic Design of Reinforced Concrete and Masonry Buildings*. 767 pages. John Wiley and Sons, New York, NY.

27

Peru

Gianfranco Ottazzi and Daniel Quiun

27.1 INTRODUCTION

From 1966 until 1970, preliminary versions of seismic design recommendations were recognized and used by designers in Peru. In 1970, the first official code was published based on the *Recommended Lateral Force Requirements and Commentary* of the Structural Engineers Association of California (SEAOC) (1967).[1]

A strong earthquake occurred on May 31, 1970; its epicenter was located 370 km NW from Lima ($M_b = 6.6$, $M_s = 7.6$, depth = 43 km).[2] Another strong earthquake occurred on October 3, 1974, with epicenter located 90 km SW from Lima ($M_b = 6.6$, $M_s = 7.5$, depth = 13 km). The damage to buildings during these earthquakes indicated that the previous code was not adequate for seismic-resistant design. The Ministry of Housing approved the current seismic code in April, 1977. This code was developed during 1976 by a committee composed of representatives of the Society of Engineers, the local universities, the Geophysical Institute, and the National Committee for Disaster Prevention.

The 1977 Code uses modern concepts developed for seismic-resistant design, recognizes the seismic areas of Peru, and considers the different construction materials and structural types commonly used. However, available research and evaluations of recent seismic effects have demonstrated the need for a new, modified code. A new version of the Seismic Code is being developed by a special committee whose members were designated by the National Institute of Research and Housing Regulations (ININVI).

The 1977 Code established the minimum require-

---

[1]SEAOC periodically has published recommendations for seismic-resistant design of structures, beginning in 1959.

[2]M_b, M_s, and also M_L are various scales for measuring the magnitudes of earthquakes. The body wave magnitude M_b is a measure of the 1-Hz amplitude of the source spectrum, the M_s (surface wave) magnitude is proportional to the source spectral amplitude of the 0.05-Hz (or 20-second-period) waves, and M_L is the original Richter scale magnitude, a measure of the 1- to 3-Hz portion of the source spectrum (See Appendix of this Handbook).

ments for earthquake-resistant design of buildings in terms of effects produced by earthquakes of various intensities. Earthquakes are classified as weak, with Modified Mercalli Intensity (MMI) of V or less, intermediate (MMI VI–VII) and strong (MMI of VIII or higher). Modern earthquake-resistant design philosophy establishes that during weak earthquakes, structures should suffer no damage, during intermediate earthquakes, slight damage is accepted, and during strong earthquakes, structural damage is allowed but buildings should not collapse.

Seismic forces are considered to act along the two main directions of the building, or in the most unfavorable direction. The analysis can be done independently in each of these directions in order to determine the total design seismic forces on the structure. That is, the 1977 Code does not require consideration of simultaneous action of two horizontal earthquake components. Elastic structural behavior is assumed in the analysis. However, energy dissipation through inelastic deformations during strong earthquakes is taken into account by reducing the equivalent elastic seismic forces. General guidelines recommend symmetry in mass and stiffness distribution, structural continuity, and general uniformity in plan and elevation in order to provide resistance to earthquakes.

This chapter presents the main provisions of the Peruvian 1977 Seismic Code for the determination of equivalent static seismic forces and corresponding response produced by those forces.

27.2 SEISMIC ZONES AND ZONE FACTOR

The locations and depths of earthquakes, magnitudes 3.4 or greater, that occurred in Peru between 1900 and 1984, are presented in Fig. 27.1 (Espinoza 1985). The 1977 Code divides Peruvian territory into three zones (Fig. 27.2) that are based on the information on the intensity and the recurrence interval provided by these historical earthquakes: Zone 1 corresponds to high seismicity, with an associated maximum intensity of VII–IX on the MMI scale; Zone 2 is of medium seismicity with an associated maximum intensity of VI–VII MMI; and Zone 3 comprises the lowest observed intensities, less than V MMI. For each of the three seismic zones, the 1977 Code assigns a seismic zone factor Z, which directly affects the base shear seismic force specified by the Code. The values for the seismic zone factor are given in Table 27.1 as well as in Fig. 27.2.

Numerical values for the Z factor assigned to Zones 2 and 3 are conservative because of the lack of information on intensities in those areas. Guidelines for more specific seismic microzonation studies are included in the Code. Such studies are required for new urban areas, vital industries, and important facilities.

27.3 CLASSIFICATION OF BUILDINGS AND BUILDING WEIGHT

In evaluating the seismic design forces, the relative importance of the building is taken into account through the use factor U and the adjustment of the percentage of live load to be considered in determining the weight of the building.

The Code considers the following four types of buildings:

Type A (Use factor $U \geq 1.3$). Includes buildings whose failure could be catastrophic—such as nuclear power plants, large smelters, and storage depots for inflammable material. For this type of building, U shall be established by the designer, but it cannot be less than 1.3.

Type B (Use factor $U = 1.3$). Includes public service facilities and those buildings where many people gather or where valuable goods are kept (e.g., hospitals, schools, water supply tanks, stadiums, churches, museums, and all the buildings that must supply vital services after an earthquake).

Type C (Use factor $U = 1.0$). Common buildings: dwellings; office, industrial and commercial buildings; warehouses.

Type D (Use factor $U \leq 1.0$). Those structures whose failure would be of less importance (such as temporary warehouses) do not require special seismic design.

The above classification of buildings also is used to evaluate the percentage of the live load to be included in the weight of the building for the determination of the seismic forces (obtained by adding the dead load and a percentage of the live load) as given in Table 27.2.

Table 27.1. Seismicity Zone Factor Z

Zone	Seismicity	Z
1	High	1.0
2	Medium	0.7
3	Low	0.3

Table 27.2. Percentage of Live Load

Building Type	Live Load Considered
A	100%
B	50%
C	25%
Warehouses	80% of capacity
Roofs	25%
Silos, tanks	100% of capacity

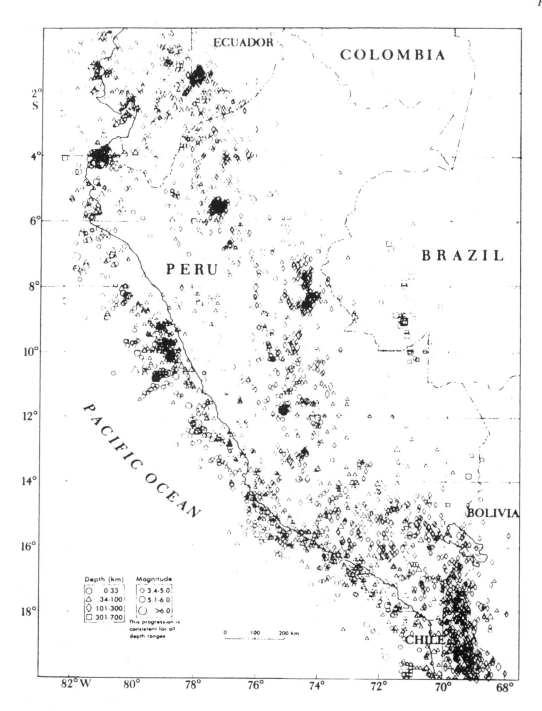

Fig. 27.1. Seismicity of Peru, 1900 through 1984

27.4 SOIL CONDITIONS

In a general sense, foundation soil conditions that directly affect the lateral seismic forces specified by the Code are considered in evaluating the characteristic soil period T_s and the soil factor S. The soil factor S predicts amplification of the earthquake effects. The S values refer to the behavior of the structure over hard soil, for which S is taken as unity. Foundation soil must be considered to a depth of one-half of the shorter of the structure base dimensions.

If no soil study is performed, the values of T_s and S indicated in Table 27.3 should be used. For intermediate foundation soil types, the Code allows interpolation between S values. If a geotechnical investigation is performed to obtain the soil period, the Code specifies a 25% increase of the value thus obtained, in order to establish the S factor.

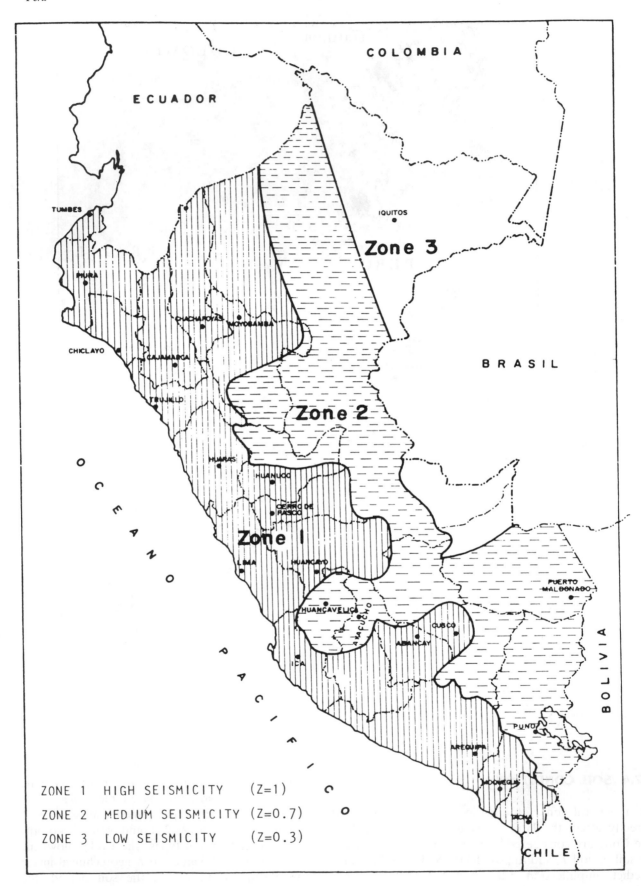

Fig. 27.2.   Seismic zones in Peru

Table 27.3. Soil Parameters for Seismic Analysis

Type	Soil Profile	T_s (sec)	S
I	Rock or dense gravel	0.3	1.0
II	Dense sand or hard cohesive soil	0.6	1.2
III	Loose sand or soft cohesive soil	0.9	1.4

27.5 BUILDING FUNDAMENTAL PERIOD

Unless a dynamic analysis is performed using the actual stiffness and mass distribution of the structure, the fundamental period of the building may be evaluated using the following empirical expressions provided by the Code:

$$T = 0.08N \quad \text{for framed structures} \tag{27.1}$$

$$T = \frac{0.09h}{\sqrt{D}} \quad \begin{array}{l}\text{for framed structures with shear} \\ \text{walls only around elevators}\end{array} \tag{27.2}$$

$$T = \frac{0.07h}{\sqrt{D}} \quad \begin{array}{l}\text{for buildings with frames and} \\ \text{shear walls}\end{array} \tag{27.3}$$

$$T = \frac{0.05h}{\sqrt{D}} \quad \begin{array}{l}\text{for structures consisting mainly} \\ \text{of shear walls}\end{array} \tag{27.4}$$

where

N = number of stories
h = height of building from ground level (m)
D = horizontal building dimension in the direction in which seismic motion is being considered (m)

Equations (27.2), (27.3), and (27.4) were obtained from wind and traffic-induced vibration tests of several buildings in Lima.

27.6 SEISMIC COEFFICIENT

The seismic coefficient C defines the elastic response acceleration spectrum. The elastic response spectra for the strongest horizontal components of the three major earthquakes recorded in Lima are shown in Fig. 27.3 for hard soil conditions (dense gravel). This figure also shows the Code spectrum for hard soil conditions corresponding to $S = 1.0$ and $T_s = 0.3$ sec.

The seismic coefficient C is specified in the Code by eq.(27.5), as a function of the building fundamental period T and of the characteristic period T_s of the soil:

$$C = \frac{0.8}{(T/T_s) + 1} \tag{27.5}$$

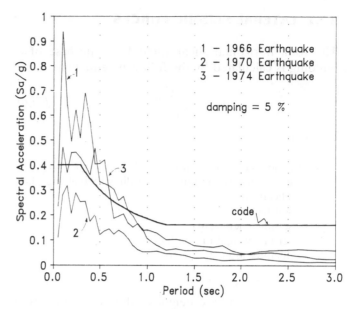

Fig. 27.3. Response spectra for the major earthquakes in Lima compared with the seismic code spectrum

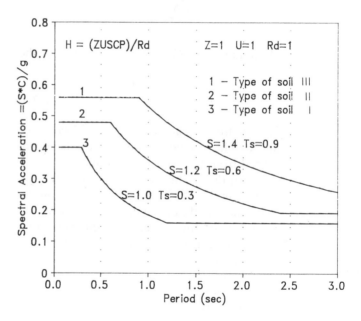

Fig. 27.4. Seismic code response spectra for different soil conditions

with the following limits for the values of C and T_s:

$$0.16 \leqslant C \leqslant 0.40 \quad \text{and} \quad 0.3 \leqslant T_s \leqslant 0.9$$

The product of the soil factor S and the seismic coefficient C, for the three soil conditions defined in Table 27.3 is depicted in Fig. 27.4. The ordinate values in Fig. 27.4 may be considered to be the response spectral acceleration for an elastic structure.

27.7 LATERAL SEISMIC FORCES

The total horizontal shear force V at the base of a building is calculated by the following equation:

$$V = \frac{ZUSC}{R_d} W \qquad (27.6)$$

in which

Z = zone factor, given in Fig. 27.2
U = use factor (determined according to building classification)
S = soil factor
C = seismic coefficient
W = seismic weight of building
R_d = ductility reduction factor.

The base shear force V, for each analyzed direction, is distributed along the height of the building according to the following expression:

$$F_i = fV \frac{W_i h_i}{\displaystyle\sum_{i=1}^{N} W_i h_i} \qquad (27.7)$$

with the additional force F_t applied at the top of the building

$$F_t = (1 - f)V \qquad (27.8)$$

where F_i denotes the lateral force acting at level i, W_i is the seismic weight of level i, and h_i is the height of level i measured from the base of the building.

Factor f in eq.(27.7) takes into account the influence of higher modes of vibration: it depends on the slenderness of the building, which is defined as the ratio of the height of the building to the horizontal dimension in the direction of analysis. Values of the factor f are as follows:

$f = 1.0$ if   slenderness $\leqslant 3$
$f = 0.85$ if   slenderness $\geqslant 6$.

For intermediate values of slenderness, linear interpolation of the factor f is allowed.

27.8 DUCTILITY REDUCTION FACTOR

The Code accounts for energy dissipation during strong earthquakes by means of the ductility reduction factor R_d, which reduces the elastic seismic forces. Basically, this factor corresponds to the global structural ductility. The Code defines ductility as the ratio between maximum inelastic displacement and yield displacement.

For steel frame buildings and ductile special reinforced concrete frame buildings, a factor $R_d = 6.0$ may be used; when special ductile frames are combined with shear walls in reinforced concrete buildings, R_d is reduced to 5.0. Because requirements for high ductility are rather difficult to attain for common reinforced concrete buildings, a value of $R_d = 4.0$ is used frequently for wall-frame structures that do not meet all the special ductility requirements, and a value $R_d = 3.0$ is used when shear walls are the main seismic-resistant elements. Reinforced masonry or masonry confined by concrete elements (collar beams and columns) is given a value of $R_d = 2.5$.

The nonlinear response of several single-degree-of-freedom (SDOF) structures, subjected to the strongest horizontal acceleration recorded in Lima during the October 17, 1966 earthquake (epicenter located 240 km NW of Lima, $M_b = 6.3$, $M_s = 7.5$, depth = 24 km), was calculated to obtain an approximate estimate of the required ductility for Peruvian earthquakes. The behavior of the structures was assumed to be perfectly elastoplastic, with a lateral yield force defined by eq.(27.6) with $Z = 1$, $U = 1$, $S = 1$, and $R_d = 4$. Two values for the structure overstrength (actual strength/code-required strength) were selected. Fig. 27.5 shows the required ductility for each structure, defined as the ratio of the maximum calculated displacement to the displacement at yield. The results in Fig. 27.5 indicate that structures with periods longer than 1 second behave elastically (required ductility = 1), whereas the required ductility is greater than the value of $R_d = 4$ (assumed by the Code) only for very stiff structures.

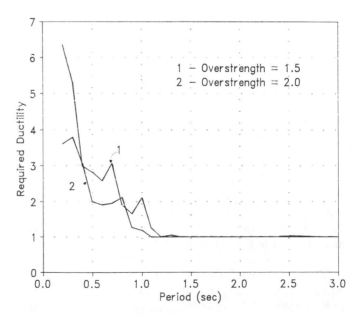

Fig. 27.5. Required ductility for elastoplastic SDOF structures designed according to the seismic code and subjected to 1966 Lima earthquake. Damping 5%

Table 27.4. Maximum Allowed Building Heights

Material	Maximum Height
Concrete, steel	No limit
Masonry	5 stories or 16 m
Wood	2 stories or 7 m
Adobe	1 story or 3 m

27.9 HEIGHT OF BUILDINGS

The Code limits the height of a structure on the basis of the main type of construction material used. Table 27.4 shows the maximum allowed height for buildings constructed of different materials.

27.10 VERTICAL SEISMIC FORCES

Vertical seismic effects have to be evaluated in all vertical elements, prestressed elements, and cantilevers. The vertical force must be considered in combination with the horizontal seismic force. The vertical force values are $0.30W_p$ for Seismic Zone 1, and $0.20W_p$ for Seismic Zone 2, whereas for Seismic Zone 3 these effects need not be considered (W_p is the weight of the element under analysis).

27.11 OVERTURNING MOMENT

The structure and its foundation should be designed to resist the overturning moment produced by the lateral forces. The overturning moment M_x at level x of the building is calculated by means of eq.(27.9), in which F_i is the lateral force at level i [eq.(27.7)], F_t is the additional force at the top [eq.(27.8)], and h_i is the height of level i measured from the base of the building.

$$M_x = F_t(h_N - h_x) + \sum_{i=x+1}^{N} F_i(h_i - h_x) \qquad (27.9)$$

where $x = 0, 1, 2, \ldots, N - 1$.

A reduction in the overturning moment is allowed for buildings 10 or more stories high; the calculated moment is reduced by 2% per story to a maximum reduction of 20%. For example, in a building 15 stories high, there would be no reduction in calculated moment for the top 10 stories, the fifth story would have a 2% reduction, the fourth story would have a 4% reduction, etc. Thus, the calculated moment at the base of a 15-story building would be reduced 10%. For inverted pendulum structures, no reduction in the overturning moment is allowed.

27.12 TORSIONAL EFFECTS

Seismic forces F_i are considered applied at the center of mass of each respective level. The torsional moment at each level is evaluated using eqs.(27.10) and (27.11), which consider the inherent eccentricity e due to the noncoincidence between the center of mass and the stiffness center, in addition to an accidental eccentricity, which is taken as 0.05 times the building plan width b, perpendicular to the direction of the seismic forces.

$$M_{ti} = V_i(1.5e + 0.05b) \qquad (27.10)$$
$$M_{ti} = V_i(e - 0.05b) \qquad (27.11)$$

where V_i denotes the shear force at level i.

Reductions in the shear forces in elements due to torsional effects are not allowed.

27.13 DYNAMIC ANALYSIS

The Code allows the use of two alternative dynamic methods for the evaluation of the seismic structural response: modal analysis and step-by-step analysis. Dynamic analysis is required for buildings higher than 25 stories, or 75 m, and for those buildings for which the static method is not sufficiently accurate.

27.13.1 Modal Analysis

The Code requires that at least three modes be considered. Higher modes for which the incremental effect does not exceed 10% of the total may be neglected in the modal superposition. If this method is used, the base shear force should not be less than 80% of the value obtained using eq.(27.6). If a lesser base shear force is obtained in the modal superposition, all the results from the modal analysis must be scaled by the ratio between the specified 80% of the static base shear force calculated with eq.(27.6) and the base shear obtained by the modal analysis.

Each modal response is associated with a spectral acceleration S_{am} expressed in units of gravitational acceleration g, and given by eq.(27.12). In this equation the coefficients Z, U, S, and R_d were defined previously, and the seismic coefficient C is evaluated from eq.(27.5) using the corresponding modal period.

$$S_{am} = \frac{ZUSC}{R_d} \qquad (27.12)$$

The design values for internal forces, moments, and displacements are obtained by the superposition of all the significant modes. This superposition is calculated as the mean value between the sum of the absolute modal responses and the Square Root of the Sum of Squared (SRSS) modal values.

27.13.2 Step-by-Step Analysis

The step-by-step method to calculate the seismic response of the structure may use true earthquake records, artificial records, or a combination of both types. The data must be typical of the intensity and general characteristics of the earthquakes that have affected the region, and a minimum of three different sets of data must be used.

27.14 LATERAL DISPLACEMENTS

When an elastic analysis is performed using the equivalent lateral forces prescribed by the Code, the resulting displacements must be multiplied by $0.75R_d$ (Section 27.8) to obtain the lateral inelastic displacements. There is no limit on the absolute lateral displacements. However, the Code stipulates that the drift between two consecutive stories shall not exceed 1.5% of the story height, but is limited to 1% of the story height for those buildings having elements that may suffer damage from such drifts. At every level, the separation between adjacent buildings shall not be less than 2/3 of the sum of the maximum displacements of the two structures at corresponding levels.

27.15 NONSTRUCTURAL ELEMENTS

Nonstructural elements, as well as their supports, shall be designed to support an equivalent horizontal force out-of-plane given by

$$V \text{ (nonstructural)} = ZUC_1W_p \qquad (27.13)$$

in which factors Z and U are similar to those defined for the main structure and W_p denotes the weight of the element. Coefficient C_1 varies from 1.0 for elements that may fall outward (such as balcony parapets) to 0.35 for inner walls and 0.25 for fences. For a floor acting as a rigid diaphragm, the force V is considered to act in the plane of the floor with the coefficient $C_1 = 0.20$.

27.16 LOAD COMBINATIONS

Load combinations are not specified in the 1977 Code, but they are included in the National Building Code (ININVI 1985). Seismic forces usually are combined with gravitational loads (dead load, live load, snow) and other load effects (earth pressure or temperature

changes) in order to obtain design internal forces. The Code does not require the simultaneous consideration of wind and seismic forces.

27.17 NUMERICAL EXAMPLE

Example 27.1

An eight-story reinforced concrete office building is presented for seismic analysis. The plan view of the building is shown in Fig. 27.6. Seismic analysis is to be performed for forces acting in the y direction. The height of the first story is 4 m; all the other stories are 3 m high. The weight calculated for each level of the building is 450 metric tons, except at the roof, where the weight is 400 metric tons. These weights were calculated as dead load plus 25% of the design live load. This is a building of Type C, thus Table 27.2 indicates use of 25% of the live load. The building is located in Lima at a site where the foundation soil is dense gravel. Perform the following analyses:

(a) Static method. According to the static method of the Peruvian Seismic Code, determine the following: (1) Fundamental building period, (2) base shear force, (3) lateral force distribution, (4) story shear forces, and (5) overturning moments.

(b) Dynamic method. Determine the stiffness and mass matrices for this structure, then use the modal superposition method to determine the following: (1) natural periods and modal shapes for all the significant modes, (2) modal spectral acceleration, (3) modal forces and modal story shears, and (4) modal overturning moments. Also, use modal combination, as stipulated in the Code, to determine the response.

Solution

(a) Static analysis

1. Fundamental period. For analysis in the y direction, the main structural seismic-resistant elements are the two shear walls parallel to that direction; by comparison the elevator walls and interior columns are of minor importance in resisting lateral seismic forces.

Because $h = 4 + (3)(7) = 25$ m (height of the building) and $D = 15$ m (dimension of the building in the y direction), the fundamental period is evaluated by eq.(27.4) as

$$T = \frac{(0.05)(25)}{\sqrt{15}} = 0.32 \text{ sec}$$

2. Base shear force. The city of Lima, in which the building is located, is an area of high seismicity (Fig. 27.2) with a seismic zone factor $Z = 1.0$, as given in Table 27.1. For an office building (Type C), $U = 1.0$

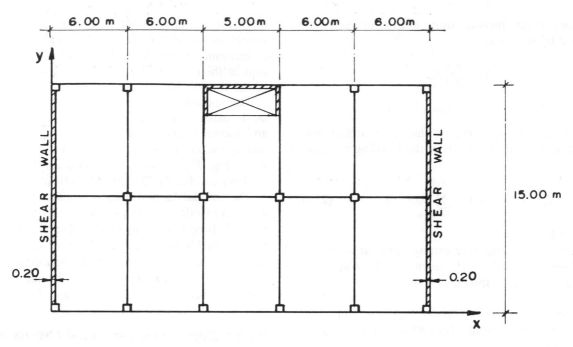

Fig. 27.6. Floor plan of the eight-story building for Example 27.1. (Seismic analysis in the *y* direction)

(Section 27.3). For dense gravel as foundation soil, Table 27.3 gives the soil factor $S = 1.0$ and the predominant period of the soil $T_s = 0.3$ sec.

The seismic coefficient C is calculated from eq.(27.5) as

$$C = \frac{0.8}{(T/T_s) + 1}$$

$$0.16 \leqslant C \leqslant 0.40 \quad \text{and} \quad 0.3 \leqslant T_s \leqslant 0.9$$

$$C = \frac{0.8}{(0.32/0.30) + 1} = 0.387$$

The total weight W of the building is calculated as

$$W = (450)(7) + 400 = 3{,}550 \text{ metric tons}$$

The base shear force is given by

$$V = \frac{ZUSC}{R_d} W \qquad \text{[eq.(27.6)]}$$

in which $R_d = 3.0$ (Section 27.8).

Hence,

$$V = \frac{(1)(1)(1)(0.387)}{3} 3{,}550 = 458 \text{ metric tons}$$

The base shear V is approximately 13% of the total weight of the building.

3. Lateral force distribution. The base shear force is distributed through the height of the building as

$$F_i = fV \frac{W_i h_i}{\displaystyle\sum_{i=1}^{N} W_i h_i} \qquad \text{[eq.(27.7)]}$$

in which the factor f depends on the slenderness of the building (slenderness = total height/plan dimension in the y direction); the slenderness is $25/15 = 1.67 < 3$, so the factor $f = 1.0$ (Section 27.7).

The calculated lateral forces F_i are shown in Table 27.5.

Table 27.5. Lateral Force Distribution, Story Shear Forces, and Overturning Moments for Example 27.1

Level i	Height h_i (m)	Weight W_i (tons)	Lateral Force F_i (tons)	Story Shear V_i (tons)	Overturning Moment M_i (ton-m)
8	25	400	89.9	89.9	——
7	22	450	89.0	178.9	270
6	19	450	76.8	255.7	806
5	16	450	64.7	320.4	1,574
4	13	450	52.6	373.0	2,535
3	10	450	40.5	413.5	3,654
2	7	450	28.3	441.8	4,894
1	4	450	16.2	458.0	6,220
Base (0)	—	—	——	——	8,052

4. Story shear forces. Story shear forces V_i are calculated by statics as

$$V_i = \sum_{j=i}^{N} F_j$$

Resulting values of V_i are shown in Table 27.5.

5. Overturning moments. The overturning moments M_i at the various levels of the building are given by

$$M_i = F_t(h_N - h_i) + \sum_{j=i+1}^{N} F_j(h_j - h_i) \qquad \text{[eq.(27.9)]}$$

where $i = 0, 1, 2, \ldots, N - 1$.

No reduction in the overturning moment is allowed for buildings less than 10 stories high. Calculated values of M_i are also shown in Table 27.5.

(b) Dynamic analysis (modal superposition). The data required to evaluate the actual stiffness of the structure are:

Modulus of elasticity: $E = 2.2 \times 10^6$ ton/m^2,

Shear modulus: $G = 0.96 \times 10^6$ ton/m^2,

Wall cross-section: 0.20 m $\times$ 15.0 m (two of these walls).

The building structure is idealized as two cantilever walls with eight lateral degrees of freedom. Use is made of a plane-frame analysis computer program to determine the stiffness matrix of the structure, including shear deformations that refer to the eight lateral displacements that correspond to the eight levels of the building. The mass, assumed lumped at the various levels of the building, produces a diagonal mass matrix indicated by

$$\lceil M \rfloor = \lceil \, 450 \ 450 \ 450 \ 450 \ 450 \ 450 \ 450 \ 400 \, \rfloor / g \ (\text{ton-sec}^2/\text{m})$$

where $g = 9.81$ m/sec^2 is the acceleration of gravity.

The first three natural periods and modal shapes were determined using a computer program that implements the subspace iteration method. Also, the effective modal weights were calculated according to eq.(4.65) of Chapter 4. These results are shown in Table 27.6. The data in this table show that the first

three modes account for 98.2% of the total effective weight in the system. This satisfies the usual practice of including a sufficient number of modes so that the sum of their effective weights is at least equal to 90% of the total weight of the building.

The spectral accelerations were calculated for each mode using eq.(27.12). The modal story shear forces and modal overturning moments were calculated by adding the effects of the modal lateral forces story by story. Fig. 27.7 depicts the lateral forces for the static analysis and for the first three modes.

The modal superposition values were obtained using the Code criteria (Section 27.13.1), which is the mean between the sum of the absolute modal values and the SRSS modal values. Figs. 27.8 and 27.9 show the story shears and the overturning moments, respectively, for the static analysis and the superposition of the first three modes.

27.18 COMPUTER PROGRAM AND EXAMPLE

A computer program has been developed to implement the main provisions of the Peruvian Seismic

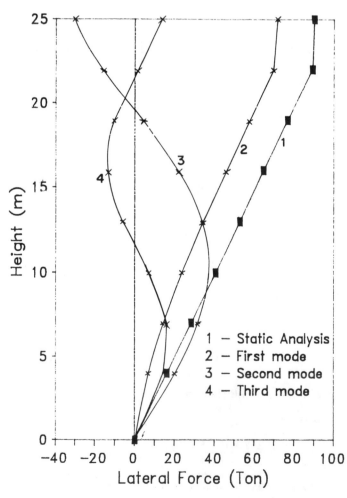

Fig. 27.7. Lateral force distribution for Example 27.1

Table 27.6. Results from Dynamic Analysis Example 27.1

Mode	Period (sec)	Spectral Acceleration (g)	Effective Weight	Weight %	Base Shear (tons)
1	0.351	0.123	2,613	73.6	321
2	0.088	0.133	735	20.7	98
3	0.044	0.133	138	3.9	18

Sum = 98.2%

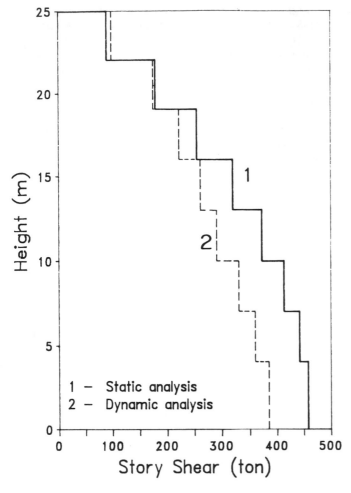

Fig. 27.8. Story shear distribution for Example 27.1

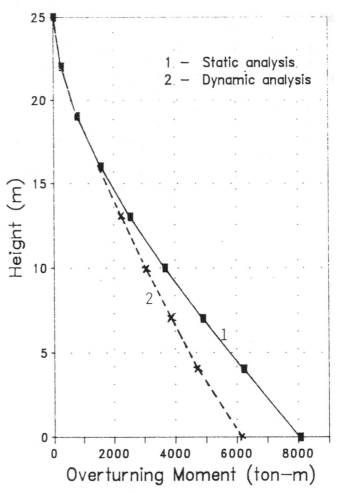

Fig. 27.9. Overturning moment distribution for Example 27.1

Resistant Design Code (1977). The program is intended to be used only for determination of static lateral forces in buildings. The program is written in BASIC for implementation on IBM PC-compatible microcomputers.

Example 27.2

Example 27.1 is solved using the computer program. Input data and output results from the computer program are provided at the end of this chapter.

```
=========================================================
              ***** SEISMIC ANALYSIS OF BUILDINGS *****
     GIANFRANCO OTTAZZI & DANIEL QUIUN, CATHOLIC UNIVERSITY OF PERU

                    SEISMIC CODE OF PERU (1977)
=========================================================

DATA FILE : a:ex232.dat

EXAMPLE 27.2  EIGHT-STORY BUILDING

Data provided for seismic analysis
-----------------------------------

SEISMIC ZONE FACTOR              Z = 1.0
BUILDING USE FACTOR              U = 1.0
SOIL FACTOR                      S = 1.0
SOIL PERIOD                      Ts = 0.30 sec
DUCTILITY FACTOR                 Rd = 3.0
NUMBER OF STORIES                Ns = 8
TOTAL HEIGHT                     HT = 25.00 m
```

```
BUILDING HORIZONTAL WIDTH IN
DIRECTION OF ANALYSIS                    D = 15.00 m
CODE EXPRESSION FOR PERIOD               Not Used
PERIOD PREVIOUSLY DETERMINED             T = 0.32 sec

STORY DATA

STORY       STORY HEIGHT      STORY WEIGHT
            meters            ton

 8          25.00             400.00
 7          22.00             450.00
 6          19.00             450.00
 5          16.00             450.00
 4          13.00             450.00
 3          10.00             450.00
 2           7.00             450.00
 1           4.00             450.00


RESULTS FROM ANALYSIS
---------------------

Fundamental Period              T = 0.32 sec
Seismic Coefficient             C = 0.387
Total building weight           W = 3550.000 ton
Base shear force                V = 458.065 ton
Slenderness                     HT/D = 1.667
Shape factor                    f = 1.00


Story    Lateral Force    Story Shear    Overt. Moment
         (ton)            (ton)          (ton-m)
=================================================================

 8        89.905           89.905          269.7
 7        89.006          178.910          806.4
 6        76.869          255.779         1573.8
 5        64.731          320.510         2535.8
 4        52.594          373.105         3654.6
 3        40.457          413.562         4895.3
 2        28.320          441.882         6221.0
 1        16.183          458.065         8053.2
```

REFERENCES

Espinoza, A. F. (1985) *Catálogo Sísmico del Perú*. Monografías del Instituto Geográfico Nacional, Madrid, Spain.

Husid, R., and Vargas, J. (1975) *Análisis Crítico del Capítulo IV, Título V, del RNC: Seguridad contra el Efecto Destructivo de los Sismos*. Pontificia Universidad Católica del Perú, Lima, Peru.

ININVI (1985) *Norma E-020*. Cargas, Normas Técnicas de Edificación, Lima, Peru.

Ministerio de Vivienda y Construcción (1977) *Normas de Diseño Sismo-Resistente*. Reglamento Nacional de Construcciones, Lima, Peru.

Ottazzi, G.; Repetto, P.; Vargas, J.; and Zegarra, L. (1980) "Bases para una Revision de las Normas Peruanas de Diseño Sismo-Resistente." Proceedings, II Seminario Latinoamericano de Ingeniería Sísmica, Lima, Peru.

Structural Engineers Association of California (SEAOC) (1990) *Recommended Lateral Force Requirements and Commentary*. 5th Edition, San Francisco, CA.

Structural Engineers Association of California (SEAOC) (1967) *Recommended Lateral Force Requirements and Commentary*. 2nd Edition, San Francisco, CA.

28

Portugal

Joao Azevedo

28.1 INTRODUCTION

Portugal is located to the north of the Azores-Gibraltar Plate boundary, in an intermediate tectonic environment, between an interplate boundary and a more stable intraplate region. The seismicity of this country thus is influenced by the seismic activity in the zone of collision between the Eurasian Plate and the African Plate and by the activity on faults in the continent.

Strong earthquakes have caused extensive damage in continental Portugal in the years 1009, 1356, 1531, 1755, 1858, 1909, and 1969, and in the Azores Islands in the years 1522, 1614, 1757, 1841, 1973, and 1980. The Lisbon earthquake of 1755, one of the largest earthquakes in world history, destroyed downtown Lisbon, caused an estimated 10,000 casualties, and greatly damaged over 50% of the buildings in Lisbon. In this century, more than 700 earthquakes of magnitude greater than 3 on the Richter scale have been recorded within a circle of 500 km around Lisbon. The most destructive of these earthquakes was the 1909 Benavente earthquake, which completely destroyed the village of Benavente, the North Atlantic earthquake of February 28, 1969, with epicenter 280 km SW of Lisbon, which was felt all over the country; and the 1980 Azores earthquake.

Following the 1755 Lisbon earthquake, extensive studies of earthquake phenomena and their effects on buildings were undertaken in what probably was the first systematic study of earthquake engineering in the world. A survey using a special questionnaire concerning the effects of the Lisbon earthquake was sent to government officials and members of the clergy throughout the country. This survey provided important information concerning levels of intensity for the different zones of the country, damage sustained by the different types of structures, and the nature of the earthquake itself. Soon after the completion of this survey, soldiers were directed to use huge hammers to test several specially designed building systems in the main square of Lisbon. This testing can be viewed as the precursor of modern-day "shaking table" experiments. According to Ferry Borges (1960), due to building provisions introduced after the 1755 earthquake, it is believed that Lisbon was the first city in the world to have a code for seismic-resistant construction. This code, prepared after the 1755 earthquake, defined the basic principles for reconstruction of the city. One of the innovations introduced in this code was the concept of wood-braced frames interacting with and supporting the masonry elements (ascribed to Carlos Mardel, the architect/engineer in charge of the reconstruction plans for Lisbon after 1955). This concept soon was adopted in other countries and was replaced later by steel frame and reinforced concrete frame structures. Finally, the code limited to three the number of stories permitted in new construction.

The fact that the return period for such catastrophic earthquakes, about 200 years, is large compared to the life of one generation helps to explain why until recently there has not been an effort to implement code regulations. Following the 1909 Benavente earthquake, a work group on seismic studies was given the task of preparing seismic specifications that could be incorporated in a code for building construction to resist earthquake loading. The Symposium on Seismic Actions [Ordem dos Engenheiros (1955)], which occurred 200 years after the great Lisbon earthquake and the reconstruction of the city, was the beginning of a new and fruitful effort in seismic studies in Portugal. This symposium alerted civil engineering researchers and design professionals to the urgent need for seismic-resistant building design provisions. As a result, a code for earthquake-resistant construction, *Regulamento de Seguranca das Construcoes contra os Sismos*, was published in 1958. This code introduced the concept of seismic risk for the several zones of Portugal and the first rules for design against seismic loads.

The 1961 version of the Code for Actions on Buildings and Bridges (*Regulamento de Solicitacoes em Edificios e Pontes*) included provisions for earthquake-resistant design of buildings and bridges. A new code, Regulation for Safety and Actions for buildings and bridges (RSA) (*Regulamento de Seguranca e Accoes para Estruturas de Edificios e Pontes*), was promulgated in 1983. Simultaneously with the publication of this code, several other codes related to the general regulation of construction in different materials were published. Two important codes published about that time were:

(a) Code for Reinforced and Prestressed Concrete Structures (*Regulamento de Estruturas de Betao Armado e Pre-Esforcado*, 1983)

(b) Code for Steel Structures (*Regulamento de Estruturas de Aco para Edificios*, 1986).

These two codes and the Code for Safety and Actions contain the seismic-design guidelines for structural engineering in Portugal. The recent Code for Safety and Actions takes into consideration seismic hazards, soil conditions, structural dynamic characteristics, ductility capacity, and indirectly, the importance-occupancy factor. The Code for Safety and Actions defines four seismic zones in Portugal, including the Azores and the Madeira Islands. These zones were defined as the result of an assessment of the level of activity for each of the possible earthquake sources. A level of seismic hazard was assigned through a *seismicity coefficient* for each one of these zones.

The code provides two methods for earthquake-resistant design: (1) the *static lateral force method* and (2) the *dynamic method*. The first method is applicable to structures regular in plan configuration and in elevation. Also, both stiffness and mass distribution should be considered when classifying structures as regular. The dynamic method uses the concepts of modal superposition and response spectra or another dynamic method of analysis such as a time-history response or a stochastic method analysis.

The effects of spatial variability of the seismic action should be considered in the analysis of structures that have any two supports at a distance greater than 100 m apart. The influence of local soil conditions also should be considered in both the static equivalent lateral force method and the dynamic method. The code also requires consideration of soil-structure interaction and of hydrodynamic effects.

The code essentially implements linear elastic analysis, but behavior coefficients designated as *q-factors*, which depend on the type of structure and its ductility characteristics are applied for different kinds of structural materials (concrete, steel, etc.). The code allows larger ductility factors if special construction details and specifications are followed. There are no special regulations in the seismic code regarding control and quality of construction. However, the general provisions for continuous supervision and control of quality by both the owner and by local municipal authorities are also enforced in the construction of structures built to resist earthquakes. Unfortunately, full compliance with these regulations has not been achieved. Control of quality of materials is performed according to Standards published by the Portuguese Laboratory for Civil Engineering (Laboratorio Nacional de Engenharia Civil, LNEC).

This chapter presents an overview of the current seismic codes for Portugal (primarily the RSA) without attempting to give extensive explanations of every provision. The presentation, examples, and computer program included with this chapter are limited to the static equivalent lateral force method.

28.2 SEISMIC ZONES

The National Territory of Portugal has been divided into seismic zones according to the historical seismicity of each region. This division was based on the level of maximum peak ground acceleration for earthquakes expected in each zone. The expected peak ground acceleration constitutes the basic parameter of the earthquake for seismic analysis. Portugal is divided, as shown in the map of Fig. 28.1, into the following four seismic zones (in decreasing order of seismic intensity): A, B, C, and D. Zone A includes the southwest region of the continental territory of the country, as well as the Azores Islands located east of the mid-Atlantic ridge. Zones B, C, and D are located progressively farther north in the continental territory. Zone D includes the Madeira archipelago and Flores and Corvo islands in the Azores archipelago, which are located on the American Plate. Each of these seismic zones has been assigned *a relative seismicity coefficient* α as indicated in Table 28.1.

28.3 SOIL CHARACTERIZATION

The code classifies soils into the following three types:

Type I. Rock or very hard cohesive soils

Type II. Very stiff to medium-stiff cohesive soils or dense cohesionless soils

Type III. Soft to very soft cohesive soils or loose cohesionless soils

Table 28.1. Relative Seismicity Coefficient α

Seismic Zone	α
A	1.0
B	0.7
C	0.5
D	0.3

Table 28.2. Fraction of Live Load Included in Seismic Mass

Use of Structure	ψ_2
Houses, hotels, small hospital rooms, stadiums, and churches.	0.2
Schools, offices, military barracks, large hospital rooms, restaurants, waiting rooms, gymnasiums, theaters, cinemas, and public spaces in buildings.	0.4
Archives, light industrial buildings, and car garages.	0.6

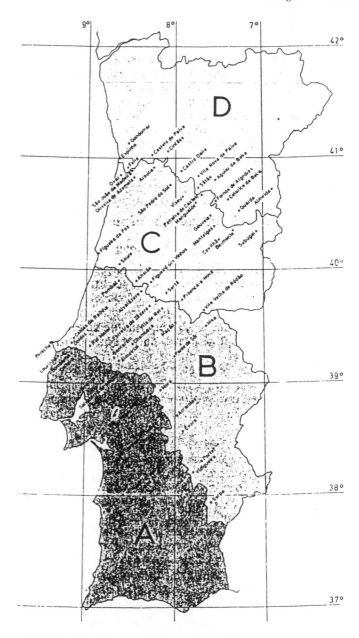

Fig. 28.1. Seismic zones of the continental Portuguese territory

28.4 SEISMIC MASS, STIFFNESS, AND DAMPING

The *seismic mass* consists of the dead load (weight of the structure plus permanent loads on the structure), as well as a fraction of the live loads. This fraction (ψ_2) depends on the use of the structure, as indicated in Table 28.2.

The stiffness of the structure is determined on the basis of average values of the stiffness of its structural elements and also of nonstructural elements, such as partitions. The damping considered in the system should include the effect of any special damping device installed in the structure.

28.5 LOAD COMBINATION

When seismic action is considered, the structure should be designed for the following load combination:

$$F_c = DL + 1.5E + \psi_2 LL \qquad (28.1)$$

where DL is the dead load, LL is the live load, E is the earthquake load, and ψ_2 is the load factor given in Table 28.2. The values of ψ_2 in Table 28.2 are applicable to occupation/use live loads. For other loads such as wind, snow, and temperature effect, the corresponding factor ψ_2 is given in Table 28.3.

For other variable loads, the values in Table 28.2 should be adopted.

28.6 METHODS OF SEISMIC DESIGN

In general, structures subjected to earthquake loading should be designed using a dynamic method of analysis. However, the simplified method known as the equivalent static method is allowed for seismic analysis of regular structures. Dynamic analysis of structures can be performed using either *acceleration response spectra* or an *acceleration power spectral density function*. Two different types of earthquake scenarios must be considered: (1) Type 1, which corresponds to earthquakes of small epicentral distances and moderate magnitude (local earthquakes), and (2) Type 2, corresponding to earthquakes of large magnitudes and large epicentral distances (earthquakes generated in the contact zone between the African and Eurasian plates). Linear elastic behavior of the structure can be assumed; however, analytical results should be corrected by appropriate behavior factors (*q*-factors) whose values depend on the type of structure and on its ductility characteristics. For structures with an orthogonal layout, seismic analysis can be performed in two orthogonal main directions of the building, thus simplifying each analysis to that of a planar system. In such a case, an additional analysis must be carried out to account for torsional effects.

28.7 EQUIVALENT STATIC METHOD

The equivalent static method for seismic analysis is applicable only to regular structures. To be considered regular, a building must satisfy the following conditions:

- It shall have an orthogonal layout and shall not be too flexible.
- In plan, the building shall not be unbalanced in its distribution of mass or stiffness.

Table 28.3. ψ_2 Values for Load Combination with Seismic Loads

Type of Load	ψ_2
Traffic loads in bridges	0
Wind	0
Snow	0
Temperature loads	0.3

- In elevation, it shall not exhibit large variations in mass or rigidity and shall not be too flexible.
- The stories shall be constructed as rigid diaphragms, undeformable in their planes.

A structure is considered to be too flexible when its natural frequency is lower than 0.5 Hz. A balanced distribution of mass and stiffness occurs when, for each story, the distance between the mass center and the stiffness center is not greater than 15% of the dimension of the building in the direction normal to the direction of the seismic action.

For buildings that do not satisfy the conditions for regularity, but have some ductility, a simplified method of static analysis still can be used. For such buildings, the equivalent static forces are obtained by multiplying the weights corresponding to the story masses by a single coefficient equal to 0.22α.

For bridges, the conditions for regularity are as follows:

- The superstructure must be supported by vertical columns.
- The longitudinal axis shall be almost straight in plan and the bridge shall not have a pronounced skew.
- The spans shall not differ greatly and the structure shall be almost symmetrical with respect to a plane perpendicular to the longitudinal axis.

28.7.1 Seismic Coefficients

When the equivalent static method of analysis is used, *seismic coefficients* are calculated in order to evaluate the equivalent static forces that simulate the seismic action. The seismic coefficient β is evaluated for each orthogonal direction by the following formula:

$$\beta = \beta_0 \frac{\alpha}{\eta} \qquad (28.2)$$

where

α = the relative seismicity coefficient (Table 28.1)
β_0 = the reference seismic coefficient (Table 28.4)
η = the behavior factor (Table 28.5).

The reference seismic coefficient β_0 depends on the local soil characteristics and on the value of the

Table 28.4. Reference Seismic Coefficient β_0

Type of Soil	Fundamental Natural Frequency of Structure f (in Hz)	β_0
I	$0.5 \leqslant f \leqslant 5.6$	$0.17\sqrt{f}$
	$f \geqslant 5.6$	0.40
II	$0.5 \leqslant f \leqslant 4.0$	$0.20\sqrt{f}$
	$f \geqslant 4.0$	0.40
III	$0.5 \leqslant f \leqslant 2.0$	$0.23\sqrt{f}$
	$f \geqslant 2.0$	0.32

fundamental natural frequency of the building. Values for β_0 are shown in Table 28.4.

The value of β_0 obtained from Table 28.4 is an approximation of the maximum acceleration response (spectral acceleration) of a single-degree-of-freedom system with a 5% damping ratio when subjected to a specific seismic action. For other damping ratio values, the value of β_0 obtained from Table 28.4 should be corrected according to the ratio of the spectral acceleration for the specified damping ratio and the spectral acceleration for a 5% damping ratio. Values of spectral acceleration are obtained from response spectral charts such as those in Figs. 28.4(a)–28.4(f).

Values of β obtained according to eq.(28.2) should not be smaller in any case than 0.04 and, if the structure has some ductility, need not be greater than 0.16α.

η is the behavior factor, which depends on the type of structure, the structural material, and the potential ductile behavior of the structure. Values of the behavior factor η (or q-factor) for concrete structures are given in Table 28.5. Normal-ductility and high-ductility structures differ in the degree of fixity detailing of joints and in the type of reinforced concrete design methodology. These differences are explained in the Code for Reinforced Concrete and Prestressed Structures. For other types of structures, appropriate values should be used because they are not yet available for steel or masonry in the Portuguese codes. It has been common practice to use values suggested by the draft versions of Eurocode regulations.

Table 28.5. η Values (q-factors) for Concrete Structures

Action	Normal Ductility	High Ductility
Internal forces in frame structures	2.5	3.5
Internal forces in mixed frame-shear wall structures	2.0	2.5
Internal forces in shear wall structures	1.5	2.0
Displacements and vertical vibrations	1.0	1.0

28.7.2 Fundamental Natural Frequency

The equivalent static analysis considers only the influence of the first mode of vibration for each orthogonal direction. The natural frequency of the structure for each orthogonal direction can be determined by analytical or experimental procedures. For buildings, the natural frequency f (in Hz) can be estimated on the basis of the following empirical expressions:

Frame structures: $f = 12/n$
Mixed shear wall-frame structures: $f = 16/n$
Shear wall structures: $f = 6b/h$

where n is the number of stories above ground level, h is the height (in meters) of the building above the ground level, and b is the horizontal dimension of the building in the direction of the seismic action.

The structure can be assigned to one of the above three categories on the basis of the ratio for each story of the stiffness of the vertical and horizontal elements; that is,

$$\rho_i = \frac{\sum_m \dfrac{I_{vmi}}{L_{vmi}}}{\sum_n \dfrac{I_{hni}}{L_{hni}}} \tag{28.3}$$

where I_{vmi} and L_{vmi} represent, respectively, the moment of inertia and the height between levels of stories above and below level i, and I_{hni} and L_{hni} represent, respectively, the moment of inertia and length of the horizontal elements existing at level i of the building.

If all the values ρ_i are less than 10, the structure can be considered a frame structure. For ρ_i values greater than 100, the structure can be considered shear wall. When the values of ρ_i fall between 10 and 100, the building is a mixed shear wall-frame structure.

An alternative method for the evaluation of the natural frequencies of a structure is the Rayleigh method. Based on this method, the natural frequency for each direction can be determined according to the expression:

$$f \text{ (in Hz)} = \frac{1}{2\pi} \sqrt{\frac{g \sum_{i=1}^n F_i d_i}{\sum_{i=1}^n F_i d_i^2}} \tag{28.4}$$

where g is the acceleration due to gravity, F_i is a force applied at level i with intensity equal to the weight of the mass at that level, and d_i is the displacement at level i due to the action of the set of forces F_i.

The evaluation of the natural frequency of the structure, for each direction, according to the Rayleigh method should be based on the assumption that the overall stiffness depends not only on the stiffness of the structural elements, but also on the stiffness conferred to the structure by nonstructural elements like partition walls. If the stiffness of these nonstructural elements is omitted, the calculated natural frequency will be too low and will yield an underestimate of the seismic coefficient.

28.7.3 Distribution of Equivalent Seismic Forces

The equivalent static forces F_i applied to the structure at each level i for each direction are calculated using the following formula:

$$F_i = \beta h_i G_i \frac{\displaystyle\sum_{i=1}^{n} G_i}{\displaystyle\sum_{i=1}^{n} h_i G_i} \tag{28.5}$$

where β is the seismic coefficient defined in eq.(28.2), h_i is the height of story i above ground level, and G_i is the weight of story i.

28.7.4 Torsional Moments

The forces obtained using eq.(28.5) shall be applied laterally at the various levels of the building, each at a corresponding eccentricity e_{1i} or each at eccentricity e_{2i}, whatever is less favorable. These eccentricities are defined, in reference to Fig. 28.2 where C_{ri} is the position of the center of rigidity at level i and C_{gi} is the position of the center of mass at that level, by the following expressions:

$$\begin{aligned} e_{1i} &= 0.5b_i + 0.05a \\ e_{2i} &= 0.05a \end{aligned} \tag{28.6}$$

where the distances a and b_i are indicated in Fig. 28.2.

The additional eccentricity given by the term $0.05a$ in eqs.(28.6), which applies even for symmetrical structures, is introduced to account for asymmetries due to nonlinear behavior and rotational movements of the soil.

Torsional moments M_{ti} at each story i of the building are calculated as the sums of the products of the equivalent static force F_i times the corresponding eccentricity e_i for levels i and above; that is,

$$M_{ti} = \sum_{j=i}^{n} F_i e_i \tag{28.7}$$

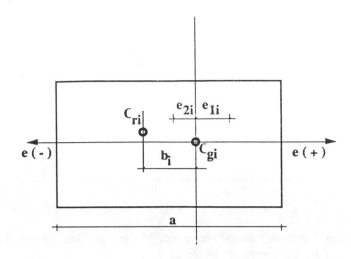

Fig. 28.2. Eccentricity between center of mass C_{gi} and center of rigidity C_{ri} at story i

If the structure is symmetrical relative to a plane containing the direction of the considered seismic action, then, as an alternative method, the forces acting on each one of the resisting structural elements (frames or walls), can be corrected for the eccentricity effects by multiplying the force on each element by a coefficient

$$\xi = 1 + 0.6\frac{x}{a} \tag{28.8}$$

where x is the distance of each element from the plane of symmetry and a is the width of the building in the direction perpendicular to the direction of the seismic forces.

28.7.5 Story Shear Force

The story shear V_i at level i is equal to the sum of the lateral forces at and above that level:

$$V_i = \sum_{j=i}^{n} F_j \tag{28.9}$$

28.7.6 Story Drift

The story drift Δ_i or relative displacement between consecutive levels of a building with rigid horizontal diaphragms at the floor levels is given by

$$\Delta_i = \frac{V_i}{k_i} \tag{28.10}$$

where k_i is the stiffness of the story, that is, the lateral force required to produce an interstory drift of one unit. The code limits the maximum permissible story drift to 1.5% of the height of the story.

28.7.7 Lateral Displacements

For a regular building with rigid horizontal diaphragms, lateral displacements at the various levels may be determined by simply adding the story drifts of the lower stories. Hence, in this case, the lateral displacement δ_i at level i is calculated by

$$\delta_i = \sum_{j=1}^{i} \Delta_j \qquad (28.11)$$

In general, the lateral displacements at the various levels of the building may be determined by static analysis of the building subjected to the equivalent lateral forces F_i.

28.7.8 Overturning Moments

The seismic lateral forces produce overturning moments that result in additional axial forces in the columns, particularly in the exterior columns of the building. The overturning moment, at any level of the building, is determined by statics as the moment produced at that level by the lateral forces above that level. Therefore, the overturning moment M_i at level i of the building is given by

$$M_i = \sum_{j=i+1}^{n} F_j(h_j - h_i) \quad (i = 0, 1, 2, \ldots, n-1)(28.12)$$

where the forces F_j are given by eq.(28.5) and h_i or h_j is the height of the ith or jth level of the building.

Example 28.1

The plane frame of Fig. 28.3 serves as a model for a six-story building that can be considered to have normal ductility. Loads of 200 kN are assigned to each level of the building, except at the roof where the load is 100 kN. The total flexural rigidity of the vertical elements in any story is 1.3×10^5 kN-m². The building is located in Seismic Zone A and its foundation is a compact sand. The building plan dimensions are 8 m × 6 m, and the structure is totally symmetrical.

Determine for the direction perpendicular to the 8-m dimension, and according to the seismic code (RSA): (1) the natural frequency (use the Rayleigh method), (2) the reference seismic coefficient and the seismic coefficient, (3) the equivalent horizontal forces at each level, (4) the shear forces at each story, (5) the overturning moment at each level, (6) the displacement at each level, (7) the effective torsional moment at each level.

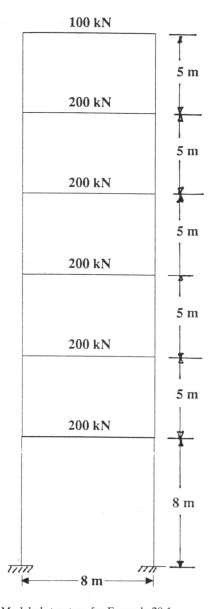

Fig. 28.3. Modeled structure for Example 28.1

Solution

$\alpha = 1.0$ (relative seismicity coefficient for Zone A, Table 28.1)

$\eta = 2.5$ (q-factor for normal concrete frame, Table 28.5)

Soil Type II (compact sand, Section 28.3)

$b = 8$ m (base dimension normal to seismic action)

$f = 0.71$ Hz [fundamental frequency using Rayleigh formula, eq.(28.4)]

$\beta_0 = 0.20\sqrt{0.71} = 0.169$ (reference seismic coefficient, Table 28.4)

$$\beta = 0.169\frac{1}{2.5} = 0.067 \qquad [\text{eq.}(28.2)]$$

Table 28.6. Lateral Forces, Shear Forces, Overturning Moments, and Lateral Displacements for Example 28.1

Story	Lateral Forces (kN)	Shear Forces (kN)	Overt. Moment (kN-m)	Displ. (cm)
1	5.6	74.2	1,682	6.09
2	9.1	68.6	1,088	7.46
3	12.5	59.6	745	8.65
4	16.0	47.0	447	9.60
5	19.5	31.0	212	10.20
6	11.5	11.5	57	10.40

Values for lateral forces, shear forces, overturning moments, and lateral displacements calculated respectively from eqs.(28.5), (28.9), (28.12), and (28.11), are shown in Table 28.6. The values in this table for the lateral displacements were calculated by multiplying the displacement produced by the lateral forces by the behavior factor η (q-factor).
Eccentricity at all levels:

$$b_i = 0$$
$$e_{1i} = 0.5b_i + 0.05a = 0.05 \times 8 = 0.4\,\text{m} \qquad [\text{eq.}(28.6)]$$
$$e_{2i} = 0.05a = 0.05 \times 8 = 0.4\,\text{m}$$

Table 28.7 shows the torsional moments at the various stories of the building, calculated by eq.(28.7).

28.8 DYNAMIC METHOD

In general, the response of structures to dynamic excitation may be obtained by application of the modal superposition method or by time-history analysis. The application of the modal superposition method requires the solution of an eigenproblem to determine natural frequencies and corresponding modal shapes. In this method of analysis, as demonstrated in Chapter 4, the problem is reduced to the solution of a single-degree-of-freedom system for each

Table 28.7. Torsional Moments

Story	e_i (m)	F_j (kN)	$F_j e_i$ (kN-m)	Torsional Moment (kN-m)
1	0.4	5.6	2.2	29.7
2	0.4	9.1	3.6	27.4
3	0.4	12.5	5.0	23.8
4	0.4	16.0	6.4	18.8
5	0.4	19.5	7.8	12.4
6	0.4	11.5	4.6	4.6

significant mode of the system. For seismic excitation, the response for each mode is obtained from spectral design charts as described in detail in Chapter 2. Finally, these modal responses are combined by a suitable method such as the Square Root of the Sum of the Squared values (SRSS).

An alternative approach to the commonly used deterministic analysis is the application of the theory of random vibrations[1] through the use of power spectral density functions. The seismic code of Portugal provides both methods of dynamic analysis: the acceleration response spectra for the deterministic analysis, and power spectral density functions for stochastic analysis.

28.8.1 Acceleration Power Spectral Density Function

Values of power spectra, provided by the code, for the horizontal components of ground acceleration in Seismic Zone A for two types of earthquakes and three soil types are shown in Table 28.8. For intermediate frequency values, linear interpolation may be used. For Seismic Zones B, C, and D, the values in

Table 28.8. Power Spectral Density Function $S(f)$ [(cm/sec²)²/Hz] for the Horizontal Components of Ground Acceleration (Zone A)

Seismic Action Type 1					
Soil Type I		Soil Type II		Soil Type III	
f	$S(f)$	f	$S(f)$	f	$S(f)$
0.04	0	0.03	0	0.02	0
1.05	250	0.90	220	0.75	190
2.10	360	1.80	300	1.50	240
4.20	360	3.60	300	3.00	240
8.40	160	7.20	130	6.00	100
16.8	50	14.4	40	12.0	35
20.0	20	20.0	16	20.0	12

Seismic Action Type 2					
Soil Type I		Soil Type II		Soil Type III	
f	$S(f)$	f	$S(f)$	f	$S(f)$
0.04	0	0.03	0	0.02	0
0.60	220	0.50	220	0.40	220
1.20	300	1.00	400	0.80	500
2.40	150	2.00	160	1.60	200
4.80	65	4.00	65	3.20	80
9.60	20	8.00	25	6.40	30
20.0	0	20.0	0	20.0	0

---

[1]Paz, Mario, *Structural Dynamics: Theory and Computation*, 3d ed., Van Nostrand Reinhold, New York, 1991, Chapter 22.

Table 28.8 should be multiplied by the square of the corresponding seismicity zone factor (α). Values of power spectra for the vertical direction are calculated by multiplying the corresponding values for the horizontal direction by 0.45. However, vertical seismic forces need be considered only when the structure is sensitive to vertical vibrations, as is the case for structures with natural frequencies lower than 10 Hz and shapes permitting large vertical deformations. Table 28.8 provides the power spectral density function for the two scenarios of seismic activity: Type 1, earthquakes of moderate magnitude and small focal distance; and Type 2, earthquakes of large magnitude and focal distance greater than for those included in Scenario 1.

In performing a time domain analysis based on power spectra, several time histories should be analyzed. Those histories should have a duration of 10 seconds for Type 1 and 30 seconds for Type 2. The average of the maximum response values should be used in design.

28.8.2 Response Spectrum Analysis

Spectra representing horizontal components of the seismic action are presented in Figs.28.4(a)–28.4(f). These figures represent the average response spectra based on the two different earthquake scenarios (Type 1 and Type 2) and three different local soil conditions (Types I, II, and III) for Seismic Zone A ($\alpha = 1.0$). Each figure contains the spectral ordinates for three damping ratio values ($\xi = 2\%, 5\%$, and 10%). For all the other seismic zones, the ordinates should be multiplied by the corresponding seismicity factor. For the vertical component, the corresponding ordinates of the horizontal spectra shall be multiplied by 0.67.

The design values for the base shear, story shear, lateral deflection, story drift, overturning moment, and torsional moment are obtained by combining corresponding modal responses. As indicated in the code, such a combination may be performed by application of the SRSS method. However, as described in Chapter 4, the SRSS method may yield relatively large errors when some of the natural frequencies are close. In recognition of this fact, the code allows the use of SRSS only when the ratio between any two frequencies of the structure ($f_i > f_j$) that contribute significantly to the response satisfies eq.(28.13).

$$f_i/f_j > 1.5 \qquad (28.13)$$

The Complete Quadratic Combination (CQC), the more refined technique presented in Section 4.8 of Chapter 4, should be used for cases in which eq.(28.13) is not satisfied. The global response can

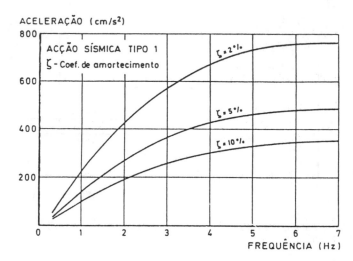

Fig. 28.4(a). Acceleration response spectra: Seismic action Type 1, Soil Type I

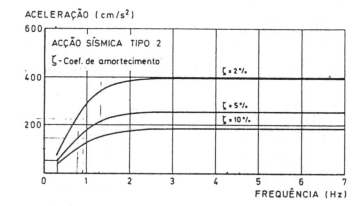

Fig. 28.4(b). Acceleration response spectra: Seismic action Type 2, Soil Type I

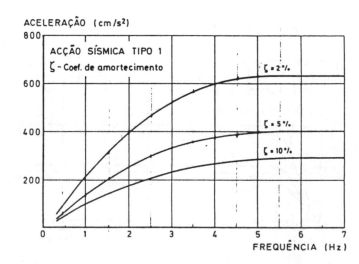

Fig. 28.4(c). Acceleration response spectra: Seismic action Type 1, Soil Type II

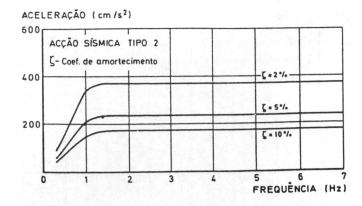

Fig. 28.4(d). Acceleration response spectra: Seismic action Type 2, Soil Type II

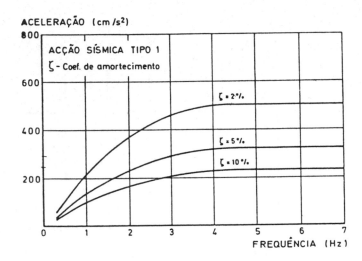

Fig. 28.4(e). Acceleration response spectra: Seismic action Type 1, Soil Type III

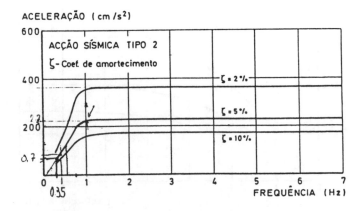

Fig. 28.4(f). Acceleration response spectra: Seismic action Type 2, Soil Type III

also be computed as the square root of the sum of the squares of the modal responses corresponding to the two main directions of the structure.

28.9 COMPUTER PROGRAM

The computer program used in the solution of examples in this chapter implements the provisions of the equivalent static method of the seismic code of Portugal. The program has been coded in BASIC for IBM-compatible microcomputers. To demonstrate the use of the program, two simple structures are analyzed in the following examples.

Example 28.2

Determine the seismic response of the structure of Example 28.1 using the computer program.

Example 28.3

A plane frame similar to the frame analyzed in Example 1 but with 20 stories (first story 5 m and other stories 3 m high) is founded on a soft clay soil and located in Zone A. A distance of 0.5 m separates the stiffness center and the center of mass for the direction perpendicular to the direction of the seismic forces.

Determine according to the seismic code (RSA) for the direction perpendicular to the 8-m dimension: (1) the natural frequency (use simplified formula), (2) the reference seismic coefficient and the seismic coefficient, (3) the equivalent horizontal forces at each level, (4) the shear forces at each story, and (5) the overturning moment at each level, (6) the displacement at each level, and (7) the effective torsional moment at each level.

28.9.1 Input Data and Output Results for Example 28.2

```
**********************************************************************

                    *** EARTHQUAKE RESISTANT DESIGN ***

        JOAO AZEVEDO and LUIS GUERREIRO, INSTITUTO SUPERIOR TECNICO
                    TECHNICAL UNIVERSITY OF LISBON

        *** USING BASIC SEISMIC CODE OF PORTUGAL (RSA-85) ***
                            PROGRAM 25


**********************************************************************


                    ***DATA FILE INFORMATION***

                    1. PREPARE NEW DATA FILE

                    2. MODIFY EXISTING DATA FILE

                    3. USE EXISTING DATA FILE

                    SELECT NUMBER ---> 3
```

```
                DRIVE USED FOR DATA FILES (A:,B:,or C:) ----> C:
                FILE NAME (OMIT DRIVE LETTER) -------------> EXAMPLE1
        INPUT DATA


    SEISMIC   ZONE                        SZ = 1
    TYPE OF SOIL                          TS = 2
    TYPE OF STRUCTURE                     TE = 1
    q FACTOR (BEHAVIOR COEFFICIENT)       Q = 2.5
    BASE DIMENSION (EARTHQUAKE DIRECTION)(m)   B = 6
    BASE DIMENSION (PERPENDICULAR DIRECTION)(m) A = 8
    NUMBER OF STORIES                     N = 6
```

STORY #	STORY HEIGHT (meter)	STORY WEIGHT (kN)	STORY EQUIVALENT EI (kN.m2)
1	8.00	200.00	0.130E+06
2	5.00	200.00	0.130E+06
3	5.00	200.00	0.130E+06
4	5.00	200.00	0.130E+06
5	5.00	200.00	0.130E+06
6	5.00	100.00	0.130E+06

```
                        <RETURN> TO CONTINUE
```

STORY DISTANCES FROM STIFFNESS TO MASS CENTERS

STORY #	DISTANCE (meter)
1	0.00
2	0.00
3	0.00
4	0.00
5	0.00
6	0.00

```
                        <RETURN> TO CONTINUE
SEISMIC RESPONSE
```

NATURAL FREQUENCY EVALUATED BASED ON RAYLEIGH METHOD - 0.71

REFERENCE SEISMIC COEFFICIENT ----- 0.169
SEISMIC COEFFICIENT -------------- 0.067

STORY #	HOR. FORCE (kN)	SHEAR (kN)	OVERT. MOMENT (kN.m)	DISPLACEMENT (m)
1	0.5572E+01	0.7418E+02	0.1682E+04	0.609E-01
2	0.9055E+01	0.6861E+02	0.1088E+04	0.746E-01
3	0.1254E+02	0.5955E+02	0.7453E+03	0.865E-01
4	0.1602E+02	0.4702E+02	0.4475E+03	0.960E-01
5	0.1950E+02	0.3100E+02	0.2124E+03	0.102E+00
6	0.1149E+02	0.1149E+02	0.5746E+02	0.104E+00

PRINTED DISPLACEMENTS INCLUDE THE BEHAVIOR FACTOR (Q-FACTOR)

```
                        <RETURN> TO CONTINUE
POSITIVE VALUES OF THE EFFECTIVE TORSIONAL MOMENTS CORRESPOND
TO SHEAR FORCES APPLIED WITH POSITIVE ECCENTRICITIES
```

A POSITIVE ECCENTRICITY IS MEASURED FROM THE STIFFNESS CENTER
IN THE DIRECTION OF THE CENTER OF MASS

STORY #	TORSIONAL MOMENT 1 (kN.m)	TORSIONAL MOMENT 2 (kN.m)
1	0.2967E+02	-.2967E+02
2	0.2744E+02	-.2744E+02
3	0.2382E+02	-.2382E+02
4	0.1881E+02	-.1881E+02
5	0.1240E+02	-.1240E+02
6	0.4597E+01	-.4597E+01

28.9.2 Input Data and Output Results for Example 28.3

```
****************************************************************

            *** EARTHQUAKE RESISTANT DESIGN ***

JOAO AZEVEDO and LUIS GUERREIRO, INSTITUTO SUPERIOR TECNICO
            TECHNICAL UNIVERSITY OF LISBON

    *** USING BASIC SEISMIC CODE OF PORTUGAL (RSA-85) ***
                        PROGRAM 25

****************************************************************

            ***DATA FILE INFORMATION***

            1. PREPARE NEW DATA FILE

            2. MODIFY EXISTING DATA FILE

            3. USE EXISTING DATA FILE

            SELECT NUMBER ---> 3

            DRIVE USED FOR DATA FILES (A:,B:,or C:) ----> C:
            FILE NAME (OMIT DRIVE LETTER) -------------> EXAMPLE2
        INPUT DATA


    SEISMIC   ZONE                        SZ = 1
    TYPE OF SOIL                          TS = 3
    TYPE OF STRUCTURE                     TE = 2
    q FACTOR (BEHAVIOR COEFFICIENT)       Q = 2.5
    BASE DIMENSION (EARTHQUAKE DIRECTION)(m)   B = 6
    BASE DIMENSION (PERPENDICULAR DIRECTION)(m) A = 8
    NUMBER OF STORIES                     N = 20
```

STORY #	STORY HEIGHT (meter)	STORY WEIGHT (kN)	STORY EQUIVALENT EI (kN.m2)
1	5.00	200.00	0.130E+06
2	3.00	200.00	0.130E+06
3	3.00	200.00	0.130E+06
4	3.00	200.00	0.130E+06
5	3.00	200.00	0.130E+06
6	3.00	200.00	0.130E+06
7	3.00	200.00	0.130E+06
8	3.00	200.00	0.130E+06
9	3.00	200.00	0.130E+06
10	3.00	200.00	0.130E+06
11	3.00	200.00	0.130E+06
12	3.00	200.00	0.130E+06
13	3.00	200.00	0.130E+06
14	3.00	200.00	0.130E+06
15	3.00	200.00	0.130E+06
16	3.00	200.00	0.130E+06
17	3.00	200.00	0.130E+06
18	3.00	200.00	0.130E+06
19	3.00	200.00	0.130E+06
20	3.00	100.00	0.130E+06

```
                        <RETURN> TO CONTINUE
```

STORY DISTANCES FROM STIFFNESS TO MASS CENTERS

STORY #	DISTANCE (meter)
1	0.50
2	0.50
3	0.50
4	0.50
5	0.50

6	0.50
7	0.50
8	0.50
9	0.50
10	0.50
11	0.50
12	0.50
13	0.50
14	0.50
15	0.50
16	0.50
17	0.50
18	0.50
19	0.50
20	0.50

<RETURN> TO CONTINUE

SEISMIC RESPONSE

NATURAL FREQUENCY EVALUATED BASED ON APROXIMATED FORMULAE (RSA) - 0.60

REFERENCE SEISMIC COEFFICIENT ----- 0.178
SEISMIC COEFFICIENT --------------- 0.071

STORY #	HOR. FORCE (kN)	SHEAR (kN)	OVERT. MOMENT (kN.m)	DISPLACEMENT (m)
1	0.2175E+01	0.2779E+03	0.1153E+05	0.557E-01
2	0.3480E+01	0.2758E+03	0.1014E+05	0.676E-01
3	0.4784E+01	0.2723E+03	0.9312E+04	0.794E-01
4	0.6089E+01	0.2675E+03	0.8496E+04	0.910E-01
5	0.7394E+01	0.2614E+03	0.7693E+04	0.102E+00
6	0.8699E+01	0.2540E+03	0.6909E+04	0.113E+00
7	0.1000E+02	0.2453E+03	0.6147E+04	0.124E+00
8	0.1131E+02	0.2353E+03	0.5411E+04	0.134E+00
9	0.1261E+02	0.2240E+03	0.4705E+04	0.144E+00
10	0.1392E+02	0.2114E+03	0.4033E+04	0.153E+00
11	0.1522E+02	0.1975E+03	0.3399E+04	0.161E+00
12	0.1653E+02	0.1822E+03	0.2807E+04	0.169E+00
13	0.1783E+02	0.1657E+03	0.2260E+04	0.176E+00
14	0.1914E+02	0.1479E+03	0.1763E+04	0.183E+00
15	0.2044E+02	0.1287E+03	0.1319E+04	0.188E+00
16	0.2175E+02	0.1083E+03	0.9329E+03	0.193E+00
17	0.2305E+02	0.8655E+02	0.6080E+03	0.197E+00
18	0.2436E+02	0.6350E+02	0.3484E+03	0.200E+00
19	0.2566E+02	0.3914E+02	0.1579E+03	0.201E+00
20	0.1348E+02	0.1348E+02	0.4045E+02	0.202E+00

PRINTED DISPLACEMENTS INCLUDE THE BEHAVIOR FACTOR (Q-FACTOR)

<RETURN> TO CONTINUE

POSITIVE VALUES OF THE EFFECTIVE TORSIONAL MOMENTS CORRESPOND
TO SHEAR FORCES APPLIED WITH POSITIVE ECCENTRICITIES

A POSITIVE ECCENTRICITY IS MEASURED FROM THE STIFFNESS CENTER
IN THE DIRECTION OF THE CENTER OF MASS

STORY #	TORSIONAL MOMENT 1 (kN.m)	TORSIONAL MOMENT 2 (kN.m)
1	0.3196E+03	0.2779E+02
2	0.3171E+03	0.2758E+02
3	0.3131E+03	0.2723E+02
4	0.3076E+03	0.2675E+02
5	0.3006E+03	0.2614E+02
6	0.2921E+03	0.2540E+02
7	0.2821E+03	0.2453E+02
8	0.2706E+03	0.2353E+02
9	0.2576E+03	0.2240E+02
10	0.2431E+03	0.2114E+02
11	0.2271E+03	0.1975E+02
12	0.2096E+03	0.1822E+02
13	0.1906E+03	0.1657E+02
14	0.1701E+03	0.1479E+02
15	0.1481E+03	0.1287E+02
16	0.1245E+03	0.1083E+02
17	0.9954E+02	0.8655E+01
18	0.7303E+02	0.6350E+01
19	0.4502E+02	0.3914E+01
20	0.1551E+02	0.1348E+01

REFERENCES

BORGES, FERRY J. (1960) "Portuguese Studies on Earthquake Resistant Structures." Proceedings of the 2nd World Conference on Earthquake Engineering, Tokyo, Japan.

IMPRENSA NACIONAL, CASA DA MOEDA (1958) *Regulamento de Seguranca das Construcoes contra os Sismos.* Lisbon, Portugal.

―― (1961) *Regulamento de Solicitacoes em Edificios e Pontes.* Lisbon, Portugal.

―― (1983) *Regulamento de Estruturas de Betao Armado e Pre-Esforcado.* Lisbon, Portugal.

―― (1983) *Regulamento de Seguranca e Accoes para Estruturas de Edificios e Pontes.* Lisbon, Portugal.

―― (1986) *Regulamento de Estruturas de Aco para Edificios.* Lisbon, Portugal.

ORDEM DOS ENGENHEIROS (1955) "Simposio sobre a accao dos sismos e sua consideracao no calculo das construcoes." Lisbon, Portugal.

29

Puerto Rico

Luis E. Suarez

29.1 INTRODUCTION AND SCOPE

Puerto Rico is an island lying at the eastern edge of the Greater Antilles near the northeastern edge of the Caribbean Plate, on a rigid block that is moving with respect to the North and South American continents (McCann 1985). The island has been stricken periodically by strong to moderate earthquakes. Although there are some records of many of these events dating back to the sixteenth century (Díaz Hernández 1987), the epicenters and magnitudes of these older seismic events are not clearly defined.

Most of the earthquakes felt in Puerto Rico have originated in one of three zones: the Puerto Rico Trench fault, 50 to 100 km to the north of the island; the Mona Canyon fault off the west coast; and the Anegada Passage fault off the southeast coast. The largest recorded shock for Puerto Rico was the May 2, 1787 earthquake whose magnitude is estimated to have been 8 to 8.25 on the Richter scale. The 1787 earthquake appears to have originated in the Puerto Rico Trench. On November 18, 1867, a strong earthquake of an estimated magnitude 7.5 on the Richter scale occurred in the Virgin Islands; it caused considerable damage in the eastern part of Puerto Rico. The 1867 earthquake originated in the Anegada Passage fault. The shock (aftershocks lasted for almost six months) was followed by the occurrence of severe tsunamis. Although of relatively smaller magnitudes, additional damaging earthquakes occurred during this century in 1918 and 1943. The October 11, 1918 earthquake, which killed more than 100 people, had a magnitude $M = 7.8$. This earthquake, which probably occurred on one of the faults bounding the Mona Canyon, was accompanied by a destructive sea wave nearly six meters (20 feet) high that ravaged the western coast of the island.

The Puerto Rican Seismic Code is contained in *Enmiendas Adoptadas al Reglamento de Edificación* (1987) (Amendments to the Building Code). The provisions of this code are based upon the recommendations of the Earthquake Commission of the Colegio de Ingenieros y Agrimensores de Puerto Rico (Puerto Rican Association of Engineers and Land Surveyors). The first three sections of the 13-section document consist of general considerations, definitions used in the code, and nomenclature. The Puerto Rican code is based partially on the U.S. Uniform Building

Code (UBC) (1985) as well as on the recommendations of the Applied Technology Council [ATC 3-06 (1978)]. However, several of the coefficients in the UBC-85 were changed to reflect the particular conditions of Puerto Rico.

Puerto Rico is included in seismic zone 3 in the ATC zone map for effective peak velocity-related acceleration coefficient. The seismic maps for the contiguous 48 states were based on historical seismicity studies and geological and tectonic evidence. Such studies were not available for Puerto Rico, according to the authors of ATC 3-06. Therefore, the seismic zoning of Puerto Rico was based on the best alternative information available from various sources.

The Puerto Rican Seismic Code does not address explicit requirements for the design and construction of structures that may require special considerations. Such structures include, but are not limited to, transmission towers and large industrial towers, power plants, dams, nuclear power plants, off-shore structures, and exceptionally slender buildings.[1] Also the Administrator of Codes and Permits determines when a building or structure shall be considered as a "special building or structure." The Administrator approves special construction systems with structural concepts not mentioned in the code, provided that evidence is submitted demonstrating that the proposed system will have ductility and energy absorption equivalent to those required by the code.

The code specifies two procedures to determine the seismic design forces, (1) *The equivalent static force method* and (2) *the dynamic method*. The first method is restricted to the seismic design of regular structures, i.e., structures with a reasonably uniform distribution of mass and stiffness. The second method is general and may be applied to the design of any structure, but must be used for irregular and special structures.

29.2 TOTAL LATERAL SEISMIC FORCE

With the exception of the cases considered in Sections 29.8 and 29.9, the code requires that all structures shall be designed and built to resist at least the total lateral seismic force or base shear force V calculated from eq.(29.1). Lateral seismic forces shall be assumed to act independently along the two main directions of the structure.

$$V = ZIKCSW \tag{29.1}$$

The following six factors define the base shear force V in eq.(29.1):

Z = zone factor which depends on the seismic zone of the country. The code assigns the value $Z = 0.6$ for the entire island of Puerto Rico.

I = occupancy importance factor which depends on the occupancy or use of the structure.

K = structural factor which depends on the type of structural system.

C = dynamic response factor which depends on the fundamental period of the structure.

S = soil-structure resonance factor that depends on the characteristics of the soil at the site.

W = total seismic load of the structure. (This load includes the total weight of the structure, partition walls, and permanent equipment.) Structures to be used as warehouses or for parking garages for heavy vehicles are required to include 25% of the design floor live load.

The 1987 amendment to the Puerto Rican Code provides an alternative method to determine the seismic design forces that can be used in lieu of eq.(29.1). Those structures that do not satisfy the ductility and energy absorption requirements implicit in the provisions of the code shall have at least the strength to resist the total lateral seismic force determined from the elastic spectrum (corresponding to the Taft Earthquake Record) shown in Fig. 29.1. In this alternative method, the total seismic force is determined as the product of the spectral acceleration $A_g(T)$ obtained from the chart in Fig. 29.1, divided by 1.4, and multiplied by the total seismic weight W, that is

$$V = \frac{A_g(T)}{1.4} W \tag{29.2}$$

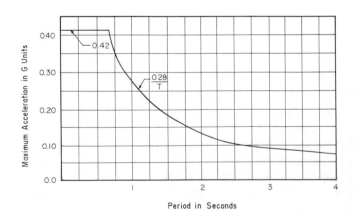

Fig. 29.1. Design spectrum for Taft S-69°-E record

---

[1]Buildings in which the ratio between the total height to the dimension at the ground level parallel to the seismic direction under consideration is equal to or greater than five shall be regarded as "exceptionally slender."

Table 29.1. Importance Factor I

Structure Function	I
Essential facilities	1.5
Buildings or structures that must remain safe and in working condition in the event of an earthquake, to preserve the health and safety of the general population. Includes hospitals, police and fire stations, government communication and emergency centers, prisons, and other facilities that are deemed vital.	
Gathering facilities	1.2
Schools, transmission towers, power transforming stations, and buildings for the gathering of more than 300 persons in a single room.	
Other structures	1.0

in which the spectral acceleration $A_g(T)$ is a function of the fundamental period T of the structure.

29.2.1 Importance Factor

The code assigns different values to the importance factor I depending on the function of the structure as shown in Table 29.1.

29.2.2 Structural Factor

The value of the horizontal structural factor K depends on the type of structural system. In the case of buildings or structures with different resisting systems along the two main directions of the building, the factor K shall be selected as the larger value obtained for either of these directions. To select the factor K, structural systems are classified into seven groups as indicated in Table 29.2.

29.2.3 Dynamic Response Factor

The dynamic response factor C represents the influence of the fundamental period in the total base shear force. This factor decreases the total base shear force for buildings with longer periods according to the following formulas:

$$C = \frac{1}{15T} \leqslant 0.10 \quad \text{for} \quad T \leqslant 1 \sec \tag{29.3}$$

$$C = \frac{1}{15T^{2/3}} \quad \text{for} \quad T \geqslant 1 \sec \tag{29.4}$$

However, the value of C need not exceed 0.10.

The code explicitly limits the maximum value for the period that can be used to determine C, which in turn limits the value of the total base shear. The value of T to be used in eq.(29.3) or (29.4) must satisfy

$$T \leqslant 1.20 T_a \tag{29.5}$$

where T_a is an approximate value for the fundamental period of the building obtained from the formulas in eq.(29.7) or (29.8).

The code provides two explicit ways to calculate the value of T. One is the Rayleigh formula:

$$T = 2\pi \sqrt{\frac{\sum\limits_{i=1}^{n} W_i \delta_i^2}{g \sum\limits_{i=1}^{n} F_i \delta_i}} \tag{29.6}$$

where g is the acceleration due to gravity, F_i are lateral forces applied at each level of the building, δ_i are the elastic lateral displacements due to the forces F_i, and W_i is the seismic weight assigned to level i. In determining the period with Rayleigh's formula, the code states that any rational distribution of lateral forces that approximately satisfies eqs.(29.12) through (29.14) can be used for the forces F_i. Equation (29.6) cannot be used initially because the lateral displacements δ_i are not yet known. However, the code provides two empirical formulas as an alternative method of estimating an approximate value T_a for the fundamental period. The other explicit way of calculating the value of T is to use empirical formulas.

For buildings with moment resisting frames in which the frames are not enclosed by, or adjacent to, more rigid elements that can impede their deflection under the seismic excitation

$$T_a = C_t h_n^{3/4} \tag{29.7}$$

In eq.(29.7) h_n is the height in feet of the upper or nth level of the main part of the structure, with respect to its base. The coefficient C_t is determined as follows:

For reinforced concrete frames: $C_t = 0.025$
For steel frames: $C_t = 0.035$

For any other type of building:

$$T_a = \frac{h_n}{20\sqrt{D}} \tag{29.8}$$

where D is the dimension in feet of the structure in the direction parallel to the applied forces.

The code specifies that the product, in eq.(29.1), of the dynamic response factor C and the soil-structure resonance factor S need not exceed 0.14.

29.2.4 Soil-Structure Resonance Factor

The factor S in eq.(29.1) takes into account the effects that the subsoil conditions have on the structural response. In those cases where the dominant period of the soil T_s has been determined, the

Table 29.2. Structural Factor K and Reduction Ductility Factor R

Structural System	K	R
(a) Bearing wall or braced frame systems:		
A structural system with bearing walls or braced frames providing support for all or the major part of the vertical loads. The seismic resistance is provided by shear walls or braced frames.		
Shear walls with reinforced concrete:	1.33	4.50
Braced frames:	1.33	4.00
Shear walls or reinforced masonry:	2.33	3.50
(b) Frame systems:		
A structural system with an essentially complete space frame providing support for the vertical loads. The seismic force resistance is provided by shear walls or braced frames capable of resisting the total prescribed force.		
Shear walls of reinforced concrete:	1.00	5.50
Braced frames:	1.00	5.00
Shear walls of reinforced masonry:	1.75	4.50
(c) Combined systems:		
A structural system with an essentially complete space frame providing support for the vertical loads. The seismic force resistance is provided by the combination of a ductile moment-resisting space frame and braced frames or shear walls in proportion to their relative rigidity.		
Shear walls of reinforced concrete:	1.00	6.50
Shear walls of reinforced masonry:	1.75	5.50
Braced frames:	1.00	5.50
(d) Dual systems:		
A structural system with an essentially complete space frame providing support for the vertical loads. A ductile moment-resisting space frame shall be provided and this shall be capable of resisting at least 25% of the prescribed seismic forces. The total seismic force resistance is provided by the combination of the ductile moment-resisting space frame and braced frames or shear walls in proportion to their relative rigidity. The shear walls or braced frames, acting independently of the ductile moment-resisting space frame, shall resist the total seismic forces.		
Shear walls of reinforced concrete:	0.80	8.00
Shear walls of reinforced masonry:	1.40	6.50
Braced frames:	0.80	6.00
(e) Moment-resisting frame systems:		
A structural system with an essentially complete space frame providing support for the vertical loads. The seismic force resistance is provided by a ductile moment-resisting space frame capable of resisting the total prescribed forces.		
Steel:	0.67	8.00
Reinforced concrete:	0.80	7.00
(f) Elevated tanks:	2.50	2.50
The minimum value of KC shall be 0.12 whereas the maximum value need not exceed 0.25. The tower must be designed for an accidental torsion of 5% according to Section 29.5. Elevated tanks supported by buildings or structures, or those that do not conform to any type of arrangement of the supporting element described in this section shall be designed according to Section 29.8 using a coefficient $C_p = 0.3$.		
(g) Other:	2.00	1.50
Structures other than buildings and those different from those listed in Section 29.8.		

coefficient S can be determined from the following expressions:

$$S = 1.0 + \frac{T}{T_s} - 0.5\left(\frac{T}{T_s}\right)^2 \quad \text{for} \quad \frac{T}{T_s} \leq 1 \quad (29.9)$$

$$S = 1.2 + 0.6\frac{T}{T_s} - 0.3\left(\frac{T}{T_s}\right)^2 \quad \text{for} \quad \frac{T}{T_s} > 1 \quad (29.10)$$

The code states that the value of S should not be less than 1.00. There are some additional restrictions. First, the value of T used in eq.(29.9) or (29.10) cannot be calculated using empirical expressions but rather should be determined from any properly substantiated analysis, such as the Rayleigh formula. Moreover, the value used for T cannot be less than

0.3 sec. However, if the value of T obtained from a properly substantiated analysis exceeds 2.5 sec, the value of S can be calculated using a soil period of $T_s = 2.5$ sec. Second, the soil-dominant period T_s must be established from the geological data of the site and its value must be within 0.5 sec and 2.5 sec. If there is a range of periods for the soils, the one closest to the fundamental period T of the structure must be used in the calculation of S.

In those cases in which in situ determination of the period of the soil deposit was not carried out, the resonance factor S can be determined according to the type of soil profile:

Profile type S_1: $S = 1.00$
    Bedrocks of any characteristic, either shale-like or crystalline in nature ($N > 100$), hard or very stiff clays

typical of the Hato Rey formations[2] ($N > 20$), very dense sands or clayey sands ($N > 60$).

Profile type S_2: $S = 1.30$
  Stiff clays ($8 \leqslant N \leqslant 20$) or medium to dense sands ($3 \leqslant N \leqslant 60$).

Profile type S_3: $S = 1.50$
  Very soft, soft, or medium-stiff clays, silty clays, organic soils, typical lake deposits and lake deposits covered by sands ($N \leqslant 7$).

Profile type S_4: $S = 1.50$
  All soil deposits where the bedrock is located 200 ft below the surface, such as in valleys of major rivers.

The value of N is the penetration blow count obtained from a standard penetration test. For soils of the S_2 or S_3 type, the value of the soil-structure resonance factor may be calculated from the following expression:

$$S = 1.5 - \frac{0.3 \log N}{1.2} \qquad (29.11)$$

Finally, if there is no data available for the soil at the site, the coefficient S shall be taken equal to 1.5.

29.3 DISTRIBUTION OF LATERAL FORCES

29.3.1 Structures with Regular Shapes

The base shear force V will be distributed in the n levels of the structure as a lateral force F_i at each level, plus an additional force F_t applied at the top of the structure. Hence, the base shear force can be written as

$$V = F_t + \sum_{i=1}^{n} F_i \qquad (29.12)$$

The force F_t applied at the upper level of the structure is defined as follows:

$$F_t = 0.07 TV \leqq 0.25V \quad \text{for} \quad T > 0.7 \sec \qquad (29.13)$$

$$F_t = 0 \quad \text{for} \quad T \leqq 0.7 \sec \qquad (29.14)$$

The remaining part of the base shear, $V - F_t$, will be distributed among the n levels of the building according to

$$F_x = \frac{(V - F_t) W_x h_x}{\sum\limits_{i=1}^{n} W_i h_i} \qquad (29.15)$$

in which F_x and h_x are, respectively, the lateral force and the height above the base at a given level x.

The code stipulates that buildings with setbacks or towers can be considered as "structures with regular shapes" provided that the plan dimension of the tower part in each main direction is at least 75% of the corresponding plan dimension of the lower part of the building.

29.3.2 Structures with Irregular Shapes

The distribution of the lateral forces in structures with highly irregular shapes, or with significant differences in the lateral resistance or stiffness between adjacent floors, or with other unusual structural characteristics shall be done taking into account the dynamic characteristics of the structural system. The code does not provide specific guidelines on how to implement this provision. However, it is common practice to determine first the dynamic properties of the building, the natural frequencies, and the modal shapes. Then, a linear elastic analysis is performed with a design spectrum appropriate for the conditions of the site. This approach is described in Chapter 4.

29.4 DISTRIBUTION OF HORIZONTAL SHEAR FORCE

The shear force V_x at a given horizontal plane x of the structure is equal to the sum of all the lateral forces acting on and above that plane. That is,

$$V_x = F_t + \sum_{i=x}^{n} F_i \qquad (29.16)$$

This force is distributed among the resisting elements at that level, proportionally to their relative stiffness with appropriate consideration for the stiffness of the horizontal diaphragm or bracing system. Any rigid elements that are not assumed to be part of the lateral resisting structural system can be incorporated in the structure, provided that their effect on the system is considered in the design.

29.5 HORIZONTAL TORSIONAL MOMENT

The code requires consideration of the shear force increment in the resisting elements due to the horizontal torsional effects generated by the eccentricity

---

[2]Hato Rey is a locality in the northeastern part of Puerto Rico.

between the mass center and stiffness center. In those cases where the vertical resisting elements depend on the diaphragm action for the distribution of shear force at any level, the shear resisting elements shall be capable of resisting a torsional moment assumed to be equivalent to the story shear acting with a minimum eccentricity of 5% of the dimension of the structure in the direction perpendicular to the applied seismic forces. However, no reduction in the shear force due to the torsional effect shall be considered. Moreover, no main vertical resisting member will be allowed to have a shear force at its base acting in the opposite direction to the seismic loads, unless a dynamic analysis as described in the next paragraph is performed.

A dynamic analysis that includes rotational degrees of freedom is required for those structures in which at any level there is an eccentricity larger than 40% of the dimension of the structure at that level perpendicular to the direction of the seismic load.

Let V_y and V_z be the story shear in the Y and Z horizontal directions, respectively, at the level under consideration, and e_y and e_z the eccentricities between the centers of mass and resistance for that story. The torsional moments due to these eccentricities are calculated as:

For seismic forces applied in the Y direction:

$$M_t = V_y e_z \qquad (29.17)$$

For seismic forces applied in the Z direction:

$$M_t = V_z e_y \qquad (29.18)$$

Let D_y and D_z be, respectively, the plan dimension of the building in the story under consideration along the Y and Z directions. The accidental torsion is defined as:

For seismic forces applied in the Y direction:

$$M_{ta} = \pm 0.05 D_z V_y \qquad (29.19)$$

For seismic forces applied in the Z direction:

$$M_{ta} = \pm 0.05 D_y V_z \qquad (29.20)$$

The accidental torsional moments are introduced to take into account the effects of several factors that were not explicitly included, such as possible differences between the computed and actual values of stiffness and deadload weights, and the rotational component of the seismic motion about the vertical axis. The accidental torsional moment M_{ta} should not be added to the actual moment M_t defined in eqs.(29.17) and (29.18), but rather the larger moment in each case must be used to calculate an adjusted design shear force, if needed.

29.6 P-DELTA EFFECT AND STORY DRIFT LIMITATIONS

The code specifies that a stability coefficient shall be used to determine whether there is a need for a P-delta (P-Δ) analysis. When the stability coefficient calculated according to eq.(29.21) is less than or equal to 0.10, it is not necessary to consider the P-delta effect in the drift or in the forces acting on the resisting elements. The coefficient θ is defined as the ratio of the P-delta moment to the story moment due to the lateral loading:

$$\theta = \frac{P_x \Delta}{V_x H_x} \qquad (29.21)$$

where

P_x = total gravity load applied above the story under consideration

$$P_x = \sum_{i=x}^{n} W_i \qquad (29.22)$$

Δ = interstory drift at the level considered due to the lateral design load and determined from a first-order analysis

V_x = shear force defined in eq.(29.16) acting between levels x and $x-1$

H_x = interstory height below the level x.

If the value of θ is larger than 0.10, the coefficient of increment a_d due to the P-delta effect is determined as

$$a_d = \frac{1}{1-\theta} \qquad (29.23)$$

To obtain the story drift and also the story shear force that includes the P-delta effect, the design story drift and the story shear force should be multiplied by the factor $(1 + a_d)$.

The story drift is the difference between the deflections at the top and bottom of the story under consideration. In calculating the total elastic drift, the Puerto Rican Seismic Code requires that the elastic deflections must be multiplied by the coefficient $(1.0/K)$:

$$\delta_x = \frac{1.0}{K} \delta_{xe}; \quad (x = 1, 2, \ldots, n) \qquad (29.24)$$

The deflections δ_{xe} are computed from an elastic analysis using the lateral forces defined in Section 29.3. The coefficient K is the structural factor used in eq.(29.1) to calculate the base shear; its value is obtained as explained in Section 29.2.2. The code specifies that the coefficient $(1.0/K)$ cannot be less than 1.0.

The deflections δ_x defined in eq.(29.24) shall be used to calculate the total elastic drift Δ_x as:

$$\Delta_x = \delta_{x+1} - \delta_x \qquad (29.25)$$

The code limits the interstory elastic drift to a maximum value of 0.005 times the story height H, unless it is demonstrated that larger displacements can be tolerated. That is,

$$\frac{\Delta_x}{K} \leqslant 0.005H \qquad (29.26)$$

with $K \leqslant 1.0$.

All the components of the structures or buildings shall be designed and built so that they work as a single unit to resist the horizontal forces, unless they are separated by a distance sufficiently large to prevent contact due to inelastic displacements induced by the seismic or wind loads. The inelastic displacements will be determined by multiplying the elastic displacements by the factor $3/K$ where the structural factor K is not less than 1.00.

29.7 OVERTURNING MOMENT

The code requires every building or structure to be designed to resist the overturning effects. The incremental changes in the design overturning moment at any level will be distributed among the various resisting elements in the floor under consideration in the same proportion as in the distribution of the shear force in the resisting system. In those cases where other vertical members capable of partially resisting the overturning moments are provided, a redistribution may be made to these members if frame members with sufficient strength and stiffness to transmit the required loads are provided.

The overturning moment at any level of the building is determined by statics to be the moment produced at that level by the lateral seismic forces applied above. Therefore the overturning moment M_x at level x of the building is given by

$$M_x = F_t(h_n - h_x) + \sum_{i=x+1}^{n} F_i(h_i - h_x) \qquad (29.27)$$

where $x = 0, 1, 2, \ldots, n-1$.

Where a vertical resisting member is not continuous between two stories, the overturning moment calculated for the lowest level of this element shall be carried as axial loads to the lower levels. The founda-

tion of the building can be designed to resist 0.75% of the overturning moment due to seismic loads.

Although some codes permit the designer to lower the overturning moment calculated from eq.(29.27) by introducing a reduction factor in the right-hand side of this equation, the Puerto Rican Seismic Code does not allow any such reduction. The only exception is for the design of the foundation, in which a reduction factor equal to 0.75 may be used. This exception is based on the fact that under the action of the overturning moments a slight uplifting of one edge of the foundation may occur for a short duration of time. In turn, this uplift produces a reduction in the seismic forces and, consequently, a reduction in the overturning moments.

29.8 LATERAL FORCES ON NONSTRUCTURAL COMPONENTS

The code provides the following expression to calculate the lateral force F_p for the design of nonstructural components and their anchorage to the main structure:

$$F_p = ZIC_pW_p \qquad (29.28)$$

where Z is the zone factor equal to 0.6 for Puerto Rico, I is the importance factor, C_p is a horizontal force factor that varies with the type of nonstructural component, and W_p is the weight of the nonstructural component.

The value of the coefficient I is the same as the one used for the main structure, except for cases in which the anchorage of equipment is essential for human safety. In these cases, the value of the coefficient I will be equal to 1.5.

For flexible or flexibly mounted equipment and machinery, the values of C_p will be determined considering the dynamic properties of the equipment and machinery as well as those of the structure where they will be located. However, their values shall not be less than those provided by the code. The equipment and machinery as well as their anchorage are an integral part of the seismic-resistant design.

The coefficient C_p can take the following two values, depending on the type of nonstructural component:

$C_p = 0.3$ for:
- Exterior bearing and nonbearing walls, partitions or interior nonbearing walls with the horizontal force acting normal to the surface, masonry or concrete fences over 6 ft high, or framing systems for suspended plafonds.
- Any of the following elements, when they are part of, connected to, or housed in a building: (a) penthouses,

anchorage and supports for chimneys, stacks and tanks (with contents included), (b) storage racks (contents included) with upper level at more than 8 ft in height, (c) all equipment or machinery.

$C_p = 0.8$ for:

- Cantilevered elements such as parapets with the force acting normal to the flat surface, chimneys and stacks with force acting in any direction. Exterior or interior ornamentation and appendages.

(For elements that are laterally supported only at ground level, C_p can be taken as equal to two-thirds of the values indicated above.)

29.9 FORCES ON DIAPHRAGMS

The code requires that the floor and roof diaphragms be designed to withstand a force at the floor level x determined in accordance with

$$F_{px} = \frac{\sum_{i=x}^{n} F_i}{\sum_{i=x}^{n} W_i} W_{px} \qquad (29.29)$$

The force F_{px} must satisfy

$$F_{px} \geq 0.14 Z I W_{px} \qquad (29.30)$$

but need not exceed

$$F_{px} \leq 0.30 Z I W_{px} \qquad (29.31)$$

In the above formulas, F_i is the lateral force applied at level i, W_i is the portion of the total weight of the building associated with level i, and W_{px} is the weight of the diaphragm and its tributary elements for level x; 25% of the live load must be included in W_{px} for storage facilities.

There are cases where the diaphragm at a given level must transfer lateral loads from resisting elements located above it to elements below it. As changes in the relative location or in the stiffness of the vertical resisting elements, these forces must be added to those calculated with eq.(29.15). The diaphragm deformation shall be considered in the design of the walls supported by the diaphragm.

29.10 ALTERNATIVE DETERMINATION OF SEISMIC FORCES

The Puerto Rican Seismic Code explicitly allows the use of other properly substantiated methods of analy-

sis to determine the lateral forces and their distribution. The dynamic characteristics of the building or structure must be considered in such analyses. The code specifies the seismic loads that must be used in any dynamic analysis by providing design spectra based on the Hollywood and Taft earthquake records, described as follows:

(a) *Hollywood record.* The effect on the building or structure based on the record obtained in the basement of the Hollywood warehouse for the S-00°-W component of the 1952 earthquake in Tehachapi, California. The design spectrum corresponding to a damping coefficient equal to 5% of the critical damping is shown in Fig. 29.1.

(b) *Taft record.* This record corresponds to the S-69°-E component obtained in Taft during the same 1952 earthquake in Tehachapi, California. The corresponding design spectrum is based on a damping ratio of 5% (see Fig. 29.2).

For a structure modeled as a shear building with one degree of freedom per floor, the maximum lateral force of the jth mode at level x of the building is given by (Paz 1991):

$$F_{xj} = \tau_j \phi_{xj} W_x S_a(T_j, \xi_j) \qquad (29.32)$$

where τ_j is the jth modal participation factor (defined in Chapter 3, Section 3.3), ϕ_{xj} is the xth row of the jth modal vector, and W_x is the weight attributed to the xth floor. The factor $S_a(T_j, \xi_j)$, referred to as the modal spectral acceleration, is the ordinate of the acceleration design spectrum in g units for a period T_j and a modal damping ratio ξ_j. Because both design spectra (Figs. 29.1 and 29.2) are defined for a 5% damping ratio, one is compelled to assume the same value for all the modes.

For an N-story building, the modal base shear associated with the jth mode is obtained as

$$V_j = \sum_{x=1}^{N} F_{xj} \qquad (29.33)$$

To determine the total base shear V, the modal shears can be combined according to the Square Root of the Sum of the Squares (SRSS) method:

$$V = \sqrt{\sum_{j=1}^{N} V_j^2} \qquad (29.34)$$

The code specifies that the design base shear V_d shall be determined as

$$V_d = \frac{SV}{R} \qquad (29.35)$$

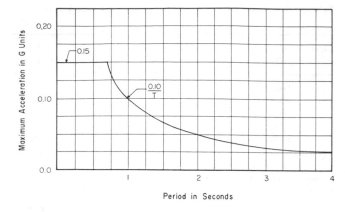

Fig. 29.2. Design spectrum for Hollywood Earthquake S-00°-W Record

where S is the soil-structure resonance factor. Unless the local soil effects have been explicitly considered in the analysis, the value of S shall be determined as explained in Section 29.2.4. The code stipulates that an importance factor I, equal to 1, shall be used in the dynamic analysis. The coefficient R is a reduction ductility factor intended to account for the ductile capacity of the structural system. The value of R should be close to 1 for a structural system of brittle material with low tolerance to deformation beyond the elastic range; R should be assigned higher values for very ductile structures that are able to tolerate considerable deformation beyond the initial yield point. The code specifies that when the spectrum corresponding to the Hollywood earthquake record (Fig. 29.2) is used to calculate S_a in eq.(29.32), the value of R in eq.(29.35) must be equal to 1.5. On the other hand, when the spectral accelerations are determined from the Taft spectrum, the value of R depends on the type of structural system. The same procedures used to determine the structural factor K are adopted in the code to define R in Table 29.2.

The code stipulates that the member forces resulting from the greater of the two seismic motions (Hollywood or Taft) shall be used in conjunction with the ultimate stress criterion to design the structural components. Moreover, under no circumstances should the total design shear force at the base V_d be less than 60% of the value obtained by multiplying the force calculated according to eq.(29.1) by 1.4; that is:

$$\frac{V_d}{1.4} \geqslant 0.60(ZIKCSW) \qquad (29.36)$$

Knowing the reduction factor R assigned to a particular structural system allows the designer to predict which of the two seismic motions will yield the higher base shear force. According to Figs. 29.1 and 29.2, the ordinates of the two spectra are analytically defined as follows:

For $T \leqslant 2/3$ sec: $SH = 0.15;$ $ST = 0.42$ (29.37)

For $T \geqslant 2/3$ sec: $SH = \dfrac{0.1}{T};$ $ST = \dfrac{0.28}{T}$ (29.38)

where SH and ST are, respectively, the spectral accelerations of the Hollywood and Taft records in g units. The ordinates of the Taft spectrum are 2.8 times those of the Hollywood spectrum. The shear base force V, or equivalently, the spectral accelerations, must be divided by $R = 1.5$ when the Hollywood spectrum is used, and by a factor R that depends on the structural system when the Taft spectrum is considered. Hence, one can equate SH divided by 1.5 to ST divided by R to obtain the particular value of R that controls the selection of one spectrum or the other. For instance, for $T \leqq 2/3$ sec,

$$\frac{0.15}{1.5} = \frac{0.42}{R} \qquad (29.39)$$

from which $R = 4.2$. Therefore, if $R > 4.2$, one must use the Hollywood spectrum. For $R \leqslant 4.2$ the Taft spectrum must be employed as it will lead to higher values of V_d.

29.11 NUMERICAL EXAMPLES

Example 29.1

To illustrate the application of the Puerto Rican Seismic Code, the five-story office building shown in Fig. 29.3 will be used. The building is regular with the floors assumed to be infinitely rigid compared to the columns. It is also assumed that at the various stories of the building there are no eccentricities between the centers of mass and the stiffness centers. In this example, the building will be analyzed using the equivalent static lateral force method. The next example will present the alternative dynamic method. The gravity loads at each level and the total lateral stiffness values of the columns are given in Fig. 29.3. The example presents an analysis in only one direction. It is assumed that there is no information about the soil below the foundation.

1. *Total lateral seismic forces*

$$V = ZIKCSW \qquad \text{[eq.(29.1)]}$$

(a) Z: Zone factor
$Z = 0.60$ (Puerto Rico)

(b) I: Occupancy importance factor
$I = 1.00$ (office building)

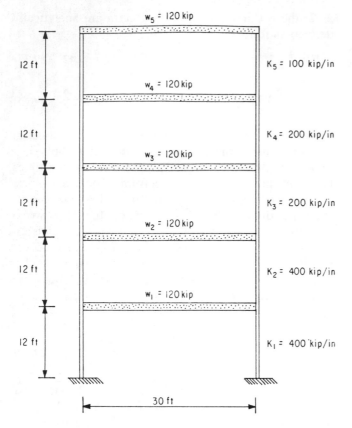

$w_5 = 120$ kip

12 ft $K_5 = 100$ kip/in

$w_4 = 120$ kip

12 ft $K_4 = 200$ kip/in

$w_3 = 120$ kip

12 ft $K_3 = 200$ kip/in

$w_2 = 120$ kip

12 ft $K_2 = 400$ kip/in

$w_1 = 120$ kip

12 ft $K_1 = 400$ kip/in

30 ft

Fig. 29.3. Modeled building for Example 29.1

(c) K: Structural factor
$K = 0.67$ (Table 29.2, Moment-resisting steel
frame)

(d) S: Soil-structure resonance factor
$S = 1.50$ (No information on soil)

(e) C: Dynamic response factor
Period: $T_a = C_t(h_n)^{3/4}$ [eq.(29.7)]
$C_t = 0.035$ (steel frame)
$h_n = 60.00$ ft
$T_a = 0.755$ sec
$C = 1/(15T_a)$ [eq.(29.3)]
   $= 0.08835 < 0.100$
Verification: $SC = 0.132 < 0.14$

(f) W: Total gravity load of structure
$W = 600.00$ kips

Thus, $V = 0.6 \times 1.00 \times 0.67 \times 0.08835 \times 1.5 \times 600$
    $= 31.97$ kips

2. Distribution of seismic lateral forces
(Additional force if $T > 0.70$ sec)

$$F_t = 0.07TV < 0.25V \quad \text{[eq.(29.13)]}$$

$$= 1.69 \text{ kips}$$

$$F_x = \frac{(V - F_t)W_x h_x}{\sum_{i=1}^{n} W_i h_i} \quad \text{[eq.(29.15)]}$$

Table 29.3. Calculation of Lateral Seismic Forces and Overturning Moments

Level i	H_i (ft)	W_i (kip)	$H_i \times W_i$ (kip-ft)	F_i (kip)	M_i (kip-ft)
5	60.00	100.00	6,000.00	10.490*	——
4	48.00	120.00	5,760.00	8.450	125.882
3	36.00	120.00	4,320.00	6.337	353.161
2	24.00	120.00	2,880.00	4.225	656.488
1	12.00	140.00	1,680.00	2.465	1,010.512
Base					1,394.111
*Including the additional force $F_t = 1.69$ kips					
SUM =			20,640.00	31.967	

Calculation results of the seismic lateral forces are shown in Table 29.3.

3. Overturning moments

$$M_x = F_n(h_n - h_x) + \sum_{i=x+1}^{n} F_i(h_i - h_x) \quad \text{[eq.(29.27)]}$$

where $x = 0, 1, 2, \ldots, n - 1$.

Calculation results of the overturning moments are given in Table 29.3.

4. Story shears and torsional moments

$$V_x = F_t + \sum_{i=x}^{N} F_i \quad \text{[eq.(29.16)]}$$

The eccentricities are assumed to be zero, therefore the torsional moments are due to the accidental torsion at the different levels of the building:

$$M_{tax} = \pm 0.05 D_z V_{yx} \quad \text{[eq.(29.19)]}$$

The values of the calculated story shears and torsional moments are shown in Table 29.4.

5. Lateral displacements and story drifts

When determining the lateral displacements, it is convenient to calculate first the story drifts by dividing

Table 29.4. Calculation of Story Shears and Torsional Moments

Level	F_x [kip]	V_x [kip]	M_{tax} [kip-ft]
5	10.490	10.490	20.980
4	8.450	18.940	37.880
3	6.337	25.277	50.554
2	4.225	29.502	59.004
1	2.465	31.967	63.934

Table 29.5. Calculation of Lateral Displacements and Story Drifts

Level	K_x [kip/in]	V_x [kip]	Δ_x [in]	δ_x [in]	Δ_x/K [in]
5	100.0	10.490	0.105	0.480	0.157
4	200.0	18.940	0.095	0.375	0.142
3	200.0	25.277	0.126	0.280	0.188
2	400.0	29.502	0.074	0.154	0.110
1	400.0	31.967	0.080	0.080	0.119

Table 29.6. Natural Frequencies and Participation Factors

Mode	Freq. (rad/sec)	Period (sec)	Participation Factor τ_j
1	8.87477	0.70798	1.09330
2	21.48793	0.29241	−0.45724
3	31.38593	0.20019	−0.33435
4	43.36510	0.14489	−0.13813
5	58.04115	0.10825	−0.13707

the shear force at each level by the corresponding flexural stiffness, that is:

$$\Delta_x = \frac{V_x}{K_x}$$

The lateral displacements then are determined as:

$$\delta_{x+1} = \Delta_x + \delta_x; \quad (x = 0, 1, \ldots, N-1)$$

starting at the base of the building with $\delta_0 = 0$. As specified in Section 29.6, the elastic drift must be multiplied by the factor $1.0/K = 1.4925$. The results of the calculations are shown in Table 29.5.

The code requires that the values of the drifts, the last column of Table 29.5, be less than 0.005 times the corresponding story height. This condition is satisfied at all levels because in this case, $0.005h = 0.72$ in.

Example 29.2

The five-story building of Example 29.1 will be considered again to illustrate seismic analysis using the dynamic method of Section 29.10. The structure is modeled as a five-degree-of-freedom shear building. A computer program was used to calculate the natural frequencies, modes of vibration, and participation factors. Table 29.6 shows the frequencies in rad/sec, and the participation factors for the structure shown in Fig. 29.3. The modes of vibration normalized with respect to the mass matrix are shown in Table 29.7. The modal spectral accelerations for the five periods in Table 29.6 that correspond to the design spectra in Figs. 29.1 and 29.2 are listed in Table 29.8. This table also shows these values divided by the response reduction factor $R = 1.5$ for the Hollywood spectrum and $R = 8$ for the Taft spectrum. The value of $R = 8.00$ corresponds to a steel moment-resisting frame, according to Table 29.2. Table 29.8 shows that the Hollywood record is the dominant seismic motion. This was expected as it was demonstrated in Section 29.10 that for $R \geqslant 4.2$ the Hollywood spectrum yields the higher spectral accelerations. Therefore, the seismic load will be defined in terms of the Hollywood spectrum for the remaining calculations.

Table 29.7. Normal Modes of Vibration

Level	Mode 1	Mode 2	Mode 3	Mode 4	Mode 5
5	1.28094	1.30035	−0.68058	0.25647	−0.01452
4	1.01963	−0.25477	1.05586	−0.99271	0.11205
3	0.76417	−0.84952	0.30771	1.28386	−0.41091
2	0.41518	−0.83469	−0.91151	−0.19158	1.21687
1	0.21528	−0.82342	−0.82342	−0.64947	−1.15451

Table 29.8. Spectral Acceleration Values

Mode	Hollywood		Taft	
	$SA[g]$	$SA[g]/1.5$	$SA[g]$	$SA[g]/R$
1	0.1412	0.0942	0.3955	0.0494
2	0.1500	0.1000	0.4200	0.0525
3	0.1500	0.1000	0.4200	0.0525
4	0.1500	0.1000	0.4200	0.0525
5	0.1500	0.1000	0.4200	0.0525

The modal lateral forces can be calculated using eq.(29.32). However, the code specifies that the design base shear shall be calculated amplifying the base shear obtained from the dynamic analysis by the soil-structure resonance factor S and reducing it by the reduction ductility factor R. In turn, this requires that all the response quantities be scaled in the same proportion. The reduction due to the factor R has already been introduced in the spectral accelerations. Therefore, if we multiply the values obtained from eq.(29.32) by the factor S ($S = 1.5$ in this example), the base shear (to be determined later) will satisfy the code requirement. The modal lateral forces obtained in this way are shown in Table 29.9a. To obtain the design values of the modal forces the values in Table 29.9a are divided by 1.4. The results are shown in Table 29.9b. The design values of the seismic lateral forces are obtained by adding the contribution of the modal forces according to a modal combination rule. The results obtained using two combination rules, the Sum of the ABSolute values (SABS) and SRSS, are shown in the last two columns of Table 29.9b.

The modal overturning moments are calculated

Table 29.9. Lateral Forces for Hollywood Spectrum
(a) Modal Lateral Forces (Ultimate Values)

Level	Mode 1	Mode 2	Mode 3	Mode 4	Mode 5
5	19.7810	8.9186	−3.4133	0.5314	−0.0299
4	18.8948	−2.0968	6.3545	−2.4682	0.2765
3	14.1609	−6.9919	1.8519	3.1921	−1.0139
2	7.6937	−6.8699	−5.4858	−0.4763	3.0024
1	4.6542	−5.0681	−5.7815	−1.8840	−3.3233

(b) Modal Lateral Forces and Design Values (Service Values) (kips)

Level	Mode 1	Mode 2	Mode 3	Mode 4	Mode 5	Design Values SABS	SRSS
5	14.1293	6.3704	−2.4381	0.3796	−0.0213	23.34	15.69
4	13.4963	−1.4977	4.5390	−1.7630	0.1975	21.49	14.43
3	10.1149	−4.9942	1.3228	2.2801	−0.7242	19.44	11.61
2	5.4955	−4.9070	−3.9184	−0.3402	2.1446	16.81	8.62
1	3.3244	−3.6201	−4.1297	−1.3457	−2.3738	14.79	6.98

Table 29.10. Modal Overturning Moments and Design Values (kip-ft)

Level	Mode 1	Mode 2	Mode 3	Mode 4	Mode 5	Design Values SABS	SRSS
5	—	—	—	—	—	—	—
4	169.5514	76.4454	−29.2569	4.5547	−0.2560	280.06	188.33
3	501.0584	134.9180	−4.0462	−12.0468	1.8578	653.93	519.06
2	953.9447	133.4605	37.0379	−1.2872	−4.7187	1,130.45	963.96
1	1,472.7774	73.1184	31.1009	5.3896	14.4400	1,596.75	1,475.00
Base	2,031.5028	−30.6646	−24.3922	−4.0818	5.1130	2,095.75	2,031.89

Table 29.11. Modal Shear Forces and Design Values (kip)

Level	Mode 1	Mode 2	Mode 3	Mode 4	Mode 5	Design Values SABS	SRSS
5	14.1293	6.3704	−2.4381	0.3796	−0.0213	23.34	15.69
4	27.6256	4.8727	2.1009	−1.3835	0.1761	36.16	28.17
3	37.7405	−0.1215	3.4237	0.8966	−0.5480	42.73	37.91
2	43.2361	−5.0285	−0.4948	0.5564	1.5966	50.91	43.56
1	46.5605	−8.6486	−4.6244	−0.7893	−0.7773	61.40	47.60

from eq.(29.27) using as F_i, $i = 1, \ldots, 5$, the values of the modal forces, and are given in Table 29.10a. The design values obtained by combining the overturning moments according to the SABS and SRSS rules are shown in the last two columns of Table 29.10. To calculate the design story shears, the modal shear forces are obtained from eq.(29.16), in which now F_i are the modal forces and F_t is set equal to 0. The modal shear forces are listed in the last two columns of Table 29.11, and the design values for the shear forces in Table 29.11. An important quantity obtained from Table 29.11 is the design base shear

$V_d = 47.60$ kip. The code demands that the base shear calculated from the dynamic analysis be larger than or equal to 60% of the base shear obtained from the equivalent static analysis, $V_s = 31.97$ kips:

$$V_d = 47.60 \text{ kips} \geq 0.6 V_s = 19.18 \text{ kips}$$

No scaling of the design response values is necessary for this example as the condition is satisfied.

The modal torsional moments are calculated using the modal story shears in eq.(29.19) with $(0.05)(D) = (0.05)(400) = 20$ ft for all floors. The

Table 29.12. Modal Torsional Moment and Design Values (kip-ft)

Level	Mode 1	Mode 2	Mode 3	Mode 4	Mode 5	Design Values SABS	Design Values SRSS
5	28.2586	12.7409	−4.8761	0.7591	−0.0427	46.68	31.39
4	55.2512	9.7454	4.2018	−2.7669	0.3523	72.32	56.33
3	75.4811	−0.2429	6.8473	1.7933	−1.0961	85.46	75.82
2	86.4721	−10.0570	−0.9895	1.1128	3.1931	101.82	87.13
1	93.1209	−17.2972	−9.2488	−1.5786	−1.5545	122.80	95.19

Table 29.13. Modal Lateral Displacements and Design Values (inches)

Level	Mode 1	Mode 2	Mode 3	Mode 4	Mode 5	Design Values SABS	Design Values SRSS
5	0.9697	0.0746	−0.0134	0.0011	−0.0000	1.0587	0.9726
4	0.7719	−0.0146	0.0208	−0.0042	0.0003	0.8117	0.7723
3	0.5785	−0.0487	0.0060	0.0055	−0.0010	0.6397	0.5806
2	0.3143	−0.0479	−0.0179	−0.0008	0.0029	0.3838	0.3184
1	0.1630	−0.0303	−0.0162	−0.0028	−0.0027	0.2149	0.1666

Table 29.14. Modal Story Drifts and Design Values (inches)

Level	Mode 1	Mode 2	Mode 3	Mode 4	Mode 5	Design Values SABS	Design Values SRSS
5	0.1978	0.0892	−0.0341	0.0053	−0.0003	0.3267	0.2197
4	0.1934	0.0341	0.0147	−0.0097	0.0012	0.2531	0.1972
3	0.2642	−0.0009	0.0240	0.0063	−0.0038	0.2991	0.2654
2	0.1513	−0.0176	−0.0017	0.0019	0.0056	0.1782	0.1525
1	0.1630	−0.0303	−0.0162	−0.0028	−0.0027	0.2149	0.1666

modal and design values of the accidental torsional moments are given in Table 29.12.

Table 29.13 shows the modal displacements for the five levels. The jth modal displacement for level x is calculated as:

$$\delta_{xj} = \frac{S\tau_j \phi_{xj} S_{aj}(T_j, \xi_j)}{\omega_j^2}$$

where

S = soil-structure resonance factor (Section 29.2.4)

τ_j = participation factor mode j

$$\tau_j = \frac{\sum\limits_{i=1}^{N} W_i \phi_{ij}}{\sum\limits_{i=1}^{N} W_i \phi_{ij}^2} \qquad \text{[eq.(4.53a)]}$$

ϕ_{xj} = jth modal displacement at coordinate x

$S_{aj}(T_j, \xi_j)$ = modal spectral acceleration for period T_j and damping ratio ξ_j corresponding to jth mode

ω_j = modal natural frequency (rad/sec).

The soil-structure resonance factor S was introduced in the above expression to scale the results in a manner consistent with eq.(29.35). The design values for lateral displacements are listed in the last two columns of Table 29.13. Finally, the modal displacements are used to calculate the modal story drifts, shown in Table 29.14. The design values for story drifts are displayed in the last two columns of Table 29.14.

Example 29.3

A computer program was written to perform the seismic analysis of buildings based on the provisions of the Puerto Rican Seismic Code. The program was written in FORTRAN 77 and it runs on any IBM-compatible personal computer. The program presents the user with three options: use the equivalent lateral static method; perform the seismic analysis using a dynamic analysis with loads defined by the response spectra in Figs 29.1 and 29.2; or use both methodologies. For the dynamic analysis, the structure is modeled as a shear building with one degree of freedom per floor. The input data (the building characteristics and other relevant information) can be provided to the program either interactively or by means of a data file. The input data required are shown in the printouts for Examples 29.3 and 29.4. The outputs of the program, for all the three options, are the design values of the lateral forces, the overturning moments, the shear forces, the torsional moments, and the lateral displacements and drifts. When a dynamic analysis is performed, the output results include the natural frequencies, the periods, the mass-normalized modes of vibration, and the participation factors. The SRSS method is the modal combination procedure adopted in the program to calculate the design response. The program automatically performs an appropriate scaling of all the design response quantities if the base shear calculated from the dynamic analysis does not verify the inequality in eq.(29.36). In this example, the program is used to verify the results obtained via hand-calculation in Examples 29.1 and 29.2. The output results[3] coincide with those of Tables 29.1–29.5 for the equivalent static analysis method and those of Tables 29.9–29.14 for the dynamic method.

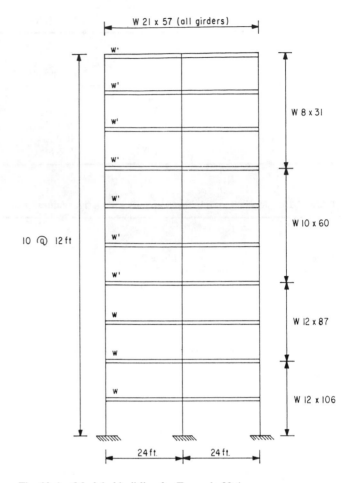

Fig. 29.4. Modeled building for Example 29.4

Example 29.4

Fig. 29.4 shows a two-bay, 10-story frame with a rectangular layout. The computer program uses both the equivalent static method and the alternative dynamic method of the Puerto Rican Code for the seismic-resistant design of this structure. The size and the essential dimensions of the steel-rolled section for the beams and columns are shown in the figure. Weights of 85 kips and 80 kips are superimposed at each of the lower seven and upper three floors, respectively. The program first performs the dynamic analysis and then calculates the base shear using the dynamic and static methods. The relation between the two shears was found to be:

$$V_d = 0.4394V_s$$

The program computes the scale factor; the response quantities were scaled by this factor (=0.6/0.4394). After the dynamic analysis is completed, the program uses the equivalent static lateral force method. However, because Option 3 was selected, the fundamental period T, determined in the previous dynamic analysis, is used in the calculations.[4]

REFERENCES

Díaz Hernández, L. E. (1987) *Temblores y Terremotos de Puerto Rico*. Impresos Comerciales, Puerto Rico.

Enmiendas Adoptadas al Reglamento de Edificación (1987) (Amendments to the Building Code). Administración de Reglamentos y Permisos (Office of Codes and Permits), Commonwealth of Puerto Rico.

---

[3]Editor's note: To save space, the computer output for this example is not reproduced.

[4]Editor's note: To save space, the computer output for this example is not reproduced. However, it is included in the diskette of programs and data files corresponding to this chapter.

McCann, W. R. (1985) "On the Earthquake Hazards of Puerto Rico and the Virgin Islands." *Bulletin of the Seismological Society of America* Vol. 75, No. 1: 251–262.

Paz, M. (1991) *Structural Dynamics: Theory and Computation.* 3d ed., Van Nostrand Reinhold, New York, NY.

Tentative Provisions for the Development of Seismic Regulations for Buildings, ATC 3-06 (1978) Applied Technology Council, Palo Alto, CA.

Uniform Building Code (UBC) (1985) International Conference of Building Officials, Whittier, CA.

30

Romania

Gelu Onu

30.1 INTRODUCTION

The first Romanian official regulations for earthquake-resistant design of buildings were published in January 1942. Before this date, on November 10, 1940, a strong seismic motion had caused much loss of life and material damage. The regulations of 1942 introduced in Romania the method of equivalent seismic forces with a constant seismic coefficient as in the contemporary Italian specifications (Beles and Ifrim 1962).

The first official map of Romania showing macroseismic zones was published in 1952. Later, in 1963, 1977, and 1991, new official maps of microseismic zones were issued. In 1963, a code designated as P13-63, with provisions for earthquake-resistant design of both industrial and civil buildings, was adopted. This code was superseded in 1970 with the publication of a new version, the P13-70 code.

A major earthquake that occurred on March 4, 1977 has been a very important source of information for the professional earthquake engineering community in Romania. On the basis of this information, in 1978 the P13-70 code was replaced with the P100-78 code, and in 1981 an improved version, the P100-81 code, was published. The code of 1981 introduced an important change regarding the dynamic coefficients that represent the spectral composition of the ground motion during an earthquake. These new coefficients represented the particular characteristics of earthquakes that originated at Vrancea's focus. The new code of 1981 also stressed the need to consider the spatial character of seismic oscillations. A comprehensive study of the 1977 earthquake and its consequences, as reflected in the official Romanian regulations for seismic design of buildings, was published by Balan, Cristescu, and Cornea (1982).

The P100-91 code for earthquake-resistant design of dwellings and buildings (socio-economical, agro-zootechnical, and industrial), was put into operation in June 1991 (Petrovici and Stanciulescu 1991). The main topics included in this code are:

- Principles of Seismic Design; Overall Construction of Buildings; Structural Analysis Under Seismic Excitation; Provisions for RC Structures (T. Postelnicu).

- Planning of Buildings and Evaluation of Building Sites; Seismic Zone Map of the National Territory; Simplified Analysis Method; Three-dimensional Analysis of Buildings Using the Modal Method (H. Sandi).
- Stipulations Regarding Steel Buildings (C. Dalban).
- Earthquake Resistant Design of Installations and Equipment (T. Teretean).
- Estimation of Seismic Protection Level of Existing Buildings and Necessary Retrofitting for Upgrading Their Seismic Integrity (L. Crainic).

Seismic provisions of the P100-91 code regarding masonry buildings and building specifications for these structures have been taken entirely from the previous version of the code. E. Titaru, C. Pavel, D. Lungu, and V. Apostolescu collaborated in the preparation of the new seismic code of 1991.

A most significant improvement in the new code in relation to previous versions is the inclusion of provisions for hazard mitigation in existing seismically deficient buildings.

Another significant improvement is the modification of old specifications for calculation of equivalent seismic forces and the addition of new provisions so that the cyclic excursions of structures into nonlinear

inelastic behavior during strong-motion earthquakes can be predicted and controlled through seismic design. The new code relies on the nonlinear dynamic response of the structure and requires an indirect estimation of this behavior, at the least. The indirect estimation entails the use of equivalent seismic forces and assumes linear elastic behavior of the structure. The cross-sectional dimensioning requires the introduction of plastic hinges at locations where failure is most probable in the event of a strong seismic motion. Moreover, criteria for resistance, stability, and deformation based on the importance of the building must be satisfied. For important edifices or large complexes of buildings, the P100-91 code requires a direct nonlinear analysis considering the inelastic behavior of the structure. The code recommends the use of computer programs for such analysis. Some approved programs for the earthquake-resistant design of buildings, as specified by the P100-81 code, are presented by Capatina (1990). Some auxiliary provisions and norms for seismic design of structures are also included in other codes and standards currently in use in Romania.

The presentation in this chapter is not intended as an extensive description of all the provisions of the P100-91 code. Therefore, only a brief presentation of the provisions for the determination of equivalent seismic forces and corresponding response for dwellings and public buildings is given.

30.2 LATERAL SEISMIC FORCES

According to the P100-91 code, weights are assumed to be applied only at the floor levels of multistory buildings. If the weight applied at level i is Q_i, then the lateral seismic force F_{ri} at this level, for vibration mode r, is given by

$$F_{ri} = S_{ri} Q_i \qquad (30.1)$$

where S_{ri}, a seismic coefficient defined as the ratio of the lateral acceleration and the acceleration of gravity, is given by eq.(30.5a). The seismic forces F_{ri} are applied at the floor levels of the building; the floors usually are assumed to be very rigid in their own planes.

Fig. 30.1 represents a typical multistory building composed of several orthogonal plane frames; three longitudinal and five transverse frames connected at each floor level by a diaphragm that is rigid in its own plane. The structural characteristics of buildings with shear walls or bracing elements may be integrated into the structural frame. As shown in the figure, the global system of reference axes is chosen parallel to the direction of the orthogonal structural frames. In

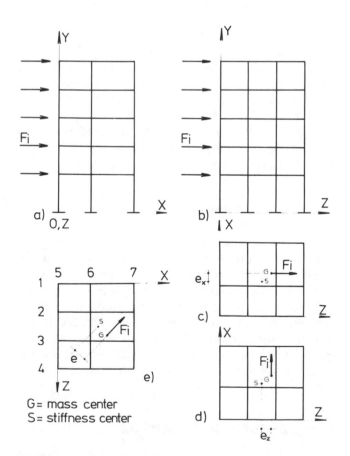

G = mass center
S = stiffness center

Fig. 30.1. Multistory building modeled as a plane structure, subjected to lateral seismic forces

cases in which the main structural resisting frames and shear walls are not oriented in two orthogonal directions, the code stipulates that the reference axes X and Z should be oriented in directions parallel to the principal inertial axes of the building.

For regular structures of the type shown in Fig. 30.1, the code permits independent plane analysis considering seismic loads first along the direction of the OX axis and then along the OZ axis (Fig. 30.1), without superposing the effects of these two analyses. However, in the case of flexible structures, the code also requires a third analysis with seismic forces acting at a 45° skew direction as shown in Fig. 30.1(e). In this latter case, the code also allows plane analysis by decomposition of the loading into loads on the orthogonal directions, but the resulting responses have to be superposed.

As shown in Figs. 30.1(c), (d), and (e), the resultant of the lateral forces does not pass through the stiffness center. In the cases of regular structures, where the stiffness centers at various stories are situated approximately on a vertical line and the mass centers of the various levels of the building exhibit a similar arrangement, the eccentricities e, e_x, and e_z are so regular that the torsional effect may be expressed in a simple manner. In this case, the eccentricities e, e_x, and e_z shown in Fig. 30.1 are calculated as follows:

$$e = e_1 \pm e_2$$
$$e_x = e_{1x} \pm \sqrt{e_2} \qquad (30.2)$$
$$e_z = e_{1z} \pm \sqrt{e_2}$$

in which e_1 is the distance between the center of stiffness and the mass center of a given floor, e_{1x} and e_{1z} are the projections of e_1 on the axes OX and OZ, and e_2 is the additional eccentricity that is added to estimate the asynchronous character of seismic motions.

For buildings with a favorable arrangement of the main structural components or resisting elements, e_2 is given by

$$e_2 = 0.05L \qquad (30.3a)$$

where L is the largest dimension of the two plane sides of the building.

When the plane configurations of the main structural components are unfavorable, the additional eccentricity e_2 is given by

$$e_2 = 0.075L \qquad (30.3b)$$

In the case of buildings having more than five levels and if $e_1 > 0.1L$, the general torsional effect has to be taken into account through a model that considers torsional oscillations. Likewise, for buildings of irregu-

lar shape with $e_1 > 0.1L$, a dynamic analysis using a spatial model is recommended.

30.3 VERTICAL SEISMIC FORCES

The seismic code requires that columns with large axial forces, beams and cantilevers with large shear forces, and slab floors supported directly on columns must be checked for the vertical action of the earthquake. The vertical seismic force N_i on a structural element i is given by

$$N_i = c_v Q_i \qquad (30.4)$$

where Q_i is the gravitational load on the element, c_v is an amplification factor in the range of $\pm 2k_s$, and k_s is the zone factor given in the seismic zone map of Fig. 30.2.

The vertical seismic forces must be superimposed on the gravitational loads or represented directly through corresponding internal forces. According to the seismic code, the effects of vertical and horizontal seismic forces should not be superimposed.

30.4 SEISMIC WEIGHTS Q_i

The seismic weight Q_i at the various levels of the building includes the contributing weight of the building and permanent equipment. It also includes the nonpermanent gravitational loads, uniformly distributed. However, when the resultant of these live loads is greater than 25% of the resultant of all the gravitational loads, it is necessary to consider uniform loads acting only on parts of the building to obtain a maximum overturning effect.

30.5 LOAD COMBINATION

The seismic forces are considered to be exceptional loads. These forces and those resulting from gravitational loads (permanent, quasi-permanent, and transient loads) form a special loading group. Other forces such as those due to wind action, temperature variation, and concrete shrinkage are not included in the special loading group. However, some buildings with special features that emphasize the effects of the lateral loads may have to be treated as exceptional cases.

30.6 SEISMIC COEFFICIENT S_{ri}

The lateral seismic forces F_{ri} given by eq.(30.1) corresponding to each level i of the building have to

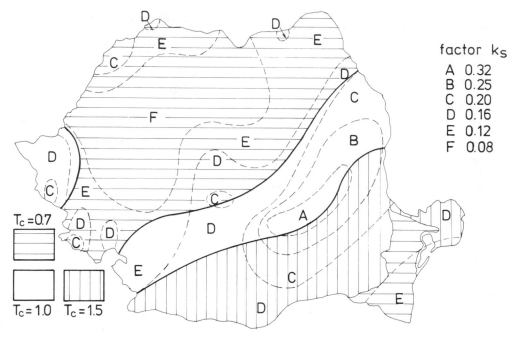

factor k_s

A 0.32
B 0.25
C 0.20
D 0.16
E 0.12
F 0.08

$T_c = 0.7$

$T_c = 1.0$ $T_c = 1.5$

Fig. 30.2. Seismic zones of the Romanian territory

be determined for all the vibration modes considered in the analysis. According to the seismic code, when the vibrations of the structure have a plane character, it is sufficient to consider only the first three modes. The seismic coefficient S_{ri} for level i in mode r is calculated as

$$S_{ri} = \alpha \psi k_s \beta_r \eta_{ri} \qquad (30.5a)$$

in which

α = building importance factor
ψ = nonlinear adjustment factor
k_s = seismic zone factor
β_r = modal amplification factor
η_{ri} = modal distribution factor for level i.

The product of the first four factors in eq.(30.5a) is designated the global seismic coefficient for vibration mode r. Hence, the global seismic coefficient k_g is given by:

$$k_g = \alpha \psi k_s \beta_r \qquad (30.5b)$$

The factors in eq.(30.5b) are described in the following sections.

30.7 BUILDING IMPORTANCE FACTOR α

The seismic code assigns different protection levels to four categories of buildings. These levels of protection are described in Table 30.1.

30.8 NONLINEAR ADJUSTMENT FACTOR ψ

This factor performs the function of reducing the effect due to the seismic forces obtained from an elastic analysis. Consequently, this factor emphasizes the nonlinear potential of the structure due to ductility, internal force redistribution, and energy dissipation through damping of nonstructural elements. Numerical values for the factor ψ for different structural types are given in Table 30.2.

30.9 SEISMIC ZONE FACTOR k_s

The seismic zone factor k_s represents the ratio of the maximum seismic acceleration of the ground and the acceleration of gravity. For earthquakes of certain

Table 30.1. Building Importance Factor α

Factor α	Description
$\alpha = 1.4$	Buildings of vital importance to society (fire stations, hospitals, ambulance services, national transmitting stations, governmental office buildings, national art galleries, etc.)
$\alpha = 1.2$	Very important buildings (schools, churches, day nurseries, trade centers, arenas, etc.)
$\alpha = 1.0$	Ordinary buildings (dwellings, hotels, dormitories, etc.)
$\alpha = 0.8$	Buildings of reduced importance (one- or two- story dwellings, warehouses, etc.)

Table 30.2. Nonlinear Adjustment Factor ψ

Factor ψ	Type of Building
$\psi = 0.17$	Multistory steel framed buildings
$\psi = 0.20$	Reinforced concrete multistory buildings without frame-shear wall interaction.
$\psi = 0.25$	Reinforced concrete multistory frame buildings with structural walls; buildings with braced masonry bearing walls.
$\psi = 0.30$	Reinforced concrete buildings with walls, columns, and slab floors but without beams; buildings with unbraced masonry bearing walls.

seismic intensity, the recurrence period is approximately 50 years for zones dominated by Vrancea's focus, and at least 100 years for other zones. On the basis of values for the k_s factor, the Romanian territory has been divided into six seismic zones, as shown in Fig. 30.2.

30.10 MODAL AMPLIFICATION FACTOR β_r

The factor β_r is a dynamic coefficient that is a function of both vibration mode and site location of the building. This factor is calculated by the following formulas:

$$\begin{aligned} \beta_r &= 2.5 - (T_r - T_c) \quad \text{for} \quad T_r > T_c \\ \beta_r &= 2.5 \quad\quad\quad\quad\;\; \text{for} \quad T_r \leqslant T_c \end{aligned} \tag{30.6}$$

in which T_r is the natural frequency for mode r, and T_c is the corner period representing the spectral composition of the seismic motion of the building foundation. (Both T_r and T_c are expressed in seconds.) Romania is divided into three zones with $T_c = 0.7$; 1.0, and 1.5 as indicated in Fig. 30.2. The seismic code stipulates that the vibration modes must be obtained using structural dynamics methods. However, for regular buildings, the code permits the fundamental period to be determined using the empirical formulas given in Section 30.11.

30.11 EMPIRICAL FORMULAS FOR THE FUNDAMENTAL PERIOD

The fundamental period T (sec) for a regular building may be calculated as follows:

1. Buildings with structural walls
   (a) Bar-type buildings

$$T = 0.045n \quad \text{(transverse direction)} \tag{30.7a}$$

$$T = 0.040n \quad \text{(longitudinal direction)} \tag{30.7b}$$

(b) Tower-type buildings

$$T = 0.065H/\sqrt{L} \tag{30.7c}$$

2. Buildings with structural walls and frames
   (a) Bar-type buildings

$$T = 0.055n \quad \text{(transverse direction)} \tag{30.7d}$$

$$T = 0.045n \quad \text{(longitudinal direction)} \tag{30.7e}$$

(b) Tower-type buildings

$$T = 0.075H/\sqrt{L} \tag{30.7f}$$

(c) Frame-type buildings

$$T = 0.3 + 0.05n \quad \text{for} \quad n < 6 \tag{30.7g}$$

$$T = 0.1n \quad\quad\quad\; \text{for} \quad 5 < n < 11 \tag{30.7h}$$

In these formulas, n is the number of floors or levels, H is the building height, and L is the plane dimension in the direction of the seismic forces (both H and L are in meters). Equations 30.7a–d are based on results of experiments conducted on instrumented buildings in Bucharest (Sandi and Serbanescu 1969).

30.12 LEVEL DISTRIBUTION FACTOR η_{ri}

η_{ri} is the coefficient of distribution of the horizontal seismic forces on the building height. The level distribution factor, corresponding to vibration mode r and level i, is calculated by eq.(30.8), in which n is the number of levels in the building.

$$\eta_{ri} = \frac{Y_{ri} \displaystyle\sum_{j=1}^{n} Q_j Y_{rj}}{\displaystyle\sum_{j=1}^{n} Q_j Y_{rj}^2} \tag{30.8}$$

Q_j = concentrated weight at level j

Y_{ri}, Y_{rj} = maximum displacement (or modal displacement) at level i, j in mode r

Since the displacements Y_{rj} are not known initially, it is necessary to use some assumed mode shapes in applying eq.(30.8). The first mode shape is usually assumed to be represented by a straight line that originates at the base of the building with the maximum displacement at the top level. For the higher modes, other assumptions have been proposed to approximate the modal shapes. The computer program presented in this chapter includes several alternatives for estimating the higher modal shapes.

For ordinary buildings, the seismic code stipulates that the fundamental mode shape may be approximated as a straight line. In this case, if the weights Q_j have roughly the same numerical value and the distances between levels are equal, the level distribution factor, for the fundamental mode, may be expressed as

$$\eta = \frac{3i}{1+2n} \tag{30.9}$$

30.13 EQUIVALENT LATERAL FORCES F_{ri}

The Romanian seismic code states that the lateral seismic forces may be calculated directly by application of eqs.(30.1) and (30.5) or indirectly as the resultant of the lateral seismic forces. The resultant of the lateral seismic forces F_r, corresponding to the mode of vibration r, is given by

$$F_r = \alpha\psi k_s\beta_r\varepsilon_r \sum_{i=1}^{n} Q_i \tag{30.10}$$

in which the factors α, ψ, k_s, and β_r have been defined previously. The global equivalence factor ε_r is a coefficient relating the actual system of n degrees of freedom vibrating in the r mode to the corresponding single-degree-of-freedom system. This coefficient is given by

$$\varepsilon_r = \frac{\left(\sum\limits_{j=1}^{n} Q_j Y_{rj}\right)^2}{G\sum\limits_{j=1}^{n} Q_j Y_{rj}^2} \tag{30.11}$$

where G is the total weight of the building and Q_j and Y_{rj} are used as in eq.(30.8). For multistory buildings, a lower limiting value of this coefficient for the fundamental mode ε_1 is stipulated by the code:

$$\varepsilon_1 \geqslant 0.65 \tag{30.12}$$

If the conditions required by eq.(30.9) are satisfied, the coefficient ε_1 then may be calculated as

$$\varepsilon_1 = \frac{3+3n}{2+4n} \tag{30.13}$$

The lateral seismic force F_{ri} at the level i of the building corresponding to the mode r is calculated by

$$F_{ri} = \frac{F_r Q_i Y_{ri}}{\sum\limits_{j=1}^{n} Q_j Y_{rj}} \tag{30.14}$$

in which the resultant seismic force F_r is given by eq.(30.10).

30.14 STORY SHEAR FORCES

For vibration mode r, the shear force V_{ri} at level i is given by

$$V_{ri} = \sum_{j=i}^{n} F_{rj} \tag{30.15}$$

in which the sum refers to lateral seismic forces F_{rj} above level i. The base shear force or resultant of the lateral seismic forces may be calculated using eq.(30.15) or eq.(30.10).

30.15 OVERTURNING MOMENTS

The evaluation of the overturning moments is needed for stability analysis of the entire structure. The overturning moment M_{ri} at level i of the building is caused by the lateral seismic forces acting above this level. For mode r at level i, the overturning moment is given by

$$M_{ri} = \sum_{i=j+1}^{n} F_{rj}(h_j - h_i) \tag{30.16}$$

in which h_j is the height of level j and F_{rj} is the lateral force calculated by eq.(30.1) or eq.(30.14).

30.16 TORSIONAL MOMENTS

The seismic code requires the consideration of torsional moments even if dynamic symmetry exists in the building. If the story stiffness centers are approximately in a vertical line and the story mass centers also are located along a vertical line, then the torsion moment Mt_{ri} at story i for mode r is given by

$$Mt_{ri} = e_i V_{ri} \tag{30.17}$$

In eq.(30.17), e_i (or e_{xi} or e_{zi}) is given by eq.(30.2), and the story shear force V_{ri} by eq.(30.15).

The torsional moments produce lateral seismic forces that are resisted by the main structural elements. Thus, the frames and structural walls that are located farther from the stiffness center will be carrying a greater load. According to the code, regular buildings may be analyzed as planar structures. In this case, when the floors are assumed to be rigid, the distribution of the lateral seismic forces caused by torsional moments is proportional to both the lateral stiffness and the distance between the resisting elements and the stiffness center. The story shear force V_{ri} is also distributed in this manner. For example, if V_{ri} is the story shear force at level i, due to the lateral seismic forces acting in the OX direction, and $Mt_{ri} = e_i V_{ri}$ is the corresponding torsional moment, then the shear force V_{rai} for resisting element a at level i (assumed to be oriented in the same OX direction) may be calculated as follows:

$$V_{rai} = V_{ri} k_a \left(1 \pm e_i d_a \frac{K_x}{K_t} \right) / K_x \qquad (30.18)$$

and

$$K_x = \sum_{OX} k_j; \qquad K_t = \sum_{j=1}^{n} k_j d_j^2 \qquad (30.19)$$

in which K_x is the lateral stiffness of the building in the OX direction, K_t the torsional stiffness of the building, k_j the lateral stiffness of the resisting element j assumed to be uniform along the height of the building, d_j the distance between element j and the stiffness center, and e_i the eccentricity given by eq.(30.2).

30.17 LATERAL DISPLACEMENTS

According to the seismic code, the lateral elastic displacement X_{ri} at level i of the building, for vibration mode r, may be determined by

$$X_{ri} = g S_{ri} \left(\frac{T_r}{2\pi} \right)^2 \qquad (30.20a)$$

where T_r is the rth natural period and the product $g S_{ri}$ represents the conventional horizontal acceleration at level i. Alternatively, the elastic displacements may be obtained by static analysis of the structure loaded by the equivalent seismic forces. Equation (30.20a) is further modified by the correction factor ν_a, which approximately accounts for the torsional effect on a structural element a. Hence,

$$X_{ri} = g \frac{F_{ri}}{Q_i} \left(\frac{T_r}{2\pi} \right)^2 \nu_a \qquad (30.20b)$$

in which the correction factor ν_a is given by

$$\nu_a = 1 + K_x d_a e_i / K_t \qquad (30.21)$$

where all the factors have been defined in relation to eq.(30.18).

To take into account the post-elastic behavior of the building, which causes an increase of the computed elastic displacements, the lateral displacements X_{ri} obtained using eq.(30.20) must be divided by the nonlinear adjustment factor ψ.

30.18 STORY DRIFT

To avoid excessive lateral displacements of the building, the seismic code imposes limitations on the relative displacement Δ_i between two consecutive levels of the building as stated in the following condition:

$$\Delta_i \leqslant \mu h_i \qquad (30.22)$$

where

$\mu = 0.0035$ for buildings with integral infilled frames
$\mu = 0.007$ for buildings with open frames
h_i = story height.

30.19 COMBINATION OF MODAL ACTIONS

The seismic code requires a distinct analysis for each vibration mode considered. For the purpose of estimating the maximum response, the modal effects are combined by the rule known as the Square Root of the Sum of the Squared modal contributions (SRSS rule). Accordingly, the shear design force V_i, the torsional design moment Mt_i, and the overturning design moment M_i are calculated as follows:

$$V_i = \left(\sum_{r=1}^{n} V_{ri}^2 \right)^{1/2}$$

$$Mt_i = \left(\sum_{r=1}^{n} Mt_{ri}^2 \right)^{1/2} \qquad (30.23)$$

$$M_i = \left(\sum_{r=1}^{n} M_{ri}^2 \right)^{1/2}$$

in which $n = 3$, corresponding to the first three vibration modes. Other quantities of interest may be calculated in a similar manner.

In the case in which two successive periods differ by less than 10%, the seismic code recommends that the absolute values of the corresponding internal forces be added, the result squared, and the squared value then used in eq.(30.23) as a single term.

Example 30.1

A four-story building with three main structural elements in both the OX and OZ directions is shown

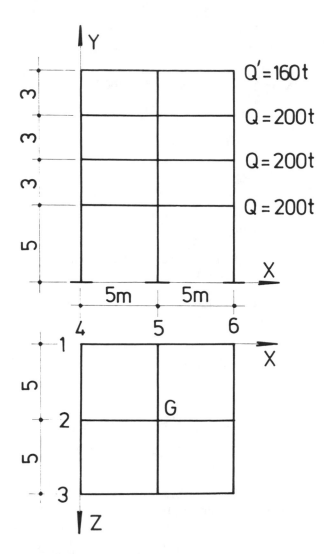

Fig. 30.3. Building model for Example 30.1

in Fig. 30.3. The weight of the building is lumped at each floor as follows: $Q_1 = Q_2 = Q_3 = 200$ tons for the first three levels and $Q_4 = 160$ tons for the top level. The seismic forces are assumed to be applied along the OX direction.

Determine, according to the Romanian seismic code, the following: (a) fundamental period, (b) distribution factors, (c) global equivalence factor, (d) global amplification factor, (e) global seismic coefficient, (f) lateral seismic forces for each story, (g) story shear forces, and (h) overturning moments. Also, for frame 1, determine (i) torsional shear force, (j) lateral displacements, and (k) relative displacements between the building levels.

Solution

(a) Fundamental period. For this type of building (tower), the seismic code considers only the fun-

damental mode. The fundamental period for this building (with structural walls and frames) is given by

$$T = 0.075 \frac{H}{\sqrt{L}} \qquad \text{[eq.(30.7)]}$$

$H = 14$ m (height of the building)
$L = 10$ m (dimension of the building in OX direction)

$$T = 0.075 \frac{14}{\sqrt{10}} = 0.332 \text{ sec}$$

(b) Distribution factors. The distribution factors are given by

$$\eta_i = \frac{Y_i \sum_{j=1}^{n} Q_j Y_j}{\sum_{j=1}^{n} Q_j Y_j^2} \qquad \text{[eq.(30.8)]}$$

Y_i, Y_j = height of levels i, j (assuming straight line for the first mode)
Q_j = weight at level j

The distribution factors for each level are:

Level:	1	2	3	4
$\eta_i =$	0.4798	0.7677	1.0556	1.3435

(c) Global equivalence factor. The global equivalence factor is given by:

$$\varepsilon_1 = \frac{\left(\sum_{j=1}^{n} Q_j Y_j \right)^2}{G \sum_{j=1}^{n} Q_j Y_j^2} \qquad \text{[eq.(30.11)]}$$

$G = 760$ tons (weight of the building)
Q_j = weight at level j
Y_j = height of level j (normalized with respect to H; i.e., $5/14 = 0.357$, $8/14 = 0.571$, etc.)

$$\varepsilon_1 = \frac{[200(0.357 + 0.571 + 0.786) + 160]^2}{760[200(0.357^2 + 0.571^2 + 0.786^2) + 160]} = 0.889$$

(d) Global amplification factor

$$\beta = 2.5 - (T - T_c) \qquad \text{[eq.(30.6)]}$$

$T = 0.332$ sec (fundamental period)
$T_c = 1.0$ sec (corner period − input data)

Because $T < T_c$ then $\beta = 2.5$ (code provision).

(e) Global seismic coefficient. The global seismic coefficient is given by

$$k_g = \alpha \psi k_s \beta_r \qquad [\text{eq.}(30.5b)]$$

According to input data,

$\alpha = 1.0$	(from Table 30.1)
$\psi = 0.25$	(from Table 30.2)
$k_s = 0.16$	(from Fig. 30.2)
$\beta_r = 2.5$	(from eq.(30.6)
$k_g = 1.0 \times 0.25 \times 0.16 \times 2.5 = 0.10$	

(f) Lateral seismic forces

$$F_i = k_g \eta_i Q_i \qquad [\text{eqs.}(30.1), (30.5)]$$

Q_i = weight at level i
k_g = global seismic coefficient
η_i = distribution factor at level i

The lateral forces are as follows:

Level:	1	2	3	4
$F_i =$	9.60 tons	15.35 tons	21.11 tons	21.50 tons

(g) Story shear forces

$$V_i = \sum_{j=i}^{n} F_j \qquad [\text{eq.}(30.15)]$$

The story shear forces are:

Level:	1	2	3	4
$V_i =$	67.56 tons	57.96 tons	42.61 tons	21.50 tons

(h) Overturning moments. The overturning moments are given by

$$M_i = \sum_{j=1+1}^{n} F_j(h_j - h_i) \qquad [\text{eq.}(30.16)]$$

The overturning moments at each level are as follows:

Level:	0	1	2	3
$M_i =$	704.0 ton-m	366.2 ton-m	192.3 ton-m	64.49 ton-m

(i) Torsional shear force for frame 1. The shear force for story i of frame 1 is given by

$$V_{1i} = V_i K_1 \left(1 + d_1 e_1 \frac{K_x}{K_t}\right)/K_x \qquad [\text{eq.}(30.18)]$$

[See Negoita, Voiculescu and Mihalache (1988).]

$K_1 = K_2 = K_3 = 1$ (normalized lateral stiffness of main structural elements)

$$K_x = \sum_{i=1}^{3} K_i = 3 (\text{lateral stiffness of the building})$$

$$K_t = \sum_{i=1}^{6} d_i^2 K_i = 5^2 \times 4 \times 1 = 100$$

where

$d_i = 5.0$ m, distance between element i and stiffness center
V_i = shear force at level i
$e_i = e_1 + e_2$ (distance between stiffness center and mass center at level i)
$e_1 = 0$
$e_2 = 0.05 \times L$ (additional eccentricity; 0.05 input data)
$L = 10$ m
$e_2 = 0.05 \times 10 = 0.5$ m

The shear torsional forces V_{ti} of frame 1 at each level are as follows:

Level:	1	2	3	4
$V_{ti} =$	24.21 tons	20.77 tons	15.27 tons	7.70 tons

(j) Lateral displacements of frame 1. The lateral displacement at level i is given by

$$X_i = g \frac{F_i}{Q_i} \left(\frac{T}{2\pi}\right)^2 \nu_a \qquad [\text{eq.}(30.20b)]$$

$\nu_a = 1 + K_x d_a e_i / K_t$
  (correction factor to account approximately for the torsional effect)

$$= 1.075$$

The displacements of frame 1 at each level are as follows:

Level:	1	2	3	4
$X_i =$	0.141 cm	0.276 cm	0.311 cm	0.396 cm

(k) Story drift. The story relative displacement or story drift ΔX_i is calculated as the difference between lateral displacements of upper and lower levels for each story. This calculation yields the following values for the relative displacements of frame 1:

Story:	1	2	3	4
$\Delta X_i =$	0.141 cm	0.085 cm	0.085 cm	0.085 cm

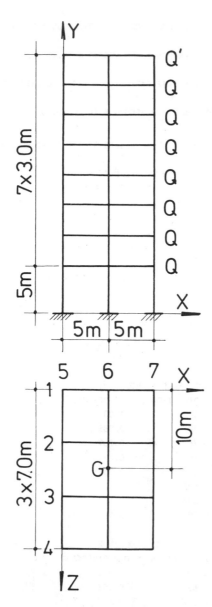

Fig. 30.4. Building model for Example 30.2 (lateral forces are $Q' = 200$ tons and $Q = 300$ tons)

30.20 COMPUTER PROGRAM AND EXAMPLES

A computer program has been developed to implement the provisions of the P100-91 code of Romania for earthquake-resistant design of buildings. The program is written in BASIC for implementation on IBM-compatible microcomputers. The following example is presented to illustrate the implementation and execution of the program.

Example 30.2

A bar-type building with seven main structural components (frames and structural walls) is shown in Fig. 30.4. This building serves here as an example of cases in which the seismic code reduces the analysis to a single mode. However, the designer may be interested to know the effects of the first three modes. The seismic code does not provide empirical formulas for mode shapes two and three. However, the computer program developed for this chapter presents some options. One of these options presented (Ifrim 1984) has been used commonly in Romania.

```
******************************************************
*                                                    *
*          EARTHQUAKE RESISTANT DESIGN               *
*                                                    *
*   PROGRAM 91 BASED ON THE SPECIFICATIONS OF        *
*                                                    *
*      SEISMIC CODE OF ROMANIA (P100/91)             *
*                                                    *
******************************************************


    DRIVE USED FOR DATA FILES (A:,B:,C:)? C

    FILE NAME WITHOUT DRIVE LETTER? EX2$

    * * DATA FILE INFORMATION * *

      1. PREPARE NEW DATA FILE

      2. MODIFY EXISTING DATA FILE

      3. USE EXISTING DATA FILE

         SELECT NUMBER? 3



    PRINT DATA
```

NUMBER OF STORIES	NIV = 8
NUMBER OF FRAMES PARALLEL TO YOX PLANE	NCX = 4
NUMBER OF FRAMES PARALLEL TO YOZ PLANE	NCZ = 3
IMPORTANCE FACTOR	CF1 = 1.2
NONLINEAR ADJUSTEMENT FACTOR	CF2 = .2
SEISMIC ZONE FACTOR	CF3 = .2
CORNER PERIOD	CF4 = 1.5
ADITIONAL ECCENTRICITY COEFFICIENT	CF5 = 0.05

STORY #	HEIGHT (METERS)	WEIGHT (TONS)	X-COOR (METERS)	Z-COOR (METERS)
8	3.00	200.0	5.00	10.00
7	3.00	300.0	5.00	10.00
6	3.00	300.0	5.00	10.00
5	3.00	300.0	5.00	10.00
4	3.00	300.0	5.00	10.00
3	3.00	300.0	5.00	10.00
2	3.00	300.0	5.00	10.00
1	5.00	300.0	5.00	10.00

FRAMES PARALLEL TO YOX PLANE

FRAME #	POSITION (METERS)	STIFFNESS (NORMALIZED)
1	0.00	3.00
2	7.00	3.00
3	14.00	3.00
4	21.00	3.00

FRAMES PARALLEL TO YOZ PLANE

FRAME #	POSITION (METERS)	STIFFNESS (NORMALIZED)
5	0.00	5.00
6	5.00	5.00
7	10.00	5.00

SEISMIC RESPONSE

STIFFNESS CENTER COORDINATES :

X-COOR = 5.00   Z-COOR = 10.50
GLOBAL STIFFNESS: STIF-X = 12.00
                     STIF-Z = 15.00
                     STIF-ROT= 985.00

WEIGHTS ABOVE LEVELS & MASS CENTERS COORDINATES

LEVEL #	WEIGHT (TONS)	X-COOR (METERS)	Z-COOR (METERS)
8	200	5.00	10.00
7	500	5.00	10.00
6	800	5.00	10.00
5	1100	5.00	10.00
4	1400	5.00	10.00
3	1700	5.00	10.00
2	2000	5.00	10.00
1	2300	5.00	10.00

ENTER DATA FOR SEISMIC RESPONSE

NUMBER OF MODES CONSIDERED (1,2 OR 3) MOD=? 3

ARE THE MODES PREVIOUSLY DETERMINATED ? (Y/N)? N

FUNDAMENTAL PERIOD MENU;

1-BAR TYPE BUILDINGS WITH STRUCTURAL WALLS

2-TOWER TYPE BUILDINGS WITH STRUCTURAL WALLS

3-BAR TYPE BUILDINGS WITH STRUCTURAL WALLS & FRAMES

4-TOWER TYPE BUILDINGS WITH STRUCTURAL WALLS & FRAMES

5-FRAME BUILDINGS

SELECT A NUMBER ? 3

MENU FOR SUPERIOR MODES OR MODES PREVIOUSLY DETERMINED

1-PROF.IFRIM'S PROPOSAL FOR SECOND & THIRD MODES

2-SPANISH CODE SPECIFICATIONS FOR SECOND & THIRD MODES

3-ANOTHER PROPOSAL FOR SECOND & THIRD MODES

4-PERIODS PREVIOUSLY DETERMINED & SIMPLIFIED MODE SHAPES

5-COMPLETE MODES PREVIOUSLY DETERMINED

SELECT A NUMBER ? 1

IT'S OK ?   -   RESPOND: A, B, C, G, OR S
                  FOR EXPLANATION RESPOND E
ENTER WA$ = ? G

VIBRATION PERIODS:

MODE	O-X DIRECTION	O-Z DIRECTION
1	0.440	0.360
2	0.169	0.138
3	0.113	0.092

VIBRATION MODE SHAPES - MODE: 1

LEVEL	O-X DIRECTION	O-Z DIRECTION
8	1.000	1.000
7	0.885	0.885
6	0.769	0.769
5	0.654	0.654
4	0.538	0.538
3	0.423	0.423
2	0.308	0.308
1	0.192	0.192

VIBRATION MODE SHAPES - MODE: 2

LEVEL	O-X DIRECTION	O-Z DIRECTION
8	1.000	1.000
7	0.653	0.653
6	0.307	0.307
5	-0.039	-0.039
4	-0.385	-0.385
3	-0.730	-0.730
2	-0.924	-0.924
1	-0.578	-0.578

VIBRATION MODE SHAPES - MODE: 3

LEVEL	O-X DIRECTION	O-Z DIRECTION
8	1.000	1.000
7	0.423	0.423
6	-0.154	-0.154
5	-0.731	-0.731
4	-0.692	-0.692
3	-0.115	-0.115
2	0.462	0.462
1	0.962	0.962

IT'S OK ?   -   RESPOND: A, B, C, G, OR S
                  FOR EXPLANATION RESPOND E
ENTER WA$ = ? G

SIMPLIFIED CALCULATION OF DISTRIBUTION FACTOR ? (Y/N) ? N

DISTRIBUTION FACTORS - MODE: 1

LEVEL #	O-X DIRECTION	O-Z DIRECTION
8	1.445	1.445
7	1.279	1.279
6	1.112	1.112
5	0.945	0.945
4	0.778	0.778
3	0.612	0.612
2	0.445	0.445
1	0.278	0.278

GLOBAL EQUIVALENCE FACTORS=   0.8363   0.8363

DISTRIBUTION FACTORS - MODE: 2

LEVEL #	O-X DIRECTION	O-Z DIRECTION
8	-0.336	-0.336
7	-0.220	-0.220
6	-0.103	-0.103
5	0.013	0.013
4	0.129	0.129
3	0.246	0.246
2	0.311	0.311
1	0.194	0.194

DISTRIBUTION FACTORS - MODE: 3

LEVEL #	O-X DIRECTION	O-Z DIRECTION
8	0.270	0.270
7	0.114	0.114
6	-0.042	-0.042
5	-0.198	-0.198
4	-0.187	-0.187
3	-0.031	-0.031
2	0.125	0.125
1	0.260	0.260

GLOBAL SEISMIC COEFFICIENTS - DIRECTION O-X
    (CF1 * CF2 * CF3 * BETA)

MODE 1	MODE 2	MODE 3
0.120	0.120	0.120

GLOBAL SEISMIC COEFFICIENTS - DIRECTION O-Z
    (CF1 * CF2 * CF3 * BETA)

MODE 1	MODE 2	MODE 3
0.120	0.120	0.120

GLOBAL LATERAL FORCES (T) - MODE 1

LEVEL #	O-X DIRECTION	O-Z DIRECTION
8	34.7	34.7
7	46.0	46.0
6	40.0	40.0
5	34.0	34.0
4	28.0	28.0
3	22.0	22.0
2	16.0	16.0
1	10.0	10.0

GLOBAL LATERAL FORCES (T) - MODE 2

LEVEL #	O-X DIRECTION	O-Z DIRECTION
8	-8.1	-8.1
7	-7.9	-7.9
6	-3.7	-3.7
5	0.5	0.5
4	4.7	4.7
3	8.8	8.8
2	11.2	11.2
1	7.0	7.0

GLOBAL LATERAL FORCES (T) - MODE 3

LEVEL #	O-X DIRECTION	O-Z DIRECTION
8	6.5	6.5
7	4.1	4.1
6	-1.5	-1.5
5	-7.1	-7.1
4	-6.7	-6.7
3	-1.1	-1.1
2	4.5	4.5
1	9.4	9.4

OVERTURNING MOMENT (TM) - DIRECTION O-X

LEVEL #	MODE 1	MODE 2	MODE 3	SUM
7	104	-24	19	109
6	346	-72	51	357
5	708	-131	79	725
4	1173	-189	85	1191
3	1721	-233	70	1738
2	2336	-250	53	2349
1	2998	-233	49	3007
0	4152	-171	89	4156

OVERTURNING MOMENT (TM) - DIRECTION O-Z

LEVEL #	MODE 1	MODE 2	MODE 3	SUM
7	104	-24	19	109
6	346	-72	51	357
5	708	-131	79	725
4	1173	-189	85	1191
3	1721	-233	70	1738
2	2336	-250	53	2349
1	2998	-233	49	3007
0	4152	-171	89	4156

GLOBAL STORY SHEAR FORCE (T) - DIRECTION O-X

LEVEL #	MODE 1	MODE 2	MODE 3	SUM
8	35	-8	6	36
7	81	-16	11	83
6	121	-20	9	123
5	155	-19	2	156
4	183	-15	-5	183
3	205	-6	-6	205
2	221	5	-1	221
1	231	12	8	231

GLOBAL STORY SHEAR FORCE (T) - DIRECTION O-Z

LEVEL #	MODE 1	MODE 2	MODE 3	SUM
8	35	-8	6	36
7	81	-16	11	83
6	121	-20	9	123
5	155	-19	2	156
4	183	-15	-5	183
3	205	-6	-6	205
2	221	5	-1	221
1	231	12	8	231

EARTHQUAKE DIRECTION MENU:

1   O-X DIRECTION

2   O-Z DIRECTION

3   SKEW DIRECTION

SELECT A NUMBER: JDIR= ? 1

* * * * SEISM DIRECTION: O-X * * * *

NUMBER OF FRAME TO ANALYSE: IFR = ? 1

   IT'S OK ?   -   RESPOND: A, B, C, G, OR S
                     FOR EXPLANATION RESPOND E
   ENTER WA$ = ? G

STORY SHEAR FORCE (T) - FRAME: 1
& AUGMENTED COORDINATES OF MASS CENTERS

LEVEL	MODE 1	MODE 2	MODE 3	SUM	ECC-X	ECC-Z
8	10.4	-2.4	1.9	10.8	6.05	8.95
7	24.2	-4.8	3.2	24.9	6.05	8.95
6	36.2	-5.9	2.7	36.8	6.05	8.95
5	46.4	-5.8	0.6	46.7	6.05	8.95
4	54.8	-4.4	-1.4	54.9	6.05	8.95
3	61.4	-1.7	-1.8	61.4	6.05	8.95
2	66.1	1.6	-0.4	66.2	6.05	8.95
1	69.1	3.7	2.4	69.3	6.05	8.95

DISPLACEMENTS (CM) OF FRAME : 1

LEVEL	MODE 1	MODE 2	MODE 3	SUM
8	1.00	-0.03	0.01	1.00
7	0.88	-0.02	0.01	0.88
6	0.77	-0.01	-0.00	0.77
5	0.65	0.00	-0.01	0.65
4	0.54	0.01	-0.01	0.54
3	0.42	0.03	-0.00	0.42
2	0.31	0.03	0.01	0.31
1	0.19	0.02	0.01	0.19

STORY DRIFT (CM) OF FRAME : 1

LEVEL	MODE 1	MODE 2	MODE 3	SUM
8	0.12	-0.01	0.01	0.12
7	0.12	-0.01	0.01	0.12
6	0.12	-0.01	0.01	0.12
5	0.12	-0.01	-0.00	0.12
4	0.12	-0.01	-0.01	0.12
3	0.12	-0.01	-0.01	0.12
2	0.12	0.01	-0.01	0.12
1	0.19	0.02	0.01	0.19

CONTINUATION MENU

1-MAINTAINING OF SEISM DIRECTION & ECCENTRICITY

2-MAINTAINING OF SEISM DIRECTION ONLY

3-MODIFYING OF SEISM DIRECTION

4-RETURN TO THE BEGINNING OF PROGRAM

5-ENDING OF PROGRAM

   SELECT A NUMBER: ? 3

EARTHQUAKE DIRECTION MENU:

1   O-X DIRECTION

2   O-Z DIRECTION

3   SKEW DIRECTION

   SELECT A NUMBER: JDIR= ? 2

* * * * SEISM DIRECTION: O-Z * * * *

NUMBER OF FRAME TO ANALYSE: IFR = ? 5

   IT'S OK ?   -   RESPOND: A, B, C, G, OR S
                     FOR EXPLANATION RESPOND E
   ENTER WA$ = ? G

STORY SHEAR FORCE (T) - FRAME: 5
& AUGMENTED COORDINATES OF MASS CENTERS

LEVEL	MODE 1	MODE 2	MODE 3	SUM	ECC-X	ECC-Z
8	12.5	-2.9	2.3	13.0	3.95	8.95
7	29.1	-5.8	3.8	29.9	3.95	8.95
6	43.5	-7.1	3.3	44.2	3.95	8.95
5	55.7	-6.9	0.7	56.1	3.95	8.95
4	65.8	-5.2	-1.7	66.0	3.95	8.95
3	73.7	-2.1	-2.1	73.8	3.95	8.95
2	79.5	2.0	-0.5	79.5	3.95	8.95
1	83.1	4.5	2.9	83.3	3.95	8.95

DISPLACEMENTS (CM) OF FRAME : 5

LEVEL	MODE 1	MODE 2	MODE 3	SUM
8	0.60	-0.02	0.01	0.60
7	0.53	-0.01	0.00	0.53
6	0.46	-0.01	-0.00	0.46
5	0.39	0.00	-0.01	0.39
4	0.32	0.01	-0.01	0.32
3	0.26	0.02	-0.00	0.26
2	0.19	0.02	0.00	0.19
1	0.12	0.01	0.01	0.12

STORY DRIFT (CM) OF FRAME : 5

LEVEL	MODE 1	MODE 2	MODE 3	SUM
8	0.07	-0.01	0.00	0.07
7	0.07	-0.01	0.00	0.07
6	0.07	-0.01	0.00	0.07
5	0.07	-0.01	-0.00	0.07
4	0.07	-0.01	-0.00	0.07
3	0.07	-0.00	-0.00	0.07
2	0.07	0.01	-0.00	0.07
1	0.12	0.01	0.01	0.12

CONTINUATION MENU

1-MAINTAINING OF SEISM DIRECTION & ECCENTRICITY

2-MAINTAINING OF SEISM DIRECTION ONLY

3-MODIFYING OF SEISM DIRECTION

4-RETURN TO THE BEGINNING OF PROGRAM

5-ENDING OF PROGRAM

SELECT A NUMBER: ? 3

EARTHQUAKE DIRECTION MENU:

1   O-X DIRECTION

2   O-Z DIRECTION

3   SKEW DIRECTION

SELECT A NUMBER: JDIR= ? 3

* * * * SEISM DIRECTION: SKEW * * * *

NUMBER OF FRAME TO ANALYSE: IFR = ? 1

IT'S OK ?   -   RESPOND: A, B, C, G, OR S
                 FOR EXPLANATION RESPOND E
ENTER WA$ = ? G

STORY SHEAR FORCE (T) - FRAME: 1
& AUGMENTED COORDINATES OF MASS CENTERS

LEVEL	MODE 1	MODE 2	MODE 3	SUM	ECC-X	ECC-Z
8	7.7	-1.5	1.4	8.0	5.74	9.26
7	17.9	-3.0	2.4	18.3	5.74	9.26
6	26.8	-3.7	2.0	27.1	5.74	9.26
5	34.3	-3.6	0.4	34.5	5.74	9.26
4	40.5	-2.7	-0.9	40.6	5.74	9.26
3	45.4	-1.1	-1.1	45.4	5.74	9.26
2	48.9	1.2	-0.3	49.0	5.74	9.26
1	51.2	2.8	1.8	51.3	5.74	9.26

DISPLACEMENTS (CM) OF FRAME : 1

LEVEL	MODE 1	MODE 2	MODE 3	SUM
8	0.82	-0.03	0.01	0.82
7	0.73	-0.02	0.00	0.73

6	0.63	-0.01	-0.00	0.63
5	0.54	0.00	-0.01	0.54
4	0.44	0.01	-0.01	0.44
3	0.35	0.02	-0.00	0.35
2	0.25	0.03	0.00	0.25
1	0.16	0.02	0.01	0.16

STORY DRIFT (CM) OF FRAME : 1

LEVEL	MODE 1	MODE 2	MODE 3	SUM
8	0.09	-0.01	0.01	0.10
7	0.09	-0.01	0.01	0.10
6	0.09	-0.01	0.01	0.10
5	0.09	-0.01	-0.00	0.10
4	0.09	-0.01	-0.01	0.10
3	0.09	-0.01	-0.01	0.10
2	0.09	0.01	-0.01	0.10
1	0.16	0.02	0.01	0.16

CONTINUATION MENU

1-MAINTAINING OF SEISM DIRECTION & ECCENTRICITY

2-MAINTAINING OF SEISM DIRECTION ONLY

3-MODIFYING OF SEISM DIRECTION

4-RETURN TO THE BEGINNING OF PROGRAM

5-ENDING OF PROGRAM

SELECT A NUMBER: ? 1

NUMBER OF FRAME TO ANALYSE: IFR = ? 5

IT'S OK ?   -   RESPOND: A, B, C, G, OR S
                 FOR EXPLANATION RESPOND E
ENTER WA$ = ? G

STORY SHEAR FORCE (T) - FRAME: 5
& AUGMENTED COORDINATES OF MASS CENTERS

LEVEL	MODE 1	MODE 2	MODE 3	SUM	ECC-X	ECC-Z
8	8.5	-1.6	1.6	8.8	5.74	9.26
7	19.7	-3.2	2.6	20.2	5.74	9.26
6	29.5	-3.9	2.2	29.9	5.74	9.26
5	37.9	-3.8	0.5	38.1	5.74	9.26
4	44.7	-2.9	-0.9	44.8	5.74	9.26
3	50.1	-1.1	-1.2	50.1	5.74	9.26
2	54.0	1.3	-0.3	54.0	5.74	9.26
1	56.5	3.0	2.0	56.6	5.74	9.26

DISPLACEMENTS (CM) OF FRAME : 5

LEVEL	MODE 1	MODE 2	MODE 3	SUM
8	0.41	-0.01	0.00	0.41
7	0.36	-0.01	0.00	0.36
6	0.31	-0.00	-0.00	0.31
5	0.27	0.00	-0.00	0.27
4	0.22	0.01	-0.00	0.22
3	0.17	0.01	-0.00	0.17

2	0.12	0.01	0.00	0.13
1	0.08	0.01	0.00	0.08

STORY DRIFT (CM) OF FRAME : 5

LEVEL	MODE 1	MODE 2	MODE 3	SUM
8	0.05	-0.00	0.00	0.05
7	0.05	-0.00	0.00	0.05
6	0.05	-0.00	0.00	0.05
5	0.05	-0.00	-0.00	0.05
4	0.05	-0.00	-0.00	0.05
3	0.05	-0.00	-0.00	0.05
2	0.05	0.00	-0.00	0.05
1	0.08	0.01	0.00	0.08

REFERENCES

BALAN, STEFAN; CRISTESCU, VALERIU; and CORNEA ION (1982) *The March 4, 1977 Earthquake in Bucharest*. Edit Academiei Romane, Romania.

BELES, AUREL, and IFRIM, MIHAI (1962) *Elements of Earthquake Engineering*. Edit Technica, Bucharest, Romania.

CAPATINA, DAN (1990) "Computer Assisted Design of Aseismic Structures." In *Applications of Earthquake Engineering (Vol. II)*, eds. Alexandru Negoita et al., pp. 273–301. Edit Technica, Bucharest, Romania.

IFRIM, MIHAI (1984) *Dynamics of Structures and Earthquake Engineering*. Edit D. and P., Bucharest, Romania.

NEGOITA, ALEXANDRU; VOICULESCU, M.; and MIHALACHE, ADR. (1988) "Torsion Problems." In *Applications of Earthquake Engineering (Vol. 1)*, eds. Alexandru Negoita et al., pp. 123–139. Edit Technica, Bucharest, Romania.

PETROVICI, RADU, and STANCIULESCU, DAN, eds. (1991) *P100-91 Code for Earthquake Resistant Design of Dwellings, Socio-Economical, Agro-Zootechnical, and Industrial Buildings*. Department of Constructions and Public Works of MLPTAT, Bucharest, Romania.

SANDI, H., and SERBANESCU, G. (1969) "Experimental Results of the Dynamic Deformation of Multi-story Buildings." Proc. of 4th World Conference on Earthquake Engineering, Santiago, Chile.

31

Spain

Alex H. Barbat and Mario Paz

31.1 HISTORY OF SEISMIC CODES OF SPAIN

Spain is not located in a region of high seismicity. However, the seismic events that occurred in the past in some parts of the country produced major damage to structures and resulted in many deaths. One such seismic event occurred in Andalusia in 1884; it damaged 17,000 buildings and resulted in 1,000 deaths. Although the need for a Spanish seismic code for the design of structures was widely recognized, there were no legal binding regulations until 1963. The first set of Spanish regulations containing a definition of seismic action, Norma M.V. 101-1962, became a law in 1963. It was limited to the methodology for computing earthquake forces on buildings and defining other actions to be considered in building design.

In 1962, an interministerial commission initiated the development of a general seismic code for Spain, P.G.S.-1 (1968), which was approved in 1969 (Presidencia del Gobierno, Comisión Interministerial 1969). The provisions of the general seismic code apply mostly to buildings, but some specific provisions govern the seismic design of other structures such as highways, railways, tunnels, bridges, dams, power plants, and nuclear power plants. A seismic hazard map for Spain was included in this code.

Although P.G.S.-1 of 1968 was approved as a provisional code, it remained in use until 1974 when an updated version (P.D.S.-1) was issued. This new code eliminated references to structures other than buildings. However, it was required that other particular codes or regulations for special structures prepared by the corresponding Ministries must define seismic actions that have at least the values prescribed in the general seismic code.

This chapter presents the main provisions of the P.D.S.-1 (1974) Seismic Code and a brief description of the new Seismic Code, Comisión Permanente de Normas Sísmicas (Draft 1991), which is in the final stages of preparation.

31.2 SPANISH SEISMIC CODE P.D.S.-1 (1974)[1]

31.2.1 System of Seismic Forces

The Spanish Seismic Code requires structures to be analyzed under the action of static lateral forces applied at discrete mass locations in the structure. In the seismic design of buildings, these locations are selected at the floor levels, as shown in Fig. 31.1. The figure shows a plane frame that models a multistory building in which the weight of the building, as well as the weight of permanent equipment, is assumed to be applied at the levels of the floors. These weights are designated in Fig. 31.1 as W_1, W_2, . . ., W_N, where N is the number of levels in the building. This figure also shows the seismic equivalent lateral forces F_1, F_2, . . ., F_N applied at the floor levels of the building.

The equivalent seismic forces F_i are calculated by

$$F_i = s_i W_i \qquad (31.1)$$

in which s_i is the seismic coefficient given by eq.(31.7) and W_i is the concentrated weight at level i of the building. The Code specifies that the forces F_i should be applied in the most unfavorable direction. Alternatively, the structure may be analyzed under the action of seismic forces applied independently in the two main orthogonal directions of the building.

In the evaluation of the concentrated weights W_i, the following loads should be considered:

- Weight of the structure and permanent loads
- Live load adjusted by a reduction factor given in Table 31.1
- Snow load equal to 50% of the snow accumulated on sites where the snow usually remains over 30 consecutive days; otherwise the weight of the snow is not considered.

The following loads should be considered, in their most unfavorable combination, in the seismic design of structures:

1. Gravitational loads
   (a) Dead load

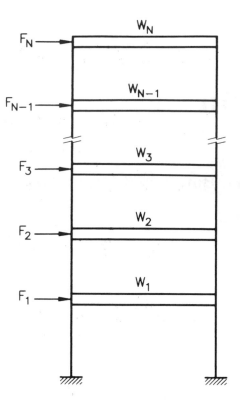

Fig. 31.1. Multistory building modeled as plane frame

   (b) Live loads affected by a reduction factor
   (c) Snow loads as specified above.
2. Dynamic actions
   (a) Wind load. This action is not considered in seismic design of structures except in cases of exposed topography. In such cases, the loads stipulated by official norms are modified by a reduction factor equal to 0.25.
   (b) Loads resulting from moving masses of permanent heavy equipment.
3. Other loads. These include earth pressure and thermal loads.

31.2.2 Seismic Zones

Spain has been divided into seismic zones based on the seismic intensity grades of the International Mac-

Table 31.1.   Reduction Coefficient for Live Load (P.D.S.-1 1974)

Case	Description	Reduction Coefficient
1	Dwellings and hotels	0.50
2	Offices, business, and garages	0.60
3	Hospitals, jails, schools, churches, buildings for public events, and hotel auditoriums	0.80

---

[1]The Seismic Code NBE-AE-88 published by Dirección General para la Vivienda y Arquitectura (1988) contains the same provisions as those in the Seismic Code P.D.S.-1 (1974) with one significant difference related to the expression that is used to evaluate the response factor (see footnote 3).

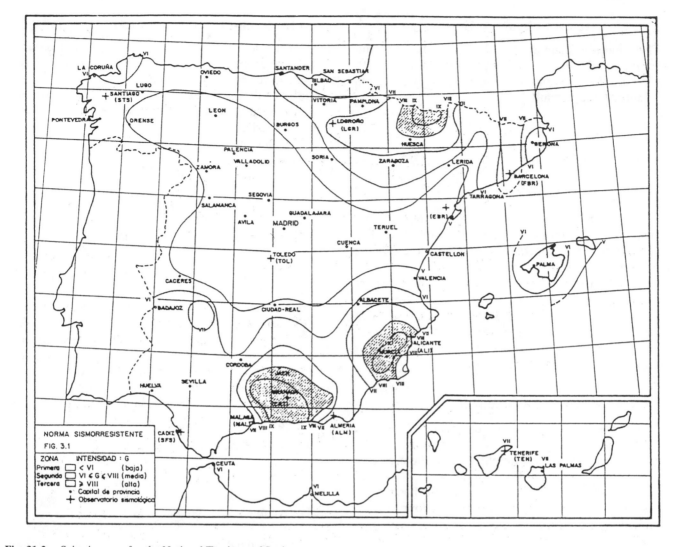

Fig. 31.2. Seismic zones for the National Territory of Spain

roseismic Intensity Scale (MSK)[2] as indicated in Fig. 31.2. The Code provides a correlation between grade G in the MSK seismic intensity scale and the spectral velocity S_V of a simple oscillator with a natural period $T = 0.5$ sec. This correlation is established by means of the following formula:

$$S_v = 1.5(2)^{G-5} \text{ (cm/sec)} \qquad (31.2)$$

The spectral displacement S_D and the spectral acceleration S_A are related to the spectral velocity through the following formulas:

$$S_D = \frac{T}{2\pi} S_V \qquad (31.3)$$

$$S_A = \frac{2\pi}{T} S_V \qquad (31.4)$$

The correspondence defined by the eqs.(31.2), (31.3), and (31.4) is indicated in Table 31.2.

Table 31.2. Correspondence Between Intensity Grades and Spectral Values for $T = 0.5$ sec (P.D.S.-1 1974)

Seismic Intensity Grade G	Spectral Values		
	Displacement S_D (cm)	Velocity S_V (cm/sec)	Acceleration S_A (cm/sec²)
V	0.12	1.5	18.9
VI	0.24	3.0	37.7
VII	0.48	6.0	75.4
VIII	0.96	12.0	150.7
IX	1.91	24.0	301.4
X	3.82	48.0	602.9

[2]The International Macroseismic Intensity Scale (MSK; Medvedev-Sponheuer-Karnik), which is similar to the Modified Mercalli Intensity (MMI) scale, provides an empirical appraisal of an earthquake based on observed damage. (See Appendix of this Handbook.)

31.2.3 Vertical Seismic Action

The Spanish Seismic Code also requires that the structure should be designed to resist vertical forces V_i applied at the locations of concentrated weights W_i according to the formula

$$V_i = v_i W_i \tag{31.5}$$

in which v_i is given by

$$v_i = \chi C \tag{31.6}$$

The coefficient χ, which depends on the seismic intensity of the earthquake, is given in Table 31.3. The other coefficient C in eq.(31.6) is the basic seismic coefficient defined in Section 31.2.4.

Generally, the Code permits the designer to neglect the effects due to vertical seismic forces. However, these effects should be considered in special cases such as beams of unusually large span, or in those cases where case-specific norms or regulations require inclusion of vertical seismic effects.

31.2.4 Seismic Coefficient s_i

The seismic coefficient s_i in eq.(31.1) is defined as the ratio of the seismic acceleration response at the point of the concentrated weight W_i and the acceleration of gravity. For buildings modeled with discrete weights concentrated at the levels of the floors, this coefficient is expressed by the product of four factors, as indicated in the following formula:

$$s_i = \alpha \cdot \beta \cdot \eta_i \cdot \delta \leq 0.20 \tag{31.7}$$

The factors in this equation are designated as follows:

α = intensity factor
β = response factor
η_i = distribution factor
δ = foundation factor

The seismic coefficient s_i depends on the natural period of the structure, through the factors α, β, η_i, and δ.

Table 31.3. Values of Coefficient χ (P.D.S.-1 1974)

Seismic Intensity Grade G	Coefficient χ
VI	2.00
VII	1.50
VIII	1.20
IX	1.00

31.2.5 Fundamental Period

The Code indicates that the fundamental period T of the structure may be determined experimentally in a structure of similar characteristics, or calculated theoretically using exact methods. The Code also allows the use of the Rayleigh formula and approximate empirical formulas that are described below.

The fundamental period T is given by the Rayleigh formula as

$$T = 2\pi \sqrt{\frac{\sum_{i=1}^{N} W_i X_i^2}{g \sum_{i=1}^{N} F_i X_i}} \tag{31.8}$$

in which W_i is the weight concentrated at level i of the building, X_i is the lateral displacement at this level produced by the equivalent forces F_i defined in eq.(31.1), and g is the acceleration of gravity. Equation (31.8) cannot be used initially because the displacements X_i are not yet known. Therefore, it is necessary to make use of empirical formulas.

The following empirical expressions are proposed in the Code to calculate the fundamental period of building structures:

For buildings with structural walls:

$$T = 0.06 \frac{H}{\sqrt{L}} \sqrt{\frac{H}{2L + H}} \geq 0.50 \sec \tag{31.9}$$

For reinforced concrete buildings:

$$T = 0.09 \frac{H}{\sqrt{L}} \geq 0.50 \sec \tag{31.10}$$

For steel buildings:

$$T = 0.10 \frac{H}{\sqrt{L}} \geq 0.50 \sec \tag{31.11}$$

For reinforced concrete buildings with structural walls or with steel bracing, the values for the fundamental period T calculated by the formulas above should be multiplied by a factor f given by

$$f = 0.85 \sqrt{\frac{1}{1 + L/H}} \tag{31.12}$$

In all these expressions H is the height of the building (in meters) and L is the plan dimension (in meters) in the direction of the seismic forces.

For modal analysis, in lieu of a more precise determination of the period of the second mode T', the following formula may be used:

$$T' = \frac{1}{3} T \geq 0.25 \tag{31.13}$$

where T is the fundamental period. Analogously, the period for the third mode T'' may be estimated as

$$T'' = \frac{1}{5} T \geqslant 0.25 \qquad (31.14)$$

31.2.6 Intensity Factor α

The intensity factor is established by

$$\alpha = CR \qquad (31.15)$$

where C is the basic seismic coefficient, equal to the maximum ground acceleration divided by the acceleration of gravity, and R is the risk coefficient.

For structures with periods equal to or less than $T = 0.5$ second, the basic seismic coefficient C is given in Table 31.4 for different values of the seismic intensity grade. For periods $T > 0.5$ second, the basic seismic coefficient C_G (corresponding to seismic intensity G) is given by

$$C_G = \frac{C}{2T} \qquad (31.16)$$

where T is the fundamental period, in seconds.

The seismic coefficients C'_G and C''_G, for the second and third modes, are determined by eq.(31.16) with $T = T'$ and $T = T''$, respectively.

The seismic risk coefficient R in eq.(31.15) is a factor that reflects the probability that an earthquake of seismic intensity G will occur during a specified period of years, as indicated in Table 31.5.

31.2.7 Response Factor[3] β

The response factor β depends both on the period of the structure and on its damping characteristics, as given by

$$\beta = \frac{B}{\sqrt[3]{T}} \geqslant 0.5 \qquad (31.17)$$

in which $B = 0.6$ for buildings with many internal partition walls, and $B = 0.8$ for other buildings. The values of the response factor β for the second and third modes also are calculated with eq.(31.17) after replacing T by T' or T'', that is the second and third periods. The product $\alpha\beta$ actually defines the seismic acceleration response spectrum, as would be obtained from a theoretical analysis.

Table 31.4. Basic Seismic Coefficient C Corresponding to Fundamental Period of the Structure $T = 0.5$ sec (P.D.S.-1 1974)

Seismic Intensity Grade G	Basic Seismic Coefficient C
V	0.02
VI	0.04
VII	0.08
VIII	0.15
IX	0.30

Table 31.5. Seismic Risk Coefficient R for Different Values of Seismic Intensity and Risk Period (P.D.S.-1 1974)

Seismic Intensity Grade G	Risk Period in Years			
	50	100	200	500
VII	1	1	1	1
VIII	0.90	0.99	1	1
IX	0.72	0.92	0.99	1
X	0.53	0.78	0.95	1

31.2.8 Distribution Factor η_i

For structures modeled with discrete concentrated weights W_i, the distribution factor η_i corresponding to the level i is given by

$$\eta_i = X_i \frac{\displaystyle\sum_{k=1}^{N} W_k X_k}{\displaystyle\sum_{k=1}^{N} W_k X_k^2} \qquad (31.18)$$

where

N = number of levels with concentrated weights
X_k = maximum displacement at level k
W_k = concentrated weight at level k.

31.2.8.1 Simplified calculation of η_i for fundamental mode of vibration. The Code provides a simplified formula to calculate the values of η_i corresponding to the fundamental mode, based on the assumption of a linear lateral displacement of the building as shown in Fig. 31.3(a). The use of proportional lines in this figure allows, instead of eq.(31.18), the following equation to be used:

$$\eta_i = Z_i \frac{S}{I} \qquad (31.19)$$

---

[3]Equation (31.17) is the expression to evaluate the response factor according to the Seismic Code P.D.S.-1 (1974). This factor is expressed as $B/\sqrt{T} \geqslant 0.5$ in the Seismic Code NBE-AE-88.

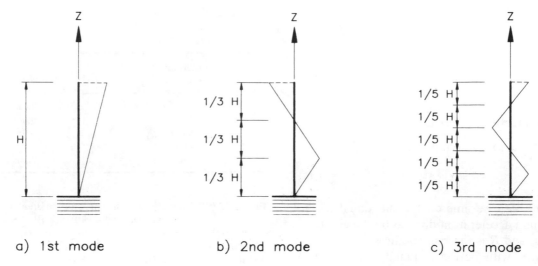

a) 1st mode   b) 2nd mode   c) 3rd mode

Fig. 31.3.   Assumed shapes for lateral displacements

In this equation, Z_i is the height of level i, S is the static moment of the weight W_i with respect to the base of the building, and I is the moment of inertia of this weight. These moments, S and I, are given by

$$S = \sum_{k=1}^{N} W_k Z_k \qquad (31.20)$$

and

$$I = \sum_{k=1}^{N} W_k Z_k^2 \qquad (31.21)$$

In buildings where the differences in height and weight between levels are not large, eq.(31.19) reduces to

$$\eta_i = \frac{3i}{2N+1} \qquad (31.22)$$

in which i is the level considered and N the total number of levels in the building.

31.2.8.2 Simplified calculation of η_i for second and third modes.

The simplified formulas to determine the distribution factors η_i', for the second mode, and η_i'' for the third mode, are given, respectively, by eqs.(31.23a) and (31.23b). These equations are obtained from assumed shapes shown in Figs. 31.3(b) and 31.3(c) for lateral displacements corresponding to the second and third modes, respectively:

$$\eta_i' = X_i \frac{\displaystyle\sum_{k=1}^{N} X_k}{\displaystyle\sum_{k=1}^{N} X_k^2} \qquad (31.23a)$$

with

$$X_i = \frac{3Z_i}{H} \quad \text{for} \quad Z_i \leqslant \frac{H}{3}$$

$$X_i = 2 - \frac{3Z_i}{H} \quad \text{for} \quad Z_i > \frac{H}{3}$$

and

$$\eta_i'' = X_i \frac{\displaystyle\sum_{k=1}^{N} X_k}{\displaystyle\sum_{k=1}^{N} X_k^2} \qquad (31.23b)$$

with

$$X_i = \frac{5Z_i}{H} \quad \text{for} \quad Z_i \leqslant \frac{H}{5}$$

$$X_i = 2 - \frac{5Z_i}{H} \quad \text{for} \quad \frac{H}{5} < Z_i \leqslant \frac{3H}{5}$$

$$X_i = -4 + \frac{5Z_i}{H} \quad \text{for} \quad Z_i > \frac{3H}{5}$$

in which H is the height of the building, and Z_i is the height of level i.

31.2.9 Foundation Factor δ

The values of the foundation factor δ are given in Table 31.6, according to the nature of the soil and type of foundation.

Table 31.6. Foundation Factor δ (P.D.S.-1 1974)

	Type of Soil				
Type of Foundation	Swamps $c \leqslant 500$	Loose Sands and Gravels $500 < c \leqslant 1,000$	Consolidated Sands and Gravels $1,000 < c \leqslant 2,000$	Compact Rocks $2,000 < c \leqslant 4,000$	Very Compact Rocks $c > 4,000$
Piles:					
Friction type	2.0	1.0	0.7	—	—
Bearing type	1.8	0.9	0.6	—	—
Shallow Foundation:					
Isolated footing	1.6	1.1	0.8	0.5	0.5
Continuous footing	1.5	1.0	0.7	0.4	0.3
Slabs	1.4	0.7	0.5	0.3	0.2

c = velocity of elastic compression wave (meter/second)

31.2.10 Evaluation of Horizontal Forces

The equivalent horizontal modal forces in each floor for the first three modes F_i, F_i', and F_i'' are determined as

$$F_i = s_i W_i$$
$$F_i' = s_i' W_i \qquad (31.24)$$
$$F_i'' = s_i'' W_i$$

where s_i, s_i', and s_i'' are the seismic coefficients for the first three modes at level i, and W_i is the weight at that level.

The modal forces given by eq.(31.24) are combined in the following expressions to obtain the effective force F_{ei} at each level of the building:

(a) For normal structures ($T \leqslant 0.75$ sec)

$$F_{ei} = s_i W_i \qquad (31.25a)$$

(b) For slender structures (0.75 sec $< T \leqslant 1.25$ sec)

$$F_{ei} = \sqrt{F_i^2 + F_i'^2} = W_i \sqrt{s_i^2 + s_i'^2} \qquad (31.25b)$$

(c) For very slender structures ($T > 1.25$ sec)

$$F_{ei} = \sqrt{F_i^2 + F_i'^2 + F_i''^2} = W_i \sqrt{s_i^2 + s_i'^2 + s_i''^2} \qquad (31.25c)$$

31.2.11 Overturning Moments, Story Shears, and Torsional Moments

31.2.11.1 Overturning moments. The lateral seismic forces produce overturning moments and axial forces in the columns, particularly on the external columns of the building. These overturning moments increase the gravity forces in the external columns alternately at one side and then at the other side of the building during the vibration of the structure. The overturning moment at any level of the building is determined as the moment produced at that level by the lateral forces at overlying levels. The overturning moment M_i at level i of height Z_i is given by

$$M_i = \sum_{k=i+1}^{N} F_k(Z_k - Z_i) \qquad (31.26)$$

$$\text{for} \quad i = 0, 1, 2, \ldots, (N-1)$$

where the lateral forces F_k for each mode are calculated by eqs.(31.24).

31.2.11.2 Story shear forces. The shear force V_i at level i, in the resisting elements of story i of a building, is given by the sum of the lateral seismic forces above that level:

$$V_i = \sum_{k=i}^{N} F_k \qquad (31.27)$$

31.2.11.3 Torsional moments. At any level of a building, the inertial forces induced by the earthquake act through the center of overlying mass, while the resultant of the resisting forces acts at the stiffness center of the resisting elements at that level. If the structure does not have dynamic symmetry, the stiffness center of the story and the center of the overlying mass do not coincide, and thus a torsional moment is developed. This torsional moment M_{ti} at each story is equal to the story shear force V_i multiplied by the eccentricity e_i, the distance between the center of the overlying mass and the stiffness center of the story. This distance is measured normal to the direction of the seismic forces. Fig. 31.4 shows the eccentricities e_x and e_y, between the center of mass G, and the stiffness center S, for earthquake forces acting respectively in the north–south (N–S) and east–west (E–W) directions.

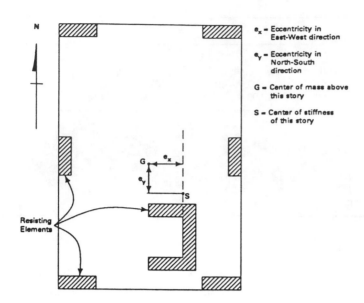

G=Center of above mass
S=Story stiffness center
e_x =North-south eccentricity
e_y =East-west eccentricity

e_x = Eccentricity in East-West direction

e_y = Eccentricity in North-South direction

G = Center of mass above this story

S = Center of stiffness of this story

Resisting Elements

Fig. 31.4. Plan of a building showing the eccentricities between the center of the above mass (G) and the story stiffness center (S) of the resisting elements

31.2.12 Lateral Displacements

The lateral displacements at the various levels of the building may be determined by static analysis of the building with regard to the equivalent lateral forces F_i. The lateral displacements also may be calculated for each mode by applying Newton's law of motion in the form

$$F_{ij} = \frac{W_i}{g} \omega_j^2 X_{ij} \qquad (31.28)$$

because the magnitude of the modal acceleration corresponding to the modal displacement X_{ij} is $\omega_j^2 X_{ij}$. Hence, from eq.(31.28)

$$X_{ij} = \frac{g}{\omega_j^2} \frac{F_{ij}}{W_i}$$

or substituting $\omega_j = 2\pi/T_j$

$$X_{ij} = \frac{g}{4\pi^2} T_j^2 \frac{F_{ij}}{W_i} \qquad (31.29)$$

where T_j is the jth natural frequency.

The story drift, the relative displacement between the upper and lower levels for the story, is given by

$$\Delta_{ij} = X_{ij} - X_{(i-1)j} \qquad (31.30)$$

with $X_{0j} = 0$, and where Δ_{ij} is the drift of story i for mode j.

31.2.13 Combination of Modal Actions

In those cases in which higher modes should be considered, it is necessary to estimate the maximum response produced by the most unfavorable combination of the modal responses. The combination of modes generally is evaluated using the method of the Square Root of the Sum of Square modal contributions (SRSS). The modal responses at level i, corresponding to the first three modes, are calculated and then combined as indicated in Table 31.7.

31.3 SPANISH SEISMIC CODE P.D.S.-1 (DRAFT 1991)

Some brief comments on the new seismic code P.D.S.-1 (Draft 1991), which is being prepared by the Spanish Interministerial Commission, are presented in this section. There are some important conceptual differences between the provisions of this draft code and those of the P.D.S.-1 (1974) Code. The seismic map in the draft code defines directly the maximum horizontal acceleration of the ground; thus, the curves in the map correspond to locations of equal ground acceleration, not to locations of equal seismic intensity as in the 1974 Code. The design seismic acceleration a_c is given by

$$a_c = \rho a_b \qquad (31.31)$$

in which a_b is the peak ground acceleration obtained from the seismic map and ρ is the risk coefficient

Table 31.7.   Maximum Values from Combination of Modal Responses

Action*	Normal Structures $T \leqslant 0.75$ sec	Slender Structures $0.75 < T \leqslant 1.25$ sec	Very Slender Structures $T > 1.25$ sec
Axial force	N_i	$\sqrt{N_i^2 + N_i'^2}$	$\sqrt{N_i^2 + N_i'^2 + N_i''^2}$
Overturning moment	M_i	$\sqrt{M_i^2 + M_i'^2}$	$\sqrt{M_i^2 + M_i'^2 + M_i''^2}$
Story shear force	V_i	$\sqrt{V_i^2 + V_i'^2}$	$\sqrt{V_i^2 + V_i'^2 + V_i''^2}$
Torsional moment	M_{ti}	$\sqrt{M_{ti}^2 + M_{ti}'^2}$	$\sqrt{M_{ti}^2 + M_{ti}'^2 + M_{ti}''^2}$
Lateral displacement	X_i	$\sqrt{X_i^2 + X_i'^2}$	$\sqrt{X_i^2 + X_i'^2 + X_i''^2}$
Relative displacement	Δ_i	$\sqrt{\Delta_i^2 + \Delta_i'^2}$	$\sqrt{\Delta_i^2 + \Delta_i'^2 + \Delta_i''^2}$

*The unprimed symbols refer to the first mode; the single-primed and double-primed symbols refer to the second and third modes, respectively.

depending on earthquake recurrence time expressed in years. The risk coefficient is given by

$$\rho = \left(\frac{t}{50}\right)^{0.37} \tag{31.32}$$

where the recurrence time t is equal to 50 years for ordinary structures and 100 years for structures considered to be very important. The equivalent static force F_{ik} corresponding to level k and mode of vibration i is determined by the following formula:

$$F_{ik} = S_{ik} W_k \tag{31.33}$$

where W_k is the weight at the level k of the building and S_{ik} is the seismic coefficient for that level corresponding to mode i. This coefficient is given by

$$S_{ik} = a_c \alpha(T_i) \beta \eta_{ik} \tag{31.34}$$

where a_c is the design acceleration previously defined by eq.(31.31), and $\alpha(T_i)$ is the spectral ordinate corresponding to the mode of vibration i and to a damping ratio of 0.05. The response spectrum $\alpha(T_i)$ is given explicitly in the draft code. It is a function of the eigenperiod T of the structure and two other coefficients, C (which depends on the characteristics of a 30-m-thick soil layer under the structure) and K (which considers the seismic hazard caused by different types of earthquakes that could occur in the future at a given location). The factor β in eq.(31.34) is the response factor, which depends on the damping ratio ν and on the ductility μ of the structure, expressed as

$$\beta = \frac{\nu}{\mu} \tag{31.35}$$

Finally, η_{ik} is the distribution factor defined by the following equation:

$$\eta_{ik} = \phi_{ik} \frac{\displaystyle\sum_{k=1}^{N} m_k \phi_{ik}}{\displaystyle\sum_{k=1}^{N} m_k \phi_{ik}^2} \tag{31.36}$$

where

N = number of levels of the structure that support concentrated weights

m_k = mass concentrated at level k of the building

ϕ_{ik} = modal ordinate corresponding to mode i and level k.

The draft of the new seismic code includes simplified formulas to calculate the periods T_i and the modal shapes ϕ_{ik} of the first three modes of vibration. The effect of the vertical ground motion on the seismic forces has been simplified by adopting an elastic response spectrum having ordinates equal to 70% of the ordinates of the horizontal response spectrum.

31.4 EXAMPLE 31.1

The plane steel frame of Fig. 31.5 shows a model for a six-story building. Loads of 20,000 kg (kilograms weight) are assigned to each level of the building except at the roof where the load is 10,000 kg. The total flexural rigidity of the columns in any story is $EI = 1.3 \times 10^{11}$ kg-cm². The risk period is 50 years and the seismic intensity $G = $ IX. The building rests on isolated footings for the columns founded on consolidated gravel and sand.

Determine according to the Spanish Seismic Code: (1) fundamental period and the second and the third natural periods; (2) intensity, response, foundation, and distribution factors; (3) seismic coefficient at each level for the first three modes; (4) equivalent forces and effective forces; (5) story shear forces; (6) overturning moments; (7) torsional moments for an assumed eccentricity of one meter at each story; (8) lateral displacements; and (9) relative displacements (drift) between levels of the building.

Solution

(1a) Fundamental period
Formula for steel building:

$$T = 0.10 \frac{H}{\sqrt{L}} \geqslant 0.5 \qquad [\text{eq.}(31.11)]$$

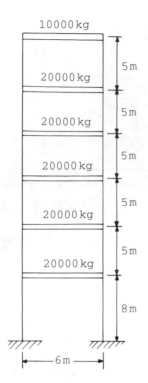

Fig. 31.5. Modeled building for Example 31.1

$H = 33\,\text{m}$ (height of the building)
$L = 6\,\text{m}$ (dimension in the direction of the seismic forces)

$$T = 0.10\,\frac{33}{\sqrt{6}} = 1.35\,\text{sec}$$

(1b) Second and third periods

$$T' = \frac{T}{3} \geqslant 0.25$$

$$\text{[eq.(31.13)]}$$

$$T' = \frac{1.35}{3} = 0.45\,\text{sec}$$

$$T'' = \frac{T}{5} \geqslant 0.25$$

$$\text{[eq.(31.14)]}$$

$$T'' = \frac{1.35}{5} = 0.27\,\text{sec}$$

(2a) Intensity factor

$$\alpha = C_G R \quad \text{(for } T > 0.5\,\text{sec)} \quad \text{[eq.(31.15)]}$$

$$C = 0.30 \quad \text{(Table 31.4 with } G = \text{IX)}$$

$$C_G = \frac{C}{2T} = \frac{0.30}{(2)(1.35)} = 0.111 \quad \text{[eq.(31.16)]}$$

$$R = 0.72 \quad \text{(Table 31.5)}$$

$$\alpha = 0.111 \times 0.72 = 0.08$$

Analogously for the second and third modes:

$$C_G' = \frac{0.30}{(2)(0.45)} = 0.333$$

$$C_G'' = \frac{0.30}{(2)(0.27)} = 0.555$$

$$\alpha' = 0.333 \times 0.72 = 0.24$$

$$\alpha'' = 0.555 \times 0.72 = 0.40$$

(2b) Response factor

$$\beta = \frac{B}{\sqrt[3]{T}} = \geqslant 0.5 \qquad \text{[eq.(31.17)]}$$

$$B = 0.8 \quad \text{(no partitions)}$$

$$\beta = \frac{0.8}{\sqrt[3]{1.35}} = 0.724$$

$$\beta' = \frac{0.8}{\sqrt[3]{0.45}} = 1.044$$

$$\beta'' = \frac{0.8}{\sqrt[3]{0.27}} = 1.238$$

(2c) Foundation factor

$$\delta = 0.8 \quad \text{(Table 31.6)}$$

(2d) Distribution factor

First mode:

$$\eta_i = Z_i \frac{S}{I} \qquad \text{[eq.(31.19)]}$$

with

$$S = \sum_{k=1}^{N} W_k Z_k \qquad \text{[eq.(31.20)]}$$

$$I = \sum_{k=1}^{N} W_k Z_k^2 \qquad \text{[eq.(31.21)]}$$

Second mode:

$$\eta_i' = X_i \frac{\displaystyle\sum_{k=1}^{N} X_k}{\displaystyle\sum_{k=1}^{N} X_k^2} \qquad \text{[eq.(31.23a)]}$$

where

$$X_i = \frac{3Z_i}{H} \quad \text{for} \quad Z_i \leq \frac{H}{3}$$

$$X_i = 2 - \frac{3Z_i}{H} \quad \text{for} \quad Z_i > \frac{H}{3}$$

Third mode:

$$\eta_i'' = X_i \frac{\sum\limits_{k=1}^{N} X_k}{\sum\limits_{k=1}^{N} X_k^2} \qquad \text{[eq.(31.23b)]}$$

where

$$X_i = \frac{5Z_i}{H} \quad \text{for} \quad Z_i \leq \frac{H}{5}$$

$$X_i = 2 - \frac{5Z_i}{H} \quad \text{for} \quad \frac{H}{5} < Z_i \leq \frac{H}{5}$$

$$X_i = -4 + \frac{5Z_i}{H} \quad \text{for} \quad Z_i > \frac{3H}{5}$$

Table 31.8 shows the results of using these equations to calculate the distribution factors η_{ij} for each level i of the building, for the modes $j = 1, 2,$ and 3.

(3) Seismic coefficient

$$s_{ij} = \alpha_j \cdot \eta_{ij} \cdot \beta_j \cdot \delta \qquad \text{[eq.(31.7)]}$$

Table 31.9 shows the values calculated for s_{ij} corresponding to the levels $i = 1, 2, \ldots, 6$ and to the modes $j = 1, 2, 3$.

(4) Lateral equivalent force and effective force

$$F_{ij} = s_{ij} W_i \qquad \text{[eq.(31.1)]}$$

Table 31.8. Distribution Factor η_{ij}

Level i	Distribution Factor η_{ij}		
	Mode 1	Mode 2	Mode 3
6	1.456	−0.103	0.331
5	1.235	−0.056	0.080
4	1.014	−0.009	−0.170
3	0.794	0.038	−0.240
2	0.573	0.085	0.010
1	0.353	0.075	0.260

Table 31.9. Seismic Coefficient s_{ij}

Level i	Seismic Coefficient s_{ij}		
	Mode 1	Mode 2	Mode 3
6	0.067	−0.021	0.131
5	0.057	−0.011	0.032
4	0.047	−0.002	−0.067
3	0.037	0.008	−0.095
2	0.027	0.017	0.004
1	0.016	0.015	0.103

Table 31.10. Lateral Equivalent Force F_{ij} and Effective Force F_{ei} (kg)

Level i	Lateral Force F_{ij}			Effective Force F_{ei}
	Mode 1	Mode 2	Mode 3	
6	675	−206	1,311	1,489
5	1,145	−225	634	1,328
4	940	−36	−1,347	1,926
3	736	152	−1,907	2,045
2	531	340	79	636
1	327	300	2,060	2,107

F_{ij} = force at level i for mode j
s_{ij} = seismic factor at level i for mode j
W_i = weight at level i

$$F_{ei} = \sqrt{F_i^2 + F_i'^2 + F_i''^2} \quad \text{for} \quad T = 1.35 > 1.25 \quad \text{[eq.(31.25c)]}$$

Table 31.10 shows calculated values for the lateral equivalent forces F_{ij} and the effective forces F_{ei} at the various levels of the building.

(5) Shear force

$$V_{ij} = \sum_{k=i}^{N} F_{kj} \qquad \text{[eq.(31.27)]}$$

V_{ij} = shear force at level i for mode j
F_{kj} = equivalent force at level k for mode j.

Table 31.11 shows the values for the shear force and for the effective shear.

Table 31.11. Shear Force V_{ij} and Effective Shear Force V_{ei} (kg)

Level i	Shear Force V_{ij}			Effective Shear V_{ei}
	Mode 1	Mode 2	Mode 3	
6	675	−206	131	1,489
5	1,820	−431	1,945	2,698
4	2,760	−467	598	2,862
3	3,496	−315	−1,304	3,745
2	4,027	25	−1,225	4,209
1	4,354	325	835	4,445

(6) Overturning moment

$$M_{ij} = \sum_{k=i+1}^{N} F_{kj}(Z_k - Z_i) \qquad \text{[eq.(31.26)]}$$

M_{ij} = overturning moment at level i for mode j
Z_k, Z_i = height at level k, i
F_{kj} = equivalent lateral force at level k for mode j.

The results are shown in Table 31.12.

(7) Torsional moment. The torsional moment is given by

$$M_{tij} = V_{ij}e_i$$

where

M_{tij} = torsional moment at level i for mode j
V_{ij} = shear force at level i for mode j
e_i = eccentricity at level i

For this building, it is assumed that the eccentricity (distance between the center of the mass of the above floors and the center of stiffness) at each level is one meter. The calculated torsional moments are shown in Table 31.13.

Table 31.12. Overturning Moment M_{ij} and Effective Overturning Moment M_{ei} (kg-m)

Level i	Overturning Moment M_{ij}			Effective Moment M_{ei}
	Mode 1	Mode 2	Mode 3	
6	—	—	—	—
5	3,375	−1,030	6,555	7,445
4	12,475	−3,185	16,280	20,756
3	26,275	−5,520	19,270	33,048
2	43,755	−7,095	12,750	46,127
1	63,890	−6,970	6,625	64,610
Base	98,722	−4,370	13,305	99,710

Table 31.13. Torsional Moment M_{tij} and Effective Torsional Moment M_{tei} (kg-m)

Level i	Torsional Moment M_{tij}			Effective Moment M_{tei}
	Mode 1	Mode 2	Mode 3	
6	675	−206	1,311	1,489
5	1,820	−431	1,945	2,698
4	2,760	−467	598	2,862
3	3,496	−315	−1,304	3,745
2	4,027	25	−1,225	4,209
1	4,354	325	835	4,445

Table 31.14. Lateral Displacement X_{ij} and Effective Displacement X_{ei} (cm)

Level i	Lateral Displacement X_{ij}			Effective Displacement X_{ei}
	Mode 1	Mode 2	Mode 3	
6	3.05	−0.10	0.24	3.06
5	2.59	−0.06	0.06	2.59
4	2.13	−0.01	−0.12	2.132
3	1.66	0.04	−0.17	1.67
2	1.20	0.09	0.01	1.20
1	0.72	0.08	0.19	0.77

(8) Lateral displacements. The lateral displacement X_{ij} at level i for mode j is calculated by

$$X_{ij} = \frac{g}{4\pi^2} \frac{T_j^2 F_{ij}}{W_i} \qquad \text{[eq.(31.29)]}$$

in which T_j is the jth natural period, F_{ij} is the lateral force at level i for mode j, W_i is the seismic weight at level i, and g is the acceleration of gravity. The lateral displacement at the first level for the first mode is

$$X_{11} = \frac{980}{4\pi^2} \frac{1.35^2 \times 320}{20,000} = 0.72$$

The calculation of the modal lateral displacements at the various levels of the building calculated by eq.(31.29), as well as the effective lateral displacements evaluated by the corresponding expression shown in Table 31.7, are given in Table 31.14.

(9) Relative displacements (story drift). The story drift is calculated as the difference between the lateral displacements of the upper and lower levels for each story. The drift for the second story in the first mode is then given by

$$\Delta_{21} = X_{31} - X_{21}$$
$$= 1.66 - 1.20 = 0.46 \text{ cm}$$

The modal story drift and effective story drift for all the stories of the building are shown in Table 31.15.

Table 31.15. Story Drift Δ_{ij} and Effective Story Drift Δ_{ei} (cm)

Level i	Story Drift Δ_{ij}			Effective Relat. Displ.
	Mode 1	Mode 2	Mode 3	
6	0.46	−0.05	0.18	0.50
5	0.46	0.05	0.18	0.50
4	0.46	0.05	0.05	0.47
3	0.46	0.05	−0.18	0.50
2	0.46	0.01	−0.18	0.50
1	0.74	0.08	0.19	0.77

31.5 COMPUTER PROGRAM

A program has been developed to implement the provisions of the Spanish Seismic Code P.D.S.-1 (1974). The program contains provisions for the calculation of the fundamental period using the empirical formulas of the Code. All the necessary factors to determine the equivalent seismic lateral forces are calculated and printed by the program. Values for seismic lateral forces, story shear forces, torsional moments, overturning moments, and lateral displacements are calculated and printed for each level of the building. All these values are calculated and printed for the first three modes and their combined values (effective values).

31.6 EXAMPLE 31.2

Solve Example 31.1 using the computer program developed for implementation of the Seismic Code P.D.S.-1 (1974).

31.6.1 Input Data and Output Results for Example 31.2

```
************************************************************************

          ***EARTHQUAKE RESISTANT DESIGN***

      MARIO PAZ,  UNIVERSITY OF LOUISVILLE

  ***USING SEISMIC BUILDING CODE OF SPAIN (PDS-1 1974)***

************************************************************************

       ***DATA FILE INFORMATION***

          1. PREPARE NEW DATA FILE

          2. MODIFY EXISTING DATA FILE

          3. USE EXISTING DATA FILE

       SELECT NUMBER? 3

       DRIVE USED FOR DATA FILES (A:,B:,or C:) ? C:
       FILE NAME (OMIT DRIVE LETTER)(SAMPLE D32B)? D32B

    INPUT DATA:


SEISMIC INTENSITY ZONE              NGI= 9
RISK PERIOD (YEARS)                 RI= 50
BASE DIMENSION (EARTHQUAKE DIRECTION) (M)   L= 6
TYPE OF SOIL                        NT= 3
FOUNDATION TYPE                     NC= 3
NUMBER OF STORIES                   ND= 6
```

STORY #	STORY HEIGHT METER	STORY WEIGHT Kg.
6	5.00	10000.00
5	5.00	20000.00
4	5.00	20000.00
3	5.00	20000.00
2	5.00	20000.00
1	8.00	20000.00

STORY TORSIONAL ECCENTRICITY

STORY #	ECCENTRICITY (METER)
6	1.00
5	1.00
4	1.00
3	1.00
2	1.00
1	1.00

FUNDAMENTAL PERIOD MENU:

  1. BUILDINGS WITH STRUCTURAL WALLS:

     T=0.06*H/SQR(L)*SQR[H/(2*L+H)]>0.5

  2. REINFORCED CONCRETE BUILDINGS:

     T=0.09*H/SQR(L)>0.5

  3. STEEL BUILDINGS:

     T=0.10*H/SQR(L)>0.5

  4. BRACED CONCRETE BUILDINGS:

     T=0.85/SQR(1+L/H)*0.09*H/SQR(L)>0.5>0.5

  5. VALUE PREVIOUSLY DETERMINED:

     SELECT A NUMBER ? 3

  RESULTS:

FACTORS OF THE SEISMIC COEFICIENT S(I,J)

FACTOR	MODE 1	MODE 2	MODE 3
INTENSITY FACTOR (ALPHA)	0.080	0.240	0.400
RESPONSE FACTOR (BETA)	0.724	1.044	1.238
FOUNDATION FACTOR (DELTA)	0.800	0.800	0.800

DISTRIBUTION FACTOR,ETA(I,J)

LEVEL	MODE 1	MODE 2	MODE 3
6	1.456	-0.103	0.331
5	1.235	-0.056	0.080
4	1.014	-0.009	-0.170
3	0.794	0.038	-0.240
2	0.573	0.085	0.010
1	0.353	0.075	0.260

SEISMIC FACTOR, S(I,J)

LEVEL	MODE 1	MODE 2	MODE 3
6	0.067	-0.021	0.131
5	0.057	-0.011	0.032
4	0.047	-0.002	-0.067
3	0.037	0.008	-0.095
2	0.027	0.017	0.004
1	0.016	0.015	0.103

SEISMIC INTENSITY	GI= .3
FUNDAMENTAL PERIOD	T1= 1.35
SECOND PERIOD	T2= .45
THIRD PERIOD	T3= .27
SEISMIC RISK	RS= .72

EQUIVALENT LATERAL FORCES: (Kg)

LEVEL	MODE 1	MODE 2	MODE 3	EFFECTIVE FORCE
6	674	-207	1309	1503
5	1144	-226	635	1343
4	940	-38	-1349	1672
3	736	151	-1905	2064
2	531	339	79	616
1	327	302	2063	2111

SHEAR FORCE (Kg):

LEVEL	MODE 1	MODE 2	MODE 3	EFFECTIVE FORCE
6	674	-207	1309	1473
5	1819	-434	1944	2662
4	2759	-471	595	2822
3	3494	-320	-1309	3731
2	4025	19	-1230	4209
1	4352	320	833	4431

OVERTURNING MOMENT (Kg-m)

LEVEL	MODE 1	MODE 2	MODE 3	EFFECTIVE MOMENT
5	3372	-1037	6547	7365
4	12465	-3205	16269	20495
3	26257	-5561	19246	32555
2	43728	-7163	12698	45535
1	63855	-7069	6547	64190
0	98675	-4505	13214	99555

LATERAL DISPLACEMENTS (cm):

LEVEL	MODE 1	MODE 2	MODE 3	EFFECTIVE DISPLACEMENT
6	3.05	-0.10	0.24	3.06
5	2.59	-0.06	0.06	2.59
4	2.13	-0.01	-0.12	2.13
3	1.66	0.04	-0.17	1.67
2	1.20	0.09	0.01	1.20
1	0.74	0.08	0.19	0.76

RELATIVE DISPLACEMENTS BETWEEN LEVELS (cm):

LEVEL	MODE 1	MODE 2	MODE 3	EFFECTIVE RELAT. DISPLACEMENT
6	0.46	-0.05	0.18	0.50
5	0.46	-0.05	0.18	0.50
4	0.46	-0.05	0.05	0.46
3	0.46	-0.05	-0.18	0.50
2	0.46	0.01	-0.18	0.50
1	0.74	0.08	0.19	0.76

TORSIONAL MOMENT (Kg-m):

LEVEL	MODE 1	MODE 2	MODE 3	EFFECTIVE MOMENT
6	674	-207	1309	1473
5	1819	-434	1944	2662
4	2759	-471	595	2822
3	3494	-320	-1309	3731
2	4025	19	-1230	4209
1	4352	320	833	4431

31.7 EXAMPLE 31.3

A 12-story moment-resisting frame serves to model a concrete building for earthquake analysis. The effective weight on all levels is 500 kg/m^2 except on the roof where it is 400 kg/m^2. The plan dimension of the building is 30 m by 50 m. All the stories in the building are 4.0 m high except for the first, which is 6.0 m high. The building site is in a zone of seismic intensity grade IX. The building is supported on firm rock with isolated footings. The risk period assigned to the building is 100 years.

Perform the seismic analysis for this building according to the Spanish Seismic Code P.D.S.-1 (1974).

31.7.1 Input Data and Output Results for Example 31.3

INPUT DATA:

SEISMIC INTENSITY ZONE	NGI= 9
RISK PERIOD (YEARS)	RI= 100
BASE DIMENSION (EARTHQUAKE DIRECTION) (M)	L= 20
TYPE OF SOIL	NT= 3
FOUNDATION TYPE	NC= 3
NUMBER OF STORIES	ND= 20

STORY #	STORY HEIGHT METER	STORY WEIGHT Kg.
20	4.00	12000.00
19	4.00	15000.00
18	4.00	15000.00
17	4.00	15000.00
16	4.00	15000.00
15	4.00	15000.00
14	4.00	15000.00
13	4.00	15000.00

12	4.00	15000.00
11	4.00	15000.00
10	4.00	15000.00
9	4.00	15000.00
8	4.00	15000.00
7	4.00	15000.00
6	4.00	15000.00
5	4.00	15000.00
4	4.00	15000.00
3	4.00	15000.00
2	4.00	15000.00
1	6.00	15000.00

STORY TORSIONAL ECCENTRICITY

STORY #	ECCENTRICITY (METER)
20	1.00
19	1.00
18	1.00
17	1.00
16	1.00
15	1.00
14	1.00
13	1.00
12	1.00
11	1.00
10	1.00
9	1.00
8	1.00
7	1.00
6	1.00
5	1.00
4	1.00
3	1.00
2	1.00
1	1.00

FUNDAMENTAL PERIOD MENU:

1. BUILDINGS WITH STRUCTURAL WALLS:

    T=0.06*H/SQR(L)*SQR[H/(2*L+H)]>0.5

2. REINFORCED CONCRETE BUILDINGS:

    T=0.09*H/SQR(L)>0.5

3. STEEL BUILDINGS:

    T=0.10*H/SQR(L)>0.5

4. BRACED CONCRETE BUILDINGS:

    T=0.85/SQR(1+L/H)*0.09*H/SQR(L)>0.5>0.5

5. VALUE PREVIOUSLY DETERMINED:

    SELECT A NUMBER ? 2

    RESULTS:

FACTORS OF THE SEISMIC COEFICIENT S(I,J)

FACTOR	MODE 1	MODE 2	MODE 3
INTENSITY FACTOR (ALPHA)	0.084	0.251	0.418
RESPONSE FACTOR (BETA)	0.508	0.732	0.868
FOUNDATION FACTOR (DELTA)	0.800	0.800	0.800

SEISMIC INTENSITY	GI=	.3
FUNDAMENTAL PERIOD	T1=	1.65
SECOND PERIOD	T2=	.55
THIRD PERIOD	T3=	.33
SEISMIC RISK	RS=	.92

EQUIVALENT LATERAL FORCES: (Kg)

LEVEL	MODE 1	MODE 2	MODE 3	EFFECTIVE FORCE
20	601	-680	1160	1320
19	715	-726	1097	1322
18	678	-601	743	1023
17	642	-477	389	773
16	605	-353	35	634
15	568	-228	-318	678
14	532	-104	-672	877
13	495	21	-1026	1154
12	458	145	-1380	1466
11	422	270	-1167	1255
10	385	394	-814	919
9	348	519	-460	606
8	312	643	-106	378
7	275	767	248	414
6	238	809	601	673
5	202	684	955	994
4	165	560	1309	1332
3	128	436	1238	1259
2	92	311	884	909
1	55	187	531	565

SHEAR FORCE (Kg):

LEVEL	MODE 1	MODE 2	MODE 3	EFFECTIVE FORCE
20	601	-680	1160	1307
19	1316	-1406	2257	2613
18	1994	-2008	3000	3602
17	2636	-2485	3389	4294
16	3241	-2837	3425	4715
15	3809	-3065	3106	4915
14	4341	-3169	2434	4977
13	4836	-3148	1408	5037
12	5294	-3003	28	5294
11	5716	-2734	-1139	5828
10	6101	-2340	-1953	6406
9	6449	-1821	-2413	6886
8	6761	-1178	-2519	7215
7	7036	-411	-2271	7393
6	7274	398	-1670	7463
5	7476	1083	-715	7510
4	7641	1643	594	7664
3	7769	2078	1833	7982
2	7861	2389	2717	8317
1	7916	2576	3248	8556

OVERTURNING MOMENT (Kg-m)

LEVEL	MODE 1	MODE 2	MODE 3	EFFECTIVE MOMENT
19	2405	-2721	4641	5228
18	7670	-8346	13670	15674
17	15648	-16377	25670	30063
16	26192	-26316	39226	47167
15	39156	-37665	52924	65834

14	54393	-49927	65349	85024
13	71757	-62603	75085	103859
12	91100	-75197	80717	121714
11	112276	-87210	80830	138345
10	135139	-98144	76273	155178
9	159542	-107502	68462	173611
8	185338	-114786	58811	194445
7	212381	-119498	48736	217901
6	240523	-121141	39651	243770
5	269619	-119548	32972	271628
4	299521	-115218	30113	301031
3	330084	-108647	32490	331679
2	361159	-100334	39821	363348
1	392601	-90777	50688	395860
0	440094	-75321	70174	445654

LATERAL DISPLACEMENTS (cm):

LEVEL	MODE 1	MODE 2	MODE 3	EFFECTIVE DISPLACEMENT
20	3.39	-0.43	0.26	3.40
19	3.22	-0.36	0.20	3.23
18	3.06	-0.30	0.13	3.06
17	2.89	-0.24	0.07	2.89
16	2.73	-0.18	0.01	2.73
15	2.56	-0.11	-0.06	2.56
14	2.40	-0.05	-0.12	2.40
13	2.23	0.01	-0.18	2.24
12	2.06	0.07	-0.25	2.08
11	1.90	0.13	-0.21	1.91
10	1.73	0.20	-0.15	1.74
9	1.57	0.26	-0.08	1.57
8	1.40	0.32	-0.02	1.40
7	1.24	0.38	0.04	1.24
6	1.07	0.40	0.11	1.08
5	0.91	0.34	0.17	0.92
4	0.74	0.28	0.24	0.78
3	0.58	0.22	0.22	0.62
2	0.41	0.16	0.16	0.44
1	0.25	0.09	0.10	0.27

RELATIVE DISPLACEMENTS BETWEEN LEVELS (cm):

LEVEL	MODE 1	MODE 2	MODE 3	EFFECTIVE RELAT. DISPLACEMENT
20	0.17	-0.06	0.06	0.18
19	0.17	-0.06	0.06	0.18
18	0.17	-0.06	0.06	0.18
17	0.17	-0.06	0.06	0.18
16	0.17	-0.06	0.06	0.18
15	0.17	-0.06	0.06	0.18
14	0.17	-0.06	0.06	0.18
13	0.17	-0.06	0.06	0.18
12	0.17	-0.06	-0.04	0.17
11	0.17	-0.06	-0.06	0.18
10	0.17	-0.06	-0.06	0.18
9	0.17	-0.06	-0.06	0.18
8	0.17	-0.06	-0.06	0.18
7	0.17	-0.02	-0.06	0.18
6	0.17	0.06	-0.06	0.18
5	0.17	0.06	-0.06	0.18
4	0.17	0.06	0.01	0.17
3	0.17	0.06	0.06	0.18
2	0.17	0.06	0.06	0.18
1	0.25	0.09	0.10	0.27

TORSIONAL MOMENT (Kg-m):

LEVEL	MODE 1	MODE 2	MODE 3	EFFECTIVE MOMENT
20	601	-680	1160	1307
19	1316	-1406	2257	2613
18	1994	-2008	3000	3602
17	2636	-2485	3389	4294
16	3241	-2837	3425	4715
15	3809	-3065	3106	4915
14	4341	-3169	2434	4977
13	4836	-3148	1408	5037
12	5294	-3003	28	5294
11	5716	-2734	-1139	5828
10	6101	-2340	-1953	6406
9	6449	-1821	-2413	6886
8	6761	-1178	-2519	7215
7	7036	-411	-2271	7393
6	7274	398	-1670	7463
5	7476	1083	-715	7510
4	7641	1643	594	7664
3	7769	2078	1833	7982
2	7861	2389	2717	8317
1	7916	2576	3248	8556

REFERENCES

Comisión Permanente de Normas Sísmicas (Draft 1991) *Norma Sismorresistente*. Madrid, Spain.

Dirección General de Arquitectura y Tecnología de la Edificación, Ministerio de la Vivienda (1973) *Norma Tecnológica de la Edificación NTE-ECS/1973*. Madrid, Spain.

Dirección General para la Vivienda y Arquitectura, Ministerio de Obras Públicas y Urbanismo (1988) *La Norma Básica de la Edificación y Acciones en la Edificación NBE-AE-88*. Madrid, Spain.

Ministerio de la Vivienda (1963) Norma M.V. 101-1962. *Acciones en la edificación*. Decreto 17 (195/1963), Madrid, Spain.

Ministerio de Obras Públicas (1962) "Instrucción para el proyecto, construcción y explotación de grandes presas." Madrid, Spain.

—— (1967), *Instrucción para el proyecto, construcción y explotación de grandes presas*. Madrid, Spain.

Presidencia del Gobierno, Comisión Interministerial (1969) *Norma Sismorresistente P.G.S.-1 (1968)*. Decreto 106, Madrid, Spain.

—— (1974) *Norma Sismorresistente P.D.S.-1 (1974)*. Decreto 3209, Madrid, Spain.

Presidencia del Gobierno, Dirección General del Instituto Geográfico Nacional, Comisión Permanente de Normas Sismorresistentes (1978) *Boletin Informativo Número 1*. Madrid, Spain.

32

Taiwan

Yohchia Chen and Julius P. Wong

32.1 INTRODUCTION

A modern version of the Taiwanese Building Code (TBC 1982) was completed in June 1982, and officially approved in July; supplementary provisions were added to the TBC in 1983. A revised code (TBC 1991) is essentially based on the following: *Uniform Building Code* (UBC 1988), *Building Code Requirements for Reinforced Concrete and Commentary* (ACI 1989), *Timber Construction Manual* (AITC 1985), *Manual of Steel Construction-Allowable Stress Design* (AISC 1989), and a few provisions based on the Japanese building code. Load factor and resistance factor designs for steel structures have not yet been considered in the revised code. The provisions for seismic-resistant design and construction of buildings in Taiwan are contained in the Building Construction section of the *Taiwanese Building Code* (TBC 1991), Section 5 of Chapter 1. These provisions are presented in this chapter.

32.2 TOTAL LATERAL FORCE (*V*)

Building structures should be designed to resist equivalent seismic forces applied along two main normal directions of the building, independently. Earthquake forces are assumed to act horizontally at the elevation of each floor above grade and at the roof. The total lateral force *V* is determined by the following expression:

$$V = ZKCIW \qquad (32.1)$$

where Z = seismic zone factor, K = framing coefficient, C = seismic force coefficient, I = occupancy importance factor, and W = total dead load. For warehouses or storage facilities, W is the total dead load and at least one-fourth of the design live load. For water towers or water tanks, W is the total dead load and the total weight of the contained water (e.g., tank completely filled).

The form of eq.(32.1) is similar to that of eq.(12.1) of the *Uniform Building Code* (UBC 1988); K is analogous to $1/R_w$. R_w is a numerical coefficient that is a function of structural system, lateral load resistance capacity, and height of the structure. The various coefficients in eq.(32.1) are described in the sections that follow.

32.2.1 Framing Coefficient (*K*)

Framing coefficient K has different values for different structural frame types as follows:

- For ductile moment-resisting frames $K = 0.67$.
- For ductile moment-resisting frames with shear walls or lateral bracing systems, $K = 0.80$ provided that one of the following conditions is met:
  1. Frames interact with shear walls or lateral braces structurally, and the total lateral force is resisted by each component proportionally to the component stiffness.
  2. Shear walls or lateral braces function separately from the ductile frames, but resist the total lateral force.
  3. Ductile frames resist at least one-fourth of the total lateral force.
- For structural systems resisting the lateral force totally by shear walls, $K = 1.33$.
- For structures other than those described above, $K = 1.00$.
- For nonstructural systems $K = 2.00$.
- For water towers and water tanks that have cross-lateral bracing and are supported by four or more legs not attached to the structural system (i.e., the legs are attached to the ground), $K = 2.50$. However, in this case, K times C must not be less than 0.23.

32.2.2 Seismic Force Coefficient (C)

The seismic force coefficient C for Taipei (the nation's capital and principal metropolitan area) is given by

$$C = \frac{0.248}{T}, \quad 0.0625 \leqslant C \leqslant 0.15 \qquad (32.2)$$

where T is the fundamental period of vibration (sec).

For other regions of the country, C can be determined from

$$C = \frac{1}{8\sqrt{T}} \leqslant 0.15 \qquad (32.3)$$

Depending on soil conditions and other related considerations, alternative seismic force coefficients may be proposed for approval by local building officials, provided those coefficients are supported by technical data and approved by the federal government. However, in no case should the coefficient C be less than 80% of that calculated by eq.(32.2) or (32.3).

32.2.3 Fundamental Period of Vibration (*T*)

For frames whose lateral resistance is not increased by stiffened members T is calculated as follows:

- For reinforced concrete structures
$$T = 0.06 h_N^{3/4} \qquad (32.4)$$
- For steel structures
$$T = 0.085 h_N^{3/4} \qquad (32.5)$$
- For structures constructed of other materials
$$T = \frac{0.09 h_N}{\sqrt{D}} \qquad (32.6)$$

where h_N in meters (m) is the height of the roof measured from the base of the building, and D (m) is the dimension of the structure parallel to the lateral force.

Alternatively, T may be determined by any suitable structural mechanics method. However, any value of T so computed should not exceed 1.4 times that given by eqs.(32.4), (32.5), or (32.6).

32.2.4 Seismic Zone Factor (*Z*)

Taiwan has been divided into three seismic zones, as shown in Fig. 32.1: (1) Seismic Zone W for weak earthquakes, (2) Seismic Zone M for moderate earthquakes, and (3) Seismic Zone S for strong earthquakes. Table 32.1 provides the value of the seismic zone factor corresponding to these three seismic zones.

32.2. Occupancy importance Factor (*I*)

The occupancy importance factor has the following values:

- $I = 1.50$ for essential structures that must remain operational at all times; e.g., fire and police stations, hospitals, power and water-supply plants, and structures containing natural gas and gasoline storage facilities.
- $I = 1.25$ for public buildings such as schools, gymnasiums, museums, galleries, libraries, conference rooms, theaters, ballrooms, auditoriums, and any other similar structures that may house more than 300 people.
- $I = 1.00$ for all other structures.

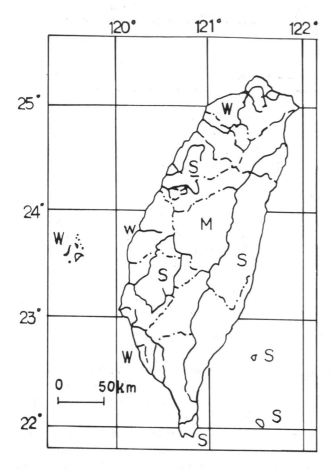

Fig. 32.1. Seismic map of Taiwan (W = weak, M = moderate, S = strong)

Table 32.1. Seismic Zone Factor (Z)

Seismic Zone	Zone Factor Z
W	0.6
M	0.8
S	1.0

32.3 VERTICAL DISTRIBUTION OF LATERAL FORCES

32.3.1 Regular Structures

The total lateral force V given by eq.(32.1) is distributed at the various levels of the building as horizontal forces F_x:

$$F_x = \frac{(V - F_t)\, W_x h_x}{\sum_{i=1}^{N} W_i h_i} \tag{32.7}$$

and F_t is the added horizontal force applied at the roof of the building:

$$F_t = 0.07TV \leqslant 0.25V \quad \text{for} \quad T \geqslant 0.7 \text{ sec} \tag{32.8}$$

$$F_t = 0 \qquad\qquad \text{for} \quad T < 0.7 \text{ sec.} \tag{32.9}$$

$$V = F_t + \sum_{i=1}^{N} F_i \tag{32.10}$$

where F_i is the lateral force for the ith level.

In eq.(32.7), W_i is the weight at level i, h_i is the height of level i measured from the base of the structure, and N is the total number of levels (stories).

32.3.2 Irregular Structures

The code requires that the dynamic characteristics of the structure be considered in the vertical distribution of lateral forces for structures that have an irregular geometry, significant differences in lateral stiffness between any two adjacent stories, or any other irregularities in plan or in elevation.

32.3.3 Structural Components

For structural components, the minimum total lateral force F_p to be resisted by the component is determined by

$$F_p = ZIC_p W_p \tag{32.11}$$

where W_p is the component weight, and C_p is the component seismic force coefficient as given in Table 32.2. For a structural component with $C_p \geqslant 1.25$, the occupancy importance factor, I, may be taken as 1.00.

For floor and roof slabs used as horizontal diaphragms, the design F_p value should be based on a C_p of 0.12 and W_i, instead of W_p. However, if F_i is greater than W_i, W_p should be replaced by F_i. The design lateral pressure should be at least 25 kgf/m² acting perpendicular to the wall plane, for interior partition walls more than 2 m high; the displacement due to the minimum lateral force should be limited to $L/240$ (L = length of wall) if the building has a brittle facade, and $L/120$ if the facade is ductile. For storage racks attached in parallel with four and more columns located along the same line, V may be calculated by eq.(32.1); W should be the dead load plus one-half the live load, and C should be taken as 0.25.

32.4 HORIZONTAL DISTRIBUTION OF LATERAL FORCES AND TORSIONAL MOMENTS

The shear due to lateral forces and the torsional moments at each story are distributed to vertical resisting members in proportion to their stiffnesses;

Table 32.2. Partial Seismic Force Coefficient C_p

Type of Partial Structure	Direction of Lateral Force	C_p
Exterior walls, interior bearing partition walls, interior nonbearing partition walls with height over 3 m, and parapet walls with height over 1.8 m.	Perpendicular to wall plane	0.35
Cantilever walls excluding retaining walls	Perpendicular to wall plane	1.25
Decorative/attached materials	Any direction	1.25
Building appurtenances:	Any direction	
• Water towers, chimneys, water tanks, and their contents		0.35
• Such appurtenances located at the top of the building with $h_N/D > 5$		0.50
• Storage racks with height over 2.4 m		0.35
• Machinery or equipment whose operation is not related directly to safety		0.35
• Safety equipment		0.90
Water tanks with contents supported on the ground	Any direction	0.20
Floor and roof slabs used as horizontal diaphragms	Any direction	0.20
Connections for precast nonstructural exterior curtain walls	Any direction	2.50
Connections for precast structural members	Any direction	0.50
Frames for suspended ceilings	Any direction	0.35

the relative ratios of vertical members and diaphragms/slabs are also taken into consideration. The torsional moment at a story of the building is due to the eccentricity of the center of the mass above that story with respect to the stiffness center of the lateral resisting elements in the story. An assumed accidental eccentricity equal to 5% of the dimension of the building perpendicular to the direction of seismic lateral forces must be added to the existing eccentricity, according to TBC 1991. Vertical members should be designed to withstand the maximum stress caused by these shears and moments.

32.5 LATERAL DISPLACEMENTS

32.5.1 Relative Lateral Displacement or Story Drift

The relative effective lateral displacement between any two adjacent stories is limited to 0.5% of the story height, or to a larger value, provided that structural analysis demonstrates that a larger story drift can be safe. The effective relative lateral displacement is calculated as the lateral displacement due to the lateral seismic forces divided by the framing coefficient K (K must not be larger than 1.0).

32.5.2 Space Between Buildings

To avoid contact of building structures deformed by earthquake or wind, structural components should be designed and constructed as a whole to resist earthquake forces. TBC 1991 requires the minimum spacing between two adjacent buildings to be 1.5% of the larger building height or 15 cm, whichever is greater.

32.6 OVERTURNING MOMENT

Structures should be designed to resist overturning moments caused by wind or earthquake. The overturning moment at each story of the building is distributed to the resisting element members on the same basis as the shear distribution ratios. For discontinuous vertical resisting members, the overturning moment taken by the lowest story is regarded as the added external load carried to the foundation. However, when the vertical resisting elements are continuous from story to story, the Code permits a 10% reduction of the overturning moment transmitted to the foundation.

32.7 BUILDING SETBACK

For a building in which upper portions are set back, the base dimensions may be used to calculate earthquake forces, if the setback dimensions are not less than 75% of the base dimensions. Otherwise, the earthquake forces on the upper setback portion should be calculated separately and considered to act at the top of the lower portion when calculating the total lateral force on the building.

32.8 STRUCTURAL REQUIREMENTS

32.8.1 Ductility Requirements

The ductility requirements are as follows:

• Ductile frames: $K = 0.67$ or 0.8 (see Section 32.2.1); the underground structural components should be designed such that lateral forces can be transferred to the foundation.

- Buildings more than 50 m high should be designed to resist at least one-fourth of the lateral earthquake force capacity of a ductile frame.

- Concrete frames that partially resist lateral forces or that are located on the perimeter of vertical supports should be designed as ductile frames. For the latter, if shear walls are designed to resist the total lateral force, the design can be based on the requirements in Section 32.8.2.

- Rigid frames not designed to resist lateral forces must be adequate to withstand gravity loads and $3/K$ times the bending moment induced by the deformation caused by the lateral force.

- Moment-resisting frames may be connected to stiffener frames without affecting lateral resistance capabilities.

- For inclined rigid frames, the minimum total lateral force for design should be 1.25 times that calculated by eq.(32.1). The connections should be designed so that all members have maximum earthquake resistance, or should be designed to resist the described minimum lateral force with no material overstrength allowance.

32.8.2 Design Requirements

The design requirements are as follows:

- If existing structures are modified, they should be able to resist the original design lateral force.

- Masonry or concrete structures should be reinforced by steel bars. For rigid masonry frames, the spacing of main reinforcement should be based on a C_p of 0.35, but spacing should not exceed 60 cm.

- Combined load or stress effects due to gravity loads (excluding roof live load) and lateral forces should be considered, whenever appropriate. The combined stress due to minimum gravity loads and lateral forces should also be considered.

- Horizontal diaphragms, or slabs supported by masonry or concrete walls, should be designed as continuous members; the design of the bearing walls should include the deformation of diaphragms or slabs.

32.8.3 Special Requirements

The following special requirements must be considered:

- Piles or caissons should be linked by connecting grade beams and should be designed so that each of the beams can resist a lateral force equal to at least one-tenth of the maximum pile/caisson load.

  Otherwise, resistance to lateral force should be guaranteed by an approved method of testing/analysis.

- Nonbearing exterior precast shear panels should be fabricated from cast-in-place concrete. The panels should meet the following requirements:
  1. Minimum space between two adjacent panels to allow relative movement: two times the lateral displacement due to wind and $3/K$ times the lateral

displacement due to an earthquake, or 6 mm, whichever is greatest.
2. Connected materials should have ductility and torsional resistance adequate to avoid concrete cracking or brittle failure of welds. Embedded bars should be spliced with main reinforcing bars so that the full bar strength is developed.
3. Acceptable devices at panel interfaces to allow relative lateral movement: sliding groove, slotted holes for bolts, bending-resistant elements, or other connectors with equivalent mobility and ductility.

32.9 EXAMPLE 32.1

The plane frame shown in Fig. 32.2 serves to model a three-story building for seismic design using the static method. The building has a reinforced concrete, moment-resisting frame with shear walls, and is located in Taipei (northeastern Taiwan). It is considered essential that this building remain functional after an earthquake. The lateral forces on the building are shown in Fig. 32.3.

Determine: (a) lateral equivalent forces, (b) overturning moments, (c) story shear forces, (d) lateral displacements, and (e) story drifts.

Solution

Natural period

$$T = 0.06 h_N^{3/4} \qquad \text{[eq.(32.4)]}$$
$$= 0.06(9.0)^{3/4} = 0.312 \text{ sec}$$

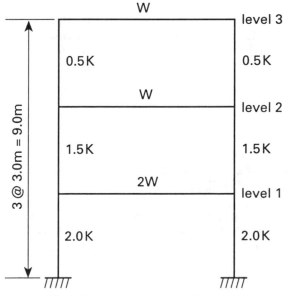

(w = 22.5 Tonnes, K = 2000 Tonne / m)

Fig. 32.2. Building system

Seismic force coefficient

$$C = \frac{0.248}{T} \qquad \text{[eq.(32.2)]}$$

$$= \frac{0.248}{0.312} = 0.795 > 0.15$$

Use $C = 0.15$ (maximum value)

Total lateral force

$$V = ZKCIW$$

$Z = 0.6$ (from Fig. 32.1 and Table 32.1; Taipei is in Seismic Zone *W*)
$K = 0.8$ (from Section 32.2.1, ductile moment with shear walls)
$I = 1.5$ (from Section 32.2.5, essential building)
$W = 4 \times 22.5 = 90$ tonne

Thus, $V = 0.6 \times 0.8 \times 0.15 \times 1.50 \times 90 = 9.72$ tonne.

Equivalent lateral forces

$$F_i = \frac{(V - F_t)W_i h_i}{\sum\limits_{j=1}^{N} W_j h_j} \qquad \text{[eq.(32.7)]}$$

Because

$$T = 0.312 < 0.7 \text{ sec}, \ F_t = 0 \qquad \text{[(eq.(32.9)]}$$

Calculation of the equivalent lateral forces is shown in Table 32.3.

Overturning moments. The overturning moments M_i at a level i of the building are calculated as the static moments developed by the lateral forces above. Hence,

$$M_i = \sum_{j=i+1}^{N} F_j(h_j - h_i) + F_t(h_N - h_i)$$

where $i = 0, 1, 2, \ldots, N-1$.

Resulting values for the overturning moments are given in Table 32.3.

Story shear force. The shear force V_i at story i of the building is equal to the sum of the equivalent lateral forces above that level. That is

$$V_i = \sum_{j=i}^{N} F_j$$

Calculated story shear forces are shown in Table 32.3.

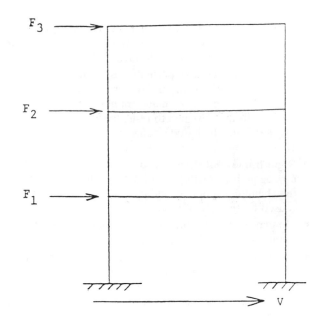

Fig. 32.3. Lateral forces

Table 32.3. Lateral Forces, Story Shear Forces, Overturning Moments, and Lateral Displacements for Example 32.1

Level	h_i (m)	W_i (tonne)	$W_i h_i$ (tonne-m)	F_i (tonne)	M_i (tonne-m)	δ_i (cm)	Δ_i (cm)
3	9.0	22.5	202.5	4.17	——	0.448	0.149
2	6.0	22.5	135.0	2.78	37.53	0.299	0.149
1	3.0	45.0	135.0	2.78	33.36	0.149	0.149
0	0	——	——	——	62.55	——	——

$$\Sigma = 472.5$$

Lateral displacements. Lateral displacements δ_i may be calculated by eq.(4.72) from Chapter 4 as

$$\delta_i = \frac{g}{4\pi^2} \frac{T^2 F_i}{W_i}$$

where

g = acceleration of gravity
T = fundamental period
F_i = equivalent lateral force at level i
W_i = seismic weight at level i.

Calculated values for the lateral displacements are shown in Table 32.3.

Story drifts. The story relative lateral displacements, or story drifts Δ_i, are determined as the difference between lateral displacements of the above and lower levels of the story; i.e.,

$$\Delta_i = \delta_i - \delta_{i-1} \qquad \text{[eq.(4.73)]}$$

with $\delta_0 = 0$.

Calculated values of story drifts are given in Table 32.3.

REFERENCES

ACI (1989) *Building Code Requirements for Reinforced Concrete and Commentary*. ACI 318-89 & 318R-89. American Concrete Institute, Detroit, MI.

AISC (1989) *Manual of Steel Construction Allowable Stress Design*. American Institute of Steel Construction. 9th ed. Chicago, IL.

AITC (1985) *Timber Construction Manual*. American Institute of Timber Construction. 3d ed. Englewood, CO.

TBC (1991) *Taiwanese Building Code*. Construction Bureau, The Ministry of Internal Affairs, Taiwan.

UBC (1988) *Uniform Building Code*. International Conference of Building Officials, Whittier, CA.

33

Thailand

Panitan Lukkunaprasit

33.1 INTRODUCTION

The Thai people were not concerned about earthquakes until the recent occurrence of several moderate earthquakes. On April 22, 1983, an earthquake of magnitude 5.9 on the Richter scale occurred near a dam site, about 200 kilometers from Bangkok, the capital of Thailand. The main tremor of this earthquake was felt all over the western part and most of the central part of the country. Five years later (November 6, 1988), an earthquake of magnitude 7.3 hit the southern part of China near the Burmese border. This earthquake was felt in Bangkok even though the epicenter was at a distance of more than 1,000 kilometers; this is a consequence of Bangkok's deep, soft-alluvial soil which tends to amplify the motion of incoming seismic waves. On September 29 and October 1 of the following year, several moderate earthquakes (5.3–5.4 on the Richter scale) hit the northern part of Thailand along the Burmese border.

In the city of Chiang Mai, about 180 kilometers from the epicenter, the intensity of ground shaking was rated as VI on the Modified Mercalli Intensity (MMI) scale.

Public concern for the safety of buildings has increased since 1983 because most structures in Thailand have not been designed to resist seismic effects even though wind load resistance is required by the building code. A subcommittee of the committee responsible for the building regulations of Thailand subsequently was organized to draft the first seismic-resistant building design code for Thailand. The first draft was presented in a workshop organized by the National Earthquake Committee of Thailand and the Southeast Asia Association of Seismology and Earthquake Engineering (Chandrangsu 1986). This first draft was resisted by some engineers and developers in the construction industry who feared that implementation of the code might produce significant extra building costs. Academicians also questioned the appropriateness of the numerical values of some parameters used in the draft code because there was a paucity of data and relevant research work concerning the seismicity of the country.

A new seismic-resistant design code has been drafted. This new code is based on evidence derived from the earthquake events of the past few decades and on recent research work (Bergado, Kim, and Yamada 1986; Lukkunaprasit, Thusneeyanont, and Kanjanakaroon 1986; Lukkunaprasit 1989, 1990). This new code recognizes the relatively low seismicity of most of the country (virtually no structural damage was caused by past earthquakes) and considers socio-economic factors. The subcommittee decided to make the code simple and easy to use, as many engineers are not familiar with the seismic-resistant design of buildings.[1]

33.2 SYSTEM OF EQUIVALENT STATIC LATERAL FORCES

The proposed new seismic code for building design in Thailand allows the designer to substitute for the dynamic system an equivalent static force system in which lateral forces are applied at discrete points of the structure that correspond to the floor levels. Fig. 33.1 shows such an equivalent lateral force system. For short-period buildings (with fundamental period of natural vibration not greater than 0.7 second), the draft code assumes a linearly varying fundamental mode shape. It follows from the theory of dynamics of structures (e.g., Chopra 1980) that

$$F_x = \frac{VW_x h_x}{\sum_{i=1}^{N} W_i h_i} \quad (33.1)$$

where F_x is the equivalent force at level x, V is the total base shear force, W_i and h_i are, respectively, the effective dead weight and height of level i, and N is the total number of stories of the building. The subscript x in eq.(33.1) denotes the level under consideration. For more flexible buildings, the fundamental mode still is the predominant contributor to the base shear. However, the higher modes increasingly contribute to the motion of the building as the fundamental period increases. The code requires that an additional force F_t be applied at the top of relatively flexible buildings to account for the effect of higher modes of motion for which the fundamental period is greater than 0.7 second. This additional force is taken as

$$F_t = 0.07TV \quad (33.2)$$

in which T (in seconds) is the fundamental period of the structure in the direction under consideration. The value of F_t is limited by the code as not greater than $0.25V$. The remaining force $(V - F_t)$ is distributed proportionately throughout the height of the building at each floor as follows:

$$F_x = \frac{(V - F_t)W_x h_x}{\sum_{i=1}^{N} W_i h_i} \quad (33.3)$$

where all the terms are as previously defined.

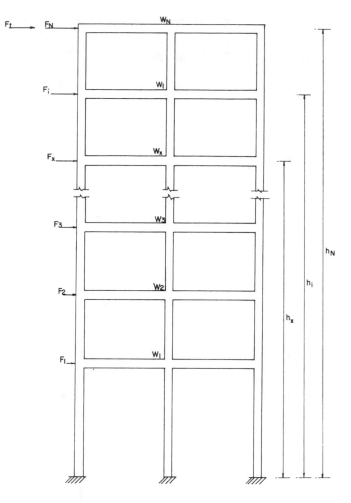

Fig. 33.1. Equivalent lateral seismic forces

33.3 BASE SHEAR

In lieu of analyses based on accepted theoretical procedures, the base shear force may be determined from the simplified formula

$$V = ZKCSIW \quad (33.4)$$

where

Z = seismic zone factor
K = structural factor
C = dynamic factor
S = site-structure resonance coefficient
I = building importance factor
W = total dead load of the building.

[1]At the time of publication of this Handbook, the draft code still had not passed through certain legislative review procedures required before regulations are promulgated. In all probability, the essence of the draft code will remain the same although slight changes may be made during review.

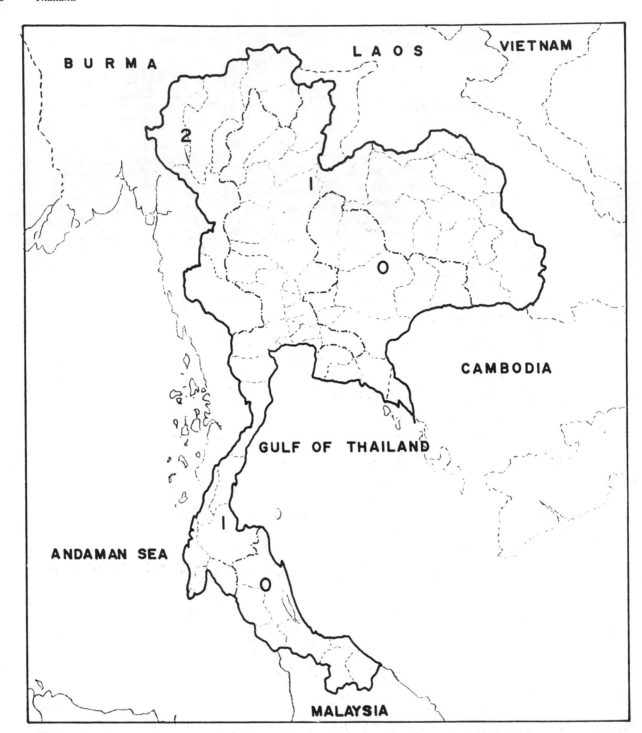

Fig. 33.2.   Seismic zone map of Thailand

These parameters are described in the sections that follow.

33.4 SEISMIC ZONES

Fig. 33.2 shows the seismic zone map of the Kingdom of Thailand (Chandrangsu 1986; Yensuang 1990). In the draft code, the provinces that are prone to seismic hazards are grouped into two zones, depending on the seismicity of the location. Table 33.1 lists the names of the provinces in each zone. The value of the seismic zone factor is set equal to 0.15 for Zone 1 and 0.25 for Zone 2. Buildings in Zone 0 need not be specifically designed for seismic motion.

Table 33.1. List of Provinces in Seismic Zones 1 and 2

Zone	Provinces
1	Bangkok, Krabi, Kalasin, Kamphaeng Phet, Khon Kaen, Chachoengsao, Chai Nat, Chumphon, Nakhon-Pathom, Nakhon Phanom, Nakhon Sawan, Nonthaburi, Nan, Pathum Thani, Prachuap Kiri Khan, Ayutthaya, Phangnga, Phichit, Phitsanulok, Phetchaburi, Phrae, Phuket, Mahasarakham, Mukdahan, Yasothon, Roi Et, Ranong, Ratchaburi, Lop Buri, Lampang, Loei, Srisaket, Sakon Nakhon, Samut Prakan, Samut Songkhram, Samut Sakhon, Sing Buri, Sukhothai, Suphan Buri, Surat Thani, Nongkhai, Ang Thong, Udon Thani, Uttaradit, Uthai Thani, and Ubon Ratchathani
2	Kanchanaburi, Chiang Rai, Chiang Mai, Tak, Phayao, Mae Hong Son, and Lamphun

33.5 STRUCTURAL FACTOR *K*

The draft code recognizes the ductility inherent in steel building frames (provided good workmanship is exercised in fabrication and construction). Accordingly, a *K* value of 0.67 is assigned to steel buildings and a *K* of 1.0 is assigned to all other structures.

33.6 FUNDAMENTAL PERIOD OF VIBRATION

In the draft code, an approximate empirical formula has been adopted for computation of the fundamental natural period of vibration *T* of the structure as follows:

$$T = \frac{0.09h_N}{\sqrt{D}} \qquad (33.5)$$

in which h_N is the height of the structure above the base and *D* is the dimension of the structure in the plane of vibration under consideration. Both h_N and *D* are measured in meters.

For cases in which a steel frame provides the sole lateral force resisting system for the building, the value of *T* may be determined from the following formula:

$$T = 0.1N \qquad (33.6)$$

in which *N* is the total number of stories of the building.

33.7 DYNAMIC FACTOR *C*

The dynamic factor *C* may be computed in accordance with eq.(33.7):

$$C = \frac{1}{15\sqrt{T}} \qquad (33.7)$$

but the code specifies a value of 0.12 set as the upper limit for *C*.

33.8 SITE-STRUCTURE RESONANCE COEFFICIENT *S*

The site-structure resonance coefficient *S* is intended to account for the amplifying effect of soft foundation soil on ground motion and thus on the response of the structure. For simplicity, the draft code stipulates that the coefficient *S* be assigned a value equal to 1.5 when the underlying soil is soft. For stiffer soil types, *S* is to be taken as 1.0. However, the product *CS* in eq.(33.4) should not exceed 0.14.

33.9 BUILDING IMPORTANCE FACTOR *I*

The severity of loss and consequence of damage caused by an earthquake differs among buildings depending on building function. For instance, buildings such as hospitals and fire stations should survive even under a strong earthquake of relatively low probability of occurrence because their functions are vital for public health and safety. From an engineering perspective, these essential buildings must be designed for higher seismic resistance. Table 33.2 lists the importance factors to be assigned to buildings of different categories.

33.10 STORY SHEARS AND OVERTURNING MOMENTS

Once the equivalent seismic lateral forces are determined, it is a matter of simple statics to obtain the story shear force V_x and overturning moment M_x at the base of the columns supporting floor level *x*. Considering that portion of the structure from above level $x-1$ up to level *N* as a free body yields the following relations for shear force and overturning moment:

$$V_x = F_t + \sum_{i=x}^{N} F_i \qquad (33.8)$$

$$M_x = M_{x+1} + (V_{x+1} + F_x)(h_x - h_{x-1}) \qquad (33.9)$$

Table 33.2. Values for Building Importance Factor *I*

Type of Building	*I*
1. Buildings essential to the public subsistence, e.g., schools, hospitals, fire stations, disaster mitigation centers, railway stations, airports.	1.5
2. Any building that can assemble more than 300 persons (in one room) at one time, or buildings of more than five stories.	1.25
3. All other buildings	1.0

where

$$1 \leqslant x \leqslant N - 1$$

and

$$M_N = (F_t + F_N)(h_N - h_{N-1}) \qquad (33.10)$$

where F_t is the higher mode force at the top of the building [eq.(33.2)] and F_i is the force at floor level i.

Table 33.3. Lumped Weight on Each Floor, per Frame

	Weight (kg)				Total Weight W_i
Level	Slabs	Beams	Columns	Walls	(kg)
4	17,280	4,030	580	3,830	25,720
3	17,280	4,030	1,150	9,050	31,510
2	17,280	4,030	1,840	9,050	32,200
1	17,280	4,030	3,360	12,530	37,200

$$\Sigma = 126,630$$

33.11 COMPUTER PROGRAM

A computer program that incorporates the provisions of the draft code for earthquake-resistant design of buildings in Thailand is presented in this chapter. Equivalent lateral forces, story shears, and overturning moments can be calculated and the results printed for each level of a building.

It is worth mentioning that the draft code makes no mention of torsion induced by seismic action. However, the equivalent story seismic forces computed by this program would be used as loads in a structural analysis that treated the building as a three-dimensional structure. The line of action of the seismic force at each floor should pass through the center of mass of that floor. In this manner, the torsional effects can be taken into account automatically.

Example 33.1

Fig. 33.3 shows a four-story reinforced concrete framed building that is typical of commercial/residential buildings (commonly called "shop houses") in Thailand. The beams are 200 mm wide by 400 mm deep. Column sections are 250 mm by 350 mm in the first two (lower) stories and 200 mm by 200 mm elsewhere. The dead load per unit area of floor slabs (100 mm thick) including screening and plastering is 360 kg/m². The weight of the walls, considered lumped on each floor level, is 12,530 kg for the first floor, 9,050 kg for the second and third floors, and 3,830 kg for the roof. The building is located in Zone 1 and the site soil is classified as soft.

Determine the seismic forces for this structure in accordance with the draft seismic code of Thailand.

Solution

(1) Effective weight at various floors. The dead weight of each floor for a frame width of 12 m is

$$360 \text{ kg/m}^2 \times 4 \text{ m} \times 12 \text{ m} = 17,280 \text{ kg per frame}$$

The weight of the beams (less the 100 mm portions that extend into the slabs) is

$$0.2 \text{ m} \times 0.3 \text{ m} \times 2,400 \text{ kg/m}^3 \times (12 \text{ m} + 16 \text{ m})$$
$$= 4,030 \text{ kg per frame}$$

The column unit weight is

$$0.25 \text{ m} \times 0.35 \text{ m} \times 2,400 \text{ kg/m}^3 = 210 \text{ kg/m}$$

in the first two stories, and

$$0.20 \text{ m} \times 0.20 \text{ m} \times 2,400 \text{ kg/m}^3 = 96 \text{ kg/m}$$

in the upper stories. Table 33.3 lists the weight of each level per frame.

(2) Fundamental period

$$T = 0.09 h_N / \sqrt{D} \qquad \text{[eq.(33.5)]}$$

$$h_N = 14 \text{ m}$$

$$D = 12 \text{ m}$$

Thus,

$$T = 0.09 \times 14 / \sqrt{12} = 0.36 \text{ sec}$$

(3) Base shear

$$V = ZKCSIW \qquad \text{[eq.(33.4)]}$$

$Z = 0.15$ (for site in Zone 1)
$S = 1.5$ (for soft soil foundation)
$K = 1.0$
$I = 1.0$

$$C = \frac{1}{15\sqrt{T}} \leqslant 0.12$$

$$= \frac{1}{15\sqrt{0.36}} = 0.110$$

$CS = 0.110 \times 1.5 = 0.165 > 0.14$; use $CS = 0.14$.

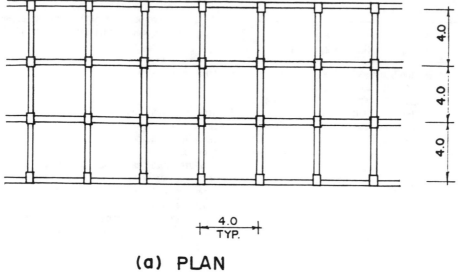

(a) PLAN

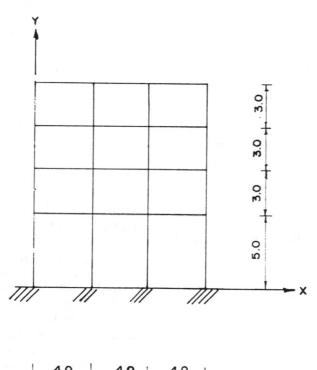

(b) ELEVATION

Fig. 33.3. Plan and elevation for the four-story building of Example 33.1

Table 33.4. Lateral Equivalent Force

Level	W_x (kg)	h_x (m)	W_xh_x (kg-m)	F_x (kg)	V_x (kg)	M_x (kg-m)
4	25,720	14	360,080	832	832	——
3	31,510	11	346,610	801	1,634	2,496
2	32,200	8	257,600	596	2,230	7,398
1	37,200	5	186,000	430	2,660	14,088
Base	——	0	——	——	——	27,388
			$\Sigma = 1,150,290$			

Thus,

$$V = 0.15 \times 1 \times 0.14 \times 1 \times 126,630$$
$$= 2,660 \text{ kg}$$

(4) Lateral seismic forces

$$F_x = \frac{(V - F_t)W_xh_x}{\sum_{i=1}^{N} W_ih_i} \qquad [\text{eq.}(33.3)]$$

$$F_t = 0 \quad \text{for} \quad T = 0.36 < 0.7 \sec$$

The calculation results are given in Table 33.4.

(5) Shear force and overturning moment at level x. The story shear force V_x and overturning moment M_x can be calculated from eqs.(33.8)–(33.10). The results are also shown in Table 33.4.

Example 33.2

Solve Example 33.1 using the computer program developed for this chapter.

```
************************************************************
*                                                          *
*         *** EARTHQUAKE  RESISTANT  DESIGN ***            *
*                                                          *
*  IN  ACCORDANCE  WITH  SEISMIC  DRAFT  CODE  OF  THAILAND *
*                                                          *
*          PANITAN  LUKKUNAPRASIT  AND  STAFF              *
*    EARTHQUAKE  ENGINEERING  AND  VIBRATION  LABORATORY    *
*              CHULALONGKORN  UNIVERSITY                   *
*                    MARCH  1992                           *
*                                                          *
************************************************************


          ***DATA FILE INFORMATION***

          1. PREPARE NEW DATA FILE

          2. MODIFY EXISTING DATA FILE

          3. USE EXISTING DATA FILE
```

```
SELECT NUMBER                              ? 3
DRIVE USED FOR DATA FILES (A:,B:,or C:)    ? C:
FILE NAME ( OMIT DRIVE LETTER )            ? EXAMFILE

=============
INPUT DATA
=============

SITE LOCATION PROVINCE IS              BANGKOK
SEISMIC ZONE FACTOR              = .15
IMPORTANCE FACTOR                = 1
K-FACTOR                         = 1
SITE-STRUCTURE RESONANCE COEFFICIENT = 1.5
TOTAL WEIGHT ( KGS )             = 126630
TOTAL HEIGHT ( M )               = 14
WIDTH OF FRAME ( M )             = 12
  STORY NO.   STORY HEIGHT   STORY WEIGHT
               ( METERS )      ( KGS )

     1           5.00          37200
     2           3.00          32200
     3           3.00          31510
     4           3.00          25720

==================================================
FINAL  RESULTS : EQUIVALENT  LATERAL  FORCES
               : SHEAR  FORCES
               : OVERTURNING  MOMENTS
==================================================

STORY NO.   LAT. FORCE     SHEAR       MOMENT
             ( KG )        ( KG )      ( KG-M )

    4          832          832         2497
    3          801         1634         7398
    2          596         2229        14086
    1          430         2659        27382
```

REFERENCES

BERGADO, D. T.; KIM, S. H.; and YAMADA, Y. (1986) "Dynamic Analysis of Ground Motion during Earthquake in Bangkok Area." pp. 105–136. Proceedings of 1st Workshop on Earthquake Engineering and Hazard Mitigation, Bangkok, Thailand.

CHANDRANGSU, S. (1986) "Earthquake Resistant Building Code for Thailand" (in Thai). pp. 25–40. Proceedings of 1st Workshop on Earthquake Engineering and Hazard Mitigation, Bangkok, Thailand.

CHOPRA, A. (1980) *Dynamics of Structures: A Primer.* Earthquake Engineering Research Institute, Berkeley, CA.

LUKKUNAPRASIT, P. (1989) "Response Spectrum for Earthquake Resistant Design of Buildings in Bangkok" (in Thai). Paper Presented at the Annual Technical Conference of the Engineering Institute of Thailand, Thailand.

—— (1990) "Considerations for the Seismic Resistant Design of Buildings" (in Thai). pp. 125–150. Proceedings of the Annual Technical Conference of the Engineering Institute of Thailand, Supplementary Volume, Thailand.

LUKKUNAPRASIT, P.; THUSNEEYANONT, S.; and KAN-JANAKAROON, W. (1986) "Assessment of Seismic Ground Motion and Response of a Building Frame in Bangkok and its Vicinity." pp. 245–255. Proceedings of 1st Workshop on Earthquake Engineering and Hazard Mitigation, Bangkok, Thailand.

YENSUANG, S. (1990) "Earthquakes" (in Thai). pp. 84–89. Proceedings of the Annual Technical Conference of the Geography Association of Thailand, Thailand.

34

Turkey

Turan Durgunoğlu*

34.1 INTRODUCTION

Turkey is located in a region that has suffered frequent and intense seismic activity.[1] Seismic-resistant design is very important in order to prevent the kind of destruction that has occurred in cities and towns of the country.

The official code for earthquake-resistant design in Turkey is entitled *Specifications for Structures to be Built in Disaster Areas* (1975), a publication of the Turkish Government Ministry of Reconstruction and Resettlement, Earthquake Research Institute. This code has been in use since 1975 and is the current code for seismic design of buildings in the country. An updated version is in preparation, but is currently available only in draft form.

The material presented in this chapter, and the accompanying program, are based on the code of 1975. Parts I and II of the 1975 code are devoted, respectively, to general provisions and to design for protection against flood and fire. Part III deals with design for resistance to earthquakes.

34.2 GENERAL CONSIDERATIONS

The analysis and design of structures to resist earthquakes are based on a set of lateral static forces applied at the various levels of the building. The lateral forces are assumed to act independently first along one main axis of the building and then along the other axis. Where the principal axes of vertical resisting elements do not coincide with the main axes of the building, the possibility of very unfavorable conditions due to eccentric loading must be investigated. The lateral static forces stipulated by the code shall be considered minimum equivalent seismic forces applied to the

---

*The author would like to thank Mr. Mutlu Koyluoglu, graduate research assistant of Bogazici University, and Mr. Fatih Kulac, project engineer of ZETAS Earth Technology Corporation, for their contribution in preparation of this chapter.

[1]A list of the major earthquakes in Turkey since 1900 is given in Appendix A34 of this chapter.

entire structure. In the design of resisting elements of the building, it is not required to assume earthquake and wind loading acting simultaneously; the more unfavorable of these two loading conditions will control the final design.

Structures are classified by the code as regular or irregular. Regular structures are those in which the vertical resisting elements, columns and shear walls, extend continuously through the height of the building down to the foundation level. The static method of analysis in the code, based on equivalent lateral forces, is applicable only to regular structures with a clear height above the base level not exceeding 75 m. Irregularity in structures could be due to mass or stiffness irregularities either in plan or along the height of the building. The code requires all other structures to be designed using an appropriate dynamic method. Such dynamic analysis shall be based on the dynamic properties of both the structure and the underlying soil. The seismic responses may be found by using the modal superposition method in conjunction with real or idealized response spectra, by time integration of the pertinent equations of motion, or through the analysis of experimental results obtained from an appropriate model. The code stipulates that the total lateral force (base shear force) obtained from a dynamic method of analysis should not be less than 70% of the lateral force obtained by using the static method of analysis.

34.3 SEISMIC LATERAL FORCES

The total equivalent static force or base shear force V is given by

$$V = CW \qquad (34.1)$$

where W is the total weight of the building calculated as the sum of the weight of the various levels; that is,

$$W = \sum_{i=1}^{N} W_i \qquad (34.2)$$

with the weight W_i at level i determined as

$$W_i = G_i + \psi P_i \qquad (34.3)$$

in which

G_i = total dead load at level i
P_i = total live load at level i
ψ = live load factor from Table 34.1

and C is the seismic coefficient defined as

$$C = C_0 KSI \qquad (34.4)$$

Table 34.1. Live Load Factor (ψ)

Type of Structure	ψ
Warehouses, depots, etc.	0.80
Schools, student housing buildings, stadiums, cinemas, concert halls, garages, restaurants, commercial establishments, etc.	0.60
Private dwellings, hotels, hospitals, office buildings, etc.	0.30

Table 34.2. Seismic Zone Coefficient (C_0)

Seismic Zone	C_0
1	0.10
2	0.08
3	0.06
4	0.03

in which

C_0 = seismic zone coefficient
K = structural coefficient
S = spectral coefficient
I = structural importance coefficient.

34.3.1 Seismic Zone Coefficient C_0

The seismic zone coefficient C_0 is obtained from Table 34.2 on the basis of Seismic Zones 1 through 4. (See the seismic zone map of Turkey, Fig. 34.1.)

34.3.2 Structural Coefficient K

Values of the structural coefficient K are given in Table 34.3 for different types of buildings. The code stipulates that the coefficient K shall be not less than 1.0 for one- or two-story buildings.

34.3.3 Structural Importance Coefficient I

The structural importance coefficient I is given in Table 34.4. The value of this coefficient is based on the use of the building and on the need to maintain the functioning of governmental and emergency facilities after a disaster.

34.3.4 Spectral Coefficient S

The spectral coefficient S in eq.(34.4) is calculated by the following formula:

$$S = \frac{1}{|0.8 + T - T_0|} \leq 1.0 \qquad (34.5)$$

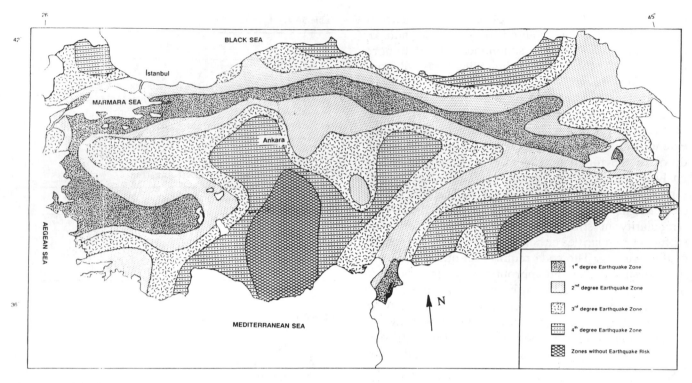

Fig. 34.1.   Seismic zone map of Turkey

Table 34.3.   Structural Coefficient (K)

Structural Type		K
All building framing systems not otherwise classified		1.00
Buildings with box systems with shear walls		1.33
Buildings with frame systems where the frame resists the total lateral force for filler wall types a, b, and c*		
1. Ductile moment resisting frames†	a)	0.60
(steel or reinforced concrete)	b)	0.80
	c)	1.00
2. Nonductile moment-resisting frames	a)	1.20
	b)	1.50
	c)	1.50
3. Steel space frames with diagonal bracing	a)	1.33
	b)	1.50
	c)	1.60
Shear wall systems with ductile frames capable of resisting at least 25% of the total lateral forces	a)	0.80
	b)	1.00
	c)	1.20
Masonry buildings		1.50
Elevated tanks not supported by a building		3.00
Structures other than buildings, towers, and chimney stacks		2.00

*Filler wall types:
 a) Reinforced concrete or partition walls of masonry blocks with horizontal and vertical reinforcement.
 b) Unreinforced masonry partition walls.
 c) Light and sparse partition walls or prefabricated concrete partition walls.
 †Ductile moment-resisting frames are those structural frames designed and constructed with the potential capacity to sustain loads and dissipate energy in the inelastic range of deformations.

where

T = fundamental period of the structure (in sec)
T_0 = predominant period of site (sec).

The code indicates that the value of S in eq.(34.4) need not be larger than 1.0, and that S shall be taken as 1.0 for masonry buildings. For different values of the predominant period of the site T_0, Fig. 34.2 shows a set of curves for the spectral coefficient S as a function of the period T [eq.(34.5)]. The value of T_0 should be selected from Table 34.5 based on soil/rock

Table 34.4.   Structural Importance Coefficient (I)

Structure Type	I
(a) Structures and buildings to be used during or immediately after an earthquake (post office, fire stations, broadcasting buildings, power stations, hospitals, stations and terminals, refineries, etc.)	1.50
(b) Buildings housing valuable and important items (museums, etc.)	1.50
(c) Buildings and structures of high occupancy (schools, stadiums, theaters, cinemas, concert halls, religious temples, etc.)	1.50
(d) Buildings and structures of low occupancy (private dwellings, hotels, office buildings, restaurants, industrial structures, etc.)	1.00

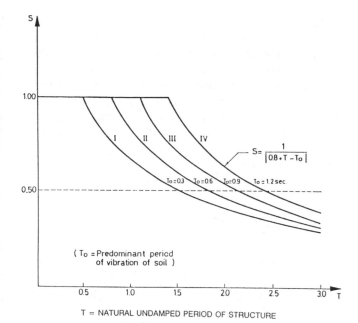

Fig. 34.2. Spectral coefficient in function of natural period for various types of soils defined by the predominant period of the soil

Table 34.5. Predominant Period T_0 for Soils Described in Table 34.6

Soil/Rock Class	Soil/Rock Types	T_0 Predominant Period of Site (sec)	T_0 Average (sec)
I	a	0.20	
	b	0.25	0.25
	c	0.30	
II	a	0.35	
	b	0.40	0.42
	c	0.50	
III	a	0.55	
	b	0.60	0.60
	c	0.65	
IV	a	0.70	
	b	0.80	0.80
	c	0.90	

class and soil/rock type as described in Table 34.6, unless it is determined by experimental, empirical, or theoretical principles based on valid assumptions and geological observations. Values obtained from Table 34.5 are valid only for the case where the layer of soil is directly above the bedrock or another rock-like formation, and has a thickness on the order of 50 m. Where the thickness of the layer of soil is greatly different from 50 m, the values of the shear wave velocity V_s of the soil stratum (in m/sec) and the thickness of the top layer soil stratum H_z (in m) shall be determined more accurately by field tests, empirical equations, or theoretical methods. For the latter

Table 34.6. Soil Classifications for Determination of Predominant Period T_0 in Table 34.5

Soil/Rock Class	Soil/Rock Types	N Blows per foot, Standard Penetration Test	D_r Relative Density (%)	q_u Unconfined Compressive Strength (kg/cm²)	V_s Shear Wave Velocity (m/sec)
I	(a) Massive and deep volcanic rocks; undecomposed, sound metamorphic rocks; very stiff cemented sedimentary rocks	NA	NA	NA	>700
	(b) Very dense sand	>50	85–100	NA	
	(c) Very stiff clay	>32	NA	>4.0	
II	(a) Loose magmatic rocks such as tuff or agglomerate, decomposed sedimentary rocks with planes of discontinuity	NA	NA	NA	400–700
	(b) Dense sand	30–50	65–85	NA	
	(c) Stiff clay	16–32	NA	2.0–4.0	
III	(a) Decomposed metamorphic rocks and soft, cemented sedimentary rocks with planes of discontinuity	NA	NA	NA	200–400
	(b) Medium-dense sand	10–30	35–64	NA	
	(c) Silty clay	8–16	NA	1.0–2.0	
IV	(a) Soft and deep alluvial layers with a high water-table, marshlands or ground recovered from sea by mud filled, all fill layers	NA	NA	NA	<200
	(b) Loose sand	0–10	≤35	NA	
	(c) Clay, silty clay	0–8	NA	≤1.0	

NA = Not Applicable.

situation, the value of T_0 shall be calculated by the equation

$$T_0 = \frac{4H_z}{V_s} \tag{34.6}$$

An accurate solution of such soil-related problems as selection of foundation type, determination of bearing capacity and settlements, as well as a realistic determination of the predominant period of vibration of the soil layer, requires an appropriate seismic exploration with laboratory experiments of the soil layer. Such an exploration shall be carried out for the following structures:

(a) Buildings having a height of more than 75 meters above foundation level
(b) Industrial structures with large spans, and buildings such as theaters, cinemas, etc.
(c) Towers, chimney stacks, elevated tanks, etc.

Where the value of V_s cannot be determined accurately for use in eq.(34.6), the values for V_s given in Table 34.6 may be used. Where the underlying soil consists of a number of layers with different values of V_s, a separate value of T_0 shall be calculated for each and every layer. Soils that have a V_s value larger than 700 m/sec shall be assumed to be very sound; layers below that very sound layer need not be considered.

34.4 FUNDAMENTAL PERIOD

Unless obtained from experiments or by theoretical methods on the basis of valid assumptions, the value of the natural period T shall be calculated by both of the approximate relations that follow. The less favorable value of T given by eq.(34.7) or eq.(34.8) shall be used in eq.(34.5).

$$T = \frac{0.09H}{\sqrt{D}} \tag{34.7}$$

$$T = \lambda N \tag{34.8}$$

in which

H = height of structure above foundation level (in m)
D = dimension of building in a direction parallel to the applied lateral forces (m)
N = number of stories above foundation level
λ = coefficient determined by interpolation between the values of 0.07 and 0.10 according to the degree of general structural flexibility of the building ($\lambda = 0.07$ pertains to very rigid buildings; $\lambda = 0.10$ pertains to very flexible structures).

Equations (34.7) and (34.8) shall not apply to structures with large spans such as industrial buildings, cinemas, sports halls, and stadiums, or to buildings with regular bearing systems with a height of more than 34.0 m above foundation level such as chimney stacks, towers, and elevated tanks. The natural periods of such structures shall be calculated by a rigorous dynamic analysis where the properties of the soil and of the structure (soil-structure interaction) are taken into consideration.

34.5 VERTICAL DISTRIBUTION OF LATERAL FORCES

The base shear force V given by eq.(34.1) shall be distributed as lateral static forces F_i applied at the various levels of the building, according to the following relationships:

$$F_i = (V - F_t) \frac{W_i h_i}{\sum\limits_{j=1}^{N} W_j h_j} \tag{34.9}$$

and

$$F_t = 0 \quad \text{for} \quad \left(\frac{H}{D}\right) \leqslant 3$$

$$F_t = 0.004V \left(\frac{H}{D}\right)^2 \leqslant 0.15V \quad \text{for} \quad \left(\frac{H}{D}\right) > 3 \tag{34.10}$$

in which

W_i = seismic weight at level i
h_i = height from the base of the building to level i
F_t = additional force applied at the top of the building
H = total height of the building
D = dimension of the building in the direction of the seismic forces.

34.6 OVERTURNING MOMENTS

Overturning moments are determined by statics as the moments resulting from the seismic design forces F_i and F_t [eqs.(34.9) and (34.10)] that act on levels above the level under consideration. Hence, the overturning moment M_i at level i of the building is given by

$$M_i = F_t(h_N - h_i) + \sum_{j=i+1}^{N} F_j(h_j - h_i) \tag{34.11}$$

where $i = 0, 1, 2, \ldots, N-1$.

34.7 TORSIONAL MOMENTS

The code requires that buildings be designed to resist horizontal torsional moments M_t due to the eccentricity or distance between the center of mass and the centers of stiffness of any floor, in addition to an accidental eccentricity e_0 of 5% of the largest plan dimension of the building perpendicular to the direction of the applied lateral forces.

34.8 APPURTENANCES AND/OR PARTS OF BUILDINGS

Earthquake loads acting on appurtenances and/or parts of buildings such as parapet walls, chimneys, cantilever elements, and balconies shall be calculated separately. In these calculations, the coefficient C as determined for the structure [eq.(34.4)], shall be increased threefold, and the lateral load V determined by eq.(34.1) shall be assumed to act at the center of mass of the appurtenance or element in the most unfavorable direction.

34.9 ALLOWABLE STRESSES

In the seismic-resistant design of members, the allowable stresses for concrete and steel may be increased by not more than 33% of the allowable values for static design. In reinforced concrete structures, an increase in bond stresses shall not be permitted. In steel structures, allowable stresses for all connections and joints shall not exceed the values for increased allowable stresses. The same requirement shall apply to the design of diagonal wind bracing and stability members. Whenever earthquake effects are considered, the allowable bearing pressures for subsoils may be increased by not more than 33% for Soil/Rock Classes I, II and III. No such increase shall be permitted for Class IV soils. Where the top foundation layer is of Class II, III, or IV, possible total settlements and/or differential settlements due to seismic vibrations should be determined, in addition to those settlements due to static loads. No increase in allowable stresses for concrete and reinforcing steel shall be permitted in foundations bearing directly on Class IV soils.

When designing retaining walls and sheet-pile walls with heights in excess of 6.00 m, the characteristics of the soil shall be determined by appropriate laboratory and field testing. In the calculation of earth pressure, the angle of shearing resistance shall be decreased by 6° in Seismic Zones 1 and 2, and by 4° in Seismic Zones 3 and 4.

34.10 EXAMPLE 34.1

A five-story symmetrical reinforced concrete building is modeled as a plane frame (Fig. 34.3) with rigid horizontal diaphragms at the floor and top levels (shear building). The dead load is 131.90 tonnes at the first floor; 121.39 tonnes on the second, third, and fourth levels; and 115.08 tonnes on the fifth level. The live load is assumed to be 44.36 tonnes on all levels of the building, except at the roof level where it is zero. The underlying soil is medium-dense sand at a site in Seismic Zone 1 (Fig. 34.1). The building is to be used for governmental offices in which post-earthquake functioning is considered necessary.

Solution

1. Seismic weights

$$W_i = G_i + \psi P_i \qquad \text{[eq.(34.3)]}$$

$\psi = 0.3$ (From Table 34.1, office building)

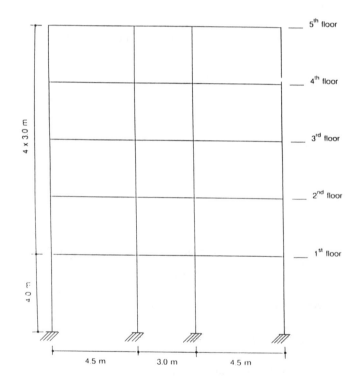

	Elevation Above Foundation Base (m.)	Total Dead Load (tons)	Total Live Load (tons)
1st floor	4.00	131.90	44.36
2nd floor	7.00	121.39	44.36
3rd floor	10.00	121.39	44.36
4th floor	13.00	121.39	44.36
5th floor	16.00	115.08	0.00

Fig. 34.3. Modeled building for Examples 34.1 and 34.2

Table 34.7. Forces, Moments and Displacements

Level i	Seismic Weight W_i (tonnes)	Force F_i (tonnes)	Story Shear V_i (tonnes)	Overturning Moment M_i (tonne-m)	Torsional Moment M_{ti} (tonne-m)	Story Drift Δ_i (mm)	Lateral Displacement δ_i (mm)
5	115.1	28.4	28.4	——	28.40	3.40	58.41
4	134.7	27.0	55.4	85.2	55.40	6.64	55.01
3	134.7	20.8	76.2	251.3	76.17	9.14	48.37
2	134.7	14.5	90.7	479.8	90.77	10.88	39.23
1	145.2	9.0	99.7	751.9	99.67	28.35	28.35
0				1,150.6			

SUM = 664.4.

Table 34.7 shows the values calculated for the seismic weights, W_i.

2. Seismic zone coefficient

$C_0 = 0.10$ (From Table 34.2, Seismic Zone 1)

3. Structural coefficient

$K = 1.0$ (From Table 34.3, framed system)

4. Structural importance coefficient

$I = 1.5$ (from Table 34.4)

5. Natural period

$$T = \frac{0.09H}{\sqrt{D}} = \frac{0.09 \times 16}{\sqrt{12}} = 0.42 \text{ sec} \quad \text{[eq.(34.7)]}$$

or

$$T = \lambda N = 0.1 \times 5 = 0.5 \text{ sec} \quad \text{[eq.(34.8)]}$$

$$\text{(Assume } \lambda = 0.1)$$

Select most unfavorable value: $T = 0.42$ sec.

6. Predominant period of soil

$T_0 = 0.6$ (from Tables 34.5 and 34.6 for Class III, Type b)

7. Spectral coefficient

$$S = \frac{1}{|0.8 + T - T_0|} \leqslant 1.0 \quad \text{[eq.(34.5)]}$$

$$S = \frac{1}{0.8 + 0.42 - 0.6} = 1.61$$

$$= 1.0$$

8. Seismic coefficient

$$C = C_0 KSI \quad \text{[eq.(34.4)]}$$

$$= 0.10 \times 1.0 \times 1.0 \times 1.5 = 0.15$$

9. Base shear force

$$V = CW \quad \text{[eq.(34.1)]}$$

$$W = 664.4 \text{ T} \quad \text{(from Table 34.7)}$$

$$V = 0.15(664.4) = 99.7 \text{ T}$$

10. Distribution of lateral seismic forces

$$F_t = 0 \quad \text{for} \quad \frac{H}{D} = \frac{16}{12} = 1.33 < 3 \quad \text{[eq.34.10)]}$$

$$F_i = (V - F_t) \frac{W_i h_i}{\sum\limits_{j=1}^{N} W_j h_j} \quad \text{[eq.(34.9)]}$$

Table 34.7 shows the calculated seismic forces F_i for the various levels of the building.

11. Story shear force

$$V_i = \sum_{j=i}^{N} F_j$$

The calculated values are given in Table 34.7.

12. Overturning moments

$$M_i = F_N(h_N - h_i) + \sum_{j=i+1}^{N} F_j(h_j - h_i) \quad \text{[eq.(34.11)]}$$

where $i = 0, 1, 2, \ldots, N-1$.

13. Torsional moments

$$M_{ti} = Ve_0 \quad \text{(accidental eccentricity)}$$

$$e_0 = 0.05D = 0.05 \times 12 \text{ m} = 0.6 \text{ m}$$

Calculated values of story shear force, overturning moment, and torsional moment are shown in Table 34.7.

14. Story drift. The interstory drift Δ_i, for a structure modeled as a shear building, is given by

$$\Delta_i = \frac{V_i}{k_i}$$

where

V_i = story shear force
k_i = story stiffness

$$k_i = \frac{12(EI)_i}{L_i^3}$$

$(EI)_i$ = total flexural stiffness of the story i
L_i = interstory height.

15. Lateral displacements. The lateral displacements δ_i of the building are calculated as

$$\delta_i = \sum_{j=1}^{i} \Delta_j$$

Table 34.7 shows the calculated values of story drift and lateral displacement.

34.11 COMPUTER PROGRAM

A computer program has been developed to implement the provisions of the Turkish seismic code of 1975. The program has provisions to either implement eqs.(34.7) and (34.8) to estimate the fundamental period of the structure or to input this value when it has been predetermined. The program calculates at each level of the building the seismic forces F_i, story shear forces V_i, overturning moments M_i, torsional moments M_{ti}, story drifts Δ_i, and lateral displacements δ_i.

Example 34.2

Use the computer program prepared for this chapter to solve Example 34.1.

34.11.1 Input Data and Output Results for Example 34.2

```
INPUT DATA:

ZONE NUMBER                          NZ= 1
OCCUPANCY IMPORTANCE FACTOR          I= 0
STRUCTURAL FACTOR                    K= 1
SOIL PROFILE TYPE                    NIS= 3
FACTOR LAMBDA                        LAMB= .1
LIVE LOAD FACTOR                     LX= .3
```

```
BUILDING DATA:

BUILDING DIMENSION, NORMAL TO FORCES (M)    DM= 20
BUILDING DIMENSION, FORCE DIRECTION (M)     D= 12
NUMBER OF STORIES                           N= 5
```

STORY # I	HIGHT (M)	FLEXURAL STIFFNESS (T-M2)	DEAD LOAD (T)	LIVE LOAD (T)
5	16.00	18750.00	115.10	0.00
4	13.00	18750.00	121.39	44.36
3	10.00	18750.00	121.39	44.36
2	7.00	18750.00	121.39	44.36
1	4.00	18750.00	131.90	44.36

```
OUTPUT RESULTS:

SEISMIC ZONE FACTOR    Z= .1
FUNDAMENTAL PERIOD     T= .4156923
SOIL PROFILE TYPE      NIS= 3
SEISMIC FACTOR         C= .15
DYNAMIC COEFFICIENT    S= 1

TOTAL BASE SHEAR       V= 99.6603

ASSUME ONLY ACCIDENTAL ECCENTRICITY (Y/N) ? Y
```

DISTRIBUTION LATERAL FORCES, MOENTS AND DISPLACEMENTS

LEVEL	SEISMIC WEIGHT (T)	LATERAL FORCE (T)	SHEAR FORCE (T)	OVERTURNNING MOMENT(T-M)	TORSIONAL MOMENT(T-M)	STORY DRIFT(mm)	LATERAL DISPL.(mm)
5	115.10	28.40	28.40	—	28.396	3.408	58.427
4	134.70	27.00	55.40	85.19	55.396	6.648	55.020
3	134.70	20.77	76.17	251.38	76.166	9.140	48.372
2	134.70	14.54	90.70	479.87	90.704	10.885	39.232
1	145.21	8.96	99.66	751.99	99.660	28.348	28.348
0				1150.63			
SUM= 664.40							

Example 34.3

Use the computer program developed from the 1975 Turkish code for seismic-resistant design to analyze the structure shown in Fig. 34.4.

34.11.2 Input Data and Output Results for Example 34.3

```
INPUT DATA:

ZONE NUMBER                          NZ= 1
OCCUPANCY IMPORTANCE FACTOR          I= 0
STRUCTURAL FACTOR                    K= 1
SOIL PROFILE TYPE                    NIS= 3
FACTOR LAMBDA                        LAMB= .1
LIVE LOAD FACTOR                     LX= .3
```

```
BUILDING DATA:

BUILDING DIMENSION, NORMAL TO FORCES (M)    DM= 26
BUILDING DIMENSION, FORCE DIRECTION (M)     D= 13
NUMBER OF STORIES                           N= 14
```

STORY # I	HIGHT (M)	FLEXURAL STIFFNESS (T-M2)	DEAD LOAD (T)	LIVE LOAD (T)
14	41.20	18750.00	40.00	18.00
13	38.30	18750.00	44.00	28.00
12	35.40	18750.00	44.00	28.00
11	32.15	18750.00	44.00	28.00

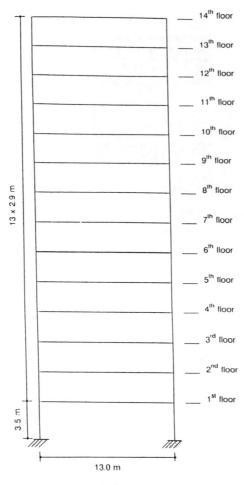

	Elevation Above Foundation Base (m.)	Total Dead Load (tons)	Total Live Load (tons)
1st floor	3.50	44.0	28.0
2nd floor	6.40	44.0	28.0
3rd floor	9.30	44.0	28.0
4th floor	12.20	44.0	28.0
5th floor	15.10	44.0	28.0
6th floor	18.00	44.0	28.0
7th floor	20.90	44.0	28.0
8th floor	23.80	44.0	28.0
9th floor	26.70	44.0	28.0
10th floor	29.60	44.0	28.0
11th floor	32.15	44.0	28.0
12th floor	35.40	44.0	28.0
13th floor	38.30	44.0	28.0
14th floor	41.20	40.0	18.0

Fig. 34.4. Modeled building for Example 34.3

```
10      29.60      18750.00      44.00      28.00
 9      26.70      18750.00      44.00      28.00
 8      23.80      18750.00      44.00      28.00
 7      20.90      18750.00      44.00      28.00
 6      18.00      18750.00      44.00      28.00
 5      15.10      18750.00      44.00      28.00
 4      12.20      18750.00      44.00      28.00
 3       9.30      18750.00      44.00      28.00
 2       6.40      18750.00      44.00      28.00
 1       3.50      18750.00      44.00      28.00


OUTPUT RESULTS:

SEISMIC ZONE FACTOR   Z= .1

FUNDAMENTAL PERIOD    T= 1.028414
SOIL PROFILE TYPE  NIS= 3
SEISMIC FACTOR        C= .1221087
DYNAMIC COEFFICIENT   S= .8140577

TOTAL BASE SHEAR      V= 88.72417

ASSUME ONLY ACCIDENTAL ECCENTRICITY (Y/N) ? Y
```

```
            DISTRIBUTION LATERAL FORCES, MOENTS AND DISPLACEMENTS

LEVEL SEISMIC    LATERAL    SHEAR    OVERTURNNING TORSIONAL   STORY      LATERAL
      WEIGHT (T) FORCE (T) FORCE (T) MOMENT(T-M)  MOMENT(T) DRIFT(mm)  DISPL.(mm)

14     45.40     13.46*     13.46        0.00       17.504    1.460     100.665
13     52.40     10.62      24.09       39.05       31.314    2.611      99.205
12     52.40      9.82      33.91      108.90       44.077    5.173      96.594
11     52.40      8.92      42.82      219.10       55.669    3.156      91.421
10     52.40      8.21      51.03      328.29       66.342    5.532      88.265
 9     52.40      7.41      58.44      476.29       75.969    6.334      82.734
 8     52.40      6.60      65.04      645.76       84.550    7.050      76.399
 7     52.40      5.80      70.84      834.37       92.086    7.678      69.350
 6     52.40      4.99      75.83     1039.79       98.576    8.219      61.671
 5     52.40      4.19      80.02     1259.69      104.020    8.673      53.452
 4     52.40      3.38      83.40     1491.73      108.419    9.320      44.779
 3     52.40      2.58      85.98     1733.59      111.772    9.320      35.739
 2     52.40      1.78      87.75     1982.93      114.079    9.512      26.419
 1     52.40      0.97      88.72     2237.41      115.341   16.907      16.907
 0                                    2547.95
_____
SUM= 726.6       * Including additional top force = 3.56
```

REFERENCES

Turkish Government Ministry of Reconstruction and Resettlement (1972) *Seismic Zone Map of Turkey*. Ankara, Turkey.

——— (1975) *Specification for Structures to be Built in Disaster Areas*. Ankara, Turkey.

APPENDIX A.34 TURKISH EARTHQUAKE CATALOG (1900–1990) (KANDILLI OBSERVATORY/BOGAZICI UNIVERSITY)

Region Bounded: lat. 33°N–45°N
long. 23°E–48°E

Min. Magnitude: 6.0
Min. Acc. % g: 0.1

No.	Day	Month	Year	Latitude	Longitude	Magnitude
1	28	4	1903	39.1	42.5	6.3
2	20	1	1905	39.6	23	6.0
3	4	12	1905	39	39	6.8
4	28	9	1906	40.5	42.7	6.2
5	28	12	1906	40.5	42	6.0
6	17	2	1908	37.4	35.8	6.0
7	17	5	1908	35.5	24	6.7
8	28	9	1908	38	44	6.0
9	9	2	1909	40	38	6.3
10	25	6	1910	41	34	6.2
11	20	4	1913	41.91	44.32	6.1
12	24	1	1916	40.27	36.83	7.1
13	20	8	1917	40.3	25.43	6.0
14	16	7	1918	36.08	26.99	6.1
15	29	9	1918	35.2	34.7	6.5
16	18	11	1919	39.26	26.71	7.0
17	11	8	1922	35.36	27.7	6.5
18	13	8	1922	35.51	27.98	6.9
19	13	9	1924	39.96	41.94	6.8
20	9	1	1925	41.33	43.41	6.0
21	1	3	1926	37.15	29.61	6.4
22	18	3	1926	35.99	30.13	6.8
23	26	6	1926	36.75	26.98	7.7
24	5	7	1926	36.52	26.69	6.2
25	22	10	1926	40.94	43.88	6.0
26	5	6	1927	36	31	6.2
27	31	3	1928	38.18	27.8	6.5
28	14	4	1928	42.34	26.02	6.8
29	2	5	1928	39.64	29.14	6.1
30	18	5	1929	40.2	37.9	6.1
31	27	3	1929	36.4	26.54	6.2
32	6	5	1930	37.98	44.48	7.6
33	8	5	1930	37.97	45	6.3
34	26	9	1932	40.39	23.81	7.1
35	29	9	1932	40.83	23.46	6.4
36	19	7	1933	38.19	29.79	6.0
37	9	11	1934	36.63	25.77	6.3
38	4	1	1935	40.4	27.49	6.4
39	4	1	1935	40.3	27.45	6.3
40	18	3	1935	35.33	26.83	6.5
41	1	5	1935	40.09	43.22	6.0
42	19	4	1938	39.44	33.79	6.6
43	22	9	1939	39.07	26.94	6.6
44	26	12	1939	39.8	39.51	7.9
45	7	5	1940	42.03	43.71	6.1
46	30	7	1940	39.64	35.25	6.2
47	20	12	1940	39.11	39.2	6.0
48	23	5	1941	37.07	28.21	6.0
49	8	11	1941	39.74	39.5	6.0
50	13	12	1941	37.13	28.06	6.5
51	21	6	1942	36.12	27.2	6.4
52	15	11	1942	39.55	28.58	6.1
53	11	12	1942	40.76	34.83	6.1
54	20	6	1943	40.85	30.51	6.5
55	16	10	1943	36.45	27.94	6.6
56	26	11	1943	41.05	33.72	7.2
57	1	2	1944	41.41	32.69	7.2
58	27	5	1944	36.23	27.25	6.2
59	25	6	1944	38.74	29	6.2
60	17	7	1944	35.91	42.55	6.1
61	6	10	1944	39.48	26.56	6.8
62	20	3	1945	37.11	35.7	6.0
63	2	9	1945	34.43	28.61	6.3
64	26	10	1945	41.54	33.29	6.0
65	16	7	1946	34.2	25.65	6.0
66	9	2	1948	35.41	27.2	7.1
67	23	7	1949	38.57	26.29	6.6
68	17	8	1949	39.57	40.62	7.0
69	13	8	1951	40.88	32.87	6.9
70	12	6	1952	34.67	26.56	6.6
71	17	12	1952	34.47	24.22	6.6
72	18	3	1953	39.99	27.36	7.2
73	7	9	1953	41.09	33.01	6.4
74	10	9	1953	34.8	32.5	6.3
75	16	7	1955	37.65	27.26	6.8
76	20	2	1956	39.89	30.49	6.4
77	9	7	1956	36.69	25.92	7.4
78	9	7	1956	36.59	25.86	6.5
79	30	7	1956	35.89	26.01	6.2
80	24	4	1957	36.43	28.63	6.8
81	25	4	1957	36.42	28.68	7.1
82	16	5	1957	40.67	31	7.1
83	15	9	1959	34.86	25.9	6.6
84	15	4	1960	40.5	42	6.0
85	23	5	1961	36.7	28.49	6.6
86	28	4	1962	36.03	26.87	6.0
87	16	7	1963	43.27	41.57	6.4
88	18	9	1963	40.77	29.12	6.3
89	14	6	1964	38.13	38.51	6.0
90	6	10	1964	40.3	28.23	7.0
91	9	3	1965	39.34	23.82	6.3
92	19	8	1966	39.17	41.56	6.9
93	20	8	1966	39.42	40.98	6.2
94	20	8	1966	39.16	40.7	6.1
95	4	3	1967	39.25	24.6	6.5
96	22	7	1967	40.67	30.69	7.2
97	26	7	1967	39.54	40.38	6.2
98	19	2	1968	39.4	24.94	7.2
99	3	9	1968	41.81	32.39	6.5
100	25	3	1969	39.25	28.44	6.0
101	28	3	1969	38.55	28.46	6.5
102	12	6	1969	34.43	25.04	6.2
103	28	3	1970	39.21	29.51	7.2
104	22	5	1971	38.85	40.52	6.8
105	4	5	1972	35.15	23.56	6.3
106	27	3	1975	40.45	26.12	6.7
107	6	9	1975	38.51	40.77	6.6
108	24	11	1976	39.05	44.04	6.1
109	11	9	1977	34.95	23.05	6.0
110	20	6	1978	40.78	23.24	6.4
111	25	2	1981	38.17	23.12	6.3
112	4	3	1981	38.24	23.26	6.4
113	19	12	1981	39.22	25.25	7.2
114	27	12	1981	38.91	24.92	6.5
115	18	1	1982	39.96	24.39	6.9
116	18	1	1982	40.03	24.56	6.8
117	6	8	1983	40.14	24.75	6.9
118	30	10	1983	40.35	42.18	6.8
119	7	12	1988	40.96	44.16	6.7

35

Union of Soviet Socialist Republics (USSR) Currently known as Commonwealth of Independent States [CIS]

Vladimir N. Alekhin

35.1 INTRODUCTION

The official code for seismic-resistant design currently used in the republics of the former USSR was promulgated in June 1981 and published in Part II of "Basic Norms and Rules for Civil and Structural Engineering" [SNIP II-7-81 (1982)]. This code includes provisions for the seismic-resistant design of buildings and other structural systems such as highway bridges and hydraulic structures. Also, it recognizes that seismic analysis and design of buildings and other structures require information on the soil condition at the site, the structural characteristics of the building, and the seismic history of the region.

Methods of seismic analysis generally are based on

dynamic analysis of the structure as an elastic system in which the magnitude of the design acceleration of the ground is set at a lower value than the peak acceleration values recorded for major earthquakes that occurred in the region. This approach provides an adequate margin of safety for buildings and occupants according to studies of structures subjected to strong-motion earthquakes. Structures successfully resist earthquakes of magnitudes much larger than those assumed for design because the ductility of the building enables the structure to dissipate energy while deforming well beyond the elastic range.

The presentation in this chapter focuses on provisions for determining equivalent seismic forces and for predicting the response of buildings subjected to seismic excitation. It is not intended as a comprehensive description of all the regulations of the SNIP II-7-81 code.

The seismic-resistant design of structures requires consideration of the following factors:

- the earthquake intensity for the region
- the soil condition at the site
- natural frequencies and corresponding modal deformation shapes
- equivalent seismic loads and their combination with other loads
- resultant seismic forces in the structural members.

The design earthquake intensity should be based on the seismic activity of the region in terms of the magnitude of strong motion of earthquakes and their expected recurrence. Based on these considerations, the SNIP II-7-81 code provides seismic zone maps for

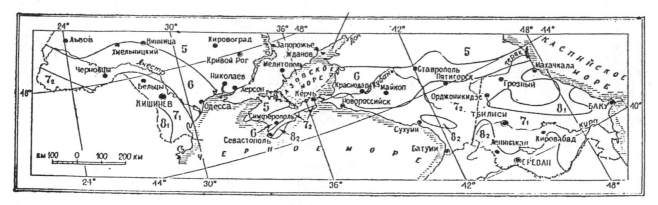

Fig. 35.1.   Seismic zone map 1 of the CIS territory (former USSR)

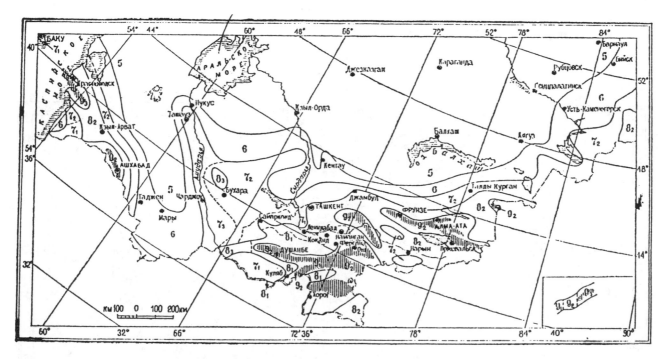

Fig. 35.2.   Seismic zone map 2 of the CIS territory (former (USSR)

the former USSR. These are reproduced in Figs. 35.1 through 35.5.

The code recommends that the building be modeled as one of the following types of structural systems:

1. Multistory frame with rigid joints in the transverse and longitudinal directions
2. Multistory frame with rigid joints in one direction and braced in the other direction
3. Multistory braced building.

The dynamic analysis may consider the building either as a cantilever structure fixed at the top of the foundation or as a two- or three-dimensional frame. The seismic code for the former USSR stipulates that the dynamic analysis must include no less than the first three modes of vibration.

35.2 LATERAL SEISMIC FORCES

The code for construction in the seismically active regions of the country establishes that the weight of the building and other loads should be applied at the floor levels of the building; equivalent lateral seismic forces for each mode of vibration should be applied at these levels, as shown in Fig. 35.6.

The lateral seismic force S_{ik} in the direction assumed for the earthquake action, corresponding to mode i at level k of the building, is calculated by the following formula:

$$S_{ik} = K_1 K_2 S_{oik} \qquad (35.1)$$

where K_1 is a coefficient that reflects the allowable damage of the building as described in Table 35.1, K_2

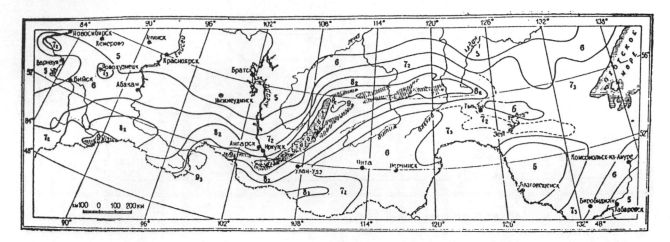

Fig. 35.3.   Seismic zone map 3 of the CIS territory (former USSR)

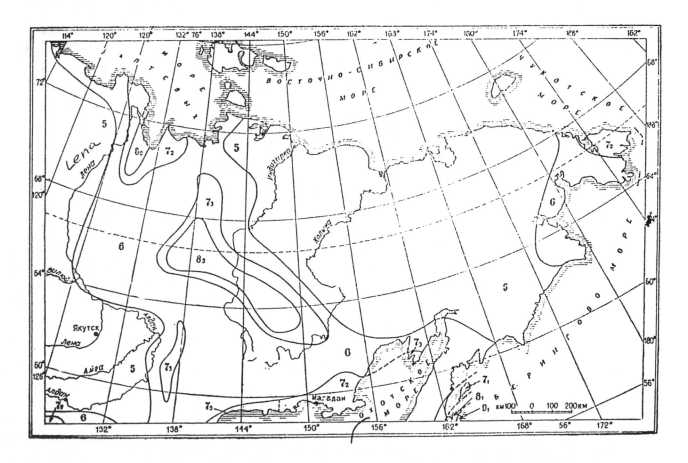

Fig. 35.4.   Seismic zone map 4 of the CIS territory (former USSR)

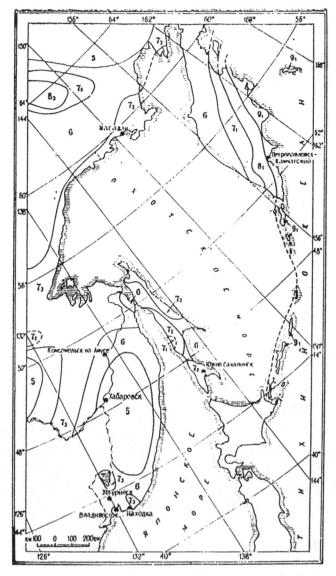

Fig. 35.5. Seismic zone map 5 of the CIS territory (former USSR)

is a structural coefficient depending on the characteristics of the system as indicated in Table 35.2, and S_{oik} is the value of the lateral seismic force corresponding to mode i at level k of the building based on the assumption that the structure deforms elastically. The lateral seismic force S_{oik} is calculated by

$$S_{oik} = Q_k A \beta_i K_\psi \eta_{ik} \qquad (35.2)$$

in which

Q_k = seismic weight at level k. (This weight includes the dead loads, permanent loads, and a fraction of the live loads as described in Section 35.4.)
A = seismic zone factor
β_i = dynamic coefficient for mode i
K_ψ = lateral stiffness coefficient
η_{ik} = distribution factor for mode i at level k of the building.

Table 35.1. Values of the Damage Coefficient K_1 (SNIP II-7-81)

Allowable Damage	K_1
1. *Structures in which no damage (residual deformations, local damages such as subsidences, cracks, etc.) is allowed.	1.00
2. Buildings and structures in which residual deformations, cracks, distress in individual members, etc., may be allowed if no hazard is created for people and equipment (dwellings; public, industrial, and agricultural buildings; hydrotechnical and transport structures; energy and water supply systems; fire stations; fire extinguisher systems; communication structures, etc.).	0.25
3. Buildings and structures in which considerable residual deformations, cracks, distress in individual members, displacements, etc., may be allowed if no hazard is created for people (single-story industrial and agricultural buildings not containing essential equipment).	0.12

*Should be coordinated with State Construction Committee of the former USSR.

Table 35.2. Values of the Structural Coefficient K_2 (SNIP II-7-81)

Structural System	K_2
1. Frame buildings with number of stories n greater than 5.	$1 + 0.1 \times (n-1)$
2. Buildings with concrete shear wall panels and up to 5 stories.	0.9
3. The same as category 2, but with more than 5 stories.	$0.9 + 0.075 \times (n-5)$
4. Buildings with several lower framed unfilled stories and shear wall upper stories.	1.5
5. Buildings with brick or masonry walls (assembled by hand).	1.3
6. Frame single-story buildings less than 8 m high to the bottom of the rafter and with spans less than 18 m.	0.8
7. Agricultural buildings with pile foundations erected on soil of Category III (in accordance with Table 35.6).	0.5
8. Buildings and structures not indicated in categories 1–7.	1.0

Notes: 1. The value of the coefficient K_2 should be not more than 1.5.
2. The values of the coefficient K_2 can be obtained more accurately in accordance with the results of experimental research confirmed by the State Construction Committee of the former USSR.

The factors on the right-hand side of eq.(35.2) are described in detail in Sections 35.4 through 35.9.

35.3 VERTICAL SEISMIC ACTION

The code establishes that vertical seismic action should be taken into consideration in the analysis of the following types of structures:

• Horizontal and inclined cantilever structures

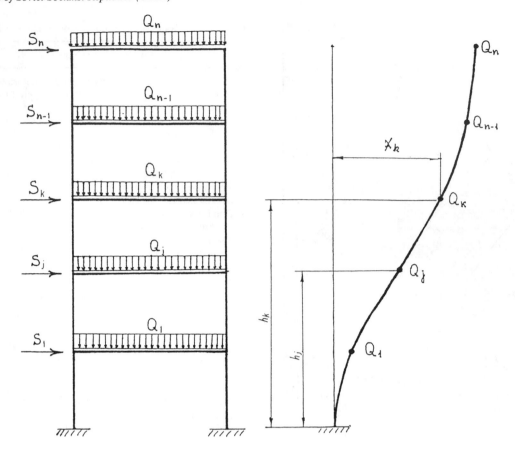

Fig. 35.6. Multistory building modeled as a plane frame

- Bridges
- Frames, arches, trusses, and space roof structures with spans equal to or greater than 24 m
- Structures analyzed for overturning stability and sliding
- Masonry structures.

Vertical forces are to be applied at appropriate points as concentrated forces corresponding to the weights Q_k; the forces should be calculated for the structures mentioned above (except masonry structures) according to eqs.(35.1) and (35.2) in which the factors K_ψ and K_2 have been taken as 1.0.

35.4 DETERMINATION OF SEISMIC WEIGHTS Q_k

The analysis of structures and their foundations for seismic regions must consider the principal and special load combinations required by the code of SNIP II-07-85 (1987). The seismic weight at level k, Q_k, is evaluated considering the following loads in their most unfavorable combination:

- Dead load

- Long-term load
- Short-term loads applied to floors and roof.

In the case of special load combinations, these loads should be multiplied by the reduction factors η_c indicated in Table 35.3.

For purposes of seismic analysis, the code does not require consideration of lateral loads caused by wind, thermal effects, moving equipment, or overhead cranes. (The code stipulates that the weight of crane and trolley, multiplied by a reduction factor of 0.8, plus 30% of the load capacity of the crane must be included in Q_k.)

The seismic forces can act in any direction. The code specifies that the seismic forces should be applied in the most unfavorable direction. However, for buildings of regular geometrical form, seismic forces

Table 35.3. Reduction Factors for Dead and Live Loads

Description of Loads	Reduction Coefficient η_c
Dead loads	0.9
Long-term live loads	0.8
Short-term live loads (on the floors and roof)	0.5

Table 35.4. Seismic Zone Factor

Seismic Intensity	Seismic Zone Factor A
7	0.1λ
8	0.2λ
9	0.4λ

Table 35.6. Seismic Category of Soils

Description of Soil	Seismic Category
Very competent rocks	I
Competent rocks, consolidated sands and gravels	II
Loose sands and gravels, soft fine-grained soils	III

may be applied horizontally in two orthogonal directions and considered to act independently.

35.5 SEISMIC ZONE FACTOR A

The seismic zone factor A depends on the seismic intensity at the site (Figs. 35.1 through 35.5) and on the function of the structure. Table 35.4 provides the values of the seismic zone factor corresponding to seismic intensity values 7, 8, and 9. The values of the factor λ in Table 35.4 is given in Table 35.5. The values of the seismic intensity given in Figs. 35.1 through 35.5 correspond to seismic Soil Category II (Table 35.6). For other soil conditions, the seismic intensity at the site must be taken from Table I of SNIP II-7-81.

35.6 DYNAMIC COEFFICIENT β_i

The dynamic coefficient β_i corresponding to mode i depends on both the natural period of the structure for that mode and the seismic category of the soil at the site. Table 35.6 provides the seismic categories for the different types of soils I, II, and III at the site.

The dynamic coefficient β_i corresponding to the

modal period T_i is given for the various seismic categories of the soil by the following expressions:

Seismic Soil Category I:

$$\beta_i = \frac{1}{T_i} = 0.159\omega_i \leqslant 3.0 \qquad (35.3)$$

Seismic Soil Category II:

$$\beta_i = \frac{1.1}{T_i} = 0.175\omega_i \leqslant 2.7 \qquad (35.4)$$

Seismic Soil Category III:

$$\beta_i = \frac{1.5}{T_i} = 0.239\omega_i \leqslant 2.0 \qquad (35.5)$$

where

$$\omega_i = 2\pi/T_i \qquad (35.6)$$

ω_i is the natural frequency for mode i in rad/sec. The code stipulates that the value of β_i must not be less than 0.8. The dynamic coefficient β_i may also be obtained from the graph shown in Fig. 35.7.

The natural frequencies ω_i and the corresponding modal shapes $\{\phi\}_i$ are obtained by solving the corres-

Table 35.5. Values of the Factor λ^*

Description and Function of Structure	Seismic Intensity	Design Seismic Intensity	Factor λ
1. Dwellings, and public and industrial buildings and structures except as indicated	7	7	1.0
below.	8	8	1.0
	9	9	1.0
2. Particularly significant buildings and structures.	7	8	1.0
	8	9	1.0
	9	9	1.5
3. Buildings and structures in which damage would cause the most unfavorable	7	7	1.5
consequences (large and medium-sized railway stations, covered stadiums, etc.).	8	8	1.5
	9	9	1.2
4. Buildings and structures that are essential for recovery after an earthquake (energy	7	7	1.2
and water supply systems, fire stations, fire extinguisher systems, some communication	8	8	1.2
structures, etc.).	9	9	1.2

*The code does not require consideration of seismic loads for buildings and structures in which damage would not cause death, destruction of valuable equipment, or interruption of production (warehouses, crane or repair trestles, small workshops, etc.) or for temporary buildings and structures.

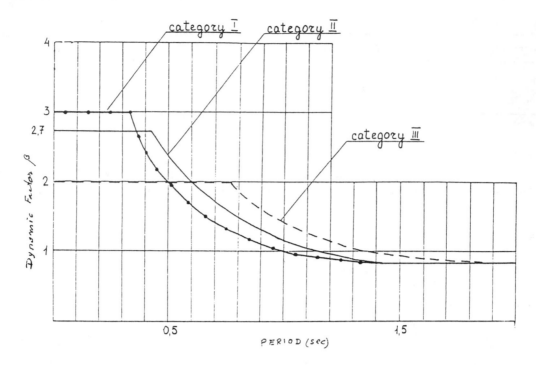

Fig. 35.7. Dynamic factor β_i for different soil categories

ponding eigenproblem for the system. The eigenproblem for a structural system, for which the stiffness and the mass matrices are, respectively, $[K]$ and $[M]$, may be written as

$$[[K] - \omega^2[M]]\{\phi\} = \{0\} \tag{35.7}$$

The computer program developed for this chapter includes an option to solve eq.(35.7), which yields the natural frequencies ω_i (or natural periods ($T_i = 2\pi/\omega_i$)) and modal shapes $\{\phi\}_i$. The fundamental period T_1 may be estimated using an empirical formula for regular buildings for which the fundamental mode predominates.

35.7 ESTIMATION OF THE FUNDAMENTAL PERIOD

The following empirical formula (Korenev and Rabinovich 1981) is based on experimental results. The fundamental period T_1 for steel buildings with no shear walls is given approximately by

$$T_1 = 0.0905 \frac{H}{\sqrt{B}} \tag{35.8}$$

where H is the height of the building and B is the width of the building (both in meters) in the direction of the applied seismic forces.

35.8 LATERAL STIFFNESS FACTOR K_ψ

The value of K_ψ depends on the dimension and the structural characteristics of the building. Table 35.7 provides values for the lateral stiffness coefficient for different types of buildings.

35.9 LEVEL DISTRIBUTION FACTOR η_{ik}

For buildings modeled as cantilever structures with concentrated weights Q_j at the various levels of the

Table 35.7. Lateral Stiffness Coefficient K_ψ (SNIP II-7-81)

Description of Structure	K_ψ
1. Tall structures of small plan dimensions (towers, masts, stacks, standalone shafts, etc.).	1.5
2. Framed buildings without frame shear walls where the ratio of the height of the struts h to the dimension of the cross-section of the struts b in the direction of the seismic loads is equal to or greater than 25.	1.5
3. The same as case 2, but where the ratio h/b is equal to or less than 15.	1.0
4. Buildings and structures not included in cases 1–3.	1.0

Notes: 1. For values of h/b between 25 and 15, the value of K_ψ should be determined by linear interpolation between 1.5 and 1.0.
  2. For buildings of irregular height, the value of K_ψ should be obtained for the average value of the ratio h/b.

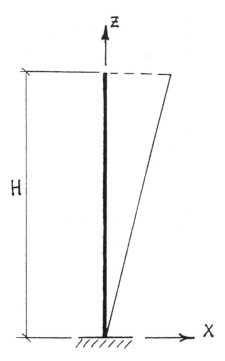

Fig. 35.8. Assumed shape for lateral displacements in the first mode of vibration

building, the level distribution factor η_{ik} for mode i at level k is given by

$$\eta_{ik} = \frac{X_{ik} \sum\limits_{j=1}^{n} Q_j X_{ij}}{\sum\limits_{j=1}^{n} Q_j X_{ij}^2} \qquad (35.9)$$

where

Q_j = seismic design weight at level j
X_{ij} = lateral displacement at level j of the building vibrating freely in mode i
n = number of levels in the building including the roof, level n.

For ordinary buildings with no more than five levels, and with fundamental period T_1 less than 0.4 sec, the seismic code provides a simplified formula to calculate the factor η_k corresponding to the fundamental mode:

$$\eta_k = \frac{h_k \sum\limits_{j=1}^{n} Q_j h_j}{\sum\limits_{j=1}^{n} Q_j h_j^2} \qquad (35.10)$$

where h_k and h_j are, respectively, the distances from the foundation to the levels k and j. Equation (35.10)

is based on the assumption that the modal shape for the first mode is represented by a straight line as shown in Fig. 35.8.

In the particular case in which the weights Q_j at the various levels of the building have approximately the same value and the distances between consecutive levels are equal, eq.(35.10) reduces to

$$\eta_k = \frac{3k}{2n+1} \qquad (35.11)$$

The code does not provide explicit expressions to calculate the distribution factors for higher vibration modes. To consider higher modes, it is necessary to determine natural frequencies and higher modal shapes by solving eq.(35.7). The computer program developed for this chapter includes an option for determining the first three modes by solving eq.(35.7).

35.10 STORY SHEAR FORCE

The shear force V_k for story k corresponding to mode i is given by the sum of the lateral seismic forces S_{ij} above that story; that is,

$$V_{ik} = \sum_{j=k}^{n} S_{ij} \qquad (35.12)$$

When the structure is modeled as a plane or space frame, the shear forces in the resisting elements can be obtained from static analysis using the computer program developed for this chapter.

35.11 OVERTURNING MOMENTS

The overturning moment M_{ik} for mode i at level k of thebuilding, can be determined from statics as the sum of the moments of the forces above that level; that is,

$$M_{ik} = \sum_{j=k+1}^{n} S_{ij}(h_j - h_k) \qquad (35.13)$$

where $k = 0, 1, 2, \ldots, n-1$.

35.12 TORSIONAL MOMENTS

According to the code, the analysis of buildings more than 30 m long in any direction, even those having regular configurations, requires the consideration of horizontal torsional moments. The torsional moment M_{tk} at story k is equal to the shear force V_k at that story, multiplied by the eccentricity e_k between the center of the overlying mass and the stiffness center of the story considered. The code requires that this distance should not be less than

$$e_k = 0.02L \qquad (35.14)$$

where L is the dimension of the building normal to the direction of the seismic lateral forces.

35.13 COMBINATION OF MODAL COMPONENTS: DESIGN FORCES, MOMENTS, AND DISPLACEMENTS

The design values for base shear, story shear, lateral deflections, and overturning moments are obtained by combining corresponding modal responses using the SRSS method (Square Root of Sum of Squared modal contributions). The design response R for a force, moment, displacement, etc., is calculated from the corresponding modal responses R_i as

$$R = \sqrt{\sum_{i=1}^{N} R_i^2} \qquad (35.15)$$

where N is the number of modes considered in the response.

35.14 LATERAL DISPLACEMENTS

The lateral displacement δ_{ik} for a multistory building at level k for mode i may be obtained by static analysis of the building subjected to the seismic lateral forces S_{ik}. Alternatively, the modal displacements δ_{ik} may be calculated from Newton's law of motion ["Manual on Design of Frame Industrial Buildings for Construction in Seismic Regions" (1984)] as follows:

$$\delta_{ik} = \frac{S_{ik} T_i^2 g}{4\pi^2 Q_k} \qquad (35.16)$$

where

S_{ik} = seismic lateral force at level k for mode i
Q_k = seismic weight at level k
T_i = period of the structure for mode i
g = acceleration of gravity.

The design displacement δ_k at level k is calculated by the SRSS method as

$$\delta_k = \sqrt{\sum_{i=1}^{n} \delta_{ik}^2} \qquad (35.17)$$

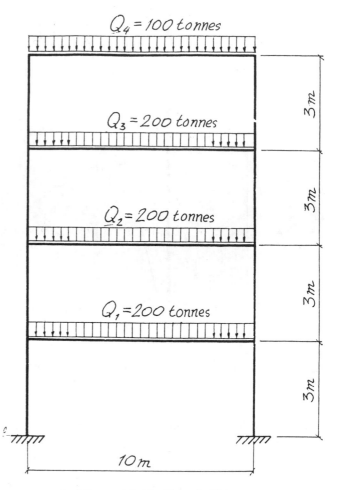

Fig. 35.9. Building modeled for Example 35.1

Example 35.1

The plane frame shown in Fig. 35.9 is chosen to model a four-story steel building. The seismic design weights at the floors are $Q_1 = Q_2 = Q_3 = 200$ tonnes and $Q_4 = 100$ tonnes at the roof. The total flexural stiffness at any story of the building, modeled as a cantilever building, is $EI = 3.5 \times 10^6$ tonne-m^2. The design seismic intensity at the site is 9, and the soil at the site is consolidated gravel.

Determine according to the CIS (former USSR) seismic design code the following: (1) fundamental period; (2) allowable damage coefficient; (3) lateral stiffness coefficient; (4) dynamic coefficient; (5) level distribution factor; (6) lateral seismic forces; (7) story shear forces; (8) overturning moments; (9) torsional moments (for an assumed eccentricity of $e = 0.02L$ at each story); (10) lateral displacements; and (11) interstory displacements.

Solution

(1) Fundamental period. The simplified formula for the fundamental period for this building with structu-

ral walls and frames, is given by

$$T_1 = 0.0905 \frac{H}{\sqrt{B}} \qquad \text{[eq.(35.8)]}$$

$H = 12$ m (height of the building)
$B = 10$ m (dimension in the direction of the seismic forces)

$$T_1 = 0.0905 \frac{12}{\sqrt{10}} = 0.343 \text{ sec}$$

The seismic code reduces the analysis to a single mode of vibration for this value of the fundamental period.

(2a) Allowed damage coefficient K_1

$$K_1 = 0.25 \qquad \text{(Table 35.1)}$$

(2b) Structural coefficient K_2

$$K_2 = 1 \qquad \text{(Table 35.2)}$$

(2c) Seismic zone factor A

$$A = 0.4 \text{ (Tables 35.4 and 35.5 for Seismic Intensity 9)}$$

(3) Lateral stiffness coefficient K_ψ

$$K_\psi = 1 \qquad \text{(Table 35.7)}$$

(4) Dynamic coefficient. The dynamic coefficient β_i is given by

$$\beta_i = \frac{1.1}{T_i} = 0.175\omega_i \leqslant 2.7 \qquad \text{[eq.(35.4)]}$$

(for Seismic Soil Category II)

$$T_1 = 0.343 \text{ sec (fundamental period)}$$

$$\beta_1 = \frac{1.1}{0.343} = 3.207 > 2.7$$

In this case, the value of the dynamic coefficient should be taken as

$$\beta_1 = 2.7 \text{ (code provision)}$$

(5) Distribution factor. The distribution factors are given by

$$\eta_k = \frac{h_k \sum\limits_{j=1}^{n} Q_j h_j}{\sum\limits_{j=1}^{n} Q_j h_j^2} \qquad \text{[eq.(35.10)]}$$

h_k, h_j = heights of levels k, j (assuming linear shape for the first mode)
Q_j = seismic weight at level j.

The distribution factors for each level are:

Level (k):	1	2	3	4
$\eta_k =$	0.3636	0.7273	1.0909	1.4545

(6) Lateral seismic forces

$$S_{ik} = K_1 K_2 S_{oik} \qquad \text{[eq.(35.1)]}$$
$$S_{oik} = Q_k A \beta_i K_\psi \eta_{ik} \qquad \text{[eq.(35.2)]}$$

K_1 = allowable damage coefficient
K_2 = structural coefficient
Q_k = seismic weight at level k
A = seismic zone factor
β_i = dynamic coefficient for vibration mode i
K_ψ = lateral stiffness coefficient
η_{ik} = level distribution factor at level k for mode i.

The lateral seismic forces for mode 1 are:

Level (k):	1	2	3	4
S_1(tonnes)	19.634	39.274	58.909	39.272

(7) Story shear forces

$$V_{ik} = \sum_{j=k}^{n} S_{ij} \qquad \text{[eq.(35.12)]}$$

The values for the story shear forces for mode 1 are:

Level (k):	1	2	3	4
V_1(tonne) =	157.089	137.454	98.180	39.272

(8) Overturning moments

$$M_{ik} = \sum_{j=k+1}^{n} S_{ij}(h_j - h_k) \qquad \text{[eq.(35.13)]}$$

where $k = 0, 1, 2, \ldots, n-1$.
The overturning moments at each level for mode 1 are:

Level (k):	0	1	2	3
M_{1k}(tonne-m) =	1,296.0	824.7	412.4	117.8

(9) Torsional moments. The torsional moment M_{tki} at level k for mode i is given by

$$M_{tki} = V_{ki} e_k$$

in which

V_{ki} = shear force at level k for mode i
e_k = eccentricity at level k

$$e_k = 0.02L \qquad \text{[eq.(35.14)]}$$

$L = 10\,\text{m}$ (dimension of the building normal to the direction of the seismic forces S_{ik}).

The torsional moments at each level are:

Level (k):	1	2	3	4
M_{t1k}(tonne-m) =	31.428	27.491	19.636	7.854

(10) Lateral displacements. The lateral displacement δ_{ik} at level k for mode i is given by

$$\delta_{ik} = \frac{S_{ik} T_i^2 g}{4\pi^2 Q_k} \qquad \text{[eq.(35.16)]}$$

The displacements at each level for mode 1 are:

Level(k)	1	2	3	4
δ_k(cm) =	0.287	0.574	0.861	1.148

(11) Relative displacements between levels of the building. The story relative displacement or story drift Δ_{ik}, for story k corresponding to mode i, is given by the difference between displacements of upper and lower levels of that story. The values of these drifts for mode 1 are as follows:

Story (k):	1	2	3	4
Δ_{1k}(cm) =	0.287	0.287	0.287	0.287

35.15 COMPUTER PROGRAM AND EXAMPLES

A computer program has been developed to implement the provisions of the II-7-81 code. The program is written in Quick BASIC for implementation on IBM-compatible personal computers. The program presents some options for calculation of the fundamental period according to the formulas given in Korenev and Rabinovich (1981) as well as an option to determine the periods and mode shapes by solving eq.(35.7). There is also an option for the static analysis of a structure of any configuration using the displacement method. Output results are calculated and printed for the natural periods and modal shapes as well as for seismic lateral forces, story shear forces, torsional moments, overturning moments, and lateral displacements. These values can be calculated and printed for the first six modes of vibration with the corresponding design values determined by the SRSS method. Example 35.2 illustrates the use of this program.

Example 35.2

Solve Example 35.1 using the computer program developed for this chapter. Consider the first three modes of vibration and use the static option for calculation of forces, moments, and displacements.

35.15.1 Input Data and Output Results for Example 35.2

```
INPUT DATA:

    SEISMIC INTENSITY              EAQ = 9

    NUMBER OF STORIES              NS = 4

    TORSIONAL ECCENTRICITY         EC = 0.2

    NUMBER OF MODES CONSIDERED     NM = 3

ALLOWED DAMAGES MENU:

    1:  STRUCTURES WITH NO DAMAGES ALLOWED

    2:  STRUCTURES WITH PARTIAL DAMAGES ALLOWED

    3:  STRUCTURES WITH EXTENSIVE DAMAGES ALLOWED

            SELECT NUMBER?  2

STRUCTURAL FORM MENU:

    1:  FRAME BUILDINGS

    2:  BUILDINGS WITH CONCRETE SHEAR WALL PANELS

    3:  BUILDINGS WITH LOWER FRAME UNFILLED STORIES AND
        SHEAR WALL UPPER STORIES

    4:  BUILDINGS WITH BRICK OR MASONRY WALLS

    5:  SMALL FRAME SINGLE STORY BUILDINGS

    6:  AGRICULTURAL BUILDINGS WITH PILE FOUNDATIONS

    7:  OTHER STRUCTURES

            SELECT NUMBER?  7

TYPE OF THE SOIL MENU:

    1:  VERY COMPACT ROCKS

    2:  COMPACT ROCKS, CONSOLIDATED SANDS AND GRAVELS

    3:  LOOSE SANDS AND GRAVELS, SWAMPS

            SELECT NUMBER?  2

BUILDING SIGNIFICANCE MENU:

    1:  IMPORTANT AND ORDINARY BUILDINGS

    2:  BUILDINGS OF VITAL IMPORTANCE TO SOCIETY

    3:  BUILDINGS OF REDUCED IMPORTANCE

            SELECT NUMBER?  1

        LATERAL STIFFNESS FACTOR Kpsi = 1
```

STORY #	STORY HEIGHT (METER)	STORY WEIGHT (TonneS)	STIFFNESS (T*M 2)
1	3	200	3500000
2	3	200	3500000
3	3	200	3500000
4	3	100	3500000

```
OUTPUT RESULTS:

VIBRATION PERIODS (SEC.):

    MODE  PERIOD
    1     0.3689
    2     0.0628
    3     0.0237

        VIBRATION MODE SHAPES
```

LEVEL	MODE 1	MODE 2	MODE 3
1	0.0959	-0.4544	1.0000
2	0.3360	-0.8282	0.2575
3	0.6541	-0.2725	-0.9174
4	1.0000	1.0000	0.8354

```
ALLOWABLE DAMAGE FACTOR        K1 = 0.25
STRUCTURAL FORM FACTOR         K2 = 1
SEISMIC ZONE FACTOR            A = 0.4
LATERAL STIFFNESS FACTOR       KPSI = 1.0
DYNAMIC COEFFICIENT (BETA):

      BETA (1) = 2.7        BETA (2) = 2.7        BETA (3) = 2.7

              LEVEL DISTRIBUTION FACTOR (ETA):

LEVEL        MODE 1          MODE 2          MODE 3

  1        0.1448157       0.3269134       0.3357422
  2        0.5074785       0.5957764       0.0864432
  3        0.9880280       0.1959988      -0.3080201
  4        1.510561       -0.719386        0.2804834
```

```
                   COMBINED MODAL RESPONSE:

           SHEAR      TORSIONAL    OVERTURNING
           FORCE      MOMENT       MOMENT       DISPLACEMENTS
LEVEL      (T)        (T-M)        (T-M)        (M)

  1        136.39     27.28        1162.77      0.0013
  2        123.84     24.77         769.74      0.0046
  3         94.98     19.00         413.57      0.0090
  4         45.80      9.16         137.41      0.0138
```

Example 35.3

A 21-story moment resisting frame shown in Fig. 35.10 serves to model a steel building for earthquake analysis. The total weight on all levels is 200 tonnes except on the roof, where it is 100 tonnes. The seismic intensity grade (G) of the earthquake is 9. The type of soil is consolidated gravel. Perform the seismic analysis for this building according to the SNIP II-07-85. Use either the cantilever model or the plane frame model for the analysis. The total flexural rigidity of the cantilever model is $EI = 3.5 \times 10^6$ (tonne-m). The total flexural rigidity of the columns and rafters in any story for the plane frame model is $EI = 1,000,000$ (tonne-m).

35.15.2 Input Data and Output Results for Example 35.3

```
INPUT DATA:

SEISMIC INTENSITY            EAQ = 9

NUMBER OF STORIES            NS = 21

TORSIONAL ECCENTRICITY       EC = 0.2

NUMBER OF MODES CONSIDERED   NM = 3

ALLOWED DAMAGES MENU:

    1:  STRUCTURES WITH NO DAMAGES ALLOWED

    2:  STRUCTURES WITH PARTIAL DAMAGES ALLOWED

    3:  STRUCTURES WITH EXTENSIVE DAMAGES ALLOWED

        SELECT NUMBER?  2

STRUCTURAL FORM MENU:

    1:  FRAME BUILDINGS

    2:  BUILDINGS WITH CONCRETE SHEAR WALLS PANELS

    3:  BUILDINGS WITH LOWER FRAME UNFILLED STORIES AND SHEAR WALL
```

```
    UPPER STORIES

4:  BUILDINGS WITH BRICK OR MASONRY WALLS

5:  SMALL FRAME SINGLE STORY BUILDINGS

6:  AGRICULTURAL BUILDINGS WITH PILE FOUNDATIONS

7:  OTHER STRUCTURES

    SELECT NUMBER?  1
```

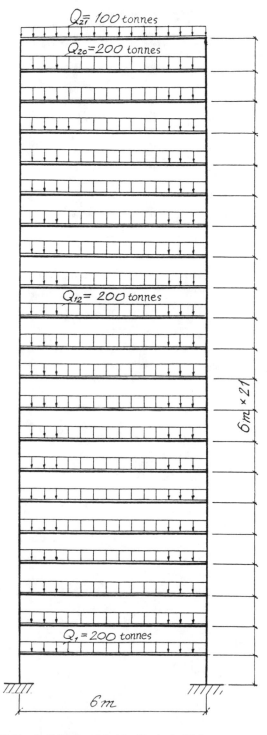

Fig. 35.10. Building modeled for Example 35.3

```
TYPE OF THE SOIL MENU:

    1:  VERY COMPACT ROCKS

    2:  COMPACT ROCKS, CONSOLIDATED SANDS AND GRAVELS

    3:  LOOSE SANDS AND GRAVELS, SWAMPS

        SELECT NUMBER?  2

BUILDING SIGNIFICANCE MENU:

    1:  IMPORTANT AND ORDINARY BUILDINGS

    2:  BUILDINGS OF VITAL IMPORTANCE TO SOCIETY

    3:  BUILDINGS OF REDUCED IMPORTANCE

        SELECT NUMBER?  1

        LATERAL STIFFNESS FACTOR Kpsi = 1
```

STORY #	STORY HEIGHT (METER)	STORY WEIGHT (Tonnes)	STIFFNESS (Tonne-M^2)
1	6	200	3500000
2	6	200	3500000
3	6	200	3500000
4	6	200	3500000
5	6	200	3500000
6	6	200	3500000
7	6	200	3500000
8	6	200	3500000
9	6	200	3500000
10	6	200	3500000
11	6	200	3500000
12	6	200	3500000
13	6	200	3500000
14	6	200	3500000
15	6	200	3500000
16	6	200	3500000
17	6	200	3500000
18	6	200	3500000
19	6	200	3500000
20	6	200	3500000
21	6	100	3500000

```
OUTPUT RESULTS

VIBRATION PERIODS (SEC.):

     MODE        PERIOD

      1         27.8577

      2          4.4763

      3          1.6025
```

ALLOWED DAMAGES FACTOR	K1 = .25
STRUCTURAL FORM FACTOR	K2 = 3
SEISMIC ZONE FACTOR	A = .4

```
LATERAL STIFFNESS FACTOR      KPSI = 1

DYNAMIC COEFFICIENT (BETA):

     BETA (1) = 0.8        BETA (2) = 0.8        BETA (3) = 0.8
```

LEVEL DISTRIBUTION FACTOR (ETA):

LEVEL	MODE 1	MODE 2	MODE 3
1	0.006097	0.019930	0.030901
2	0.023843	0.073195	0.106174
3	0.052242	0.150065	0.200368
4	0.091019	0.240997	0.2901433
5	0.138822	0.336833	0.3560267
6	0.195027	0.429009	0.3840759
7	0.258837	0.509794	0.3671141
8	0.329466	0.572520	0.3052487
9	0.406148	0.611771	0.2055363
10	0.488136	0.623541	0.0808368
11	0.574709	0.605332	−0.0521130
12	0.665179	0.556173	−0.1750478
13	0.758893	0.476592	−0.2706526
14	0.855240	0.368471	−0.3249256
15	0.953655	0.234862	−0.3289792
16	1.053635	0.079731	−0.2800042
17	1.154732	0.092395	−0.1814286
18	1.256570	0.276786	−0.0419380
19	1.358839	0.468959	0.1264421
20	1.461315	0.665109	0.3109251
21	1.563855	0.862606	0.5012654

REFERENCES

KORENEV, B. G., and RABINOVICH, I. M. (1981) *Design Handbook: Dynamic Analysis of Structures Subjected to Special Loads*, Stroiizdat, Moscow.

SNIP II-7-81 (1982) Basic Norms and Rules for Civil and Structural Engineering, Chapter 7, *Construction in Seismic Regions*, Stroiizdat, Moscow.

SNIP II-07-85 (1987) Basic Norms and Rules for Civil Engineering Loads and Actions, Gosstroi USSR, Moscow.

"The Manual on Design of Frame Industrial Buildings for Construction in Seismic Regions" (1984) (a contribution to SNIP II-7-81), Stroiizdat, Moscow.

36

United States of America

Mario Paz

36.1 INTRODUCTION

Several seismic building codes are currently in use in different regions of the United States of America (U.S.A.). The *Uniform Building Code* (UBC), published by the International Conference of Building

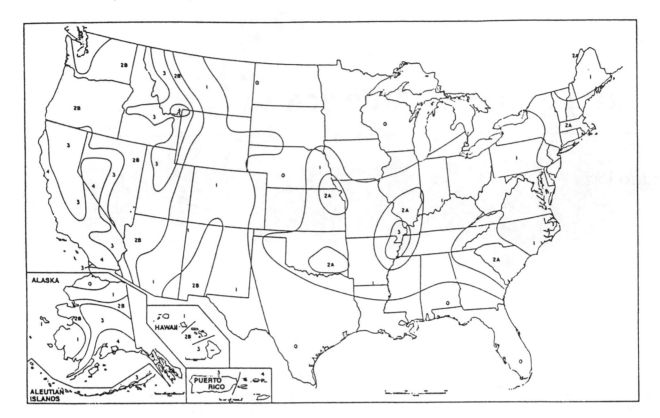

Fig. 36.1.   Seismic map of the United States of America (UBC-91) (Reproduced from *Uniform Building Code*, 1991, with permission of the publisher, the International Conference of Building Officials)

Officials (1991), is the building code most extensively used, particularly in the western part of the country. In addition to the UBC, three other major building codes are used: (1) The *BOCA*, or *Basic Building Code* [Building Officials and Code Administrators International (1990)]; (2) *The National Building Code*, published by the American Insurance Association (1976); and (3) the *Standard Building Code*, of the Southern Building Code Congress International (1991).

In addition to the preceding codes, there is the ASCE *Standard Minimum Design Loads for Buildings and Other Structures* (1988) (Revision of ANSI A58.1). Several organizations involved in earthquake-resistant design have published recommendations that form the basis for requirements in the official codes. These organizations include (1) the Structural Engineers Association of California (SEAOC 1990), (2) the Applied Technology Council (1978), (3) the Building Seismic Safety Council (BSSC 1988), and (4) The Federal Emergency Management Agency (FEMA) which leads the National Earthquake Hazard Reduction Program (NEHRP) with publications issued by Building Seismic Safety Council (BSSC 1988). These organizations periodically issue recommendations and requirements for earthquake-resistant design of struc-

tures that are based on a combination of theory, experiment, and practical observation.

36.2 IMPLEMENTATION OF THE UBC-91: EQUIVALENT STATIC LATERAL FORCE METHOD

The earthquake-resistant design regulations of the Uniform Building Code of 1991, UBC-91 (International Conference of Building Officials 1991) are based mainly on the recommendations of the Structural Engineers Association of California entitled, *Recommended Lateral Force Requirements and Tentative Commentary* (SEAOC 1990). This in turn is based partly on the provisions of the Applied Technology Council recommendations (ATC 3-06-1978) and partly on the Building Seismic Safety Council guidelines (BSSC 1988). The key UBC-91 provisions for earthquake-resistant design are presented in this chapter.

In the UBC-91, the United States of America is divided into Seismic Zones 0, 1, 2A, 2B, 3, and 4 as shown in Fig. 36.1. Two methods for earthquake-resistant design are provided in the Code: (1) the static lateral force method and (2) the dynamic method.

Table 36.1. Structural Factor

Lateral Load-Resisting System	R_W	$H_{\max}$*
A. BEARING WALL SYSTEM		
1. Light-framed walls with shear panels		
(a) Plywood walls for structures of three stories or less	8	65
(b) All other framed walls	6	65
2. Shear Walls		
(a) Concrete	6	160
(b) Masonry	6	160
3. Light steel-framed bearing walls with tension bracing	4	65
4. Braced frames where bracing carries gravity loads		
(a) Steel	6	160
(b) Concrete**	4	—
(c) Heavy timber	4	65
B. BUILDING FRAME SYSTEM		
1. Steel braced frame (EBF)°	10	240
2. Light-framed walls with shear panels		
(a) Plywood walls for structure three stories or less	9	65
(b) All other light-framed walls	7	65
3. Shear walls		
(a) Concrete	8	240
(b) Masonry	8	160
4. Concentrated braced frames		
(a) Steel	8	160
(b) Concrete**	8	—
(c) Heavy timber	8	65
C. MOMENT-RESISTING FRAME SYSTEM		
1. Special moment-resisting frames (SMRF)°		
(a) Steel	12	N.L.
(b) Concrete	12	N.L.
2. Concrete intermediate moment-resisting frames (IMRF)****,°	8	—
3. Ordinary moment-resisting frames (OMRF)°		
(a) Steel	6	160
(b) Concrete***	5	—
D. DUAL SYSTEM		
1. Shear walls		
(a) Concrete with SMRF°	12	N.L.
(b) Concrete with steel OMRF°	6	160
(c) Concrete with concrete IMRF****,°	9	160
(d) Masonry with SMRF°	8	160
(e) Masonry with steel OMRF°	6	160
(f) Masonry with concrete IMRF**,°	7	—
2. Steel EBF°		
(a) With steel SMRF°	12	N.L.
(b) With steel OMRF°	6	160
3. Concentrically braced frames		
(a) Steel with steel SMRF°	10	N.L.
(b) Steel with steel OMRF°	6	160
(c) Concrete with concrete SMRF**,°	9	—
(d) Concrete with concrete IMRF**,°	6	—

*$H_{\max}$ = height limit applicable to Seismic Zones 3 and 4

**Prohibited in Seismic Zones 3 and 4

***Prohibited in Seismic Zones 2, 3, and 4

****Prohibited in Seismic Zones 3 and 4 except as permitted by the Code [Section 2338(b)]

°Structural systems defined in Section 1233(f) of the Code:

    EBF – Eccentrically Braced Frame is a steel braced frame designed in conformance with Section 2710(i) of the Code.

    IMRF – Intermediate Moment-Resisting Frame is a concrete frame designed in accordance with Section 2625(k) of the Code.

    OMRF – Ordinary Moment-Resisting Frame is a moment-resisting frame not meeting special design requirements for ductile behavior.

    SMRF – Special Moment-Resisting Frames is a moment-resisting frame specially detailed to provide ductile behavior and comply with the requirements given in 2601 through 2617, and proportioned to satisfy Sections 2625(c), (d), (e), (g), and (h) of the Code.

N.L. = No Limit

(Reprinted from *Uniform Building Code*, 1991, with permission of the publisher, the International Conference of Building Officials.)

The static lateral force method may be used for structures with the following characteristics:

(a) All regular or irregular[1] in Seismic Zone 1

(b) Regular under 240 feet in height with lateral force resistance provided by systems listed in Table 36.1, except for structures located in Soil Profile Type S_4 (Table 36.2) that have a period greater than 0.7 second

(c) Irregular of no more than five stories or 65 feet (ft) in height

(d) With a flexible upper portion on a rigid lower portion where both portions of the structure considered separately can be classified as being regular (the average story stiffness of the lower portion is at least 10 times the average stiffness of the upper portion, and the period of the entire structure is not greater than 1.1 times the period of the upper portion considered as a separate structure fixed at the base).

The dynamic method may be used for any structure. However, it must be used for structures over 240 ft in height, irregular structures of over five stories or 65 ft in height, and structures with dissimilar structural systems located in Seismic Zones 3 and 4. The UBC-91 stipulates that the structure should be designed for a total base shear force given by the following formula:

$$V = \frac{ZIC}{R_W} W \qquad (36.1)$$

in which

$$C = \frac{1.25S}{T^{2/3}} \leqslant 2.75 \qquad (36.2)$$

The Code also stipulates a minimum value for the ratio C/R_W of 0.075 except where scaling of forces by $3R_W/8$ is prescribed.

The factors in eqs.(36.1) and (36.2) are defined as follows:

Z is the seismic zone factor related to the seismic zones of Fig. 36.1. It is equal to 0.075 for Zone 1, 0.15 for Zone 2A, 0.20 for Zone 2B, 0.30 for Zone 3, and 0.40 for Zone 4. Buildings in Zone 0 are not required to be specifically designed for earthquake. The values

Table 36.2. Site Coefficient

Site Factor S^*	Description
$S_1 = 1.0$	A soil profile with either: (a) A rock-like material characterized by a shear wave velocity greater than 2,500 feet per second or by other suitable means of classification, or (b) Stiff or dense soil conditions where the soil depth is less than 200 feet.
$S_2 = 1.2$	A soil profile with dense or stiff soil conditions, where the soil depth exceeds 200 feet.
$S_3 = 1.5$	A soil profile 70 feet or more in depth and containing more than 20 feet of soft to medium-stiff clay but not more than 40 feet of soft clay.
$S_4 = 2.0$	A soil profile containing more than 40 feet of soft clay characterized by a shear wave velocity less than 500 feet per second.

*The site factor shall be established from properly substantiated geotechnical data. In locations where the soil properties are not known in sufficient detail to determine the soil profile type, soil profile S_3 shall be used. Soil profile S_4 need not be assumed unless the building official determines that soil profile S_4 may be present at the site, or in the event that soil profile S_4 is established by geotechnical data.

(Reprinted from *Uniform Building Code*, 1991, with permission of the publisher, the International Conference of Building Officials.)

of this coefficient can be considered to represent the effective peak ground acceleration (associated with an earthquake that has a 10% probability of being exceeded in 50 years) expressed as a fraction of the acceleration of gravity.

S is the site coefficient depending on the characteristics of the soil at the site as described in Table 36.2.

I is the occupancy importance factor related to the anticipated use of the structures as classified in Table 36.3. The importance factor I is equal to 1.25 for essential and hazardous facilities as listed, and equal to 1.00 for all other structures.

R_W is the structural factor ranging from 4 to 12 as given in Table 36.1. It is a measure of the capacity of the structural system to absorb energy in the inelastic range through ductility and redundancy. Its value is based primarily on the performance of similar systems in past earthquakes.

W is the seismic weight which includes the dead

---

[1]Irregular structures are those that have significant physical discontinuities in configuration or in their lateral force resisting systems. Specific features for vertical structural irregularities are described in Table 23-M of UBC-91; features for horizontal irregularities are in Table 23-N.

[2]A moment resisting frame is a structural frame in which the members and joints are capable of resisting forces primarily by flexure.

[3]Alternatively, the value of C_t for structures with concrete or masonry structural walls (shear walls) may be taken as $0.1/\sqrt{A_c}$. The value of A_c is determined from the following formula:

$$A_c = \Sigma A_e[0.2 + (D_e/h_N)^2]$$

where A_e is the minimum cross-sectional area in any horizontal plane in the first story of a structural wall (in square feet), D_e is the length (feet) of a structural wall in the first story (in the direction parallel to the applied forces), and h_N is the height of the building (feet). The value of D_e/h_N shall not exceed 0.9.

Table 36.3.   Occupancy Importance Factor

Occupancy Factor I	Occupancy Type or Functions of Structure
I. Essential Facilities* $I = 1.25$	Hospitals and other medical facilities having surgery and emergency treatment areas. Fire and police stations. Tanks or other structures containing, housing, or supporting water or other fire-suppression materials or equipment required for the protection of essential or hazardous facilities, or special occupancy structures. Emergency vehicle shelters and garages. Structures and equipment in emergency-preparedness centers. Standby power-generating equipment for essential facilities. Structures and equipment in government communication centers and other facilities required for emergency response.
II. Hazardous Facilities $I = 1.25$	Structures housing, supporting, or containing sufficient quantities of toxic or explosive substances to be dangerous to the safety of the general public if released.
III. Special Occupancy $I = 1.0$	Covered structures whose primary occupancy is public assembly—capacity > 300 persons. Buildings for schools through secondary level, or day-care centers—capacity > 250 students. Buildings for colleges or adult education schools—capacity > 500 students. Medical facilities with 50 or more resident incapacitated patients, included above. Jails and detention facilities. All structures with occupancy > 5,000 persons. Structures and equipment in power-generating stations and other public utility facilities not included above, and required for continued operation.
IV. Standard Occupancy Structure $I = 1.0$	All structures having occupancies or functions not listed above.

*Essential facilities are those structures which are necessary for emergency operations subsequent to a natural disaster.

(Reprinted from *Uniform Building Code*, 1991, with permission of the publisher, the International Conference of Building Officials.)

weight of the building and applicable percentage of other loads:

1.   In storage and warehouse occupancies, a minimum of 25% of the floor live load shall be applicable.
2.   Where a partition load is used in floor design, a load of not less than 10% per square foot (psf) shall be included.
3.   Where the design snow load is greater than 30 psf, the snow load shall be included in W, although it could be reduced up to 75% when approved by the building official.
4.   Total weight of permanent equipment shall be included.

T is the fundamental period of the building, which may be approximated from the following formula (Method A):

$$T = C_t[h_N^{3/4}] \tag{36.3}$$

where

h_N = total height of the building in feet
C_t = 0.035 for steel moment-resisting frames[2]
C_t = 0.030 for reinforced concrete moment-resisting frames and eccentrically braced frames
C_t = 0.020 for all other buildings.[3]

Alternatively, the fundamental period of the structure may be determined by appropriate dynamic analysis or from Rayleigh's formula (Method B) as

$$T = 2\pi \sqrt{\frac{\sum_{i=1}^{N} W_i \delta_i^2}{g \sum_{i=1}^{N} f_i \delta_i}} \tag{36.4}$$

where f_i represents any lateral force distribution (applied at the various levels of the building) approximately consistent with results obtained using eqs.(36.5) and (36.6), W_i is the seismic weight attributed to level i, and δ_i are the elastic lateral displacements produced by the lateral forces f_i.

The value of C in eq.(36.2) resulting from the use of T given by eq.(36.4) shall not be less than 80% of the value obtained by using T from eq.(36.3). This provision of the Code was adopted to avoid the possibility of using an excessively long calculated period to justify an unreasonably low base shear.

36.2.1 Distribution of Lateral Forces

The base shear force V calculated from eq.(36.1) is distributed at the various levels of the building according to the following formulas:

$$F_x = \frac{(V - F_t) W_x h_x}{\sum\limits_{i=1}^{N} W_i h_i} \qquad (36.5)$$

in which

$$F_t = 0.07TV \leqslant 0.25V \quad \text{for} \quad T > 0.7 \sec$$
$$F_t = 0 \qquad\qquad\quad \text{for} \quad T \leqslant 0.7 \sec \qquad (36.6)$$

and

$$V = F_t + \sum_{i=1}^{N} F_i \qquad (36.7)$$

where

N = total number of stories above the base of the building

F_x, F_i, F_N = lateral force applied at level x, i, or N

F_t = portion of the base force V at the top of the structure in addition to F_N

h_x, h_i = height of level x or i above the base

W_x, W_i = seismic weight of xth or ith level.

The distribution of the lateral forces F_x and F_t for a uniform multistory building is shown in Fig. 36.2. The Code stipulates that the force F_x at level x be applied over the area of the building according to the mass distribution at that level.

36.2.2 Story Shear Force

The shear force V_x at any story x is given by the sum of the lateral seismic forces above that story, that is,

$$V_x = F_t + \sum_{i=x}^{N} F_i \qquad (36.8)$$

36.2.3 Horizontal Torsional Moment

The UBC-91 states that provisions should be made for the increased shear force resulting from horizontal torsion where diaphragms are not flexible. Diaphragms are considered flexible when the maximum lateral deformation of the diaphragm is more than twice the average story drift of the associated story. This condition may be implemented by comparing the computed midpoint in-plane deflection of the diaphragm under a lateral load with the story drift of adjoining vertical resisting elements under an equivalent tributary lateral load. The torsional design moment at a given story shall be the moment resulting from eccentricities between applied design lateral

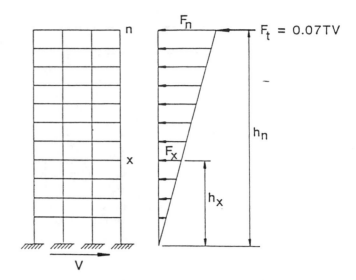

Fig. 36.2. Distribution of lateral forces in a multistory building

forces (applied through each story mass center) at levels above that story and the center of stiffness of the vertical resisting elements of the story. The Code also requires that an accidental torsional moment be added to the actual torsional moment. The story accidental torsional moment is determined by displacing the calculated mass center in each direction to a distance that is equal to 5% of the building dimension perpendicular to the direction of the seismic forces.

Torsional irregularities in the building are considered by increasing the accidental torsion by an amplification factor A_x, given by

$$A_x = \left(\frac{\delta_{\max}}{1.2\delta_{\text{avg}}} \right)^2 \leqslant 3.0 \qquad (36.9)$$

where

$\delta_{\max}$ = the maximum displacement at level x

δ_{avg} = the average displacement at the extreme points of the structure at level x.

36.2.4 Overturning Moment

The code requires that overturning moments shall be determined at each level of the structure. The overturning moment is determined using the seismic design forces F_x and F_t [eqs.(36.5) and (36.6)] which act on levels above the level under consideration. Hence, the overturning moment M_x at level x of the building is given by

$$M_x = F_t(h_N - h_x) + \sum_{i=x+1}^{N} F_i(h_i - h_x) \qquad (36.10)$$

where $x = 0, 1, 2, \ldots, N-1$.

36.2.5 Story Drift Limitation

Story drift, the relative displacement between consecutive floor levels produced by the design lateral forces, shall include calculated translational and torsional deflections. The calculated story drift shall not exceed $0.04/R_W$ times the story height, nor 0.005 times the story height for structures having a fundamental period of less than 0.7 second. For structures having a period of 0.7 second or greater, the calculated story drift shall not exceed $0.03/R_W$ times the story height, nor 0.004 times the story height. In addition, the drift limitations are not subject to the 80% limit mentioned in the calculation of the period T, nor the limitation imposed on the ratio C/R_W cited in relation to the use of eq.(36.2).

36.2.6 P-Delta Effects (P-Δ)

As stated in Section 4.9 of Chapter 4, the P-Δ effect refers to the additional moment produced by the vertical loads and the lateral displacements of columns and other resisting elements. The Code specifies that the resulting member forces and moments, as well as the story drift induced by the P-Δ effect, shall be considered in the evaluation of overall structural frame stability. The P-Δ effect need not be considered when the ratio of the secondary moment (resulting from the story drift) to the primary moment (due to the seismic lateral forces), for any story, does not exceed 0.10. At any story, this ratio may be evaluated as the product of the total seismic loads above the story times the seismic story drift, divided by the product of the seismic shear force at that story times the height of the story. That is, the ratio θ_x at level x of the secondary moment M_{xs}, resulting from the P-Δ effect, and the primary moment M_{xp}, due to the seismic lateral forces, may be calculated from the following formula:

$$\theta_x = \frac{M_{xs}}{M_{xp}} = \frac{P_x \Delta_x}{V_x H_x} \qquad (36.11)$$

where

P_x = total seismic weight at level x and above
Δ_x = drift of story x
V_x = shear force of story x
H_x = height of story x

In Seismic Zones 3 and 4, the P-Δ effect need not be considered when the story drift does not exceed $0.02/R_W$.

---

[4]1 kip = 1,000 pounds

36.2.7 Diaphragm Design Force

The Code stipulates that floor and roof diaphragms should be designed to resist the forces determined by the following formula:

$$F_{px} = \frac{F_t + \sum_{i=x}^{N} F_i}{\sum_{i=x}^{N} W_i} W_{px} \qquad (36.12)$$

in which F_i is the seismic lateral force, W_i is the seismic weight at each level, and W_{px} is the weight of the diaphragm and attached parts of the building. The Code states that the force F_{px} calculated by eq.(36.12) need not exceed $0.75ZIW_{px}$, but shall not be less than $0.35ZIW_{px}$.

36.2.8 Example 36.1

A four-story reinforced concrete framed building has the dimensions shown in Fig. 36.3. The sizes of the exterior columns (nine each on lines A and C) are 12 in. × 20 in. The sizes of the interior columns (nine on line B) are 12 in. × 24 in. for the bottom two stories, and, respectively, 12 in. × 16 in. × 20 in. for the top and second-highest stories. The height between floors is 12 ft. The dead load per unit area of floor is estimated to be 140 psf; it consists of floor slab, beam, half the weight of columns above and below the floor, partition walls, etc. The normal live load is assumed as 125 psf. The soil below the foundation is assumed to be hard rock. The building site is in Seismic Zone 3. The building is intended to be used as a warehouse.

Perform the seismic analysis for this structure (in the direction normal to lines A, B, and C) in accordance with the UBC-91.

Solution

1. Seismic weight at various floors. For a warehouse, the design load should include 25% of the live load. No live load needs to be considered in the roof. Hence, the effective weight at all floors, except at the roof, will be $140 + 0.25 \times 125 = 171.25$ psf, and the effective weight for the roof will be 140 psf. The plan area is $48\,\text{ft} \times 96\,\text{ft} = 4,608\,\text{ft}^2$. Hence, the seismic weights of various levels are:

$$W_1 = W_2 = W_3 = 4,608 \times 0.17125 = 789.1\,\text{kip}^4$$

$$W_4 = 4,608 \times 0.140 = 645.1\,\text{kip}$$

The total seismic weight of the building is then

$$W = 789.1 \times 3 + 645.1 = 3,012.4\,\text{kip}$$

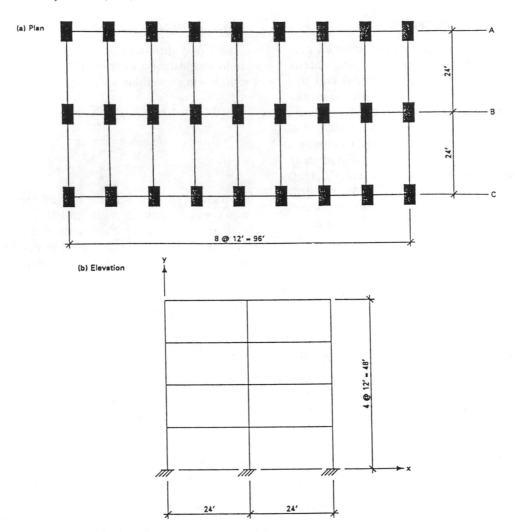

Fig. 36.3. Plan and elevation for a four-story building for Example 36.1.

2. Fundamental period

$$T = C_t h^{3/4} \qquad \text{[eq.(36.3)]}$$

where

$C_t = 0.030$ (for reinforced concrete moment-resisting frame)

$h_N = 48$ ft (total height of the building)

$$T = 0.030 \times 48^{3/4} = 0.55 \text{ sec}$$

3. Base shear force

$$V = \frac{ZICW}{R_W} \qquad \text{[eq.(36.1)]}$$

$Z = 0.3$ (site in Zone 3)

$I = 1.0$ [warehouse (Table 36.3)]

$R_W = 12$ [special moment-resisting space frame (Table 36.1)]

$S = 1.0$ [rock (Table 36.2]

$$C = \frac{1.25S}{T^{2/3}} \leqslant 2.75$$

$$C = \frac{1.25 \times 1.0}{(0.55)^{2/3}} = 1.862 \qquad \text{[eq.36.2)]}$$

and

$$C/R_W = 0.155 > 0.075$$

Therefore,

$$V = \frac{0.3 \times 1.0 \times 1.862}{12} \, 3,012.4 = 140.23 \text{ kip}$$

4. Lateral forces

$$F_x = \frac{(V - F_t) W_x h_x}{\sum\limits_{i=1}^{N} W_i h_i} \qquad \text{[eq.(36.5)]}$$

$$F_t = 0 \quad \text{for} \quad T = 0.55 < 0.7 \qquad \text{[eq.(36.6)]}$$

Calculated lateral forces are shown in Table 36.4.

Table 36.4. Lateral Forces

Level	W_i (kip)	h_x (ft)	$W_x h_x$ (kip-ft)	F_x (kip)	V_x (kip)
4	645.12	48	30,965	49.5	49.5
3	789.12	36	28,408	45.4	94.9
2	789.12	24	18,939	30.3	125.2
1	798.12	12	9,469	15.1	140.3
			$\Sigma = 87{,}781$		

5. Shear force at story x

$$V_x = F_t + \sum_{i=x}^{N} F_x \qquad [\text{eq.}(36.8)]$$

Calculated shear forces are shown in Table 36.4.

6. Story lateral displacement and drift. For this structure modeled as a shear building, the flexural stiffness calculated for the first and second stories of the building is:

$$K_1 = K_2 = 3{,}236.0 \text{ kip/in.}$$

and for the third and fourth stories

$$K_3 = K_4 = 1{,}757.7 \text{ kip/in.}$$

The drift for the stories is calculated by

$$\Delta_x = V_x/K_i$$

The lateral displacement at any level is given by the sum of the drifts of the overlying stories. The results of these calculations to obtain the story drifts and lateral displacements are shown in Table 36.5.

The Code stipulates that the story drift Δ_x should not exceed $(0.04/R_W)$ times the story height, nor 0.005 times the story height. Hence, the maximum permissible drift Δ_{max} is given by the smallest of the following results:

$$\Delta_{max} = (0.04/12)144 = 0.48 \text{ in.}$$
$$\Delta_{max} = 0.005 \times 144 = 0.72 \text{ in.}$$

We observe in Table 36.5 that values of the drift Δ_x for all stories are well below the maximum permissible drift.

7. Natural period using Rayleigh's formula

$$T = 2\pi \sqrt{\sum_{i=1}^{N} W_i \delta_i^2 / g \sum_{i=1}^{N} f_i \delta_i} \qquad [\text{eq.}(36.4)]$$

$$T = 2\pi \sqrt{\frac{38.72}{386 \times 17.42}} = 0.48 \text{ sec}$$

Table 36.5. Story Drifts and Lateral Displacements

Level	Story Shear V_x (kip)	Story Stiffness K_i (kip/in.)	Story Drift Δ_x (in.)	Lateral Displacement δ_x (in.)
4	49.5	1,757.7	0.028	0.164
3	94.9	1,757.7	0.054	0.136
2	125.2	3,236.0	0.039	0.082
1	140.3	3,236.0	0.043	0.043

Table 36.6. Calculations for Rayleigh's Formula

Level	Weight W_i (kip)	Displacement δ_i (in.)	Lateral Force F_i (kip)	$W_i \delta_i^2$	$F_i \delta_i$
4	645.12	0.164	49.5	17.35	8.12
3	789.12	0.136	45.4	14.60	6.17
2	789.12	0.082	30.3	5.31	2.48
1	789.12	0.043	15.1	1.46	0.65
				$\Sigma = 38.72$	$\Sigma = 17.42$

The results of the necessary calculations are shown in Table 36.6.

Because Rayleigh's formula yields a value for the fundamental period ($T = 0.48$ sec) somewhat lower than the approximate value used in the calculations ($T = 0.55$ sec), the seismic analysis could be recalculated using $T = 0.48$ sec.

8. Overturning moments. The overturning moment at each level of the building is given by eq.(36.10) as

$$M_x = F_t(h_N - h_x) + \sum_{i=x+1}^{N} F_i(h_i - h_x)$$

where $x = 0, 1, 2, 3, \ldots, N-1$.

Table 36.7 shows the results of the necessary calculations for this example.

9. Accidental torsional moments

$$T_x = 0.05 \times D \times V_x$$
$$= 0.05 \times 96 \times V_x$$

Table 36.7. Overturning Moments and Torsional Moments

Level x	Lateral Force F_x (kip)	Story Height H_x (ft)	Overturning Moment M_x (kip-ft)	Shear Force V_x (kip)	Torsional Moment T_x (kip-ft)
4	49.5	12	—	49.5	238
3	45.4	12	594	94.9	456
2	30.3	12	1,724	125.2	601
1	15.1	12	3,235	140.3	673
Base	—	—	4,919	—	—

Table 36.8.   Ratio of Secondary Moment to Primary Moment

Level x	Story Weight (kip)	Overlying Weight P_x(kip)	Story Drift Δ_x(in.)	Story Shear V_x(kip)	Story Height H_x(in.)	$M_{xs}/M_{xp} = P_x\Delta_x/V_xH_x$
4	645.1	645	0.028	49.5	144	0.002
3	789.1	1,434	0.054	94.9	144	0.006
2	789.1	2,223	0.039	125.2	144	0.005
1	789.1	3,012	0.043	140.3	144	0.008

Table 36.7 shows the necessary calculations for the torsional moments considering only the accidental torsion at the various levels of the building.

10. The P-Δ effect. The code stipulates that the P-Δ effect need not be evaluated if the ratio of the secondary moment M_{xs} to the primary moment M_{xp} at each level of the building is less than 0.10. The results of the necessary calculations to evaluate this ratio [eq.(36.11)] are shown in Table 36.8. This table shows that the largest P-Δ moment ratio is 0.008, which is well below the Code stipulation of 0.1. Consequently, there is no need to account for the P-Δ effect.

11. Diaphragm design force. The Code requires that horizontal diaphragms (floors and roof) be designed to resist the force

$$F_{px} = \frac{F_t + \sum\limits_{i=x}^{N} F_i}{\sum\limits_{i=x}^{N} W_i} W_{px} \qquad \text{[eq.(36.12)]}$$

which need not be greater than $0.75ZIW_{px}$, but shall not be less than $0.35ZIW_{px}$. Table 36.9 contains the results of the necessary calculations to determine the diaphragm design forces at the various levels of the building. The minimum values of F_{px}, shown in the last column, in this case, are the design forces for the diaphragms of the building.

36.2.9 Program 36A UBC-91: Equivalent Static Lateral Force Method

Program 36A implements the provisions of the Uniform Building Code of 1991 using the static lateral

Table 36.9.   Diaphragm Forces

Level x	F_i (kip)	W_i (kip)	ΣF_i (kip)	ΣW_j (kip)	F_{px}[eq.(36.12)] (kip)	F_{px}min. (kip)
4	49.5	645.1	49.5	645.1	49.5	67.7
3	45.4	789.1	94.9	1,434.2	52.2	82.8
2	30.3	789.1	125.2	2,223.3	44.4	82.8
1	15.1	789.1	140.3	3,012.4	36.8	82.8

force method. The implementation of the program estimates the fundamental period T of the building from eq.(36.3), and calculates the total base shear force V from eq.(36.1) and the lateral forces F_x and F_t from eqs.(36.5) and (36.6). The program calculates at each level of the building the shear force V_x, the overturning moment M_x, the torsional moment T_x, the lateral displacement δ_x, the diaphragm design force F_{px}, and the story drift Δ_x. The secondary moment M_{xs}, the primary moment M_{xp}, and their ratio θ_x using eq.(36.11) are also calculated. Finally, the fundamental period T is recalculated using Rayleigh's formula, eq.(36.4).

Example 36.2

Use Program 36A to solve Example 36.1 after modeling the building as a plane orthogonal frame (Fig. 36.4). Assume all horizontal members to be 12 in. by 24 in. of reinforced concrete beams.

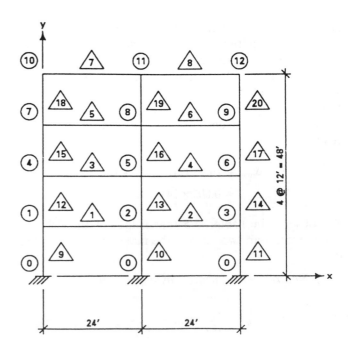

Fig. 36.4.   Plane frame modeling the four-story building of Example 36.2

Solution

Program 29 (Paz 1991), which is included in the set of programs for this chapter, is designed to model the structure as a plane frame. The program uses the stiffness method and static condensation method to determine the reduced stiffness matrix that corresponds to the lateral displacement coordinates at the various levels of the building. Next, Program 36A is executed to implement the provisions of UBC-91. The implementation of this program requires numbering the joints of the frame consecutively excluding the fixed joints at the foundation (labeled zero) as shown in Fig. 36.4. In this model each vertical or horizontal member represents all nine elements or members along the building. The flexural stiffness values EI for the members of the plane frame are calculated as follows:

Horizontal members

$$EI = 9 \times 3,000 \times 12 \times 24^3/12$$
$$= 0.37325 \times 10^9 \, (\text{kip-in}^2)$$

Exterior columns

First and second stories: $EI = 0.216 \times 10^9$ (kip-in^2)
Third and fourth stories: $EI = 0.111 \times 10^9$

Interior columns

First and second stories: $EI = 0.373 \times 10^9$ (kip-in^2)
Third and fourth stories: $EI = 0.216 \times 10^9$

The numerical results given by the computer for this example are the same as the values obtained by hand calculation in Example 36.1, except for the values of the story drifts and lateral displacements. For the building modeled as a plane frame, the story drifts and lateral displacements are slightly more than twice the values obtained in Example 36.1, for the structure modeled as a shear building.

36.2.9.1 Input Data and Output Results

```
PROGRAM 36A: UBC-91 (STATIC METHOD)          DATA FILE: D'36A

          INPUT DATA:

NUMBER OF STORIES                  N= 4
SEISMIC ZONE                       NZ$= 3
OCCUPANCY IMPORTANCE FACTOR        I= 1
STRUCTURAL SYSTEM FACTOR           RW= 12
SOIL COEFFICIENT                   S= 1
NORMAL BUILDING DIMENSION (FT)     D= 96
```

STORY #	STORY HIGHT FEET	STORY WEIGHT KIP
4	12.00	645.12
3	12.00	789.12
2	12.00	789.12
1	12.00	789.12

	COORDINATE MASS CENTER	COORDINATE STIFFNESS CENTER
1	0.00	0.00
2	0.00	0.00
3	0.00	0.00
4	0.00	0.00

```
ESTIMATE NATURAL PERIOD MENU:

1. USE  CT=0.035 FOR STEEL MOMENT-RESISTING FRAMES

2. USE  CT=0.030 FOR REINFORCED CONCRETE MOMENT RESISTING FRAMES
                 AND ECCENTRIC BRACED FRAMES

3. USE  CT=.020 FOR ALL OTHER BUILDINGS

4. USE ESTIMATED VALUE FOR THE PERIOD

        SELECT NUMBER ? 2


        OUTPUT RESULTS:

SEISMIC ZONE FACTOR        Z= .3
FUNDAMENTAL PERIOD         T= .547  SEC
SOIL FACTOR                S= 1
DYNAMIC FACTOR             C= 1.8689
TOTAL BASE SHEAR           V= 140.75  KIP
```

```
DISTRIBUTION LATERAL FORCES:
```

LEVEL	LATERAL FORCE (KIP)	SHEAR FORCE (KIP)
4	49.65	49.65
3	45.55	95.20
2	30.37	125.57
1	15.18	140.75

```
FLEXIBLE DIAPHRAGMS (Y/N) ? N

CONSIDER ONLY ACCIDENTAL ECCENTRICITY (Y/N) ? Y

TORSIONAL AND OVERTURNING MOMENTS:
```

LEVEL	OVERTURNING MOMENT (KIP-FT)	TORSIONAL MOMENT (KIP-FT)
4	0.00	238.32
3	595.80	456.96
2	1738.21	602.72
1	3245.01	675.60
0	4934.01	0.00

```
DISPLACEMENT AND STORY DRIFT:
```

LEVEL	DRIFT (IN)	DISPLACEMENT (IN)	MAX. PERMITTED DRIFT (IN)
4	0.062	0.369	0.480
3	0.114	0.307	0.480
2	0.112	0.193	0.480
1	0.081	0.081	0.480

```
                (KIP)

PERIOD BY RAYLEIGH'S FORMULA:    T= .714  SEC
```

DIAPHRAGM FORCES:

LEVEL	DIAPHRAGM FORCE
4	67.74
3	82.86
2	82.86
1	82.86

P-DELTA EFFECT:

LEVEL	LATERAL FORCE DISPL.(IN)	P-DELTA DISPL.(IN)	TOTAL DISPL.(IN)	RATIO OF MOMENTS
4	0.369	0.004	0.373	0.006
3	0.307	0.004	0.311	0.012
2	0.193	0.003	0.196	0.014
1	0.081	0.001	0.082	0.012

36.3 IMPLEMENTATION OF THE UBC-91: DYNAMIC METHOD

36.3.1 Introduction

The Uniform Building Code of 1991 stipulates that when the dynamic method is used, it shall be implemented using accepted principles of dynamics and it shall conform to criteria established in the Code. This stipulation is generally implemented by the application of the modal superposition method, described in Chapter 4, Section 4.7. The code criteria include the following provisions:

1. The ground motion representation shall, as a minimum, be one having a 10% probability of being exceeded in 50 years, and may be one of the following:

(a) The normalized response spectra given in the Code and reproduced in this chapter as Fig. 36.5.[5]

(b) A site-specific spectrum based on the geologic, tectonic, seismologic, and soil characteristics associated with the specific site. Such a spectrum shall be developed for a damping ratio of 5% unless a different value can be justified.

(c) Ground motion time histories developed for the specific site shall be representative of actual earthquake motions. Response spectra developed from time histories, either individually or in combination, shall approximate the site design spectrum as indicated in the preceding point.

(d) For structures in Soil Profile Type S_4 that have a period greater than 0.7 second, the following shall apply:

   (i) The ground motion representation shall be developed in accordance with items (b) and (c).

   (ii) Possible amplification of building response due to the effects of soil-structure interaction and lengthening of the building period caused by inelastic behavior shall be considered.

   (iii) The base shear determined by these procedures may be reduced to a design base shear by dividing by a factor not greater than the appropriate R_W factor for the structure.

(e) The vertical component of the ground motion may be defined by scaling the corresponding

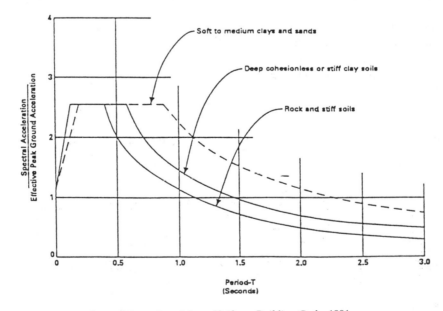

Fig. 36.5. Normalized response spectra shapes (Reproduced from *Uniform Building Code*, 1991, with permission of the publisher, the International Conference of Building Officials)

---

[5]Values obtained from the UBC-91 response spectra should be multiplied by the modification factor Z/R_W where Z is the seismic zone coefficient and R_W the structural factor. Also, the response spectrum curves provided by UBC-91 have a 5% damping as stated in the Commentary of *Recommended Lateral Force Requirements and Tentative Commentary* (SEAOC 1990).

horizontal acceleration by a factor of 2/3. Alternative factors may be used when substantiated by site-specific data.

2. The adopted mathematical model should represent the spatial distribution of the mass and stiffness of the structure to an extent adequate for the calculation of the significant features of its dynamic response.

3. The Code indicates that the dynamic analysis may be performed using (a) an elastic response spectrum in which all the significant maximum modal contributions are combined in a statistical manner to obtain approximately the total structural response,[6] or (b) a time-history response in which the dynamic response of the structure to a specific ground motion is obtained at each increment of time.

36.3.2 Scaling of Results

The base shear calculated by dynamic analysis shall be scaled up to match 100% of the base shear determined by the static lateral force procedure when the base shear force calculated by the dynamic method is less than that determined by the static lateral force method, for irregular buildings. For regular buildings, the base shear force shall be scaled up to match 90% of the base shear determined by the static lateral force procedure, except that it shall not be less than 80% of the value obtained when eq.(36.3) is used to determine the fundamental period of the structure. All corresponding response parameters, including member forces and moments, shall be adjusted proportionally.

The Code also stipulates that the base shear for a given direction, determined using dynamic analysis, need not exceed the value obtained by the static lateral force method. Therefore, in this case, the base shear force and all the response parameters may be scaled down to the value obtained by the static method.

36.3.3 Example 36.3

Consider again the four-story reinforced concrete building of Example 36.1. Model this building as a plane orthogonal frame and perform the seismic analysis by the UBC-91 dynamic method. Use the response spectra provided by the code (Fig. 36.5).

Problem Data (from Example 36.1)
Seismic weights:

Total seismic weight: $W = 3,012.4$ kip
Seismic zone factor: $Z = 0.3$ (for Zone 3)
Occupancy importance factor: $I = 1.0$ (warehouse)
Site coefficient: $S = 1.0$ (rock)

1. Modeling the structure. This structure is modeled as a plane orthogonal frame for which an appropriate computer program may be used such as Program 29 in Paz (1991). With reference to the four lateral displacement coordinates of the building, the reduced stiffness matrix is then given by

$$[K] = \begin{bmatrix} 5,611 & -2,916 & 458 & -45 \\ -2,916 & 3,863 & -1,885 & 269 \\ 458 & -1,885 & 2,893 & -1,398 \\ -45 & 259 & -1,398 & 1,168 \end{bmatrix} \text{kip/in.}$$

and the mass matrix (W_i/g) by

$$[M] = \begin{bmatrix} 2.036 & 0 & 0 & 0 \\ 0 & 2.036 & 0 & 0 \\ 0 & 0 & 2.036 & 0 \\ 0 & 0 & 0 & 1.671 \end{bmatrix} \text{kip·sec}^2\text{/in.}$$

2. Natural periods and modal shapes. The natural frequencies and the normalized modal shapes are obtained by solving the eigenproblem

$$[[K] - \omega^2[M]]\{\phi\} = \{0\} \tag{e}$$

The roots of the corresponding characteristic equation

$$[[K] - \omega^2[M]] = 0 \tag{f}$$

are

$$\omega_1^2 = 77.22, \quad \omega_2^2 = 1,939.75$$
$$\omega_3^2 = 678.60, \quad \omega_4^2 = 4,052.78 \tag{g}$$

resulting in the natural frequencies $(f = \omega/2\pi)$

$$f_1 = 1.40 \text{ cps}, \quad f_3 = 7.01 \text{ cps}$$
$$f_2 = 4.14 \text{ cps}, \quad f_4 = 10.13 \text{ cps} \tag{h}$$

or natural periods $(T = 1/f)$

$$T_1 = 0.715 \text{ sec}, \quad T_3 = 0.143 \text{ sec}$$
$$T_2 = 0.241 \text{ sec}, \quad T_4 = 0.099 \text{ sec} \tag{i}$$

and the corresponding mode shapes arranged in the columns of the modal matrix are

$$[\Phi] = \begin{bmatrix} 0.11277 & -0.31074 & -0.35308 & 0.50518 \\ 0.27075 & -0.46790 & -0.11497 & -0.42860 \\ 0.43123 & -0.06907 & 0.50310 & 0.21286 \\ 0.51540 & 0.45452 & -0.34652 & -0.07766 \end{bmatrix} \tag{j}$$

---

[6]As indicated in Chapter 4, Section 4.7.2, this requirement is readily satisfied by including a sufficient number of modes with a total effective modal weight equal to 90% or more of the seismic weight.

Table 36.10. Modal Effective Weight and Modal Base Shear

Mode m	Modal Effective Weight		Modal Base Shear V_m (kip)
	W_m (kip)	(%)	
1	2,465	82	84
2	366	12	23
3	99	3	6
4	82	3	4

Total Weight = 3,012 kips

3. Spectral accelerations. The spectral accelerations (scaled down by the effective peak ground acceleration of 0.3*g* and divided by the structural factor $R_W = 12$ for the natural periods above obtained from the spectral chart for rock or stiff soils in Fig. 36.5) are

$$S_{a1} = 0.034g, \quad S_{a3} = 0.061g$$

$$S_{a2} = 0.063g, \quad S_{a4} = 0.050g$$

4. Effective modal weights. The effective modal weight is given by eq.(4.65) of Chapter 4 as

$$W_m = \frac{\left[\sum_{i=1}^{N} \phi_{im} W_i \right]^2}{\sum_{i=1}^{N} \phi_{im}^2 W_i}$$

Values obtained for W_m ($m = 1, 2, 3, 4$) are shown in Table 36.10. This table also shows the effective modal weight as a percentage of the total seismic weight of the building.

5. Modal base shear. The modal base shear is defined by eq.(4.63) as

$$V_m = W_m S_{am}$$

Numerical values of V_m are also given in Table 36.10. The total base shear force is then calculated, using the Square Root of the Sum of the Squares (SRSS) rule, from values in the last column of Table 36.10 as

$$V = \sqrt{(84)^2 + (23)^2 + (6)^2 + (4)^2} = 87.4 \text{ kip}$$

6. Scaling modal effective weight and modal base shear. The modal values for the effective weight and for the base shear force in Table 36.10 are scaled up by the ratio *r* of 90% of the base shear. This ratio is determined by the equivalent static lateral force method and the value for the base shear force calculated by the dynamic method. The base shear determined in Example 36.1 using the equivalent static lateral force method was equal to 140.75 kip, while the

value calculated in this example using the dynamic method is equal to 87.4 kip. Therefore, the scaling ratio is

$$r = \frac{0.90 \times 140.75}{87.4} = 1.45$$

Table 36.11 shows the result of scaling (by the factor $r = 1.45$) the values in Table 36.10 for the modal effective weight W_m and the modal base shear V_m.

7. Modal seismic force. The numerical values for the seismic coefficients C_{xm} and the seismic lateral forces F_{xm} calculated from eqs.(4.68) and (4.67) are shown respectively in Tables 36.12 and 36.13. The seismic forces calculated by using the Square Root of Sum of Squares (SRSS) rule [eq.(4.77), Chapter 4] are shown in the last column of Table 36.13.

8. Modal shear force. Values for the modal shear force V_{xm} at level *x* calculated using eq.(4.60) and results from Table 36.13 are given in Table 36.14.

Table 36.11. Scaled Values for Modal Effective Weight and Modal Base Shear

Mode m	Modal Effective Weight		Modal Base Shear V_m (kip)
	W_m (kip)	(%)	
1	3,584	82	122
2	531	12	33
3	144	3	8
4	119	3	5

Σ = 4,378 kips

Table 36.12. Modal Seismic Coefficient C_{xm}

Level	Mode 1	Mode 2	Mode 3	Mode 4
4	0.340	−0.780	1.141	−0.281
3	0.348	0.145	−2.027	0.942
2	0.219	0.982	0.463	−1.897
1	0.091	0.652	1.422	2.235

Table 36.13. Modal Seismic Force F_{xm} (kip)

Level	Mode 1	Mode 2	Mode 3	Mode 4	Design F_x Values
4	41	−26	9	−1	50
3	42	5	−16	5	46
2	27	32	4	−9	43
1	11	22	11	11	29

Table 36.14. Modal Shear Force V_{xm} (kip)

Level x	Mode 1	Mode 2	Mode 3	Mode 4	Design V_x Values
4	41	−26	9	−1	49
3	84	−21	−7	3	86
2	111	11	−3	−6	111
1	122	33	8	5	126

Table 36.15. Modal Lateral Displacement δ_{xm} (in.)

Level x	Mode 1	Mode 2	Mode 3	Mode 4	Design δ_x Values
4	0.321	−0.23	0.002	0.001	0.320
3	0.268	0.003	−0.005	0.000	0.266
2	0.169	0.023	0.000	−0.002	0.167
1	0.070	0.015	0.002	0.001	0.063

Table 36.16. Modal Story Drift Δ_{xm} (in.)

Story	Mode 1	Mode 2	Mode 3	Mode 4	Design Δ_x Values
4	0.053	−0.026	0.007	−0.001	0.059
3	0.099	−0.020	−0.005	0.002	0.101
2	0.099	0.008	−0.002	−0.003	0.099
1	0.070	0.015	0.003	0.001	0.071

Table 36.17. Modal Overturning Moments M_{xm} (kip-ft)

Level x	Mode 1	Mode 2	Mode 3	Mode 4	Design M_x Values
4	—	—	—	—	—
3	497	−310	109	−17	596
2	1,504	−562	24	22	1,605
1	2,832	−425	−17	−52	2,864
Base	4,293	−30	79	8	4,293

Table 36.18. Modal Torsional Moments T_{xm} (kip-ft)

Level x	Mode 1	Mode 2	Mode 3	Mode 4	Design T_x Values
4	199	−124	43	−7	238
3	402	−101	−35	15	416
2	531	54	−17	−30	534
1	584	−157	38	23	606

Table 36.19. Calculation of Ratio of Secondary to Primary Moments

Level x	Story Weight W_x(kip)	Above Weight P_x(kip)	Story Drift Δ_x(in.)	Story Shear V_x(kip)	Story Height H_x(in.)	$M_s/M_p = P_x\Delta_x/V_xH_x$
4	645.1	645	0.059	49	144	0.005
3	789.1	1,434	0.101	86	144	0.012
2	789.1	2,223	0.099	111	144	0.015
1	789.1	3,012	0.071	126	144	0.013

Design values for the story shear forces V_x calculated using the SRSS rule, are given in the last column of Table 36.14. Note that the values shown in Table 36.14 for the modal shear force at level $x = 1$ are precisely the values for the base shear force calculated by eq.(4.63) and shown in Table 36.11.

9. Modal lateral displacement. Values for the modal lateral displacement δ_{xm} at level x, calculated from eq.(4.72), are shown in Table 36.15, which also shows, in the last column, the design values for lateral displacements δ_x, calculated using the SRSS rule.

10. Modal story drift. Table 36.16 shows the values for modal story drift Δ_{xm} calculated by eq.(4.73). Design values for story drift Δ_x obtained using the SRSS rule, are given in the last column of this table. The maximum story drift permitted by the UBC-91 should not exceed $(0.04/R_W)H_x = (0.04/12)144 = 0.48$ in. or $0.005H_x$ $(0.005 \times 144 = 0.72$ in.), where H_x is the story height and R_W is the structural factor from Table 36.3. The design values δ_x calculated in Table 36.16 for this example are well below the maximum story drift permitted by the UBC-91.

11. Modal overturning moments. Table 36.17 shows the values for modal overturning moment calculated using eq.(4.74). The last column of this table gives the design values for overturning moments M_x calculated by the SRSS rule.

12. Modal torsional moments. Table 36.18 shows the values calculated from eq.(4.76) for the modal torsional moments assuming only an accidental eccentricity e_x of 5% at each level x of the building (for this example $e_x = 0.05 \times 96$ ft = 4.8 ft). Design values for torsional moments T_x calculated by the SRSS rule are shown in the last column of Table 36.18.

13. The P-Δ effect. The UBC-91 specifies that the P-Δ effect does not need to be considered when the ratio of the secondary moment $M_s = P_x\Delta_x$ to the overturning or primary moment M_p is less than 0.10, calculated for each level x of the building. The results of the necessary calculations to evaluate this ratio [eq.(36.11)] are shown in Table 36.19. The largest moment ratio in Table 36.19 is 0.013, which is well below the code limit of 0.1. Consequently, there is no need to account for the P-Δ effect.

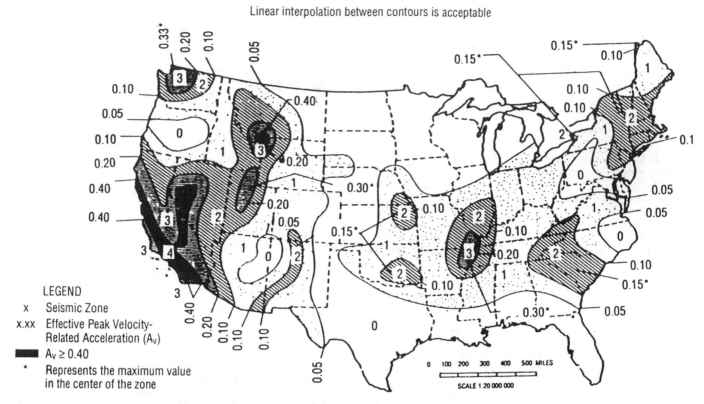

Fig. 36.6. Map of seismic zones and effective peak velocity-related acceleration (A_v) for United States of America (Reproduced from *BOCA National Basic Building Code*, 1990, publisher Building Officials and Code Administrators International)

36.3.4 Program 36B UBC-91: Dynamic Lateral Force Method

The determination of the seismic response of a building by modal superposition and response spectra requires the calculation of the natural frequencies (or natural periods) and the modal shapes. This calculation can be obtained by determining the stiffness and mass matrices of the structure and solving the corresponding eigenproblem.

The stiffness and mass matrices may be obtained by the execution of one of the following programs for modeling buildings: (1) Program 27 for modeling structures as shear buildings, (2) Program 28 for modeling structures as cantilever buildings, or (3) Program 29 for modeling structures as plane orthogonal frames. The solution of the corresponding eigenproblem can then be obtained by the execution of Program 30 (Jacobi method) or Program 31 (subspace iteration method) to determine natural frequencies and normal modes. Program 36B uses the following analytical expressions[7] of the response spectra provided by the UBC-91 in Fig. 36.5:

Soil Type S1:

$$S_a = 1 + 10T \quad \text{for} \quad 0 < T \le 0.15 \text{ sec}$$
$$S_a = 2.5 \quad \text{for} \quad 0.15 < T \le 0.39 \text{ sec}$$
$$S_a = 0.975/T \quad \text{for} \quad T > 0.39 \text{ sec}$$

Soil Type S2:

$$S_a = 1 + 10T \quad \text{for} \quad 0 < T \le 0.15 \text{ sec}$$
$$S_a = 2.5 \quad \text{for} \quad 0.15 < T \le 0.585 \text{ sec}$$
$$S_a = 1.463/T \quad \text{for} \quad T > 0.585 \text{ sec}$$

Soil Type S3:

$$S_a = 1 + 75T \quad \text{for} \quad 0 < T \le 0.2 \text{ sec}$$
$$S_a = 2.5 \quad \text{for} \quad 0.2 < T \le 0.915 \text{ sec}$$
$$S_a = 2.288/T \quad \text{for} \quad T > 0.915 \text{ sec}$$

where S_a is the spectral acceleration normalized to a peak ground acceleration of $1.0g$ and T is the fundamental period of the building.

The program calculates the effective modal weights W_m from eq.(4.65) and the modal base shear force V_m

---

[7]*Recommended Lateral Force Requirements and Tentative Commentary*, SEAOC 1990.

from eq.(4.63). Then it requests information for scaling these effective values when the base shear calculated by the equivalent lateral force method differs from the value calculated by the dynamic method.

The output from Program 36B includes the modal values for the lowest four modes and the design values for each level of the building as follows:

1. Seismic lateral forces F_x from eqs.(4.67), (4.68), and (4.77)
2. Shear forces V_x from eqs.(4.60) and (4.77)
3. Lateral displacements δ_x from eqs.(4.69) and (4.77)
4. Story drifts Δ_x from eqs.(4.73) and (4.77).

Example 36.4

Solve Example 36.3 using (a) Program 29 to model the structure as a plane orthogonal frame, (b) Program 30 to calculate the natural periods and normal modes, and (c) Program 36B for earthquake-resistant design by the dynamic method of the UBC-91.

36.3.4.1 Input Data and Output Results

```
PROGRAM 36B: UBC-91 (DYNAMIC METHOD)        DATA FILE: D36B

          RESULTS FROM PREVIOUS PROGRAM:

NUMBER OF STORIES:          ND= 4

  STORY DATA:
STORY #  STORY HEIGHT  STORY WEIGHT
           (FT)          (KIP)
   4      12.00          645.12
   3      12.00          789.12
   2      12.00          789.12
   1      12.00          789.12

STORY        COORDINATE         COORDINATE
            MASS CENTER       STIFFNESS CENTER
  1            0.00               0.00
  2            0.00               0.00
  3            0.00               0.00
  4            0.00               0.00

  THE NATURAL PERIODS (SEC) ARE:

 0.715      0.241      0.143      0.099
            INPUT DATA

SEISMIC FACTOR                         Z= .3
IMPORTANCE FACTOR                      I= 1
STRUCTURAL FACTOR                      RW= 12
BUILDING LENGTH NORMAL TO FORCES (FEET) LBN= 96

         SOIL PROFILE MENU

  1.- ROCK AND STIFF SOILS
  2.- DEEP COHESIONLESS OR STIFF CLAY SOILS
  3.- SOFT MEDIUM CLAYS OR SANDS
       SELECT NUMBER ? 1

NUMBER OF MODES IN THE RESPONSE = ? 4
```

```
OUTPUT RESULTS:

SPECTRAL ACCELERATION:

  0.034     0.063     0.061      0.050

MODAL EFFECTIVE WEIGHT Wm AND EFFECTIVE BASE SHEAR Vm:

MODE    MODAL EFFECTIVE WEIGHT    MODAL BASE SHEAR
           (KIP)        %            (KIP)
  1      2464.98      81.83          84.03
  2       365.73      12.14          22.86
  3        99.41       3.30           6.03
  4        82.37       2.73           4.09


        DESIGN BASE SHEAR (KIP)= 87.39186
SCALE THE RESPONSE, AS INDICATED BY THE UBC-91 (Y/N)? Y

BASE SHEAR FORCE DETERMINED BY STATIC METHOD: ? 126.7

      SCALE FACTOR FOR BASE SHEAR = 1.449792

MODAL EFFECTIVE WEIGHT Wm AND EFFECTIVE BASE SHEAR Vm:

MODE    MODAL EFFECTIVE WEIGHT    MODAL BASE SHEAR
           (KIP)        %            (KIP)
  1      3573.71      81.83         121.83
  2       530.23      12.14          33.14
  3       144.12       3.30           8.74
  4       119.41       2.73           5.93

      DESIGN BASE SHEAR (KIP)= 126.7

    FLEXIBLE DIAPHRAGMS (Y/N) ? N

**DIAPHRAGMS NOT FLEXIBLE**: CALCULATE TORSION

DESIGN FOR ONLY ACCIDENTAL ECCENTRICITY (Y/N) ? Y

        ***DESIGN VALUES***

LEVEL  LATERAL FORCE  STORY SHEAR  LAT. DISPL.  STORY DRIFT  MAX.DRIFT
         (KIP)          (KIP)        (IN)         (IN)        (IN)
  4      49.96          49.96        0.323        0.059        0.48
  3      46.63          87.06        0.269        0.102        0.48
  2      43.77         111.62        0.171        0.099        0.48
  1      30.36         126.70        0.072        0.072        0.48


LEVEL   OVERTURNING    TORSIONAL   DIAPHRAGM  P-DELTA EFFECT
       MOMENT(K-FT)  MOMENT(K-FT) FORCE(KIP) MOMENT RATIO
  4       599.50        239.80       67.74       0.005
  3      1608.93        417.90       82.86       0.012
  2      2867.68        535.76       82.86       0.015
  1      4298.30        608.16       82.86       0.013
```

36.4 IMPLEMENTATION OF BOCA-90 AND STANDARD CODE-91

36.4.1 Introduction

The provisions of the 1990 BOCA (Building Officials and Code Administrators) National *Basic Building Code* published by the American Insurance Association (1976) are virtually identical to those of the 1991 Standard Building Code and the ASCE *Standard Minimum Design Loads for Buildings and Other Structures* (1988). The ASCE Standard is a revision of ANSI A58.1-1982. The difference in the provisions of these three documents lies in the description of the

Table 36.20. Occupancy Importance Factor, *I*

Nature of Occupancy	*I* Factor
All buildings and structures except those listed below	1.00
Buildings and structures of Use Group A in which more than 300 people congregate in one area	1.25
Buildings and structures designated as essential facilities including, but not limited to:	1.50
1. Hospitals having surgery or emergency treatment areas	
2. Fire or rescue and police stations	
3. Primary communication facilities and disaster operation centers	
4. Power stations and other utilities required in an emergency	
5. Structures having critical national defense capabilities	
6. Designated shelters for hurricanes	

various coefficients in tables for occupancy importance factor *I*, soil profile coefficient *S*, and horizontal force factor *K*. The tables for these coefficients reproduced in this section are taken from BOCA-1990.

36.4.2 Base Shear Force

The following formulas are equally valid for the codes BOCA-90 and STANDARD-91. These codes stipulate that all structures (with some exceptions) shall be designed and constructed to resist minimum lateral forces assumed to act nonconcurrently in the direction of each of the main axes of the structure. The resulting total base shear force *V* is given by the following formula:

$$V = 2.5A_v IKCSW \qquad (36.13a)$$

or by the equivalent formula as given in the ASCE-1988 Standard

$$V = ZIKCSW \qquad (36.13b)$$

The various factors in eqs.(35.13) are defined as follows:

A_v factor. The factor A_v is the Effective Peak Velocity-related acceleration determined from the map of seismic zones and the effective peak velocity A_v as indicated in Fig. 36.6. The factor A_v shall be determined between areas of contour lines by linear interpolation; A_v shall not be less than 0.05, and is not required to exceed 0.40.

I factor. The factor *I* is the Occupancy Importance Factor. Numerical values for this factor are given in Table 36.20.

Table 36.21. Horizontal Force Factor, *K*

Arrangement of Lateral Force-Resisting Elements	Value of *K*
Bearing wall system: A structural system with bearing walls providing support for all or major portions of the vertical loads. Seismic force resistance is provided by shear walls or braced frames in accordance with Section 1113.9.1, utilizing:	
Seismically reinforced masonry shear walls or braced frames[a]	1.33
Other masonry shear walls or braced frames[b]	
For Seismic Zones 0 and 1	1.33
For Seismic Zone 2	4.00
Reinforced concrete shear walls or braced frames	1.33
One-, two-, or three-story light wood or metal-frame walls	1.00
Building framing system: A structural system with an essentially complete space frame providing support for vertical loads. Seismic force resistance is provided by shear walls or braced frames in accordance with Section 1113.9.2	1.00
Moment-resisting frame system: A structural system with an essentially complete space frame providing support for vertical loads. Seismic force resistance is provided by moment-resisting frame systems in conformance with:	
Requirements for ordinary concrete frames	1.50
Sections 1113.9.3.1 and 1113.9.3.2 for ordinary steel frames	1.00
Section 1113.9.3.4 for semiductile concrete frames	1.00
Sections 1113.9.3.1 and 1113.9.3.3 for special frames	0.67
Dual system: A structural system with an essentially complete space frame providing support for vertical loads. Seismic force resistance is provided by a combination of:	
Special moment-resisting frame systems and shear walls or braced frames in accordance with Section 1113.9.4.1; or	0.80
Semiductile moment-resisting frames and shear walls or braced frames in accordance with Section 1113.9.4.2	1.00
Elevated tanks: Tanks plus full contents, where tanks are supported on four or more cross-braced legs and not supported by a building	2.50[c]
Structures other than buildings: Structures other than buildings and other than those set forth in Table 1113.10	2.00

Note a. Defined in Section 1113.2.
Note b. Not permitted in Seismic Zones 3 and 4.
Note c. The minimum value of KC shall be 0.12. The maximum value of KCS is not required to exceed 0.29 (or 0.23 for Soil Profile 3 in Seismic Zones 3 and 4). The tower shall be designed for an accidental torsion of 5% as specified in Section *1113.5.5*. Elevated tanks that are supported by buildings or do not conform to the type or arrangement of supporting elements as described herein shall be designed in accordance with Section *1113.10* using $C_p = 0.3$.

K factor. Values for the horizontal force factor *K* are shown in Table 36.21.

C factor. The value of the factor *C* shall be determined as

$$C = \frac{1}{15\sqrt{T}} \leqslant 0.12 \tag{36.14}$$

where *T* is the fundamental period of the building. The product *CS*, where *S* is the soil-profile coefficient, is not required to exceed 0.14; this product need not exceed 0.11 for soil profile S_3 in Seismic Zones 3 and 4.

The seismic code provides the Rayleigh formula to determine the fundamental period *T*:

$$T = 2\pi \sqrt{\frac{\sum_{i=1}^{N} W_i \sigma_i^2}{g \sum_{i=1}^{N} f_i \sigma_i}} \tag{36.15}$$

In this equation, σ_i ($i = 1, 2, \ldots, N$) are the lateral displacements at the floor levels of the building produced by statically applied forces f_i such as those calculated by eqs.(36.17) and (36.18), and W_i are seismic weights at the various levels of the building. The value of *C*, calculated using the period *T* determined by eq.(36.15), shall be not less than 80% of the value of *C* based on the period *T*, estimated by the appropriate equation [see eqs.(36.16a), (36.16b), and (36.16c)]. As the lateral displacements are not yet known, eq.(36.15) cannot be used initially. However, the code provides the following empirical formulas to estimate the fundamental period *T*:

1. For buildings with shear walls or exterior concrete frames utilizing deep beams or wide piers, or both

    $$T = \frac{0.05 h_N}{\sqrt{D}} \tag{36.16a}$$

    where *D* is the dimension of the building (in ft) parallel to the direction of the applied lateral forces and h_N is the height of the building.

2. For buildings with braced or isolated shear walls not interconnected by frames, or for braced frames

    $$T = \frac{0.05 h_N}{\sqrt{D_s}} \tag{36.16b}$$

    where D_s is the longest dimension of a shear wall or braced frame (ft) parallel to the applied forces.

3. For buildings in which the lateral force-resisting system consists of moment-resisting frames capable of resisting 100% of the required lateral forces, such that the frames are not closed by or adjoined by

Table 36.22. Soil Profile Coefficient, *S*

Soil Profile	Description
$S_1 = 1.0$	(a) Rock of any characteristic that is either shalelike or crystalline in nature. Such material is characterized by a shear wave velocity greater than 2,500 feet/sec (762 m/sec); or
	(b) Stiff soil conditions where the soil depth is less than 200 feet (61 m) and the soil types overlying rock are stable deposits of sands, gravels, or stiff clays.
$S_2 = 1.2$	A profile with deep cohesionless deposits or stiff clay conditions, including sites where the soil depth exceeds 200 feet (61 m) and the soil types overlying rock are stable deposits of sands, gravels, or stiff clays.
$S_3 = 1.5$	A profile with soft to medium-stiff clays and sands, characterized by 30 feet (9.1 m) or more of soft- to medium-stiff clays without intervening layers of sand or other cohesionless soils. In locations where the soil properties are not known in sufficient detail to determine the soil profile type, or the profile does not fit any of the three types, soil profile S_2 or S_3 shall be used, whichever gives the larger value of *CS*.

more rigid elements tending to prevent the frame from resisting lateral forces:

$$T = C_T h_N^{3/4} \tag{36.16c}$$

where $C_T = 0.035$ for steel frames and 0.030 for concrete frames.

S factor. The values for *S*, the soil profile coefficient, are given in Table 36.22.

Z factor. The seismic zone factor *Z* is given on the map of seismic zones, Fig. 36.6, and is equal to 2.5 times the effective peak velocity-related acceleration A_v; this relationship renders eqs.(36.13a) and (36.13b) identical.

W factor. The factor *W* is equal to the total dead load of the building, including partition loading, plus 25% of the floor live load. The snow load shall be included in the value of *W* when the design snow load is greater than 30 lb/ft². However, a reduction of the snow load up to 75% is permitted in areas subjected to short durations of snow loads.

36.4.3 Distribution of Lateral Forces

For structures having regular shapes or framing systems, the total base shear force *V* calculated from eq.(36.13) is distributed over the height of the structure as a force at each level F_x, plus an additional force F_t (at the top of the building), as follows:

$$F_x = \frac{(V - F_t) W_x h_x}{\sum_{i=1}^{N} W_i h_i} \tag{36.17}$$

$$F_t = 0.07TV \leqslant 0.25V \quad \text{for} \quad T > 0.7 \text{ sec}$$
$$F_t = 0 \qquad\qquad\qquad \text{for} \quad T < 0.7 \text{ sec} \quad (36.18)$$

$$V = F_t + \sum_{i=1}^{N} F_i \quad (36.19)$$

In these equations, N is the number of stories above the base, h_x or h_i the height of level x or level i, and W_x or W_i the weight of level x or level i.

At each level, the force F_x is applied over the area of the building according to the mass distribution on that level. The lateral forces given by eqs.(36.17) and (36.18) are applicable only to regular structures. For structures having irregular shapes, large differences in lateral resistance (or stiffness) between adjacent stories, or other unusual structural features, lateral forces must be determined considering the dynamic characteristics of the structure.

36.4.4 Horizontal Torsional Moments

The design shall include both the torsional moment resulting from the location of the building masses and the torsional moments caused by assumed displacement of the mass by a distance, in each way, equal to 5% of the dimension of the building perpendicular to the direction of the applied forces.

36.4.5 Horizontal Shear Force

The story shear force V_x at level x is the sum of the seismic lateral forces at and above that level:

$$V_x = F_t + \sum_{i=x}^{N} F_i \quad (36.20)$$

The shear force V_x in any horizontal plane shall be distributed to the various elements of the lateral force-resisting system in proportion to their relative stiffness.

36.4.6 Diaphragms

Floor and roof diaphragms shall be designed to resist forces determined in accordance with the following formula:

$$F_{px} = \frac{F_t + \sum_{i=x}^{N} F_i}{\sum_{i=x}^{N} W_i} W_{px} \quad (36.21)$$

F_i and F_t are the lateral seismic forces determined by eqs.(36.17) and (36.18), W_i is the seismic weight, and

W_{px} is the weight of the diaphragm and attached parts at the various levels of the building. The force F_{px} is not required to exceed the product $0.75A_v I W_{px}$.

36.4.7 Overturning Moments

The overturning moment M_x at each level due to the lateral forces F_t and F_i is calculated as follows:

$$M_x = F_t(h_N - h_x) + \sum_{i=x+1}^{N} F_i(h_i - h_x) \quad (36.22)$$

where $x = 0, 1, 2, \ldots, N-1$.

The forces resulting from the overturning moment shall be distributed to the resisting elements of the story in the same proportion as the distribution of the horizontal shear forces. In tall buildings, the maximum overturning moment in any element shall be determined by multiplying M_{x-1} in eq.(36.22) by a factor k, depending on the location of the element, as follows:

$k = 1.0$ for the top 10 stories
$k = 0.8$ for the 20th story from the top and those below
$k = $ a value between 1.0 and 0.8 determined by linear interpolation for stories between the 10th and the 20th stories below the top.

36.4.8 Story Drift and Building Separation

Relative lateral deflections between consecutive levels of the building, that is, story drift, shall not exceed 0.005 times the story height (0.0025 times the story height in buildings with unreinforced masonry) unless it can be demonstrated that greater deformation can be tolerated. Also, an amplification factor $1/K$ must be applied to the calculated displacements when the horizontal force factor K is less than 1. It must be recognized that, regardless of the reduced loads permitted for ductile structures, the actual deformation may be produced by forces larger than the design forces, that is, forces not reduced by the K factor.

36.4.9 Dynamic Method

To establish the lateral forces and their distribution, the Code permits the design of the structure by either elastic or inelastic dynamic analysis. In such an analysis, the dynamic characteristics of the structure shall be considered and the following principles shall be observed:

1. The base shear shall not be less than 90% of that determined by eq.(36.13) using the equivalent static lateral force method.
2. Values for the horizontal force factor ($K = 0.67$ to 2.5), given in Table 36.21, are applicable only if the

structure is designed and detailed with special requirements to provide the necessary level of ductility. Otherwise, the structure shall be designed for a base shear consistent with its ability to dissipate energy by inelastic cyclic straining, which will generally require a value of *K* from 2.5 to 4.0 or greater.

3. Story drift and building separation shall be satisfied using the forces determined by eqs.(36.17) and (36.18).

4. The input to the dynamic analysis shall be either a smoothed response spectrum or a suitable ground motion time history (one that is approved by the code official and reflects the characteristics of the structure site). In either case, the input shall be scaled in accordance with the preceding three principles.

36.4.10 Computer Program 36C

Computer program 36C implements the provisions of the U.S.A. seismic code, BOCA-90, using the equivalent static lateral force method.

Example 36.5

Use Program 36C to solve Example 36.1 according to the seismic code BOCA-90.

36.4.10.1 Input Data and Output Results

```
        INPUT DATA:


ZONE NUMBER                              NZ= 3
EFFECTIVE VELOCITY RELATED ACCELERATION  AV= .3
  OCCUPANCY IMPORTANCE FACTOR            I= 1
  HORIZONTAL FORCE FACTOR                K= .67
  SOIL PROFILE FACTOR                    S= 1


    BUILDING DATA:


LARGEST SHEAR WALL OR BRACED FRAME  (FEET)  DM= 0
BUILDING DIMENSION, FORCE DIRECTION (FEET)  D= 48
BUILDING DIMENSION NORMAL TO FORCES (FEET)  DN= 96
NUMBER OF STORIES                           N= 4

STORY #   STORY HIGHT   STORY WEIGHT
            (FEET)        (KIP)
   4        12.00         645.12
   3        12.00         789.12
   2        12.00         789.12
   1        12.00         789.12


   ESTIMATE NATURAL PERIOD MENU:

 1. USE  T=0.05*H/SQR(D) (BUILDINGS WITH SHEAR WALLS,ETC.)

 2. USE  T=0.05*H/SQR(DS) (BLD. WITH ISOLATE SHEAR WALLS, ETC)

 3. USE  T=0.035*H^3/4  (STEEL FRAMES)

 4. USE  T=0.030*H^3/4  (CONCRETE FRAMES)

 5. USE ASSUMED VALUE FOR T
```

```
SELECT NUMBER ? 4

  OUTPUT RESULTS:

SEISMIC ZONE FACTOR    Z= .75
FUNDAMENTAL PERIOD     T= .5470818
SOIL FACTOR            S= 1
DYNAMIC FACTOR         C= 9.013275E-02
TOTAL BASE SHEAR       V= 136.4404

      DISTRIBUTION LATERAL FORCES

LEVEL     LATERAL FORCE      SHEAR FORCE
             (KIP)             (KIP)
  4         48.13             48.13
  3         44.16             92.29
  2         29.44            121.72
  1         14.72            136.44


ASSUME ONLY ACCIDENTAL ECCENTRICITY (Y/N) ? Y


    TORSIONAL AND OVERTURNING MOMENTS

LEVEL   ECCENTRICITY   TORSIONAL MOMENT   OVERTURNING MOMENT
          (FT)           (KIP-FT)            (KIP-FT)
  4       0.00           231.02              577.56
  3       0.00           442.97             1684.98
  2       0.00           584.27             3145.65
  1       0.00           654.91             3587.20


      LATERAL DISPLACEMENTS


LEVEL    DRIFT     DISPLACEMENTS   MAX. DRIFT PERMITTED
         (IN)          (IN)              (IN)
  4      0.089        0.534            0.720
  3      0.165        0.444            0.720
  2      0.162        0.280            0.720
  1      0.117        0.117            0.720

PERIOD BY RAYLEIGH'S FORMULA:  T= .714 SEC

     DIAPHRAGM FORCES

LEVEL     DIAPHRAGM FORCE
             (KIP)
  4         67.74
  3         82.86
  2         82.86
  1         82.86
```

36.5 SIMPLIFIED THREE-DIMENSIONAL EARTHQUAKE-RESISTANT DESIGN OF BUILDINGS

36.5.1 Modeling of Buildings

The general provisions of most seismic codes allow the modeling of a building by concentrating its weight at the various horizontal levels (floor diaphragms) which are assumed to be rigid in their planes. At each level of the building three degrees of freedom are considered, two translational degrees of freedom on a horizontal plane along the main axes *X* and *Y* of the building and one torsional degree of freedom along the vertical axis *Z*.

The main lateral resisting structural elements are provided by columns, structural walls (shear walls), and bracing elements. The horizontal diaphragms at the various levels of the building may provide different degrees of fixity at the ends of columns and structural walls depending on the relative stiffness of the structural walls and columns and the horizontal diaphragms.

In general, the lateral stiffness, K, for a lateral resisting element may be expressed by the following formula:

$$K = \frac{1}{(L^3/\lambda EI) + (1.2L/GA)} \qquad (36.23)$$

with

$$G = \frac{E}{2(1 + \nu)}$$

where

E = modulus of elasticity
G = modulus of rigidity
I = cross-sectional moment of inertia
A = cross-sectional area
L = length of the element
ν = Poisson's ratio

The coefficient λ in eq.(36.23) has a theoretical value of 12 for a structural element that is completely fixed against rotation about both ends, and 3 for a cantilever element fixed in one end and free to rotate in the other end. Investigating studies [Blume, J. A. et al (1961), MacLeod, I. A. (1971)] on the seismic response behavior of multistory reinforced-concrete buildings have shown that the coefficient λ in

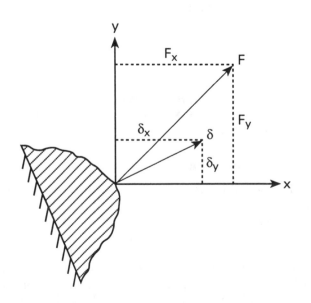

Fig. 36.7.   Schematic diagram of a linear biaxial stiffness element

eq.(36.23) has a value from 1.30 to 6.91 depending on the relative rigidity of columns and the floor slab. In practice, a numerical value of 3 is usually assumed for the coefficient λ.

36.5.2 Transformation of Stiffness Coefficients

The use of eq.(36.23) provides the values of the stiffness coefficients of the structural elements in reference to principal cross-sectional axes (x, y) of the element. Given a general layout of structural elements (columns, walls) at any story of the buildings, it is then necessary to transform the stiffness coefficients referred to local or element axes (x, y) to the global or system axes (X, Y).

Figure 36.7 shows a schematic diagram of linear biaxial stiffness element acted upon by a force F with components F_x and F_y, in the local axes, producing a displacement δ with components δ_x and δ_y along these axes. The forces f_{xx} and f_{yx}, in the x and y directions, required to produce the displacement δ_x are given by

$$f_{xx} = k_x \delta_x \quad \text{and} \quad f_{yx} = k_{yx} \delta_x \qquad (36.24)$$

Analogously, the forces f_{xy} and f_{yy} required to produce the displacement δ_y are

$$f_{xy} = k_{xy} \delta_y \quad \text{and} \quad f_{yy} = k_y \delta_y \qquad (36.25)$$

where k_x and k_y are the necessary forces, respectively, in the x and y directions resulting from a unit displacement in the corresponding directions; and k_{xy} ($k_{yx} = k_{xy}$ by Maxwell's reciprocity principle) is the force in the x direction resulting from a unit displacement in the y direction.

The force components f_x and f_y required to produce the displacements having components δ_x and δ_y are then the sums from eqs.(36.24) and (36.25):

$$f_x = k_x \delta_x + k_{xy} \delta_y \qquad (36.26)$$
$$f_y = k_{yx} \delta_x + k_y \delta_y$$

or

$$\{f\} = [k]\{\delta\} \qquad (36.27)$$

in which

$$[f] = \begin{bmatrix} f_x \\ f_y \end{bmatrix}, \quad [\delta] = \begin{bmatrix} \delta_x \\ \delta_y \end{bmatrix}, \quad [k] = \begin{bmatrix} k_x & k_{xy} \\ k_{yx} & k_y \end{bmatrix} \qquad (36.28)$$

The transformation of the stiffness coefficients k_x, k_y and $k_{xy} = k_{yx}$ from local axes x, y to the global axes

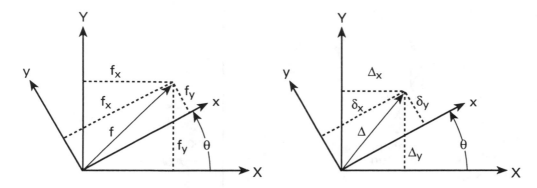

Fig. 36.8. Force and displacement diagrams for a linear biaxial stiffness element

X, Y (Fig. 36.8), requires the substitution into eq.(36.27) of

$$\{f\} = [T]\{F\} \quad \text{and} \quad \{\delta\} = [T]\{\Delta\} \quad (36.29)$$

which gives

$$\{F\} = [K]\{\Delta\} \quad (36.30)$$

with

$$[F] = \begin{bmatrix} F_X \\ F_Y \end{bmatrix}, \quad [\Delta] = \begin{bmatrix} \Delta_X \\ \Delta_Y \end{bmatrix}, \quad [K] = \begin{bmatrix} k_X & k_{XY} \\ k_{YX} & k_Y \end{bmatrix},$$

$$[T] = \begin{bmatrix} \cos\theta & \sin\theta \\ -\sin\theta & \cos\theta \end{bmatrix}$$

where

F_X, F_Y = force components in the X, Y direction
Δ_X, Δ_Y = displacement components in the X, Y direction
θ = angle between the global axis X and the element local axis x

and

$$[K] = [T]^T[k][T] \quad (36.31)$$

or

$$k_X = k_x \cos^2\theta + k_y \sin^2\theta - 2k_{xy}\sin\theta\cos\theta$$
$$k_Y = k_x \sin^2\theta + k_y \cos^2\theta + 2k_{xy}\sin\theta\cos\theta \quad (36.32)$$
$$k_{XY} = k_{YX} = (k_x - k_y)\sin\theta\cos\theta + k_{xy}$$

36.5.3 Center of Rigidity

The center of rigidity, or center of resistance, at a story of a building is defined as the point where the total stiffness $K_X = \Sigma k_X$ and $K_Y = \Sigma k_Y$ and $K_{XY} = K_{YX} = \Sigma k_{YX}$ of the story may be assumed to

be concentrated; these summations include all the structural elements in the story. The coordinates (X_r, Y_r) of the center of rigidity at a story are determined as the solution of eq.(36.33) and eq.(36.34) expressing that the sum of the moments of the forces (k_X, k_{YX}) in the structural elements due to a unit displacement in the X direction, or the sum of the moments of the forces (k_Y, k_{XY}) in the structural elements due to a unit displacement in the Y direction, are equal to the moment of the resultant forces (K_X, K_{YX}) or to the moment of the resultant forces (K_Y, K_{XY}), respectively produced by these displacements and shown in Figs. 36.9(a) and 36.9(b).

$$-K_{YX}X_r + K_X Y_r = \Sigma(k_X Y) - \Sigma(k_{YX} X_r) \quad (36.33)$$
$$-K_Y X_r + K_{XY} Y_r = \Sigma(k_{XY} Y) - \Sigma(k_Y X) \quad (36.34)$$

In equations (36.33) and (36.34) the summations include all the lateral resisting structural elements in the story.

36.5.4 Story Eccentricity

The story eccentricity is defined as the distance (measured in the direction normal to the assumed seismic force direction) between the center of rigidity and the center of the applied design lateral forces above the story considered. The center of the lateral seismic forces (X_{Fi}, Y_{Fi}) of the stories above the story i is calculated as

$$X_{Fi} = \frac{\sum\limits_{j=i}^{N} F_j x_{ij}}{\sum\limits_{j=i}^{N} F_j} \quad (36.35)$$

$$Y_{Fi} = \frac{\sum\limits_{j=i}^{N} F_j Y_{ij}}{\sum\limits_{j=i}^{N} F_j}$$

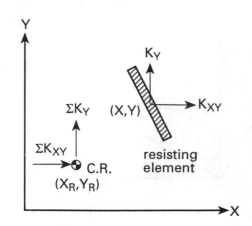 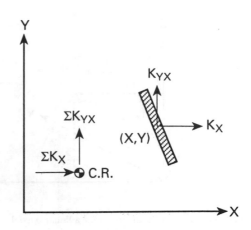

Fig. 36.9. Forces at the center of rigidity and on a resisting element due to a unit displacement (a) in the Y-direction; (b) in the X-direction

where

F_j = lateral force applied at level j

X_j, Y_j = centroidal coordinates of the mass center for story j (point of application of the seismic force F_j)

N = number of levels in the building.

The eccentricities e_{xi} and e_{yi}, at story i, in X and Y directions are then given by

$$e_{xi} = X_{ri} - X_{Fi}$$
$$e_{yi} = Y_{ri} - Y_{Fi} \qquad (36.36)$$

in which X_{ri}, Y_{ri} are the coordinates of the center of rigidity for story i [solution of eqs.(36.33) and (36.34)].

36.5.5 Rotational Stiffness

The rotational stiffness at story i is defined as the torsional moment necessary, at this story, to produce a unit rotation of the story. The rotational stiffness, K_{ti}, at story i is then given by

$$K_{ti} = \sum_j (k_{Xij}a_{2j}^2) + \sum_j (k_{Yji}a_{1j}^2) - 2\sum_j (k_{XYij}a_{1j}a_{2j}) \quad (36.37)$$

in which k_{Xij}, k_{Yij}, and k_{XYij} are the j-resisting element stiffness coefficients of the i story evaluated by eq.(36.32); and

$$a_{1j} = X_{ij} - X_{ri}$$
$$a_{2j} = Y_{ij} - Y_{ri} \qquad (36.38)$$

where X_{ij} and Y_{ij} are the centroidal coordinates of the resisting element j of story i.

The story rotation θ_i is then given by

$$\theta_i = M_{ti}/K_{ti} \qquad (36.39)$$

where the torsional moment M_{ti} is calculated by eq.(36.52), and the torsional stiffness K_{ti} by eq.(36.37).

36.5.6 Fundamental Period

Empirical formulas

1. $T = aN$ (36.40)

2. $T = bH/L$ (36.41)

3. $T = C_t H^x$ (36.42)

where the value of the coefficients a, b, C_t, and of the exponent x are prescribed by seismic codes, and H and L are respectively the total height and the length of the building as defined by the seismic code.

Rayleigh's formula

$$T = 2\pi \sqrt{\frac{\sum_{i=1}^{N} W_i \delta_i^2}{g \sum_{i=1}^{N} F_i \delta_i}} \qquad (36.43)$$

where

W_i = weight at level i

δ_i = lateral displacement at level i

F_i = lateral force at level i prescribed by the seismic code

g = acceleration of gravity

36.5.7 Seismic Factors

The following factors are generally required in the implementation of seismic codes:

1. Z_a = peak ground acceleration
2. Z_v = peak ground velocity
3. I = occupancy importance factor
4. K = structural type
5. R = structural reduction factor
6. S = soil profile factor
7. C = seismic coefficient given by

$$C_1 = \frac{1}{pT^x}; \quad C_2 = \frac{qS}{T^x} \qquad (36.44)$$

in which the numerical values of p, q, and x are prescribed by the seismic code; T and S are, respectively, the fundamental period of the building and the soil profile factor whose value is specified by the Code for different types of soils.

36.5.8 Base Shear Force

The base shear force V is determined in different seismic codes by formulas of the form

$$V = Z_a IKCSW \qquad (36.45)$$

or

$$V = \frac{Z_a IC}{R} \qquad (36.46)$$

where the values of the seismic factors Z_a, I, K, C, S are prescribed by the seismic code and W is the total seismic weight of the building which is calculated by

$$W = D + \lambda L \qquad (36.47)$$

where

D = dead load
L = design live load
λ = fraction of the live load prescribed by the seismic code

36.5.9 Equivalent Lateral Seismic Forces

The equivalent seismic forces F_i acting horizontally at the various levels of the building are given by

$$F_i = \frac{(V - F_t) W_i h_i^x}{\sum\limits_{j=1}^{N} W_j h_j^x} \qquad (36.48)$$

where

W_i = seismic weight at level i
h_i, h_j = height of level i, j from the base of the building
x = exponent prescribed by the seismic code (usually, $x = 1$)
F_t = additional force at the top of the building given by

$$F_t = 0.07TV \leqslant 0.25V \quad \text{for} \quad T > 0.7 \, \text{sec}$$
$$F_t = 0 \qquad\qquad \text{for} \quad T \leqslant 0.7 \, \text{sec} \qquad (36.49)$$

in which

T = fundamental period of the building
V = base shear force [eq.(36.45) or eq.(36.46)].

36.5.10 Overturning Moments

The overturning moment M_i at the level i of the building is calculated as

$$M_i = F_t(h_N - h_i) + \sum_{j=i+1}^{N} F_j(h_j - h_i) \qquad (36.50)$$

where

$i = 0, 1, 2, \ldots, N-1$
F_j = equivalent seismic force at level j
F_t = additional force at the top of the building
h_i, h_j, h_N = height to levels i, j, N from the base of the building
N = total number of levels in the building.

36.5.11 Story Shear Force

The story shear force V_i at level i is given by

$$V_i = F_i + \sum_{j=i}^{N} F_j \qquad (36.51)$$

36.5.12 Torsional Moments

Torsional moments M_{ti} are given at each story of the building as the product of the story shear force [eq.(36.51)] and the eccentricity at the story [eq.(36.36)] usually increased by a factor a and by an additive term e_0 to account for accidental eccentricity. Hence,

$$M_{ti} = (ae_i + e_0) V_{xi} \qquad (36.52)$$

36.5.13 Story Drift and Lateral Displacements

The elastic story drift or interstory displacement at story i, Δ_{ie}, due to the lateral shear force only, may be calculated by

$$\Delta_{ie} = \frac{V_i}{K_i} \qquad (36.53)$$

in which K_i is the total stiffness of story i and V_i is the story shear force. Most seismic codes specify an amplification of the story drift by dividing the calculated elastic story drift by the structural type factor K when $K < 1.0$ or multiplying the elastic story drift by the reduction factor R. Hence the amplified or inelastic story drift Δ_i is calculated by

$$\Delta_i = \frac{\Delta_{ie}}{K} \quad (K < 1.0) \tag{36.54}$$

or by

$$\Delta_i = R\Delta_{ie} \tag{36.55}$$

The lateral displacement δ_i at level i due to the lateral shear force only is then given by the summation of the story drifts of the above stories. That is,

$$\delta_i = \sum_{j=i}^{N} \Delta_i \tag{36.56}$$

The total displacements including the effect of story rotation δ_{PX} and δ_{PY}, in the X- and Y-directions, at a selected point $P(X_P, Y_P)$, at level i, is then given by

$$\delta_{PX} = \delta_{iX} - \theta Y_P \tag{36.57}$$

$$\delta_{PY} = \delta_{iY} + \theta X_P \tag{36.58}$$

where δ_{iX} and δ_{iY} are the lateral displacements due to the seismic lateral forces applied either in the X- or in the Y-direction; and θ is the story rotation given by eq.(36.39). The total displacement δ_P of a selected point P is then calculated as

$$\delta_P = \sqrt{\delta_{PX}^2 + \delta_{PY}^2} \tag{36.59}$$

36.5.14 Forces and Moments on Structural Elements

The story shear force V_i [eq.(36.51)] and the overturning moment M_i [eq.(36.50)], as well as the shear force resulting from the torsional moment M_{ti} [eq.(36.52)], are distributed among the lateral resisting elements in proportion to their relative stiffness as presented in the following section.

36.5.14.1 Element shear force resulting from the story shear force. The shear force f_{sYj} on element j, assuming the seismic forces in the Y-direction, is given by

$$f_{sYj} = k_{Yj}\Delta_{Yi} \tag{36.60}$$

where

k_{Yj} = stiffness in the Y-direction of element j
Δ_{Yi} = elastic drift in the Y-direction for story i.

36.5.14.2 Element shear force resulting from the torsional moment. The shear force f_{tYj} in the Y-direction on element j resulting from the torsional moment is calculated by

$$f_{tYj} = [k_{XYj}(Y_{ij} - Y_{ri}) - k_{Yj}(X_{ij} - X_{ri})]\theta_i \tag{36.61}$$

where

k_{Xj}, k_{XYj} = stiffness coefficients in the X-direction of element j, due to displacements of this element, respectively, in the X- and Y-directions
X_{ij}, Y_{ij} = centroidal coordinates of a resisting element j in the i story
X_{ri}, Y_{ri} = coordinates of the center of rotation of story i.

36.5.14.3 Element total shear force. The total shear force in the Y-direction f_{Yj} on element j resulting from the story shear force and from the torsional moment is then given by

$$f_{Yj} = f_{sYj} + f_{tYj} \tag{36.62}$$

Analogously for the X-direction, the element shear forces f_{sXj}, f_{tXj}, and f_{Xj} are calculated by

$$f_{sXj} = k_{XYj}\Delta_{Yi} \tag{36.63}$$

$$f_{tXj} = [k_{Xj}(Y_{ij} - Y_{ri}) - k_{XYj}(X_{ij} - X_{ri})]\theta_i \tag{36.64}$$

and

$$f_{Xj} = f_{sXj} + f_{tXj} \tag{36.65}$$

36.5.14.4 Element moments resulting from the story overturning moment. The story overturning moment M_i is distributed among the resisting elements in the story. The incremental moment Δm_{ij} distributed to the resisting element j of story i is given by

$$\Delta m_{ij} = \frac{k_{Yj}}{K_i}\Delta M_i \tag{36.66}$$

where

ΔM_i = incremental moment at story i from story $(i-1)$
k_{Yj} = stiffness of resisting element j in the Y-direction
K_i = total stiffness of story i in the y-direction.

The total moment for the resisting element at the story is then

$$m_{ij} = m_{(i+1)j} + \Delta m_{ij}$$

The components m_{ijx} and m_{ijy} of the moment m_{ij} along the local coordinates x, y of the resisting element are then calculated as

$$\begin{aligned} m_{ijx} &= m_{ij} \cos \theta \\ m_{ijy} &= m_{ij} \sin \theta \end{aligned} \qquad (36.67)$$

where θ is the angle between the global axis X and the element local axis x.

36.5.15 Computer Program

A general computer program has been developed to implement the provisions of different seismic codes. The program is written in QUICK BASIC for execution on microcomputers. It may be executed either interactively (with the data supplied and stored in a file during the interactive session) or in a batch mode (with the data being input from a file previously prepared).

The program algorithm assumes that all lateral load-resisting elements extend the entire height of the building. However, structural systems with discontinuous elements may be analyzed by modeling a discontinuous element as a continuous element having zero lateral stiffness in those stories in which it does not exist.

A lateral load-resisting element may be oriented in an arbitrary direction on a floor plan. The orientation of the resisting element is defined by the angle which is measured counterclockwise from the positive global X-axis to the positive local element x-axis. The lateral stiffnesses of the resisting elements are described in their local element coordinate systems. These coordinates are specified as input data for the analysis.

The program calculates the seismic design base shear and the corresponding lateral seismic forces distributed to the stories. The program also determines the centers of rigidity at each of the stories, the distribution of the lateral shear among the resisting elements at each story, the effect of the torsional moments corresponding to the design eccentricities specified in the Code, the story drift resulting from the lateral story shear and the rotations due to the story torsional moment.

36.5.15.1 Example 36.6.
A 20-story steel building, with a lateral load-resisting system composed of a moment-resisting space frame, is shown in Fig. 36.10. A gravity load of 1,320 kips is attributed to each floor level of the building. Lateral stiffness of each column

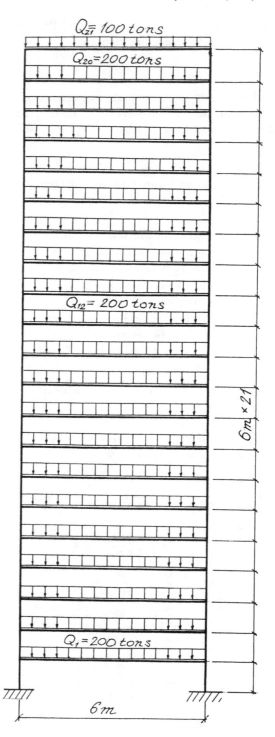

Fig. 36.10. Building modeled for Example 36.6

at all floor levels is 250 kip/in; it is assumed to be the same in both principal directions. The building is located in a seismic region for which the design peak ground acceleration $Z_A = 0.2g$, occupancy importance factor $I = 1.0$, soil site factor $S = 1.0$, and structural reduction factor $R = 12$. The analysis of building using the computer program is to be performed in the

N–S direction according to the UBC-1991 Seismic Code of the United States of America.

Solution

The input data and complete results are given by the computer output; however, to save space only selected portions of the results are reproduced for this example.

36.5.15.2 Computer Input Data and Output Results for Example 36.6

```
******************************************************************

            ***EARTHQUAKE RESISTANT DESIGN***

         MARIO PAZ, UNIVERSITY OF LOUISVILLE

         ***USING UNIFORM BUILDING CODE (UBC-91)***

******************************************************************

         ***DATA FILE INFORMATION***

            1. PREPARE NEW DATA FILE

            2. USE EXISTING DATA FILE

         SELECT NUMBER? 2

         DRIVE USED FOR DATA FILES (A:,B:,or C:) ? C:
         FILE NAME (OMIT DRIVE LETTER)(SAMPLE Q3)? Q3

   INPUT DATA:

      GENERAL DATA:

      PEAK GROUND ACCELERATION        ZA = .2
      OCCUPANCY IMPORTANCE FACTOR     I = 1
      STRUCTURAL REDUCTION FACTOR     R = 12
      SOIL SITE COEFFICIENT           S = 1

      BUILDING DATA:

   BUILDING DIMENSION NORMAL TO FORCES          DN= 1600
   BUILDING DIMENSION, SEISMIC FORCE DIRECTION  DF= 1600
   NUMBER OF STORIES                            N= 20
```

LEVEL #	HEIGHT	WEIGHT	X-MASS CENTER	Y-MASS CENTER	LAYOUT #
1	144.00	1320.00	0.00	0.00	1
2	144.00	1320.00	0.00	0.00	1
3	144.00	1320.00	0.00	0.00	1
4	144.00	1320.00	0.00	0.00	1
5	144.00	1320.00	0.00	0.00	1
6	144.00	1320.00	0.00	0.00	1
7	144.00	1320.00	0.00	0.00	1
8	144.00	1320.00	0.00	0.00	1
9	144.00	1320.00	0.00	0.00	1
10	144.00	1320.00	0.00	0.00	1
11	144.00	1320.00	0.00	0.00	1
12	144.00	1320.00	0.00	0.00	1
13	144.00	1320.00	0.00	0.00	1
14	144.00	1320.00	0.00	0.00	1
15	144.00	1320.00	0.00	0.00	1
16	144.00	1320.00	0.00	0.00	1
17	144.00	1320.00	0.00	0.00	1
18	144.00	1320.00	0.00	0.00	1
19	144.00	1320.00	0.00	0.00	1
20	144.00	1320.00	0.00	0.00	1

```
      STRUCTURAL LAYOUT

   NUMBER OF DIFFERENT STRUCTURAL LAYOUTS = 1
   NUMBER OF STRUCTURAL ELEMENTS IN LAYOUT # 1
                                    20
      STRUCTURAL ELEMENTS OF LAYOUT # 1
```

ELEM #	X-COORD.	Y-COORD.	ANG	KX	KY	KXY
1	-800.00	-800.00	0.00	.25000E+03	.25000E+03	.00000E+00
2	-480.00	-800.00	0.00	.25000E+03	.25000E+03	.00000E+00
3	-160.00	-800.00	0.00	.25000E+03	.25000E+03	.00000E+00
4	160.00	-800.00	0.00	.25000E+03	.25000E+03	.00000E+00
5	480.00	-800.00	0.00	.25000E+03	.25000E+03	.00000E+00
6	800.00	-800.00	0.00	.25000E+03	.25000E+03	.00000E+00
7	-800.00	800.00	0.00	.25000E+03	.25000E+03	.00000E+00
8	-480.00	800.00	0.00	.25000E+03	.25000E+03	.00000E+00
9	-160.00	800.00	0.00	.25000E+03	.25000E+03	.00000E+00
10	160.00	800.00	0.00	.25000E+03	.25000E+03	.00000E+00
11	480.00	800.00	0.00	.25000E+03	.25000E+03	.00000E+00
12	800.00	800.00	0.00	.25000E+03	.25000E+03	.00000E+00
13	-800.00	-480.00	0.00	.25000E+03	.25000E+03	.00000E+00
14	-800.00	-160.00	0.00	.25000E+03	.25000E+03	.00000E+00
15	-800.00	160.00	0.00	.25000E+03	.25000E+03	.00000E+00
16	-800.00	480.00	0.00	.25000E+03	.25000E+03	.00000E+00
17	800.00	-480.00	0.00	.25000E+03	.25000E+03	.00000E+00
18	800.00	-160.00	0.00	.25000E+03	.25000E+03	.00000E+00
19	800.00	160.00	0.00	.25000E+03	.25000E+03	.00000E+00
20	800.00	480.00	0.00	.25000E+03	.25000E+03	.00000E+00

```
   FUNDAMENTAL PERIOD MENU

      1.- USE   T = C1 * H**E

      2.- USE   T = C2 * H/SQR(L)

      3.- USE   T = 0.1*N

      4.- USE PREDETERMINED VALUE FOR T

            SELECT NUMBER =? 1
   INPUT COEFFICIENT  C1=? 0.35035
   INPUT EXPONENT  E=? 0.75

   OUTPUT RESULTS:

   FUNDAMENTAL PERIOD T = 1.14665
```

STORY #	CENTER OF RIGIDITY X-Coord.	Y-Coord.	TOTAL STIFFNESS KX	KY	KXY
1	0.0000	0.0000	.50000E+04	.50000E+04	.00000E+00
2	0.0000	0.0000	.50000E+04	.50000E+04	.00000E+00
3	0.0000	0.0000	.50000E+04	.50000E+04	.00000E+00
4	0.0000	0.0000	.50000E+04	.50000E+04	.00000E+00
5	0.0000	0.0000	.50000E+04	.50000E+04	.00000E+00
6	0.0000	0.0000	.50000E+04	.50000E+04	.00000E+00
7	0.0000	0.0000	.50000E+04	.50000E+04	.00000E+00
8	0.0000	0.0000	.50000E+04	.50000E+04	.00000E+00
9	0.0000	0.0000	.50000E+04	.50000E+04	.00000E+00
10	0.0000	0.0000	.50000E+04	.50000E+04	.00000E+00
11	0.0000	0.0000	.50000E+04	.50000E+04	.00000E+00
12	0.0000	0.0000	.50000E+04	.50000E+04	.00000E+00
13	0.0000	0.0000	.50000E+04	.50000E+04	.00000E+00
14	0.0000	0.0000	.50000E+04	.50000E+04	.00000E+00
15	0.0000	0.0000	.50000E+04	.50000E+04	.00000E+00
16	0.0000	0.0000	.50000E+04	.50000E+04	.00000E+00
17	0.0000	0.0000	.50000E+04	.50000E+04	.00000E+00
18	0.0000	0.0000	.50000E+04	.50000E+04	.00000E+00
19	0.0000	0.0000	.50000E+04	.50000E+04	.00000E+00
20	0.0000	0.0000	.50000E+04	.50000E+04	.00000E+00

```
   BASE SHEAR FORCE

      UBC-91:

         V = Z*I*C*W/R

         C = 1.25 S/T**2/3

   DESIGN BASE SHEAR  V= 502.0445
```

STORY #	LATERAL FORCE	STORY SHEAR	OVERTURNING MOMENT
20	84.27	84.27	0.00
19	41.78	126.05	12135.29
18	39.58	165.63	29727.83
17	37.38	203.01	52239.55
16	35.18	238.19	79209.50
15	32.98	271.17	110242.80
14	30.78	301.95	145000.20
13	28.58	330.54	183188.30
12	26.39	356.92	224550.20
11	24.19	381.11	268857.00
10	21.99	403.10	315900.00
9	19.79	422.89	365482.90
8	17.59	440.48	417415.20
7	15.39	455.87	471506.60
6	13.19	469.06	527560.20
5	10.99	480.06	585369.30
4	8.80	488.85	644711.50
3	6.60	495.45	705346.10
2	4.40	499.85	767010.70
1	2.20	502.04	829418.30
0			892256.10

STORY	TORSIONAL STIFFNESS
1	0.435E+10
2	0.435E+10
3	0.435E+10
4	0.435E+10
5	0.435E+10
6	0.435E+10
7	0.435E+10
8	0.435E+10
9	0.435E+10
10	0.435E+10
11	0.435E+10
12	0.435E+10
13	0.435E+10
14	0.435E+10
15	0.435E+10
16	0.435E+10
17	0.435E+10
18	0.435E+10
19	0.435E+10
20	0.435E+10

```
      ECCENTRICITY

         EC1 = EC + 0.05 DN

         EC2 = EC - 0.05 DN
```

TORSIONAL MOMENTS AND STORY ROTATIONS

STORY #	COMPUTED ECC	DESIGN ECC1	DESIGN ECC2	TORSIONAL MOMENTS MT1	MT2	STORY ROTATION RT1	RT2
20	0.000	80.000	-80.000	6741.83	-6741.83	0.186E-04	-.186E-04
19	0.000	80.000	-80.000	10084.00	-10084.00	0.278E-04	-.278E-04
18	0.000	80.000	-80.000	13250.27	-13250.27	0.365E-04	-.365E-04
17	0.000	80.000	-80.000	16240.63	-16240.63	0.448E-04	-.448E-04
16	0.000	80.000	-80.000	19055.09	-19055.09	0.525E-04	-.525E-04
15	0.000	80.000	-80.000	21693.65	-21693.65	0.598E-04	-.598E-04
14	0.000	80.000	-80.000	24156.31	-24156.31	0.666E-04	-.666E-04
13	0.000	80.000	-80.000	26443.06	-26443.06	0.729E-04	-.729E-04
12	0.000	80.000	-80.000	28553.91	-28553.91	0.787E-04	-.787E-04
11	0.000	80.000	-80.000	30488.85	-30488.85	0.841E-04	-.841E-04
10	0.000	80.000	-80.000	32247.88	-32247.88	0.889E-04	-.889E-04
9	0.000	80.000	-80.000	33831.02	-33831.02	0.933E-04	-.933E-04
8	0.000	80.000	-80.000	35238.25	-35238.25	0.972E-04	-.972E-04
7	0.000	80.000	-80.000	36469.58	-36469.58	0.101E-03	-.101E-03
6	0.000	80.000	-80.000	37525.00	-37525.00	0.103E-03	-.103E-03
5	0.000	80.000	-80.000	38404.52	-38404.52	0.106E-03	-.106E-03
4	0.000	80.000	-80.000	39108.13	-39108.13	0.108E-03	-.108E-03
3	0.000	80.000	-80.000	39635.85	-39635.85	0.109E-03	-.109E-03
2	0.000	80.000	-80.000	39987.65	-39987.65	0.110E-03	-.110E-03
1	0.000	80.000	-80.000	40163.56	-40163.56	0.111E-03	-.111E-03

STORY DRIFT AND LATERAL DISPLACEMENTS (DIRECT SHEAR)

LEVEL	DRIFT	DISPL.
20	0.202	17.080
19	0.303	16.877
18	0.398	16.575
17	0.487	16.177
16	0.572	15.690
15	0.651	15.118
14	0.725	14.468
13	0.793	13.743
12	0.857	12.950
11	0.915	12.093
10	0.967	11.178
9	1.015	10.211
8	1.057	9.196
7	1.094	8.139
6	1.126	7.045
5	1.152	5.919
4	1.173	4.767
3	1.189	3.594
2	1.200	2.405
1	1.205	1.205

STRUCTURAL ELEMENT'S LATERAL SHEAR AND OVERTURNING MOMENT[*]

STORY # 20   [DESIGN ECCENTRICITY ECC= 80]

ELMENT #	LATERAL SHEAR COMPONENTS x-LOCAL	y-LOCAL	OVERTURNING MOMENT x-LOCAL	y-LOCAL
1	-.30983E+00	0.45235E+01	0.65138E+03	0.44615E+02
2	-.30983E+00	0.43995E+01	0.63353E+03	0.44615E+02
3	-.30983E+00	0.42756E+01	0.61569E+03	0.44615E+02
4	-.30983E+00	0.41517E+01	0.59784E+03	0.44615E+02
5	-.30983E+00	0.40277E+01	0.58000E+03	0.44615E+02
6	-.30983E+00	0.39038E+01	0.56215E+03	0.44615E+02
7	0.30983E+00	0.45235E+01	0.65138E+03	-.44615E+02
8	0.30983E+00	0.43995E+01	0.63353E+03	-.44615E+02
9	0.30983E+00	0.42756E+01	0.61569E+03	-.44615E+02
10	0.30983E+00	0.41517E+01	0.59784E+03	-.44615E+02
11	0.30983E+00	0.40277E+01	0.58000E+03	-.44615E+02
12	0.30983E+00	0.39038E+01	0.56215E+03	-.44615E+02
13	-.18590E+00	0.45235E+01	0.65138E+03	0.26769E+02
14	-.61965E-01	0.45235E+01	0.65138E+03	0.89230E+01
15	0.61965E-01	0.45235E+01	0.65138E+03	-.89230E+01
16	0.18590E+00	0.45235E+01	0.65138E+03	-.26769E+02
17	-.18590E+00	0.39038E+01	0.56215E+03	0.26769E+02
18	-.61965E-01	0.39038E+01	0.56215E+03	0.89230E+01
19	0.61965E-01	0.39038E+01	0.56215E+03	-.89230E+01
20	0.18590E+00	0.39038E+01	0.56215E+03	-.26769E+02

*To save space only results for STORY-20 corresponding to design eccentricity ECC-80 are reproduced.

REFERENCES

Structural Dynamics

BATHE, K. J. (1982) *Finite Element Procedures in Engineering Analysis*. Prentice Hall, Englewood Cliffs, NJ.

BERG, GLEN V. (1989) *Elements of Structural Dynamics*. Prentice Hall, Englewood Cliffs, NJ.

BIGGS, J. M. (1964) *Introduction to Structural Dynamics*. McGraw-Hill, New York, NY.

BLEVINS, R. D. (1979) *Formulas for Natural Frequency and Mode Shape*. Van Nostrand Reinhold, New York, NY.

CHOPRA, A. (1981) *Dynamics of Structures: A Primer*. Earthquake Engineering Research Institute. Berkeley, CA.

CLOUGH, R. W., and PENZIEN, J. (1975) *Dynamics of Structures*. McGraw-Hill, New York, NY.

DER KIUREGHIAN, A. (1980) *A Response Spectrum Method for Random Vibration*. Report No. UCB/EERC-80/15, Earthquake Engineering Research Center, University of California, Berkeley, CA.

GALLAGHER, R. H. (1975) *Finite Element Analysis*. p. 115. Prentice Hall, Englewood Cliffs, NJ.

GUYAN, R. J. (1965) "Reduction of Stiffness and Mass Matrices." *AIAA J*. 13: 380.

HARRIS, CYRIL M. (1987) *Shock and Vibration Handbook*. 3d ed. McGraw-Hill, New York, NY.

NASHIF, A. D.; JONES, D. I. C.; and HENDERSON, J. P. (1985) *Vibration Damping*. Wiley, New York, NY.

NEWMARK, N. M. (1959) "A Method of Computation for Structural Dynamics." *Trans. ASCE* 127: 1406–35.

PAZ, MARIO (1973) "Mathematical Observations in Structural Dynamics." *Int. J. Comput. Struct*. 3: 385–396.

——— (1983) "Practical Reduction of Structural Problems." *J. Struct. Eng., ASCE* 109(111): 2590–2599.

——— (1984a) "Dynamic Condensation." *AIAA J*. 22(5): 732–737.

——— (1984b) In Structural Mechanics Software Series. Vol. V, pp. 271–286. The University Press of Virginia, Charlottesville, VA.

——— (1985) *Micro-Computer Aided Engineering: Structural Dynamics*. Van Nostrand Reinhold, New York, NY.

——— (1989) "Modified Dynamic Condensation Method." *J. Struct. Eng., ASCE* 115(1): 234–238.

——— (1991) *Structural Dynamics: Theory and Computation*. Van Nostrand Reinhold, New York, NY.

PAZ, M., and DUNG, L. (1975) "Power Series Expansion of the General Stiffness Matrix for Beam Elements." *Int. J. Numer. Methods Eng*. 9: 449–459.

WILSON, E. L.; DER KIUREGHIAN, A.; and BAYO, E. P. (1981) "A Replacement for the SRSS Method in Seismic Analysis." *Int. J. Earthquake Eng. Struct. Dyn*. 9: 187–194.

WILSON, E. L.; FARHOOMAND, I.; and BATHE, K. J. (1973) "Nonlinear Dynamic Analysis of Complex Structures." *Int. J. Earthquake and Structural Dynamics*. Vol. 1: pp. 241–252.

Earthquake Engineering

BLUME, J. A.; NEWMARK, N. M.; and CORNING, L. (1961) *Design of Multi-story Reinforced Concrete Buildings for Earthquake Motions*. Portland Cement Association, Chicago, IL.

HART, GARY C., and ENGLEKIRK, ROBERT E. (1982) *Earthquake Design of Concrete Masonry Buildings: Response Spectra Analysis and General Earthquake Modeling Considerations*. Prentice Hall, Englewood Cliffs, NJ.

HOUSNER, G. W. (1970) "Design Spectrum." In *Earthquake Engineering*, R. L. Weigel, ed. Prentice Hall, Englewood Cliffs, NJ.

HOUSNER, G. W., and JENNINGS, P. C. (1982) *Earthquake Design Criteria*. Earthquake Engineering Institute, Berkeley, CA.

HUDSON, D. E. (1970) "Dynamic Tests of Full Scale Structures." In *Earthquake Engineering*, R. L. Weigel, ed. Prentice Hall, Englewood Cliffs, NJ.

MACLEOD, I. A. (1971) *A Design Aid with Commentary*, Portland Cement Association, Skokie, IL.

NAEIM, FARZAD (1989) *The Seismic Design Handbook*. Van Nostrand Reinhold, New York, NY.

NEWMARK, N. M., and HALL, W. J. (1973) *Procedures and Criteria for Earthquake Resistant Design: Building Practices for Disaster Mitigation*. pp. 209–23. Building Science Series 46, National Bureau of Standards, Washington, D.C.

—— (1982) *Earthquake Spectra and Design*. Earthquake Engineering Research Institute, Berkeley, CA.

NEWMARK, N. M., and RIDDELL, R. (1980) In "Inelastic Spectra for Seismic Design." Vol. 4, pp. 129–136. Proceedings of Seventh World Conference on Earthquake Engineering, Istanbul, Turkey.

NEWMARK, N. M., and ROSENBLUETH, E. (1971) *Fundamentals of Earthquake Engineering*. Prentice Hall, Englewood Cliffs, NJ.

POPOV, E. P., and BERTERO, V. V. (1980) "Seismic Analysis of Some Steel Building Frames." *J. Eng. Mech., ASCE* 106: 75–93.

STEINBRUGGE, KARL V. (1970) "Earthquake Damage and Structural Performance in the United States." In *Earthquake Engineering*, R. L. Weigel, ed. Prentice Hall, Englewood Cliffs, NJ.

WASABAYASHI, MINORU (1986) *Design of Earthquake-Resistant Buildings*. McGraw-Hill, New York, NY.

Building Codes

American Insurance Association (1976) *The National Building Code*. New York, NY.

American Society of Civil Engineers (ASCE) (1988) *Standard Minimum Design Loads for Buildings and Other Structures*. (Revision of American National Standard Institute (ANSI) A58.1-1982.) New York, NY.

Applied Technology Council (1978) *Tentative Provisions for the Development of Seismic Regulations for Buildings*. ATC 3-06, National Bureau of Standards, Special Publication 510, U.S. Government Printing Office, Washington, D.C.

Building Officials and Code Administrators International (1990) *BOCA National Basic Building Code*. Homewood, IL.

Building Seismic Safety Council (BSSC) (1988) *Recommended Provisions for the Development of Seismic Regulations for New Buildings*. NEHRP (National Earthquake Hazard Reduction Program). Washington, D.C.

International Conference of Building Officials (1991) *Uniform Building Code (UBC)*. Whittier, CA.

Southern Building Code Congress International (1991) *Standard Building Code*. Birmingham, AL.

Structural Engineers Association of California (SEAOC) (1990) *Recommended Lateral Force Requirements and Tentative Commentary*. San Francisco, CA.

37

Venezuela

William Lobo Quintero and Edward D. Thomson

37.1 INTRODUCTION

Venezuela is located at the intersection of the Caribbean Plate and the South American Plate and has a system of faults that originate in Colombia in the west and extend under the Caribbean Sea to the east. This system includes the Bocono, Oca, Moron, La Victoria, and El Pilar faults, which seismically have been active, causing severe earthquakes in 1812, 1894, and 1967. Steep mountain ranges cross the country in an east–west (E–W) direction; erosion from these mountains has filled intervening valleys with mixed alluvial strata, which may exacerbate the effects of seismic shocks. Many severe earthquakes have been recorded in Venezuela in the last 400 years.

In Venezuela, earthquake-resistant structural design, as taught at universities and used in professional practice, has been based on seismic codes published by the Ministry of Public Works (Ministerio de Obras Públicas MOP). The first Venezuelan seismic code was published in 1939, and revised in 1947 (MOP 1947). The 1947 version of the code followed the provisions of the Static Method of Analysis Uniform Building Code UBC (Grases et al. 1984). It used seismic coefficients based on the bearing capacity of the soil. The seismic coefficient was assigned a value of 0.10 for soils with a bearing capacity of less than 2 kg/cm^2 (200 kPa) and 0.05 for soils with a greater bearing capacity of more than 2 kg/cm^2. The code of 1947 divided the country into three seismic zones; the eastern zone was assigned the greatest seismic risk. The 1947 code also provided seismic coefficients for nonstructural elements and stipulated that the overturning moments due to the equivalent seismic forces should not exceed 75% of the stabilizing moments produced by the gravity loads.

A new version of the seismic code (MOP 1955) required the dynamic method of analysis for special and irregular structures and made the static method applicable only to regular buildings. Earthquake-resistant design was not required for buildings of one or two stories.

The seismic coefficient C_s at a given story of the building was calculated by the following equation:

$$C_s = \frac{0.30}{n + 4.5} \qquad (37.1)$$

where n is the number of levels above the level under consideration. Equation (37.1) was applicable to Zones A or B. For Seismic Zone C, the seismic coefficient was set as twice the value given by eq.(37.1). This equation was based on the expression for the design spectral acceleration S_a given by the UBC (Housner 1962) as

$$S_a = \frac{0.06g}{T + 0.35} \qquad (37.2)$$

where g is the acceleration due to gravity and T is the fundamental period of the structure, in seconds.

The Seismic Code of 1955 included improvements in the analysis of masonry buildings, by specifying the minimum thickness of floor slabs and detailing the requirements for intermediate infill columns, horizontal and vertical bars for reinforced concrete walls, and special reinforcement around doors and windows. These masonry design requirements and other requirements for baked clay blocks are still in force. The MOP (1955) code was not revised until the Caracas earthquake of 1967.

On July 29, 1967 at 8:02 P.M., an earthquake of a magnitude [1]$M_s = 6.3$ and a maximum intensity of $I = VIII$ on the Modified Mercalli Intensity scale occurred. The epicenter was in the Caribbean Sea 70 km from Caracas, and the earthquake lasted 35 seconds. In Caracas and in the central coastal region of the country, the earthquake caused 245 deaths, structural damage to 206 buildings, nonstructural damage to 563 buildings, and the complete collapse of 4 buildings. Additionally, it caused soil liquefaction and ground subsidence in the neighborhood of Guigue, Carabobo state, and lesser damage in nearby towns. According to the report of the Presidential Committee for Study of the Earthquake (Comisión Presidencial para el Estudio del Sismo 1978), the damage resulting from the effects of the 1967 Caracas earthquake can be attributed to several causes as it is described in Table 37.1.

Table 37.1   Relative Number of Buildings with Indicated Defect in Design

Defects in Design:	*Relative Number	Percentage
1. Subestimation of the dead loads.	72/121	59.5
2. Seismic analysis without determination of maximum combination values.	63/117	53.4
3. Torsion effects neglected.	58/129	46.0
4. Overturning effects neglected.	116/129	89.9
5. Inadequate details: in columns	72/119	60.5
in beams	60/108	55.6
6. Design did not include all effects.	36/61	60.0
7. Comparison with wind effects not made.	116/128	90.6
Structural Geometry:		
8. Slenderness of building H/B: $5 < H/B < 10$.	33/128	25.8
9. Symmetry with respect to one or two axes.	72/130	55.4
10. Length/width ratio of building ($L/B > 3$).	15/21	71.0
11. Asymmetric rectangular ground plan.	7/11	64.0
12. Asymmetric "H" ground plan.	6/11	55.0
13. Number of floors (10 or 13 floors).	46/129	36.7
Soil-Structure Interaction:		
14. Type of soil (alluvial deposits).	76/80	95.0
Code Requirements:		
15. Hyperbolic seismic distribution (MOP 55).	74/115	63.8
16. Construction period (1963–1967).	59/129	45.7

*Number of damaged buildings with the cited characteristics (i.e., defects in design) from a sample of 130 buildings selected in the most affected zones.

---

[1]M_s is a scale for measuring the magnitude of earthquakes. Specifically, the M_s (surface wave) magnitude is proportional to the source spectral amplitude of the 20-second-period wave. (See Appendix of this Handbook.)

The report by Hanson and Degenkolb (1969) summarized the critical factors identified from the Caracas earthquake of 1967 as follows: influence of nonstructural elements; asymmetry of masonry; failure of columns due to overturning effect; incomplete design for different load cases; inadequate reinforcement in beams and slabs; lack of confinement at ends of columns; lack of confinement at beam-column connections; shear failure in beams and flat slabs; stress concentrations due to interrupted diaphragms; damage due to pounding of adjacent buildings; defects in the design of stairs, beams, and water tanks; and interaction between foundation sand and gravel deposits and supported structures. The Caracas earthquake illustrated the effects of faults in design, in construction, in structural shape, and in detailing of connections.

Three months after the Caracas earthquake of 1967, the government adopted provisions for a new seismic code, the Provisional Code for Earthquake Resistant Buildings (NP-MOP 1967). This code prescribed the static method of analysis for buildings up to 20 floors or no more than 60 meters high, and the dynamic analysis for higher buildings, with the provision that the value for the base shear force used by the dynamic analysis should not be less than 60% of the value obtained by the static method. The current seismic code for Venezuela was issued in 1987 under the title *Norma Venezolana de Edificación Antisísmica* (1987) (Venezuelan Rules for Seismic Construction). This code is usually referred to as the COVENIN 1756-87 Code.

37.2 PHILOSOPHY AND VALIDITY OF THE 1987 SEISMIC CODE

The Venezuelan Seismic Code of 1987, COVENIN 1756-87, is conceptually based on expected inelastic behavior of structural resisting elements when subjected to the effects of strong motion earthquakes with expected low probability of recurrence. The fundamental concepts are as follows:

(a) The structure should be capable of responding elastically to earthquakes of small magnitude without damage.

(b) The nonstructural damage caused by intermediate or moderate earthquakes should be able to be repaired economically.

(c) The design may allow a risk of serious structural damage under the action of strong earthquakes, but it should prevent collapse of the structure.

The requirements of the 1987 Code for the seismic design of structures have an impact on the usual

Table 37.2. Seismic Coefficient A_0

Zone:	1	2	3	4
$A_0 =$	0.08	0.15	0.22	0.30

design of typical buildings, not including prefabricated structures, bridges, and special buildings. COVENIN 1756-87 gives detailed information on the application of the requirements for the seismic design of structures, including assignment of seismic zones and classification of structures according to use, structural type, and foundation soil. The Code also provides design spectra, methods of analysis, specifications on the quality of materials, design requirements for elements and connections, and limits on maximum displacements.

37.3 SEISMIC ZONES

The Code provides the seismic zone map reproduced in Fig. 37.1, in which the territory of Venezuela has been divided into five distinct seismic zones with corresponding values for the seismic coefficient (peak ground acceleration) A_0 given in Table 37.2. The seismic zones in Fig. 37.1 are based on an average probability of exceedance of the ground motion of 10% in 50 years, as described by Grases (1985). The delimitation between seismic zones was adjusted to conform to boundaries of districts, territories, or states.

37.4 BUILDING USE GROUPS

Buildings are classified according to their use as follows:

Group A. Includes essential facilities such as structures used for emergencies or to provide important services (e.g., hospitals, health centers, government buildings, monuments and temples, museums and libraries, educational institutes, fire stations, police stations and barracks, power stations, telephone exchanges, telegraph offices, radio and television stations, and warehouses for toxic, explosive, or radioactive substances).

Group B. Buildings for private or public use (detached dwellings, apartments, offices, hotels, banks, movie theaters, theaters, warehouses, depots, and industrial installations, and buildings whose collapse would affect buildings in Group A or Group B).

Group C. Structures not classified in Group A or B.

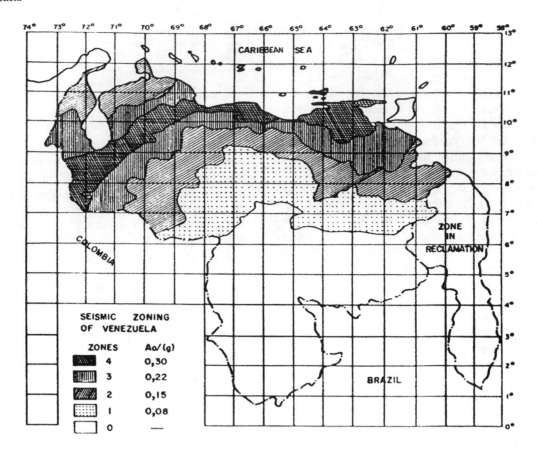

Fig. 37.1.   Seismic zoning of Venezuela

The Code stipulates group use factors $\alpha = 1.25$ and $\alpha = 1.00$, respectively, for buildings in Group A and in Group B. No specific value for the use factor is given in the Code for structures in Group C.

37.5 DESIGN LEVELS ND

Design levels ND are established according to the use of the building and the seismic zone of the site as shown in Table 37.3 and defined in Chapter 18 of the Code for Reinforced Concrete Design (MDU 1985).

No seismic design is required at the ND1 level. At the ND2 level, an intermediate condition of design is required. At level ND3, all the special stipulations of earthquake-resistant design are required. At this level, the structure should be designed to deform inelastically, with great potential for energy dissipation, without suffering significant loss in strength.

37.6 STRUCTURAL CLASSIFICATION

CONVENIN 1756-87 classifies structures into the following types:

> *Type I.* Moment-resisting frames
> *Type II.* Combined frame-wall structures where the frames must resist at least 25% of the seismic forces
> *Type III.* Walls, or braced frames (structures of Type II with frames resisting less than 25% of the seismic forces)
> *Type IV.* Inverted pendulum type.

When two or more types of structural systems are used in a building, the type with the smallest ductility should be used in classifying the structure, unless that system part resists less than 10% of the seismic forces.

Table 37.3.   Design Levels ND

Group Use Factor	Seismic Zone			
	1	2	3	4
A	ND2, ND3	ND2, ND3	ND3*	ND3
B	ND1, ND2, ND3	ND2, ND3	ND2, ND3	ND3

*When the simplified static method is used, design level ND2 can be applied.

37.7 DUCTILITY FACTOR D_f AND RESPONSE MODIFICATION FACTOR R

The ductility factors D_f corresponding to the structural type and design level are indicated in Table 37.4.

The response modification factor R depends on the expected ductility, as given by the following relationships:

$$R = 1 + (T/0.15)(D_f - 1) \quad T < 0.15 \text{ sec} \quad (37.3)$$

$$R = D_f \quad\quad\quad\quad\quad\quad T \geqslant 0.15 \text{ sec} \quad (37.4)$$

where T is the fundamental period of the building (Section 37.11) and D_f is the ductility factor (Table 37.4).

37.8 STRUCTURAL REGULARITY

Regular buildings are considered to be those that have the following characteristics: there are no great variations in mass, strength, or stiffness between adjacent levels; the eccentricities between the center of mass and the center of stiffness at each story of the building do not exceed 8% of the larger plan dimension; and the plan dimensions do not vary substantially with height. Buildings are considered to be *irregular* when these conditions are not satisfied.

37.9 TYPICAL SOIL PROFILES AND LIQUEFACTION

The Code classifies soils into three profiles:

Profile S_1. Rock; hard and/or dense soil with a depth of less than 50 meters above the rock base

Profile S_2. Medium-dense to dense sands and gravels, stiff to very stiff silts or clays, or a mixture of these soils

Profile S_3. Loose granular soils and/or soft to medium-cohesive soils with depths greater than 10 meters.

When the classification of the soil profile is doubtful, the most unfavorable condition should be adopted. In Seismic Zones 2, 3, and 4, the potential liquefaction of soil should be evaluated for the first 20 meters of saturated fine sands or silty sands. Liquefiable soils should not be used as foundation layers; they should be made dense so that they are no longer liquefiable.

37.10 DESIGN ACCELERATION SPECTRA

The design spectral acceleration A_d is defined by the following relationships:

$$A_d = \frac{\alpha A_0}{R}\left(1 + \frac{T}{0.15}(\beta - 1)\right) \quad \text{for} \quad T < 0.15 \text{ sec} \quad (37.5)$$

$$A_d = \frac{\alpha \beta A_0}{R} \quad \text{for} \quad 0.15 \leqslant T < T^* \quad (37.6)$$

$$A_d = \frac{\alpha \beta A_0}{R}\left(\frac{T^*}{T}\right)^p \quad \text{for} \quad T \geqslant T^* \quad (37.7)$$

where

α = group use factor (Section 37.4)
β = soil amplification factor (Table 37.5)
T^* = characteristic period of the soil at the site (Table 37.5)
A_0 = seismic coefficient (Table 37.2)
R = response modification factor [eqs.(37.3) and (37.4)]
p = exponent of descending branch of the spectrum (Table 37.5).

Table 37.5 provides numerical values for β, T^*, and p for soil profiles S_1, S_2, and S_3. Fig. 37.2 shows the plot of eqs.(37.5), (37.6), and (37.7) for both the elastic spectrum ($R = 1$) and the inelastic spectrum ($R > 1$).

Table 37.5. Values of β, T^*, and p

Soil Profile	β	T^*	p
S_1	2.2	0.4	0.8
S_2	2.2	0.6	0.7
S_3	2.0	1.0	0.6

Table 37.4. Ductility Factors D_f

Design Level	D_f for Structural Type			
	I	II	III	IV
ND3	6.0	5.0	4.0	1.5
ND2	4.5	3.75	3.0	1.25
ND1	2.5	2.0	1.5	1.0

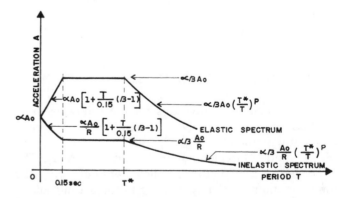

Fig. 37.2. Elastic and reduced spectral forms

37.11 FUNDAMENTAL PERIOD

An approximate value for the fundamental period T_a may be determined by an empirical formula according to the structural type as a function of the height of the building h_n and the greater plan dimension L. The Code gives the following empirical formulas to estimate the fundamental period of a building:

$$T_a = 0.061 h_n^{3/4} \quad \text{(for Type I structures)} \quad (37.8a)$$

$$T_a = 0.09 h_n / \sqrt{L} \quad \text{(for Type II and III structures)} \quad (37.8b)$$

The fundamental period can be calculated more accurately by the Rayleigh formula:

$$T = 2\pi \sqrt{\frac{\sum_{i=1}^{n} W_i \delta_i^2}{g\left[\sum_{i=1}^{n-1} F_i \delta_i + (F_t + F_n)\delta_n \right]}} \quad (37.9)$$

where

W_i = seismic weight at level i
δ_i, δ_n = lateral displacement at level i, n
F_i, F_n = equivalent seismic forces at level i, n
F_t = additional seismic force at the top level
g = acceleration of gravity
n = number of stories

The seismic weight includes the dead weight plus the weight of permanent machinery and equipment, and the prorated live load (100% of fully loaded tanks, 80% of floor loads in stores and warehouses, 50% of floor loads in parking lots, and 25% of interstory floor loads).

The lateral displacements δ_i in eq.(37.9) are not known until the structure is analyzed; therefore the Rayleigh formula cannot be used initially. After the structure has been analyzed, the value for the fundamental period obtained with the Rayleigh formula may be used to reanalyze the structure, if this value is significantly different than the approximate value obtained from eq.(37.8a) or (37.8b). However, the value obtained from eq.(37.9) cannot exceed $1.2T_a$ with T_a obtained from eqs.(37.8).

37.12 METHODS OF ANALYSIS

The Code contains provisions for the following three methods of seismic analysis of buildings: (1) equivalent static method (ESM), (2) simplified static method (SSM), and (3) dynamic modal analysis (DMA). The equivalent static method, which considers only the fundamental mode, may be used in the seismic analysis of regular buildings of no more than 20 stories nor

Table 37.6.   Method of Analysis for Irregular Buildings

Relative Eccentricity	Method of Analysis
$e/B \leq 0.08$	DMA plus EST or DA3D
$0.08 < e/B \leq 0.12$	DA3D or DMA plus EST
$e/B > 0.12$	DA3D

Notes:
e = eccentricity between the center of mass and the center of stiffness.
B = dimension of the building perpendicular to direction of seismic forces.
EST = equivalent static torsion due to torsional moments from the action of the seismic forces (Section 37.17).
DMA = dynamic modal analysis with one degree of freedom per level (Section 37.16).
DA3D = dynamic analysis with three degrees of freedom per level (Section 37.18).

higher than 60 m. The equivalent static method may be used for regular buildings of more than 20 stories or higher than 60 m provided that all the significant modes of vibration are considered; thus, requiring a dynamic analysis of the building to determine natural periods and corresponding modal shapes. For irregular buildings, the method permitted for seismic analysis of buildings will depend on the relative eccentricity e/B of the building, as indicated in Table 37.6.

The simplified static method (SSM) may be used for the seismic analysis of buildings in occupancy Group B of no more than three stories and not more than 10.5 m in height. The DA3D method of analysis may be used for any structure, but it *must* be used for irregular buildings when the relative eccentricity $e/B > 0.12$, as indicated in Table 37.6.

37.13 EQUIVALENT STATIC METHOD (ESM)

The base shear force V_0 is determined as

$$V_0 = \mu A_d W \quad (37.10)$$

where

W = total seismic weight of the building
A_d = design spectral acceleration [eqs.(37.5)–(37.7)]
μ = shear modification factor

The shear modification factor μ must be selected as the largest value given by the following formulas:

$$\mu_1 = \frac{3}{2} \frac{(n+1)}{(2n+1)} \quad (37.11)$$

or

$$\mu_2 = 0.80 + \frac{1}{20}\left(\frac{T_a}{T^*} - 1\right) \quad (37.12)$$

where

- n = number of stories in the building
- T_a = approximate value for the fundamental period (Section 37.11)
- T^* = characteristic period of site (Table 37.5).

37.13.1 Equivalent Static Forces

The base shear force V_0 is distributed into forces F_i applied at the various levels of the building in addition to a force F_t at the top of the building, according to the following formulas:

$$V_0 = F_t + \sum_{i=1}^{n} F_i \tag{37.13}$$

$$F_t = \left(0.06\frac{T}{T^*} - 0.02\right)V_0 \tag{37.14}$$

with

$$0.04V_0 \leq F_t \leq 0.10V_0 \tag{37.15}$$

and

$$F_i = \frac{W_i h_i}{\sum_{j=1}^{n} W_j h_j}(V_0 - F_t) \tag{37.16}$$

in which

- h_i, h_j = heights at levels i and j measured from the base of the building
- W_i, W_j = seismic weights at levels i and j.

37.13.2 Overturning Moments

The overturning moment M_i at level i of the building is determined by statics as the moment resulting from the equivalent static forces above that level; that is,

$$M_i = F_t(h_n - h_i) + \sum_{j=1+i}^{n} F_j(h_j - h_i) \tag{37.17}$$

where $i = 0, 1, 2, \ldots, n-1$.

37.13.3 Simplified Static Method (SSM)

In the simplified static method, the base shear force V_0 is calculated by

$$V_0 = \alpha A_0 W \frac{\beta+1}{D+1} \tag{37.18}$$

and the equivalent lateral forces F_i by

$$F_i = \frac{W_i h_i}{\sum_{i=1}^{n} W_i h_i} V_0 \tag{37.19}$$

in which

- α = group use factor (Section 37.4)
- A_0 = seismic coefficient (Table 37.2)
- W = total seismic weight of the building
- β = soil amplification factor (Table 37.5)
- h_i = height of level i measured from the base of the building
- W_i = seismic weight at level i.

When the simplified static method is applicable, the Code relies on the judgment of the designer to consider torsional effects and to verify that the story drift at each level does not exceed Code limits.

37.14 NONSTRUCTURAL ELEMENTS

Nonstructural elements should be designed as integral parts of the main structure, or as subjected to the force F_p given by eq.(37.20), whichever condition is more severe.

$$F_p = \alpha C_p W_p \tag{37.20}$$

where

- α = group use factor (Section 37.4)
- W_p = weight of the element considered
- C_p = seismic coefficient (Table 37.7).

Table 37.7. Seismic Coefficients for Nonstructural Parts

Element	Direction	C_p
Walls	Perpendicular to seismic forces	0.3
Railings or parapets	Perpendicular to seismic forces	1.0
Ornaments and balconies	Vertical	1.8(D + L)
Eaves and cantilevers		−0.2(D + L)
Appurtenances	Any	1.0
Diaphragms	Any	>0.15W_p
Connections in prefabricated walls	Any	2.0
Machine rooms and water tanks	Any	1.8/R

37.15 P-Δ EFFECTS

So-called P-Δ (P-delta) effects must be calculated and taken into account at any level of the structure where the stability factor $\theta_i \geqslant 0.08$. The stability factor θ_i is given by

$$\theta_i = \frac{\Delta_i \sum_{j=i}^{n} W_j}{V_i D_f (h_i - h_{i-1})} \tag{37.21}$$

in which

  Δ_i = drift for story i
  W_j = seismic weight at level j
  h_i = height of level i measured from the base of the building
  D_f = ductility factor (Table 37.4)
  V_i = shear force at level i.

37.16 DYNAMIC MODAL ANALYSIS (DMA)

In a dynamic modal analysis, the structure is modeled with a concentrated mass at each floor of the building, with one degree of freedom per floor. Modal shapes of vibration and corresponding natural periods are determined. The minimum number of modes N_1 that must be considered in each direction depends on the number of floors and on the fundamental period T_1:

(a) For $n < 20$ floors:
$$N_1 = 1/2[(T_1/T^*) - 1.5] + 3 \geqslant 3 \tag{37.22}$$

(b) For $n \geqslant 20$ floors:
$$N_1 = 2/3[(T1/T^*) - 1.5] + 4 \geqslant 4 \tag{37.23}$$

The modal base shear force V_{0m} which corresponds to mode m is determined as:

$$V_{0m} = W_m A_{dm} \tag{37.24}$$

where the effective modal mass W_m is given by

$$W_m = \frac{\left[\sum_{i=1}^{n} W_i \phi_{im}\right]^2}{\sum_{i=1}^{n} W_i \phi_{im}^2} \tag{37.25}$$

and

  A_{dm} = design spectral acceleration for mode (Section 37.10)
  ϕ_{im} = modal coordinate of level i and mode m.

The modal forces F_{im} are determined from

$$F_{im} = C_{im} V_{0m} \tag{37.26}$$

where

$$C_{im} = \frac{W_i \phi_{im}}{\sum_{j=1}^{n} W_j \phi_{jm}} \tag{37.27}$$

The design values are obtained taking the Square Root of the Sum of the Squares of each modal value (SRSS method). The design value V_0 obtained for the base shear force is compared to the base shear V_{ESM} obtained with the ESM method using $T = 1.4 T_a$. If $V_0 < V_{\text{ESM}}$ the modal values are scaled up by multiplying by V_{ESM}/V_0.

37.17 TORSIONAL MOMENTS

At each level and in each direction, a torsional moment is calculated as follows:

$$M_{ti} = V_i (\tau e_i + 0.10 B_i) \tag{37.28}$$
$$M_{ti} = V_i (e_i - 0.10 B_i) \tag{37.29}$$

where

  e_i = eccentricity at level i between the center of mass and the center of stiffness
  B_i = width of the plan dimension perpendicular to the direction of the seismic forces
  τ = torsional dynamic amplification factor.

The values of τ are:

  $\tau = 1.5$ when the lateral resisting elements are located around the perimeter of the building
  $\tau = 5$ when the lateral resisting elements are concentrated in the center of the building
  $\tau = 3$ for intermediate conditions.

In the commentary of the Code, there is a graph (proposed by Kan and Chopra 1976) to obtain a more precise value for τ.

37.18 DYNAMIC ANALYSIS WITH THREE DEGREES OF FREEDOM PER LEVEL (DA3D)

This method takes into account the coupling of translational and torsional vibrations, considering three degrees of freedom per level, and assuming horizontal diaphragms at the floor levels that are infinitely rigid in their plane. The minimum number of modes that should be considered in the analysis is $3 \times N1$, where $N1$ is the value given by eqs.(37.22) and (37.23). For each direction considered in the analysis, an accidental eccentricity equal to 10% of the width of the plan perpendicular to the direction of analysis should be included. The design value V_0 obtained for the base shear force is compared to the base shear V_{ESM} obtained with the ESM method using $T = 1.4T_a$. If $V_0 < V_{ESM}$, the modal values are scaled up by multiplying by V_{ESM}/V_0.

37.19 LATERAL DISPLACEMENT CONTROL

The total inelastic lateral displacement δ_i at level i is calculated by the following expression:

$$\delta_i = D_f \delta_{ei} \qquad (37.30)$$

where

D_f = ductility factor from Table 37.4
δ_{ei} = lateral elastic displacement at level i calculated for the design forces. This displacement should be increased for P-Δ effects according to eq.(37.21).

To limit damage to nonstructural elements, stairs, joints, and other features due to excessive lateral movement, the lateral drift calculated with eq.(37.30) must not exceed stipulated limits:

$$\varepsilon_i = \frac{\delta_i - \delta_{i-1}}{h_i - h_{i-1}} \qquad (37.31)$$

where

$\delta_i - \delta_{i-1}$ = relative displacement between two consecutive floors
$h_i - h_{i-1}$ = interstory height.

The limiting values for lateral drift are shown in Table 37.8 for different groups of buildings and according to the possibility of damage to nonstructural elements.

Table 37.8. Limiting Values for Lateral Drift (ε_i)

Type and Disposition of Nonstructural Elements	ε_i for Buildings	
	Group A	Group B
Can be damaged	0.015	0.018
Cannot be damaged	0.020	0.024

Buildings must be separated by a distance greater than s determined by the following expression:

$$\begin{aligned} s = 0.5\delta_{en}(D_f+1) &\geqslant 35 \text{ mm} && \text{if } h \leqslant 6\text{ m} \\ s = 0.5\delta_{en}(D_f+1) &\geqslant 35 + 0.004(h-6) && \text{if } h > 6\text{ m} \end{aligned} \qquad (37.32)$$

where

δ_{en} = elastic lateral displacement at level n
h = total height of building
D_f = ductility factor (Table 37.4).

Equation (37.32) takes into account the fact that the maximum inelastic deformations of each level do not occur simultaneously. When the minimum separation between two adjoining buildings must be determined, the square root of the sum of the squares of the values obtained from eq.(37.31) for each building should be used. The Code permits the construction of two buildings in contact with each other if the floor slabs are aligned at each level and if it can be shown that their interaction does not cause unfavorable effects.

37.20 FOUNDATIONS

Foundations preferably should be of one type (e.g., footings or piles). If by necessity mixed foundations are used, the behavior of the group under seismic action should be checked. The loads Q to be used under service conditions are the following:

$$Q = D + L \pm S \qquad (37.33\text{a})$$
$$Q = D \pm S \qquad (37.33\text{b})$$

where

D = permanent load
L = variable load
S = earthquake effects.

Foundation elements should be connected by reinforced concrete beams designed for tension and compression, with axial forces equal to at least 10% of the loads applied on the columns.

The contact pressures applied to the ground by overlying foundations should not exceed the following values:

(a) 50% of the ultimate load capacity of the soil under static loading
(b) Twice the permissible load capacity.

For deep foundations, 75% of the ultimate load capacity may be applied, but the pile stresses must not exceed 50% of the capacity of the piles.

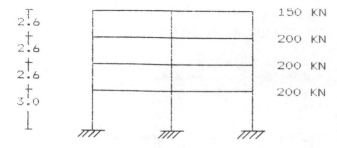

2.6
2.6
2.6
3.0

150 KN
200 KN
200 KN
200 KN

Fig. 37.3.   Frame for Example 37.1

37.21 EXAMPLE 37.1

The plane frame of Fig. 37.3 is to be part of a Type I building located in Seismic Zone 4, built on a foundation of dense gravel and sand. The building is an essential structure, so it is classified in Group A. The weights and stiffnesses associated with each level are:

Level	Weight (kN)	Stiffness (kN/m)
4	150	10,000
3	200	15,000
2	200	20,000
1	200	25,000

Determine the seismic forces and shears used for the design of this frame according to the Venezuelan Seismic Code, and check lateral displacements. The equivalent static method is used.

Estimated fundamental period for frame (Type I) buildings

$$T_a = 0.061 h_n^{3/4}$$

$$h_n = 10.8 \, \text{m}, \quad \text{then} \quad T_a = 0.363 \, \text{sec}$$

Shear modification factor

$$\mu_1 = \frac{3}{2} \frac{(n+1)}{(2n+1)} = \frac{3}{2} \frac{(4+1)}{(2 \times 4 + 1)} = 0.833$$

$$\mu_2 = 0.80 + \frac{1}{20} \left(\frac{T_a}{T^*} - 1 \right)$$

For dense soil (S1 profile) $T^* = 0.40 \, \text{sec}$ (Table 37.5), so

$$\mu_2 = 0.80 + \frac{1}{20} \left(\frac{0.363}{0.40} - 1 \right) = 0.795$$

$$\mu = \max(\mu_1, \mu_2) = 0.833$$

Design spectral acceleration (A_d)

$$\text{for } 0.15 \leqslant T_a \leqslant T^*, \quad A_d = \frac{\alpha \beta A_0}{R}$$

where

$\alpha = 1.25$ for essential structures
$\beta = 2.2$ for S_1 soil (Table 37.5)
$A_0 = 0.30$ for Seismic Zone 4 (Table 37.2)
$R = D_f = 6$ for $T > 0.15 \, \text{sec}$, ND3 and structural type 1 (frames) (Table 37.4)

Thus,

$$A_d = (1.25)(2.2)(0.30)/6 = 0.1375$$

Seismic coefficient

$$C_s = \mu A_d$$
$$= 0.833 \times 0.1375 = 0.115$$

$$(C_s)_{\min} = (\alpha A_0)/6 = (1.25)(0.30)/6 = 0.0625$$

Base shear

$$V_0 = C_s W = 0.115(750) = 86.25 \, \text{kN}$$

Top force

$$F_t = \left(0.06 \frac{T}{T^*} - 0.02 \right)$$

$$V_0 = \left(0.06 \left(\frac{0.363}{0.40} - 0.02 \right) (86.25) \right) = 2.97 \, \text{kN}$$

$$F_{t\min} = 0.04(86.25) = 3.45 \, \text{kN}$$

therefore,

$$F_t = 3.45 \, \text{kN}$$

$$V_0 - F_t = 86.25 - 3.45 = 82.80 \, \text{kN}$$

Vertical distribution of seismic forces and shears. (Calculation results are given in Table 37.9.)

$$F_i = \frac{W_i h_i}{\sum\limits_{j=1}^{n} W_j h_j} (V_0 - F_t) \qquad [\text{eq.}(37.16)]$$

Table 37.9.   Calculations of Vertical Distribution of Seismic Forces and Shears

Level	W_i (kN)	h_i (m)	$W_i h_i$	F_i (kN)	V_i (kN)
4	150	10.8	1,620	26.93 + 3.45 = 30.38	30.38
3	200	8.2	1,640	27.27	57.65
2	200	5.6	1,120	18.62	76.27
1	200	3.0	600	9.98	86.25

$$\Sigma W_i h_i = 4,980$$

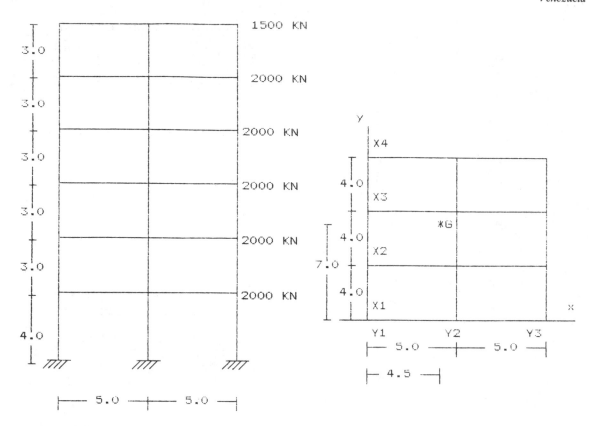

Fig. 37.4. Building for Example 37.2

Lateral displacements. (Calculation results are shown in Table 37.10.)

The maximum permissible inelastic lateral drift for essential structures with nonstructural elements connected to the structure is equal to 0.015 (Table 37.8). In this example, all levels of the building comply with the code limitation on drift (see last column of Table 37.10).

Table 37.10. Calculations of Lateral Displacements

Level	h_i (m)	V_i (kN)	K_j (kN/m)	$\delta_{ei} - \delta_{ei-1}$ (m)	$\dfrac{(\delta_{ei} - \delta_{ei-1})}{(h_i - h_{i-1})}$
4	10.8	30.38	5,000	0.0061	0.014
3	8.2	57.65	9,000	0.0064	0.015
2	5.6	76.27	12,000	0.0064	0.015
1	3.0	86.25	14,000	0.0062	0.012

37.22 EXAMPLE 37.2

The building of Fig. 37.4 is to be located in Seismic Zone 4 on a foundation of firm soil (S_1 profile). The building belongs to the group of nonessential frame-type structures to be designed for level ND3 require-ments. Determine the seismic forces according to the Venezuelan Seismic Code using the computer program developed for this purpose, with the equivalent static method (ESM).

```
Program printout:

Seismic forces according to the Venezuelan seismic code:

Building data:

Seismic zone.........................: 4
Maximum probable ground acceleration: 0.30g
Structural group.....................: B
Design level.........................: ND3
Structural type......................: 1
Ductility factor.....................: 6.00
Foundation soil type.................: S1

Number of levels.....................:  6
Number of frames in X direction.....:  4
Number of frames in Y direction.....:  3

ANALYSIS USING THE EQUIVALENT STATIC METHOD

Seismic coefficient for X earthquake = 0.0693
Estimated fundamental period    Tax = 0.555 sec
Seismic coefficient for Y earthquake = 0.0693
Estimated fundamental period    Tay = 0.555 sec
Ftx = 50.456 kN        Fty = 50.456 kN

Translational seismic forces and shears:

                                          X earthquake        Y earthquake
Level  Wi(kN)   hi(m)    Wihi         Fi(kN)   Vi(kN)     Fi(kN)   Vi(kN)

  1    2000.00   4.00    8000.00      46.51    797.47     46.51    797.47
  2    2000.00   7.00   14000.00      81.39    750.96     81.39    750.96
  3    2000.00  10.00   20000.00     116.27    669.58    116.27    669.58
  4    2000.00  13.00   26000.00     151.15    553.31    151.15    553.31
  5    2000.00  16.00   32000.00     186.03    402.16    186.03    402.16
  6    1500.00  19.00   28500.00     216.14    216.14    216.14    216.14
```

Distribution of translational shears to frames:

Frame X	Stiffness(kN/m)	Level	Vi(kN)
1	40000	1	177.22
1	40000	2	166.88
1	40000	3	148.79
1	30000	4	127.69
1	30000	5	92.81
1	10000	6	43.23
2	50000	1	221.52
2	50000	2	208.60
2	50000	3	185.99
2	35000	4	148.97
2	35000	5	108.27
2	15000	6	64.84
3	50000	1	221.52
3	50000	2	208.60
3	50000	3.	185.99
3	35000	4	148.97
3	35000	5	108.27
3	15000	6	64.84
4	40000	1	177.22
4	40000	2	166.88
4	40000	3	148.79
4	30000	4	127.69
4	30000	5	92.81
4	10000	6	43.23

Frame Y	Stiffness(kN/m)	Level	Vi(kN)
1	50000	1	221.52
1	50000	2	208.60
1	50000	3	185.99
1	40000	4	170.25
1	40000	5	123.74
1	15000	6	64.84
2	80000	1	354.43
2	80000	2	333.76
2	80000	3	297.59
2	50000	4	212.81
2	50000	5	154.68
2	20000	6	86.45
3	50000	1	221.52
3	50000	2	208.60
3	50000	3	185.99
3	40000	4	170.25
3	40000	5	123.74
3	15000	6	64.84

Torsion moments:

Dynamic amplification factor for torsion = 3.00

Level	ey(m)	X earthquake		ex(m)	Y earthquake	
		T1(kNm)	T2(kNm)		T1(kNm)	T2(kNm)
1	-1.000	3349.37	-159.49	-0.500	1993.67	-398.73
2	-1.000	3154.04	-150.19	-0.500	1877.41	-375.48
3	-1.000	2812.22	-133.92	-0.500	1673.94	-334.79
4	-1.000	2323.90	-110.66	-0.500	1383.27	-276.65
5	-1.000	1689.08	-80.43	-0.500	1005.41	-201.08
6	-1.000	907.77	-43.23	-0.500	540.34	-108.07

Torsional shears:

Level 1

Frame	Position(m)	V1(kN)	V2(kN)	V3(kN)	V4(kN)
X 1	0.000	139.07	-6.62	82.78	-16.56
X 2	4.000	57.95	-2.76	34.49	-6.90
X 3	8.000	-57.95	2.76	-34.49	6.90
X 4	12.000	-139.07	6.62	-82.78	16.56
Y 1	0.000	144.87	-6.90	86.23	-17.25
Y 2	5.000	0.00	0.00	0.00	0.00
Y 3	10.000	-144.87	6.90	-86.23	17.25

Level 2

Frame	Position(m)	V1(kN)	V2(kN)	V3(kN)	V4(kN)
X 1	0.000	130.96	-6.24	77.95	-15.59
X 2	4.000	54.57	-2.60	32.48	-6.50
X 3	8.000	-54.57	2.60	-32.48	6.50
X 4	12.000	-130.96	6.24	-77.95	15.59
Y 1	0.000	136.42	-6.50	81.20	-16.24
Y 2	5.000	0.00	0.00	0.00	0.00
Y 3	10.000	-136.42	6.50	-81.20	16.24

Level 3

Frame	Position(m)	V1(kN)	V2(kN)	V3(kN)	V4(kN)
X 1	0.000	116.77	-5.56	69.51	-13.90
X 2	4.000	48.65	-2.32	28.96	-5.79
X 3	8.000	-48.65	2.32	-28.96	5.79
X 4	12.000	-116.77	5.56	-69.51	13.90
Y 1	0.000	121.64	-5.79	72.40	-14.48
Y 2	5.000	0.00	0.00	0.00	0.00
Y 3	10.000	-121.64	5.79	-72.40	14.48

Level 4

Frame	Position(m)	V1(kN)	V2(kN)	V3(kN)	V4(kN)
X 1	0.000	94.21	-4.49	56.08	-11.22
X 2	4.000	36.64	-1.74	21.81	-4.36
X 3	8.000	-36.64	1.74	-21.81	4.36
X 4	12.000	-94.21	4.49	-56.08	11.22
Y 1	0.000	104.68	-4.98	62.31	-12.46
Y 2	5.000	0.00	0.00	0.00	0.00
Y 3	10.000	-104.68	4.98	-62.31	12.46

Level 5

Frame	Position(m)	V1(kN)	V2(kN)	V3(kN)	V4(kN)
X 1	0.000	68.48	-3.26	40.76	-8.15
X 2	4.000	26.63	-1.27	15.85	-3.17
X 3	8.000	-26.63	1.27	-15.85	3.17
X 4	12.000	-68.48	3.26	-40.76	8.15
Y 1	0.000	76.08	-3.62	45.29	-9.06
Y 2	5.000	0.00	0.00	0.00	0.00
Y 3	10.000	-76.08	3.62	-45.29	9.06

Level 6

Frame	Position(m)	V1(kN)	V2(kN)	V3(kN)	V4(kN)
X 1	0.000	34.26	-1.63	20.39	-4.08
X 2	4.000	17.13	-0.82	10.20	-2.04
X 3	8.000	-17.13	0.82	-10.20	2.04
X 4	12.000	-34.26	1.63	-20.39	4.08
Y 1	0.000	42.82	-2.04	25.49	-5.10
Y 2	5.000	0.00	0.00	0.00	0.00
Y 3	10.000	-42.82	2.04	-25.49	5.10

Design shears:

Level 1:

Frame	X earthquake		Y earthquake		V(kNm)
	Vxd(kNm)	Vxt(kNm)	Vyd(kNm)	Vyt(kNm)	
X 1	177.22	139.07	0.00	82.78	316.29
X 2	221.52	57.95	0.00	34.49	279.47
X 3	221.52	2.76	0.00	34.49	224.28
X 4	177.22	6.62	0.00	82.78	183.84
Y 1	0.00	139.07	221.52	82.78	304.30
Y 2	0.00	57.95	354.43	34.49	388.92
Y 3	0.00	57.95	221.52	6.90	228.42

Level 2:

Frame	X earthquake		Y earthquake		V(kNm)
	Vxd(kNm)	Vxt(kNm)	Vyd(kNm)	Vyt(kNm)	
X 1	166.88	130.96	0.00	77.95	297.84
X 2	208.60	54.57	0.00	32.48	263.17
X 3	208.60	2.60	0.00	32.48	211.20
X 4	166.88	6.24	0.00	77.95	173.12
Y 1	0.00	130.96	208.60	77.95	286.56
Y 2	0.00	54.57	333.76	32.48	366.24
Y 3	0.00	54.57	208.60	6.50	215.10

Level 3:

Frame	X earthquake		Y earthquake		V(kNm)
	Vxd(kNm)	Vxt(kNm)	Vyd(kNm)	Vyt(kNm)	
X 1	148.79	116.77	0.00	69.51	265.56
X 2	185.99	48.65	0.00	28.96	234.65
X 3	185.99	2.32	0.00	28.96	188.31
X 4	148.79	5.56	0.00	69.51	154.36
Y 1	0.00	116.77	185.99	69.51	255.50
Y 2	0.00	48.65	297.59	28.96	326.55
Y 3	0.00	48.65	185.99	5.79	191.79

Level 4:

Frame	X earthquake		Y earthquake		V(kNm)
	Vxd(kNm)	Vxt(kNm)	Vyd(kNm)	Vyt(kNm)	
X 1	127.69	94.21	0.00	56.08	221.90
X 2	148.97	36.64	0.00	21.81	185.61
X 3	148.97	1.74	0.00	21.81	150.71
X 4	127.69	4.49	0.00	56.08	132.17
Y 1	0.00	94.21	170.25	56.08	226.33
Y 2	0.00	36.64	212.81	21.81	234.62
Y 3	0.00	36.64	170.25	4.36	174.61

Level 5:

Frame	X earthquake		Y earthquake		V(kNm)
	Vxd(kNm)	Vxt(kNm)	Vyd(kNm)	Vyt(kNm)	
X 1	92.81	68.48	0.00	40.76	161.28
X 2	108.27	26.63	0.00	15.85	134.90
X 3	108.27	1.27	0.00	15.85	109.54
X 4	92.81	3.26	0.00	40.76	96.07
Y 1	0.00	68.48	123.74	40.76	164.50
Y 2	0.00	26.63	154.68	15.85	170.53
Y 3	0.00	26.63	123.74	3.17	126.91

Level 6:

Frame	X earthquake		Y earthquake		V(kNm)
	Vxd(kNm)	Vxt(kNm)	Vyd(kNm)	Vyt(kNm)	
X 1	43.23	34.26	0.00	20.39	77.48
X 2	64.84	17.13	0.00	10.20	81.97
X 3	64.84	0.82	0.00	10.20	65.66
X 4	43.23	1.63	0.00	20.39	44.86
Y 1	0.00	34.26	64.84	20.39	85.23
Y 2	0.00	17.13	86.45	10.20	96.65
Y 3	0.00	17.13	64.84	2.04	66.88

Lateral displacements of frames:

Level 1:

Frame	hn(m)	Vi(kN)	Ki(kN/m)	Rel.disp. Δe(m)	Inel.Drift (Δe*Duct/hn)
X 1	4.00	316.29	40000	0.0079	0.0119
X 2	4.00	279.47	50000	0.0056	0.0084
X 3	4.00	224.28	50000	0.0045	0.0067
X 4	4.00	183.84	40000	0.0046	0.0069
Y 1	4.00	304.30	50000	0.0061	0.0091
Y 2	4.00	388.92	80000	0.0049	0.0073
Y 3	4.00	228.42	50000	0.0046	0.0069

Level 2:

Frame	hn(m)	Vi(kN)	Ki(kN/m)	Rel.disp. Δe(m)	Inel.Drift (Δe*Duct/hn)
X 1	3.00	297.84	40000	0.0074	0.0149
X 2	3.00	263.17	50000	0.0053	0.0105
X 3	3.00	211.20	50000	0.0042	0.0084
X 4	3.00	173.12	40000	0.0043	0.0087
Y 1	3.00	286.56	50000	0.0057	0.0115
Y 2	3.00	366.24	80000	0.0046	0.0092
Y 3	3.00	215.10	50000	0.0043	0.0086

Level 3:

Frame	hn(m)	Vi(kN)	Ki(kN/m)	Rel.disp. Δe(m)	Inel.Drift (Δe*Duct/hn)
X 1	3.00	265.56	40000	0.0066	0.0133
X 2	3.00	234.65	50000	0.0047	0.0094
X 3	3.00	188.31	50000	0.0038	0.0075
X 4	3.00	154.36	40000	0.0039	0.0077
Y 1	3.00	255.50	50000	0.0051	0.0102
Y 2	3.00	326.55	80000	0.0041	0.0082
Y 3	3.00	191.79	50000	0.0038	0.0077

Level 4:

Frame	hn(m)	Vi(kN)	Ki(kN/m)	Rel.disp. Δe(m)	Inel.Drift (Δe*Duct/hn)
X 1	3.00	221.90	30000	0.0074	0.0148
X 2	3.00	185.61	35000	0.0053	0.0106
X 3	3.00	150.71	35000	0.0043	0.0086
X 4	3.00	132.17	30000	0.0044	0.0088
Y 1	3.00	226.33	40000	0.0057	0.0113
Y 2	3.00	234.62	50000	0.0047	0.0094
Y 3	3.00	174.61	40000	0.0044	0.0087

Level 5:

Frame	hn(m)	Vi(kN)	Ki(kN/m)	Rel.disp. Δe(m)	Inel.Drift (Δe*Duct/hn)
X 1	3.00	161.28	30000	0.0054	0.0108
X 2	3.00	134.90	35000	0.0039	0.0077
X 3	3.00	109.54	35000	0.0031	0.0063
X 4	3.00	96.07	30000	0.0032	0.0064
Y 1	3.00	164.50	40000	0.0041	0.0082
Y 2	3.00	170.53	50000	0.0034	0.0068
Y 3	3.00	126.91	40000	0.0032	0.0063

Level 6:

Frame	hn(m)	Vi(kN)	Ki(kN/m)	Rel.disp. Δe(m)	Inel.Drift (Δe*Duct/hn)
X 1	3.00	77.48	10000	0.0077	0.0155
X 2	3.00	81.97	15000	0.0055	0.0109
X 3	3.00	65.66	15000	0.0044	0.0088
X 4	3.00	44.86	10000	0.0045	0.0090
Y 1	3.00	85.23	15000	0.0057	0.0114
Y 2	3.00	96.65	20000	0.0048	0.0097
Y 3	3.00	66.88	15000	0.0045	0.0089

Maximum lateral inelastic drift permitted by the code = 0.018.
All frames and levels fulfil the code requirements.

REFERENCES

Comisión Presidencial para el Estudio del Sismo (1978) *Segunda Fase del Estudio del Sismo Occurrido en Caracas el 29 de Julio de 1967.* MOP-FUNVISIS, Vols. A and B, Caracas, Venezuela.

GRASES, J. (1985) "Fundamentos para la Elaboración del Nuevo Mapa de Zonificación Sísmica de Venezuela con Fines de Ingeniería." *Funvisis, Serie Técnica 05-84*, Caracas, Venezuela.

GRASES, J. et al. (1984) *Edificaciones Sismorresistentes: Manual de Aplicación de las Normas.* Fondur, Caracas, Venezuela.

HANSON, R. D., and DEGENKOLB, H. J. (1969) *The Venezuelan Earthquake, July 29, 1967.* AISC, New York, NY.

HOUSNER, G. M. (1962) "Fundamentos de Ingeniería Sísmica." *Revista Ingeniería.* U.N.A.M., pp. 25–55, Mexico.

KAN, C., and CHOPRA, A. (1976) *Coupled lateral torsional response of buildings to ground shaking.* Pp. 76–130. University of California, EERC. Berkeley, CA.

Ministerio del Desarrollo Urbano [MDU] (1985) *Normas Venezolanas para Estructuras de Concreto Armado: Análisis y Diseño.* COVENIN 1753-85, Caracas, Venezuela.

——— (1987) *Norma Venezolana de Edificación Antisísmica.* Dirección General Sectorial de Equipamiento Urbano, Funvisis-COVENIN 1756-87, Caracas, Venezuela.

Ministerio de Obras Públicas (MOP) (1947, 1955, 1967) *Normas para el Cálculo de Edificios.* Dirección de Edificios e Instalaciones Industriales, Imprenta Nacional, Caracas, Venezuela.

Uniform Building Code (UBC) (1927–1991) International Conference of Building Code Officials, Whittier, CA.

38

(Former) Yugoslavia

Dimitar Jurukovski and Predrag Gavrilovic

38.1 INTRODUCTION

One year after the 1963 Skopje earthquake, the first Yugoslavian Code for Construction in Seismic Regions (temporary code 1964), was prepared by a committee consisting of international and national experts. In spite of the fact that seismic activity had been severe in Yugoslavia, earthquake forces had not been considered in the design or construction of structures. The first seismic map of Yugoslavia was prepared in 1950 by Professor J. Mihailovic, who used historical earthquakes that had occurred on the Balkan Peninsula.

The temporary code of 1964 was used until 1981 when the current Code, *Technical Regulations for Design and Construction of Buildings in Seismic Regions*, was published. Minor modifications were introduced into the Code during 1982 and 1983. In 1989, the Yugoslavian Association of Earthquake Engineering evaluated the need for a new version of the Yugoslav Code. Concurrently, a special national committee prepared a Code for seismic design of structures other than buildings (dams, bridges, embankments, pipelines, supporting structures, etc.). This Code was completed in 1989, but has not been enforced because of political problems in Yugoslavia.

The number of seismological monitoring stations in Yugoslavia was continually increased during the period 1965–1972. By 1972 about one hundred strong-motion instruments (SMA-1) had been installed on bedrock and alluvial deposits; a total of 1,500 strong-motion records had been obtained by 1991. New maps based on these records of seismic activity in Yugoslavia were published by the Association of Seismology (1987) for return periods of 50, 100, 200, 500, 1,000, and 10,000 years. These new probability maps are not incorporated in the existing earthquake Code although they are published officially by the Government.

After the implementation of the new Yugoslav Code in 1981, several national subcommittees were established for preparation of national standards for various activities. These subcommittees are responsible for such areas as strong-motion instrumentation of structures, data processing, repair and strengthening of damaged structures, microzonation, vulnerability estimation, and other pertinent subjects. The current

Seismic Code implements the most recent advances for design and construction of seismic-resistant structures.

38.2 EARTHQUAKE FORCES

According to the Yugoslav Code, *Technical Regulations for Design and Construction of Buildings in Seismic Regions* (1981), two approaches are considered for earthquake-resistant design of buildings: (1) the equivalent static force method applicable to buildings in Categories I, II, III or IV, as described in Section 38.3.3; and (2) the dynamic analysis method, which may be used for any structure but must be used for buildings not included in Categories I through IV.

38.3 EQUIVALENT STATIC FORCE METHOD

In the application of the equivalent static force method, the structure is analyzed under the action of static lateral forces applied at each level of the building where the mass of the structure is concentrated, as illustrated in Fig. 38.1. This figure shows a plane frame representing a multistory building in which the weights of the building are assumed to be applied at various levels of the building. These weights are designated in Fig. 38.1 as $W_1, W_2, \ldots, W_n$, where n is the total number of stories in the building.

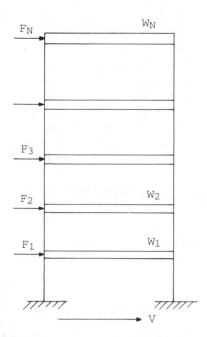

Fig 38.1. Building modeled as a plane frame

38.3.1 Base Seismic Force

The total horizontal seismic force or base seismic force V on the building is calculated by the formula

$$V = KW \qquad (38.1)$$

where K is the total seismic coefficient for the horizontal direction, and W is the total weight of the building and its permanent equipment.

38.3.2 Seismic Zones

Except for those structures classified in Building Category I (see Section 38.3.3), the seismic coefficient for each region is estimated on the basis of the seismic map of Yugoslavia, which is a constituent part of the Code (Fig. 38.2). This map was developed with the use of seismological data on earthquakes that have occurred in Yugoslavia. Seismic hazard and other seismic parameters for a specific region are determined on the basis of macrozoning and microzoning procedures.

When structures are classified as Building Category I or as out of category, the evaluation of seismic coefficients, as well as seismic design parameters, must be performed for each particular site using appropriate procedures for seismic microzonation.

38.3.3 Seismic Coefficient *K*

The total seismic coefficient K is calculated from the expression:

$$K = K_0 K_s K_d K_p \qquad (38.2)$$

where

K_0 = coefficient based on building category
K_s = coefficient based on seismic intensity
K_d = coefficient based on dynamic response
K_p = coefficient based on ductility and damping.

The total seismic coefficient K shall have a minimum value of 0.02. The coefficient based on building category, K_0, has the following values:

- *Building Category I*. Buildings in which people gather (cinemas, theaters, exhibition halls), university buildings, schools, hospitals, fire stations, telecommunication, radio, and television buildings, and some important industrial buildings ($K_0 = 1.5$)
- *Building Category II*. Residential buildings, hotels, restaurants, public buildings, and all other dwellings not classified in Category I ($K_0 = 1.0$)
- *Building Category III*. Auxiliary industrial buildings, agro-industrial buildings ($K_0 = 0.75$)

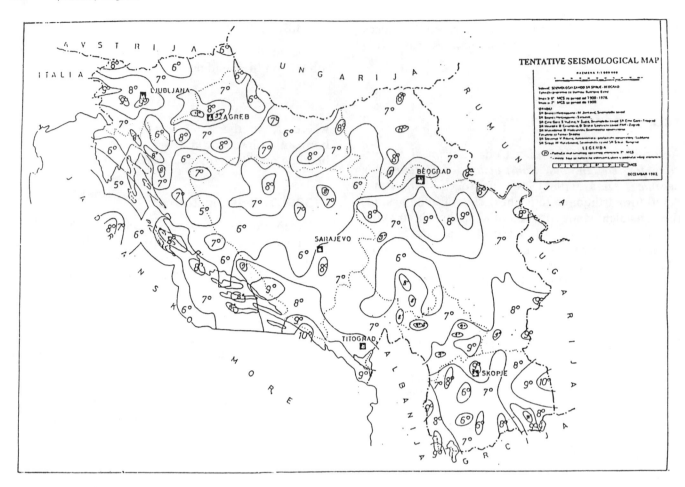

Fig. 38.2. Tentative seismic map of Yugoslavia

- *Building Category IV*. Temporary buildings and structures that could collapse without danger to human lives, and therefore do not require earthquake protection ($K_0 = 0$).

Values of the seismic intensity coefficient K_s shown in Table 38.1 are given in terms of the seismicity zone for the site, obtained from Fig. 38.2.

The coefficient based on dynamic response, K_d, is determined from Table 38.2 as a function of the foundation category and the fundamental period of free vibration of the structure. Alternatively, the dynamic response coefficient may be obtained from Fig. 38.3.

Foundation Category I includes rock and rock-like soil and dense hard soils, such as dense sand or stiff clay on top of a firm bedrock formation. Foundation Category II is represented by dense and medium-dense soils. Foundation Category III includes soils of low density and soft consistency. More precise definitions of the foundation categories are given in Table 38.1 of the Yugoslav Seismic Code (1981).

The natural periods of the structure are determined either by the methods of structural dynamics, or by approximate or empirical formulas, which are based on the principles of structural dynamics.

The coefficient of ductility and damping, K_p, depends on the type of structure as follows:

- For all modern reinforced concrete, steel, and wooden structures, $K_p = 1.0$.

Table 38.1. Seismic Coefficient K_s

Seismicity Zone	K_s
7^0	0.025
8^0	0.050
9^0	0.100

Table 38.2. Dynamic Response Coefficient K_d

Foundation Category	K_d	Limit Values for K_d
I	$0.50/T$	$0.33 \leq K_d < 1.0$
II	$0.70/T$	$0.47 \leq K_d < 1.0$
III	$0.90/T$	$0.60 \leq K_d < 1.0$

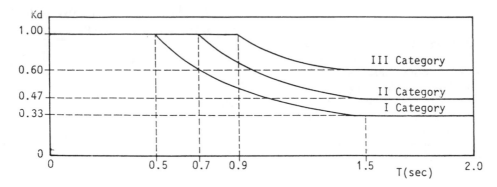

Fig. 38.3. Dynamic coefficient for Foundation Categories I, II, and III as a function of the natural period

- For reinforced masonry and braced steel structures, $K_p = 1.3$.
- For masonry and reinforced concrete buildings with shear walls and for very high and slender structures with a fundamental period $T \geqslant 2.0$ sec, $K_p = 1.6$.
- For structures with a flexible story, or with an abrupt change in stiffness between adjacent floors, and for ordinary masonry structures, $K_p = 2.0$.

38.3.4 Distribution of the Base Shear Force

The distribution of seismic forces over the height of a structure must be obtained by using structural dynamics methods. However, for buildings of no more than five stories, the distribution of seismic forces may be determined as

$$F_i = 0.85V \frac{W_i H_i}{\sum_{j=1}^{n} W_j H_j} \tag{38.3}$$

with the additional force F_t ($F_t = 0.15V$) applied at the top of the building, where

V = base seismic force calculated by eq.(38.1)
F_i = seismic force at level i
W_i = weight at level i
H_i = height of level i measured from the base of the building.

38.3.5 Vertical Seismic Action

The vertical seismic action S on the building is determined in accordance with the formula

$$S = K_v W \tag{38.4}$$

where K_v is the total seismic coefficient for the vertical

direction, and W is the total weight of the building. The total seismic coefficient K_v is calculated from the formula

$$K_v = 0.7K \tag{38.5}$$

where K is the total seismic coefficient for the horizontal direction given by eq.(38.2).

In general, the Code allows the designer to ignore the effects due to vertical seismic forces, except in structures containing elements with unusually large spans.

38.3.6 Determination of the Weights W_i

The total weight of a building W or concentrated weights W_i at the various levels of the building include the dead load, probable live load, and snow load.

The probable live load is considered to be 50% of the specified design loads. If the live load is significant (such as for warehouses, silos, libraries, archives, and storage buildings), the seismic loads shall be determined for the most unfavorable case of maximum, or minimum, actual loading. Loads caused by wind, or live loads caused by cranes, need not be taken into consideration when calculating the seismic behavior of buildings. The full weights of permanently installed equipment shall be included in the calculation of the weight W_i.

38.3.7 Torsional Effects

The inertial forces induced by the earthquake develop a torsional moment when the center of mass and the center of the stiffness at any story of the building are not at the same point. The torsional moment M_{ti} at any story i is equal to

$$M_{ti} = K_t V_i e_i \tag{38.6}$$

where

V_i = the story shear force at the ith story chosen for analysis

e_i = the eccentricity of the center of stiffness with respect to the center of mass in the ith story

K_t = the amplification factor (by which the eccentricity is increased due to coupling between lateral and torsional vibrations and nonuniform movement of the foundations). If no value of the coefficient K_t is calculated, then a value of $K_t = 1.5$ is used.

38.3.8 Lateral Displacements

The lateral displacement x_i at each floor level of the building is calculated by static analysis of the building subjected to equivalent earthquake forces F_i applied at the various levels of the building. Alternatively, the simplified procedure described in Chapter 4, eq.(4.72), which is based on Newton's law of motion, may be used. That is

$$x_i = \frac{g}{4\pi^2} \cdot \frac{T^2 F_i}{W_i} \qquad (38.7)$$

where

g = acceleration of gravity
T = fundamental period
F_i = equivalent seismic force at level i
W_i = weight at level i.

The maximum permissible displacement x_{tmax} at the top of the building produced by the seismic forces F_i, without considering soil-structure interaction, is limited to

$$x_{\text{tmax}} = \frac{H}{600} \qquad (38.8)$$

where H is the total height of the building.

For linear behavior of the structure, the story drift Δ_i is limited to $h_i/350$; it is limited to $h_i/150$ for nonlinear behavior. The term h_i is the height of the ith story in cm.

38.3.9 Overturning Moments

The seismic forces produce overturning moments at the various levels of the building; additional axial forces are generated in the columns, particularly in the external columns of the building. These additional axial forces in the external columns, combined with the gravitational forces, produce the total axial forces on the columns. Structural elements must be designed

to resist these forces. The overturning moment M_i at any floor level i is calculated by statics as

$$M_i = \sum_{k=i+1}^{n} F_k(h_k - h_i) \qquad (38.9)$$

where $i = 0, 1, 2, \ldots, n - 1$ and

F_k = seismic force acting at level k
h_i, h_k = height from the base of the building of levels i and k.

Note: F_n includes the additional force F_t at the top of the building.

38.3.10 Story Shear Force

The story shear force V_i at level i of the building is given by the sum of the lateral seismic forces above that level

$$V_i = \sum_{k=i}^{n} F_k \qquad (38.10)$$

38.3.11 Seismic Design of the Structure

The design of the structural elements for given seismic effects (shear forces, bending moments, and axial forces) is based on the limit state theory. According to the existing Yugoslav Code (1981), the following safety factors must be used:

- For reinforced and prestressed concrete structures, 1.30
- For steel structures, 1.15
- For masonry structures, 1.50.

38.4 DYNAMIC ANALYSIS METHOD

According to the Code, all out-of-category buildings are assumed to be subject to the earthquake ground motion that is expected at the proposed construction site. Such buildings must be designed using the dynamic analysis method. Also, dynamic analysis is required for prototypes of prefabricated buildings or structures constructed by industrialized methods. Wooden structures are subject to this requirement irrespective of the type of construction method used. The stresses and deformations in the structure can be predicted for the maximum expected earthquakes and the design requirements can be established by the use of dynamic analysis. This type of analysis can also show how to limit damage (to the structural and nonstructural elements of the building) to acceptable levels.

Definitions of the design and predictions of the maximum expected earthquakes are based on geo-

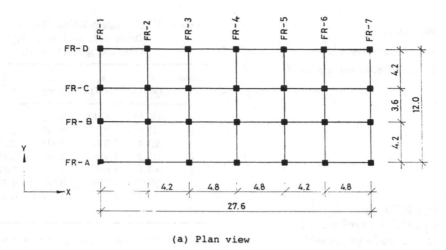

(a) Plan view

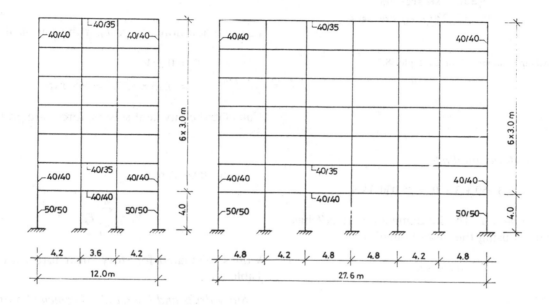

(b) Elevation views

Fig. 38.4. Plan and elevation views for the building in Example 38.1

logical, geotechnical, seismological, and geophysical data, the life expectancy of the building, and the acceptable seismic risk level. There is no written standard for definition of such earthquake parameters. Therefore, any method for microzonation accepted in engineering seismology could be used.

The total horizontal seismic force F obtained by the dynamic analysis method shall be not less than 75% nor less than $0.02W$ of the force obtained by the equivalent static method.

Example 38.1. *Reinforced concrete frame structure*

Figs. 38.4(a) and 38.4(b) show the plan and the elevation of a seven-story reinforced concrete frame structure. This structure serves as an example of

analysis applying the Yugoslav Code. The structure, $27.6\,\text{m} \times 12.0\,\text{m}$ in plan, is a residential building with a serviceability period of 100 years. The structure, within a 9° seismic intensity zone, is founded on a gravel and sand soil of moderate bearing capacity. The weight at the first level is $3,150\,\text{kN}$; at all the other levels of the building the weight is $3,650\,\text{kN}$.

The building is modeled as a shear building (Fig. 38.5) for which the calculated story stiffnesses are as follows:

$$k_1 = 4,909\,\text{kN/cm}, \quad k_2 = 3,517\,\text{kN/cm},$$
$$k_3 = 3,059\,\text{kN/cm}, \quad k_4 = 2,976\,\text{kN/cm},$$
$$k_5 = 2,936\,\text{kN/cm}, \quad k_6 = 2,902\,\text{kN/cm},$$
$$k_7 = 2,865\,\text{kN/cm}$$

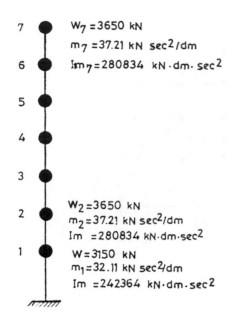

7    $W_7 = 3650$ kN

    $m_7 = 37.21$ kN sec^2/dm

6    $Im_7 = 280834$ kN·dm·sec^2

5

4

3

2    $W_2 = 3650$ kN
    $m_2 = 37.21$ kN sec^2/dm
    $Im = 280834$ kN·dm·sec^2

1    $W = 3150$ kN
    $m_1 = 32.11$ kN sec^2/dm
    $Im = 242364$ kN·dm·sec^2

Fig. 38.5. Mathematical model for Example 38.1

Solution

Total weight of the building

$$W = 6 \times 3,650 + 3,150 = 25,050.0 \text{ kN}$$

Fundamental period. The fundamental period T has been determined using the Stodola method:

$$T = 0.893 \text{ sec}$$

Seismic coefficient

$$K = K_0 K_s K_d K_p \qquad \text{[eq.(38.2)]}$$

$K_0 = 1.0$ for Building Category II
$K_s = 0.10$ for Seismic Intensity Zone 9° (Table 38.1)
$K_p = 1.00$ for a modern reinforced concrete building
$K_d = 0.7T = 0.71$ for Foundation Category II (Table 38.2).

$$K = 1.0 \times 0.10 \times 0.71 \times 1.0 = 0.071$$

Base seismic force

$$V = KW \qquad \text{[eq.(38.1)]}$$

$$= 0.071 \times 25,050 = 1,778.0 \text{ kN}$$

Equivalent seismic forces

$$F_i = 0.85V \frac{W_i H_i}{\sum\limits_{j=1}^{n} W_j H_j} \qquad \text{[eq.(38.3)]}$$

Table 38.3. Equivalent Seismic Forces F_i, Shear Forces V_i, Story Drifts Δ_i, Lateral Story Displacements δ_i, and Overturning Moments M_i

Level i	F_i (kN)	V_i (kN)	Δ_i (cm)	δ_i (cm)	M_i (kN-cm)
7	634.5*	634.5	0.221	2.822	—
6	317.6	951.1	0.328	2.601	1,903
5	267.4	1,219.5	0.415	2.273	4,760
4	217.3	1,436.8	0.483	1.858	8,418
3	167.1	1,603.9	0.524	1.375	12,729
2	117.0	1,720.9	0.489	0.851	17,540
1	57.7	1,778.6	0.362	0.362	22,703
0	—	—	—	—	29,818

*Includes additional force F_t at the top of the building.

with the additional force F_t at the top of the building:

$$F_t = 0.15V$$

$$= 0.15 \times 1,778 = 266.0 \text{ kN}$$

Calculated equivalent seismic forces are given in Table 38.3.

Story shear force

$$V_i = \sum_{k=i}^{n} F_k \qquad \text{[eq.(38.10)]}$$

Values calculated for story shear forces are shown in Table 38.3.

Story drifts and lateral displacements. Story drifts Δ_i for a building modeled as a shear building are given by

$$\Delta_i = \frac{V_i}{k_i}$$

where V_i is the story shear force and k_i is the story stiffness. Lateral displacements δi are given by

$$\delta_i = \sum_{k=1}^{i} \Delta_k$$

Calculated values for story drifts and lateral displacements are given in Table 38.3.

Check maximum permissible lateral displacement at the top of the building:

$$x_{t\max} = H/600 \qquad \text{[(eq. 38.8)]}$$

$$= 2,200/600 = 3.67 \text{ cm} > 2.82 \text{ cm}$$

from Table 38.3 (acceptable).

Overturning moments

$$M_i = \sum_{k=i+1}^{n} F_k(h_k - h_i) \qquad \text{[eq.(38.9)]}$$

where $i = 0, 1, 2, \ldots, n-1$.

Values calculated for the overturning moments are shown in Table 38.3.

Example 38.2

Use the computer program developed for this chapter to solve Example 38.1.

```
INPUT DATA AND OUTPUT RESULTS FOR EXAMPLE 38.2[1]

INPUT DATA:

NUMBER OF STORIES N = 7

BUILDING CATEGORY   KO = 1.0

SEISMIC INTENSITY   KS = 0.10

DYNAMIC RESPONSE    KD = 0.71

DUCTILITY DAMPING   KP = 1.00


STORY   WEIGHT  STIFFNESS  HEIGHT

  7      3650     2865      300
  6      3650     2902      300
  5      3650     2936      300
  4      3650     2976      300
  3      3650     3059      300
  2      3650     3517      300
  1      3150     4909      400


OUTPUT RESULTS:

SEISMIC COEFFICIENT:   K = 0.071
```

FUNDAMENTAL PERIOD: T = 0.984

STORY	EQUIVALENT FORCES	STORY SHEAR	OVERTURNING MOMENT	STORY DRIFT	STORY DISPLACEMENT
7	635.54	635.54	----	0.222	2.828
6	318.09	953.54	1906.6	0.328	2.606
5	267.86	1221.50	4767.5	0.416	2.278
4	217.64	1439.10	8432.0	0.484	1.862
3	167.41	1606.5	12749.4	0.525	1.378
2	117.19	1723.7	17569.0	0.490	0.853
1	57.79	1781.5	22740.2	0.363	0.363
0	----	----	29866.3	---	---

[1]Units used in this example are: kN, cm, and sec.

REFERENCES

Applied Technology Council (1978) *Tentative Provisions for the Development of Seismic Regulations for Buildings, ATC 3-06*. National Bureau of Standards Special Publication 510, U.S. Government Printing Office, Washington, D.C.

Association of Seismology (SFR) (1987) *Seismic Maps of Yugoslavia* (Maps and guidelines are available at the Institute of Earthquake Engineering and Engineering Seismology) (IZIIS), Skopje, P.O. Box 101, 91000 Skopje, Republic of Macedonia.

Building Construction Under Seismic Conditions in the Balkan Region (1983) UNDP/UNIDO RER/79/015, Vols. 1–7, Vienna, Austria.

Code for Repair and Strengthening of Damaged Buildings in the Republic of Montenegro (1979) Institute of Earthquake Engineering and Engineering Seismology (IZIIS), Skopje, P.O. Box 101, 91000, Skopje, Republic of Macedonia.

Comite Euro-International an Beton (CEB) (1978) *Model Code for Seismic Design of Concrete Structures*. Bulletin d'Information No. 160, Bureau de Paris, 6 Rue Lariston F75116, Paris, France.

Technical Regulations for Design and Construction of Buildings in Seismic Regions (1981) Yugoslav Code, Yugoslav Official Register No. 31, Belgrade, Yugoslavia.

Appendix
Magnitude and Intensity of Earthquakes*

Alberto Sarria M.

A.1 INTRODUCTION

The concepts of magnitude and intensity are used in seismology and in earthquake engineering to compare seismic events and to quantify the energy that is released during an earthquake. It is important that the engineer have a precise understanding of the terminology that is employed in order to have a basis for uniformly evaluating and comparing different seismic events.

A.2 MAGNITUDE

The original concept of earthquake magnitude was introduced in the year 1935 by C. F. Richter (1958) with the objective of comparing the energy liberated by different earthquakes. Seismologists have found many limitations in the use of this concept of magnitude. Therefore, it may be possible that the use of Richter's concept of magnitude will decrease in the years to come, even though it currently continues to be the most used parameter for the estimation of energy released by earthquakes. Richter expressed the magnitude, M, of an earthquake by the following formula:

$$M = \log(A/T) + f(\Delta, h) + C_s + C_n \qquad (A.1)$$

in which

A = maximum amplitude of vibration registered by a seismograph, expressed in thousandths of millimeters

T = period of the seismic wave, in seconds
Δ = distance to the epicenter, in grades
h = focal depth, in kilometers
C_s = correction factor for the seismological station
C_r = regional correction factor.

The determination of $f(\Delta, h)$ as a function of distance and depth is based on a combination of analytical and empirical studies, in which the attenuation is considered as well as the particular type of wave. The magnitude of an earthquake determined by eq.(A.1) should provide a unique value for a specific seismic event. Although the magnitude M does not have an upper limit, natural seismic events have a maximum value of $M = 8$ or slightly higher.

Since the publication in 1935 of the original Richter magnitude [eq.(A.1)], several other scales have been proposed that consider the different types of waves propagating from the same seismic source. The following paragraphs recapitulate the different scales for earthquake magnitude that are in use at the present time.

A.2.1 Local Magnitude M_L

The local magnitude M_L corresponds to the original formulation proposed by Richter in 1935 for local seismic events in southern California. The magnitude M_L is defined as the logarithm of the maximum amplitude that is obtained from the record of a seismic event using a Wood-Anderson torsional seismograph located at 100 kilometers from the epicenter of the earthquake. This seismograph must have a natural period of 0.8 second, magnification of 2,800, and

---

*Translated from Ingeniería Sísmica (Alberto Sarria M., 1990)

damping coefficient of 80% of critical damping. The magnitude for seismic events registered at locations other than 100 kilometers from the epicenter may be determined on the basis of the variation of the amplitude as a function of distance.

A.2.2 Surface Magnitude M_s

The surface magnitude M_s was proposed by Gutenberg and Richter in the year 1945 as a result of detailed studies. It is currently the magnitude scale most widely used for great epicentral distances, but it is valid for any epicentral distance and for any type of seismograph. The magnitude M_s requires a more precise knowledge of the variation of the wave amplitude as a function of distance. In order to utilize different seismographs, the amplitude of vibration of the soil should be used, not the amplitude recorded. M_s may be evaluated for surface waves with periods in the order of 20 seconds by the following expression, known as the Praga formula:

$$M_s = \log(A/T) + 1.66\log\Delta + 3.3 \tag{A.2}$$

where

A = spectral amplitude, the horizontal component of the Rayleigh wave, with a period of 20 seconds, measured on the ground surface in microns
T = period of the seismic wave in seconds
Δ = epicentral distance in degrees.

After a detailed study of the Praga formula, Okal (1989) concluded that it could be extended to longer periods at an appropriate epicentral distance, for which eq.(A.2) would require some modifications.

A.2.3 Magnitude M_b

Gutenberg has proposed another method to measure the magnitude of an earthquake based on the amplitude of internal waves with periods in the order of one second. The following expressions (Bath 1973) relate the value of M_b to that of M_s or M_L:

$$M_s = 1.7 + 0.8M_L - 0.01(M_L)^2 \tag{A.3}$$
$$M_b = 0.56M_s + 2.9 \tag{A.4}$$

Special expressions have been developed for specific geographic locations.
For northwest South America:

$$M_s = 1.51M_b - 2.96 \tag{A.5}$$

and for any location in South America:

$$M_s = 2.18M_b - 6.44 \tag{A.6}$$

A.3 SEISMIC MOMENT

A concept known as seismic moment has been introduced recently to quantify and compare seismic events; it is evaluated by the expression

$$M_0 = ADG \text{ (ton-m)} \tag{A.7}$$

in which

M_0 = seismic moment
A = fault area (length × depth) (m^2)
D = longitudinal displacement of the fault (m)
G = modulus of rigidity (approximately 3×10^6 ton/m^2 = 3×10^{11} dyna/cm^2)

A.3.1 Seismic Moment Magnitude M_w

The magnitude M_w (seismic moment) of an earthquake is expressed as (Kanamori and Anderson 1975):

$$M_w = 2/3\log M_0 - 10.7 \tag{A.8}$$

According to Bullen and Bolt (1985), the Alaska earthquake of 1964 had the following magnitude:

$$M_s = 8.4 \quad \text{or} \quad M_w = 9.2$$

A.4 RELEASED ENERGY

The energy E, in joules, released by an earthquake can be estimated from the magnitude M_s as [Newmark and Rosenblueth (1971)]

$$E = 10^{4.8+1.5M_s} \text{ (joules)} \tag{A.9}$$

According to Newmark and Rosenblueth (1971), the energy released by an earthquake is comparable to that of a nuclear explosion. A nuclear explosion of one megaton releases 5×10^{15} joules; thus, by eq.(A.9), to release the equivalent amount of energy released by an earthquake of magnitude $M_s = 7.3$[1] would require a nuclear explosion of 50 megatons. This brief calculation helps to demonstrate the enormous destructive power of a strong earthquake; the vast amount of energy that accumulates along a fault for many years, even centuries, is suddenly released in a short interval of time, about 60 seconds or less.

---

[1]Earthquakes of magnitude M_s = 7.3 or greater occur worldwide at an average of seven per year.

Table A.1. Modified Mercalli Intensity (MMI) Scale (Abbreviated Version)

Intensity	Evaluation	Description
I	Insignificant	Only detected by instruments
II	Very light	Only felt by sensitive persons; oscillation of hanging objects
III	Light	Small vibratory motion
IV	Moderate	Felt inside buildings; noise produced by moving objects
V	Slightly strong	Felt by most persons; some panic; minor damages
VI	Strong	Damage to nonseismic-resistant structures
VII	Very strong	People running; some damages in seismic-resistant structures and serious damage to unreinforced masonry structures
VIII	Destructive	Serious damage to structures in general
IX	Ruinous	Serious damage to well-built structures; almost total destruction of nonseismic-resistant structures
X	Disastrous	Only seismic-resistant structures remain standing
XI	Disastrous in extreme	General panic; almost total destruction; the ground cracks and opens
XII	Catastrophic	Total destruction

Table A.2. International Macroseismic Intensity Scale (MSK) (Abbreviated Version)

Intensity	Description
I	Not perceptible by humans, only detected by seismographs.
II	Perceptible only by some people at rest, particularly in the upper stories of buildings.
III	Perceptible by some people inside buildings, but only under favorable conditions.
IV	Perceptible by many people inside buildings and some outside. Hanging objects swing.
V	Perceptible by most people inside buildings and by many outside. Movement of objects. Some light damage to Type A structures.
VI	Perceptible by most people inside buildings and outside. Moderate damage to Type A structures and damage to some Type B structures.
VII	Most people running in the street. Many Type A structures highly damaged. Moderate damage to Type B structures.
VIII	Damage to some Type C structures. General panic. Many Type A structures are destroyed and many Type B structures experience serious damage. Also, many Type C structures suffer moderate damage.
IX	General panic. Many Type C structures collapse. Earth displacements and large waves in lakes and reservoirs occur.
X	Most Type A structures and many Type B structures suffer complete collapse. Many Type C structures experience destruction. Ground opens; displacements of sand on sea cause change in the water levels of wells.
XI	Important damage to all kinds of structures including well-designed ones.
XII	Virtually complete destruction of all structures including those underground.

A.5 INTENSITY

Although the magnitude of an earthquake provides a measure of the total energy released, it does not describe the damaging effects caused by the earthquake at a particular locality. Therefore, it is necessary to distinguish between the magnitude and intensity of an earthquake. Magnitude refers to the energy liberated, while intensity measures the effect that an earthquake produces at a given geographic location.

Several intensity scales have been proposed. The one most widely used is the intensity scale of Mercalli-Cancani, modified by Wood-Newman (Newman 1954), now known as the Modified Mercalli Intensity scale or MMI scale.

A.5.1 Modified Mercalli Intensity (MMI) Scale

The value assigned to the MMI scale at a particular location is based on the observations of the damage produced by the earthquake. The MMI scale has a total of 12 intensity values usually expressed in Roman numerals as described in Table A.1. Intensities up to VI usually do not produce damage, while intensities from VI to XII result in progressively greater damage to buildings and other structures.

A.5.2 International Macroseismic Intensity Scale (MSK)

The International Macroseismic Intensity Scale (MSK) proposed by Medvedev-Sponheuer-Karnik, which is similar to the Modified Mercalli Intensity (MMI) scale, provides an empirical appraisal of the effects of an earthquake based on observed damage. The degrees in the MSK are defined by: (a) effects felt by people, (b) damage produced in different types of structures, and (c) changes observed in nature after the earthquake. Damages are observed in three types of structures not designed to resist seismic action:

Type A: with masonry walls of adobe or similar materials;

Type B: with brick walls, concrete blocks, or masonry with wood reinforcement; and

Type C: steel or reinforced concrete structures.

Table A.2 presents an abbreviated description of the 12 degrees of the International Macroseismic Intensity Scale (MSK).

REFERENCES

BATH, MARKUS (1973) *Introduction to Seismology*. John Wiley and Sons, New York, NY.

BULLEN, K. E., and BOLT, BRUCE (1985) *An Introduction to the Theory of Seismology*. 4th ed. Cambridge University Press, London and New York.

KANAMORI, H., and ANDERSON, DON L. (1975) "Theoretical Basis of Some Empirical Relations in Seismology." *Bulletin of the Seismological Society of America*, Vol. 65, #5.

NEWMAN, F. (1954) *Earthquake Intensities and Related Ground Motions*. University of Washington Press, Seattle, WA.

NEWMARK, NATHAN, and ROSENBLUETH, EMILIO (1971) *Fundamentals of Earthquake Engineering*. Prentice Hall, Englewood Cliffs, NJ.

OKAL, EMILE (1989) "A Theoretical Discussion of Time Domain Magnitude: The Praga Formula for M_s and the Mantle Magnitude M_m." *Journal of Geophysical Research*, Vol. 94.

RICHTER, CHARLES F. (1958) *Elementary Seismology*. W. H. Freeman and Company, San Francisco, CA.

SARRIA M., ALBERTO (1990) *Ingeniería Sísmica*. Ediciones Uniandes, Universidad de Los Andes, Bogotá, Colombia.

Diskette Order Form

Professor Mario Paz
P.O. Box 35101
Louisville, KY 40232
USA

Date _____

Please send to:

Name _____

Street Address _____

City, State, Zip code _____

Country _____

Earthquake Resistant Design Programs:

Implementing Seismic Codes for 25 Countries in this Handbook

Set of Programs, source and compiled versions (menu driven) $240
20% discount to a purchaser of the Handbook   48
                                                                 ————
                                                                 $192
Shipping charge
                                                                 ————
Total (check enclosed)

USA shipping and handling, add $5
Overseas shipping, add $15
Canada shipping, add $10

Index